HANDBOOK OF APPLIED ECONOMIC STATISTICS

T0203652

STATISTICS: Textbooks and Monographs

A Series Edited by

D. B. Owen, Founding Editor, 1972–1991

W. R. Schucany, Coordinating Editor
Department of Statistics
Southern Methodist University
Dallas, Texas

S. C. Chow, Associate Editor
for Biostatistics
Bristol-Myers Squibb Company

W. J. Kennedy, Associate Editor
for Statistical Computing
Iowa State University

A. M. Kshirsagar, Associate Editor
for Multivariate Analysis and
Experimental Design
University of Michigan

E. G. Schilling, Associate Editor
for Statistical Quality Control
Rochester Institute of Technology

Additional Volumes in Preparation

HANDBOOK OF APPLIED ECONOMIC STATISTICS

edited by

Aman Ullah
University of California
Riverside, California

David E. A. Giles
University of Victoria
Victoria, British Columbia, Canada

CRC Press
Taylor & Francis Group
Boca Raton London New York

CRC Press is an imprint of the
Taylor & Francis Group, an **informa** business

First published in 1998 by Marcel Dekker

Published in 2020 by CRC Press
Taylor & Francis Group
6000 Broken Sound Parkway NW, Suite 300
Boca Raton, FL 3487-2742

First issued in paperback 2020

© 1998 by Taylor & Francis Group, LLC
CRC Press is an imprint of Taylor & Francis Group, an Informa business

No claim to original U.S. Government works

ISBN-13: 978-0-367-57937-1 (pbk)
ISBN-13: 978-0-8247-0129-1 (hbk)

Visit the Taylor & Francis Web site at
http://www.taylorandfrancis.com

and the CRC Press Web site at
http://www.crcpress.com

Library of Congress Cataloging-in-Publication Data

Handbook of applied economic statistics/edited by Aman Ullah, David E. A. Giles.
    p. cm.—(Statistics, textbooks, and monographs; v.155)
    Includes bibliographical references and index.
    ISBN 0-8247-0129-1
    1. Economics—Statistical methods. I. Ullah, Aman. II. Giles, David E. A. III. Series.
HB137.H36   1998
330'.01'5195—dc21
                                                97–47379
                                                   CIP

Preface

Many applied subjects, including economic statistics, deal with the collection of data, measurement of variables, and the statistical analysis of key relationships and hypotheses. The attempts to analyze economic data go back to the late eighteenth century, when the first examinations of the wages of the poor were done in the United Kingdom, followed by the the mid-nineteenth century research by Engle on food expenditure and income (or total expenditure). These investigations led to the early twentieth-century growth of empirical studies on demand, production, and cost functions, price determination, and macroeconomic models. During this period the statistical theory was developed through the seminal works of Legendre, Gauss, and Pearson. Finally, the works of Fisher and Neyman and Pearson laid the foundations of modern statistical inference in the form of classical estimation theory and hypothesis testing. These developments in statistical theory, along with the growth of data collections and economic theory, generated a demand for more rigorous research in the metholodogy of economic data analysis and the establishment of the International Statistical Institute and the Econometric Society.

The post–World War II period saw significant advances in statistical science, and the transformation of economic statistics into a broader subject: econometrics, which is the application of mathematical and statistical methods to the analysis of economic data. During the last four decades, significant works have appeared on econometric techniques of estimation and hypothesis testing, leading to the application of econometrics not only in economics but also in sociology, psychology, history, political science, and medicine, among others. We also witnessed major developments in the literature associated with the research at the interface between econometrics and statistics, especially in the areas of censored models, panel (longitudinal) data models, the analysis of nonstationary time series, cointegration and volatility, and finite sample and asymptotic theories, among others. These common grounds are of considerable importance for researchers, practitioners, and students of both of these disciplines and are of direct interest to those working in other areas of applied statistics.

The most important objective of this volume is to cover the developments in both applied economics statistics and the econometric techniques of estimation and hypothesis testing. It is in this respect that our book differs from other publications in which the emphasis is on econometric methodology. With the above purpose in view, we deal with the material that is of direct interest to researchers, practitioners, and graduate students in many applied fields, especially economics and statistics. It covers reasonably comprehensive and up-to-date reviews of developments in various aspects of economic statistics and econometrics, and also contains papers with new results and scopes for future research. The objective behind all this was to produce a handbook that could be used by professionals in economics, sociology, econometrics, and statistics, and by teachers of graduate courses.

The Handbook consists of eighteen chapters that can be broadly classified into the following three groups:

1. Applied Economic Statistics
2. Econometric Methodology and Data Issues
3. Model Specification and Simulation

Chapters 1–5 belong to Part 1 and they are applied papers dealing with important statistical issues in development economics and microeconomics. The chapter by Davies, Green, and Paarsch reviews the literature on using economics statistics, such as income inequality and other aggregate poverty indices, and they make a strong case for the use of disaggregated dominance criteria to make social welfare comparisons. They also discuss some statistical issues related to parametric and nonparametric inference concerning Lorenz Curves, with reference to stochastic dominance. In contrast, Krämer's chapter develops two ways of looking at inequality measurement: the first, a preordering based on majorization defined over income vectors, and the second, an axiomatic-based approach in which axioms are defined over inequality measurements. The chapter also includes the empirical application of inequality measurement primarily focused on dealing with the fact that data is usually grouped by quantile. Ravallion's chapter addresses an important issue of persistence in the geography of poverty. It proposes a methodology for empirically testing the validity of two competing explanations of poverty: an individualistic model and a geographic model. His proposed approach contributes to our understanding of the determinants of poverty and provides information for policymakers regarding which policy interventions are likely to be most effective for its alleviation. The chapter by Deolalikar explores another dimension of the poverty issue in developing countries, that is, whether decreased spending on government health programs will reduce the demand for public health services by the poor and hence will adversely affect the health status of the poor. This question is analyzed using data from research conducted in Indonesia. The chapter also attempts to address the shortcomings of the existing literature. Finally, the chapter on mobility by Maasoumi reviews two different, but related, approaches to testing for income mobility and shows that the two ways converge to the same ordering of states. The relationship between this ordering and the partial ordering given by Lorenz dominance is shown.

Chapters 6–10 and 15–17 deal with econometric methodologies related to different kinds of data used in empirical research. Chapter 6 by Russell, Breunig and Chiu is perhaps the first comprehensive treatment of the problem of aggregation as it relates to empirical estimation of aggregate relationships. It is well known that the analysis of individual behavior based on aggregate data is justified if the estimated aggregate relationships can be consistently disaggregated to the individual relationships and vice versa. Most empirical studies have ignored this problem; those that have not are reviewed in this chapter. Anselin and Bera's chapter details another data problem ignored in the econometric analysis of regression model: the problem of spatial autocorrelation and the correlation in cross-sectional data. This chapter reviews the methodological issues related to the treatment of spatial dependence in linear models. Another data issue often ignored in empirical development economics and labor economics is related to the fact that most of the survey data is based on complex sampling from a finite population, such as stratified, cluster, and systematic sampling. However, the econometric analysis is carried out under the assumption of random sampling from an infinite population. The chapter by Ullah and Breunig reviews the literature on complex sampling and indicates that the effect of misspecifying or ignoring true sampling schemes on the econometric inference can be quite serious.

Panel data is the multiple time series observations on the same set of cross-sectional survey units (e.g., households). Baltagi's chapter reviews the extensive existing literature on econometric inference in linear and nonlinear parametric panel data models. In a related chapter, Ullah and Roy develop the nonparametric kernel estimation of panel data models without assuming their functional forms. The chapter by Golan, Judge, and Miller proposes a maximum-entropy approach to the estimation of simultaneous equations models when the economic data is partially incomplete. In Chapter 15 Teräsvirta looks into the modeling of time series data that exhibit nonlinear relationships due to discrete or smooth transitions and to regimes' switching. He proposes and develops a smooth transition regression analysis for such situations. Finally, the chapter by Franses surveys econometric issues concerning seasonality in economic time series data due to weather or other institutional factors. He discusses the statistical models that can describe forecasts of economic time series with seasonal variations encountered in macroeconomics, marketing and finance.

Chapters 12 and 18 are related to the simulation procedures and 11, 13, and 14 to the model and selection procedures in econometrics. The chapter by DeBenedictis and Giles surveys the diagnostic tests for the model misspecifications that can have serious consequences on the sampling properties of both estimators and tests. In a related chapter, Hadi and Son look into diagnostic procedures for revealing outliers (influential observations) in the data which, if present, could also affect the estimators and tests. They also propose a methodology of estimating linear models with outliers, which is an alternative to computer-intensive quantile estimation techniques used in practice. Next, the chapter by Dufour and Torrès systematically develops the general theory of union-intersection and sample split methods in various specification testing problems in econometrics. They apply their results for testing problems in the SURE model and a model with MA(1) errors. In contrast to the analytical approaches of specification testing, the chapter by Veall provides a survey of bootstrap simulation procedures that is especially useful in small samples. The book concludes with the chapter by Pagan, in which he debates about the calibration methodology of estimation and specification analysis. Several thought-provoking questions are raised and discussed.

In summary, this volume brings together survey material and new methodological results which are vitally important to modern developments in applied economic statistics and econometrics. The emphasis is on data problems, methodological issues, and inferential techniques that arise in practice in a wide range of situations that are frequently encountered by researchers in many related disciplines. Accordingly, the contents of the book should have wide appeal and application. We are very pleased with the end product and would like to thank all the authors for their contributions, and for their cooperation during the preparation of this volume. We are also most grateful to Benicia Chatman, University of California at Riverside, for the efficient assistance that she has provided, and to the editorial and production staff at Marcel Dekker, especially Maria Allegra and Lia Pelosi, for their patience, guidance, and expertise.

Aman Ullah
David E. A. Giles

Contents

Part 3   Model Specification and Simulation

Contributors

Luc Anselin, Ph.D.   Research Professor, Regional Research Institute and Department of Economics, West Virginia University, Morgantown, West Virginia

Badi H. Baltagi, Ph.D.   Professor, Department of Economics, Texas A&M University, College Station, Texas

Anil K. Bera, Ph.D.   Professor, Department of Economics, University of Illinois, Champaign, Illinois

Robert V. Breunig   Graduate Student, Department of Economics, University of California at Riverside, Riverside, California

Chia-Hui Chiu   Graduate Student, Department of Economics, University of California at Riverside, Riverside, California

James B. Davies, Ph.D.   Professor and Chair, Department of Economics, University of Western Ontario, London, Ontario, Canada

Linda F. DeBenedictis, M.A.   Senior Policy Analyst, Policy and Research Division, Ministry of Human Resources, Victoria, British Columbia, Canada

Anil B. Deolalikar, Ph.D.   Professor, Department of Economics, University of Washington, Seattle, Washington

Jean-Marie Dufour, Ph.D.   Professor, C.R.D.E. and Department of Economic Sciences, University of Montreal, Montreal, Quebec, Canada

Philip Hans Franses, Ph.D.   Associate Professor, Department of Econometrics, Erasmus University Rotterdam, Rotterdam, The Netherlands

David E. A. Giles, Ph.D.   Professor, Department of Economics, University of Victoria, Victoria, British Columbia, Canada

Amos Golan, Ph.D.   Visiting Associate Professor, Department of Agricultural and Resource Economics, University of California at Berkeley, Berkeley, California, and Department of Economics, American University, Washington, D.C.

David A. Green, Ph.D.   Department of Economics, University of British Columbia, Vancouver, British Columbia, Canada

Ali S. Hadi, Ph.D.   Professor, Department of Statistics, Cornell University, Ithaca, New York

George Judge, Ph.D.   Professor, University of California at Berkeley, Berkeley, California

Walter Krämer, Dr. rer. pol.   Professor, Department of Statistics, University of Dortmund, Dortmund, Germany

Esfandiar Maasoumi, Ph.D. (FRS)   Professor, Department of Economics, Southern Methodist University, Dallas, Texas

Douglas Miller, Ph.D.   Assistant Professor, Department of Economics, Iowa State University, Ames, Iowa

Harry J. Paarsch, Ph.D.   Associate Professor, Department of Economics, University of Iowa, Iowa City, Iowa

Adrian Rodney Pagan, Ph.D.   Professor, Economics Program, The Australian National University, Canberra, Australia

Martin Ravallion, Ph.D.   Lead Economist, Development Research Group, World Bank, Washington, D.C.

Nilanjana Roy, Ph.D.   Assistant Professor, Department of Economics, University of Victoria, Victoria, British Columbia, Canada

R. Robert Russell, Ph.D.   Professor, Department of Economics, University of California at Riverside, Riverside, California

Mun S. Son, Ph.D.   Associate Professor, Department of Mathematics and Statistics, University of Vermont, Burlington, Vermont

Timo Teräsvirta, Ph.D. Professor, Department of Economic Statistics, Stockholm School of Economics, Stockholm, Sweden

Olivier Torrès, Ph.D. Maître de Conférences, U.F.R. Mathématiques, Sciences Économiques et Sociales, Université de Lille, Villeneuve d'Ascq, France

Aman Ullah, Ph.D. Professor, Department of Economics, University of California at Riverside, Riverside, California

Michael R. Veall, Ph.D. Professor, Department of Economics, McMaster University, Hamilton, Ontario, Canada

HANDBOOK OF APPLIED
ECONOMIC STATISTICS

1

Economic Statistics and Social Welfare Comparisons
A Review

James B. Davies
University of Western Ontario, London, Ontario, Canada

David A. Green
University of British Columbia, Vancouver, British Columbia, Canada

Harry J. Paarsch
University of Iowa, Iowa City, Iowa

I. INTRODUCTION

We present a selective survey of the literature concerned with using economic statistics to make social welfare comparisons. First, we define a number of different summary income inequality measures and social welfare indices as well as functional summary measures associated with disaggregated dominance criteria. Next, we describe the theoretical basis for the use of such measures. Finally, we outline how data from conventional sample surveys can be used to estimate the functionals that can be interpreted in terms of social welfare and compared in decision-theoretic terms with other functionals.

While we define and discuss some of the properties of popular summary inequality indices, our main focus is on functional summary measures associated with disaggregated dominance criteria. In particular, we examine the estimation and comparison of Lorenz and generalized Lorenz curves as well as indicators of third-degree stochastic dominance.* In line with a growing body of opinion, we believe that the

---

*The estimation of summary indices is discussed, for example, by Cowell (1989); Cowell and Mehta (1982); and Cowell and Victoria-Feser (1996).

partial-ordering approach using disaggregated dominance criteria is more attractive than the complete-ordering approach using summary indices, since the former requires only widely appealing restrictions on social preferences, whereas the latter requires the (explicit or implicit) choice of a specific form for the social welfare function.

The issues in making applied welfare comparisons are well reviewed in several places; see, for example, Atkinson (1975) and Cowell (1979). One of the most important of these issues concerns the definition of income. Applied researchers face choices between money income versus broader definitions including imputations; before- versus after-tax income; measuring over a short versus a long horizon; and even the choice between income and alternative measures of welfare, such as consumption. Other important conceptual issues concern the choice of unit (individual? family? household?) and whether one should examine total income, income per capita, or perhaps income per adult equivalent. In what follows, we assume that an appropriate definition of income has already been chosen.

Practical difficulties in making welfare comparisons center around measurement and related problems. Official data, for example from tax records, omit income components, are contaminated by avoidance and evasion, and sometimes do not allow the researcher to use the desired family unit. Survey data are affected by differential response according to income and other characteristics, and misreporting. Also, both of these sources tend to neglect in-kind income. These problems should be borne in mind by applied researchers working with the techniques discussed here.

In Section II we define the bulk of the notation used and describe different summary inequality indices as well as functional summary measures which are often used to relate economic statistics to social welfare comparisons. Section III provides an axiomatic foundation for many of the measures listed in Section II. We show how data from conventional sample surveys can be used to estimate the functionals of interest in Section IV, and in Section V we discuss how observed covariates can be introduced into this framework. We summarize and present our conclusions in Section VI. Section VII, an appendix, describes how to access several programs designed to carry out the analysis described here. These programs reside in the Econometrics Laboratory Software Archive (ELSA) at the University of California, Berkeley; this archive can be accessed easily by using a number of different browsers (e.g., Netscape Navigator) via the Internet.

II. DIFFERENT ECONOMIC MEASURES OF SOCIAL WELFARE

In this section we set out notation and define several scalar summary indices of economic inequality and social welfare as well as two functional summary measures associated with disaggregated dominance criteria. We confine ourselves to the positive evaluation of the behavior of the various measures. The normative properties of the measures are discussed in Section III. Nonetheless, it is useful from the outset to note some of the motivation for the various indices. In this section we provide a heuristic discussion, an approach that mimics how the field has developed historically.

Economists have always agreed that increases in everyone's income raise welfare, so there is natural interest in measures of central location (the mean, the median), which would reflect such changes. But there has also always been a view that increases in relative inequality make society worse off. The latter view has sometimes been based on, but is logically separate from, utilitarianism. Given the interest in inequality, it was natural that economists would like to measure it. Historically, economists have proposed a number of essentially ad hoc methods of measuring inequality, and found that the proposed indices were not always consistent in their rankings. This gave rise to an interest in and the systematic study of the normative foundations of inequality measurement. Some of the results of that study are surveyed in Section III.

One central concern in empirical studies of inequality has been the ability to allocate overall inequality for a population to inequality between and within specific subpopulations. Thus, for example, one might like to know how much of the overall income inequality in a country is due to inequality for females, how much is due to inequality for males, and how much is due to inequality between males and females. One reason for the popularity of particular scalar measures of inequality (e.g., the variance of the logarithm of income) is that they are additively decomposable into these various between- and within-group effects. The decompositions are often created by dividing the population into subpopulations and applying the inequality measure to each of the subpopulations (to get within-group inequality measures) and to the "sample" consisting of the means of all the subpopulations (to get a between-group measure). The more subpopulations one wants to examine, the more unwieldy this becomes. Furthermore, in some instances, one is interested in answering the question, how would inequality change if the proportion of the population who are unionized increases, holding constant all other worker characteristics? The decompositions allowed in the standard inequality measures do not provide a clear answer to this question. This point will be raised again in Section IV. In the following section, the reader should keep in mind that indices like the generalized entropy family of indices possess the decomposability property. For a more detailed discussion of the decompositions of various indices, see Shorrocks (1980, 1982, 1984).

We consider a population each member of which has nonnegative income Y distributed according to the probability density (or mass) function $f(y)$ with corresponding cumulative distribution function $F(y)$.

A. Location Measures

Under some conditions, economists and others attempt comparisons of economic welfare which neglect how income is distributed. It has been argued, for example by Harberger (1971), that gains and losses should generally be summed in an un-weighted fashion in applied welfare economics. This procedure may identify poten-tial Pareto improvements; i.e., situations where gainers could hypothetically com-pensate losers. Furthermore, it is possible to conceive of changes that would affect all individuals' incomes uniformly, so that distributional changes would be absent. For these reasons, measures of central location are a natural starting point in any discussion of economic measurement of social welfare.

1. Per Capita Income

Perhaps the most common measure of welfare is aggregate or per capital income. Focusing on the latter, we have mean income

$$\mu = \mathcal{E}[Y] = \int_0^\infty yf(y)\, dy$$

in the continuous case or

$$\mu = \mathcal{E}[Y] = \sum yf(y)$$

in the discrete case.* Such a measure is easy to calculate and has considerable intu-itive appeal, but it is sensitive to outliers in the tails of the distributions. For example, an allocation in which 99 people each have an income of \$1 per annum, while one person has an income of \$999,901 would be considered to be equivalent in welfare terms to an allocation in which each person has \$10,000 per annum. Researchers are frequently attracted to alternative measures that are relatively insensitive to be-havior in the tails of the distribution. One measure of location that is robust to tail behavior is the median.

2. Median Income

The median is defined as that point at which half of the population is above and the other half is below. In terms of the probability density and cumulative distribution functions, the median solves the following:

$$0.5 = \int_0^{\xi(0.5)} f(y)\, dy = F(\xi(0.5))$$

Alternatively,

$$\xi(0.5) = F^{-1}(0.5)$$

---

*Hereafter, without loss of generality, we shall focus almost exclusively on the continuous case.

where $F^{-1}(\cdot)$ is the inverse function of the cumulative distribution function. Clearly, other quantiles could also be entertained: for example, the lower and upper quartiles, which solve

$$\xi(0.25) = F^{-1}(0.25) \quad \text{and} \quad \xi(0.75) = F^{-1}(0.75)$$

respectively.

Using the example considered above, the median of the first allocation would be $1, while that of the second would be $10,000. Clearly, tail behavior (or dispersion) is important in ranking allocations. Accordingly, a natural progression is to consider measures of the *scale* of the distributions being considered, such as the variance. However, alternative income distributions differ in more than just scale. They also differ in *shape*, and both scale and shape can affect the degree of inequality which observers perceive in a distribution. This leads to an array of different possible inequality measures, each member of which is an acceptable measure of scale, but each of which reacts differently to shape.

B. Scale Measures and Inequality Indices

Consider two alternative income distributions which are related in the following way:

$$Y' = a + bY$$

If $a \neq 0$ then these distributions differ in location, and if $b \neq 1$ they differ in scale. Any index which rises monotonically in b, but which is invariant to a, is a measure of the scale of a distribution. Standard examples include the variance and standard deviation, but a wide variety of other scale measures is possible.

How do measures of scale relate to measures of inequality? Some scale measures, such as the standard deviation, make sensible inequality measures under some circumstances. Other scale measures, such as the interquartile range discussed here, are woefully inadequate inequality measures. The reason is that real-world distributions differ in ways other than simple differences in location and scale. An adequate inequality measure must, at a minimum, increase when income is transferred from any poorer person to any richer person. Not all scale measures satisfy this criterion. Moreover, there is considerable interest in relative inequality measures, which are defined on incomes normalized by the mean, and are therefore independent of scale.

In this subsection we first look at three types of measures of scale: quantile-based measures, the variance, and the standard deviation. We then go on to a few popular relative inequality indices: the coefficient of variation, the variance of logarithms, and the Gini coefficient. Holding mean income constant, these indices are all inverse measures of social welfare. Finally, we examine Atkinson's index and the generalized entropy family of indices.

1. Quantile-Based Measures of Income

Quantile differences or quantile ratios have been employed to provide rough-and-ready descriptions of the degree of income dispersion. Attention is often paid, for example, to the interquartile range $\xi(0.75)-\xi(0.25)$ or the 90–10 percentile ratio $\xi(0.90)/\xi(0.10)$. While such differences or ratios respond to some changes in scale, they may remain invariant in the face of major changes in the distribution of income. Each is completely insensitive, for example, to redistributions of income occurring exclusively in a wide, middle range of incomes.

2. Variance of Income

A standard measure of scale is the variance of Y, defined by

$$\sigma^2 = \mathcal{V}[Y] = \int_0^\infty (y - \mu)^2 f(y) \, dy$$

An important characteristic of the variance, which carries through to the related measures like the standard deviation and the coefficient of variation, is that it is highly sensitive to the tails of the distribution. Given that distributions of income and related variables are typically skewed positively, in practice the sensitivity of these measures is generally greatest to the length of the upper tail. Another important property of the variance is that it is additively decomposable both in terms of income components and population subgroups; see Shorrocks (1980, 1982, 1984). The decomposability of the variance also carries through to convenient decompositions of the standard deviation and the coefficient of variation.

3. Standard Deviation of Income

A drawback of the variance is that it is in the units squared of income. Another measure of scale, which has the same units as the mean, is the standard deviation, which is defined as

$$\sigma = \sqrt{\sigma^2}$$

4. Coefficient of Variation of Income

We now turn to some measures of relative inequality—i.e., to indices which are defined over income normalized by the mean. These measures are invariant to a particular kind of change in scale, one where all incomes change equiproportionally. Such measures are, of course, also insensitive to the choice of units of measurement (e.g., dollars versus thousands of dollars).

The first measure of relative inequality we define is the coefficient of variation τ which is simply the standard deviation divided by the mean:

$$\tau = \frac{\sigma}{\mu}$$

Like the variance and the standard deviation, in comparison to other popular measures, τ is especially sensitive to changes in the tails of the distribution.

5. Variance of the Logarithm of Income

An apparently attractive measure of income inequality, which is often employed, is the variance of the logarithm of Y. One reason for the frequent use of this index may lie in the popularity of "log-earnings" or "log-income" regressions. R^2 gives an immediate measure of the proportion of inequality explained by the regressors if the variance of the logarithm of Y is accepted as an appropriate inequality measure, and inequality can be decomposed into components contributed by the various factors.

In the case of the variance, and its related measures, we have seen that the use of a linear scale gives great weight to the right tail of the distribution. Applying a logarithmic transformation reduces this effect.* Thus, introducing the transformation

$$Z = \log Y$$

we define the variance of the logarithm of Y by

$$\sigma_Z^2 = \int_{-\infty}^{\infty} (z - \mu_Z)^2 f(\exp(z)) \exp(z) \, dz$$

where

$$\mu_Z = \int_{-\infty}^{\infty} z f(\exp(z)) \exp(z) \, dz$$

is the average value of $\log Y$.[†] Note that the variance of the logarithm of Y is independent of scale, as is the standard deviation of the logarithm of income σ_Z, with which it may be used interchangeably.[‡]

---

*In some cases, the transformation can actually go too far as is discussed in Section III.

[†] Here, we have used the fact that the probability density function of Z is

$$f_Z(z) = f(y)\frac{dy}{dz} = f(\exp(z)) \exp(z)$$

since

$$y = \exp(z) \quad \text{and} \quad \frac{dy}{dz} = \exp(z)$$

[‡] The coefficient of variation of the logarithm of income τ_Z is not used since τ_Z depends on scale and would fall, for example, with an equiproportionate increase in all incomes. Thus, changing from measurement in dollars to cents would cause τ_Z to fall.

6. Gini Coefficient

Perhaps the most popular summary inequality measure is the Gini coefficient κ. Several quite intuitive alternative interpretations of this index exist. We highlight two of them. The Gini coefficient has a well-known geometric interpretation related to the functional summary measure of relative inequality, the Lorenz curve, which is discussed in the next subsection. Here, we note another interpretation of the index, which may be defined as

$$\kappa = \frac{1}{2\mu} \int_0^\infty \int_0^\infty |u - v| f(u) f(v) \, du \, dv$$

In words, the Gini coefficient is one half the expected difference between the incomes of two individuals drawn independently from the distribution, divided by the mean μ.

One virtue of writing κ this way is that it draws attention to the contrasting weights that are placed on income differences in different portions of the distribution. The weight placed on the difference $|u - v|$ is relatively small in the tails of the distribution, where $f(u) f(v)$ is small, but relatively large near the mode. This means that, in practice, κ is dramatically more sensitive to changes in the middle of the income distribution than it is to changes in the tails. This contrasts sharply with the behavior of many other popular inequality indices which are most sensitive to either or both tails of the distribution.

7. Atkinson's Index

Atkinson (1970) defined a useful and popular inequality index which is based on the family of additive social welfare functions (SWFs) of the form

$$W(Y) = \begin{cases} \int_0^\infty \left[A + B \frac{y^{(1-\varepsilon)}}{(1 - \varepsilon)} \right] f(y) \, dy, & \varepsilon \neq 1 \\[2em] \int_0^\infty \log y f(y) \, dy, & \varepsilon = 1 \end{cases} \tag{1}$$

The parameter ε governs the concavity, and therefore the degree of inequality aversion, shown by the function. Note also that for ε equal to zero the function merely aggregates all incomes, and therefore ranks the same as the mean, given a constant population.

Atkinson's index ψ is defined as

$$\psi = 1 - \frac{y_{\text{EDE}}}{\mu}$$

where y_{EDE} is "equally distributed equivalent income"—the income such that if all individuals had income equal to y_{EDE}, then W would have the same value as the actual income distribution. With the SWF given by (1), this yields

$$\psi_E = 1 - \left[\int_0^\infty \left(\frac{y}{\mu}\right)^{1-\varepsilon} f(y)\, dy \right]^{1/(1-\varepsilon)}$$

The parameter ε plays a dual role. As it rises, inequality aversion increases, but, in addition, the degree of sensitivity to inequality at lower income levels also rises with ε. In the limit, as ε goes to infinity, the index is overwhelmingly concerned with inequality at the bottom of the distribution. While the sensitivity of this index to inequality at different levels can be varied by changing ε, it is always more sensitive to inequality that occurs lower in the distribution.

8. Generalized Entropy Family of Indices

An important class of inequality measures is the generalized entropy family. These measures are defined by

$$\rho_c = \frac{1}{c(c-1)} \int_0^\infty \left[\left(\frac{y}{\mu}\right)^c - 1 \right] f(y)\, dy, \qquad c \neq 0, 1$$

$$\rho_0 = \int_0^\infty \log\left(\frac{\mu}{y}\right) f(y)\, dy, \qquad c = 0$$

$$\rho_1 = \int_0^\infty \left(\frac{y}{\mu}\right) \log\left(\frac{y}{\mu}\right) f(y)\, dy, \qquad c = 1$$

ρ_2 is defined as one half of the square of the coefficient of variation ($\tau^2/2$). The first and second entropy measures of Theil (1967) are ρ_1 and ρ_0, respectively. Letting c equal $1 - \varepsilon$, one obtains

$$\psi_E = \begin{cases} 1 - [c(c-1)\rho_c + 1]^{1/c}, & c < 1, \quad c \neq 0 \\ 1 - \exp(-\rho_c), & c = 0 \end{cases}$$

Hence, for the particular form of Atkinson's index given by some value of ε, there is always a corresponding, ordinally equivalent, generalized entropy measure.

All members of the generalized entropy family of inequality indices are based on some notion of the average distance between relative incomes. These indices do not take into account rank in the income distribution in performing this averaging, which makes them fundamentally different from the Gini coefficient, for which rank is very important.

The attraction of the generalized entropy family is enhanced by the fact that it comprises all of the scale-independent inequality indices satisfying anonymity

and the strong principle of transfers which are also additively decomposable into inequality within and between population subgroups; for more on this, see Shorrocks
(1980). Furthermore, all indices which are decomposable (i.e., not necessarily additively) must be some positive transformation of a member of the generalized entropy
family; for more on this, see Shorrocks (1984).

C.   Disaggregated Summary Measures

Each of the above measures summarizes the information contained in $F(y)$ in a single number and, in the process, discards a great deal of information and can mask
important features of the distribution. As an alternative, researchers have sought to
implement disaggregated summary measures which provide more information concerning the shape of the distribution, but which are still convenient to use. We shall
examine two.[*]

1.   Lorenz Curve

The Lorenz curve (LC) is the plot of the cumulative distribution function q on the
abscissa (x-axis) versus the proportion of aggregate income held by the quantile $\xi(q)$
and below on the ordinate (y-axis). The qth ordinate of the LC is defined as

$$\mathcal{L}(q) = \frac{\int_0^{\xi(q)} yf(y)\,dy}{\int_0^{\infty} yf(y)\,dy} = \frac{\int_0^{\xi(q)} yf(y)\,dy}{\mathcal{E}[Y]} = \frac{\int_0^{\xi(q)} yf(y)\,dy}{\mu}$$

Note that $\mathcal{L}(0)$ is zero, while $\mathcal{L}(1)$ is one. A graph of a representative LC for income
is provided in Figure 1. The 45° line denotes the LC of perfect income equality. The
further is the LC bowed from this 45° line, the more unequal is the distribution of
income. The Gini coefficient κ can also be defined as twice the area between the 45°
line and the LC $\mathcal{L}(q)$. Thus,

$$\kappa = 2 \int_0^1 (q - \mathcal{L}(q))\,dq$$

2.   Generalized Lorenz Curve

Like the coefficient of variation, the variance of logarithms, and the Gini coefficient,
the LC is invariant to the mean. Thus, while it is the indicator of relative inequality
par excellence, it does not provide a complete basis for making social welfare comparisons. Shorrocks (1983) has shown, however, that a closely related indicator is a

---

[*]Howes (forthcoming) provides an up-to-date and more technical discussion of other measures related to
these.

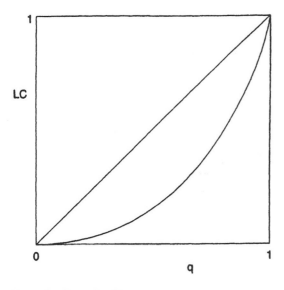

Figure 1   Example of Lorenz curve.

valid social welfare measure. This indicator is the generalized Lorenz curve (GLC) which has ordinate

$$\mathcal{G}(q) = \mathcal{E}[Y]\mathcal{L}(q) = \mu\mathcal{L}(q) = \int_0^{\xi(q)} yf(y)\,dy$$

$$= \Pr[Y \leq \xi(q)]\mathcal{E}[Y|Y \leq \xi(q)] = q\gamma(q)$$

and abscissa q as another functional summary measure. Note that $\mathcal{G}(q)$ equals $\mathcal{L}(q)$ times μ. Thus, $\mathcal{G}(0)$ is zero, while $\mathcal{G}(1)$ equals the mean $\mathcal{E}[Y]$. A graph of a representative GLC for income is provided in Figure 2.

As discussed in the next section, when the LC for one distribution lies above that for another (a situation of LC dominance), then the distribution with the higher LC has unambiguously less relative inequality. A distribution with a dominating (higher) GLC, on the other hand, provides greater welfare according to all social welfare functions defined on and increasing in individual incomes and having appropriate concavity.

D. Stochastic Dominance

In some of the discussion we shall use stochastic dominance concepts. These were first defined in the risk-measurement literature, but it was soon found that they paralleled concepts in inequality and social welfare measurement. We introduce here

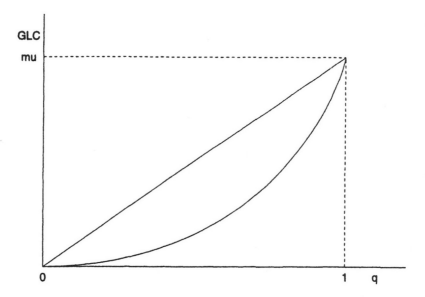

Figure 2 Example of generalized Lorenz curve.

the notions of first-, second-, and third-degree stochastic dominance for two random variables Y_1 and Y_2, each having the respective cumulative distribution functions $F_1(y)$ and $F_2(y)$. First-degree stochastic dominance holds in situations where one distribution provides a Pareto improvement compared to another. As discussed in the next section, second-degree stochastic dominance corresponds to GLC dominance. Finally, third-degree stochastic dominance may be important in the ranking of distributions whose LCs or GLCs intersect.

1. First-Degree Stochastic Dominance

The random variable Y_1 is said to dominate the random variable Y_2 stochastically in the first-degree sense (FSD) if

$$F_2(y) \geq F_1(y) \quad \forall y$$

and

$$F_2(y) > F_1(y) \qquad \text{for some } y$$

In words, strict FSD means that the cumulative distribution function of Y_1 is everywhere to the right of that for Y_2.

2.   Second-Degree Stochastic Dominance

The random variable Y_1 is said to dominate the random variable Y_2 stochastically in the second-degree sense (SSD) if

$$\int_0^y [F_2(u) - F_1(u)]\, du \geq 0 \quad \forall y$$

and

$$\int_0^y [F_2(u) - F_1(u)]\, du > 0 \qquad \text{for some } y$$

Note that FSD implies SSD. If the means of Y_1 and Y_2 are equal (i.e., $\mathcal{E}[Y_1] = \mathcal{E}[Y_2] = \mu$), then Y_1 SSD Y_2 implies that Y_1 is more concentrated about μ than is Y_2.

3.   Third-Degree Stochastic Dominance

The random variable Y_1 is said to dominate the random variable Y_2 stochastically in the third-degree sense (TSD) if

$$\int_0^y \int_0^v [F_2(u) - F_1(u)]\, du\, dv \geq 0 \quad \forall y$$

and

$$\int_0^y \int_0^v [F_2(u) - F_1(u)]\, du\, dv > 0 \qquad \text{for some } y$$

with the following endpoint condition:*

$$\int_0^\infty [F_2(y) - F_1(y)]\, dy \geq 0$$

III.   SOCIAL WELFARE AND INCOME INEQUALITY

It is possible that everyone may be better off in one distribution than in another. For example, this may happen as a result of rapid economic growth raising all incomes. In this case, there is an actual Pareto improvement. As a result, there will be fewer individuals with incomes less than any given real income level Y as time goes on.

---

*Corresponding endpoint conditions can be stated for FSD and SSD, but they are satisfied trivially.

In other words, we will have a situation of first-degree stochastic dominance. More generally, we may have cases where some individuals become worse off, but the fraction of the population below any income cutoff again declines. Under a suitable anonymity axiom, the change would be equivalent in welfare terms to a Pareto improvement. Thus, it appears that first-degree stochastic dominance may sometimes be useful in making social welfare comparisons. This is confirmed by Beach et al. (1994).

In situations where first-degree stochastic dominance does not hold, we need to consult some of the measures of inequality and welfare defined in the previous section. But on which of the wide variety of popular indicators of inequality or welfare should we focus? To answer this question, Atkinson (1970) argued that we must recognize that each indicator either maps into a specific social welfare function (SWF) or restricts the class of admissible SWFs. (This argument was spelled out precisely by Blackorby and Donaldson 1978.) In order to decide which inequality or welfare indicators are of greatest interest, we must examine the preferences embodied in the associated SWFs.

A. Principle of Transfers, Second-Degree Stochastic Dominance, and Lorenz Curves

Atkinson (1970) provided the first investigation of the SWF approach, using the additive class of SWFs:

$$W(Y) = \int_0^\infty U(y) f(y) \, dy \tag{2}$$

where $U'(y)$ is positive and $U''(y)$ is nonpositive. (If $U(y)$ is thought of as a utility function, then this is a utilitarian SWF. However, $U(y)$ may be regarded, alternatively, as a "social evaluation function," not necessarily corresponding to an individual utility function.) Because $U(y)$ is concave, W embodies the property of inequality aversion. Of course, when $U(0)$ equals zero and $U'(y)$ is constant for all y,

$$W(Y) = \mathcal{E}[Y] = \mu \tag{3}$$

This SWF corresponds to the use of per capita income to evaluate income distribution. Note that by using the mean one shows indifference to income inequality.

Atkinson (1970) pointed out that strictly concave SWFs obeyed what has come to be known as the *Pigou-Dalton principle of transfers*. This principle states that income transfers from poorer to richer individuals (i.e., *regressive transfers*) reduce social welfare. Atkinson showed that this principle corresponds formally to that of risk aversion and that a regressive transfer is the analogue of a *mean-preserving spread* (MPS) introduced into the risk-measurement literature by Rothschild and Stiglitz (1970).

Atkinson also noted that a distribution $F_1(y)$ is preferred to another $F_2(y)$ according to all additive utilitarian SWFs if and only if the criterion for second-degree stochastic dominance (SSD) is satisfied. Finally, he showed that a distribution $F_1(y)$ would have $W(Y_1)$ which weakly exceeds $W(Y_2)$ for all $U(y)$ with $U'(y)$ positive and $U''(y)$ nonpositive, if and only if $\mathcal{L}_1(q)$, the LC of $F_1(y)$, lies weakly above $\mathcal{L}_2(q)$, the LC of $F_2(y)$, for all q. In summary, Atkinson (1970) showed the following theorem.

Theorem 3.1. The following conditions are equivalent (where $F_1(y)$ and $F_2(y)$ have the same mean):

(i) $W(Y_1) \geq W(Y_2)$ for all $U(y)$ with $U'(y) > 0$ and $U''(y) \leq 0$.
(ii) $F_2(y)$ can be obtained from $F_1(y)$ by a series of MPSs, regressive transfers.
(iii) $F_1(y)$ dominates $F_2(y)$ by SSD.
(iv) $\mathcal{L}_1(q) \geq \mathcal{L}_2(q)$ for all q.

Dasgupta, Sen, and Starrett (1973) generalized this result. First, they showed that it is unnecessary for $W(Y)$ to be additive. A parallel result is true for any concave $W(Y)$. Dasgupta, Sen, and Starrett also showed that the concavity of W could be weakened to Schur or "S" concavity. These generalizations are important since they indicate that Lorenz dominance is equivalent to unanimous ranking by a very broad class of inequality measures.

To a large extent, in the remainder of this chapter we shall be concerned with stochastic dominance relations and their empirical implementation. The standard definitions of stochastic dominance, which were set out in the last section, are stated with reference to an additive objective function, reflecting their origin in the risk-measurement literature. Therefore, for the sake of exposition, it is convenient to continue to refer to the additive class of SWFs in the treatment that follows. This does not involve any loss of generality since we are studying dominance relations rather than the properties of individual inequality measures.* Dominance requires the agreement of *all* SWFs in a particular class. As the results of Dasgupta, Sen, and Starrett show for the case of SSD, in situations where all additive SWFs agree on an inequality ranking (i.e., there is dominance), all members of a much broader class of SWFs also may agree on the ranking.

---

*In the study of individual inequality indices, it would be a serious restriction to confine ones attention to those which are associated with additive SWFs. This would eliminate the coefficient of variation, the Gini coefficient, and many members of the generalized entropy family from consideration. Nonadditive SWFs may be thought of as allowing interdependence of social preferences toward individual incomes. For additional discussion of these topics, see Maasoumi (1997) and the references cited therein.

B. Some Welfare Properties of Summary Inequality and Welfare Indices

As implied above, the mean ranks distributions in the same way as any additive SWF when $U''(y)$ equals zero. In contrast, the median is generally inconsistent with the SWF framework. This is because the median concerns itself only with the welfare of the median individual. Another measure with a similar property is the Rawlsian SWF; in that case, the income of the worst-off member of society is treated as a sufficient statistic for welfare.

Note that the mean violates the strong form of the principle of transfers: it does not change in response to *any* regressive transfer. Note, however, that the mean obeys the weak form of this principle since it never increases as a result of such a transfer. In contrast, the median may well rise when a regressive transfer occurs. This would be the case, for example, if a small amount of income is transferred from people in the bottom half of the income distribution to persons around the median, without altering the ordering of individual incomes.

The variance of the logarithm of income σ_Z^2 may also violate the weak form of the principle of transfers. This is because σ_Z^2 is not convex at high levels of income; see Sen (1973, p. 29) for more on this. While this flaw calls into question the uncritical use of σ_Z^2, its use is justified where a suitable parametric form of $F(y)$ is assumed. For example, when $F(y)$ is lognormal, σ_Z^2 is a sufficient statistic for inequality.

As noted in Section II, the variance and standard deviation are dependent on scale. Therefore, they are questionable as inequality measures. The coefficient of variation τ does not suffer from this property. It also always obeys the strong form of the principle of transfers.

It is evident from the definition of τ that there is no additive SWF of the form given in (2) to which it corresponds. For more on this, see Blackorby and Donaldson (1978). This is also true for the Gini coefficient κ. Like a number of other popular summary measures (e.g., Theil's index, see Sen 1973, p. 35), however, these measures correspond to SWFs within the S-concave family.

C. Second-Degree Stochastic Dominance and Generalized Lorenz Curves

It is traditional in much of the inequality measurement literature to focus attention on *relative* inequality, i.e., to examine the distribution of income normalized by its mean. This reflects the early development of the LC as a central tool of inequality measurement, the fact that most popular inequality indices are relative, and perhaps also the influence of Atkinson (1970).

Inequality comparisons are, of course, only a part of welfare comparisons. Therefore, Atkinson (1970) argued for supplementing LCs by comparisons of means. He noted the following result.

Theorem 3.2. If $\mathcal{E}[Y_1] \geq \mathcal{E}[Y_2]$ and $\mathcal{L}_1(q) \geq \mathcal{L}_2(q)$ for all q, then $F_1(y)$ is preferred to $F_2(y)$ according to second-degree stochastic dominance.

Shorrocks (1983) went beyond this sufficient condition and established the equivalence of GLC dominance and SSD. Using his results, we can state

Theorem 3.3. The following conditions are equivalent (where $F_1(y)$ and $F_2(y)$ may have different means):

  (i) $W(Y_1) \geq W(Y_2)$ for all $U(y)$ with $U'(y) > 0$ and $U''(y) \leq 0$.
  (ii) $F_1(y)$ dominates $F_2(y)$ by SSD.
  (iii) $\mathcal{G}_1(q) \geq \mathcal{G}_2(q)$ for all q.

Theorem 3.3 is of great practical importance in making welfare comparisons because the means of real-world distributions being compared are seldom equal. Also, Shorrocks (1983) and others have found that, in many cases when LCs cross, GLCs do not. Thus, using GLCs greatly increases the number of cases where unambiguous welfare comparisons can be made in practice.

D. Aversion to Downside Inequality and Third-Degree Stochastic Dominance

Another approach to extending the range of cases where unambiguous welfare comparisons can be made, beyond those which could be managed with the techniques of Atkinson (1970), has been to continue to restrict attention to relative inequality, but to adopt a normative axiom stronger than the Pigou-Dalton principle of transfers. This axiom has been given at least two different names. Shorrocks and Foster (1987) referred to it as *transfer sensitivity*. Here, we follow Davies and Hoy (1994, 1995) who referred to it as *aversion to downside inequality* (ADI).

ADI implies that a regressive transfer should be considered to reduce welfare more if it occurs lower in the income distribution. In the context of additive SWFs, this clearly restricts $U(y)$ to have $U'''(y)$, which is nonnegative in addition to $U'(y)$'s being positive and $U''(y)$'s being nonpositive.

The ordering induced by the requirement that $W(Y_1)$ be greater than or equal to $W(Y_2)$ for all $U(y)$ such that $U'(y)$ is positive, $U''(y)$ is nonpositive, and $U'''(y)$ is nonnegative corresponds to the notion of *third-degree stochastic dominance* (TSD) introduced in the risk-measurement literature by Whitmore (1970). TSD initially received far less attention in inequality and welfare measurement than SSD, despite its embodiment of the attractive ADI axiom. This was, in part, due to the lack of a readily available indicator of when TSD held in practice. Shorrocks and Foster (1987) worked to fill this gap.

When LCs do not intersect, unambiguous rankings of relative inequality can be made under SSD, provided the means are equal. When GLCs do not intersect, unambiguous welfare rankings can also be made under SSD. When either LCs or GLCs intersect difficulties arise.

Suppose that $\mathcal{E}[Y_1]$ equals $\mathcal{E}[Y_2]$ and that the LCs $\mathcal{L}_1(q)$ and $\mathcal{L}_2(q)$ intersect once. A necessary condition for $F_1(y)$ to dominate $F_2(y)$ by TSD is that $\mathcal{L}_1(q)$ is greater than or equal to $\mathcal{L}_2(q)$ at lower incomes and that $\mathcal{L}_1(q)$ is less than or equal to $\mathcal{L}_2(q)$ at higher incomes. In other words, given that the LCs cross once, the only possible candidate for the "more equal" label is the distribution which is better for the poor. This is because the strength of aversion to downside inequality may be so high that no amount of greater equality at high income levels can repair the damage done by greater inequality at low incomes.

Shorrocks and Foster (1987) added the sufficient condition in the case of singly intersecting LCs, proving the equivalent of the following theorem for discrete distributions:

Theorem 3.4.   If $\mathcal{E}[Y_1] = \mathcal{E}[Y_2]$ and the LCs for $F_1(y)$ and $F_2(y)$ have a single intersection, then $F_1(y)$ is preferred to $F_2(y)$ by TSD if and only if

   (i)   The LC $\mathcal{L}_1(q)$ cuts the LC $\mathcal{L}_2(q)$ from above.
   (ii) $\mathcal{V}[Y_1] \leq \mathcal{V}[Y_2]$.

This theorem provides a new and important role for the variance in inequality measurement. It has been extended by Davies and Hoy (1995) to the case where the two LCs in question intersect any (finite) number of times n. Lambert (1989) has also extended the analysis to the case where $\mathcal{E}[Y_1]$ does not equal $\mathcal{E}[Y_2]$ by investigating rankings of distributions when GLCs intersect. Davies and Hoy (1995) proved:

Theorem 3.5.   If $\mathcal{E}[Y_1] = \mathcal{E}[Y_2]$, $F_1(y)$ dominates $F_2(y)$ by TSD if and only if, for all Lorenz crossover points $i = 1, 2, \ldots, (n + 1)$, $\lambda_1^2(q_i) \leq \lambda_2^2(q_i)$, with $\lambda_j^2(q_i)$ denoting the cumulative variance for incomes up to the ith crossover point $\xi(q_i)$ for distribution j.

Since multiple intersections of LCs are far from rare in applied work, this result has considerable practical value. Its implementation has been studied by Beach, Davidson, and Slotsve (1994) and is discussed in the next section.

IV.   FROM POPULATIONS TO SAMPLES: UNIVARIATE CASE

Typically, it is far too expensive and impractical to sample the entire population to construct $F(y)$. Thus, researchers usually take random samples from the population, and then attempt to estimate $F(y)$ as well as functionals that can be derived from $F(y)$, such as LCs and GLCs. Often, researchers are interested in comparing LCs (or GLCs) across countries or, for a given country, across time. To carry out this sort of analysis, one needs estimators of LCs and GLCs as well as a distribution theory for these estimators.

In this section, we show how to estimate LCs and GLCs using the kinds of microdata that are typically available from sample surveys. We consider the case where the researcher has a sample $\{Y_1, Y_2, \ldots, Y_N\}$ of N observations, each of which represents an independent and identical random draw from the distribution $F(y)$.*

A. Parametric Methods

Because parametric methods of estimation and inference are the most well known to researchers, we begin with them.[†] Using the parametric approach, the researcher assumes that $F(y)$ comes from a particular family of distributions (exponential, Pareto, lognormal, etc.) which is known up to some unknown parameter or vector of parameters. For example, the researcher may assume that

$$F(y; \mu) = 1 - \exp\left(-\frac{y}{\mu}\right), \qquad \mu > 0, \quad y > 0$$

when Y is from the exponential family with μ unknown, or

$$F(y; \theta_0, \theta_1) = 1 - \left(\frac{\theta_0}{y}\right)^{\theta_1}, \qquad \theta_0 > 0, \quad \theta_1 > 0, y > \theta_0$$

when Y is from the Pareto family with θ_0 and θ_1 unknown, or

$$f(y; \mu, \sigma) = \frac{1}{y\sqrt{2\pi\sigma^2}} \exp\left(-\frac{(\log y - \mu)^2}{2\sigma^2}\right),$$

$$-\infty < \mu < \infty, \qquad \sigma^2 > 0, y > 0$$

when Y is from the lognormal family with μ and σ^2 unknown.

Given a random sample of size N drawn from $F(y)$, a number of estimation strategies exist for recovering estimates of the unknown parameters. The most efficient is Fisher's method of maximum likelihood. The maximum likelihood estimator M of μ in the exponential case is

$$M = \frac{\sum_{i=1}^{N} Y_i}{N}$$

---

*Many large, cross-sectional surveys have weighted observations. For expositional reasons, we avoid this complication, but direct the interested reader to, among others, the work of Beach and Kaliski (1986) for extensions developed to handle this sort of complication. We also avoid complications introduced by dependence in the data, but direct the interested reader to, among others, the work of Davidson and Duclos (1995).

[†] A useful reference for the epistemology of distribution functions is Chipman (1985).

while the maximum likelihood estimators of θ_0 and θ_1 in the Pareto case are

$$T_0 = \min[Y_1, Y_2, \ldots, Y_N]$$

$$T_1 = \frac{N}{\sum_{i=1}^{N} \log(Y_i/T_0)}$$

and those of μ and σ^2 in the lognormal case are

$$M = \frac{\sum_{i=1}^{N} \log Y_i}{N}$$

$$S^2 = \frac{\sum_{i=1}^{N} (\log Y_i - M)^2}{N}$$

In each of these cases (except that of T_0), the maximum likelihood estimators are consistent and distributed normally, asymptotically.* One can also derive the pointwise asymptotic distribution of both the LC and the GLC. This is easiest to do in the exponential case where the GLC is linear in the parameter to be estimated. Note that

$$\xi(q; u) = -\mu \log(1 - q)$$

so

$$\mathcal{G}(q; \mu) = \int_0^{\xi(q)} \left(\frac{1}{\mu}\right) y \exp\left(-\frac{y}{\mu}\right) \, dy$$
$$= \mu[q + (1 - q) \log(1 - q)]$$

An unbiased estimator $G(q; M)$ of $\mathcal{G}(q; \mu)$ is then

$$G(q; M) = M[q + (1 - q) \log(1 - q)]$$

Now

$$\sqrt{N}(M - \mu^0) \xrightarrow{\text{d}} \mathcal{N}(0, (\mu^0)^2)$$

by a central limit theorem, where μ^0 is the true value of μ, so one can show that for each q, a pointwise characterization of the asymptotic distribution is

---

*The estimator T_0 of θ_0 is superconsistent and converges at rate N rather than $\sqrt{N}$; its asymptotic distribution is exponential. This is a property of extreme order statistics when the density is positive at the boundary of the support; see Galambos (1987). Because T_0 converges at a rate N, which is faster than the rate $\sqrt{N}$ for the other estimator T_1, it can be treated as if it actually equals θ_0, and thus ignored in an asymptotic expansion of T_1.

$$\sqrt{N}(G(q; M) - \mathcal{G}(q; \mu^0)) \xrightarrow{d} \mathcal{N}(0, (\mu^0)^2[q + (1 - q)\log(1 - q)]^2)$$

Using $\hat{m}$, a realization of the maximum likelihood estimator M, one can estimate $\mathcal{G}(q; \mu)$ at q by

$$\hat{g}(q; \hat{m}) = \hat{m}[q + (1 - q)\log(1 - q)]$$

and calculate its standard error $\mathcal{SE}[g(q; \hat{m})]$ by substituting $\hat{m}$ for μ^0 to get

$$\mathcal{SE}[g(q; \hat{m})] = \frac{\hat{m}[q + (1 - q)\log(1 - q)]}{\sqrt{N}}$$

Note that in the exponential case, the LC is invariant to the parameter μ since

$$\mathcal{L}(q; \mu) = \frac{\mathcal{G}(q)}{\mathcal{E}[Y]} = \frac{\mathcal{G}(q)}{\mu} = [q + (1 - q)\log(1 - q)]$$

so no estimation is required.

In the other two examples, the LCs and GLCs are nonlinear functions of the parameters. We find it useful to consider these examples further, since they illustrate well the class of technical problems faced by researchers. In the Pareto case, where

$$\xi(q; \theta_0, \theta_1) = \frac{\theta_0}{(1 - q)^{1/\theta_1}}$$

the GLC solves

$$\mathcal{G}(q; \theta_0, \theta_1) = \int_0^{\xi(q;\theta_0,\theta_1)} \frac{\theta_1 \theta_0^{\theta_1}}{y^{\theta_1}} \, dy$$

$$= \frac{\theta_0 \theta_1}{\theta_1 - 1}[1 - (1 - q)^{(\theta_1 - 1)/\theta_1}]$$

Now

$$\mathcal{E}[Y] = \frac{\theta_0 \theta_1}{\theta_1 - 1}$$

so the LC is

$$\mathcal{L}(q; \theta_1) = [1 - (1 - q)^{(\theta_1 - 1)/\theta_1}]$$

Because $\mathcal{G}(q; \theta_0, \theta_1)$ and $\mathcal{L}(q; \theta_1)$ are continuous functions of θ_0 and θ_1, the maximum likelihood estimators of them are

$$G(q; T_0, T_1) = \frac{T_0 T_1}{T_1 - 1}[1 - (1 - q)^{(T_1-1)/T_1}]$$

$$L(q; T_1) = [1 - (1 - q)^{(T_1-1)/T_1}]$$

respectively. These estimators are also consistent.*

To find the asymptotic distributions of $G(q; T_0, T_1)$ and $L(q; T_1)$, one needs first to know the asymptotic distribution of T_1. Now

$$T_1 = \left[\frac{\sum_{i=1}^{N} \log(Y_i/T_0)}{N} \right]^{-1}$$

where one can treat T_0 as if it were θ_0 because T_0 is a superconsistent estimator of θ_0. The random variable

$$Z_i = \log\left(\frac{Y_i}{\theta_0}\right)$$

is distributed exponentially with unknown parameter θ_1, so its probability density function is

$$f_Z(z) = \theta_1 \exp(-\theta_1 z), \qquad \theta_1 > 0, \quad z > 0$$

Thus, T_1 is a function of the sample mean $\overline{Z}_N$ of N independently and identically distributed exponential random variables $\{Z_1, Z_2, \ldots, Z_N\}$ where, by a central limit theorem (such as Lindeberg-Levy), $\overline{Z}_N$ has the following asymptotic distribution:

$$\sqrt{N}(\overline{Z}_N - \mathcal{E}[\overline{Z}_N]) \xrightarrow{d} \mathcal{N}\left(0, \left(\frac{1}{\theta_1^0}\right)^2\right)$$

with θ_1^0 being the true value of θ_1. Because T_1 is a continuous and differentiable function of the sample mean $\overline{Z}_N$, we can use the delta method (see, for example, Rao 1965) to derive T_1's asymptotic distribution. We proceed by expanding the function $T_1(\overline{Z}_N)$ in a Taylor's series expansion about $\mathcal{E}[\overline{Z}_N]$, which equals $1/\theta_1^0$. Thus,

$$T_1 = T_1(\mathcal{E}[\overline{Z}_N]) + \frac{dT_1(\mathcal{E}[\overline{Z}_N])}{d\overline{Z}_N}(\overline{Z}_N - \mathcal{E}[\overline{Z}_N]) + R_2^{T_1}$$

---

*When T_1 is a consistent estimator of θ_1; the true value of θ_1, then

$$\plim_{N \to \infty} T_1 = \theta_1^0$$

Also, if, for example, $L(T_1)$ is a continuous function of T_1, then

$$\plim_{N \to \infty} L(T_1) = L\left(\plim_{N \to \infty} T_1\right) = L(\theta_1^0)$$

by Slutsky's theorem; see, for example, Rao (1965).

Noting that T_1 evaluated at $\mathcal{E}[\bar{Z}_N]$ equals θ_1^0 and that $dT_1(\mathcal{E}[\bar{Z}_N])/d\bar{Z}_N$ equals $(\theta_1^0)^2$ and ignoring the remainder term $R_2^{T_1}$ since it will be negligible in a neighborhood of $\mathcal{E}[\bar{Z}_N]$, which equals $1/\theta_1^0$, one obtains

$$\sqrt{N}(T_1 - \theta_1^0) \overset{d}{\to} \mathcal{N}(0, (\theta_1^0)^2)$$

Thus, to find the asymptotic distribution of $L(q; T_1)$, for example, expand $L(q; T_1)$ in a first-order Taylor's series about the point θ_1^0, with R_2^L being the remainder, to get

$$L(q; T_1) = L(q; \theta_1^0) + \frac{dL(q; \theta_1^0)}{dT_1}(T_1 - \theta_1^0) + R_2^L$$

With a minor amount of manipulation, one then obtains

$$\sqrt{N}(L(q; T_1) - \mathcal{L}(q; \theta_1^0)) \overset{d}{\to} \mathcal{N}\left(0, \left[\frac{dL(q; \theta_1^0)}{dT_1}\right]^2 [\theta_1^0]^2\right)$$

Similar calculations can be performed for the estimator $G(q; T_0, T_1)$ of $\mathcal{G}(q; \theta_0, \theta_1)$.[*]

In the lognormal case, the GLC solves

$$\mathcal{G}(q; \mu, \sigma^2) = \int_0^{\xi(q;\mu,\sigma^2)} \frac{1}{\sqrt{2\pi\sigma^2}} \exp\left(-\frac{(\log y - \mu)^2}{2\sigma^2}\right) dy \tag{4}$$

where $\xi(q; \mu, \sigma^2)$ is implicitly defined by

$$q = \int_0^{\xi(q;\mu,\sigma^2)} \frac{1}{y\sqrt{2\pi\sigma^2}} \exp\left(-\frac{(\log y - \mu)^2}{2\sigma^2}\right) dy \tag{5}$$

In this case, both $\xi(q; \mu, \sigma^2)$ and $\mathcal{G}(q; \mu, \sigma^2)$ are only defined numerically. For a specific q, conditional on some estimates $\hat{m}$ and $\hat{s}^2$, one can solve the quantile equation (5) numerically and then the GLC equation (4). To apply the delta method, one would have to use Leibniz's rule to find the effect of changes in M and S^2 on the asymptotic distribution of $\mathcal{G}(q; \mu, \sigma^2)$ at q. A similar analysis could also be performed to find the asymptotic distribution of $\mathcal{L}(q; \mu, \sigma^2)$. As one can see, except in a few simple cases, the technical demands can increase when parametric methods are used because the quantiles are often only defined implicitly and the LCs and GLCs can typically only be calculated numerically, in the continuous case. Moreover, the

---

[*]Note that one could perform a similar large-sample analysis to find the asymptotic distribution for estimators of such summary measures as the Gini coefficient, but this is beyond the focus of this chapter.

calculations required to characterize the asymptotic distribution of the pointwise estimator of either the LC or the GLC can be long and tedious. This, of course, is not the main drawback of this approach.

The main drawback of the parametric approach is that it requires the researcher to impose considerable structure on the data by assuming $F(y)$ comes from a particular family of distributions. Some researchers, such as McDonald (1984), have sought richer parametric specifications and have provided specification tests for these models, while Harrison (1982) has demonstrated the value of a careful empirical investigation using this parametric approach. Because the parametric approach is not an entirely satisfactory solution, other researchers have sought to relax assumptions concerning structure and have pursued nonparametric methods.

B.   Nonparametric Estimators of Lorenz Curves and Generalized Lorenz Curves

The first researcher to propose a nonparametric estimator of the LC was Sendler (1979).* Unfortunately, Sendler did not provide a complete characterization of the asymptotic distribution of the LC. This task was carried out by Beach and Davidson (1983), who, in the process, also derived the asymptotic distribution of the GLC.

Beach and Davidson were interested in conducting nonparametric estimation and inference concerning the set of population ordinates $\{\mathcal{L}(q_i)|i = 1, \ldots, J\}$ corresponding to the abscissae $\{q_i|i = 1, \ldots, J\}$. When J is nine and the q_is are $\{0.1, 0.2, \ldots, 0.9\}$, for example, Beach and Davidson would be interested in estimating the population LC vector of the deciles

$$\mathcal{L} = (\mathcal{L}(0.1), \mathcal{L}(0.2), \ldots, \mathcal{L}(0.9))'$$

where

$$\mathcal{L}(q_i) = q_i \frac{\gamma_i}{\mu}$$

with $\gamma_i = \gamma(q_i)$. To carry out this sort of analysis, one must first order the sample $\{Y_1, Y_2, \ldots, Y_N\}$ so that $Y_{(1)} \le Y_{(2)} \le \cdots \le Y_{(N)}$. Beach and Davidson defined $E(q)$, an estimator of the qth population quantile $\xi(q)$, to be the rth-order statistic $Y_{(r)}$ where r denotes the greatest integer less than or equal to qN. Thus,

---

*McFadden (1989) and Klecan, McFadden, and McFadden (1991) have developed nonparametric procedures for examining SSD, while Anderson (1996) has employed nonparametric procedures and FSD, SSD, and TSD principles to income distributions. Xu, Fisher, and Wilson (1995) have also developed similar work. Although these procedures are related to estimation and inference concerning LCs and GLCs, for space reasons we do not discuss this research here.

$\mathbf{E} = (E(q_1), E(q_2), \ldots, E(q_J))'$ is an estimator of the vector of quantiles $\boldsymbol{\xi} = (\xi(q_1), \xi(q_2), \ldots, \xi(q_J))'$ and has the asymptotic distribution

$$\sqrt{N}(\mathbf{E} - \boldsymbol{\xi}^0) \overset{d}{\to} \mathcal{N}(\mathbf{0}, \boldsymbol{\Lambda}^0)$$

where $\boldsymbol{\xi}^0$ denotes the true value of the vector $\boldsymbol{\xi}$ and where

$$\boldsymbol{\Lambda}^0 = \begin{pmatrix} \dfrac{q_1(1 - q_1)}{f(\xi^0(q_1))^2} & \cdots & \dfrac{q_1(1 - q_J)}{f(\xi^0(q_1))f(\xi^0(q_J))} \\ \vdots & & \vdots \\ \dfrac{q_1(1 - q_J)}{f(\xi^0(q_1))f(\xi^0(q_J))} & \cdots & \dfrac{q_J(1 - q_J)}{f(\xi^0(q_J))^2} \end{pmatrix}$$

The asymptotic distribution of the LC estimator

$$\mathbf{L} = (L(q_1), L(q_2), \ldots, L(q_J))'$$

depends on the joint asymptotic distribution of the estimator

$$\mathbf{G} = (q_1 C_1, q_2 C_2, \ldots, q_J C_J, M)' = (G(q_1), G(q_2), \ldots, G(q_J), M)'$$

of the parameter vector

$$\mathcal{G} = (q_1 \gamma_1, q_2 \gamma_2, \ldots, q_J \gamma_J, \mu)' = (\mathcal{G}(q_1), \mathcal{G}(q_2), \ldots, \mathcal{G}(q_J), \mathcal{G}(1))'$$

where

$$\mathcal{G}(q_i) = q_i \gamma_i \qquad \text{and} \qquad C_i = \frac{\sum_{j=1}^{r_i} Y_{(j)}}{r_i}$$

with r_i being the greatest integer less than or equal to $q_i N$. Note that $\mathcal{G}$ is, in fact, the GLC and that $\mathbf{G}$ is an estimator of the GLC. Beach and Davidson showed that the distribution of $\sqrt{N}(\mathbf{G} - \mathcal{G}^0)$ is asymptotically normal, centered about the $(J + 1)$ zero vector, and has variance-covariance matrix $\boldsymbol{\Omega}^0$ where

$$\omega_{ij}^0 = q_i[(\lambda_i^0)^2 + (1 - q_j)(\xi^0(q_j) - \gamma_j^0) + (\xi^0(q_i) - \gamma_i^0)(\gamma_j^0 - \gamma_i^0)],$$
$$i \leq j$$

with $(\lambda_i^0)^2$ being $\lambda^2(q_i)$, the variance of Y given that Y is less than $\xi^0(q_i)$.

Now, $\mathbf{L}$ is a nonlinear function of $\mathbf{G}$. By the delta method, Beach and Davidson showed that it has the asymptotic distribution

$$\sqrt{N}(\mathbf{L} - \mathcal{L}^0) \overset{d}{\to} \mathcal{N}(\mathbf{0}, \boldsymbol{\Phi}^0 \boldsymbol{\Omega}^0 \boldsymbol{\Phi}^{0'})$$

where

$$\Phi^0 = \begin{pmatrix} \nabla_{\mathcal{G}} \mathcal{L}^0(q_1)' \\ \nabla_{\mathcal{G}} \mathcal{L}^0(q_2)' \\ \vdots \\ \nabla_{\mathcal{G}} \mathcal{L}^0(q_J)' \end{pmatrix}$$

where $\nabla_{\mathcal{G}}$ denotes the gradient vector of the function to follow with respect to the vector $\mathcal{G}$.

Beach and Davidson (1983) appear to have ignored the implications of their results of GLCs. These notions were first applied by Bishop, Chakraborti, and Thistle (1989).

C. Testing for (Generalized) Lorenz Curve Equivalence and Dominance

Once a researcher has estimated either the LCs or the GLCs for two populations, it is natural to think about testing the null hypothesis of LC (or GLC) equivalence or that of LC (or GLC) dominance. Testing the null hypothesis of equivalence turns out to be much easier than testing the null hypothesis of dominance, so we outline first how tests of equivalence can be carried out.

Suppose that there are two independent samples, 1 and 2, of size N_1 and N_2 respectively, and one seeks to test whether $\mathcal{L}_1$ equals $\mathcal{L}_2$ (or $\mathcal{G}_1$ equals $\mathcal{G}_2$) at a countable number of points J. (This null hypothesis can also be written as $\Pi = \mathcal{L}_1 - \mathcal{L}_2 = 0$.) Assume that consistent estimators of the variance-covariance matrices of L_1 and L_2 (or G_1 and G_2) exist, and denote them V_1 and V_2, respectively. A test of LC (or GLC) equivalence is referred to by Engle (1984) as a Wald test. It involves calculating the quadratic form

$$(L_1 - L_2)' \left[\frac{V_1}{N_1} + \frac{V_2}{N_2} \right]^{-1} (L_1 - L_2) \equiv P' V^{-1} P$$

This statistic has an asymptotic distribution which is χ^2 with J degrees of freedom.

If LC (or GLC) equivalence is rejected, then one may want to know further which particular ordinates differ from one another. One way of providing such information is through multiple comparisons. Beach and Richmond (1985), building on the work of Richmond (1982), have derived the formulae for multiple comparison intervals (Savin 1984).

Alternatively, a researcher may wish to test the null hypothesis of LC (or GLC) dominance. LC dominance implies that

$$\mathcal{L}_1(q) \geq \mathcal{L}_2(q) \quad \forall q$$

with

$$\mathcal{L}_1(q) > \mathcal{L}_2(q) \qquad \text{for some } q$$

at a countable number of points J, so the null hypothesis can also be written as $\mathbf{\Pi} \geq \mathbf{0}$. Testing this hypothesis involves deciding whether the random variable $\mathbf{L}_1 - \mathbf{L}_2$ (or $\mathbf{G}_1 - \mathbf{G}_2$ for GLC dominance) is in the nonnegative orthant of $\mathbf{R}^J$ (or $\mathbf{R}^{J+1}$). A number of authors have built on the work of Perlman (1969) to provide solutions to this problem; see, for example, Kodde and Palm (1986) and Wolak (1987, 1989, 1991). To calculate the test statistic LR for realizations $\hat{\mathbf{l}}_1$ and $\hat{\mathbf{l}}_2$ of $\mathbf{L}_1$ and $\mathbf{L}_2$ and estimates $\hat{\mathbf{V}}_1$ and $\hat{\mathbf{V}}_2$ of $\mathbf{V}_1$ and $\mathbf{V}_2$, requires one to solve the following constrained, quadratic programming problem:

$$\min_{(\mathbf{p})} \left((\hat{\mathbf{l}}_1 - \hat{\mathbf{l}}_2) - \mathbf{p}\right)' \left[\frac{\hat{\mathbf{V}}_1}{N_1} + \frac{\hat{\mathbf{V}}_2}{N_2}\right]^{-1} \left((\hat{\mathbf{l}}_1 - \hat{\mathbf{l}}_2) - \mathbf{p}\right) \quad \text{subject to } \mathbf{p} \geq \mathbf{0}$$

which we rewrite as

$$\min_{(\mathbf{p})} (\hat{\mathbf{p}} - \mathbf{p})' \hat{\mathbf{V}}^{-1} (\hat{\mathbf{p}} - \mathbf{p}) \quad \text{subject to } \mathbf{p} \geq \mathbf{0}$$

Letting

$$\tilde{\mathbf{p}} = \text{argmin}(\hat{\mathbf{p}} - \mathbf{p})' \hat{\mathbf{V}}^{-1} (\hat{\mathbf{p}} - \mathbf{p}) \quad \text{subject to } \mathbf{p} \geq \mathbf{0}$$

the test statistic is

$$LR = (\hat{\mathbf{p}} - \tilde{\mathbf{p}})' \hat{\mathbf{V}}^{-1} (\hat{\mathbf{p}} - \tilde{\mathbf{p}})$$

which has asymptotic distribution

$$\lim_{N \to \infty} \text{Pr}[LR \geq c] = \sum_{j=1}^{J} \text{Pr}[\chi^2(j) \geq c] w(J, J - j, \mathbf{V})$$

where $\text{Pr}[\chi^2(j) \geq c]$ denotes the probability that a χ^2 random variable with j degrees of freedom exceeds some constant c, and $w(J, J - j, \mathbf{V})$ is a weighting function.

D. Nonparametric Analysis when Lorenz Curves Cross

Sometimes in empirical studies, LCs cross. Extending the work of Shorrocks and Foster (1987), in which they investigated but one crossing, Davies and Hoy (1995) have considered how to make welfare statements when there are n crossings. One can summarize the results of Davies and Hoy (1995) in terms of the cumulative coefficients of variation (CCVs):

$$\mathcal{T}(q_i) = \frac{\lambda(q_i)}{\gamma(q_i)}$$

In particular, if $\mathcal{T}_1(q)$ for population 1 is always less than $\mathcal{T}_2(q)$ for population 2, then population 1 dominates population 2 in the social welfare sense.

Beach, Davidson, and Slotsve (1994) derived an estimator of the vector

$$\mathcal{T} = (\mathcal{T}(q_1), \mathcal{T}(q_2), \ldots, \mathcal{T}(q_{J+1}))'$$

as well as the asymptotic theory required to test for dominance of the cumulative coefficients of variation. In particular, they derived the asymptotic distribution of the estimator

$$\mathbf{T} = (T(q_1), T(q_2), \ldots, T(q_{J+1}))'$$

of $\mathcal{T}$ where

$$T(q_i) = \frac{\sqrt{\sum_{j=1}^{r_i} Y_{(j)}^2 / r_i - [\sum_{j=1}^{r_i} Y_{(j)} / r_i]^2}}{\sum_{j=1}^{r_i} Y_{(j)} / r_i}$$

with r_i again being the largest integer less than or equal to $q_i N$. As with the test of LC (or GLC) dominance, a test of the null hypothesis of CCV dominance for two populations involves deciding whether $\mathbf{T}_1 - \mathbf{T}_2$ is in the nonnegative orthant of $\mathbf{R}^{J+1}$.

V. FROM POPULATIONS TO SAMPLES WITH OBSERVED COVARIATES

In many empirical applications, considerable observed covariate heterogeneity exists. This sort of heterogeneity is of interest for a number of reasons. One important reason is that the sources of differences in welfare between populations arise from this covariate heterogeneity. To the extent that some covariate heterogeneity can be manipulated by policy, some scope exists for improving welfare.

Thus, consider the random variable Y which is observed conditional on a $1 \times k$ random covariate vector $\mathbf{X}$. Denote the probability density function of Y conditional on $\mathbf{X}$ equaling $\mathbf{x}$ by $f(y|\mathbf{x})$. As above, one can define the conditional qth population quantile of Y conditional on $\mathbf{X}$ equaling $\mathbf{x}$ by

$$q = F(\xi(q|\mathbf{x})|\mathbf{x}) = \int_0^{\xi(q|\mathbf{x})} f(y|\mathbf{x}) \, dy$$

One can also define the conditional LC ordinate by

$$\mathcal{L}(q|\mathbf{x}) = \frac{\int_0^{\xi(q|\mathbf{x})} yf(y|\mathbf{x}) \, dy}{\int_0^{\infty} yf(y|\mathbf{x}) \, dy} = \frac{\int_0^{\xi(q|\mathbf{x})} yf(y|\mathbf{x}) \, dy}{\mathcal{E}[Y|\mathbf{x}]}$$

A LC of Y conditional on $\mathbf{X}$ equaling $\mathbf{x}$ is the plot of $(q, \mathcal{L}(q|\mathbf{x}))$. The conditional GLC is the plot of $(q, \mathcal{E}[Y|\mathbf{x}]\mathcal{L}(q|\mathbf{x}))$, which we denote by $(q, \mathcal{G}(q|\mathbf{x}))$.

A. Parametric Methods

As in Section IV.A one could assume that $f(y|\mathbf{x})$ comes from a particular parametric family. For example, suppose

$$\log Y = \mathbf{x}\boldsymbol{\beta} + V$$

where V is distributed independently and identically normal having mean zero and variance σ^2. One could then use the method of maximum likelihood to get estimates of $\boldsymbol{\beta}$ and σ^2, and these estimates in turn could be used to estimate $\mathcal{G}(q|\mathbf{x})$ or $\mathcal{L}(q|\mathbf{x})$. Since the maximum likelihood estimators of $\boldsymbol{\beta}$ and σ^2 are asymptotically normal and since both the LC and the GLC are functionals of $f(y|\mathbf{x})$, one can then use the delta method to find the pointwise asymptotic distribution of $\mathcal{G}(q|x)$ and $\mathcal{L}(q|\mathbf{x})$. Of course, as in the univariate case, the major criticism of the parametric approach is that it imposes too much structure on the data.

B. Adapting Nonparametric Methods

At the other end of the spectrum is nonparametric analysis. To estimate LCs (or GLCs) nonparametrically in the presence of observed covariates, one would simply break up the observed covariates into cells, and implement, for example, the methods of Beach and Davidson (1983) on each cell. This approach is feasible when the number of covariates is small and if the covariates are discrete. When the number of covariates is large or if covariates are continuous, then the methods described in Beach and Davidson (1983) are impractical. A natural solution to this quandary is a method that imposes some structure on $f(y|\mathbf{x})$, but which is not wholly parametric (i.e., a semiparametric method).

C. Semiparametric Methods

The idea behind any semiparametric method is to put enough structure on the distribution of Y given $\mathbf{X}$ equals $\mathbf{x}$ so as to reduce the curse of dimensionality that arises in nonparametric methods. In this section, we focus on one particularly useful approach (that of Donald, Green, and Paarsch 1995) to estimating GLCs semiparametrically. These methods can be adapted to recover estimates of LCs, but in the interest of space we omit the explicit parallel development of methods for LCs.

Donald et al. (1995) translated techniques developed for estimating spell-duration distributions (Kalbfleisch and Prentice 1980) to the estimation of income distributions. The main building block in their approach is the hazard function. For a nonnegative random variable Y with associated probability density function $f(y)$ and cumulative distribution function $F(y)$, the hazard function $h(y)$ is defined by the conditional probability

$$h(y) = \frac{f(y)}{1 - F(y)} \equiv \frac{f(y)}{S(y)} \tag{6}$$

where $S(y)$ is the survivor function. From (6) one can see that the hazard function is simply a transformation of the probability density (or mass) function. One key result from the literature on spell-duration estimation is that the conditional nature of $h(y)$ makes it easy to introduce flexible functions of the covariates and to entertain complex shapes for the hazard function.

Donald et al. introduced covariates using an extension of a proportional hazard model in which the range of Y is partitioned into P subintervals $\Upsilon_p = [y_L^p, y_U^p)$, where $\Upsilon_p \cap \Upsilon_q = \emptyset$ for all $p \neq q$ with $\bigcup_{p=1}^P \Upsilon_p = [0, \infty)$, and they allow the covariate effects to vary over these subintervals.* They referred to these subintervals as covariate segments. In particular, following Gritz and MaCurdy (1992), they replaced $x_i\beta$ in Cox's model with

$$g(\mathbf{x}_i(\Upsilon_p), \boldsymbol{\beta}) = \sum_{p=1}^P \mathbf{1}[\Upsilon_p]\mathbf{x}_i(\Upsilon_p)\boldsymbol{\beta}^p \tag{7}$$

where $\mathbf{1}[\Upsilon_p]$ is an indicator function equaling one if Y is contained in the set Υ_p and zero otherwise, $\mathbf{x}_i(\Upsilon_p)$ is a $1 \times k$ vector of covariates defined on the set Υ_p, $\boldsymbol{\beta}^p$ is a $k \times 1$ vector of unknown parameters, and $\boldsymbol{\beta} = (\boldsymbol{\beta}^{1'}, \boldsymbol{\beta}^{2'}, \ldots, \boldsymbol{\beta}^{P'})'$ is a $K \times 1$ parameter vector with $K = P \times k$. Within this specification, the covariates can shift the hazard function up over some regions and down over others, providing the possibility of quite different shapes for the hazard function for individuals with different covariate vectors.

There are several options for specifying the baseline hazard $h_0(y)$. One could assume a particular density function, such as the Weibull, and use the associated functional form for the hazard. This may aid in tractability, but imposes unwarranted shape restrictions on the density. Donald et al. approximated $h_0(y)$ using a step function. In particular, they created a set of dummy variables corresponding to each of the segments $[y_j, y_{j+1})$ for $j = 1, \ldots, J$ where J is finite and estimated a parameter associated with each of these "baseline" segments. The properties of a proportional

---

*This class of models was first introduced by Cox (1972, 1975). Specifically, the hazard rate for person i conditional on $\mathbf{x}_i$, a particular realization of the covariate vector $\mathbf{X}$, is

$$h(y|\mathbf{x}_i) = \exp(\mathbf{x}_i\boldsymbol{\beta})h_0(y)$$

where $h_0(y)$ is the baseline hazard function common to all individuals and $\boldsymbol{\beta}$ is a vector of unknown parameters. An important shortcoming of this specification is the restriction that individuals with very different covariate vectors have hazard functions with the same basic shape, and that any particular covariate shifts the entire hazard function up or down relative to the baseline specification. It can be shown that if a particular element of the vector is negative, then $F(y|\mathbf{x})$ for a person with a positive value for the associated covariate and zero values for all other covariates first-degree stochastically dominates $F(y|\mathbf{0})$, the cumulative distribution function for a person with zero values for all covariates. This is quite a strong restriction.

hazard model with this form of baseline specification are discussed in Meyer (1990). The major advantage of this approach is that, with a sufficiently large value for J, it can capture complicated shapes for the hazard, while allowing for very straightforward transformations from hazard to density estimates and then to estimates of GLC ordinates. This latter occurs because sums rather than integrals are involved. The main disadvantage is that density estimates are very "spikey," including spikes such as those induced by focal-point income reporting, e.g., integer multiples of $1000. The latter spikes may prove distracting when they are not the main focus of the analysis.

For a sample of size N, the logarithm of the likelihood function specification is

$$\sum_{i=1}^{N} \{1[Y_i < y_J]\log[1 - \exp(-\exp(\alpha_{j_i^*} + g(\mathbf{X}_i(\Upsilon_{j(j_i^*)}), \boldsymbol{\beta})))]$$

$$- \sum_{j=1}^{j_i^*-1} \exp(\alpha_j + g(\mathbf{X}_i(\Upsilon_{p(j)}), \boldsymbol{\beta}))\} \tag{8}$$

where the dependent variable for individual i falls in the j_i^*th baseline segment, Υ_j is the set of Y values corresponding to the jth baseline segment, $\Upsilon_{p(j)}$ is the set of Y values corresponding to the pth covariate segment which itself is associated with the jth baseline segment, and Y_i is less than y_J if Y_i is not right-censored. The vector α contains the J baseline parameters and the notation α_j indicates the element of α corresponding to the jth baseline segment. Note that

$$\alpha_j = \log\left[\int_{y_j}^{y_{j+1}} h_0(u) \, du\right]$$

For consistent estimates, Donald et al. required that the covariate values not change within the baseline segments; they may, however, vary across baseline segments.

For the jth baseline segment, an estimate of the hazard rate is

$$\hat{h}(y_j|\mathbf{x}) = 1 - \exp\left(-\exp\left(\hat{a}_j + \sum_{p=1}^{P} 1[\Upsilon_{p(j)}]\mathbf{x}\hat{b}^p\right)\right)$$

where $\hat{a}_j$ is the estimate of the jth element of the baseline parameter vector and $\hat{b}^p$ is an estimate of $\boldsymbol{\beta}^p$. An estimate of the survivor function is

$$\hat{S}(y_j|\mathbf{x}) = \exp\left(-\sum_{i=1}^{j-1} \hat{h}(y_i|\mathbf{x})\right)$$

The discrete form of the baseline hazard makes estimating the survivor function very simple. An estimate of the probability mass function is then

$$\hat{f}(y_j|\mathbf{x}) = \hat{S}(y_j|\mathbf{x}) - \hat{S}(y_{j+1}|\mathbf{x})$$

Define $\boldsymbol{\delta}$, a $(K + J)$ vector of unknown parameters with true value $\boldsymbol{\delta}^0$. Donald et al. assumed that the maximum likelihood estimator $\mathbf{D}$ satisfies

$$\sqrt{N}(\mathbf{D} - \boldsymbol{\delta}^0) \xrightarrow{d} \mathcal{N}(\mathbf{0}, \boldsymbol{\Sigma}^0)$$

and that a consistent estimate of $\boldsymbol{\Sigma}$ exists.* For segment j, they defined

$$\boldsymbol{\Gamma}_j = \nabla_\delta S(y_j|\mathbf{x})$$

which is a $(K + J)$ column vector with elements

$$\nabla_{\beta^k} S(y_j|\mathbf{x}) = -S(y_j|\mathbf{x}) \sum_{i=1}^{j-1} \mathbf{1}[\Upsilon_{p(i)}] \exp(\alpha_i) \exp\left(\sum_{p=1}^{P} \mathbf{1}[\Upsilon_{p(i)}]\mathbf{x}\boldsymbol{\beta}^p\right) \mathbf{x}'$$

and

$$\frac{\partial}{\partial \alpha_l} S(y_i|\mathbf{x}) = \begin{cases} -S(y_j|\mathbf{x})h(y_l|\mathbf{x}) & \text{for } l < j \\ 0 & \text{otherwise} \end{cases}$$

By the delta method, they showed that

$$\sqrt{N}(\hat{S}(y_j|\mathbf{x}) - S(y_j|\mathbf{x})) \xrightarrow{d} \mathcal{N}(0, \boldsymbol{\Gamma}_j^{0'}\boldsymbol{\Sigma}^0\boldsymbol{\Gamma}_j^0)$$

and

$$\sqrt{N}(\hat{f}(y_j|\mathbf{x}) - f(y_j|\mathbf{x})) \xrightarrow{d} \mathcal{N}(0, \boldsymbol{\Theta}_j^{0'}\boldsymbol{\Sigma}^0\boldsymbol{\Theta}_j^0)$$

where $\boldsymbol{\Theta}_j = \boldsymbol{\Gamma}_j - \boldsymbol{\Gamma}_{j+1}$. Thus, the standard errors of the density estimate at each baseline segment are easy to construct given $\hat{\mathbf{d}}$ and an estimate of $\boldsymbol{\Sigma}$. The discrete form of the baseline hazard is helpful in this construction since one requires only summations rather than integration.

One can also recover estimates of quantiles conditional on the covariates. The qth quantile of Y conditional on $\mathbf{x}$ is defined by

$$q = \int_0^{\xi(q|\mathbf{x})} f(y|\mathbf{x}) \, dy = F(\xi(q|\mathbf{x}))$$

---

*In practice, Donald et al. use the inverse of the Hessian matrix of the logarithm of the likelihood function (divided by the sample size) to estimate $\boldsymbol{\Sigma}$.

In the discrete case, however, estimates of the quantiles are easily found by $E(q|\mathbf{x}) = y_j$ if

$$\hat{S}(y_j|\mathbf{x}) \geq (1 - q) > \hat{S}(y_{j+1}|\mathbf{x})$$

and $E(q|\mathbf{x}) = y_J$ if

$$\hat{S}(y_J|\mathbf{x}) \geq (1 - q)$$

Donald et al. characterized the limiting distribution of $E(q|\mathbf{x})$ for different values of q.

The main advantage of the approach of Donald et al. is its combination of flexibility and tractability. Convergence for their likelihood function can be obtained relatively quickly and easily. The transformations from $\hat{\mathbf{d}}$ to estimates of the hazard and then to the density functions described above are straightforward. At the same time, the specification imposes few restrictions on the shape of the density function and permits quite different shapes for different covariate vectors. In addition, the approach admits the examination of both decomposability of subgroups and marginal effects of covariates. Finally, this approach provides a consistent means of addressing top-coded data since top-coded values are just right-censored "spells."*

Once a consistent estimate of $f(y|\mathbf{x})$ is obtained, one can then estimate the GLC ordinates conditional on the covariates. When the Ys are discrete, the ordinate of the GLC at q is

$$\mathcal{G}(q|\mathbf{x}) = \sum_{j=1}^{J} \mathbf{1}[y_j \leq \xi(q|\mathbf{x})]y_j f(y_j|\mathbf{x})$$

where

$$f(y_j|\mathbf{x}) = S(y_j|\mathbf{x}) - S(y_{j+1}|\mathbf{x})$$

and where $\xi(q|\mathbf{x})$ is the conditional qth quantile of Y:

$$\xi(q|\mathbf{x}) = \inf\{\xi : S(\xi|\mathbf{x}) \leq (1 - q)\}$$

for $0 \leq q \leq 1$. Donald et al. (1995) derived the asymptotic distribution of

$$G(q|\mathbf{x}) = \sum_{j=1}^{J} \mathbf{1}[y_j \leq E(q|\mathbf{x})]y_j \hat{f}(y_j|\mathbf{x})$$

---

*In sample surveys, individuals with high incomes can often be identified by answers to other questions on the survey instrument. For example, the richest man in a particular geographical region might have a wife and six children. Thus, an interested individual might infer the man's income from information concerning where he lived and his household characteristics. To provide confidentiality, those who conduct surveys often list high incomes as greater than some value, say $100,000 and higher. This practice is often referred to as "top-coding."

to be

$$\sqrt{N}(G(q|\mathbf{x}) - \mathcal{G}^0(q|\mathbf{x})) \overset{d}{\to} \mathcal{N}(0, \Psi^0(q|\mathbf{x})'\Sigma^0\Psi^0(q|\mathbf{x}))$$

where

$$\Psi(q|\mathbf{x}) = \sum_{j=1}^{J} \mathbf{1}[y_j \le \xi(q|\mathbf{x})]y_j\Theta_j$$

VI.   SUMMARY AND CONCLUSIONS

We have presented a selective survey of the literature concerned with interpreting economic statistics in terms of social welfare. We started by defining and reviewing the properties of location measures and the most popular summary inequality indices. Recently, use of summary indices has been augmented, and increasingly replaced, in applied work by the use of disaggregated dominance criteria. A summary inequality index must correspond to a single, specific, social welfare function or narrow family of such functions. In contrast, dominance criteria require only that social preferences should satisfy certain widely appealing axioms of social choice.

The principal dominance criteria examined here have been LC and GLC dominance. In both cases, the main distributional judgement required is that one should object to a regressive income transfer; i.e., these criteria embody the Pigou-Dalton principle of transfers. Both criteria are equivalent to SSD when the mean is invariant. However, the LC departs from SSD in that it is unconcerned with changes that merely raise or lower mean income without affecting individuals' income shares. Therefore, strictly speaking, the Lorenz criterion is an indicator of *relative inequality* rather than of social welfare. A further criterion we have discussed briefly, and which is receiving increasing use in practice, is ADI (equivalent to TSD when means are the same). ADI comes into play in the comparison of distributions whose LCs intersect. If the cumulative coefficients of variation for one distribution are less than those of another, cumulating up to each successive point of LC intersection, then there is dominance from the viewpoint of all individuals who are more concerned with inequality lower in the distribution, i.e., from the viewpoint of everyone averse to downside inequality.

We have surveyed results on the implementation of these dominance criteria, beginning with the parametric and nonparametric estimation of LCs and GLCs. We have also reviewed the results on tests for LC and GLC equivalence, which is relatively easy, and dominance, which is more difficult. We have also reviewed results on the nonparametric implementation of the ADI criterion.

A recent development in the literature has been the estimation of GLCs which depend on covariates. We have discussed limitations of parametric and nonparametric approaches, and how these limitations can be avoided through the use of tractable semiparametric methods. The latter can be made sufficiently flexible to reflect complex effects of covariates. A further advantage of the approach is that it provides a consistent means of addressing top-coded data.

As the review conducted in this chapter shows, recent literature has provided powerful theoretical and statistical tools which can be applied in social welfare comparisons. At this point there is an important challenge to workers on income distribution topics to make greater use of these techniques in their analysis of the data.

VII. APPENDIX

In this appendix, we describe how to access several programs designed to carry out the analysis described above. These programs reside in the Econometrics Laboratory Software Archive (ELSA) at the University of California, Berkeley; this archive can be accessed easily using a variety of different browsers via the Internet. To contact the staff at ELSA, simply send e-mail to

`elsa@econ.berkeley.edu`

For those with access to an Internet browser like Netscape Navigator, simply click on the Open icon and then enter

`http://econ.berkeley.edu`

This will put you into the HomePage of the Department of Economics at the University of California, Berkeley. A number of options exist. Click the option

`[Research Facilities]`

You will then be on the page for which the icon

`•ELSA (Econometrics Laboratory Software Archive)`

exists. Click on either the icon denoted Code or the icon denoted Doc and scroll through until you find the entries you desire.

The following describes the entries:

a. A Nonparametric Estimator of Lorenz Curves
    FORTRAN code for implementing Beach and Davidson (1983)
b. A Nonparametric Estimator of Generalized Lorenz Curves
    FORTRAN code for implementing Bishop, Chakraborti, and Thistle (1989)
c. A Semiparametric Estimator of Distribution Functions with Covariates
    Gauss code for implementing Donald, Green, and Paarsch (1995)

ACKNOWLEDGMENTS

Green and Paarsch wish to thank the SSHRC of Canada for research support and their co-author, Stephen G. Donald, for allowing them to borrow from their unpublished work. Most of this paper was written while Paarsch was the Arch W. Shaw National Fellow at the Hoover Institution, Stanford, California. Paarsch would like to thank the Hoover Institution for its hospitality and support. The authors are also grateful to Gordon Anderson and Charles M. Beach as well as two anonymous referees for useful comments and helpful suggestions.

REFERENCES

Anderson, G. (1996), Nonparametric Tests for Stochastic Dominance in Income Distributions, *Econometrica*, 64, 1183–1193.

Atkinson, A. (1970), On the Measurement of Inequality, *Journal of Economic Theory*, 2, 244–263.

Atkinson, A. (1975), *The Economics of Inequality*, Oxford University Press, Oxford.

Beach, C., K. Chow, J. Formby, and G. Slotsve, (1994), Statistical Inference for Decile Means, *Economics Letters*, 45, 161–167.

Beach, C. and R. Davidson (1983), Distribution-Free Statistical Inference with Lorenz Curves and Income Shares, *Review of Economic Studies*, 50, 723–735.

Beach, C., R. Davidson, and G. Slotsve (1994), Distribution-Free Statistical Inference for Inequality Dominance with Crossing Lorenz Curves, typescript, Department of Economics, Queen's University, Kingston, Canada.

Beach, C. and S. Kaliski (1986), Lorenz Curve Inference with Sample Weights: An Application to the Distribution of Unemployment Experience, *Journal of the Royal Statistical Society*, Series C, 35, 38–45.

Beach, C. and J. Richmond (1985), Joint Confidence Intervals for Income Shares and Lorenz Curves, *International Economic Review*, 26, 439–450.

Bishop, J., S. Chakraborti, and P. Thistle (1989), Asymptotically Distribution-Free Statistical Inference for Generalized Lorenz Curves, *Review of Economics and Statistics*, 71, 725–727.

Blackorby, C. and D. Donaldson (1978), Measures of Relative Equality and their Meaning in Terms of Social Welfare, *Journal of Economic Theory*, 18, 59–80.

Chipman, J. (1985), The Theory and Measurement of Income Distributions, in R. Basmann and G. Rhodes, Jr. (eds.), *Advances in Econometrics*, vol. 4, JAI Press, Greenwich, CT.

Cowell, F. (1979), *Measuring Inequality*, Philip Allan, Oxford.

Cowell, F. (1989), Sampling Variance and Decomposable Inequality Measures, *Journal of Econometrics*, 42, 27–41.

Cowell, F. and F. Mehta (1982), The Estimation and Interpolation of Inequality Measures, *Review of Economic Studies*, 49, 273–290.

Cowell, F. and M. Victoria-Feser (1996), Robustness Properties of Inequality Measures, *Econometrica*, 64, 77–101.

Cox, D. (1972), Regression Models and Life-Tables, *Journal of the Royal Statistical Society*, Series B, 34, 187–202.

Cox, D. (1975), Partial Likelihood, *Biometrika*, 62, 269–276.

Dasgupta, P., A. Sen, and D. Starrett (1973), Notes on the Measurement of Inequality, *Journal of Economic Theory*, 6, 180–187.

Davidson, R. and J.-Y. Duclos (1995), Statistical Inference for the Measurement of the Incidence of Taxes and Transfers, typescript, Department of Economics, Queen's University, Kingston, Canada.

Davies, J. and M. Hoy (1994), The Normative Significance of Using Third-Degree Stochastic Dominance in Comparing Income Distributions, *Journal of Economic Theory*, 64, 520–530.

Davies, J. and M. Hoy (1995), Making Inequality Comparisons when Lorenz Curves Intersect, *American Economic Review*, 85, 980–986.

Donald, S., D. Green, and H. Paarsch (1995), Differences in Earnings and Wage Distributions between Canada and the United States: An Application of a Semi-Parametric Estimator of Distribution Functions with Covariates, typescript, Department of Economics, University of Western Ontario, London, Canada.

Engle, R. (1984), Wald, Likelihood Ratio, and Lagrange Multiplier Tests in Economics, in Z. Griliches and M. Intriligator (eds.), *Handbook of Econometrics*, vol. II, North-Holland, Amsterdam.

Galambos, J. (1987), *The Asymptotic Theory of Extreme Order Statistics*, 2nd ed., Krieger, Malabar, FL.

Gritz, M. and T. MaCurdy (1992), Unemployment Compensation and Episodes of Nonemployment, *Empirical Economics*, 17, 183–204.

Harberger, A. (1971), Three Basic Postulates of Applied Welfare Economics, *Journal of Economic Literature*, 9, 785–797.

Harrison, A. (1982), A Tail of Two Distributions, *Review of Economic Studies*, 48, 621–631.

Howes, S. (forthcoming), Asymptotic Properties of Four Fundamental Curves of Distributional Analysis, *Journal of Econometrics*.

Kalbfleisch, J. and R. Prentice (1980), *The Statistical Analysis of Failure Time Data*, Wiley, New York.

Klecan, L., R. McFadden, and D. McFadden (1991), A Robust Test for Stochastic Dominance, typescript, Department of Economics, MIT, Cambridge, MA.

Kodde, D. and F. Palm (1986), Wald Criteria for Jointly Testing Equality and Inequality Restrictions, *Econometrica*, 54, 1243–1248.

Lambert, P. (1989), *The Distribution and Redistribution of Income, A Mathematical Analysis*, Basil Blackwell, Cambridge, MA.

Maasoumi, E. (1997), Empirical Analysis of Inequality and Welfare, in M. Pesaran and P. Schmidt (eds.), *Handbook of Applied Microeconometrics*, Basil Blackwell, London.

McDonald, J. (1984), Some Generalized Functions for the Size Distribution of Income, *Econometrica*, 52, 647–663.

McFadden, D. (1989), Testing for Stochastic Dominance, in T. Fomby and T. Seo (eds.), *Studies in the Economics of Uncertainty: In Honor of Josef Hadar*, Springer-Verlag, New York.

Meyer, B. (1990), Unemployment Insurance and Unemployment Spells, *Econometrica*, 58, 757–782.

Perlman, M. (1969), One Sided Problems in Multivariate Analysis, *Annals of Mathematical Statistics*, 40, 549–567.

Rao, C. (1965), *Linear Statistical Inference and Its Applications*, Wiley, New York.

Richmond, J. (1982), A General Method for Constructing Simultaneous Confidence Intervals, *Journal of the American Statistical Association*, 77, 455-460.

Rothschild, M. and J. Stiglitz (1970), Increasing Risk. I: A Definition, *Journal of Economic Theory*, 2, 225–243.

Savin, N. (1984), Multiple Hypothesis Testing, in Z. Griliches and M. Intriligator (eds.), *Handbook of Econometrics*, vol. II, North-Holland, Amsterdam.

Sen, A. (1973), *On Economic Inequality*, Oxford University Press, London.

Sendler, W. (1979), On Statistical Inference in Concentration Measurement, *Metrika*, 26, 109–122.

Shorrocks, A. (1980), The Class of Additively Decomposable Inequality Measures, *Econometrica*, 48, 613–625.

Shorrocks, A. (1982), Inequality Decomposition by Factor Components, *Econometrica*, 50, 193–211.

Shorrocks, A. (1983), Ranking Income Distributions, *Economica*, 50, 3–17.

Shorrocks, A. (1984), Inequality Decomposition by Population Subgroups, *Econometrica*, 52, 1369–1385.

Shorrocks, A. and J. Foster (1987), Transfer Sensitive Inequality Measures, *Review of Economic Studies*, 54, 485–497.

Theil, H. (1967), *Economics and Information Theory*, North-Holland, Amsterdam.

Whitmore, G. (1970), Third-Degree Stochastic Dominance, *American Economic Review*, 60, 457–459.

Wolak, F. (1987), An Exact Test for Multiple Inequality and Equality Constraints in the Linear Regression Model, *Journal of the American Statistical Association*, 82, 782–793.

Wolak, F. (1989), Local and Global Testing of Linear and Nonlinear Inequality Constraints in Nonlinear Econometric Models, *Econometric Theory*, 5, 1–35.

Wolak, F. (1991), The Local Nature of Hypothesis Tests Involving Inequality Constraints in Nonlinear Models, *Econometrica*, 59, 981–995.

Xu, K., G. Fisher, and D. Willson (1995), New Distribution-Free Tests for Stochastic Dominance, Department of Economics, Dalhousie University, Working Paper 9502.

2
Measurement of Inequality

Walter Krämer
University of Dortmund, Dortmund, Germany

I. THE PROBLEM

This chapter surveys the statistical measurement of inequality, where "inequality" is a property of the elements of the set

$$D := \bigcup_{n=2}^{\infty} \mathbf{R}_+^n \tag{1}$$

where $\mathbf{R}_+^n = \{x = (x_1, \ldots, x_n) | x_i \in \mathbf{R}, x_i \geq 0, \sum x_i > 0\}$.

For concreteness, I will often argue in terms of income distributions, but most of what follows also applies to variables other than income. The discussion is in terms of vectors rather than distribution functions, as a reliance on the latter would automatically restrict attention to measures which are symmetric and population invariant (see Section IV.A.2 and IV.A.8), and it is important for many problems to be more permissive here.

Given some vectors x and y from D, the problem is to rank these vectors according to inequality, whatever "inequality" may mean. In a sense this is similar to comparing distributions with respect to location and dispersion (scale), only less intuitive: while everybody "knows" what is meant be scale and location, and (to a lesser extent) by skewness and kurtosis, the inequality dimension commands much less intuitive appeal, and, in fact, allows for views which are outright contradictory.

Accordingly, this survey first discusses ordinal approaches to inequality; i.e., it considers some widely accepted concepts of "inequality" that allow at least some distributions x and y to be ranked. Section III considers cardinal measures of inequality, Section IV discusses the axiomatic foundation of these measures, Section V addresses various empirical issues that arise when data are grouped or otherwise

39

not completely available, and Section VI concludes with some recent developments in multidimensional inequality.

As the focus here is on mathematical aspects of the ordering of given vectors x and y, it is only fair to point out that in most applications these problems are dwarfed by the complications and ambiguities involved in the definition and measurement of the basic vectors x and y themselves. This is perhaps most obvious with income inequality, where the outcome of any analysis depends much more on how income is defined, on the unit (individuals versus households), and on the accounting period than on the statistical procedures applied afterward. However, such problems of basic definition and measurement differ so much across applications that there is no hope of an adequate treatment in a general survey such as this.

II. ORDINAL COMPARISONS OF INEQUALITY

There is wide agreement across value systems and applications that if some vector is subjected to a progressive transfer, then inequality should be reduced.* Formally, let $y = xT$, where both y and x are of dimension n and where T is an $n \times n$ matrix of the form

$$T = \begin{bmatrix} 1 & & \cdots & & 0 \\ & p & \cdots & 1-p & \\ \vdots & \vdots & \ddots & \vdots & \vdots \\ & 1-p & \cdots & p & \\ 0 & & \cdots & & 1 \end{bmatrix} \tag{2}$$

with $0 \le p \le 1$. Then $\sum y_i = \sum x_i$ and y differs from x only in two components, say i and j, who are each replaced by their weighted arithmetic mean. In the context of income inequality, such a transfer means that the richer of the persons i and j gives some or all of the surplus to the poorer (but no more; i.e., the richer does not thereby become poorer as the poorer has been before; both incomes stay the same if $x_i = x_j$). It was first suggested as an unequivocal reduction in inequality by Pigou (1912) and Dalton (1920) and is often called a Pigou-Dalton transfer, but some prefer Arnold's (1987) more evocative "Robin Hood transfer" instead.

Obviously, a sequence of successive Robin Hood transfers should lead to successive reductions in inequality, which in turn leads naturally to the majorization or-

---

*Or at least: should not increase. In what follows, "reduction" excludes the limiting case of "no change." If "no change" is included, I say "weak reduction" (and similarly "increase" and "weak increase").

dering in D.* Let $x = (x_{(1)}, \ldots, x_{(n)})$ and $y = (y_{(1)}, \ldots, y_{(n)})$, with $\sum x_{(i)} = \sum y_{(i)}$, be ordered such that $x_{(1)} \leq x_{(2)} \leq \cdots \leq x_{(n)}$ and $y_{(1)} \leq y_{(2)} \leq \cdots \leq y_{(n)}$. Then, by definition, x majorizes y (i.e., x is more unequal than y: $x \geq_M y$) if

$$\sum_{i=1}^{k} x_{(i)} \leq \sum_{i=1}^{k} y_{(i)} \qquad (k = 1, 2, \ldots, n) \tag{3}$$

and a central result in majorization theory, due to Hardy, Littlewood, and Polya (1934), states that x majorizes y if and only if $y = xT_1 T_2 \cdots T_L$ with finitely many matrices T_i of the form (2).

The obvious drawback of this (the only) universally accepted preordering is that only very few pairs of elements x and y from D can be compared. Both vectors must have the same dimension and the same sum of elements, which renders this ordering almost useless in empirical applications.

One way out of this dilemma is to base a comparison on one of the scalar-valued measures of inequality to be discussed in Section III. As these measures all involve value judgments of one sort or the other, one might compromise here and declare one vector y more equal than some other vector x if $I(x) \geq I(y)$ for all inequality measures $I(\cdot)$ in some conveniently chosen class. For instance, almost by definition, the majorization order on D is the one implied by the set of all Schur-convex functions $I: D \to R$ (a function $I(\cdot)$ is Schur convex if it respects the majorization ordering), and by restricting this set of functions, one might hope to obtain an order which is richer (i.e., allows more pairs of vectors to be ranked).

Quite surprisingly, however, if we restrict $I(\cdot)$ to be symmetric and convex in the ordinary sense (i.e., $I(\lambda x + (1-\lambda)y) \geq \lambda I(x) + (1-\lambda)I(y)$ for all $0 \leq \lambda \leq 1$; this implies Schur convexity), or even further to be of the form

$$I(x) = \sum_{i=1}^{n} g(x_i) \tag{4}$$

with a convex function $g: R \to R$, we still end up with the very restrictive majorization ordering $\geq_M$ (Mosler 1994, Proposition 2.2).

A major breakthrough in this respect was the discovery by Atkinson (1970) and Dasgupta, Sen, and Starrett (1973) that the preordering on D becomes much richer, and in fact identical to the familiar Lorenz order, if we require in addition to (4) that the measure I be population invariant:

$$I(x) = I(x, x, \ldots, x) \qquad (x \text{ repeated } m \text{ times}) \tag{5}$$

---

*Actually a *preordering*, to be precise. A reorder is a *partial order* if $x \leq y$ and $y \leq x$ implies $x = y$. In what follows, "order" is short for "preorder" and a preorder is an order if either $x \leq y$ or $y \leq x$.

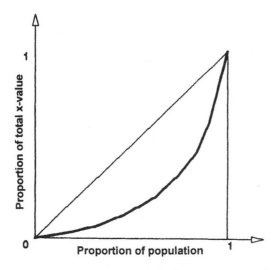

Figure 1   A typical Lorenz curve: convex, monotone, passing through (0, 0) and (1, 1).

and homogeneous of degree zero:

$$I(\lambda x) = I(x) \qquad (\lambda > 0) \tag{6}$$

Then y is more equal than x with respect to *all* inequality indices in that class if and only if the Lorenz curve of y lies everywhere above the Lorenz curve of x, where the Lorenz curve of x (and similarly for y) is obtained in the usual way by joining the points (0, 0) and

$$\left(\frac{i}{n}, \sum_{j=1}^{i} x_{(j)} \Big/ \sum_{j=1}^{n} x_{(j)} \right) \qquad (i = 1, \ldots, n) \tag{7}$$

by line segments in a two-dimensional diagram. This preordering, which I denote by "$\geq_L$," allows comparisons of vectors with different means and dimensions. When applied to vectors with identical means and dimensions, it reduces to "$\geq_M$." (Note that "$\geq_L$" and "$\geq_M$" implies in increase in *in*equality.)

A typical Lorenz curve is shown in Figure 1: it is convex, with slope $x_{(i)}/\bar{x}$ in the interval $((i - 1)/n, i/n)$, passing through (0, 0) and (1, 1).

The proof that the Lorenz order in D is implied by inequality indices with properties (4), (5) and (6) is an immediate consequence of the fact that with (5) and (6), we can without loss of generality assume x and y to have identical means and dimensions: if $\bar{x} = \frac{1}{n} \sum x_i \neq \bar{y}$ and/or dimension $(x) = n \neq m =$ dimension (y), consider

$$\tilde{x} = (x, x, \ldots, x) \qquad x \text{ repeated } m \text{ times}$$

$$\tilde{y} = \frac{\overline{x}}{\overline{y}}(y, y, \ldots, y) \qquad y \text{ repeated } n \text{ times} \tag{8}$$

Then $\tilde{x}$ has the same Lorenz curve as x, and $\tilde{y}$ has the same Lorenz curve as y, and $f(x) = f(\tilde{x})$ and $f(y) = f(\tilde{y})$ for all functions with properties (5) and (6) and the equivalence of the Lorenz order with the ordering implied by functions with properties (4), (5), and (6) follows from the fact that the majorization ordering is implied by functions with property (4) alone.

If one does not want to subscribe to (6), but requires instead that inequality remains unaffected if the same positive amount is added to all components of x:

$$I(x + \lambda e) = I(x) \qquad (e = (1, \ldots, 1), \lambda > 0) \tag{9}$$

one obtains a different ordering which turns out to be equivalent to absolute Lorenz domination as introduced by Moyes (1987): With $L_x(p)$ the conventional Lorenz curve of x, the absolute Lorenz curve is

$$LA_x(p) = \overline{x}(L_x(p) - p) \tag{10}$$

and x is more unequal than y in the absolute Lorenz sense (in symbols: $x \geq_{LA} y$) if $LA_x(p) \leq LA_y(p)$. This gives another preordering on D which reduces to majorization whenever the latter applies, but it incorporates a more leftist ideal of equality, as first pronounced by Kolm (1976), for pairs of income vectors which cannot be ranked by majorization.

There is an infinity of additional preorderings on D which are induced by families of "compromise inequality indices" (Eichhorn 1988, Ebert 1988, Bossert and Pfingsten 1989), i.e., indices with the property

$$I(x + \tau(\lambda x + (1 - \lambda)e)) = I(x) \tag{11}$$

where $0 \leq \lambda \leq 1$ and τ is any scalar such that $x + \tau(\lambda x + (1-\lambda)e) \in \mathbf{R}_+^n$. Obviously, this requirement reduces to (9) for $\lambda = 0$ and to (6) for $\lambda = 1$, but it allows some "intermediate" value judgments too: while (6) incorporates the "rightest" view that an equiproportional increases in all incomes leaves inequality the same, whereas an equal absolute increase reduces it (this is not immediate, but follows from (6) in conjunction with some minor regularity conditions) and (9) incorporates the "leftist" view that an equal absolute increase in income leaves inequality the same, whereas an equiproportional increase increases it (which can likewise be shown to follow from (9) plus some regularity conditions), the requirement (11) implies that inequality is increased both by an equiproportional increase of all incomes and by an equal absolute reduction, with the λ parameter governing the closeness to the pure leftist and rightist view, respectively: As λ tends to zero, one tends to take the leftist point of view, and the rightist point of view is taken as λ tends to one.

Given λ, the preordering "$\geq_\lambda$" induced by Schur-convex compromise indices satisfying (11) amounts to

$$x \geq_\lambda y \Leftrightarrow \sum_{j=1}^{i} \frac{x_{(j)} - \bar{x}}{\lambda\bar{x} + 1 - \lambda} \leq \sum_{j=1}^{i} \frac{y_{(j)} - \bar{y}}{\lambda\bar{y} + 1 - \lambda} \qquad (i = 1, \ldots, n) \qquad (12)$$

which in turn is equivalent to $\tilde{x} \geq_M \tilde{y}$, where $\tilde{x} = (x - \bar{x}e)/(\lambda\bar{x} + 1 - \lambda)$ and $\tilde{y} = (y - \bar{y}e)/(\lambda\bar{y} + 1 - \lambda)$, with $e = (1, \ldots, 1)$ a vector of ones (Pfingsten 1988a).

Given the multitude of pairs of vectors x and y in D which still cannot be ranked, the primary purpose of any such preordering is not the ranking as such (since this is often a rather disappointing exercise), but rather to provide a benchmark test for various procedures that affect inequality, like taxation in the case of income inequality. Given any such preordering, it is both sensible and relevant to ask whether some tax functions will always (i.e., irrespective of the pretax income vector x) reduce inequality.

For the Lorenz order, the answer is well known (see Eichhorn et al. 1984): A tax function $t(z)$ will produce an after-tax income vector y which is majorized by any pretax income vector x if and only if $t(z)$ is incentive preserving and progressive (i.e., if both residual income $z - t(z)$ and average tax rates $t(z)/z$ are increasing functions of z).[*] For the ordering implied by (11), the tax functions compatible with that ordering are derived in Pfingsten (1988b). Lambert (1993) gives a detailed discussion of such issues, and for the effects on data transformations on Lorenz curves in general see Fellman (1976).

III. CARDINAL COMPARISONS OF INEQUALITY

As Lorenz curves, whether ordinary, absolute or generalized,[†] do often cross, there remain many pairs of vectors in D which cannot be ranked by any of the methods from Section II. In an empirical comparison involving 72 countries, Kakwani (1984) for instance found more than 700 Lorenz crossings among the 2556 possible pairwise comparisons, hence the desire to attach some cardinal measure of inequality to a given distribution.

The best known of these cardinal, scalar-valued measures is the Gini index $G(x)$, defined as twice the area between the Lorenz curve and the 45° line of equality

---

[*]Note that progressiveness by itself does not guarantee a reduction in inequality. It is easy to find examples where income inequality in the Lorenz sense *increases* after a progressive income tax: take two incomes of 5 and 10 and a progressive tax rate of 0% up to 6 and 100% afterwards.

[†]A generalized Lorenz curve is an ordinary Lorenz curve multiplied by $\bar{x}$. It is most useful in welfare rankings of income distributions (Shorrocks 1983, Thistle 1989).

(the concentration area). It perfectly captures Lorenz's original intuition that, as the bow is bent, so inequality increases, taking a minimum value 0 when all x_i are equal (i.e., when the Lorenz curve is equal to the 45° line), and a maximum value of $(n-1)/n$ whenever all x_i except one are zero. It depends on the underlying vectors x only via the Lorenz curve of x, so it is obviously symmetric, homogeneous of degree zero, and, by construction, compatible with the Lorenz ordering on D. The history of this "mother of all indices of inequality" is nicely summarized in Giorgi (1990).

There are various equivalent expressions for the Gini coefficient, like Gini's own expression

$$G(x) = \frac{\Delta(x)}{2\bar{x}} \tag{13}$$

where

$$\Delta(x) = \frac{1}{n^2} \sum_{i,j=1}^{n} |x_i - x_j| \tag{14}$$

is Gini's mean difference. Rewriting the double sum in (14) as

$$\sum_{i,j=1}^{n} |x_i - x_j| = 2 \sum_{i>j=1}^{n} (x_{(i)} - x_{(j)}) = 2 \sum_{i=1}^{n} (2i - n - 1)x_{(i)} \tag{15}$$

one easily sees that

$$G(x) = \frac{\Delta(x)}{2\bar{x}} = \left(\frac{2}{n \sum x_i} \sum_{i=1}^{n} i x_{(i)} \right) - 1 - \frac{1}{n} \tag{16}$$

where the latter expression is most useful when discussing sensitivity issues (see below), or for showing that $\Delta(x)/2\bar{x}$ is indeed equal to twice the concentration area.

Yet another algebraic identity was unearthed by Lerman and Yitzhaki (1984), and independently by Berrebi and Silber (1987), who show that twice the Gini index equals the covariance between the "rank gaps" $r_i := i - (n - i + 1)$ and the "share gaps" $s_i = (x_{(i)} - x_{(n-i+1)})/\sum x_{(i)}$ of the data. This identity, for instance, implies that the Gini index is never smaller than $1/2$ for symmetric distributions and that a necessary condition for $G(x) < 1/2$ is that x is skewed to the right. It can also be used to facilitate the empirical computation of the Gini index from large sets of individual data, where competing measures were often preferred to the Gini index only because of computational convenience.

In the context of income equality, an interesting intuitive foundation of the Gini index based on (13) which does not rely on the geometry of the Lorenz curve was suggested by Pyatt (1976): Choose randomly some unit, say i, with income x_i, and choose independently and randomly another unit x_j. Then, if i keeps his income if $x_j \leq x_i$ and receives the difference if $x_j > x_i$, the Gini index is the expected profit from this game, divided by average income.

A similar idea has led Blackburn (1989) to a nice interpretation of *changes* in the Gini index: Let x and y be income vectors in $\mathbf{R}_+^n$ such that $G(x) > G(y)$. Then $G(x) - G(y)$ equals the percentage of $\bar{x}$ that units in x below the median have to transfer to units above the median to make $G(x)$ equal to $G(y)$.

There are various generalizations of the Gini index, such at Yitzhaki's (1983) generalized Gini coefficients or Mehran's (1976) linear measures of income inequality, which like the Gini index can be expressed as functions of the Lorenz curve, but which avoid certain shortcomings of the Gini index such as its lack of transfer sensitivity (see below).

The Yitzhaki family of inequality indices is defined as

$$G_\nu(x) = 1 - \nu(\nu - 1) \int_0^1 (1 - p)^{\nu-2} L_x(p) \, dp \tag{17}$$

with one coefficient for each value of a parameter $\nu > 1$. For $\nu = 2$, this is the standard Gini index $G(x)$, but as ν increases, higher weights are attached to small incomes; the limit as $\nu \to \infty$ of $G_\nu(x)$ is $1 - x_{(1)}/\bar{x}$, so in the limit inequality depends only on the lowest income (given $\bar{x}$), expressing the familiar value judgment introduced by Rawls, that social welfare (viewed as a function of inequality) depends only on the poorest member of society.

Mehran's (1976) linear measures of inequality are defined as

$$I(x) = \int_0^1 (p - L_x(p)) \, dW(p) \tag{18}$$

where $W(p)$ is some increasing "score function" not depending on x, which allows one to incorporate value judgments as to the magnitude of a decrease in inequality that is engendered by a Robin Hood transfer involving individuals a given number of ranks apart: As is shown in Mehran (1976), the decrease in inequality decreases as the rank of the recipient increases (i.e., transfers among the rich are less important than transfers among the poor) if and only if the score functions has a strictly decreasing derivative. We will return to this issue of transfer sensitivity in Section IV.

Again, the family defined by (18) includes the Gini index as a special case (take $W(p) = 2p$), but it covers various other indices as well, allowing for a unified discussion of their properties. For instance, it is easily seen that an index defined by (18) is always compatible with the majorization order, but strictly compatible (i.e., strict majorization always leads to a strict reduction in inequality) only if $W(p)$ is a strictly increasing function of p. Taking $W(p) = 0$ if p is less than the proportion of the x_i's below $\bar{x}$ and $W(p) = 1$ if p is equal to or larger than the proportion of the x_i's above $\bar{x}$, we obtain the maximum vertical distance between the egalitarion 45° line and the Lorenz curve, and the well known result that this distance is not affected by transfers either below or above $\bar{x}$.

Historically, this maximum vertical distance between the 45° line and the Lorenz curve has come in various guises; it is also known as the "maximum equalization percentage" (the relative amount of income transfers necessary for total equalization), and it is also algebraically identical to one half of the relative mean deviation:

$$\frac{\sum\limits_{i=1}^{n} |x_i - \bar{x}|}{2\bar{x}} \tag{19}$$

This measure was suggested, more or less independently, by Bresciani-Turroni (1910), Ricci (1916), von Bortkiewicz (1931), Schutz (1951), Kuznets (1959), and Elteto and Frigyes (1968), to name a few of the authors who have given it their attention. For simplicity, I will refer to it as the Pietra index $P(x)$, in honor of Pietra (1914/15).*

In the context of income inequality, it has long been viewed as a major drawback of both the Gini and Pietra indices and of related indices derived from them that they are not explicitly linked to some social welfare function $W(x)$.[†] This uneasiness was first expressed in Dalton's (1920) proposal to use the "ratio of the total economic welfare attainable under an equal distribution to the total welfare attained under the given distribution" as a measure of income inequality; it has led Atkinson (1970) to define inequality as $A(x) = 1 - y/\bar{x}$, where the scalar y is the income that, if possessed by everybody, would induce the same social welfare as the actual income vector x.

Given an additively separable welfare function $W(x) = \sum_{i=1}^{n} U(x_i)/n$, where $U(\cdot)$ is utility, and a utility function

$$U_\varepsilon(z) = \begin{cases} \dfrac{z^{1-\varepsilon} - 1}{1 - \varepsilon}, & 0 \le \varepsilon < \infty, \quad \varepsilon \ne 1 \\[2ex] \ln(z) & \varepsilon = 1 \end{cases} \tag{20}$$

this approach leads to a family of inequality measures given by

$$A_\varepsilon(x) = \begin{cases} 1 - \left[\dfrac{1}{n} \sum\limits_{i=1}^{n} \left(\dfrac{x_i}{\bar{x}} \right)^{1-\varepsilon} \right]^{\frac{1}{1-\varepsilon}} & (\varepsilon \ge 0, \varepsilon \ne 1) \\[3ex] 1 - \dfrac{\text{geom. mean of } x}{\text{arith. mean of } x} & (\varepsilon = 1) \end{cases} \tag{21}$$

---

*See Kondor (1971) or Chipman (1985, pp. 142–143) for a brief sketch of its rather long and winding genesis.

[†]Of course, an *implicit* link can in most cases easily be constructed, as shown by Sheshinski (1972) for the Gini index. In view of Newbery (1970), who shows that the Gini index is incompatible with additive separable welfare functions, these implied welfare functions are sometimes rather odd.

where the parameter ε measures aversion to inequality: As $\varepsilon \to \infty$, $A_\varepsilon(x)$ approaches $1 - x_{(1)}/\bar{x}$, similar to Yitzhaki's index $G_\nu(x)$. Also, with $\varepsilon = 1$ the Atkinson index is one minus the ratio of the geometric to the arithmetic mean, and with $\varepsilon = 2$ it equals one minus the ratio of the harmonic to the arithmetic mean.

IV. AXIOMATIC APPROACHES TO THE MEASUREMENT OF INEQUALITY

A different approach to avoid the ad hockery implicit in much of the measurement of inequality, other than the welfare-based development above, is to specify a set of axioms or minimal requirements that an index should obey. Ideally, this leaves one with a single index formula, which is then characterized by these initial axioms.

The first and most obvious axiom is that, as a result of a Robin Hood transfer, the inequality index decreases:

(A1) $I(xT) \leq I(x)$ for a matrix T of the form (2).

This minimal requirement is violated by at least one well-known measure of inequality, the empirical variance of the logs of the data: It is easy to construct examples where income is shifted from rich to poor and inequality *increases* (Cowell 1988).*

Another seemingly obvious axiom is symmetry: For any permutation matrix P, we should have

(A2) $I(x) = I(xP)$.

Together, (A1) and (A2) amount to Schur convexity (and vice versa):

(A3) $x \leq_M y \Rightarrow I(x) \leq I(y)$.

These axioms, though disputable, are widely accepted. The next two constitute a watershed, separating leftist from rightist measures of inequality:

(A4) $I(\lambda x) = I(x)$ for all $\lambda > 0$ (rightist).
(A5) $I(x + \lambda e) = I(x)$ for all $\lambda > 0$ (leftist).

It is easily seen that the only functions $I: D \to \mathbf{R}$ satisfying (A1)–(A5) are constants, so if these are excluded there is no inequality measure that is both leftist and rightist at the same time.

As a compromise, one might in this context also settle for the Eichhorn-Bossert-Pfingsten axiom:

(A6) $I(x + \tau(\lambda x + (1 - \lambda)e)) = I(x)$ (see Eq. (11)).

---

*As this can happen only for very high incomes, where the very rich give to the not so very rich, the significance of this aberration is in practice much disputed, and this measure continues to be widely used.

Indices which satisfy (A6) remain constant when an income vector x is replaced by a weighted mean of x and $e = (1, \ldots, 1)$, so by adjusting weights one can fine-tune one's degree of social rightism or leftism as closely as desired.

Another, independent watershed is provided by the next two axioms, which separate measures of industry concentration from measures of the inequality of income and wealth:

(A7) $I(x, 0) = I(x)$ (i.e., appending zeros does not affect inequality).

(A8) $I(x, x, \ldots, x) = I(x)$ (the "population invariance": replicating a population m times does not affect inequality).

A prominent index satisfying (A7) but not (A8) is the Herfindahl coefficient

$$H(x) = \sum_{i=1}^{n} \left(\frac{x_i}{\sum(x_j)} \right)^2 \tag{22}$$

It is often used in legal disputes about merger activities (see, e.g., Finkelstein and Friedberg 1967), where (relatively) small firms do not matter for the degree of concentration in the market, and where the axiom (A8) would in fact constitute a liability.

Unlike (A4) and (A5), (A7) and (A8) are nontrivially compatible with each other and (A1)–(A4): consider the well-known Theil index

$$T(x) = \frac{1}{n} \sum_{i=1}^{n} \left(\frac{x_i}{\bar{x}} \right) ln \left(\frac{x_i}{\bar{x}} \right) \tag{23}$$

where $0 \cdot ln(0)$ is taken to be 0. However, most other measures of inequality satisfy either (A7) or (A8), but not both.

Another important set of axioms concerns decomposability: If we partition the x-vector into $x = (x^{(1)}, \ldots, x^{(m)})$, where $x^{(i)}$ is $n_i \times 1$ and $\Sigma n_i = n$, it appears desirable that we have

(A9) $I(x)$ is uniquely determined by $I(x^{(i)})$, n_i, and $\bar{x}^{(i)}$.

Or in words: Knowing inequality in the subgroups and the subgroups' size and mean suffices to determine overall inequality.

Many well-known inequality measures such as the Gini index, the Pietra index, or the logarithmic variance fail this test, sometimes by wide margins. For the Gini index, it can be shown that decomposability obtains if the subgroups do not overlap, but other indices are not decomposable even then.* For the logarithmic variance and the Pietra index, Cowell (1988) gives an example where, even when all

---

*At least not in the sense defined by (A9). However, as shown by Kakwani (1980, pp. 178–181), the Gini index can be computed from information about factor incomes. This issue of decomposition by factors versus decomposition by subgroups has engendered its own literature, which we do not have space to cover here.

subgroups are strictly ordered by income, it is possible to find a change in income such that (a) mean income in every group is constant, (b) inequality in *every* group goes up, but still (c) overall inequality goes down.

A stronger version of (A9) is additive decomposability, defined as

$$(A10) \quad I(x^{(1)}, \ldots, x^{(m)}) = \sum_{i=1}^{m} w_i I(x^{(i)}) + I(\bar{x}^{(1)} e^{(1)}, \ldots, \bar{x}^{(m)} e^{(m)})$$

where $e^{(i)}$ is an n_i-vector of ones, $w_i \geq 0$, and $\sum_{i=1}^{m} w_i = 1$. This means that overall inequality can be expressed as a weighted sum of within-group inequalities, plus between-group inequality, defined as the overall inequality that would obtain if there were no inequality within the groups.

Still stronger is what Foster (1983) calls Theil decomposability:

$$(A11) \quad w_i = \frac{n_i \bar{x}^{(i)}}{n \bar{x}};$$

i.e., the weights are the income shares of the groups. Not surprisingly, the Theil coefficient $T(x)$ from (23) is decomposable that way.

A final set of axioms refers to "transfer sensitivity" (Shorrocks and Foster 1987): These principles strengthen the Pigou-Dalton axiom (A3) by requiring that the reduction in inequality resulting from a Robin Hood transfer should ceteris paribus be larger, the poorer the recipient. Relying solely on (A3), one could have "a situation in which a millionaire made a small (regressive) transfer to a more affluent millionaire and a simultaneous large (progressive) transfer to the poorest person in society" (Shorrocks and Foster 1987, p. 485), but where the combined effect of these transfers is that inequality *increases*: while the regressive transfer increases inequality and the progressive transfer reduces inequality, the axiom (A3) by itself puts no restraint on the relative magnitude of these effects, so one needs an additional axiom to prohibit such eccentric behavior.

(A12) $I(y') \leq I(y)$ whenever y' and y differ from x by a Robin Hood transfer of the same size, with spender and recipient equal amounts apart, but where the recipient in y' is poorer.

This notion can also be formalized by defining the *sensitivity* of an inequality index $I(\cdot)$, evaluated at the components $x_{(i)}$ and $x_{(j)}$ of some vector x, as

$$S_I(x, i, j) = \lim_{\delta \to 0} \frac{I(x) - I(x_{\delta,i,j})}{\delta} \tag{24}$$

whenever this limit exists, where

$$x_{\delta,i,j} = (x_{(1)}, \ldots, x_{(i)} + \delta, \ldots, x_{(j)} - \delta, \ldots, x_{(n)})$$

For the Gini index, using (16), this is easily seen to be

$$S_G(x, i, j) = \frac{2}{n \sum (x_i)} (j - i) \tag{25}$$

i.e., the sensitivity depends only on the *rank* of the units involved in the transfer, not upon their incomes. In particular, this implies that equal transfers of incomes have most effect where the population is densest, which is usually in the center.

On the other hand, for the Atkinson family, sensitivity is easily seen to be proportional to $1/x_{(i)}^{\varepsilon} - 1/x_{(j)}^{\varepsilon}$, so sensitivity increases both with ε and with $x_{(i)}$, given $|x_{(j)} - x_{(i)}|$ and with $x_{(j)} - x_{(i)}$ given $x_{(j)}$: ceteris paribus the decrease in inequality is larger, the poorer the recipient and the larger the income gap between spender and recipient.

Given some set of axioms (often augmented by normalization restrictions on the range, or requirements concerning continuity and differentiability), the following questions arise: (1) Are the axioms consistent with each other? (2) Are all the axioms really necessary? (3) What do indices which satisfy these axioms look like?

The first question is usually answered by exhibiting some specific measure that satisfies all requirements. The second question is trickier. While some axioms are easily seen to be implied by others ((A3) by (A1) and (A2), (A9) by (A10), (A8) by (A11)), others are not: As Russell (1985) demonstrates, at least two of the requirements that Cowell (1980) imposes to characterize the CES class of inequality indices are already implied by the others, and this implication is anything but trivial to see. Usually, such questions of minimality are settled by exhibiting, for every axiom, at least one index that fails this test but satisfies the others (Eichhorn and Gehrig 1982).

The third question has generated a minor industry, producing results of the type: Any function $I: D \rightarrow \mathbf{R}$ with continuous first-order derivatives satisfying (A2), (A4), (A8) and (A10), plus $I(e) = 0$, can be expressed as a positive scalar multiple of some function

$$I_c(x) = \frac{1}{nc(c - 1)} \sum_{i=1}^{n} \left[\left(\frac{x_i}{\bar{x}} \right)^c - 1 \right] \tag{26}$$

with some $c \in \mathbf{R}$ (Shorrocks 1980). This class of indices has become known as the generalized entropy family; as special cases corresponding to $c = 1$ and $c = 0$ (defined as the limit of $I_c(x)$ as $c \rightarrow 1$ or $c \rightarrow 0$), it includes the Theil index $T(x)$ defined in (23) and another index proposed by Theil (1967):

$$T^*(x) = \frac{1}{n} \sum_{i=1}^{n} [\ln(\bar{x}) - \ln(x_i)] \tag{27}$$

In the same vein, Foster (1983) shows that an inequality index satisfies (A3), (A4), and (A11) if and only if it is a positive multiple of the Theil coefficient $T(x)$, and

Eichhorn (1988) shows that the only functions $I: D \to \mathbf{R}$ satisfying (A3), (A6), and $I(e) = 0$ are

$$I(x) = f\left(\frac{\lambda x + (1 - \lambda)e}{\lambda \bar{x} + (1 - \lambda)}\right) \tag{28}$$

with some Schur-convex function f such that $f(e) = 0$.

V. EMPIRICAL IMPLEMENTATION

The empirical application of measures of inequality raises various issues which set this branch of statistics apart from others. A first and minor problem is the proper inference from a sample to a larger population. As samples are typically large, or nonrandom, or populations rather small as in the context of industrial concentration, relatively little work has been done on this.*

Much more important, in particular in the context of income inequality, is the incompleteness of the data: Typically, income figures are available only for certain quantiles of the population, and the rather voluminous literature on inequality measurement with incomplete data can be classified according to the amount of additional information available.

For the case where only selected points of the Lorenz curve are given (i.e., fractions of total income received by fractions of the total (ordered) population), Mehran (1975) gives the most extreme Lorenz curves that are compatible with these points, in the sense that the resulting concentration areas are the smallest and the largest possible.

Obviously, the upper bound is attained by joining the observed points by straight lines, as in Figure 2. This is at the same time an upper bound for the true underlying Lorenz curve. Likewise, an obvious lower bound to the true underlying Lorenz curve is given by extending these lines, as again in Figure 2. However, as these line segments do not form a Lorenz curve, Mehran (1975) proposes tangents to the true Lorenz curve at the observed points such that the concentration area and thus the Gini index are maximized, and he gives a recursive algorithm to compute the tangents' slopes. Although this curve does not necessarily bound from below the true Lorenz curve, it gives the Lorenz curve which represents (in the sense of the Gini index) the most unequal distribution compatible with the data.

If, in addition to selected points of the Lorenz curve, we are given the interval means $\bar{x}^{(i)}$, the interval endpoints a_{i-1} and a_i, and the fraction of the population in

---

*See, for instance, Sendler (1979) on the asymptotic distribution of the Gini index and the Lorenz curve, or McDonald and Ransom (1981) for the effects of sampling variability on the bounds below.

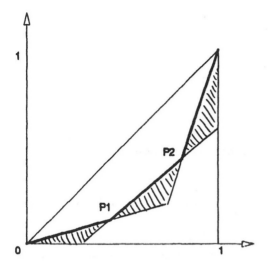

Figure 2 Given 2 nontrivial points P1 and P2, the true underlying Lorenz curve must pass through the shaded area.

the various nonoverlapping subgroups that compose the vector $x = (x^{(1)}, \ldots, x^{(m)})$, the bounds on the Gini index derived by Mehran can be sharpened as shown by Gastwirth (1972). The famous Gastwirth bounds[*] are based on the expression $G(x) = \Delta(x)/2\bar{x}$ and on the decomposition

$$\Delta(x) = \frac{1}{n^2} \sum_{i,j=1}^{m} n_i n_j |\bar{x}^{(i)} - \bar{x}^{(j)}| + \frac{1}{n^2} \sum_{i=1}^{m} n_i^2 \Delta^{(i)} \tag{29}$$

where $\Delta^{(i)}$ is Gini's mean difference for group i. In conjunction with

$$\Delta^{(i)} \leq \frac{2(\bar{x}^{(i)} - a_{i-1})(a_i - \bar{x}^{(i)})}{a_i - a_{i-1}} =: \tilde{\Delta}^{(i)} \tag{30}$$

this decomposition provides immediate bounds for the Gini index, with the lower one being attained for $\Delta^{(i)} = 0, (i = 1, \ldots, m)$, i.e., when there is no inequality within the groups, and the upper bound being attained when the observations in a given group are placed at both ends of the interval in a proportion such that the group mean is $\bar{x}^{(i)}$:

---

[*]In part already in Pizzetti (1955); see Giorgi and Pallini (1987).

$$G_L(x) = \frac{1}{2n^2\overline{x}} \sum_{i,j=1}^{m} n_i n_j |\overline{x}^{(i)} - \overline{x}^{(j)}|$$

$$\leq G(x) \leq G_U(x) = G_L(x) + \frac{1}{2n^2\overline{x}} \sum_{i=1}^{m} n_i^2 \tilde{\Delta}^{(i)} \tag{31}$$

Still sharper bounds can be obtained from information on symmetry or skewness of the distribution within groups: for instance, if distances between observations are increasing, as often happens in the higher brackets of income distributions, it can be shown that

$$\frac{2}{3}(\overline{x}^{(i)} - a_{i-1}) \leq \Delta^{(i)} \leq \frac{2(\overline{x}^{(i)} - a_{i-1})}{a_i - a_{i-1}} \left[a_i - \overline{x}^{(i)} - \frac{1}{3}(\overline{x}^{(i)} - a_{i-1}) \right] \tag{32}$$

A useful rule of thumb when no such extra information is available is provided by the "$\frac{1}{3}$–$\frac{2}{3}$" rule, as proposed by Champernowne in an unpublished manuscript and as formally derived in Cowell and Mehta (1982): Approximate the unknown true Gini index by $\frac{1}{3}G_L(x) + \frac{2}{3}G_U(x)$. As shown by Cowell and Mehta, this rule gives the *exact* Gini index when the observations within each group are split at $\overline{x}^{(i)}$ and spread evenly across the subintervals $(a_{i-1}, \overline{x}^{(i)})$ and $(\overline{x}^{(i)}, a_i)$ such that the group mean is preserved (the "split-histogram technique"), but it gives a good approximation to the true Gini index also for many other disturbances.

In the same vein, one obtains upper and lower bounds also for other measures of inequality, which, like the Gini index, are decomposable by population subgroups when the subgroups are nonoverlapping (basically the Gini index plus the generalized entropy family from (26); see Shorrocks 1984): The lower bound is given by a within-group distribution where everybody receives $\overline{x}^{(i)}$, and the upper bound is given by placing the observations at the ends of the group interval as above. However, if the measure is not the Gini index, the $\frac{1}{3}$–$\frac{2}{3}$ rule has to be adapted: the optimal approximation—in some sense—to the true inequality index is now $\frac{2}{3}$ lower bound $+\frac{1}{3}$ upper bound (Cowell and Mehta 1982).

Using similar techniques, Gastwirth (1975) extends the bounds above to arbitrary inequality measures that can be expressed as ratios of a measure of spread to $\overline{x}$ (which in view of (13) and (19) comprises the Gini and Pietra indices a special cases, but covers many other indices as well, possibly after suitable transformations).

As an alternative to bounds, it is often suggested to fit a theoretical Lorenz curve to selected values $L_i(p)$ either by interpolation, such that the fitted curve passes through the points $(p_i, L(p_i))$, or by some least-squares technique, where the curve is chosen from some given family, and to compute the desired measure of inequality from the fitted Lorenz curve.

When doing interpolation, Gastwirth and Glauberman (1976) propose cubic polynomials for the respective intervals whose derivatives at the endpoints match those of the true underlying Lorenz curve* (Hermite interpolation). The resulting function need not be convex, i.e., need not be a Lorenz curve itself, but it is easily seen (see, e.g., Schrag and Krämer 1993) that the Hermite interpolator is convex and monotone if in each interval

$$\frac{1}{3} \le \frac{\bar{x}^{(i)} - a_{i-1}}{a_i - a_{i-1}} \le \frac{2}{3} \tag{33}$$

i.e., if the group means stay clear of the endpoints of the intervals.

As to fitting parametric Lorenz curves, various families have been proposed, whose usefulness however seems much to be in doubt as the implied indices of inequality often violate the Gastwirth bounds. One such popular family, suggested by Rasche et al. (1980) is

$$L_{\alpha,\beta}(p) = (1 - (1 - p)^{\alpha})^{1/\beta} \tag{34}$$

where $0 < \alpha \le 1, 0 < \beta \le 1$. The implied Gini coefficient is

$$G(x) = 1 - \frac{2}{\alpha} B\left(\frac{1}{\alpha}, \frac{1}{\beta} - 1\right) \tag{35}$$

where $B(\cdot, \cdot)$ is the familiar Beta function, so this approach amounts to first fitting (34) to the observed points of a true Lorenz curve via some nonlinear least-squares technique and approximating the true Gini coefficient by (35).

However, as shown by Schader and Schmid (1994), the approximations thus obtained often fall outside the Gastwirth bounds (for curves of the form (34) and for other families as well), so the consensus is that such approximations do not work (see also Slottje 1990).

VI. MULTIDIMENSIONAL INEQUALITY AND POVERTY

There are various additional research areas which impinge on inequality, such as the welfare ranking of income distributions (Shorrocks 1983, Thistle 1989), the measurement of interdistributional inequality (Butler and McDonald 1987, Dagum 1987), or the parametric modeling of income distributions and the inequality orderings that are implied by the parameters (Chipman 1985, Wilfling and Krämer 1993), which I do not cover here. Rather, I pick the issues of multidimensional inequality and poverty to show how the concepts introduced above can be extended.

Take poverty. Although among certain sociologists, this is taken to be almost synonymous to inequality ("relative deprivation"), the majority consensus is that

---

*Remember that the Lorenz curve has the slope $x_{(i)}/\bar{x}$ in the interval $((i-1)/n, i/n)$, so its (left-hand) slope in p_i is $a_i/\bar{x}$.

poverty is something different, and needs different measures for quantification. Following Sen (1976), this is usually done by first identifying a poverty line z below which an income unit is considered poor, and then combining the poverty characteristics of different units into an overall measure of poverty. Two obvious candidates are the "head-count ratio"

$$H(x) = \frac{n^*}{n} \tag{36}$$

where n^* is the number of units below the poverty line, and the "income gap ratio"

$$I(x) = \frac{\sum_{i=1}^{n^*}(z - x_{(i)})}{n^* z} \tag{37}$$

i.e., the normalized per unit percentage shortfall of the poor. Individually, both measures have serious shortcomings—the head-count ratio is completely insensitive to the extent of poverty, and might even fall when income is transferred from a poor person to somebody not so poor, who thereby moves above the poverty line, and the income gap ratio takes no account of the numbers of the poor—but they can be combined into a measure with various desirable properties (Sen 1976):

$$S(x) = H(x)[I(x) + (1 - I(x))\tilde{G}(x)] \tag{38}$$

where $\tilde{G}(x) = G(x_{(1)}, \ldots, x_{(n^*)})$ is the Gini index of the incomes of the poor. This measure reduces to $H(x) \cdot I(x)$ if all the poor have the same income, and it increases whenever an income of a poor person is reduced (while incomes above the poverty line always receive uniform weights, entering only via the head-count ratio $H(x)$).

For large n^*, the coefficient $S(x)$ can be shown to be almost identical to

$$\tilde{S}(x) = \frac{2}{(n^* + 1)nz} \sum_{i=1}^{n^*}(z - x_{(i)})(n^* + 1 - i) \tag{39}$$

i.e., it is a weighted sum of the income gaps of the poor, with weights increasing as income gaps increase, which at the same time points to a serious shortfall of this measure of poverty; it is based solely on the *ranks*, not on the distances of the incomes of the poor, so there has been an enormous literature in the wake of the seminal paper by Sen (1976), surveyed in Seidl (1988), which goes on from here.

Comparatively little work has been done on the second important issue of multidimensional inequality. While it is widely recognized that a single attribute such as income is often not sufficient to capture the phenomenon whose inquality is to be determined, the statistician's toolbox is almost empty here.

There is a short chapter on multivariate majorization in Marshall and Olkin (1979), where majorization among n-vectors x and y is extended to majorization among $m \times n$ matrices X and Y: By definition, $X \geq_M Y$ if and only if $Y = XP$ with some doubly stochastic matrix P. When $m = 1$, this is equivalent to $Y =$

$XT_1, T_2, \ldots, T_L$ with finitely many matrices of the form (2), but when $m \geq 2$ and $n \geq 3$, the latter condition is more restrictive, as simple examples show (Marshall and Olkin 1979, p. 431). Therefore I denote the latter preordering by "$\geq_T$."

As shown by Rinott (1973), a differentiable function $f: \mathbf{R}^{m \times n} \rightarrow \mathbf{R}$ respects the preordering $\geq_T$ if and only if $f(x) = f(x\tilde{P})$ for all permutation matrices $\tilde{P}$, and if for all $j, k = 1, \ldots, n$,

$$\sum_{i=1}^{m} (x_{ik} - x_{ij})[f_{ik}(x) - f_{ij}(x)] \geq 0 \tag{40}$$

where $f_{ik}(x) = \partial f / \partial x_{ik}$ evaluated at X. Similar to the univariate case, this condition could presumably be used to suggest new indices and screen old indices of inequality. However, as the rationale for the $\geq_T$ preorder is much less compelling for $m > 1$ as it is for $m = 1$ (with data matrices, even fewer pairs will in general be ordered in a given application than if we had only vectors), the condition (40) does not immediately suggest a convenient index of multivariate inequality.

Similar to the univariate case, these majorization orderings can be embedded in a more general order, which in turn is based on a multivariate generalization of the Lorenz curve (Koshevoy and Mosler 1996). For $X \in \mathbf{R}_+^{m \times n}$, define the *Lorenz zonoid* of X as the convex hull of the set

$$LZ(X) := \left\{ \sum_{i=1}^{n} \delta_i \left(\frac{1}{n}, \frac{x_{1i}}{\sum_{j=1}^{n}(x_{1j})}, \ldots, \frac{x_{mi}}{\sum_{j=1}^{n}(x_{mj})} \right) \mid 0 \leq \delta_i \leq 1 \right\} \tag{41}$$

and define

$$X \geq_L Y \Leftrightarrow LZ(x) \supseteq LZ(Y) \tag{42}$$

For $m = 1$, this boils down to the univariate Lorenz order, as $LZ(X)$ is then the subset of $\mathbf{R}^2$ enclosed by the Lorenz curve $L_X(p)$ and the "dual Lorenz curve" $\tilde{L}_X(p) := 1 - L_X(p - 1)$ (the area of which is equal to the Gini index $G(x)$). For $m > 1$, the Lorenz zonoid is a convex subset of the unit cube in $\mathbf{R}^{m+1}$.

The multivariate Lorenz ordering defined by (42) allows matrices of different means and row dimensions to be compared. It is further related to univariate Lorenz dominance by the fact that $X \geq_L Y$ if and only if the generalized Lorenz curve of $d'X$ is below the generalized Lorenz curve of $d'Y$ for all coefficient vectors d such that $d_i \geq 0$ and $\sum(d_i) = 1$ (Koshevoy and Mosler 1996, Theorem 3.1).

Other generalizations from univariate to multivariate concepts of inequality exploit the parallel between inequality and choice under uncertainty (Kolm 1977). Atkinson and Bourguignon 1982), or the obvious relationship between multivariate inequality and the decomposition of inequality by factor components (Shorrocks 1988, Maasoumi 1986, Rietveld 1990), but as this field has not yet reached the maturity for a useful survey, I had better close my survey here.

ACKNOWLEDGMENTS

I am grateful to Christian Kleiber, Sonja Michels, Karl Mosler, Andreas Pfingsten, Friedrich Schmid, Andreas Stich, Mark Trede, Thorsten Ziebach, and an unknown referee for helpful discussion and comments.

REFERENCES

Arnold, B. C. (1987), *Majorization and the Lorenz Order: A Brief Introduction*, Springer, Berlin.
Atkinson, A. B. (1970), On the Measurement of Inequality, *Journal of Economic Theory*, 21, 244–263.
Atkinson, A. B. and F. Bourgignon (1982), The Comparison of Multi-Dimensional Distribution of Economic Status, *Review of Economic Studies*, 49, 183–201.
Berrbi, Z. M. and J. Silber (1987), Dispersion, Asymmetry and the Gini Index of Inequality, *International Economic Review*, 28, 331–338.
Blackburn, M. L. (1989), Interpreting the Magnitude of Changes in Measures of Income Inequality, *Journal of Econometrics*, 42, 21–25.
Bossert, W. and A. Pfingsten (1989), Intermediate Inequality: Concepts, Indices and Welfare Implications, *Mathematical Social Sciences*, 19, 117–134.
Bresciani-Turroni, C. (1910), Di un Indice Misuratore Della Disugualianza Nella Distribuzione Della Richezza, *Studi in Onore di Biagio Brugi*, Palermo, 793–812.
Butler, R. J. and J. B. McDonald (1987), Interdistributional Income Inequality, *Journal of Business and Economic Statistics*, 5, 13–18.
Chipman, J. (1985), The Theory and Measurement of Income Distribution, *Advances in Econometrics*, 4, 135–165.
Cowell, F. A. (1980), On the Structure of Additive Inequality Measure, *Review of Economic Studies*, 47, 521–531.
Cowell, F. A. (1988), Inequality Decomposition: Three Bad Measures, *Bulletin of Economic Research*, 40, 309–312.
Cowell, F. A. and F. Mehta (1982), The Estimation and Interpolation of Inequality Measures, *Review of Economic Studies*, 44, 273–290.
Dagum, C. (1987), Measuring the Economic Affluence between Populations of Income Receivers, *Journal of Business and Economic Statistics*, 5, 5–12.
Dalton, H. (1920), The Measurement of Inequality of Incomes, *Economic Journal*, 30, 348–361.
Dasgupta, P., A. K. Sen, and D. Starrett (1973), Notes on the Measurement of Inequality, *Journal of Economic Theory*, 180–187.
Ebert, U. (1988), A Family of Aggregative Compromise Inequality Measures, *International Economic Review*, 29, 363–376.
Eichhorn, W. (1988), On a Class of Inequality Measures, *Social Choice and Welfare*, 5, 171–177.
Eichhorn, W., H. Funke, and W. Richter (1984), Tax Progression and Inequality of Income Distribution, *Journal of Mathematical Economics*, 13, 127–131.

Eichhorn, W. and W. Gehrig (1982), Measurement of Inequality in Economics, in B. Korte (ed.), *Modern Applied Mathematics*, North-Holland, Amsterdam, 657–693.

Elteto, O. and E. Frigyes (1968), New Income Inequality Measures as Efficient Tools for Causal Analysis and Planning, *Econometrica*, 35, 383–396.

Fellman, J. (1976), The Effect of Transformations on the Lorenz Curve, *Econometrica*, 44, 823–824.

Finkelstein, M. O. and R. M. Friedberg (1967), The Application of an Entropy Theory of Concentration to the Clayton Act, *Yale Law Journal*, 76, 677–717.

Foster, J. E. (1983), An Axiomatic Characterization of the Theil Measure of Income Inequality, *Journal of Economic Theory*, 31, 105–121.

Gastwirth, J. L. (1972), The Estimation of the Lorenz Curve and Gini Index, *Review of Economics and Statistics*, 54, 306–316.

Gastwirth, J. L. (1975), On Estimating a Family of Measures of Economic Inequality, *Journal of Econometrics*, 3, 61–70.

Gastwirth, J. L. and M. Glauberman (1976), The Interpolation of the Lorenz Curve and Gini Index From Grouped Data, *Econometrica*, 44, 479–483.

Giogi, G. M. (1990), Bibliographic Portrait of the Gini Concentration Ration, *Metron*, 48, 183–221.

Giogi, G. M. and A. Pallini (1987), About a General Method for the Lower and Upper Distribution-Free Bounds on Gini's Concentration Ratio From Grouped Data, *Statistica*, 47, 171–184.

Hardy, G. H., J. E. Littlewood, and G. Polya (1934), *Inequalities*, Cambridge University Press, London.

Kakwani, N. C. (1980), *Income, Inequality and Poverty*, Oxford University Press, Oxford.

Kakwani, N. C. (1984), Welfare Ranking of Income Distributions, *Advances in Econometrics*, 3, 191–213.

Kolm, S. Ch. (1976), Unequal Inequalities I, *Journal of Economic Theory*, 12, 416–442.

Kolm, S. Ch. (1977), Multidimensional Egalitarianism, *Quarterly Journal of Economics*, 91, 1–13.

Kondor, Y. (1971), An Old-New Measure of Income Inequality, *Econometrica*, 39, 1041–1042.

Koshevoy, G. and K. Mosler (1996), The Lorenz Zonoid of a Multivariate Distribution, *Journal of the American Statistical Association*, 91, 873–882.

Kuznets, S. (1955), *Six Lectures on Economic Growth*, The Free Press, Glencoe.

Lambert, P. J. (1993), *The Distribution and Redistribution of Income*, Manchester University Press, Manchester.

Lerman, L. and S. Yitzhaki (1984), A Note on the Calculation and Interpretation of the Gini-Index, *Economics Letters*, 15, 363–368.

Maasoumi, E. (1986), The Measurement and Decomposition of Multi-Dimensional Inequality, *Econometrica*, 54, 991–997.

Marshall, A. W. and I. Olkin (1979), *Inequalities: Theory of Majorization and Its Applications*, Academic Press, New York.

McDonald, J. B. and M. Ransom (1981), An Analysis of the Bounds for the Gini Coefficient, *Journal of Econometrics*, 17, 177–188.

Mehran, F. (1975), Bounds on the Gini-Index Based on Observed Points of the Lorenz Curve, *Journal of the American Statistical Association*, 70, 64–66.

Mehran, F. (1976), Linear Measures of Income Inequality, *Econometrica*, 44, 805–809.

Mosler, K. (1994), Majorization in Economic Disparity Measures, *Linear Algebra and Its Applications*, 199, 91–114.

Moyes, P. (1987), A New Concept of Lorenz Domination, *Economics Letters*, 23, 203–207.

Newbery, D.A. (1970), Theorem on the Measurement of Inequality, *Journal of Economic Theory*, 2, 264–266.

Pfingsten, A. (1988a), New Concepts of Lorenz Domination and Risk Aversion, *Methods of Operations Research*, 59, 75–85.

Pfingsten, A. (1988b), Progressive Taxation and Redistributive Taxation: Different Labels for the Same Product?, *Social Choice and Welfare*, 5, 235–246.

Pietra, G. (1914/1915), Della Relazione Tra Gli Indice di Variabilita, *Atti del Realo Istituto Veneto di Scienze*, 775–792; 793–804.

Pigou, A. C. (1912), *Wealth and Welfare*, Macmillan, London.

Pizzetti, E. (1955), Osservazioni Sul Calculo Aritmetica del Rapporto di Concentrazione, *Studi in onore di Gaetano Pietra*, Capelli, Bologna.

Pyatt, G. (1976), On the Interpretation and Disaggregation of Gini-Coefficients, *Economic Journal*, 86, 243–255.

Rasche, R. H., J. Gaffney, A. Y. C. Koo, and N. Obst (1980), Functional Forms for Estimating the Lorenz Curve, *Econometrica*, 48, 1061–1062.

Ricci, U. (1916), L'indice di Variabilità e la Curva de Redditi, *Giornali Degli Economisti e Reivsta de Statistica*, 3,177–228.

Rietveld, P. (1990), Multidimensional Inequality Comparisons, *Economics Letters*, 32, 187–192.

Rinott, Y. (1973), Multivariate Majorization and Rearrangement Inequalities with Some Applications to Probability and Statistics, *Israel Journal of Mathematics*, 15, 60–67.

Russell, R. (1985), A Note on Decomposable Inequality Measures, *Review of Economic Studies*, 52, 347–352.

Schader, M. and F. Schmid (1994), Fitting Parametric Lorenz Curves to Grouped Income Distributions—A Critical Note, *Empirical Economics*, 19, 361–370.

Schrag, H. and W. Krämer (1993), A Simple Necessary and Sufficient Condition for the Convexity of Interpolated Lorenz Curves, *Statistica*, 53, 167–170.

Schutz, R. R. (1951), On the Measurement of Income Inequality, *American Economic Review*, 41, 107–122.

Seidl, Ch. (1988), Poverty Measurement: A Survey, Welfare and Efficiency, in D. Bös, M. Rose, and Ch. Seidl (eds.), *Public Economics*, Springer, Berlin.

Sen, A. (1976), Poverty: An Ordinal Approach to Measurement, *Econometrica*, 44, 219–231.

Sendler, W. (1979), On Statistical Inference in Concentration Measurement, *Metrika* 26, 109–122.

Sheshinski, E. (1972), Relation between a Social Welfare Function and the Gini Index of Income Inequality, *Journal of Economic Theory*, 4, 98–100.

Shorrocks, A. F. (1980), The Class of Additively Decomposable Inequality Measures, *Econometrica*, 48, 613–625.

Shorrocks, A. F. (1983), Ranking Income Distributions, *Economica*, 50, 3–17.

Shorrocks, A. F. (1984), Inequality Decomposition by Population Subgroups, *Econometrica*, 52, 369–1385.

Shorrocks, A. F. (1988), Aggregation Issues in Inequality Measurement, in W. Eichhorn (ed.), *Economics*, Physica, Heidelberg.

Shorrocks, A. F. and J. E. Foster (1987), Transfer Sensitive Inequality Measures, *Review of Economic Studies*, 54, 486–497.

Slottje, D. J. (1990), Using Grouped Data for Constructing Inequality Indices—Parametric vs. Nonparametric Methods, *Economics Letters*, 32, 193–197.

Theil, H. (1967), *Economics and Information Theory*, North-Holland, Amsterdam.

Thistle, P. D. (1989), Ranking Distributions with Generalized Lorenz Curves, *Southern Economic Journal*, 56, 1–2.

von Bortkiewicz, L. (1931), Die Disparitätsmaße der Einkommensstatistik, *Bulletin de l'Institut International de Statistique*, 25, 189–298.

Wilfling, B. and W. Krämer (1993), The Lorenz Ordering of Singh-Maddala Income Distributions, *Economics Letters*, 43, 53–57.

Yitzhaki, S. (1983), On an Extension of the Gini-Index, *International Economic Review*, 24, 617–628.

3
Poor Areas

Martin Ravallion
World Bank, Washington, D.C.

> As China's economic miracle continues to leave millions behind, more and more Chinese are expressing anger over the economic disparities between the flourishing provinces of China's coastal plain and the impoverished inland, where 70 million to 80 million people cannot feed or clothe themselves and hundreds of millions of others are only spectators to China's economic transformation. *The New York Times*, December 27, 1995, p. 1.

China is not unusual; almost all countries have their well-recognized "poor areas," in which the incidence of absolute poverty is unusually high by national standards. In China, there is high poverty incidence in rural areas of the southwest and northwest (the "inland" areas referred to in the quotation). Similar examples in other countries include some of the eastern Outer Islands of Indonesia, parts of northeastern India, northwestern and southern rural areas of Bangladesh, much of northern Nigeria, the rural Savannah in Ghana, the northeast of Brazil, and many other places.

We would hope, and under certain conditions expect, that the growth process will help these poor areas catch up. But that does not appear to be happening in some countries. Figure 1 illustrates the divergence over time between the relatively well off *and* more rapidly growing coastal areas of China and the lagging inland areas. The figure plots the aggregate rate of consumption growth at county level in southern China 1985–1990 against the initial county mean wealth. The data cover 119 counties spanning a region from the booming coastal province of Guangdong through to the poor inland areas of Guizhou.* There is a positive regression coefficient, sug-

---

*The figure is reproduced from Ravallion and Jalan (1996). It is based on a panel of farm-household level data for rural areas in four provinces of southern China. The data cover 4700 households living in 119 counties. The consumption measure is comprehensive, in that it includes imputed values (at local market

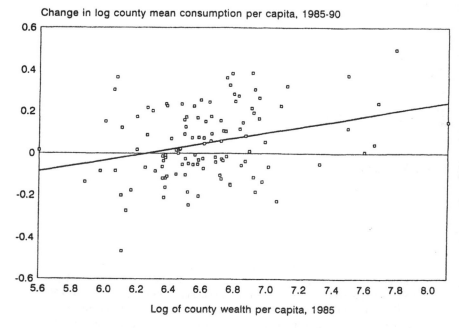

Figure I   Consumption growth by county in rural South China.

gesting divergence, and it is significant (at the 1% level). Initially wealthier counties tended to have higher subsequent rates of consumption growth.

Nor is China the only country in which poor areas appear to persist in spite of robust economic growth; for example, the eastern Outer Islands of Indonesia appear to have shared rather little in that country's sustained (and generally pro-poor) economic growth since 1970. It seems that there is a degree of persistence in the economic geography of poverty; indeed, a generation or more ago, the above list of "poor areas" by country would probably have looked pretty similar.

As the opening extract suggests, there are widespread concerns about poor areas, particularly when they persist amidst robust aggregate economic growth. In assessing the social impact of economic growth or growth-oriented economic reform, economists have traditionally focused on the impact on one or more measures of

---

prices) of consumption from own production plus the current service flows from housing and consumer durables. The data also include a seemingly complete accounting of all wealth including valuations of all fixed productive assets, cash, deposits, housing, grain stock, and consumer durables. The data are discussed at length in Chen and Ravallion (1996).

social welfare, including various measures of aggregate poverty. Yet, it appears that impacts are typically diverse among the poor, and in the society as a whole; some lose and some gain from economy-wide changes. This can be important to know, if only to better understand the political economy of growth and reform, though policymakers may well also make the (normative) judgment that a premium should be attached to more "balanced" growth. Studying these diverse impacts may also hold important keys to our understanding of the growth process itself and to a variety of questions often asked by policymakers, such as what is the best policy response to the problem of why some subgroups are lagging.

So why do some people, and in particular some regions, do so much better than others in a growing economy? It turns out that most of the standard tools of analysis used in studying poverty, distribution, and growth are ill-equipped to answer this question. After reviewing those tools, this chapter suggests some new tools of empirical analysis that may offer a better chance of answering it. We look at the dynamics of the geography of poverty from a microlevel to help understand the way various initial conditions and exogenous shocks impinge on *household-level* prospects of escaping poverty over time. While we note the links to various strands of theoretical and empirical economics, the chapter is not a survey. Rather it tries to be forward-looking on a set of seemingly important research questions, to explore how future research might better address them.

This is also an issue of considerable relevance, as the chapter will emphasize. The empirical approach outlined here would appear to entail a substantial expansion in the number of policy-relevant variables which are included in microempirical models of poverty. Past interventions in poor areas are amongst those variables. Faced with lagging regions amidst overall growth, governments and donors are regularly called upon to do something about these lagging poor areas. Area-based interventions are now found in most countries.* How much impact do such interventions have on living standards? To answer this we must be able to assess what would have happened to living standards in the absence of the interventions. It should not be as-

---

*For example, on recognizing the problem of lagging rural areas, China introduced a large antipoverty program in 1986 which declared that 272 (rural) counties were "national-poor counties," and targeted substantial aid to those counties. The extra aid took the form of subsidized credit for village-level projects (provided at well below market rates of interest), funding for public works projects (under "food-for-work" programs), and direct budgetary support to the county government. This national poor area program is the main direct intervention in the government of China's current poverty reduction policy (Leading Group 1988, World Bank 1992, Riskin 1994). Again China is not unusual. The World Bank has assisted over 300 area development projects since the early 1950s spread over all regions; most of these projects were designed to develop a selected rural area for the benefit of poor people. Other agencies, such as the International Fund for Agricultural Development, also provide substantial support for such programs (Jazairy et al. 1992). There has been a recent resurgence of interest in such programs in the World Bank and elsewhere.

sumed that such schemes will even entail net *gains* to poor people; by acting against the flow of factor mobility from low to high productivity areas, it could be argued that such interventions actually make matters worse in the longer term. Depending on how the economy works in the absence of intervention—the nature of the technology, preferences, and any constraints on factor mobility—a poor area program could entail either a net benefit or net cost to poor people.

The paper argues that the geographic variation in both initial conditions and the evolution of living standards over time offers scope for disentangling the effects of poor-area programs from other factors. Even within poor countries, geographic areas differ widely in their endowments of various aspects of "geographic capital," including locally provided public services and access to area-specific subsidies. These differences are both geoclimatic and the outcomes of past policies and projects. There is typically also a spatial variance in poverty indicators. The spatial variation in both the incidence of poverty and in area characteristics offers hope of better understanding why we see poor areas, what can be done to help them, and how well past efforts have performed. By exploiting this spatial variation, we should be in a better position to understand what role the lack of geographic capital plays in creating poor areas, versus other factors including residential differentiation, whereby people who lack "personal capital" end up being spatially concentrated.

The following section explains the motivation for studying the problem of "poor areas." Section II discusses the "standard" empirical tools found in practice, relying on either static micromodels or aggregate dynamic models. Section III discusses a micromodeling approach and its links with recent work in economics on the determinants of economic growth. Some potential lessons for policy are described in Section IV.

I.   MOTIVATION

A.   Why Do We See Unusually Poor Areas?

The starting point for assessing the pros and cons of poor-area policies is an understanding of why we see poor areas in the first place. Among economists and policymakers, a common explanation of poverty is based on an *individualistic model* in which poverty arises from low household-level endowments of privately held productive resources, including human capital, albeit with important links to the regional and macroeconomy, notably through wages and prices. This view is epitomized in the familiar human-capital earnings functions. The dynamic version is the standard neoclassical growth model, the microfoundation of which assumes atomistic agents linked only through trade at common prices. If one believes this model, then poor areas presumably arise because people with poor endowments tend to live together. Area differences in access to (for example) local public goods might still be allowed in such a model, but as long as there is free mobility they will not matter to the welfare of an individual household, which will depend on its a-spatial exogenous attributes. The types of antipoverty policies influenced by this model emphasize raising the endowments of poor people, such as by enhanced access to schooling.

Regional divergence is still possible in such a model. If there are increasing returns to scale in private production inputs, then initially better-off areas, with better endowments of private capital, will tend to see subsequently higher rates of growth. This is the essence of the view of regional growth that one finds in the writings of Myrdal (1957), Hirschman (1958), and others since (see the review in Richardson and Townroe 1986). By this interpretation, persistently poor areas, and divergence from wealthier areas, reflect the nature of the technology and the geography of natural resource endowments. There may still be a case for targeting poor areas, but it would be a *redistributive* case, and it would imply a trade-off with the overall rate of economic growth.

The individualistic model does not attach any causal significance to man-made spatial inequalities in geographic capital—the set of physical and social infrastructure endowments held by specific areas. Indeed, with free mobility, the individualistic model predicts that household welfare will *only* depend on private, mobile, endowments, and other exogenous attributes of the household. Against this view, one can postulate a *geographic model* in which individual poverty depends heavily on geographic capital and mobility is limited. By this view, the marginal returns to a given level of schooling, or a loan, depend substantially on where one lives, and limited factor mobility entails that these differences persist. Relevant geographic factors might include local agroclimatic conditions, local physical infrastructure, access to social services, and the stock of shared local knowledge about agroclimatic conditions and about the technologies appropriate to those conditions. It is not implausible that some or all of these geographic factors alter the returns to investments in private capital. As I argue later, it is likely that they will also entail *increasing* returns to geographic capital when there are *nonincreasing* returns to private production inputs. Thus it might well be that people are being left behind by China's growth process precisely because they live in poor areas; given their private endowments, they would do better in China's coastal areas.

If this model is right, then the policies called for will entail either public investment in geographic capital or (under certain conditions, discussed below) proactive efforts to encourage migration, and such policies need not entail a trade-off with the overall rate of growth. That will depend on the precise way in which differences in geographic capital impact on the marginal products of private capital and, hence, the rate of growth. That is an empirical question.

Neither model provides a complete explanation for poor areas. The individualistic model begs the questions of why individual endowments differ persistently and why residential differentiation occurs. The geographic model begs the questions of why community endowments differ and why mobility is restricted. But, as I will argue, knowing which model is right, or what the right hybrid model looks like, can provide valuable information for policy.

B.   Past Work

What do we already know about poor areas that might throw light on which of these models is most relevant? There has been a vast amount of empirical research testing the individualistic model, including human-capital earnings functions and similarly motivated income and consumption determination models estimated on microdata (Section II.B). This research has often assumed that the individual, "private capital," model holds. While sometimes spatial variables are added, this is done in an ad hoc way. At the same time, there is also a large, but mostly independent, literature on economic geography and regional science which has emphasized the importance of spatial effects on the growth process (for a survey see Richardson and Townroe 1986). The individualistic model has not been tested rigorously against the geographic model, in an encompassing framework which would allow the two models to fight it out.

But there is evidence of spatial effects in the processes relevant to creating and perpetuating poverty. The evidence of spatial effects comes from a variety of sources, including the following.

1.   "Poverty profiles" (decompositions of aggregate poverty measures by subgroups of a population, including area of residence) typically contain evidence of seemingly significant spatial differences in poverty incidence or severity. However, typical poverty profiles do not allow one to say whether it is the individualistic model or the geographic model that is producing these spatial effects. In the (far fewer) cases in which suitable controls were used, spatial effects did appear to persist (van de Walle 1995, Jalan and Ravallion 1996, Ravallion and Wodon 1997).

2.   In some of the settings in which there are persistently poor rural areas there does not appear to be much mobility among rural areas (though more so from rural to urban areas). In some cases (such as China) mobility has been deliberately restricted, but intrarural mobility seems uncommon elsewhere (such as in much of South Asia, though exceptions exist, such as seasonal migration of agricultural labor). Then the individualistic model immediately seems implausible; for how did the residential differentiation come about with rather little mobility?

3.   The literature on the diffusion process for new farm technologies has emphasized local community factors, including the demonstration effect of the presence of early adopters in an area, and there is some supportive evidence for India in Foster and Rosenzweig (1995).

4.   There is also evidence for India that areas with better rural infrastructure grow faster and that infrastructure investments tend to flow to areas with good agroclimatic conditions (Binswanger et al. 1993). The type of data used has not, however, allowed identification of external effects (discussed further in Section III).

5.   Gains from the geographic concentration of some industries (arising from scale economies and limited factor mobility; see Krugman 1991) could also entail spatial effects in a growth process. This could be magnified by intersectoral linkages. For example, there is evidence for China (Sengupta and Lin 1995) that higher non-farm-sector growth—possibly policy induced—brings external benefits to the traditional farm sector through improved technologies and management. There is evidence of positive external effects of higher density of economic activity on productivity across states of the United States (Ciccone and Hall 1996).

6.   There is also some evidence of human-capital spillover effects. In the United States, the neighborhood where a child was raised appears to influence her schooling performance and adult wages (Borjas 1995, Datcher 1982, Wilson 1987, Case and Katz 1991).

7.   Informal risk-sharing arrangements within poor communities entail that individual consumption depends in part on the community's aggregate consumption; there is supportive evidence for India in Townsend (1991), though also see Ravallion and Chaudhuri (1997).

8.   Preferences, including discount rates, may well be formed within communities, entailing spatial effects on (among other things) the growth process. This is hard to test, but there is evidence for Indonesia of spatial autocorrelation in consumer demand behavior (Case 1991).

9.   There is also some evidence of spatial autocorrelation in growth processes at the country level; the higher the average growth rate of a country's neighbors the higher is its own growth rate ceteris paribus (e.g., Easterly and Levine 1995).

All this is suggestive, but we are still a long way from a good understanding of why poor areas exist and persist. The answers could have great bearing on development policy. If the process of escaping poverty involves strong spatial effects then there may be large benefits from policies and projects which are targeted to poor areas, even if they are not targeted to households with poor endowments per se. It may also mean that, without (possibly substantial) extra resources, or greater mobility, the poor may be caught in a spatial poverty trap. To have any chance of success, an antipoverty policy may have to break the community-level constraints on escaping poverty, by public investment or encouraging migration.

C.   Poor-Area Policies

What are the policy options in assisting poor areas? Two broad types of area-based policy intervention can be identified which are aimed (explicitly or implicitly) at poverty reduction: one is *geographic targeting* of subsidies, taxes, or public investments, and the other is *migration policy.*

Geographic targeting of antipoverty schemes has been popular, though so far the assessments of poverty impacts have largely ignored dynamic effects. The attraction of this policy option for targeting stems from the existence of seemingly substantial regional disparities in living standards in many developing countries. Place of residence may thus be a useful indicator of poverty. Local governments provide an administrative apparatus. However, some assessments of the potential for this policy instrument to have more than a minor impact on aggregate poverty have not been encouraging. While regional poverty profiles for LDCs typically show large geographic disparities, regional targeting still entails a leakage of benefits to the nonpoor in poor regions, and a cost to the poor in rich regions. And even with marked regional disparities, these effects can wipe out a large share of the aggregate gains to the poor.*

A deeper analysis of why poor areas exist could have a number of implications for all such geographically targeted development policies and projects. The case for poor-area interventions depends on precisely why we observe poor areas. Here it can be important to know just how much of the poverty one sees is attributable to area-specific attributes versus personal attributes which may best be dealt with through a-spatial programs. If poor areas arise from residential differentiation through mobility (as the individualistic model would suggest), then such mobility will clearly limit the scope for targeting on the basis of where people live. On the other hand, if place of residence does matter even when one controls for personal characteristics then the case for area-based programs and investments could be greatly strengthened.

Past assessments of geographic targeting have been essentially static, though the limitations of this view have been noted by Ravallion (1993). Yet without successful intervention, the competitive equilibrium in the geographic-capital model would be unlikely to achieve a Pareto optimum, given the pervasive externalities. The geographic model could thus imply dynamic efficiency gains from investing in geographic capital; that will depend on the way in which geographic capital affects the marginal returns to investment in private capital. If borne out by the evidence, the geographic model may thus lead one to question any presumption that targeting poor areas would necessarily have an aggregate growth cost.

A deeper understanding of why we see poor areas is also needed to inform choices about the specific types of poor-area programs needed. For example, a common debating point in formulating poor-area programs is the priority to be given to basic health and education versus credit and physical infrastructure. Should policy be focusing on education or should it be roads? Is a package on interventions called for, as in the "Integrated Rural Development Programs"? This is also a matter of the

---

*For India, Datt and Ravallion (1993) consider the effects on poverty of pure (nondistortionary) transfers among states, and between rural and urban areas. They find that the *qualitative* effect of reducing regional/sectoral disparities in average living standards generally favors the poor. However, the *quantitative* gains are small. For example, the elimination of regional disparities in the means while holding *intra*regional inequalities constant, would yield only a small reduction in the proportion of persons below the poverty line, from an initial 33% to 32%. Also see Ravallion (1993) for Indonesia.

nature and extent of the *interaction effects* among area characteristics as they affect living standards and their evolution over time.

There are also potentially important implications for economic evaluations of the dynamic gains from area-specific interventions to reduce poverty. With little mobility, living in a designated poor area can be taken as exogenous to household choices. However, the existence of spatial externalities may well entail that the *growth path* of future household living standards is dependent on the same area characteristics which influence the public decision to declare the community poor. The problem is essentially one of omitted-variable bias when there is state dependence in the growth process. For example, a low endowment of local public goods may simultaneously induce a lower rate of growth *and* a higher probability of the community being declared poor. Unless this is accounted for, the value to households of living in an area which is targeted under a poor-area program will be underestimated (Jalan and Ravallion 1996).

Assessing the case for all such interventions requires a deeper understanding of how poor areas came to exist. That understanding can also inform other areas of policy. The case for proactive migration policies may be strengthened if one finds that there are strong geographic factors in the creation and perpetuation of poor areas. That will depend in part on the precise nature of those factors. If the geographic effect is largely explicable in terms of physical infrastructure endowments (as at least the proximate cause), then the case for migration policies will be strengthened; migrants who go to better-endowed areas will gain, and those left behind will also gain if there is less crowding of the existing infrastructure in the poor area. But if the geographic factors are largely *social* (to do with social capital, or the spillover effects of local endowments of human capital), then the migration policy may make matters even worse for those left behind. (This is often said about the effects of suburbanization or inner-city areas in the United States.)

Motivated by the above discussion, the following sections will discuss various approaches to empirical modeling which might prove fruitful in understanding the economic geography of poverty so as to inform these difficult policy issues.

II. STATIC MICROMODELS AND DYNAMIC AGGREGATE MODELS

A. Static Poverty Profiles

Standard empirical models of poverty are estimated on single cross-sectional sample surveys. To illustrate, suppose we have a single cross-sectioned sample of households giving consumption C_{it} for household i at date t. If one regresses this (or its log) against a set of location dummy variables $d_1, d_s, \ldots, d_m$ for m regions, then one will retrieve an estimate of what can be termed the "unconditional geographic poverty profile":

$$C_{it} = \alpha + \sum_j \beta_{jt} d_{ij} + u_{it} \tag{1}$$

where α and β_{jt} are parameters to be estimated and u_{it} is an unobserved error term. If we were to repeat this for other years, then we could also see how the geographic poverty profile changes over time.

This type of geographic poverty profile is found in (for example) almost every poverty profile found in the World Bank's Country Poverty Assessments (the regression may not be run, but in this case a bivariate cross-tab of the poverty measure by region is just another way of running the regression). While useful for some purposes (including geographic targeting), this type of poverty profile tells us nothing about why there is more poverty in one place than another. It may be, for example, that people with little education tend to live in certain places. Then if one controlled for education, the regional dummy variables would become insignificant.

Extending this logic, it is becoming common practice to estimate more complex multivariate models in which a set of household characteristics are added, represented by the vector $\mathbf{x}_i$, giving the augmented regression

$$C_{it} = \alpha + \sum_j \beta_{jt} d_{ij} + \gamma' \mathbf{x}_{it} + u_{it} \tag{2}$$

If the geographic dummy variables remain significant in this augmented regression, and one has controlled for all relevant household characteristics, then we can conclude that there are location-specific factors at work independently of personal characteristics. Of course, the personal characteristics may well also be a function of where one lives, and if one "solves out" this effect one would be back to (1). The interest in (2) is in testing if there are locational effects which appear to have little or nothing to do with private endowments; people in some area may tend to be less well educated (a personal characteristic) or (because of poor local infrastructure, for example) they may be unable to obtain a good return to their education, even when they are as well educated as people elsewhere. The aim in estimating (2) is to identify the latter effect (Ravallion and Wodon 1997).

If one also has community-level data, giving the stocks of physical and human infrastructure and locations of any area-based interventions, then one can use these data to try to explain the (conditional and unconditional) regional poverty maps implied by the above regressions. This can either be done using a two-step estimator or (probably more efficiently) in one step, by replacing the area-dummy variables in (1) and (2) by geographic characteristics to give

$$C_{it} = \alpha + \gamma' \mathbf{x}_{it} + \theta' \mathbf{z}_i + u_{it} \tag{3}$$

The specification in (2) and (3) imposes additive separability between the regional effects and household-level effects. However, this can be readily relaxed by adding interaction effects (so, for example, returns to education may depend on where one lives). All parameters can be allowed to vary locationally by estimating a separate regression on household characteristics for each region.

If one has access to repeated cross-section samples representing the same population then these static models can be repeated to see how the identifiable regional effects (both conditional and unconditional) have evolved over time. Does one find, for example, that poor regions are catching up over time, or are they diverging? This can be addressed by studying how the (conditional and unconditional) data-specific coefficients on the area dummies evolve.

So far the discussion has focused on a single welfare indicator. But this is almost surely too restrictive. More generally one can postulate a set of indicators aiming to capture both "income" and "non-income" dimensions of well-being. In addition to consumption of market goods and services one could include indicators of attainments in terms of basic capabilities, such as being healthy and well nourished. The aim here is not to make a long and overlapping list of such indicators but to capture the aspects of welfare that may not be convincingly captured by consumption or income as conventionally defined (Ravallion 1996). So indicators of child nutritional status or morbidity would be compelling since conventional household-level aggregates may be weak indicators of distribution within households.

B. Interpreting Static Micromodels

The interpretation of these regressions is often difficult, given that it is often unclear how one would motivate them from economic theory. Under special conditions (perfect credit markets, rational expectations, quadratic instantaneous utility function) one can interpret these equations as models of permanent income, in which case the right-hand-side variables are the various lifetime assets and their rates of return, or determinants of these. More generally there is a more complex intertemporal model, probably involving liquidity constraints such that current income and holdings of liquid assets are the key explanatory variables one should be looking to account for with whatever data can be observed in the current cross-sectional survey. Fuzziness in the link from observed household and geographic data to the relevant assets adds to the difficulties of interpretation. This discussion will focus on some of the special problems posed by the geographic variables.

Insignificant regression coefficients on the regional dummy variables (β_{jt}s in (1) and (2)) would suggest that area characteristics have no independent effect on living standards, either because there is free mobility—so that households with the same x can achieve the same standard of living everywhere—or there is no mobility, but area characteristics are fundamentally irrelevant to welfare. Provided one has fully captured the relevant household characteristics, significant values for the β_{jt}s, on the other hand, are inconsistent with free mobility. If, however, there are important omitted household characteristics—a possibility that one would be unwise to dismiss—and the equilibrium under free mobility entails residential differentiation according to those characteristics, then this could produce significant β_{jt}s.

The interpretation of the coefficients on area dummy variables or area characteristics as "pure" geographic effects thus depends critically on having a complete set of the **x** variables to control for household characteristics. The geographic dummies might be picking up an omitted a-spatial household characteristic in a residentially differentiated location equilibrium. As always, it is a matter of judgment to what extent omitted variable bias of this sort is a problem. However, this concern does speak to the need for rich integrated microdata sets in which a very wide range of data on individual and household characteristics are collected for the same sampled households.

Consistent estimation by OLS requires that the regional dummies and the household characteristics are uncorrelated with the error term. Another source of bias is migration. There are ways of dealing with this, such as by estimating a switching regression which determines which region or sector the household is located. The estimated probabilities of being in a given region or sector can then be used to correct for selectivity bias (Maddala 1986 surveys this class of econometric models). As always, the problem of identification will arise in that the same variables determining levels of living at a given location will presumably also influence location choice. A regional switching regression could also be plagued by endogeneity problems of its own; is being better educated the cause or the effect of living in urban areas, say, where schools are better and more accessible? This is a case where the econometric cure could be worse than the disease.

The importance of these considerations should be judged on a case-by-case basis. The extent to which endogeneity of *household* location is a serious concern in many developing-country applications is unclear. As a stylized fact, the cost of moving the whole household can be considerable; a rural farm-household typically does not abandon its land to move in its entirety to urban areas (say) but exports surplus workers, who retain the right (or even obligation) to return. Intrarural migration of whole farm-households also appears to be uncommon. In many settings it may be plausible to identify locational effects for a (possibly large) subset of households, such as farm households but not others (relatively mobile landless laborers and urban workers, for example.)

C. Estimating Static Micromodels in the Form of a "Poverty Regression"

Static "poverty regressions" have become a standard tool in poverty analysis. The most common approach assumes that the poverty measure is the headcount index, given by probability of living below the poverty line. One postulates that real consumption or income C_i is a function of a (column) vector of observed household characteristics $\mathbf{x}_i$, namely $C_i = \beta \mathbf{x}_i + \varepsilon_i$, where β is a (row) vector of parameters and ε_i

is an error term; this can be termed the "levels regression." A now common method in poverty analysis is not to estimate the levels regression but to define the binary variable $h_i = 1$ if $z \geq C_i$, and $h_i = 0$ otherwise. The method then pretends not to observe the y_is, acting as if only h_i and the vector of characteristics x_i is observed. The probability that a household will be poor is $P = \text{Prob}[C < z \mid x] = \text{Prob}[\varepsilon < z - \beta x] = F(z - \beta x)$, where F is the cumulative distribution function specified for the residuals in the levels regression. A probit or logit is then usually estimated, depending on the assumption one makes about the distribution of the error term ε_i.[*] (One could also use a semiparametric estimator which allows the distribution of the error to be data determined.) One can also generalize this procedure to other ("higher order") poverty measures and estimate censored regression models.

However, this common practice is difficult to defend since—unlike the usual binary response model—here the "latent" variable is fully observed. So there is no need for a binary response estimator if one wants to test impacts on poverty of household characteristics. The parameters of interest can be estimated directly by regressing C_i on x_i. The relevant information is already contained in the levels regression which is consistently estimable under weaker assumptions about the errors. Measurement errors at extreme Cs may prompt the use of probits, though there are almost certainly better ways of dealing with such problems, which do not entail the same loss of information, such as by using more robust estimation methods for the levels regression.

Nor is the "poverty regression" method necessary if one is interested in calculating poverty measures conditional on certain household characteristics. Subject to data availability, $\text{Prob}[C < z \mid x]$ can be estimated directly from sample data. When the number of sampled households with a specific vector of characteristics of interest, x_1 say, is too small to reliably estimate $\text{Prob}[C < z \mid x_1]$ from a subsample, one can also turn to regression methods for out-of-sample predictions. But these predictions can also be retrieved from the levels regression, though one must then know the distribution of the errors. (For example, if the errors are normally distributed with zero mean and a variance σ^2, then the probability of being poor is $F[(z - \beta x)/\sigma]$, where F is standard normal.) There is nothing gained from using a binary-response estimator, so the econometric sophistication of "probits" and so on buys us very little in this case.

Poverty regressions may make more sense if one wants to test the stability of the model for poverty across a range of potential poverty lines. Suppose, for example, that one of the regressors is the price of food, and that very poor people tend to be net

---

[*]The earliest example that I know of is Bardhan (1984), who used a logit regression of the probability of a household being poor against a range of household and community characteristics using sample survey data for rural West Bengal. Other examples include Gaiha (1988), Grootaert (1994), Foley (1995), and World Bank (1995).

consumers of food, while those who are somewhat better off tend to be net producers. Then the distributional shift with a change in the price of food will not entail first-order dominance, as assumed by the standard levels regression. Instead one might want to specify a set of regression functions the parameters of which vary according to the segment of the distribution one is considering. One way of estimating such a model is by assuming that the segment-specific error terms are of the logit form, entailing a multinomial logit model (Diamond et al. 1990).

All the above models are essentially static; some welfare indicator at a single date is modeled as a function of a range of individual and geographic data. Such models cannot distinguish effects on the growth rates of consumption (or other welfare metric) from effects on its level. The true model could be $C_{it} = (\beta + \gamma t)x_i + \varepsilon_{it}$, implying that x influences the growth rate of consumption as well as its level. However, static data cannot distinguish β from γ.

D. Aggregate Models

Another increasingly common tool of analysis is a regional (or country)-level time-series model. Here the dynamics can be readily introduced, by allowing for the effects of past outcomes and other variables on current outcomes. Among the models that might be postulated, regional growth regressions are being seen increasingly; by this approach, the growth rate over time for the region as a whole is modeled as a function of initial conditions, exogenous shocks and policy changes over the period.[*]

These aggregate dynamic models also have their limitations for understanding the economic geography of poverty. Clearly the distributional analysis of policy and other impacts on living standards which is impossible with only aggregate data.[†] Working from the micromodel allows one to better understand the distributional implications of the aggregate growth process.

When attempting to assess the welfare impacts of area-based interventions, aggregate models can also be particularly vulnerable to bias arising from endogeneity of program placement. The political decision on program placement may itself be a function of observed aggregate poverty levels. Knowledge of how programs were assigned and the history of area characteristics can help avoid this problem (as in Pitt et al. 1995). The problem is not of course confined to aggregate models. With micro (household-level) data it may be plausible that program placement is exogenous to the extent that no individual household has much influence on the placement de-

---

[*]See Barro and Sala-i-Martin (1995) for an overview of these models, with developed country applications; examples for developing countries include Cashin and Sahay (1995) and Datt and Ravallion (1996).

[†]Datt and Ravallion (1996) use a long time series of repeated cross-sectional surveys (for India) to relax this attribute of standard growth models.

cision. However, bias can still arise from either migration (discussed above) or omitted variables which simultaneously influence household-level welfare outcomes and program placement.

Another disadvantage of aggregate models is that they do not allow one to distinguish internal and external effects on production and welfare. This can matter to policy. By using the household as the unit of observation one can identify external effects of geographic capital, including local public goods, on production processes at household level (Ravallion and Jalan 1996). Consider Figure 1, based on aggregate (county-level) data for China. Such evidence tells us nothing per se about the spatial effects in a growth process. The highly aggregated form of such data does not allow one to distinguish two possible ways in which initial conditions may influence the growth process at the microlevel. One way is through effects of individual conditions on the individual growth process, and this is a common interpretation given to nonzero values of the regression coefficient on initial income in a growth regression; declining marginal product of capital would suggest a tendency for convergence; by this interpretation, the type of divergence depicted in Figure 1 suggests increasing returns to *private* capital. If this is right, then regional divergence, and the existence of persistently poor areas, is to be expected when the rate of growth is at its maximum. Conversely, under these conditions, governmental attempts to shift the allocation of investment in favor of poor areas will entail a growth cost, though a policymaker may still be willing to pay that cost to achieve a more balanced (and possibly more sustainable) growth path.

But there is another way in which the divergence in Figure 1 can arise, even with declining marginal products with respect to own capital at the microlevel and constant returns to scale in private inputs. The microgrowth process might be driven by *intraregional externalities;* individual growth prospects may be better in an initially better-off region through positive local spillover effects. Quite generally, the marginal product of capital will depend on area characteristics. (Only with rather special separability assumptions will this not be true.) There may well be declining marginal products with respect to "own capital" but increasing marginal products to geographic capital. Indeed, if there is constant returns to scale in the private inputs, and geographic capital is productive, then there must be increasing returns overall.* That may well be why we see the aggregate divergence in Figure 1, with the external effect dominating. But the aggregation hides the difference. If in fact the regional divergence is really due to the external effect of differences in geographic capital,

---

*Let output be $F(\mathbf{K}, \mathbf{G})$, where $\mathbf{K}$ is a vector of private inputs ("own capital") and $\mathbf{G}$ is a vector of public inputs ("geographic capital"), and consider any $\lambda > 1$. By constant returns to $\mathbf{K}$, $F(\lambda\mathbf{K}, \lambda\mathbf{G}) = \lambda F(\mathbf{K}, \lambda\mathbf{G}) > \lambda F(\mathbf{K}, \mathbf{G})$, since geographic capital has a positive marginal product. Thus F exhibits increasing returns to scale overall.

then successful interventions to reduce the inequality in geographic capital need not entail any cost to the overall rate of growth.

This second "external" channel through which area characteristics can alter a growth process has received relatively little attention in empirical work on the determinants of economic growth, though the possibility has been recognized in some of the theoretical literature (notably Romer 1986 and Lucas 1988). The reason why this external channel has been relatively neglected in growth empirics is undoubtedly that the level of aggregation in past work has meant that—even if one was aware of the possibility—the genuinely spatial effects of intraregional spillover effects could not possibly be identified empirically.

To encompass both the "internal" (individualistic) and "external" (geographic) channels through which initial conditions can affect a growth process one needs to model that process at the microlevel. The growth rate for each household will be a function of both its own initial conditions, characteristics of the area in which the household lives, and external shocks during the period. The areawide growth relationships (such as depicted in Figure 1) can then be interpreted as (approximate) averages formed over the underlying microgrowth processes; but in the averaging one loses the ability to distinguish the internal from the external effects (Ravallion and Jalan 1996).

The recurrent problem in aggregate models is that the economic theory which motivates them is typically a microeconomic model. So tests using aggregate data always beg the question of whether one is testing the micromodel or the aggregation assumptions.

III. THEORY AND ESTIMATION FOR DYNAMIC MICROMODELS

A. Theoretical Foundations for a Microempirical Model of Growth

A theoretical model capable of motivating an empirical analysis of a number of the issues raised above can be formulated by a reinterpretation of the Romer (1986) model of endogenous growth under increasing returns to scale.* Analogously to the distinction between firm-specific knowledge and economy-wide knowledge in Romer's model, one can conjecture that output of the farm household is a concave function

---

*A number of versions of the classic Ramsey model—in which an intertemporal utility integral is maximized subject to flow constraints and production functions—have been proposed which can yield a nonzero solution for the rate of consumption growth which will be a function of initial human and physical assets as well as preference and production parameters. For surveys of the theories of endogenous growth see Grossman and Helpman (1991), Hammond and Rodriguez-Clare (1993), and Barro and Sala-i-Martin (1995).

of various privately provided inputs, but that output also depends positively on the level of geographic capital. On fully accounting for all private inputs (all profits being reckoned as payments for those inputs), there will then be constant returns to scale to the privately provided inputs, but increasing returns to scale over all inputs, including geographic capital. With the farm-household maximizing an intertemporal utility sum—with instantaneous utility depending on current consumption, which must be partly forgone to ensure future output—one can derive an endogenous consumption growth rate which depends on the initial endowments of both private capital and geographic capital. With this reinterpretation, the results on existence and welfare properties of equilibrium in Romer (1986) model can be applied to the present problem.

The key intertemporal equilibrium condition from such a model equates the intertemporal marginal rate of substitution with the marginal product of "own capital," which is a decreasing function of the initial endowment of own capital and increasing in the amount of geographic capital, taken as exogenous at the microlevel. With appropriate functional forms, the farm-household's consumption growth rate over any period is then a decreasing function of its endowment of private capital and an increasing function of the level of geographic capital.

Past growth empirics have relied on country or regional aggregates. The translation of this approach to the microlevel is straightforward; one is simply undoing the aggregation conditions used to go from the microgrowth theory to the aggregate regional or country data. The translation is even more straightforward when it is noted that many of the households in the world's poor areas are farm households who jointly produce and consume, rather than economies in which separate consumers and producers interact through trade. But this is largely a matter of interpretation; the separation of an economy into households (which consume) and firms (which produce) is not essential in theoretical growth models.* In the present setting, the farm-household can be thought of as a small open economy, trading with those around it.

It should also be recognized that a poor area may have become poor due to a location-specific transient shock (a local drought, for example). There may also be lags in the growth process of consumption. By explicitly modeling these features of the data generation process, it should be possible to identify longer-term impacts in panels of sufficient length. (Averaging prior to estimation is not an efficient way of dealing with these data features.)

Motivated by the above considerations, an empirical approach can be suggested which entails consistently estimating a dynamic model of consumption growth

---

*Standard endogenous growth models postulate separate households and firms, but an equivalent formulation is possible in which households both consume and produce (Barro and Sala-i-Martin 1995).

at the household level using panel data. The model allows one to test the dynamic impact over the length of the panel of a wide range of initial conditions at both household and community levels. The proposed approach differs from the usual "fixed-effects" method. While it is common to model the variables of interest in first-difference form, or as deviations from their time means, this is typically done in the context of a static model in levels, for which the time slope is a constant and there are unobserved fixed effects. Clearly such a formulation is of little interest here since it does not allow initial conditions—including area-specific policies and projects—to affect the growth path of the variable of interest.

B.   Toward an Estimable Model

Unlike standard single cross-sectional sample surveys of households, here one needs panel data in which the same households are observed over time. (Later I will discuss possible approaches using repeated cross-sectional surveys.) Such data sets are, however, becoming more common, and panels of varying lengths are available for a number of developing and developed countries. Notice also that estimating a microgrowth model will require less data in the second survey than the first. At a minimum, a second reading of the household consumption or income level will do, though information on relevant demographic or other "shocks" would be desirable. So this approach is already feasible and will probably become more widely applicable in the future.

To see how an empirical model capable of addressing the questions posed here can be constructed, let us assume (following the discussion above) that the long-run household-level consumption growth rate is determined in part by a vector of exogenous initial conditions, comprising both *internal* (within the household) and *external* (community) endowments of physical and human assets, as well as any exogenous household characteristics influencing the discount rate, liquidity constraints, tastes, and production functions. The importance of the internal factors can arise from the dependence of equilibrium growth rates on the initial human and physical capital stocks in household-operated production processes. The external effects can arise from the existence of local public goods or differing agroclimatic conditions. The direction of the effects of all such initial conditions are difficult to predict on a priori grounds, and will depend on the nature of the technology (for example, if there are increasing returns then divergent effects are possible whereby higher initial wealth can result in higher future growth), how well markets work (credit market imperfections, for example, can entail that liquidity-constrained households cannot realize the same growth potential as others) and the political economy of local public policy (it has been argued that higher initial inequality, for example, may promote policy choices which inhibit growth).

In carrying such a formulation to data, it would be unrealistic to assume that the growth rate actually observed at any date is the steady-state value as implied by a standard growth-theoretic model. One would want to allow for deviations from the underlying steady-state solution, due to shocks and/or adjustment costs. It is thus better to postulate an autoregressive distributed lag structure for the growth process, augmented by exogenous shocks and unobserved effects. This will permit a more powerful test of the impact of initial conditions on the evolution of living standards than is possible by only modeling the long-run average growth rates.

Thus one can postulate an econometric model of the growth rate in living standards at the household level as a function of (1) initial conditions at the household level, (2) initial conditions at the local community level, and (3) exogenous time-varying factors ("shocks") at both levels. The variance in household-level growth rates due to the second set of variables could in principle be "explained" by a complete set of area-dummy variables. However, by collating the micro (household-level) data with geographic data bases on agroclimatic variables, and stocks of physical and social infrastructure, it will be possible to obtain a far more illuminating specification in which specific attributes of the local area enter explicitly. As a check for omitted-variable bias, one can then compare the results with a model in which the geographic variation is picked up entirely by dummy variables. This may also suggest idiosyncratic regional effects, such as due to local political factors, that might be best studied on an ad hoc (case study) basis.

On introducing dynamics and both time-invariant and time-varying unobserved effects, a suitable dynamic model could take the form

$$C_{it} = \alpha_0 + \alpha_1 t + \beta C_{it-1} + \gamma_0' x_{it} + \gamma_1' x_{it-1} + \theta' z_i \cdot t + \eta_i + u_{it} \tag{4}$$

for household i ($= 1, \ldots, N$) at date t ($= 1, \ldots, T$), where C_{it} is consumption by i at date t, x_{it} is a $1 \times k$ vector of time-varying explanatory variables, z_i is a p-dimensional vector of initial conditions, and η_i is a time-invariant household-level fixed effect. The vector z_i comprises both area-specific factors (such as initial values of indicators of physical and social infrastructure) and household-specific characteristics (such as age of the head of the household and education levels). Both x_{it} and z_i include interaction effects (including between individual and areas characteristics).*

The problem of estimating this model is different from the usual "within" estimator for panel data. It is known that the ordinary least-squares estimator of an autoregressive fixed effects model is not consistent for a typical panel where the number of periods is small and where the asymptotics are driven by the number of cross sections going to infinity (Hsiao 1986). The inconsistency arises because of the potential correlation between the lagged endogenous variables and the residuals in the transformed model.

---

*One might also hypothesize that area-mean consumption enters (4), but this effect is generally not identifiable; see Manski (1993).

Thus consistent estimation of the above model does present a more difficult problem than either the static micro- or dynamic macromodels. But a solution is available (Jalan and Ravallion 1996). First notice that the error term in (4) has two components: an unobserved individual specific time-invariant fixed effect, η_i, and the standard innovation error term, u_{it}. Let us assume that the unobserved individual-specific effect η_i is correlated with the regressors, i.e., $E(\eta_i z_i)$, $E(\eta_i, x_{it})$, and $E(\eta_i C_{it-1})$ are nonzero.* The error u_{it} is however serially uncorrelated and thus satisfies the orthogonality conditions:

$$E(C_{is}u_{it}) = E(x_{is}u_{it}) = 0 \qquad \text{for } s < t \tag{5}$$

These conditions ensure that suitably lagged values of C_{it} and x_{it} can be used as instruments. In order to get consistent estimators, the unobserved fixed effects η_i need to be eliminated. This can be done by taking the first differences of (4) to obtain the transformed "growth model":[†]

$$\Delta C_{it} = \alpha_1 + \beta\,\Delta C_{it-1} + \gamma_0'\,\Delta x_{it} + \gamma_1'\,\Delta x_{it-1} + \theta' z_i + \Delta u_{it} \tag{6}$$

There are various options for estimating such a model. GMM methods appear to offer the best approach (Arellano and Bond 1991). Given that the u_{it}s are serially uncorrelated, the GMM estimator is the most efficient one within the class of instrumental variable (IV) estimators. In estimating (6), C_{it-2} or higher lagged values (wherever feasible) are valid instrumental variables. Heteroscedasticity-consistent standard errors can be computed using the residuals from a first-stage regression to correct for any kind of general heteroscedasticity. Inferences on the estimated parameter vector are appropriate provided the moment conditions used are valid. Tests for overidentifying restrictions can be implemented to test the null hypothesis that the instruments are optimal (i.e., the instruments and the error term are orthogonal); see Sargan (1958, 1988) and Hansen (1982). In addition, a second-order serial correlation test (the test statistic will be normally distributed) can be constructed given that the consistency of the GMM estimators for the first-differenced model depends on the assumption that $E(\Delta u_{it}\,\Delta u_{it-2}) = 0$.[‡] Tests for spatial correlation in the errors—arising from omitted geographic effects—can also be performed (following Frees 1995), though they will need to be adapted to the present problem.

---

*Bhargava and Sargan (1983) offer a dynamic random-effects model where it is assumed that some of the regressors are uncorrelated with the unobserved individual specific effect.

[†] Various transformations can be used to eliminate the nuisance parameters, though the estimation procedures used are similar to the one proposed here.

[‡] There may be some first-order serial correlation; i.e., $E(\Delta u_{it}\,\Delta u_{it-1})$ may not be equal to zero since Δu_{it} are the first differences of serially uncorrelated errors. Alternatively, if u_{it} is a random walk, then there should not be any serial correlation in the first differenced Δu_{it}.

If corrective action is called for, then one can try introducing more geographic data, or more geographic structure to the error process.*

C. Further Specification Issues

Questions are often asked about the prioritization of physical versus social infrastructure development. Should one even assign a priority, or is a balanced "integrated" approach needed? Does the answer depend on the "stage" of development? It has often been argued that it is the *combination* of certain physical and/or human infrastructure endowments that matters. One at a time they may not help much. The returns to irrigation for example may depend on education (as van de Walle 1995 found for Vietnam). There are other possible interaction effects such as the possibility that greater ethnic cohesion in an area increases the chances of cooperation and hence the returns to investments in geographic capital.

In principle, all the explanatory variables in the models described above could enter the model in highly nonlinear ways, and this should be tested. However, the above considerations suggest that identifying interaction effects could be of special interest in this context.

Another issue concerns *mobility* and the possible endogeneity of a household's area characteristics. The plausibility of a free-mobility equilibrium in the settings considered here is questionable; even with no governmental restrictions on mobility, migration within India over a long period has responded little to regional disparities (Datt and Ravallion 1996). Nonetheless, the existence of even limited mobility raises questions about the possible *endogeneity* of area characteristics in the microgrowth process. How might one test for effects of area characteristics on the spatial distribution of the population?

The significance of area fixed effects in the *levels* of living standards, after controlling for a-spatial household characteristics, would be suggestive that mobility is imperfect; if there was free mobility then any two households with the same personal (mobile) characteristics should be able to achieve the same level of welfare. This is directly testable.

There are other tests that may also help. A theoretically consistent and empirically tractable approach to introducing mobility, allowing for adjustment costs, can be proposed following the approach outlined in Ravallion (1982, 1984) in the related context of local public finance in which (under certain conditions) mobility can reveal preferences over local public goods. This entails modeling the spatial distribution of specific population subgroups as functions of, inter alia, area characteristics, allowing for dynamic effects in the adjustment process to a free-mobility

---

*Potential approaches include Froot (1989) and Conley (1996); also see the special issue of *Regional Science and Urban Economics*, September 1992, on "Space and Applied Econometrics."

equilibrium. In both estimating the poverty regressions (with "social externalities") and in deriving aggregate welfare (including poverty) impacts, this model could then be used to endogenize the population shares.

The dynamic model estimated on panel data suggests a further test. One of the main sources of attrition in a panel is outmigration. Thus standard tests for attrition bias and corrective actions (see, for example, Hsiao 1986, Chap. 8) can also be interpreted as a means of dealing with migration responses to area characteristics.

Household mobility is not the only way in which endogeneity of area-based interventions can bias results from the types of models described above. Another problem is that program placement may be a function of variables which influence welfare impacts (Pitt et al. 1995). A version of this problem arises in the dynamic micromodels described above when program placement is determined by initial conditions which also influence the future evolution of living standards (Jalan and Ravallion 1996). The best solution to this problem is to find out what area characteristics influenced program placement and include those characteristics as explanatory variables. Since program placement must have been a function of observable area attributes suitable controls should be available in practice.

D. Alternative Methods Using Repeated Cross Sections

The attractions of panel data for the types of analysis described above are clear. Though more of these data sets are emerging, panel data sets are still far less common than repeated cross-sectional surveys. Is there any way of estimating spatial effects on the microgrowth process using repeated cross sections? If the answer is yes, then this would open up a wide range of potential applications in setting in which panel data are unavailable.

One approach is by the analysis of demographic cohorts.* For each wave of a set of cross-sectional surveys one can calculate mean household consumption or income for persons in a given cohort defined by initial age and (in this case) place of residence. One can also do this for other household characteristics. One can then construct a model that looks like an individual-level model but is for cohorts. In effect, one takes cohort averages of the household-level model. Thus one can still identify the *internal* (cohort averages of household characteristics) and *external* (area characteristics) effects on the evolution of the poverty indicator.

However, there may well be a better approach. We want to see how (say) consumption growth from time t to $t + 1$ is affected by time t characteristics of the household and its area of residence. With panel data we simply regress the change in (log) consumption from t to $t + 1$ on (inter alia) household is characteristics at t.

---

*This approach has showed promise in research on other topics, such as intertemporal consumption behavior and inequality (Deaton and Paxson 1994).

With cross sections we do not know consumption at $t + 1$. But if we know the future values of one or more predictors of consumption then these can be used as instruments. One first models time $t + 1$ consumption as a function of variables observed in time $t + 1$ but also at time t. Then one uses that model to predict the consumption at time $t + 1$ of each household in the time t sample and, hence, estimate its rate of consumption growth from t to $t + 1$. This can then be regressed on the individual and area characteristics at time t.

Many cross-sectional surveys do obtain information about likely future characteristics which can be used as instruments. For example, the Rural Household Surveys for China collect both beginning and end-of-year data on financial and physical wealth in each round. So the end-of-year data can be used to predict the next period's consumption along with other time-invariant variables. There are other potential instruments; the next period's demographic composition (number of persons by age groups) of the household can be predicted from the current period's composition. R^2 will be lower, but consistent estimates should still be possible under regular conditions. Estimators are available for dynamic models of this sort using repeated cross sections (see Moffitt 1993 and references therein). The performance of these methods could be studied using the panel data, but treating it as repeated cross sections; it would also be of interest to try the method out on the original samples (prior to panel construction) so as to assess effects of panel attrition.

IV. CONCLUSIONS AND POTENTIAL LESSONS FOR POLICY

When confronted with the reality of extreme poverty in remote rural areas with poor natural resources, observers often ask: "Why don't these miserably poor people just move out?" Those who claim that outmigration is the answer often also argue against public investment in these areas. "This will just be at the expense of more profitable investments elsewhere" the argument goes.

This chapter has questioned this reasoning, but it certainly has not refuted it. That will be a matter of future empirical research. But some points can be made now. One is that we should understand the nature of the incentives and constraints on outmigration from these areas. You need some money to start up elsewhere, you need some basic skills, and you need information. All are generally lacking, but more so for some people than others. The reasons they are lacking can be traced to market imperfections of one sort or another and how these interact with poverty. The outside non-farm-labor-market options are typically thin or nonexistent for someone who is illiterate, reflecting a lack of substitution possibilities with moderately educated labor in even quite labor-intensive manufacturing. Credit market failures mean that there is little chance of borrowing to finance the move. There is highly imperfect information about prospects elsewhere, and sizable uninsured risk.

The process of outmigration may be a mixed blessing for a poor area at least initially. Those who have the money, skills, and information will naturally tend to be the relatively better off. Their departure is likely to put upward pressure on the incidence of poverty in the poor area. This comes about in various ways. As a purely statistical proposition, most measures of poverty will rise when the nonpoor leave. But there are more subtle dynamic effects through "ghettoization." The local skill base is likely to have external effects on the local growth process. It follows that the outmigration of the better educated workers entails an erosion of local resource base with adverse longer-term growth consequences. Results from research on poor areas of southwest China have suggested that there exist strong external effects of physical and human infrastructure on the returns to private investment and (hence) the prospects of escaping poverty there (Jalan and Ravallion 1997). These effects will be mitigated to some extent by remittances and reduced pressure on the land.

All this suggest that one of the best ways that government can help is by investing in the schooling, health, and nutrition of the children of the poor in these areas. Public assistance with credit (to cover search costs for poor outmigrants), and information will complete the package.

Should we also be investing in the land and physical capital of these areas? What should be the balance between those investments and human resource development? Here one could proceed on an ad hoc basis; if the investment passes a standard (distribution-unweighted) cost-benefit test, then it should be done. But the "anti-investment" argument would maintain that private capital flows would already have found such opportunities.

One response is that, unless there is perfect factor mobility (which nobody seems to consider plausible), there may still be an equity case for such investments up to some point. Then they are part of a redistributive policy, exploiting the possibilities for geographic targeting (Lipton and Ravallion 1995). That is fine. However, for the same reasons that there may be too little outmigration, there may also be too little investment in these areas from an efficiency point of view as well. Credit market imperfections can entail that there are unexploited opportunities for investing in the land and physical capital of these areas. The liquidity constraints that make it hard to finance outmigration will also make it hard to finance otherwise profitable local investments. And asymmetric information and supervision costs deter outside investors.

The argument that investing in poor areas would entail lower overall growth in the economy also breaks down as soon as one introduces local public goods and other forms of "geographic capital" into the analysis, i.e., goods which cannot be supplied efficiently by markets and which alter the rate of return to private investment. Poor rural infrastructure in these areas could then be the underlying reason for low private investment; better infrastructure would then encourage private capital inflow.

However, much of this is conjecture, based on little more than casual observations and common sense. The chapter has suggested some econometric methods which might be used to address these issues more rigorously. But is it an ambitious research agenda, both in terms of the types of data needed, and the level of econometric sophistication needed to convincingly disentangle these effects. So it is important to ask: What will we have learned from such research that can reliably inform the above policy choices about poor areas? Three types of potential lessons for policy can be identified.

The first set of policy lessons concern the economic case for area-based interventions. Should such interventions only be viewed as a specific kind of redistributive policy with probable costs to the overall rate of growth? This view derives from a growth model in which the existence of persistently poor areas, and regional divergence more generally, are traced to the natural resource endowments and technologies, notably the (claimed) existence of increasing returns to private production inputs. Private investment flows to the areas with the highest returns which are also (according to this model) the initially richer areas. This still begs a number of policy questions. For example, we need to know more about what the best indicators are for this type of redistributive policy when the aim is to help individuals escape poverty in the future.

But maybe we will find that the empirical results from the type of research proposed here will reject this model at a fundamental level in favor of one which says, in effect, that poor areas and divergence reflect spatial inequalities in access to credit, and publicly provided social and physical infrastructure, and have rather little to do with increasing returns to private capital, residential differentiation, and so on. That conclusion could well dramatically alter the policy dialogue on poor-area interventions and shift the emphasis to the task of redressing these preventable spatial inequalities. If that conclusion is borne out by the data, then such policies will be good for growth and good for equity. Or the results may point to a more complex and mixed picture, possibly with a degree of country, and even regional, specificity.

A second set of broad policy lessons stem from the fact that the approach proposed here allows one to measure spatial externalities. This can throw light on, for example, how much of the welfare gain from schooling is transmitted though the internal effects on earnings and so on, and how much is external, arising from the (presumably positive) neighborhood effects of better education. This will have implications for the priority one attaches to efforts at finely targeting education subsidies and for the policy arguments often made about how much basic education needs to be subsidized on the grounds of its external benefits.

A third set of policy implications will be more specific to the types of projects that should be recommended for dealing with persistently poor areas. In the process of addressing these broad questions, empirical models can include explanatory variables of more or less direct policy relevance. One set of such variables is the very existence of poor-area interventions. Is the subsequent rate of growth in living standards of poor people higher when a poor-area program is in place than would otherwise have been the case, controlling for both household and community-level initial conditions and time-varying exogenous shocks? What were the longer-term welfare gains? How do they compare to the budgetary outlays on such programs? There will be other explanatory variables of policy relevance, such as the initial stocks of various components of publicly provided social and physical infrastructure, for which all of the same questions apply, though here of course it may not always be easy to account fully for their historical costs (though costs of new facilities will often be known). What priority should be attached to social services versus physical infrastructure or credit, and how does this vary with other factors? This should allow a deeper understanding of what the complementarities are among these various types of publicly provided inputs; we may learn, for example, how much access to one type of infrastructure alters returns to another, or how much poor agroclimatic conditions affect returns to different types of publicly provided inputs.

This chapter has argued that the long-standing problem of lagging poor areas in growing economies, and more generally the diversity in prospects of escaping poverty that one finds, are explicable with the right empirical tools and data. This offers hope for better informing a number of difficult public choices on appropriate responses to poor areas.

ACKNOWLEDGMENTS

I have had many useful discussions on this topic with Jyotsna Jalan. For their helpful comments on an earlier draft, I am also grateful to the Handbook's referee, and to Hans Binswanger, Ken Chomitz, Klaus Deininger, Lionel Demery, Bill Easterly, Paul Glewwe, Emmanuel Jimenez, Aart Kraay, Valerie Kozel, Peter Lanjouw, Andy Mason, Branko Milanovic, Lant Pritchett, Martin Rama, Zmarak Shalizi, Lyn Squire, Dominique van de Walle, Mike Walton, and Quentin Wodon.

REFERENCES

Arellano, M. and S. Bond (1991), Some Tests of Specification for Panel Data: Monte-Carlo Evidence and An Application to Employment Equation, *Review of Economic Studies*, 58, 277–298.
Bardhan, P. K. (1984), *Land, Labor and Rural Poverty: Essays in Development Economics*, Columbia University Press, New York.
Barro, R. and X. Sala-i-Martin (1995), *Economic Growth*, McGraw-Hill, New York.
Bhargava, A. and J. D. Sargan (1983), Estimating Dynamic Random Effects Models from Panel Data Covering Short Time Periods, *Econometrica*, 51, 1635–1659.
Binswanger, H. S. R. Khandker, and M. Rosenzweig (1993), How Infrastructure and Financial Institutions Affect Agricultural Output and Investment in India, *Journal of Development Economics*, 41, 337–366.

Borjas, G. J. (1995), Ethnicity, Neighborhoods, and Human-Capital Externalities, *American Economic Review*, 85, 365–390.

Case, A. (1991), Spatial Patterns in Household Demand, *Econometrica*, 59, 953–65.

Case, A. and L. F. Katz (1991), Company You Keep: The Effects of Family and Neighborhood on Disadvantaged Youths, Working Paper 3705, National Bureau of Economic Research, Washington, D.C.

Cashin, P. and R. Sahay (1995), Internal Migration, Center-State Grants and Economic Growth in the States of India, IMF Working Paper WP/95/66.

Chen, S. and M. Ravallion (1996), Data in Transition: Assessing Rural Living Standards in Southern China, *China Economic Review*, 7, 23–56.

Ciccone, A. and R. E. Hall (1996), Productivity and the Density of Economic Activity, *American Economic Review*, 86, 54–70.

Conley, T. (1996), Econometric Modelling of Cross-Sectional Dependence, mimeo, Northwestern University.

Datcher, L. (1982), Effects of Community and Family Background on Achievement, *Review of Economics and Statistics*, 64, 32–41.

Datt, G. and M. Ravallion (1993), Regional Disparities, Targeting, and Poverty in India, in M. Lipton and J. van der Gaag (eds.), *Including the Poor*, World Bank, Washington, D.C.

Datt, G. and M. Ravallion (1996), Why Have Some States of India Performed Better than Others in Reducing Absolute Poverty?, *Economica*, forthcoming.

Deaton, A. and C. Paxson (1994), Intertemporal Choice and Inequality, *Journal of Political Economy*, 102, 437–467.

Diamond, C. A., C. J. Simon, and J. T. Warner (1990), A Multinomial Probability Model of the Size Distribution of Income, *Journal of Econometrics*, 43, 43–61.

Easterly, W. and R. Levine (1995), Africa's Growth Tragedy. A Retrospective, 1960–89, Policy Research Working Paper 1503, World Bank, Washington, D.C.

Foley, M. C. (1995), Poverty in Russia: Static and Dynamic Analyses, in J. Klugman (ed.), *Poverty in Russia during the Transition*, World Bank, Washington, D.C.

Foster, A. D. and M. Rosenzweig (1995), Learning by Doing and Learning from Others: Human Capital and Technical Change in Agriculture, *Journal of Political Economy*, 103, 1176–1209.

Frees, E. (1995), Assessing Cross-Sectional Correlation in Panel Data, *Journal of Econometrics*, 69, 393–414.

Froot, K. (1989), Consistent Covariance Matrix Estimation with Cross-Sectional Dependence and Heteroskedasticity in Financial Data, *Journal of Financial and Quantitative Analysis*.

Gaiha, R. (1988), On Measuring the Risk of Poverty in Rural India, in T. N. Srinivasan, and P. K. Bardhan (eds.), *Rural Poverty in South Asia*, Oxford University Press, Oxford.

Grootaert, C. (1994), The Determinants of Poverty in Côte d'Ivoire, mimeo, World Bank, Washington D.C.

Grossman, G. M. and E. Helpman (1991), *Innovation and Growth in the Global Economy*, MIT Press, Cambridge, MA.

Hammond, P. J. and A. Rodriguez-Clare (1993), On Endogenizing Long-Run Growth, *Scandinavian Journal of Economics*, 95, 391–425.

Hansen, L. P. (1982), Large Sample Properties of Generalized Method of Moments Estimators, *Econometrica*, 50, 1029–1054.

Hirschman, A. O. (1958), *The Strategy of Economic Development*, Yale University Press, New Haven, CT.

Hsiao, C. (1986), *Analysis of Panel Data*, Cambridge University Press, New York.

Jalan, J. and M. Ravallion (1996), Are There Dynamic Gains from a Poor Area Development Program?, *Journal of Public Economics*, forthcoming.

Jalan, J. and M. Ravallion (1997). Spatial Poverty Traps? Policy Research Working Paper, World Bank, Washington, D.C.

Jazairy, I., M. Alamgir, and T. Panuccio (1992), *The State of World Rural Poverty: An Inquiry into its Causes and Consequences*, New York University Press for the International Fund for Agricultural Development, New York.

Krugman, P. (1991), *Geography and Trade*, MIT Press, Cambridge MA.

Leading Group (1988), *Outlines of Economic Development in China's Poor Areas*, Office of the Leading Group of Economic Development in Poor Areas under the State Council, Agricultural Publishing House, Beijing.

Lipton, M. (1995), and M. Ravallion, Poverty and Policy, in J. Behrman and T. N. Srinivasan (eds.), *Handbook of Development Economics*, vol. 3, North-Holland, Amsterdam.

Lucas, R. E. (1988), On the Mechanics of Economic Development, *Journal of Monetary Economics*, 12, 3–42.

Maddala, G. S. (1986), Disequilibrium, Self-Selection, and Switching Models, in Z. Griliches, and M. D. Intriligator (eds.), *Handbook of Econometrics*, North-Holland, Amsterdam.

Manski, C. F. (1993), Identification of Endogenous Social Effects: The Reflection Problem, *Review of Economic Studies*, 60, 531–542.

Moffitt, R. (1993), Identification and Estimation of Dynamic Models with a Time Series of Repeated Cross-Sections, *Journal of Econometrics*, 59, 99–123.

Myrdal, G. (1957), *Economic Theory and Underdeveloped Regions*, Duckworth, London.

Pitt, M., M. Rosenzweig, and D. Gibbons (1995), The Determinants and Consequences of the Placement of Government Programs in Indonesia, in D. van de Walle and K. Nead (eds.), *Public Spending and the Poor Theory and Evidence*, Johns Hopkins University Press.

Ravallion, M. (1982), The Welfare Economics of Local Public Spending: An Empirical Approach, *Economica*, 49, 49–61.

Ravallion, M. (1984), The Social Appraisal of Local Public Spending Using Revealed Fiscal Preferences, *Journal of Urban Economics*, 16, 46–64.

Ravallion, M. (1993), Poverty Alleviation through Regional Targeting: A Case Study for Indonesia, in K. Hoff, A. Braverman, and J. Stiglitz (eds.) *The Economics of Rural Organization*, Oxford University Press, Oxford.

Ravallion, M. (1996), Issues in Measuring and Modelling Poverty, *Economic Journal*, 106, 1328–44.

Ravallion, M. and S. Chaudhuri (1997), Risk and Insurance in Village India: Comment, *Econometrica*, 65, 171–184.

Ravallion, M. and J. Jalan (1996), Growth Divergence due to Spatial Externalities, *Economics Letters*, 53, 227–232.

Ravallion, M. and Q. Wodon (1997), Poor Areas, or Only Poor People? Policy Research Working Paper 1363, World Bank, Washington, D.C.

Richardson, H. W. and P. M. Townroe (1986), Regional Policies in Developing Countries, in P. Nijkamp (ed.), *Handbook of Regional and Urban Economics*, North-Holland, Amsterdam.

Riskin, C. (1994), Chinese Rural Poverty: Marginalized or Dispersed?, *American Economic Review, Papers and Proceedings*, 84, 281–284.

Romer, P. (1986), Increasing Returns and Long-Run Growth, *Journal of Political Economy*, 94, 1002–1037.

Sargan, J. D. (1958), The Estimation of Economic Relationships Using Instrumental Variables, *Econometrica*, 26, 393–415.

Sargan, J. D. (1988), Testing for Misspecification after Using Instrumental Variables, in E. Massoumi (ed.), *Contributions to Econometrics: John Denis Sargan*, vol. I, Cambridge University Press, Cambridge.

Sengupta, J. L. and Bo Q. Lin (1995), Recent Rural Growth in China: The Performance of the Rural Small-Scale Enterprises, mimeo, University of California, Santa Barbara, CA.

Townsend, R. (1991), Risk and Insurance in Village India, *Econometrica*, 62, 539–591.

van de Walle, D. (1995), Rural Poverty in an Emerging Market Economy: Is Diversification into Non-Farm Activities in Rural Viet Nam the Solution?, mimeo, Public Economics Division, Policy Research Department, World Bank, Washington, D.C.

Wilson, W. J. (1987), *The Truly Disadvantaged*, University of Chicago Press, Chicago, IL.

World Bank (1992), *China: Strategies for Reducing Poverty in the 1990s*, World Bank, Washington, D.C.

World Bank (1995), *Understanding Poverty in Poland*, World Bank, Washington, D.C.

4

The Demand for Health Services in a Developing Country
The Role of Prices, Service Quality, and Reporting of Illnesses

Anil B. Deolalikar
University of Washington, Seattle, Washington

I. INTRODUCTION

As a result of the debt crises of the 1980s and the ensuing structural adjustment and stabilization programs, many less-developed countries (LDCs) have had to cut back social spending, including spending on government health programs (Cornea, Jolly, and Stewart 1987). As a result, these countries have been forced to explore alternative means of financing health services, including greater recovery of (re-current) costs in the government health sector via user fees. Proponents of greater cost recovery base their recommendations on the findings of several empirical studies that suggest that the demand for health care in LDCs is price inelastic (Akin et al. 1987, Jimenez 1987, World Bank 1987). On the other hand, opponents of the cost recovery argument contend that raising fees will reduce access to care, especially by the poor, and consequently adversely affect health status (Cornea, Jolly, and Stewart 1987, Gilson 1989).

Unfortunately, the empirical bases on which both arguments are made are weak. The relatively few empirical studies of health-care demand for LDCs are flawed, largely because of their failure to recognize (1) the role of quality of health services in influencing demand and (2) the effect of health-care prices on utilization of health services via their effect on the reporting of illnesses by individuals. The most obvious reason for the lack of control for quality is that observable and quantifiable data on quality are rarely available. But, since the price charged for medical care often reflects the quality of care provided, the lack of control for quality confounds quality with price effects and biases estimated price effects toward zero (as price and quality influence demand in opposite directions). In addition, health-care demand functions that are conditioned on reported morbidity can greatly understate the total effect of health-care prices on the utilization of health services, since they ignore the potentially adverse effect that these prices can have on the reporting of morbidity.

II. PREVIOUS STUDIES

A number of studies have previously attempted to estimate the demand for health services in LDCs. Unfortunately, the existing literature in this area offers confusing evidence regarding the price response of health-services utilization to user fees. One strand of literature suggests that prices are not important determinants of health-care utilization. Heller (1981), Akin et al. (1984, 1986), Birdsall and Chuhan (1986), and Schwartz et al. (1988) all report very small and sometimes positive price effects, most of which are statistically insignificant. Another strand of work by Mwabu (1986), Gertler et al. (1987), Alderman and Gertler (1988), and Gertler and van der Gaag (1990) conclude that prices are important. The results of the first group of studies contrast sharply with most studies on the demand for medical care in developed countries which report price elasticities ranging from −0.2 to as high as −2.1 (Rosset and Huang 1973, Goldman and Grossman 1978, Newhouse and Phelps 1974, Manning et al. 1987). This divergence between the literature on developed and developing countries is paradoxical, since one would expect prices to be more important in determining utilization in developing than in developed countries for two reasons: first, income levels are substantially lower in the developing countries; second, medical insurance, which is almost universal in developed countries, is virtually nonexistent in most developing countries.

The paradox may be explained by the fact that most previous studies on health-services utilization in developing countries are flawed in three respects. First, the treatment of the price of health services in much of the previous work has been far from satisfactory. While some studies have used expenditures per medical visit reported by consumers as the relevant price, other studies have used standard fee schedules, as reported by providers. Both methods are incorrect and can cause misleading results. The amount paid by a consumer per provider visit (namely, the "unit value") depends not only on the price charged by that provider for a standard treatment but also on the type of treatment and quality of service chosen by the patient. For example, a visit for a common cold will necessarily cost less than a visit for a more serious problem. In addition, health providers, like other suppliers of goods and services, can typically provide a range of treatments of varying quality (and price) for the same ailment. To calculate the true price of health services, the disease-specific technological effect and the consumer-chosen quality effect need to be purged from observed unit values. Much of the previous work on health-care demand has confounded these price, quality, and disease-specific variations.*

The use of established or official fees as price constructs does not solve the problem either. Indeed, this procedure introduces another set of biases in the estimates of demand functions. For example, in estimating the demand for health services, Gertler and van der Gaag (1990) assume that the price of obtaining health services from government medical establishments in Peru and the Ivory Coast is zero, since such establishments do not have user fees in principle.[†] However, a number of recent surveys in developing countries suggest that there may be a wide discrepancy between officially established fees for medical visits and payments actually made by patients (World Bank 1992a, Deolalikar and Vashishta 1992). Individuals may be able to obtain speedier service and higher-quality treatments by paying for services, even when such services are officially free of charge. Imposing the assumption that prices do not vary in the sample (when, if fact, they do) can reduce the efficiency of price elasticity estimates and incorrectly lead to the result that the price elasticity of demand for health services is not significantly different from zero.

The second major problem with previous studies is that they estimate the demand for health services, *conditional on an illness episode being reported* by an individual or household. To the extent that health-care prices can affect morbidity (i.e., the probability of an individual experiencing an illness episode) and the reporting of morbidity by individual respondents, the price effects obtained from a conditional health-care demand model are partial. To complicate matters, health-care prices are likely to have opposing effects on morbidity (positive), reporting of morbidity (negative), and health-care utilization (negative), so it is impossible to infer the *total* effect of prices on health-services utilization from the conditional (and partial) demand estimates.

The third major problem in studies of health-care demand is the omission of food prices. Within a general behavioral model of health determination, the demand for food and medical care are jointly determined, since nutrition and medical care are (possibly substitutable) inputs in the "production" of health status. This means that the demand for health care is influenced not only by the price of health services but also by food prices, in much the same way as the demand for different foods is determined by food and health-care prices.[‡] Of course, the omission of food prices from health-care demand functions will not necessarily bias the estimated effects of health-care prices on health-care demand unless food prices are correlated with the price of health services.

---

*See Deaton (1988) for a discussion of a somewhat similar problem in the analysis of food demand in LDCs.
†They assume that all of the price variation occurs in the form of variation in distance traveled to providers.
‡See Behrman and Deolalikar (1988).

This chapter attempts to address all of these shortcomings of existing research. A model is developed which separates interhousehold and spatial variation in household-reported unit values of health services. Under the assumption that interhousehold variations in unit values within a geographical area or cluster reflect quality variations and spatial variations reflect true price variation, cluster-specific prices of health services that are purged of quality and disease-specific variations are derived and used in a multinomial logit treatment choice model. In order to assess the effect of health-care prices on the reporting of illness episodes, the utilization of curative health services is estimated both conditionally and unconditionally on the probability of an illness being reported. Finally, prices of important foods (also purged of quality variations) are included as determinants of the demand for health services.

III. THE MODEL

Since the theory of demand for medical care is well-developed,[*] there is no need here to develop an elaborate model of individual health determination. If it is assumed that individuals maximize a utility function having health status and other consumption as its argument, subject to a budget constraint and a health production function that includes food and medical care as inputs, the resulting reduced-form derived demand functions for food and medical care will include as their arguments the prices of food and medical care, household income, and socio-demographic individual and household characteristics. As noted earlier, the major empirical problem in estimating such a reduced-form demand system is that health-care prices are not directly observed; what are observed instead are the (endogenous) unit values. The latter need to be purged of disease-specific technological and household-specific quality variations before they can be treated as health-care prices. This is a problem that Deaton (1988) has dealt with in the context of food prices.

We assume that (1) for a given type of health provider (e.g., private physician versus a public health clinic), interhousehold variation in expenditure per illness episode that is explained by individual and household characteristics, such as sex, age, marital status, household income, household size and composition, and traits (e.g., age, schooling and occupation) of the household head, reflects variation in the quality of health services, and (2) that the spatial (intercluster) component of the *unexplained* variation in unit values reflects true (quality-constant) price variation. In other words, it is assumed that when individuals with high income and better schooling spend a larger amount on treating the same ailment from the same type of health provider, they are in effect buying higher quality of health care. However,

---

[*]See Grossman (1972). Behrman and Deolalikar (1988) also develop a generic model of health determination for an LDC.

even after controlling for income, education, and other characteristics, if individuals in one location spend more than consumers in another location for treating the same ailment from the same type of provider, that difference reflects true price variation in the cost of health services across the two locations. While this is a strong assumption, it does not appear to be unreasonable.* Given that the quality of health care selected by households is unobserved, an assumption of this type is required to identify prices from observed unit values.

Controlling for disease-specific effects in the estimation of health-care prices is much more straightforward, since the ailments for which individuals obtain medical care are observed in the data.

An individual's decision not to seek treatment for an illness or to seek treatment from a traditional healer or modern provider is modeled as a multinomial logit problem. The probability of seeking no, tradition, or modern care is

$$P(M_i = k) = \frac{\exp(a_k \ln p_j^M + b_k \ln p_j^F + d_k \ln Y_{ij} + e_k Z_{ij} + \mu_{ij})}{\sum_z \exp(a_z \ln p_j^M + b_z \ln p_j^F + d_z \ln Y_{ij} + e_z Z_{ij} + \mu_{ij})} \tag{1}$$

$$k = 0, 2, \quad z = 1, 2$$

where i = index of the individual

j = index of the location or cluster of residence

M = choice of medical care

k = 0 for no treatment, 1 for a traditional healer, and 2 for a "modern" provider (private physician, health clinic, or hospital)

p^M = vector of health-care prices (i.e., the price of services obtained from traditional healers and modern health providers), derived below

p^F = vector of food prices, derived below

Y = household income

Z = vector of individual and household characteristics, including age, education, family size and composition, etc.

μ = disturbance term

The only problem in estimating relation (1) is that quality-constant health-care prices, p^M, are unobserved. What are observed instead are the unit values of health services obtained from various providers. The derivation of prices from unit values is based on the assumption, stated earlier, that interhousehold variations in unit values reflect quality variations, while spatial variations in unit values reflect true price variation. The unit value can be thought of as the product of the true (quality-constant) price of health services and the quality of health services purchased, i.e.,

$$q_{ij}^M = p_j^M Q_{ij}^M \tag{2}$$

---

*This assumption is similar to that made by Deaton (1988) to separate the effect of true prices from the effect of quality variations on consumer food demand.

where q^M is the unit value and Q is the quality of care obtained from a particular provider.* Note that the true price charged by that provider, p^M, does not have an i subscript, since it is assumed to vary only spatially. Taking the logs of both sides of (2), we have

$$\ln q_j^M = \ln p_j^M + \ln Q_{ij}^M \tag{3}$$

If it is further assumed that the quality of care purchased, Q^M, is a function of household income and a vector of household characteristics (Z), namely

$$\ln Q_{ij}^M = \beta^M \ln Y_{ij} + \gamma^M Z_{ij} + v_{ij}^M \dagger \tag{4}$$

where v is a disturbance term, then (3) can be rewritten as

$$\ln q_{ij}^M = \ln_j^M + \beta^M \ln Y_{ij} + \gamma^M Z_{ij} + v_{ij}^M \tag{5}$$

Equation (5) is a standard cluster fixed-effects model of unit values, in which $\ln p_j^M$ are the cluster fixed effects (or intercepts).

In order to be consistent, food prices, p^F, are derived in an identical manner, even though the control for variations in food quality is not central to this chapter.

IV.   BACKGROUND, DATA, AND EMPIRICAL MODEL

With a total population estimated at 178.2 million in mid-1990, Indonesia is the fourth most populous country in the world (World Bank 1992b). The Indonesian economy has enjoyed rapid economic growth during the last two decades; for example, between 1965 and 1990, Indonesia achieved an annual growth rate of per capita GNP of 4.5%, a rate that few developing countries could match. Available estimates also suggest an impressive reduction in infant mortality in Indonesia during the same period, from 128 infant deaths per 1000 live births in 1965 to 61 in 1990 (World Bank 1992b). Despite this performance, Indonesia has one of the highest infant and maternal mortality rates among Southeast Asian countries. For example, Indonesia has higher infant and maternal mortality levels than the Philippines, Thailand, or even Vietnam. Anemia is the major cause of maternal mortality (Gopalan

---

*The true price and the quality of service are easier to interpret if the unit value (or observed price) is modeled as a product (as opposed to another function) of the true price and quality. In this case, the true price, p_j^M, can be thought of as the amount (in Indonesian Rupiahs) paid per provider visit *of standardized quality*, while the quality variable, Q_{ij}^M, can be regarded as the ratio of a visit of *standardized quality* to an actual visit.

†In the absence of any priors on the functional form for the quality-of-service function, I have chosen the log-linear form for reason of convenience and tractability.

1988), while immunizable diseases (particularly tetanus), diarrhea, and acute respiratory infections are thought to be the leading causes of infant and child mortality (Government of Indonesia–UNICEF 1989, p. 40).

The data for this study come from the 1987 round of the National Socioeconomic Survey (SUSENAS), which is a nationally representative survey of Indonesia that is undertaken periodically. The 1987 round, conducted in January, covered roughly 250,000 individuals residing in 50,000 households. While focusing on the health status of individuals and the choice of health providers for curative care, the 1987 SUSENAS survey obtained detailed information on household consumption expenditures and income as well. Using both one-week and three-month recall techniques, the health module collected data on perceived illnesses (occurrence, type, and length) and the choice of provider for any treatment obtained.

The other data source used is the Village Potential (*Potensi Desa*) module of the Economic Census 1986, a census of all the villages in Indonesia. The Economic Census reports, among other things, information on the environmental hygiene conditions of villages. Although, in principle, it is possible to merge the SUSENAS household data with the village-level information from the Economic Census, the SUSENAS data tapes identify only the district (*kabupaten*)—not the village—of residence of households in order to protect their confidentiality. Hence, the Economic Census Information has been aggregated at the level of *kabupatens* before merging it with the household-level SUSENAS data.

For the purposes of this study, there are two major drawbacks of the 1987 SUSENAS health module. First, as no clinical diagnosis was performed in assessing morbidity, all measures of morbidity are respondent-reported and as such subject to measurement error and respondent biases in illness perception.* For this reason, as previously mentioned, it is even more important not to condition the demand for health services on an illness episode being reported, as there is a strong likelihood of health-care prices influencing the reporting of morbidity by individual respondents. The second problem of the 1987 SUSENAS data is that information was collected for each illness episode, not for each visit to a health provider.† Much of the discussion in the previous section is based on survey information being available on expenditures per visit. To the extent that individuals may have made multiple visits to the same or different health providers for treating an illness episode, observed household expenditures on medical care are unlikely to be unit values. In turn, this would contaminate the derivation of health-care prices from expenditure data. Fortunately, however, the 1987 SUSENAS obtained information on health-care visits and expenditures for two reference periods—one week and three months preceding an

---

*Of course, most household health surveys are plagued by this problem, since the collection of objective indicators of ill health is difficult and expensive.

† Indeed, no information is available on how many visits an individual made to the same or different health providers in treating an illness episode.

interview. It is much less likely that individuals would have made multiple visits to a health provider within a one-week period than in a three-month period. Therefore, the one-week recall data are likely to provide more meaningful and reliable results. In addition, the one-week recall data are more dependable because three months is generally too long a period to recall an illness episode with much accuracy. Therefore, only the one-week recall data are used in the analysis that follows.

The 1987 SUSENAS reports eight treatment alternatives for illnesses reported during the past week and the past three months: no treatment, self-treatment, traditional healer, private physician, hospital, public health center, polyclinic, and paramedic. Since each treatment alternative adds a total of 34 parameters to be estimated in the multinomial logit model, the eight treatment choices have been collapsed into three broad alternatives: no treatment (with the dependent variable assuming a value of zero), self-treatment, or treatment from a traditional healer (value of one), and treatment from a modern health provider (which includes all the remaining choices) (value of 2). With the one-week recall data, the percentage of adults reporting an illness episode who sought these treatment methods were 31.3%, 5.4%, and 63.3%, respectively (Table 1). For children below five years of age, the corresponding figures are 27.6%, 4.3%, and 68.1%. The illness episodes of children aged 5–17 years appear to receive the least attention. For this age group, nearly 37.2% of reported illness episodes were not treated, and fewer than 60% were treated by a modern health provider.

The *household characteristics* (Z) included in the unit value equations and the treatment choice model are size and demographic composition of the household, age, and schooling years of the household head, whether the household head is a salaried employee, and whether the household is covered by any health insurance. The presence of health insurance can often dramatically affect both the choice of quality of health care as well as the choice of health providers for curative services. In Indonesia, the only households typically covered by health insurance are those with a member working in the public sector; in the current sample, only 1.5% of the households belong to this category. Two *cluster characteristics* are included in the unit value and treatment choice models: the proportion of villages in the *kabupaten* having organized garbage collection and those having piped (and hence presumably clean) drinking water. Since both of these environmental hygiene variables affect the production of health status from food and medical care inputs, they are expected to influence the demand for medical care. Finally, three *individual characteristics* are included in the treatment choice model; these are sex, age, and schooling years.*

There are two econometric problems that warrant discussion. First, since household income includes labor earnings, which are likely to be affected adversely

---

*Since the unit values are observed only at the household level, the unit-value equations do not include individual characteristics as explanatory variables.

by an illness in the family and by the treatment choices (including no treatment) made for that illness, there may be a simultaneous-equations bias introduced by the inclusion of household income in the multinomial logit treatment choice model. To address this problem, I use an instrumental variable for log of household income, with the instruments being all of the variables in vector Z (see next paragraph) and nonlabor income.* The latter variable thus provides identification, since the vector Z already appears in the treatment choice model.

Second, the estimates of the cluster fixed effects ($\ln p_j^M$ in (5)) contain error, which can lead to biased estimates of the price effects in the treatment choice model. Fortunately, the *Potensi Desa* data provide *kabupaten*-level information on a large number of environmental and physic 1 .nfrastructure variables, which can be used as instruments for the set of estimate 1 fixed effects. Thus, the parameters of (1) are estimated using instrumental-variable methods to correct for errors in variables in the measurement of $\ln p_j^M$ and $\ln p_j^F$, with *kabupaten* characteristics as identifying instruments. These characteristics include the proportion of villages in a *kabupaten* having a primary school, a lower secondary school, a higher secondary school, a college, a resident health worker, a physician, a hospital, another health facility, a library, piped water, organized garbage collection, and an all-weather road; the proportion of villages in a *kabupaten* that are urban centers, coastal, and accessible only by water; average village population, population density, cultivated area per person, and the proportion of cultivated area that is double-cropped and that is irrigated; average age and schooling of the village head; and the proportion of village heads (in a *kabupaten*) that are female.

V. EMPIRICAL RESULTS

A. Unit Value Regressions

Cluster fixed-effects estimates of the unit value regressions are reported in Table 2.[†] Since the unit value regression for traditional providers is estimated with very few observations, only the empirical results relating to modern-provider unit values providers are discussed here. The estimates indicate that these unit values depend

---

*Usually, the problem in using nonlabor income in most developing-country data sets is that it provides very little additional information, since few households report any asset income. However, this is not the case with the 1987 SUSENAS survey, in which over 94% of households reported nonzero values of annual nonlabor income. Transfers comprise a large fraction of nonlabor income.

[†] Since the true prices of medical care are derived from the estimated cluster effects, it is not possible to control additionally for household fixed effects in the unit value regressions. If individuals within a household have correlated error, not taking into account the error structure may lead to incorrect standard errors. However, this is not a major problem in the current sample. Fewer than 10% of the sample observations for the unit value regressions are accounted for by multiple individuals residing in the same household.

Table 1 Descriptive Statistics, Indonesia, 1987

Variable	Household statistics		Children under 5 years		Children 5–17 years		Adults over 17 years	
	Mean	Std. dev.	Mean	Std. dev.	Mean	Std. dev.	Mean	Std. dev.
Whether illness episode reported during preceding week?			0.114	0.318	0.054	0.227	0.074	0.262
Proportion of reported illness episodes treated by								
no one or self			0.276	0.504	0.372	0.533	0.313	0.547
traditional health provider			0.043	0.204	0.036	0.189	0.054	0.238
modern health provider			0.681	0.520	0.593	0.531	0.633	0.556
Percentage distribution of reported illness episodes								
tuberculosis-type symptoms			0.7%	8.3%	1.1%	10.4%	4.0%	19.7%
diarrhea-type symptoms			6.3%	24.2%	2.4%	15.2%	1.7%	13.1%
malaria-type symptoms			14.8%	35.6%	25.2%	43.4%	16.3%	36.9%
measles-type symptoms			6.3%	24.2%	3.6%	18.7%	0.7%	8.1%
liver-related ailments			5.1%	22.0%	6.3%	24.3%	13.9%	34.6%
other single symptoms			61.4%	48.7%	57.3%	49.5%	57.0%	49.5%
multiple symptoms			5.5%	22.8%	4.1%	19.9%	6.4%	24.5%
"Unit value" (in Rupiahs per reported illness episode) of								
Traditional health services	568							
Modern health services	1.218							
Log total household income	11.514	0.725						
Log nonlabor household income	8.547	3.432						
Age of household head	43.012	13.802						
Schooling of household head	5.032	3.978						

	2.018	0.513
	1.366	0.500
	10.671	0.516
	3.713	0.500
Whether household head employee?	0.386	0.487
Whether health insurance coverage?	0.015	0.120
Age of individual	37.040	14.930
Whether individual male?	0.488	0.500
Schooling of individual	5.219	3.978
Household size	4.788	2.235
Proportion in total household size of:[a]		
Males 0–4 years	0.055	0.108
Males 5–14 years	0.118	0.153
Males 15–24 years	0.090	0.160
Males 25–34 years	0.085	0.143
Males 35–44 years	0.052	0.106
Males 45–54 years	0.039	0.098
Males 55–59 years	0.016	0.069
Males 60 years and over	0.035	0.111
Females 0–4 years	0.052	0.105
Females 5–14 years	0.110	0.146
Females 15–24 years	0.104	0.159
Females 25–34 years	0.081	0.125
Females 35–44 years	0.051	0.108
Females 45–54 years	0.049	0.124
Females 55–59 years	0.017	0.084
Proportion of villages in district (kabupaten) having		
Organized trash collection	0.159	0.186
Piped drinking water	0.089	0.170

[a] Excluded category is females 60 years of age and over.

Table 2 Cluster *(kabupaten)* Fixed-Effects Estimates of Health-Care Log "Unit Values" (i.e., expenditures per illness episode), Indonesian Households, 1987

	Unit values for:			
	Traditional providers		Modern providers	
Independent Variable	Parameter	T ratio	Parameter	T ratio
Whether household head employee?	−0.222	−1.4	0.030	0.8
Household size	0.023	0.4	−0.074	−5.8
Log total household income[a]	0.550	1.7	0.804	11.3
Age of household head	−0.014	−1.4	0.003	1.5
Schooling of household head	−0.012	−0.3	0.021	2.8
Whether health insurance coverage?	−0.716	−0.7	0.434	3.1
Proportion in total household size of[b]				
Males 0–4 years	−0.526	−0.5	−0.597	−2.7
Males 5–14 years	−1.353	−1.5	−0.357	−1.9
Males 15–24 years	−2.467	−2.6	−0.117	−0.6
Males 25–34 years	−2.069	−1.9	−0.265	−1.1
Males 35–44 years	−1.035	−0.9	−0.213	−0.8
Males 45–54 years	−2.791	−2.3	−0.225	−0.9
Males 55–59 years	−1.634	−1.1	−0.105	−0.3
Males 60 years and over	0.050	0.0	−0.249	−1.0
Females 0–4 years	−1.151	−1.2	−0.424	−1.9
Females 5–14 years	−1.526	−1.9	−0.209	−1.1
Females 15–24 years	−1.060	−1.3	−0.200	−1.0
Females 25–34 years	−2.289	−2.2	−0.053	−0.2
Females 35–44 years	−0.978	−0.8	−0.135	−0.6
Females 45–54 years	−0.046	−0.1	−0.081	−0.4
Females 55–59 years	−1.118	−0.7	0.318	1.2
Distribution of illness episodes in household during past week[c]				
tuberculosis-type symptoms	−1.744	−4.0	0.524	6.8
diarrhea-type symptoms	−0.736	−1.8	−0.102	−1.3
malaria-type symptoms	−0.045	−0.2	−0.208	−4.6
measles-type symptoms	−0.057	−0.2	−0.126	−1.4
liver-related ailments	0.254	1.3	0.197	4.4
multiple symptoms	0.094	0.3	0.092	1.4
F-ratio	1.750		2.107	
R-square	0.239		0.535	
No. of obs.	178		51,248	

[a]Instrumental variable. Identifying instrument is log nonlabor household income.
[b]Excluded category is females aged 60 years and over.
[c]Excluded category is "other symptoms."

significantly on the type of illness and symptoms for which treatment is sought. For example, treatment for liver-related diseases costs 19.7% more, and that for tuberculosis costs 52.4% more, than treatment for the excluded category of "other illnesses." In contrast, malaria treatment costs 20.8% less than the excluded category.* Controlling for the disease-specific and the spatial (or cluster) effects, household income has a strong positive impact on both modern- and traditional-provider unit values, with the income elasticity for the former being 0.80 and the latter, 0.55. As would be expected, individuals with health insurance coverage spend significantly (43.4%) more per illness episode than individuals without insurance coverage. The unit value regressions also imply that individuals from large households buy lower levels of health-care quality (i.e., spend less per illness episode), while those from households with better-schooled heads purchase higher quality. Finally, the household composition effects indicate that additional numbers of children aged 0–4 years and boys aged 5–14 years significantly reduce household expenditure per illness episode.

B.   Treatment Choice Model

Table 3 shows the conditional[†] and unconditional multinomial logit estimates of the treatment choice model for children under 5 years of age, while Table 4 and 5 show the corresponding estimates for children aged 5–17 years and adults aged 18 and over. Ten major empirical findings are discussed.

    1.   The conditional estimates, which are comparable to the estimates reported by most previous studies, indicate that while the demand for traditional healer services is significantly responsive to price (with own price elasticities of −0.10, −0.45, and 0.60[‡] for individuals aged 0–4, 5–17, and 18 and over, respectively), the demand for health care from modern providers is not significantly influenced by price. However, the unconditional estimates imply significant own-price responsiveness of demand for modern health services, with estimated elasticities of −0.13, −1.64, and −0.82 (all significantly different from zero at the 5% level) for the three age groups. The estimated price responses are large, especially for children aged 5–17 but also for adults. Indeed, such large own-price elasticities have rarely been reported in the literature. The difference in own price elasticity estimates for modern health care between the conditional and unconditional provider choice models suggests that the price of care in the modern sector strongly reduces the reporting of morbidity by

---

*The large coefficient on the tuberculosis dummy variable in the unit-value regression for traditional providers is not reliable, since only a couple of individuals in the sample obtained treatment for tuberculosis from a traditional provider.

†These are conditioned on an individual reporting an illness during the one-week reference period.

‡Unless otherwise stated, all elasticities reported are evaluated at sample means.

Table 3 Multinomial Logit Estimates of the Probability of Seeking Treatment from a Traditional or Modern Health Provider for a Reported Illness Episode Experienced during the Preceding Week, Children under 5 Years of Age, Indonesia, 1987

	Conditioned on an illness being reported						Unconditioned on an illness being reported					
	Traditional health provider			Modern health provider			Traditional health provider			Modern health provider		
Independent variable	Parameter	T ratio	Elasticity	Parameter	T ratio	Elasticity	Parameter	T ratio	Elasticity	Parameter	T ratio	Elasticity
Intercept	-19.369	-1.5		2.461	0.5		-23.597	-2.1		-6.148	-2.1	
Whether household head employee?	-0.786	-2.8	-0.128	0.248	2.5	0.008	-0.778	-3.0	-3.088	0.226	4.1	0.188
Household size	-0.028	-0.3	0.045	-0.092	-2.4	-0.003	-0.026	-0.3	1.173	-0.089	-4.2	-0.071
Log total household income[a]	0.150	0.3	-0.007	0.205	1.0	0.000	-0.023	0.0	-0.643	0.186	1.6	0.039
Age of household head	-0.026	-1.4	-0.131	0.003	0.4	0.008	-0.033	-1.9	-3.349	-0.004	-0.9	0.203
Schooling of household head	0.020	0.4	-0.023	0.054	2.7	0.001	-0.010	-0.2	0.094	-0.015	-1.4	-0.006
Whether health insurance coverage?	-32.234	0.0	-3.962	-0.181	-0.5	0.240	-16.746	0.0	-51.979	0.146	0.7	3.157
Age of child	-0.171	-2.1	-0.036	-0.015	-0.4	0.002	-0.229	-3.2	-0.939	-0.077	-4.2	0.057
Whether child male?	-0.095	-0.3	0.015	-0.213	-1.5	-0.001	0.076	0.3	0.417	-0.059	-0.7	-0.025
Price of												
Traditional health services[b]	-0.899	-4.0	-0.103	-0.067	-0.7	0.006	-0.688	-3.3	-2.238	0.040	0.9	0.136
Modern health services[b]	0.812	2.1	0.129	-0.235	-1.5	-0.008	0.305	0.9	2.186	-0.406	-4.8	-0.133
Domestic rice	1.755	1.4	0.016	1.625	3.1	-0.001	1.039	1.0	1.659	0.500	1.8	-0.101
Cassava	-0.252	-0.6	-0.025	-0.049	-0.3	0.002	-0.035	-0.1	-0.276	0.055	0.5	0.017
Onion	-0.505	-0.6	0.006	-0.350	-1.7	0.000	-0.220	-0.5	-0.549	-0.041	-0.4	0.033
Canned milk	-1.365	-1.8	-0.173	0.032	0.1	0.010	-0.883	-1.4	-3.200	0.157	0.9	0.194
Meat soya	0.945	2.0	0.096	0.171	0.8	-0.006	0.807	1.9	1.243	0.403	3.4	-0.076
Sugar	1.327	0.7	0.463	-2.420	-2.8	-0.028	2.344	1.3	9.150	-0.630	-1.3	-0.556

Proportion in total household size of

Males 0–4 years	6.225	1.8	0.146	−1.510	−1.3	−0.009	6.247	1.8	3.001	−0.541	−0.9	−0.182
Males 5–14 years	4.748	1.4	0.079	−1.597	−1.5	−0.005	4.618	1.4	2.027	−1.052	−1.8	−0.123
Males 15–24 years	3.453	1.0	0.026	−1.210	−1.1	−0.002	3.386	1.0	0.558	−0.650	−1.0	−0.034
Males 25–34 years	0.954	0.3	0.039	−1.787	−1.5	−0.002	1.658	0.5	0.721	−0.622	−1.0	−0.044
Males 35–44 years	1.937	0.5	0.033	−3.561	−2.9	−0.002	2.854	0.8	0.736	−1.640	−2.4	−0.045
Males 45–54 years	6.922	1.7	0.021	−2.626	−1.8	−0.001	7.797	2.0	0.550	−1.435	−1.8	−0.033
Males 55–59 years	−396.746	0.0	−0.201	1.693	0.7	0.012	−178.350	0.0	−2.513	−0.597	−0.5	0.153
Males 60 years and over	8.985	1.9	0.013	−2.019	−1.1	−0.001	9.377	2.2	0.312	−0.829	−0.9	−0.019
Females 0–4 years	6.246	1.8	0.140	−1.763	−1.5	−0.008	6.392	1.9	3.047	−0.795	−1.3	−0.185
Females 5–14 years	4.836	1.4	0.076	−1.285	−1.2	−0.005	5.151	1.6	1.989	−0.689	−1.2	−0.121
Females 15–24 years	3.682	1.0	0.057	−1.070	−0.9	−0.003	4.217	1.2	1.128	−0.078	−0.1	−0.069
Females 25–34 years	1.284	0.3	0.025	−0.656	−0.5	−0.002	1.570	0.4	0.542	−0.146	−0.2	−0.033
Females 35–44 years	2.840	0.7	0.015	−1.789	−1.3	−0.001	3.115	0.8	0.366	−0.713	−1.0	−0.022
Females 45–54 years	4.981	1.2	0.010	−0.202	−0.1	−0.001	5.635	1.5	0.219	0.578	0.8	−0.013
Females 55–59 years	−1.619	−0.2	−0.002	2.319	1.2	0.000	−2.661	−0.4	−0.046	0.342	0.4	0.003

Proportion of villages in district having

Organized trash collection	−0.852	−0.7	−0.023	0.374	0.8	0.001	−1.914	−1.8	−0.641	−0.679	−3.0	0.039
Piped drinking water	−0.451	−0.3	−0.017	1.023	2.2	0.001	0.148	0.1	−0.292	1.138	5.2	0.018
No. of obs.	2,767						24,202					
Log-likelihood ratio	−1.986						−7,010					

All elasticities evaluated at sample means.

[a] Instrumental variable. Identifying instrument is log nonlabor household income.

[b] Instrumental variable. Identifying instruments are various district (*kabupaten*) characteristics. See text for a list of these characteristics.

Table 4 Multinomial Logit Estimates of the Probability of Seeking Treatment from a Traditional or Modern Health Provider for a Reported Illness Episode Experienced during the Preceding Week, Children Aged 5–17 Years of Age, Indonesia, 1987

| | Conditioned on an illness being reported | | | | | | Unconditioned on an illness being reported | | | | | |
| | Traditional health provider | | | Modern health provider | | | Traditional health provider | | | Modern health provider | | |
Independent variable	Parameter	T ratio	Elasticity	Parameter	T ratio	Elasticity	Parameter	T ratio	Elasticity	Parameter	T ratio	Elasticity
Intercept	-27.732	-2.8		-4.503	-1.2		-35.104	-4.0		-13.391	-5.5	
Whether household head employee?	-0.348	-1.6	-0.266	0.056	0.7	0.017	-0.197	-1.0	-5.492	0.115	2.4	0.354
Household size	-0.015	-0.2	0.112	-0.041	-1.7	-0.007	-0.043	-0.7	3.213	-0.071	-4.4	-0.207
Log total household income[a]	-0.244	-0.6	-0.310	0.226	1.6	0.020	-0.063	0.2	-6.473	0.305	3.3	0.418
Age of household head	-0.022	-1.6	-0.797	0.007	1.4	0.052	-0.034	-2.5	-24.440	-0.002	-0.7	1.577
Schooling of household head	0.004	0.1	-0.201	0.062	4.2	0.013	-0.066	-1.7	-6.871	0.006	0.7	0.443
Whether health insurance coverage?	-35.475	0.0	-23.329	-0.042	-0.2	1.512	-15.222	0.0	-271.515	0.208	1.4	17.517
Age	-0.029	-1.1	-0.044	-0.022	-2.1	0.003	-0.067	-2.6	-0.204	-0.066	-9.9	0.013
Whether male?	0.085	0.4	0.056	-0.001	0.0	-0.004	0.076	0.4	1.270	0.004	0.1	-0.082
Price of												
Traditional health services[b]	-0.584	-3.2	-0.448	0.097	1.4	0.029	-0.471	2.7	-9.720	0.081	2.1	0.627
Modern health services[b]	1.351	4.1	0.972	-0.126	-1.1	-0.063	0.751	2.3	25.433	-0.695	-9.7	-1.641
Domestic rice	2.811	2.6	0.401	2.202	5.5	-0.026	2.058	2.2	15.510	1.177	5.1	-1.001
Cassava	-0.743	-2.3	-0.500	0.015	0.1	0.032	-0.610	-2.1	-13.260	0.144	1.6	0.855
Onion	-0.617	-1.5	-0.290	-0.176	-1.2	0.019	-0.492	-1.3	-10.701	0.116	1.2	0.690
Canned milk	-1.322	-2.2	-0.790	-0.122	-0.5	0.051	-0.742	-1.3	-20.678	0.433	3.1	1.334
Meat soya	-0.007	0.0	-0.218	0.324	2.2	0.014	-0.496	-1.4	-13.283	0.259	2.6	0.857
Sugar	5.602	3.6	5.369	-2.554	-4.1	-0.348	6.962	4.8	139.206	-0.949	-2.4	-8.981

Proportion in total household size of												
Males 0–4 years	0.626	0.3	−0.018	1.217	1.5	0.001	0.868	0.4	0.011	0.852	1.6	−0.001
Males 5–14 years	−0.061	0.0	−0.165	1.182	1.6	0.011	−0.310	−0.2	−2.047	0.269	0.5	0.132
Males 15–24 years	1.107	0.5	0.004	1.038	1.4	0.000	1.095	0.6	0.916	0.526	1.0	−0.059
Males 25–34 years	0.057	0.0	−0.072	2.331	2.7	0.005	−0.662	−0.3	−1.099	0.866	1.5	0.071
Males 35–44 years	0.385	0.2	−0.067	2.120	2.5	0.004	−0.367	−0.2	−0.887	0.495	0.9	0.057
Males 45–54 years	2.158	0.9	0.015	1.599	1.9	−0.001	2.127	1.0	1.158	0.572	1.0	−0.075
Males 55–59 years	2.362	0.8	0.004	1.742	1.6	0.000	2.253	0.8	0.315	0.601	0.8	−0.020
Males 60 years and over	2.225	0.7	0.010	1.072	1.0	−0.001	2.184	0.7	0.538	0.169	0.2	−0.035
Females 0–4 years	1.191	0.5	0.016	0.636	0.8	−0.001	1.531	0.8	0.784	0.388	0.7	−0.051
Females 5–14 years	0.049	0.0	−0.171	1.418	1.9	0.011	−0.209	−0.1	−2.266	0.481	1.0	0.146
Females 15–24 years	0.074	0.0	−0.071	1.528	2.0	0.005	−0.147	−0.1	−0.967	0.515	1.0	0.062
Females 25–34 years	0.040	0.0	−0.047	0.976	1.1	0.003	0.236	0.1	−0.653	0.794	1.4	0.042
Females 35–44 years	1.580	0.7	0.030	0.796	0.9	−0.002	1.518	0.7	0.935	0.652	1.2	−0.060
Females 45–54 years	−0.668	−0.3	−0.024	0.521	0.6	0.002	−0.185	−0.1	−0.423	0.535	0.9	0.027
Females 55–59 years	−1.651	−0.5	−0.003	−0.993	−0.9	0.000	−0.178	−0.1	0.059	−0.654	−0.8	−0.004
Proportion of villages in district having												
Organized trash collection	−0.226	−0.3	−0.034	0.130	0.4	0.002	−0.721	−0.9	−1.145	−0.338	−1.8	0.074
Piped drinking water	−3.434	−2.4	−0.190	0.163	0.5	0.012	−2.337	−1.8	−4.846	0.522	2.7	0.313
No. of obs.	4,218						77,573					
Log-likelihood ratio	−3,262						−11,150					

All elasticities evaluated at sample means.

[a] Instrumental variable. Identifying instrument is log nonlabor household income.

[b] Instrumental variable. Identifying instruments are various district (*kabupaten*) characteristics. See text for a list of these characteristics.

Table 5 Multinomial Logit Estimates of the Probability of Seeking Treatment from a Traditional or Modern Health Provider for a Reported Illness Episode Experienced during the Preceding Week, Adults 18 Years of Age and Over, Indonesia, 1987

| | Conditioned on an illness being reported | | | | | | Unconditioned on an illness being reported | | | | | |
| | Traditional health provider | | | Modern health provider | | | Traditional health provider | | | Modern health provider | | |
Independent variable	Parameter	T ratio	Elasticity	Parameter	T ratio	Elasticity	Parameter	T ratio	Elasticity	Parameter	T ratio	Elasticity
Intercept	-37.223	-6.4		-4.617	-1.7		-45.595	-8.9		-14.149	-8.8	
Whether household head employee?	-0.158	-1.3	-0.171	-0.012	-0.2	0.014	-0.061	-0.5	-1.818	0.066	2.0	0.149
Household size	-0.033	-0.8	-0.225	0.02	0.1	0.019	-0.053	-1.3	-2.955	-0.016	-1.5	0.243
Log total household income[a]	0.245	1.0	0.006	0.239	2.4	-0.001	0.129	0.6	0.534	0.092	1.5	-0.044
Age of household head	0.002	0.4	-0.079	0.004	1.3	0.007	-0.005	-0.8	-0.661	-0.004	-2.2	0.054
Schooling of household head	-0.044	-1.4	-0.392	0.030	2.3	0.033	-0.072	-2.5	-5.221	-0.003	-0.3	0.429
Whether health insurance coverage?	-0.947	-0.9	-0.931	-0.155	-0.6	0.078	-1.333	-1.3	-13.953	-0.358	-2.6	1.147
Age of individual	-0.002	-0.4	-0.165	0.001	0.7	0.014	0.021	4.7	-2.738	0.027	20.6	0.225
Whether individual male?	0.225	1.9	0.421	-0.133	-2.5	-0.035	0.369	3.4	5.656	-0.026	-0.8	-0.465
Schooling (years) of individual	-0.078	-2.7	-0.556	0.036	3.2	0.046	-0.121	-4.5	-8.800	-0.003	-0.4	0.724
Price of												
Traditional health services[b]	-0.438	-4.0	-0.597	0.070	1.4	0.050	-0.369	-3.8	-6.152	0.060	2.3	0.506
Modern health services[b]	0.840	4.5	0.915	0.062	0.8	-0.076	0.147	0.8	10.018	-0.553	-11.2	-0.824
Domestic rice	2.871	4.3	1.687	1.437	5.1	-0.141	2.438	4.2	26.345	0.597	3.6	-2.166
Cassava	-0.192	-0.8	-0.156	-0.059	-0.6	0.013	0.098	0.5	-0.768	0.152	2.5	0.063
Onion	-0.559	-2.2	-0.043	-0.523	-4.9	0.004	-0.159	-0.7	-2.556	0.019	0.3	0.210
Canned milk	0.568	1.6	0.511	0.134	0.8	-0.043	0.697	2.2	6.713	0.228	2.4	-0.552
Meat soya	0.501	2.0	0.487	0.087	0.8	-0.041	0.403	1.7	3.534	0.156	2.3	-0.291
Sugar	2.104	2.2	3.754	-1.088	-2.4	-0.313	3.497	4.0	38.382	0.816	3.0	-3.156

Proportion in total household size of

Males 0–4 years	0.268	0.3	0.018	−0.067	−0.2	−0.002	0.098	0.1	0.202	−0.187	−0.9	−0.017
Males 5–14 years	0.427	0.7	0.057	0.006	0.0	−0.005	0.104	0.2	0.657	−0.284	−1.7	−0.054
Males 15–24 years	0.208	0.3	0.054	−0.254	−0.9	−0.004	0.099	0.2	0.686	−0.341	−1.9	−0.056
Males 25–34 years	1.877	2.7	0.157	−0.060	−0.2	−0.013	1.612	2.4	2.277	−0.436	−2.1	−0.187
Males 35–44 years	1.311	1.5	0.053	0.267	0.7	−0.004	0.604	0.8	0.697	−0.432	−1.8	−0.057
Males 45–54 years	1.849	2.5	0.082	0.268	0.8	−0.007	1.303	1.9	0.922	−0.207	−0.9	−0.076
Males 55–59 years	1.521	1.7	0.028	0.220	0.5	−0.002	0.685	0.8	0.249	−0.336	−1.3	−0.020
Males 60 years and over	0.480	0.7	0.024	0.095	0.3	−0.002	0.449	0.7	0.227	−0.003	0.0	−0.019
Females 0–4 years	1.177	1.6	0.069	−0.118	−0.3	−0.006	1.154	1.6	0.751	0.031	0.1	−0.062
Females 5–14 years	0.260	0.4	0.071	−0.321	−1.2	−0.006	−0.059	−0.1	0.730	−0.520	−3.0	−0.060
Females 15–24 years	0.543	0.9	0.079	−0.088	−0.3	−0.007	0.574	1.0	0.978	−0.024	−0.1	−0.080
Females 25–34 years	−0.647	−0.9	−0.053	−0.006	0.0	0.004	−0.423	−0.6	−0.667	0.206	1.0	0.055
Females 35–44 years	0.191	0.3	0.017	−0.102	−0.3	−0.001	0.229	0.3	0.265	−0.137	−0.7	−0.022
Females 45–54 years	0.374	0.6	0.009	0.245	0.9	−0.001	0.399	0.7	0.121	0.232	1.4	−0.010
Females 55–59 years	0.862	1.3	0.015	0.189	0.6	−0.001	0.575	0.9	0.204	−0.292	−1.3	−0.017

Proportion of villages in district having

Organized trash collection	0.466	1.0	0.007	0.430	2.0	−0.001	0.106	0.2	0.418	−0.055	−0.4	−0.034
Piped drinking water	−1.315	−2.3	−0.128	−0.089	−0.4	0.011	−0.598	−1.1	−1.608	0.445	3.5	0.132
No. of obs.	8,512						114,559					
Log-likelihood ratio	−6,778						−22,922					

All elasticities evaluated at sample means.

[a]Instrumental variable. Identifying instrument is log nonlabor household income.

[b]Instrumental variable. Identifying instruments are various district (*kabupaten*) characteristics. See text for a list of these characteristics.

respondents, with the result that health-care demand models estimated only on the sample of individuals self-reporting an illness episode are likely to significantly understate the adverse effect of health-care prices on the demand for care in the modern health sector.

The fact that the unconditional own-price elasticity of demand for curative treatment is small and insignificant for the 0–5 years age group suggests that very young children in households are relatively more protected than other members against price changes. On the other hand, children aged 5–17 years are most vulnerable to price increases, with their modern provider visits falling by 1.6% for a 1% increase in the price of modern care.

2.   The estimated own-price elasticities of demand for traditional health services also differ significantly between the conditional and the unconditional model. However, all elasticities for traditional health services tend to be unusually large because the sample mean of the unconditional probability of using traditional health services, which is the point at which the elasticities are evaluated, is extremely small.

3.   Household size has a strong negative effect on both the conditional and unconditional demand for modern care among children (0–17 years) but not among adults. Surprisingly, household income is generally not significant in determining the demand for either traditional or modern care. The only case where income is significant is in influencing the demand for modern care among children aged 5–17 years (with an income elasticity of 0.42). However, two facts are relevant in putting the nonsignificance of income in perspective: (1) the significant negative effect of household size implies that the demand for modern health providers does increase with household income per capita, especially for children aged 0–4 and 5–17 years of age; and (2) as noted earlier, household income has strong positive effects on the demand for *quality* (i.e., unit values) of both modern and traditional health services.

4.   The cross-price elasticities of demand for health care—estimated by few previous studies for developing countries—are positive for all three age groups (although not significantly so for children under 5), indicating that traditional and modern health services are (gross) substitutes for each other. The estimated elasticity of demand for modern care with respect to the price charged by traditional health providers is 0.63 for children aged 5–17 years and 0.51 for adults. The corresponding cross-price elasticities for traditional health services are extremely large (25.43 and 10.02 for persons aged 5–17 and 18 and over, respectively), again reflecting the extremely low levels of usage of traditional healers. These estimates indicate that an increase in the price of modern health services can cause a significant shift away from the modern to the traditional health sector.

5.   The effects of schooling of the individual* and that of the household head on health-care demand also differ between the conditional and unconditional mod-

---

*An individual's own schooling was included in the demand equations only for adults over 17 years of age, since it is likely to be endogenous for children.

els. While the former suggest that the demand for modern health care is strongly (positively) influenced by an individual's schooling or the schooling of the household head, the unconditional estimates shown neither or these effects to be significant.* The latter result reflects the fact that the unconditional estimates confound the effects of schooling on morbidity (which are likely to be negative), the reporting of morbidity (positive), and the treatment of illness episodes from modern health providers (positive).

6.   Food prices, which have not been included in the demand for health care by previous studies, are observed to generally have very strong effects on the demand for both traditional and modern health care. Most of the estimated food price effects on the demand for traditional care are positive, implying that food and traditional health care inputs are viewed as substitutes. However, the probability of choosing a modern provider declines with most food prices, implying that food and modern health-care inputs are complementary. Thus, the effect of rising food prices, especially the price of rice and sugar, appears to be to shift people from modern to traditional health providers.

7.   Among all three age groups, age is associated with an increased (unconditional) probability of using modern providers. Since age has generally no effect on the conditional probability of choosing a modern provider,[†] this suggests that either true morbidity or the reporting of morbidity increases with age. Among adults, morbidity is likely to increase with age; however, it is likely that, among children under 5, it is the perception and reporting of morbidity—not morbidity itself—that increases with age.

8.   Health insurance coverage and employee/self-employed status of the household head have strong influences on the demand for modern medical care. The magnitudes of the estimated health insurance coverage effects are especially large (with insured individuals aged 0–4, 5–17, and 18 and over having 315.7%, 1751.7%, and 114.7%, respectively, greater demand for modern care than uninsured individuals). However, only the estimate for adults over 17 is significant. A possible reason for the nonsignificance of the insurance variable is that it is highly correlated with the employee status variable, since all government employees in Indonesia as well as most salaried individuals employed in the formal business sector are covered by health insurance schemes. Likelihood-ratio tests showed that the employee status and insurance coverage variables were jointly significant in determining the choice of modern health providers for all three age groups.

---

*For adults, both own schooling and the household head's schooling do significantly reduce the conditional and unconditional probability of using a traditional health provider. For children aged 5–17 years, schooling of the household head is associated with a significantly lower unconditional probability of choosing a traditional healer.
[†] For children aged 5–17 years, age does have a significant, although numerically small, effect on the conditional probability of using a modern provider.

9.   Relatively few of the household composition effects are significant in influencing the choice of providers. An increase in the proportion of males aged 5–14, 35–44, and 45–54 in the household significantly reduces the use of modern health providers for children aged 0–5 years. For children aged 5–14, there are no significant household composition effects. Finally, an increase in the population of males aged 5–44 and females aged 5–14 years in the household significantly reduces the use of modern health providers by adults.

10.   The environmental hygiene variables generally have significant and positive effects on the (unconditional) probability of using modern providers. Both the availability of safe (piped) drinking water and organized trash collection serve to increase the demand for modern health providers among persons of all ages,* implying that environmental hygiene and modern medical care are viewed by households as complementary inputs in the production and maintenance of health status.

VI. CONCLUDING REMARKS

The empirical results in this chapter clearly show that the demand for both traditional and modern health services in Indonesia is highly responsive to own price when it is not *conditioned on the reporting of morbidity* and when quality variations are purged from the price of health services. Indeed, unlike most previous studies for LDCs that show very small price effects on health-care demand, the price elasticities estimated here are large in magnitude (-0.13, -1.64, and -0.82 for modern care and -2.2, -9.7, and -6.2 for traditional care for children under 5, children aged 5–17 and adults over 17, respectively).[†] The fact that the estimated own-price elasticities for modern health providers are large and significant in the unconditioned probability model but small and insignificant in the conditioned probability model suggests that the price of care in the modern sector strongly reduces the reporting of morbidity by respondents, so that health-care demand models estimated only on the sample of individuals self-reporting an illness episode are likely to significantly understate the total effect of health-care prices on the demand for care in the modern health sector.

Thus, these findings cast doubt on the assumption, commonly maintained in the literature, that increasing user fees at government health facilities will have limited effects on the utilization of health services and thus enable governments to raise health-sector revenues. The empirical results clearly indicate that the utilization of modern health services, especially by children aged 5–17 and adults, will decline

---

*The single exception is the effect of garbage collection on the demand for modern providers among adults, which is not statistically significant.

[†]The demand elasticities for traditional healer services are particularly large because the sample mean level of usage of traditional healers is very low.

sharply if prices in the modern health sector are raised *without simultaneously improving the quality of health services.*

The empirical results also indicate that a number of other prices—in particular, the price for health services charged by traditional healers and the price of food staples (namely rice and sugar)—have significant effects on the demand for modern health services. An increase in user fees charged by traditional healers is associated positively, while an increase in the price of rice and sugar is associated inversely, with the utilization of modern health services. Interestingly, the availability of piped drinking water and organized trash collection in the community, both of which are health-improving public interventions, serves to increase the use of modern health providers. These results thus imply that individuals view traditional and modern health services as substitutes, modern medical care and food inputs as complements, and modern health care and environmental hygiene interventions as complements, in the production and maintenance of health status. There are obvious implications here for food price policy. Likewise, the empirical findings indicate that there may be important positive externalities (in the form of increased use of modern health services) to environmental hygiene and sanitation interventions.

The empirical estimates also imply a massive increase in the demand for modern health services with an expansion in health insurance coverage in the Indonesian population. Currently, a very small fraction of households—only those having a member in government or formal-sector wage employment—are covered by health insurance schemes. However, like many other LDCs, Indonesia has been experimenting with expanded health insurance coverage for a much larger proportion of the population.

Finally, the empirical results suggest that income growth alone is unlikely to increase the utilization of modern health services in Indonesia. With increasing incomes, however, individuals are likely to purchase higher quality of health services (i.e., spend more per illness episode).

ACKNOWLEDGMENTS

I would like to thank Jeffrey Hammer and Martin Ravallion, with whom I have had numerous discussions on this topic. This chapter is a product of a World Bank research project, "Determinants of Nutritional and Health Outcomes in Indonesia and Implications for Health Policy Reforms," in which both of them participated. I am grateful to the World Bank Research Committee for funding the research. I would also like to acknowledge the useful comments of an anonymous referee and of seminar participants at Harvard, Yale, and the University of Pennsylvania, where an earlier version of this work was presented. The responsibility for all errors rests entirely with me.

REFERENCES

Akin, J. S., N. Birdsall, and D. de Ferranti (1987), *Financing Health Services in Developing Countries*, The World Bank, Washington, D.C.

Akin, J. S., C. Griffin, D. K. Guilkey, and B. M. Popkin (1984), *The Demand for Primary Health Care in the Third World*, Littlefield, Adams, Totowa, NJ.

Akin, J. S., C. Griffin, D. K. Guilkey, and B. M. Popkin (1986), The Demand for Primary Health Care in the Bicol Region of the Philippines, *Economic Development and Cultural Change*, 34, 755–782.

Alderman, H. and P. Gertler (1988), The Substitutability of Public and Private Medical Care Providers for the Treatment of Childrens' Illnesses in Urban Pakistan, Living Standards Measurement Study Working Paper, World Bank, Washington, D.C.

Behrman, J. R. and A. B. Deolalikar (1988), Health and Nutrition, in Hollis Chenery and R. N. Srinivasan (eds.), *Handbook of Development Economics*, vol. 1, North-Holland, Amsterdam.

Birdsall, N. and P. Chuhan (1986), Client Choice of Health Treatment in Rural Mali, mimeo, Population, Health and Nutrition Department, World Bank, Washington, D.C.

Cornea, A. P., R. Jolly, and F. Stewart (eds.) (1987), *Adjustment with a Human Face*, vol. I, Clarendon Press for UNICEF, Oxford.

Deaton, A. (1988), Quality, Quantity and Spatial Variation of Price, *American Economic Review*, 78, 418–430.

Deolalikar, A. B. and P. Vashishta (1992), The Utilization of Government and Private Health Services in India, mimeo, University of Washington, Seattle.

Gertler, P., L. Locay and W. Sanderson (1987), Are User Fees Regressive? The Welfare Implications of Health Care Financing Proposals in Peru, *Journal of Econometrics*, 33, 67–88.

Gertler, P. and J. van der Gaag (1990), *The Willingness to Pay for Medical Care: Evidence from Two Developing Countries*, Johns Hopkins University Press, Baltimore.

Gilson, L. (1988), Government Health Care Charges: Is Equity Being Abandoned? EPC Publication No. 15, London School of Hygiene and Tropical Medicine.

Goldman, F. and M. Grossman (1978), The Demand for Paediatric Care: A Hedonic Appraisal, *Journal of Political Economy*, 86, 259–280.

Government of Indonesia and UNICEF (1989), *Situation Analysis of Children and Women in Indonesia*, Jakarta.

Grossman, M. (1972), On the Concept of Health Capital and the Demand for Health, *Journal of Political Economy*, 80.

Heller, P. (1981), A Model of the Demand for Medical and Health Services in Peninsular Malaysia, *Social Science and Medicine*, 16, 267–284.

Jimenez, E. (1987), *Pricing Policy in the Social Sectors: Cost Recovery for Education and Health in Developing Countries*, Johns Hopkins for The World Bank, Baltimore.

Manning, W. G., J. P. Newhouse, N. Daan, E. Keeler, B. Benjamin, A. Leibowitz, M. S. Marquis, and J. Zwanziger (1987), Health Insurance and the Demand for Medical Care, mimeo, The Rand Corporation, Santa Monica.

Mwabu, G. (1986), Health Care Decisions at the Household Level: Results of Health Survey in Kenya, *Social Science and Medicine*, 22, 313–319.

Newhouse, J. P. and C. E. Phelps (1974), Price and Income Elasticities for Medical Services, in M. Perlman (ed.), *The Economics of Health and Medical Care*, Wiley, New York.

Rosset, R. N. and L. F. Huang (1973), The Effect of Health Insurance on the Demand for Medical Care, *Journal of Political Economy*, 81, 281–305.

Schwartz, J. G., J. S. Akin, and B. M. Popkin (1988), Price and Income Elasticities of Demand for Modern Health Care: The Case of Infant Delivery in Indonesia, *The World Bank Economic Review*, 2, 49–76.

World Bank (1987), Health Center Expenditure in Indonesia, Background Paper III for *Health Planning and Budgeting*, Washington, D.C.

World Bank (1992a), Viet Nam: Population, Health and Nutrition Sector Review, Report No. 10289-VN, Country Department I, East Asia Regional Office, The World Bank.

World Bank (1992b), *World Development Report* 1992, Oxford University Press for The World Bank, New York.

5
On Mobility

Esfandiar Maasoumi
Southern Methodist University, Dallas, Texas

I. INTRODUCTION

Mobility in any social hierarchy is an indication of opportunity. It is therefore a measure of fairness in any economic system. In a real sense, mobility should be of greater concern to policymakers and analysts than such other important concerns as inequality. It may be understandable and socially tolerable for the young, say, to have less assets or income, but the likelihood and the opportunity to move up and earn in relation to effort is surely a decidedly desirable social goal. What matters in this context is "lifetime equity" rather than instantaneous equality.

There are essentially two types of movements that constitute mobility. One is a "growth" or "structural" mobility which may arise from a general economic movement up, or down. The other is the "lateral type," or "exchange mobility" which obtains when individuals, households, or groups of such units move from one state to another. Since there are at least some temporal limits to such upward movements, mobility, particularly of the lateral type, must be "equalizing." Thus welfare comparison between any two mobility states necessarily requires functionals that reflect degrees of social preference for equity, exchange mobility, and growth (new opportunities).

There are two important questions that need attending to. One is how does one characterize economic status? The other is, how does one evaluate and compare different mobility situations? The latter question poses the issue of why is mobility socially desirable and, if it is, what kinds of social welfare functions (SWFs) represent our preferences for mobility? The first question is all too often resolved by using "income" as a proxy. Such a proxy is evidently more meaningful the more market oriented an economy and a culture, and the less is the level of in-kind and other transfer payments and entitlements.

The reliance on SWFs for evaluating and comparing welfare situations has provided a scientifically useful tool that provides for economy of thought, as well as discipline, since it forces a declaration of principles that are too often implicit. In the instant case, the desirability of both types of mobility suggests functionals that are increasing in incomes (for the growth component), and *ultimately* equality preferring (for the exchange component). While changes in earnings and incomes have been and should be studied with the aim of identifying significant "causal" factors, the evaluation of an existing degree of mobility requires welfare comparisons.

There are currently at least two complementary lines of analyzing mobility. The older approach requires specification of transition probabilities between social states, and a welfare evaluation of the transition matrices that are estimated from the existing data. This clearly requires detailed data, with large panel data being the best source for sufficient cell repetitions which is necessary for reliable statistical inference. There are numerous mobility indices which are mappings from the transition matrices to scalars. Shorrocks (1978a) and Geweke et al. (1986) provide systematic discussions of criteria for sensible mobility indices.

In line with a general direction toward unanimous partial ordering in the literature on inequality, useful welfare ranking relations have been developed for transition matrices which rekindle the essential role played by Lorenz and generalized Lorenz criteria. While the SWF evaluation has begun to be emphasized in this approach, the task of devising consensus SWFs over matrices remains a challenge reminiscent of that faced in the multidimensional analysis of inequality. I provide an account of this line of inquiry in Section III.

The second approach was initiated by Shorrocks (1978b) and generalized by recasting mobility as a multidimensionality question as in Maasoumi (1986); e.g., see Maasoumi and Zandvakili (1986, 1989, 1990). In this approach inequality indices are computed for multiperiod incomes, that is, a type of "permanent income" measured over more than a single period, and mobility indices with a profile of equalization over time are obtained. The latter are directly related to and interpretable by the familiar classes of increasing and equality preferring (Schur concave) SWFs. It will be seen that the more recent welfare theoretic development of the transition matrices approach is converging to the same welfare comparisons, similar notion of long-run or "permanent" income, and thus similar preferred mobility indices! Quite general mobility indices are proposed and empirically implemented by Maasoumi and Zandvakili (1989, 1990). This approach is also demanding of data and, like the first approach, it is ideally implemented with plentiful micro data. But the Maasoumi-Shorrocks-Zandvakili (MSZ) measures may be adequately estimated with data grouped by age, income, education, etc. This is useful since much data is made available in this aggregated form, and the approach is particularly focused on "equalization" between income groups in a way that allows controlling for sources of heterogeneity among individuals and households. Such controls with aggregate data appeal to some who may wish to tolerate some diversity due to such things as age or education. I present an account of this approach in Section II.

Our survey makes clear that both approaches share a concern for the distribution of a welfare attribute as well as its evolution. And the convergence in both approaches to rankings by Lorenz-type curves is of econometric significance. There is now a well developed asymptotic inference theory for empirically testing for order relations such as stochastic dominance, Lorenz dominance, and concentration curves. This type of testing is a crucial first step since comparing mobility indices (statistically or otherwise) is of questionable value when, for instance, generalized Lorenz curves cross. Section IV contains a brief account of the statistical tools that are available for inference on both the indices of mobility and order relations.

A loosely related strand of econometric research seeks to specify statistical models of earnings or income "mobility." While it is true that such models are focused on explaining "change" and "variation" which are not as meaningful as "mobility," they can shed light on significant *explanations* of earnings changes, as well as account for heterogeneity. This is consequential for policy analysis. Further, there are econometric models that seek to fit transition probabilities. Such studies are directly useful for not only estimating transition matrices, but for explaining the estimated probabilities. We do not delve into this empirically substantive area in this survey. Section V concludes with several empirical applications of the "inequality reducing" mobility measures. An insightful survey of income mobility concepts is given in Fields and Ok (1995). Absolute measurement of income mobility and partial ordering of absolute mobility states is treated in Mitra and Ok (1995).

I

II. INEQUALITY REDUCING MOBILITY INDICES

A.   The MSZ Indices

The most common approach to comparing welfare situations is based on indices. For example, two income vectors depicting the respective distribution of households at two points in time, or in different regions, are explicitly and implicitly ranked by scalar measures of such things as inequality, mobility, or poverty. The welfare theoretic underpinnings and limitations of index-based comparisons are now well understood. Notwithstanding these serious limitations, the need for "measurement" has led to the development of many "sensible" indices, supplemented with statistical tools which are based upon the asymptotic distributions of the indices; see Maasoumi (1996) for a recent survey. This "index-based" approach is limited because of a lack of unanimity on the acceptable index even when there is broad consensus about certain normative principles. But the main difficulty is a consequence of the difficulties of welfare "comparability" as well as the appropriate cardinalization of the class of admissible criterion functions.

In the case of mobility indices, Shorrocks (1978b) and Maasoumi and Zand-vakili (1986) argued that a more reliable measure of individual welfare and the distribution of incomes is to be obtained by considering the individual or household's "long-run" income. They argued that such a measure of income computed over increasingly longer periods of time would remove transitory and some other life cycle related movements which are picked up by year-to-year comparisons of income distributions. The annual "snapshots" are incapable of accounting for mobility and returns to investments and/or human capital. The effects of seniority alone may make the notion of income inequality meaningless.

These authors are therefore concerned with dynamics of income distribution, and are thus accounting for, and challenged by, a fundamental lack of *homogeneity* among households. The natural labelling of individuals at different points in their life cycle is an essential form of heterogeneity which contradicts the common assumption of symmetry (anonymity) which plays a crucial role in much of the welfare theory that underlies inequality analysis. Shorrocks (1978b) proposed the simple sum of incomes over T periods as the aggregate income. Maasoumi and Zandvakili (1986, 1989, 1990) proposed more general measure of "permanent income" encompassing the simple average and sum. This author recognized the essential multidimensionality of the mobility analysis and proposed its treatment on the basis of the techniques and the concepts developed in Maasoumi (1986). Consequently, general functions for "permanent incomes" where developed which are maximum entropy (ME) aggregators.

The next step in this development is to analyze the inequality in the permanent incomes and compare with single period inequalities. A weighted average of the latter was used by these authors to represent "short-run" inequality over any desired number of periods. Clearly, a distribution of permanent incomes is being compared and ranked with a reference short run income distribution. In view of this, all of the rich welfare theory supporting Lorenz and generalized Lorenz (second order stochastic dominance) dominance relations, as well as the convex inequality measures consistent with such relations, comes at the disposal of the analyst for evaluating mobility profiles and dynamic evolution of the income distribution. At first sight, this appears an unnecessarily restrictive setting for defining the welfare value of mobility. Further reflection suggests that this is not so. Indeed, we will see that the recent development of a welfare theory basis for the alternative of transition matrices has made clear a certain inevitability for the role of the same welfare criteria and, therefore, the same types of mobility measures as the Maasoumi-Shorrocks-Zandvakili indices!

Let X_{it} be the income of the ith individual in the tth state (period, say); $i \in [1, n]$, and $t \in [1, T]$. Let $S_i(X_i; \alpha, \beta)$ denote the ith individual's *permanent* income (living standard?) over a number of periods $k = 1, 2, \ldots, T$, and $S = (S_1, S_2, \ldots, S_n)$ denote the vector of such incomes for the n households or individuals. The inequality measures which are consistent with a set of axioms to be described below are represented by $I_\gamma(\cdot)$. Let X denote the welfare matrix with the typical element X_{it} and denote its ith row by X_i and its tth column by X^t. The latter is the income vector in the tth period/state.

The k-*period* long-run inequality is given by $I_\gamma(S)$, and short run inequality may be represented by $I_\gamma^k = \sum_{t=1}^{k} \alpha_t I_\gamma(X^t)$, for $k = 1, 2, \ldots, T$. The vector $\alpha = (\alpha_1, \alpha_2, \ldots, \alpha_T)$ represents the weights given to the income distribution in different periods, such as the ratio of the mean income in the period, μ_j, and the overall mean of incomes in all the k periods under analysis.

Shorrocks (1978b) proposed $S_i = \sum_{t=1}^{k} X_{it}$, and μ_t/μ, the ratio of means just described, as weights, and Maasoumi and Zandvakili (1986) generalized S_i and the weight functions, suggesting the following index of mobility:

$$M_\gamma^k = 1 - \frac{I_\gamma(S)}{I_\gamma^k} \tag{1}$$

with

$$R_\gamma^k = \frac{I_\gamma(S)}{I_\gamma^k}, \quad \text{as a measure of stability} \tag{2}$$

where it is to be noted that $S(\cdot)$ is measured over the same k periods. When S is quasi-concave $R \in [0, 1]$ follows from the convexity of the inequality measures considered. A "mobility profile" is generated by depicting R as k increases over its range. Both Shorrocks (1978b) and Maasoumi and Zandvakili (1986, 1989, 1990) report empirical studies of the US data from the PSID. Section V describes some of these studies.

Bounds on R_γ^k may be established using the decompositions of multidimensional measures of inequality given in Maasoumi (1986, Propositions 1 and 2). $P_T = (R_1, R_2, \ldots, R_T)$ is a stability profile which can also reveal the effect of increasing smoothing of the income variable starting from $R_1 = 1$. Other than this smoothing out of the short-run effects, R_γ is capable of revealing durable "mobility" toward equalization in a way that may be obscured by looking at each $I_\gamma(X^t)$, $t = 1, \ldots, T$. This is most clearly seen by considering a situation in which only a permutation of the income vector has occurred between two periods. As is well known, this leaves our *anonymous* (symmetric), relative inequality measures unchanged, but the distribution of the *aggregated* incomes over the two periods will be changed, possibly dramatically (unless there is perfect equality to begin with!). Consequently, $I_\gamma(S)$ and R_γ will measure any mobility over the two periods.

Given that the indices $I_\gamma(.)$ are supported by Schur-concave SWFs, ordering mobility states by the class of measures in R is consistent with unanimous partial orderings conducted with generalized Lorenz curves and related orderings. For this reason, and to help in the discussion of Section III, it is useful to have a brief account of the relevant welfare theory.

1. The Fundamental Welfare Axioms

Kolm (1969) and Atkinson (1970) provided the modern and influential formalizations of the relationship between SWFs and inequality measures. The need for establishing this important relationship is now recognized widely when analyzing poverty and mobility. However, the axiomatic SWF approach does not by itself identify unique indices even when a particular set of normative properties are consented to by a majority. To appreciate this point, consider the Atkinson family of inequality measures for income x having mean μ_x:

$$A_v = 1 - \frac{1}{\mu_x} \left[\int_0^\infty x^{1-v} \, dF \right] 1/(1 - v), \qquad v > 0, \quad v \neq 1 \tag{3}$$

$$= 1 - \exp \left[\int \log \left(\frac{x}{\mu_x} \right) dF \right], \qquad v = 1 \tag{4}$$

where F is the c.d.f. of income. Similarly, the generalized entropy (GE) family of indices is

$$I_\gamma(X) = \frac{1}{\gamma(1 + \gamma)} \int_0^\infty \frac{x}{\mu_x} \left[\left(\frac{x}{\mu_x} \right)^\gamma - 1 \right] dF, \qquad \gamma \text{ real} \tag{5}$$

Ordinally I_γ is equivalent to the coefficient of variation and the Herfindahl index, and includes the variance of logarithms and Theil's first and second measures, I_0 and I_{-1}, respectively. Also, up to a monotonic transformation, there is a unique member of GE corresponding to each member of the Atkinson family. $v = -\gamma$ is the degree of aversion to relative inequality; the higher its absolute value the greater is the sensitivity of the measure to inequality (transfers) in the tail areas of the distribution.

Employing the functional theory first developed in "information theory" for identifying appropriate measures of divergence between distributions, (Maasoumi 1993), and noting that inequality and many other indices are similar measures of divergence, one puts down an *explicit* set of normative properties (axioms) which inequality indices and/or SWFs must satisfy. Using these axioms as explicit constraints on the function space one then obtains the appropriate inequality index. To exemplify, let us follow Bourguignon (1979) or Shorrocks (1980, 1984) in their discussion of the "fundamental welfare axioms" of symmetry (anonymity), continuity, principle of transfers, and additive decomposability. Combined with homogeneity, these axioms identify members of the GE as the desirable family of relative inequality measures.

Axiom 1 (Anonymity).   The inequality index (function) is symmetric in incomes.

Axiom 2 (Homogeneity).   Invariance to scalar multiplication.

Axiom 3.   Continuity.

Axiom 4.   Principle of transfers.

This requires that inequality decrease if we redistribute from a single richer individual to a poorer one, leaving their respective ranking and all the other individuals' incomes unchanged.

Axiom 2 is a serious limitation as it restricts attention to "relative" inequality. This is so since this requirement implies mean invariance—doubling everyone's income would leave inequality unchanged.

The class of functionals satisfying Axioms 1–4 is large! Also, any further axioms are less likely to command unanimity. In fact, any further requirements must be justified by plausible considerations of such things as, heterogeneity amongst households and/or individuals, policy, empirical necessity, and practical interest. The most commonly invoked of such requirements is:

Axiom 5.   Additive decomposability.

This requirement, later strengthened as an "aggregation consistency" axiom by Shorrocks (1984), says that total inequality must be the sum of a "between-group" component, obtained over group means, and an additive component which is a weighted sum of "within-group" inequalities. This kind of decomposability is very useful for controlling and dealing with heterogeneity of populations, and as a means of unambiguously identifying the sources of inequality and those that are affected by it. In the context of mobility analysis, this property further serves to identify the contributions of each time interval under consideration.

If the additive decomposability requirement of Axiom 5 is imposed, such measures as Gini, the correlation coefficient, and variance of logarithms must be excluded. The latter measures provide ambiguous decompositions of overall inequality by population subgroups (Shorrocks 1984). For the GE family, for instance, a discretized (estimation) formula that helps to demonstrate its decomposability is

$$I_\gamma = \sum_{r=1}^{R} \left[X_r \bigg/ \sum_{j=1}^{n} x_j \right]^{\gamma+1} \left(\frac{n_r}{n}\right)^{-\gamma} I^r + I_\gamma^b \tag{6}$$

where I_γ^b is the between group inequality computed over the group mean incomes, I^r is the inequality within the rth group, $r = 1, 2, \ldots, R$, n_r is the number of units in group r, and X_r is total income of the rth group.

2.   Aggregate or Permanent Income

As in Maasoumi (1986), we consider the following weighted generalized entropy measure of divergence between S, on the one hand, and X^t, $t = 1, 2, \ldots, T$, we have

$$D_\beta(S, X; \alpha) = \sum_{t=1}^{T} \alpha_t \sum_{i=1}^{n} \frac{S_i[(S_i/X_{it})^\beta - 1]}{\beta(\beta + 1)} \tag{7}$$

where $\alpha_j s$ are the weights attached to each period. Minimizing D_β with respect to S_i such that $\sum S_i = 1$, produces the following "optimal" aggregate income functions:

$$S_i \propto \left(\sum_t^T \alpha_t X_{it}^{-\beta}\right)^{-1/\beta}, \qquad \beta \neq 0, -1 \tag{8}$$

$$S_i \propto \prod_t^T X_{it}^{\alpha_t}, \qquad \beta = 0 \tag{9}$$

$$S_i \propto \sum_t \alpha_t X_{it}, \qquad \beta = -1 \tag{10}$$

These are, respectively, the hyperbolic, the generalized geometric, and the weighted means of the incomes over time. Noting that the "constant elasticity of substitution" $-\sigma = 1/(1 + \beta)$, these functional solutions include many of the well-known utility functions in economics, as well as some arbitrarily proposed aggregates in empirical applications. For instance, the weighted arithmatic mean subsumes the simple total income discussed earlier, and a popular "composite welfare indicator" based on the principal components of X, when α_t's are the elements of the first eigenvector of the $X'X$ matrix (Maasoumi 1989a). The "divergence measure" $D_\gamma(\cdot)$ forces a choice of an aggregate income vector $S = (S_1, S_2, \ldots, S_n)$ with a distribution that is closest to the distributions of its constituent variables. This is desirable when the goal of our analysis is the assessment of income *distribution* and its dynamic evolution. The entropy principle establishes that any other S would be extra distorting of the objective information in the data matrix X. The distribution of the data reflect the outcome of all optimal allocative decisions of all agents in the economy (Maasoumi 1986).

The next step in constructing general mobility indices as proposed by Maasoumi and Zandvakili (1986) is the selection of a measure of inequality. The GE index described above was computed for the S_i functions just obtained. It is instructive to analyze this measure in the discrete case:

$$I_\gamma(S) = \sum_{i=1}^n \frac{p_i[(S_i^*/p_i)^{1+\gamma} - 1]}{\gamma(1+\gamma)}, \qquad \gamma \neq 0, -1 \tag{11}$$

$$I_0(S) = \sum S_i^* \log\left(\frac{S_i^*}{p_i}\right), \qquad \text{Theil's first index}$$

$$I_{-1}(S) = \sum p_i \log\left(\frac{p_i}{S_i^*}\right), \qquad \text{Theil's second index} \tag{12}$$

where p_i is the ith unit's population share (typically $= 1/n$), and S_i^* is S_i divided by the total $K + \sum_{j=1}^n S_j$.

These inequality indices are normalized isoelastic transformations of the aggregate income functions S_i. As such they are "symmetric," "homogeneous," and consistent with the Lorenz criterion. They will be homogeneous with respect to every X^t, the tth column/period in X, if in all the above one works with the matrix of shares, $x = (x_{it})$. While this will not change the functional solutions given above, it requires a rather unusual assertion that individual well-being depends on relative incomes. Alternatively, one can impose a general form of scale invariance at the outset (Maasoumi 1996b).

Useful decomposability properties are possessed by these measures both in population groups and in the time directions.

Theorem 1 (Decomposability of GE).   Let $x_{it} = X_{it}/T_t$, $T_t = \sum_i X_{it}$, $W_t = T_t/K$, $I_\gamma(X^t) = $ the GE inequality in the tth period, and $\delta_t = \alpha_t/\sum_{f=1}^T \alpha_f$. Then
    (i)   *If $1 + \gamma = -\beta$, we have*

$$I_\gamma(S) = \sum_{t=1}^T \delta_t W_t^{1+\gamma} I_\gamma(X^t) + \frac{\sum_t \delta_t W_t^{1+\gamma} - 1}{\gamma(1+\gamma)} \tag{13}$$

    (ii)   *If the marginal distributions are identical—i.e., $x^t = x^k$, $\forall t$ and k—we have*

$$I_\gamma(S) = I_\gamma(X^t), \qquad any\ t \in [1, T] \tag{14}$$

    (iii)   *For Theil's first and second measures ($\dot\gamma = 0, -1$), we have*

$$I_0(S) = \sum_{j=1}^m C_t I_0(X^t) - D_{-1}(x, S^*; C) \tag{15}$$

where $C_t = \delta_t T_t/\sum_k \delta_k T_k$, and

$$I_{-1}(S) = \sum_{t=1}^T \delta_t I_{-1}(X^t) - D_0(x, S^*; \delta) \tag{16}$$

where by application of L'Hôpital's rule to $D_\beta(\cdot)$ defined earlier, we have

$$D_0(S^*, x; \delta) = \sum_t \delta_t \sum_i S_i^* \log\left(\frac{S_i^*}{x_{it}}\right) \geq 0 \tag{17}$$

and S_i defined at $\beta = 0$ and

$$D_{-1}(S^*, x; C) = \sum_t C_t \sum_i x_{it} \log\left(\frac{x_{it}}{S_i^*}\right) \geq 0 \tag{18}$$

and S_i defined at $\beta = -1$.

Proof.   See Maasoumi (1986, Propositions 1–2).

In view of the nonnegativity of the $D(\cdot)$ terms in part (iii) it is clear that multi-period inequality is no more than the weighted average of inequalities in the single periods. This is due to the intertemporal "substitution" effects and a consequence of the convexity of the inequality measures.

B.   Welfare Properties of R_y Indices

In the spirit of the "axiomatic approach" discussed above for inequality measures, Shorrocks (1992) lays down a set of desirable properties that any mobility index may satisfy. These properties are quite suggestive and also useful in any discussion including that of mobility indices based on transition matrices. The desirability of some of the properties may be evident while that of others may be less obvious or compelling. In fact some of the proposed properties are not consistent, so one or more has to be abandoned.

The domain of income structures X over which the measure is well defined is given as follows.

(A1) Universal domain.   Suppose the feasible set of n units and T periods for positive incomes is denoted

$$\mathcal{X}_{nT} = \{X \mid \dim X = n \times T; \, X_{it} > 0\}$$

A mobility measure should be well defined for all $X \in \mathcal{X}$, where

$$\mathcal{X} = \bigcup_{n=2}^{\infty} \bigcup_{T=2}^{\infty} \mathcal{X}_{nT}$$

However, there are some types of income structures that are eliminated from this set. An example is situations in which, in every period t, all individuals receive the same income μ_t.

The MSZ indices satisfy this requirement. Other indices may not; see below.

(A2) Continuity.   The degree of mobility varies continuously with the incomes in X.

The MSZ indices satisfy this property.

(A3) Population symmetry.   If $X' = \Pi X$ for some permutation matrix Π, then X and X' are equally mobile. This requires "rank invariance" to be acceptable since X and X' may not given the same ranking of individuals.

The term "population symmetry" is used because in the analysis of mobility, unlike inequality, we can consider permutation of the time period distributions (the columns of X) as well as permutations of the individual income profiles (the rows of X). One may thus define time symmetry separately:

(A3') Time symmetry.   A mobility measure is *time symmetric* if X and X' are equally mobile wherever $X = X\Pi$ for some permutation matrix Π.

Broadly speaking, time symmetry suggest that we care about the distribution of an individual's income receipts over time, but not about the time sequence of those receipts. This is more than time symmetry implies, however, since a permutation Π swaps all the incomes in two periods s and t, not just X_{is} with X_{it} for a single person i. But even the weaker idea of time symmetry, as stated, is objectionable since we may not be indifferent to the time sequence of incomes as, for instance, between the situation in which incomes are originally different and then become equal, and the time symmetric equivalent structure in which incomes are initially equal, and then become different.

The MSZ indices satisfy time symmetry, but this may or may not be a desirable quality.

(A4) Population replication invariance. X and X' are equally mobile whenever X' is a population replication of X.

(A4') Time replication invariance.   A mobility measure is time replication invariant if X and X' are equally mobile whenever X' is a time replication of X.

$X' \in \mathcal{X}_{rn,T}$ is a *population replication* of $X \in \mathcal{X}_{nT}$ if r is a positive integer and $X'_{jt} = X_{it}$ wherever $j = kn + i$ for some integer $k \geq 0$. That is, X' is the aggregate income structure for subpopulations each having the income structure X. Similarly, $X' \in \mathcal{X}_{n,kT}$ is a *time replication* of $X \in \mathcal{X}_{nT}$ if k is a positive integer and $X' = [X, X, \ldots, X]$. Invariance with respect to population replication is the assumption typically used in inequality measurement to compare income distributions for different sized populations.

The MSZ index with additive aggregate income functions exhibit replication invariance with respect to time and the population. However, replication invariance with respect to time may in any case be regarded as a suspect property, for much the same reason as time symmetry. It implies that the degree of mobility is unchanged if the pattern of incomes received in the first T period, say, is exactly repeated for all individuals in the next T periods, and so on. *"But this does not take into account the fact that the distribution of income in period T may be radically different from that in period 1, so moving from period T to period $T + 1$ (and hence back to period 1 incomes again) may be quite a jolt. The desirability of time replication invariance is therefore less than transparent"* (Shorrocks 1992).

(A5) Normalization.   Mobility is a minimum whenever X is completely immobile.

Definition.   A structure X is completely immobile if and only if $X_{is}/\mu_s = X_{it}/\mu_t$ for all i, s, and t.

Completely immobile income structures perform a role similar to that played by completely equal distributions in inequality measurement. But it is implicit in the above definition that *relative* mobility is measured since one is only looking at changes in *relative* incomes. This rules out pure exchange mobility. It is a type of rank invariance property that goes beyond requiring that a structure in which all individuals incomes are constant over time is completely immobile.

The MSZ indices satisfy this property. They also satisfy the following stronger normalization property:

(A6) Strong normalization.   Mobility is a minimum if and only if X is completely immobile.

The other benchmark of interest is perfect mobility. Ideally we would define such a state as one in which the probability of achieving an income level in period $t + 1$ is independent of the income received in period t. The concept of perfect mobility is difficult to formulate in terms of *observed income structures* X. A plausible parallel that falls short of "independence" is to require that not only are X_s and X_t uncorrelated, but also any, arbitrarily transformed vectors $\phi_s(X_s)$ and $\phi_t(X_t)$, where $\phi_t(X_t) \equiv (\phi_t(X_{1t}), \phi_t(X_{2t}), \ldots, \phi_t(X_{nT}))'$ for some real-valued function ϕ_t. Thus:

Definition.   An income structure $X = [X_1, \ldots, X_T]$ is *perfectly mobile* if and only if $\phi_s(X_s)$ and $\phi_t(X_t)$ are uncorrelated for all s, t, and all real functions ϕ_s, ϕ_t.

An example is given by Shorrocks (1992). Suppose there are J income levels $y_{11}, y_{21}, \ldots, y_{J1}$ at a time 1 and K income levels $y_{12}, y_{22}, \ldots, y_{K2}$ at time 2, and let X consist of the JK income profiles (y_{j1}, y_{k2}) for $j = 1, \ldots, J$ and $k = 1, \ldots, K$. Now consider

$$\overline{\phi}_1 = \frac{1}{J} \sum_{j=1}^{J} \phi_1(y_{j1}) \quad \text{and} \quad \overline{\phi}_2 = \frac{1}{K} \sum_{k=1}^{K} \phi_2(y_{k2}) \tag{19}$$

Then the sample covariance of the observed income profile can be shown to be zero.

The associated property is defined as follows:

(A7) Perfect mobility.   Mobility is a maximum whenever X is perfectly mobile.

Note that this requires that all perfectly mobile structures have the same index value as well as this common value being a maximum. Not all members of the MSZ family satisfy this requirement. See below for a member that partially satisfies (A7). But none of the members satisfy a strong version of (A7) defined as follows:

(A8) Strong perfect mobility.   Mobility is a maximum if and only if X is perfectly mobile.

(A9) Unit interval range.   Mobility index should be in [0, 1].
This is convenient and satisfied by MSZ and many other indices.

(A10) *Scale invariance.* X and X' are equally mobile if $X' = \lambda X$, for any scalar $\lambda > 0$.

And,

(A11) *Intertemporal scale invariance.* X and X' are equally mobile whenever $X' = X\Lambda$, for any positive diagonal matrix Λ.

Shorrocks' index based on sum of incomes satisfies (A10) but not (A11). But a generalization to a weighted sum aggregate proposed in Maasoumi and Zandvakili (1986, 1990) does satisfy both (A10) and (A11).

How do mobility indices (welfare functions) rank intermediate situations between perfect mobility and total immobility? This requires a careful consideration of transfer sensitivity of mobility measures and has implications for the class of welfare functions. Smoothing transfers that are generally considered as equalizing (therefore preferred) conflict with normalization and change the mean of incomes at the point of transfer. It is difficult to consider them as mobility reducing. Compensating smoothing transfers, on the other hand, preserve the mean incomes (μ_t) but can change cross section distribution (Shorrocks 1992). Thus the following type of "switches" are considered which can be seen to be mobility enhancing:

Definition.   The income structure X' is obtained for X by a "simple switch" if, for some i and j,

$$(X_{i1} - X_{j1})(X_{i2} - X_{j2}) > 0, \qquad \text{and} \qquad (20)$$

$$X'_{it} = X_{jt}; \qquad X_{jt} = X_{it}$$

$$X'_{ks} = X_{ks} \qquad \text{for } s \neq t \text{ and all } k$$

$$X'_{ks} = X_{ks} \qquad \text{for all } s \text{ and all } k \neq i, j$$

Then

(A12) *Atkinson-Bourguignon condition (for two periods).*   The income structure X' is more mobile than X whenever X' is obtained from X by a simple switch.

This condition implies that if income profiles of the two persons i and j are initially rank correlated, then a switch of incomes in either period enhances mobility. This condition has not been generalized to $T \geq 2$. The MSZ indices satisfy this condition when income aggregates are weighted averages.

Shorrocks (1992) studied a particular member of the MSZ family which satisfies all of the above properties except (A8), strong perfect mobility. This member is obtained from the measures defined by Maasoumi and Zandvakili (1986, 1989) where aggregate income is $\sum_t \alpha_t X_{it}$, $\alpha_t = 1/\mu_t$, and the short run inequality is represented by $\sum \alpha_t \mu_t I_\gamma(X^t)$, and $I_\gamma(\cdot)$ is the coefficient of variation, σ^2/μ. He refers to this as an "ideal" index. Shorrocks (1992) also looked at another mobility index attributed to Hart. The latter index is more conveniently described in relation to indices defined over transition matrices to which we now turn.

III. TRANSITION MATRICES

As was argued before, income distributions change over time under the effect of different transition mechanisms. Transition mechanisms affect social welfare by changing the income distribution. Two societies with the same income distribution at a point in time may have different levels of social welfare depending on the mobility of the populations. This requires welfare functions defined over an expanding time *dimension*.

In a Markov chain model of income generation, Dardanoni (1993) considers how economic mobility influences social welfare by following the approaches of Atkinson (1970), Markandya (1984), and Kanbur and Stiglitz (1986). He considers the welfare prospects of individuals in society by deriving the discounted stream of income distributions which obtain under different mobility structures. He proposes a class of SWFs *over the aggregates of these welfare prospects*, and derives some necessary and sufficient conditions for unambiguous welfare rankings. Since these aggregates are the discounted stream of incomes, a special case of the aggregates proposed by this author and described in the previous section, the two approaches of this section and the previous one converge when the same welfare functions and the same measures of "permanent" income are used.

The fundamental inequality theorem states that the Lorenz curve gives the normatively significant ordering of *equal mean* income distributions. Inequality indices are difficult to interpret when Lorenz curves cross. In a similar vein, Dardanoni (1993) derives a partial order of social mobility matrices which can be considered as the natural extension of the Lorenz ordering to mobility measurement. The derived ordering may provide conditions for an unambiguous welfare recommendation without employing a specific mobility measure.

Summary mobility measures induce a complete order on the set of mobility matrices and have the advantage of providing intuitive measurements and firm rankings. However, it is clear that there are substantial problems in trying to reduce a matrix of transition probabilities into a single number. This is very much the problem of multidimensional inequality measurement addressed by Maasoumi (1986), Ebert (1995a), and Shorrocks (1995). Dardanoni (1993) offers the following example of three mobility matrices

$$P_1 = \begin{bmatrix} .6 & .35 & .05 \\ .35 & .4 & .25 \\ .05 & .25 & .7 \end{bmatrix} ; \quad P_2 = \begin{bmatrix} .6 & .3 & .1 \\ .3 & .5 & .2 \\ .1 & .2 & .7 \end{bmatrix} ; \quad P_3 = \begin{bmatrix} .6 & .4 & 0 \\ .3 & .4 & .3 \\ .1 & .2 & .7 \end{bmatrix}$$

The rows denote current state and columns denote future state. Suppose we use some common summary immobility measures as proposed and discussed by, for instance, Bartholomew (1982) and Conlisk (1990). Consider the second largest eigenvalue modulus, the trace, the determinant, the mean first passage time, and Bartholomew's measure. These indices are defined below. Any of the three matrices may be considered the most mobile depending on which immobility index is chosen. This is illustrated in the following table, which shows the most mobile mobility matrix according to the different indices.

Indices	Eigenvalue	Trace	Determinant	Mean first passage	Bartholomew
Most Mobile	P_2	P_1, P_3	P_1	P_3	P_1, P_2, P_3

Different mobility rankings obtain depending on the mobility measure adopted. The welfare-based partial order is similar to the axiomatic welfare analysis exemplified above as it aims to clarify the situation depicted in this example in a sound fashion.

A.   The Welfare Ranking of Mobility Matrices

Consider a discrete Markov chain process for income. Let there be n income states. Let $P = [p_{ij}]$, such that $p_{ij} \geq 0$ and $\sum_j p_{ij} = 1$, be the $n \times n$ transition matrix, assumed regular (i.e., P^k is strictly positive for sufficiently large integer k), so that the strictly positive steady state probability vector π exists and is the unique solution to $\pi' = \pi' P$. The element p_{ij} is the probability that an individual in state i will be in state j in the following period. $\pi_{t+1} = \pi_t P$ denotes the vector whose ith element gives the fraction of the population which is in state I at time $t + 1$. It is assumed that transitions are independent across individuals and P is constant over time. Strictly speaking, P need not be square. Restricting attention to this case exploits the properties of bistochastic matrices. Lastly, income states are in ascending order.

For a given transition matrix, P, we may derive the implied distribution of expected lifetime welfare for the individuals who live in the society whose mobility is governed by P. Consider a society, assumed in equilibrium, consisting of identical individuals who are born simultaneously and live exactly for τ periods. The transition mechanism may be either intra-generational or intergenerational.

Let $u = (u_1, u_2, \dots, u_n)'$ denote a vector of instantaneous utilities, where u_i denotes the utility value of income in state i, and $V^P = (V_1, V_2, \dots, V_n)'$ denotes a vector of expected discounted lifetime utilities. The typical element V_i^P denotes the expected lifetime discounted utility of an individual beginning life in income class i, given by the ith element of the vector $V^P = u + \rho P u + \rho^2 P^2 u + \cdots + \rho^\tau P^\tau u$, where $0 \leq \rho < 1$ denotes the discount factor. This typical element is comparable with the S_i functions of the previous section. V^P will in general depend on the vector u, on the transition matrix P, on the discount factor ρ, and on the time period τ. As $\tau \to \infty$, $V^P = [I - \rho P]^{-1} u$ which, for convenience, may be normalized as

$$V^P = (1 - \rho)[I - \rho P]^{-1} u = P(\rho) u \qquad (21)$$

say. The typical element of $P(\rho)$, $P_{ij}(p)$, may be interpreted as the average discounted "lifetime" probability of moving from the initial state i to state j.

Dardanoni suggests that transition matrices be ranked according to real-valued SWFs defined over the vector of lifetime expected utilities V^p. This is comparable with the "inequality reducing" rankings of the previous section where Schur-concave welfare functions correspond to inequality measures. There is more a priori structure imposed on the data here by assuming a Markov transition process.

Mobility can occur through general growth in equilibrium income distributions. This is known as "structural mobility" or the "growth component." But there is some intertemporal movement of individuals among the different social classes, for a given equilibrium distribution of the number of individuals in each class; this latter effect is defined as "exchange" or "pure" mobility. Dardanoni (1993) isolates the pure mobility effect by assuming societies with identical steady-state income distributions. In other words two societies have within each period an identical spot income distribution, but individuals may move between income states differently under the two transition mechanisms. Note that under the stated assumptions the distribution of individuals in each state will be given by the equilibrium vector π, with the typical element π_i indicating the proportion of people in income state i. This procedure is the dynamic counterpart to the usual static inequality analysis (e.g., Atkinson 1970), where to isolate the pure inequality effect on social welfare one considers societies with equal mean incomes. Lorenz rankings would suffice in that situation.

Allowing for different mean incomes requires consideration of growth. This would be similar to the analysis of inequality on the basis of generalized Lorenz (GL) curves, proposed by Shorrocks (1983), and second order stochastic dominance. This is the extension considered by in the context of mobility matrices.

Following Dardanoni (1993), take two regular transition matrices P and Q with equal steady-state income distribution vector $\pi' = \pi'P = \pi'Q$. He considers the class of symmetric and additively separable (i.e., linear) SWFs $\sum_i \pi_i V_i^P$ which adds up, for a given u and ρ, the expected lifetime utility of the individuals in the population. This is equivalent to the SWFs considered by Atkinson (1970) and Kolm (1969) for the inequality ranking of income distributions. Noting that $\pi'P(\rho) = \pi'Q(\rho)$, we have

$$\sum_i \pi_i V_i^p = \pi'V^p = \pi'P(\rho)u = \pi'u = \pi'Q(\rho)u = \pi'V^Q = \sum_i \pi_i V_i^Q \quad (22)$$

Therefore, given a vector u and a ρ, any two transition matrices with equal steady-state income distributions will be indifferent. This result is given by Atkinson (1983) and Kanbur and Stiglitz (1986) and indicates that we are not ranking mobility as such, but the social welfare implications of each mobility matrix. The symmetric additive social welfare functional implies that movement between income states is irrelevant. What is important is the spot distribution at each period since additive separable lifetime welfares remove any influence that exchange mobility may have on intertemporal social welfare. Thus additive SWFs take inadequate account of fairness considerations. Under the stated assumptions, the equilibrium Lorenz curve of the distribution of income will look identical each period under any transition matrix with equal steady-state distribution, so that any (additive or otherwise) symmetric ex post SWF defined on the vector of realized utilities will rank the matrices as indifferent. Yet, under different transition matrices the composition of people in each income state will be different in each time period. For example, under the identity transition matrix each individual in the population remains in the same income group as in the initial situation; on the other hand, if transition is governed by a matrix in which each entry is equal to $1/n$, each individual will have the same probability of belonging to any of the n income groups regardless of the initial state. Therefore, though the equilibrium ex post Lorenz curves associated with each of these matrices could look identical for each period, social welfare may well be considered different if we take account of several periods in terms of the position of each individual in the past.

Clearly the natural labelling of welfare units in the context of mobility requires a relaxation of the "symmetry" assumption, such as (A3), which are replaced by additional assumptions on "comparability." These assumptions are discussed in, for example, Sen (1970) and Atkinson and Bourguignon (1987). Here the natural "label" for each individual is his/her starting position in the income ranking. Thus one restricted SWF would be the weighted sum of the expected welfares of the individuals, *with greater weights to the individuals who start with a lower position in the society.* That is, $W(V^p, \lambda) = \sum_i \pi_i \lambda_i V_i^P$, where $\lambda = (\lambda_1 \lambda_2, \ldots, \lambda_n)'$ denotes a nonincreasing nonnegative vector of weights. This is a step toward cardinalization (Maasoumi 1996b). Furthermore, this asymmetric treatment makes sense only if it is a disadvantage to start at a lower position. With no restriction on the mobility matrices, this is not necessarily a disadvantage. There could be a transition matrix such that the lower states are the preferred starting point in terms of lifetime expected utility. Therefore the additional assumption:

Assumption.   Transition matrices are *monotone.*

A transition matrix is called monotone if each row stochastically dominates the row above it (Conlisk 1990). In an intergenerational mobility context, a monotone mobility matrix implies that each child at time t is better off, in terms of stochastic dominance, by having a parent from state $i + 1$ than a parent from state i. In an intragenerational mobility context, a monotone mobility matrix implies that an individual who at time t is in state $i + 1$ faces a better lottery, in terms of stochastic dominance, than an individual who is in state i. If we let y be a $n \times 1$ vector it may be shown that Py is nondecreasing for all nondecreasing y if and only if P is monotone. Since $P(\rho)$ is monotone when P is monotone, the expected lifetime utility vector will be nondecreasing. Estimated transition matrices are often either exactly monotone or within sampling errors from being monotone.

Considering two extreme cases, $\lambda_1 = 1$ and $\lambda_i = 0$ for all $i > 1$, which is "Rawlsian," and $\lambda_i = 1$ for all i, which is the symmetric case, it is seen that there is a need for exploring necessary and sufficient conditions on transition matrices for the unanimous ranking of $W(V^P, \lambda) = \sum_i \pi_i \lambda_i V_i^P$ for *all* nonincreasing positive λ.

Theorem 2. Let P and Q be two monotone regular transition matrices such that $\pi' = \pi'P = \pi'Q$, and for a given ρ, the following conditions are equivalent:

(1i) $W(V^P, \lambda) \geq W(V^Q, \lambda)$.

(1ii) $T'\Pi[P(p) - Q(p)]T \leq 0$.

Proof. See Dardanoni (1993, Theorem 1).

T is the triangular summation matrix with its inverse, T^{-1}, as the first "differencing" matrix. For instance, PT transforms each row of P to the cumulative distribution function. Π is a block-diagonal matrix with the typical block being the π-vector.

This result is an extended horizon version of the first order stochastic dominance relations obtained by Atkinson (1983). Further, if we denote by $\mathfrak{M}(\pi)$ the set of regular monotone transition matrices, condition (1ii) induces (iff) a partial ordering $\succeq_M$, that is reflexive, antisymmetric, and transitive.

If one further assumes monotonicity of the reverse chain—i.e., at each time t an individual in state i has faced a stochastically dominant lottery than an individual in state $i - 1$, the above result would hold for *all* $0 \leq \rho < 1$.

An interesting result concerns the effect of transfers such as smoothing and "simple switches" considered in an earlier section. Dardanoni considers the following dynamic Pigou-Dalton (DPD) transfers: Given integers $0 < i, j, s, k < n$, with $i + k < n$ and $j + s < n$, let us decrease the probabilities of the event, "initial state i/lifetime state j," and the event, "initial state $(i + k)$/lifetime state $(j + s)$," by a quantity $0 \leq h \leq 1$. Simultaneously, increase by the same h the probabilities of the events, "initial class i/lifetime class $(j + s)$," and the "initial class $(i + k)$/lifetime class j," in such a manner as to not violate monotonicity. This transfer is mobility enhancing as it would leave the row and column sums unchanged and improve the lifetime status of a poorer individual. Finally noting that for a more mobile situation there will be smaller covariance between the initial and the lifetime status, the following general result is established:

Theorem 3. Let P and Q be two transition matrices in $\mathfrak{M}(\pi)$ and let ρ be given. Then the following conditions are equivalent:

(i) $P(\rho) \succeq_M Q(\rho)$.

(ii)   *The Lorenz curve of permanent income for P lies nowhere below that of Q $\forall$ nondecreasing income vectors.*

(iii)   *The covariance between initial status and lifetime status is greater under Q for any nondecreasing score (rank) vectors.*

(iv) $P(\cdot)$ *can be derived from* $Q(\cdot)$ *by a finite sequence of DPD exchanges.*

Formby et al. (1995) extend this result by relaxing the assumption of identical steady states. They note that $T'\Pi P(\rho)y$ is the generalized Lorenz vector of "permanent incomes," and show the following result:

Theorem 4.   Let P and Q be two monotone transition matrices with a given discount factor ρ. *Then the following conditions are equivalent:*

(1i) $W(V^P, \lambda) \geq W(V^Q, \lambda)$.

(1ii) $T'\Pi[(P(p) - Q(p)]T \leq 0$.

(1iii)   *The generalized Lorenz curve of permanent income for P lies nowhere below that for Q for all nondecreasing income vectors y.*

(1iv) $P(p)$ *can be derived from* $Q(p)$ *by a finite sequence of DPD exchanges and simple increments.*

Proof.   By noting that the assumption of steady-state income distribution is not crucial in proving Dardanoi's results as well as in Theorem 1, Formby et al. (1995) prove the equivalence among conditions (1i), (1ii), and (1iii). Also, (1iv) implies (1ii). The converse can be similarly proved. Note that each DPD or simple increment leaves all elements other than the (i, j)th of $T'\Pi[(P(p) - Q(p)]T$ unchanged.

Formby et al. (1995) demonstrate a further result which can be useful in empirical testing based on the generalized concentration curves:

Theorem 5.   Let P and Q be two monotone transition matrices. Then the following conditions are equivalent:

(1i) $P(p) \succ_M Q(p)$.

(1ii) $T'\Pi Py \leq T'\Pi Qy$ *for all nondecreasing income vectors y, i.e., the "snapshot" generalized concentration curve for P lies nowhere below that for Q for all nondecreasing income vectors, if the condition of the monotonicity of the reverse Markov chain is further assumed.*

This result establishes that if the generalized concentration curve for P lies nowhere below that for Q, then the generalized Lorenz curve of the "lifetime" for P also lies above that for Q. If one adopts Dardanoni's assumption of identical steady states (no growth mobility), the above result reduces to the ordinary concentration curve dominance which can be used as an intuitive tool in ranking "pure" mobility.

This result is empirically significant as it suggests the concentration curve as the dynamic counterpart of the standard Lorenz curve in terms of social welfare. If the generalized concentration curve of P dominates that of Q, the social welfare of *permanent income* under P will be no less than under Q for all *Schur*-concave symmetric SWFs.

Two main points emerge. One is that, both the inequality reducing and transition matrix measures are supported by the same type of welfare functions in terms of "lifetime incomes." Second, the empirical techniques for testing Lorenz type dominance, or stochastic dominance, are all made available. For instance, Bishop, Chow, and Formby (1994) show that matched pairs of estimates of generalized Lorenz and concentration curve ordinates have a joint normal distribution and this sampling property is distribution free. In general they have to be applied to the same statistics which are, however, defined over the types of aggregate income functions proposed by Maasoumi and Zandvakili (1986, 1989, 1990). Further description of some statistical techniques is given in the next section.

B.   Some Mobility Indices

We close this section with a definition of some measures of immobility based on transition matrices. The first is attributed to Prais and is called the trace index:

$$\text{Trace} = \frac{\text{trace}(P) - 1}{n - 1} \tag{23}$$

This index has been criticized for ignoring the off-diagonal transition probabilities. It also violates transfer properties such as (A12), for instance when DPD transfers increase any element along the diagonal of P.

$$\text{Determinant} = |P|^{1/(n-1)} \tag{24}$$

This is an index of immobility such as $1 - R_\gamma$. Unfortunately, this index obtains its perfectly mobile limit whenever *any* two rows or columns of P are identical! It also violates the ranking condition of the above theorems.

$$\text{Second largest eigenvalue modulus } |\lambda_2| \tag{25}$$

This measure of immobility is in the unit interval (A9) and measures the speed of escape from the initial state (Theil 1972). But it too is incoherent with the ranking conditions for DPD-type exchanges:

$$\text{Bartholomew's} = \frac{1}{n - 1} \sum_i \sum_j \pi_i p_{ij} |i - j| \tag{26}$$

This immobility index measures the expected number of crossings between periods in the steady state. It satisfies the normalization and the unit interval conditions. This measure is also coherent with DPD rankings.

$$\text{The mean first passage time} = \pi' M^P \pi \tag{27}$$

where M^P is the mean first passage matrix. This index measures the expected number of periods before the "first" individual reaches the state of the second individual. This measure of immobility is not coherent with DPD rankings.

As we have seen, all of the MSZ measures in the previous section, or indeed any reasonable measure corresponding to Schur-concave SWFs, and based on "permanent income" is coherent with DPD rankings and the results of the above theorems. Shorrocks (1992) considered an interesting measure due to Hart (1976a). Let the Galtonian model of income evolution be written in terms of the geometric mean incomes, m_t at time t, as follows:

$$\ln\left(\frac{x_{t+1}}{m_{t+1}}\right) = \beta_t \ln\left(\frac{x_t}{m_t}\right) + \varepsilon_t \tag{28}$$

This model can be used to analyze both income movements over time as well as the effect of mobility on the distribution of income. β_t measure the extent to which incomes regress toward the geometric mean. The case of a "unit root" corresponds to Gibrat's law of proportional effect: changes in relative incomes are independent of current income. This simple model of mobility may be extended by further modeling the ε_t in terms of individual specific characteristics and/or time varying effects. An example is Lillard and Willis (1978) where panel data are used. Of course, using the techniques of limited dependent variable models, transition probabilities can be similarly modeled in terms of individual specific and time varying components. A survey of several applications is given in Creedy (1985). Alternative models of diffusion describing the evolution of income have been proposed which derive its steady-state distribution forms. An example is Sargan (1957). Interestingly, the focus seems to have shifted to analyzing the properties of the equilibrium distribution and instantaneous inequality and poverty, rather than the dynamic welfare implications of the evolution mechanism. Mobility analysis is thus a return to first principles.

Shorrocks defined the following function of Hart's mobility index (Hart 1976a, 1976b, 1981, 1983), which is derived from the Galtonian model above:

$$H(x_t, x_{t+1}) = 1 - \frac{\text{cov}(\ln x_t, \ln x_{t+1})}{\sigma(\ln x_t)\sigma(\ln x_{t+1})} \tag{29}$$

This is related to the simple correlation coefficient. In the Galtonian model, $\beta_t < 1$ will reduce inequality while $\sigma(\varepsilon_t)$ would increase it. Shorrocks (1992) shows that $H(\cdot)$ is consistent with (A2)–(A5), (A10)–(A12), and partially with (A7) and (A9). It fails the universal domain, strong normalization, and strong perfect mobility properties. Also, it satisfies time symmetry but not necessarily time replication invariance. It is coherent with DPD-type rankings. As noted above, the MSZ indices are consistent with almost all of (A1)–(A12) properties and thus are superior to $H(\cdot)$.

Recently, Chakravarty (1995) proposed the Kullback's minimum discrimination statistic as a measure of mobility. This is given by

$$K = \sum_i \pi_i^t \ln\left(\frac{\pi_i^t}{\pi_i^{t+1}}\right) \tag{30}$$

where π_i^t stands for proportion of population at time t in state i.

This is a natural entropy measure of divergence between two distributions. Indeed it may be extended so that we may measure the divergence between the distribution of the "permanent income" and the distribution of income at any desired point. Examples of the latter are the short-run income distribution of MSZ indices, perfect mobility distribution, and complete immobility distributions.

The first property to note about K is that it is not a metric as it violates the triangularity rule. There are other entropy measures that have similar properties and are "metric" (Maasoumi 1993). Chakravarty (1995) notes that K satisfies many of the useful properties discussed in Shorrocks (1978a) and summarized above. But it fails to satisfy the "monotonicity" property. We note that the ordering relation discussed by Dardanoni (1993) is coherent with perfect mobility, as is K, but does not imply monotonicity.

Chakravarty (1995) points out that the well-known asymptotic χ^2 distribution of $2K$ may be used to test some very interesting hypotheses about mobility.

Shorrocks (1976) gives a good account of the properties of mobility indices based on transition matrices. Dardanoni's (1993) account of the same also leaves one with the conclusion that at least some members of the MSZ family are "ideal."

IV. STATISTICAL INFERENCE

There have been several significant advances in the development of statistical inference tools in the area of income inequality. These are generally applicable to inference on mobility indices and on ranking distributions.

For mobility indices such as the MSZ the connection is rather immediate. Inequality indices are estimated by the method of moments (MM) estimators since they are functions of population moments. Explicit formulae are derived for derivatives that are required in the delta method which extends the well-known theory of MM asymptotic distributions to that of inequality indices. This is surveyed in Maasoumi (1996a) which contains an extensive citation to original sources. The extension to mobility indices requires thinking in terms of long-run incomes and the inequality in their distributions. Trede (1995) gives the details for the asymptotic distribution of some of the MSZ measures, such as those based on the Atkinson family and Theil's inequality indices, but where aggregate income is the simple sum of incomes analyzed by Shorrocks (1978b). Extension to weighted sum function is immediate, but the statistical theory for the more complicated aggregate functions is developed in Maasoumi and Trede (1997). Trede (1995) also gives the asymptotic distributions of the mobility indices that are based on transition matrices. Some of these measures were discussed earlier. An application to German data is reported in Trede (1995) which analyzes earnings mobility for different sexes. A program written in GAUSS code is made available by Trede.

But inference about indices derived in either of the two approaches described in this paper may be inconclusive, if not bewildering, when Lorenz-type curves cross. Therefore it is desirable to test for rank relations of the type described above and in Maasoumi (1996a, 1996b). This type of testing is now possible and is inspired by testing for inequality restrictions in econometrics and statistics. A brief account of some of the available techniques for stochastic dominance would be helpful and follows.

A. Tests for Stochastic Dominance

Stochastic dominance (SD) relations and comparisons of distributions on the basis of their Lorenz and GL curves are intended to avoid the "index choice" problem. As we have seen, the SWF rankings of mobility structures can essentially follow the same path as that of static inequality analysis but in terms of lifetime incomes. In practice, however, numerical SD rankings often encounter a predictable difficulty since many distributions and (Lorenz) curves cross, making it impossible to be decisive. But the realization that all such comparisons are based on sample-based estimates of distribution functions (or curves) suggests that such comparisons should be conducted statistically and tested accordingly. The statistical approach is both sound and able to deliver more clear-cut statistical decisions!

The basic characteristic of tests for rankings is that of ordered populations and inequality restrictions. Starting with the work of Lehmann (1959) and Bartholomew (1959), likelihood ratio and Wald-type tests have been and are being developed for such hypotheses. These tests supplement other well-known procedures based on one-sided Wilcoxon rank, and the multivariate versions of the Kolmogorov-Smirnov tests. See Maasoumi (1996a) for a recent selective survey.

In the area of income distributions and tax analysis, initial developments focused on tests for Lorenz curve comparisons as in Beach and Davidson (1983), Bishop, Formby, and Thistle (1989). In practice, a finite number of ordinates of the desired curves or functions are compared. These ordinates are typically represented by quantiles and/or conditional interval means. Thus, the distribution theory of the proposed tests are typically derived from the existing asymptotic theory for ordered statistics and quantiles. Recently Beach, Davidson, and Slotsve (1995) outlined the asymptotic distribution theory for cumulative/conditional means and variances which are useful for statistically comparing Lorenz and GL curves. This theory is particularly useful for third-order stochastic dominance (TSD) ranking of crossing GL curves when a "transfer sensitivity" condition is assumed; see the definition of TSD below. To control for the size of a sequence of tests at several points the union intersection (UI) test and Studentized maximum modulus technique for multiple comparisons is generally favored in this area.

Some alternatives to these multiple comparison techniques have been suggested which are typically based on Wald-type joint tests of equality of the same ordinates, see Bishop et al. (1994) and Anderson (1994). These alternatives are sometimes problematic when their implicit null and alternative hypotheses are not a satisfactory representation of the inequality (order) relations that need to be tested. Xu et al. (1995) and Xu (1995) take proper account of the inequality nature of such hypotheses and adapt econometric tests for inequality restrictions to testing for FSD and SSD, and to GL dominance, respectively. Their tests follow the work in econometrics of Gouieroux et al. (1982), Kodde and Palm (1986), and Wolak (1988, 1989), which complements the work in statistics exemplified by Perlman (1969), Robertson and Wright (1981), and Shapiro (1988). The asymptotic distributions of these $\bar{\chi}^2$ tests are mixtures of chi-squared variates with probability weights which are generally difficult to compute. This leads to bounds tests involving inconclusive regions and conservative inferences. In addition, the computation of the $\bar{\chi}^2$ statistic requires Monte Carlo or bootstrap estimates of covariance matrices, as well as inequality restricted estimation which requires optimization with quadratic linear programming.

In contrast, Maasoumi et al. (1996) propose a direct bootstrap approach that bypasses many of these complexities while making less restrictive assumptions about the underlying processes. They offer an empirical application for ranking U.S. income distributions from the CPS and the PSID data. Their chosen statistic is the Kolmogorov-Smirnov (KS) as characterized by McFadden (1989), Klecan et al. (1991), and Kaur et al. (1994).

McFadden (1989) and Klecan, McFadden, and McFadden (1991) have proposed tests of first- and second-order "maximality" for stochastic dominance which are extensions of the Kolmogorov-Smirnov statistic. McFadden (1989) assumes i.i.d. observations and independent variates, allowing him to derive the asymptotic distribution of his test, in general, and its exact distribution in some cases (Durbin 1973, 1985). Klecan et al. generalize this earlier test by allowing for weak dependence in the processes both across variables and observations. They demonstrate with an application for ranking investment portfolios. The asymptotic distribution of these tests cannot be fully characterized, however, prompting Monte Carlo and bootstrap methods for evaluating critical levels. In the following section some definitions and results are summarized which help to describe these tests.

1. Definitions and Tests

Let X and Y be two income variables at either two different points in time, or two lifetime income vectors. Let $X_1, X_2, \ldots, X_n$ be the not necessarily i.i.d observations on X, and $Y_1, Y_2, \ldots, Y_m$ be similar observations on Y. Let U_1 denote the class of all utility functions u such that $u' \geq 0$ (increasing). Also, let U_2 denote the class of all utility functions in U_1 for which $u'' \leq 0$ (strict concavity). Let $X_{(i)}$ and $Y_{(i)}$ denote the ith order statistics, and assume $F(x)$ and $G(x)$ are continuous and monotonic cumulative distribution functions (cdfs) of X and Y, respectively.

Quantiles $q_x(p)$ and $q_y(p)$ are implicitly defined by, for example, $F[X \leq q_x(p)] = p$.

Definition (FSD). X first-order stochastic dominates Y, denoted X FSD Y, if and only if any one of the following equivalent conditions holds:

(1) $E[u(X)] \geq E[u(Y)]$ for all $u \in U_1$, with strict inequality for some u.
(2) $F(x) \leq G(x)$ for all x in the support of X, with strict inequality for some x.
(3) $q_x(p) \geq q_y(p)$ for all $0 \leq p \leq 1$.

Definition (SSD). X second-order stochastic dominates Y, denoted X SSD Y, if and only if any of the following equivalent conditions holds:

(1) $E[u(X)] \geq E[u(Y)]$ for all $u \in U_2$, with strict inequality for some u.
(2) $\int_{-\infty}^{x} F(t) \, dt \leq \int_{-\infty}^{x} G(t) \, dt$ for all x in the support of X and Y, with strict inequality for some x.
(3) $\int_0^p q_x(t) \, dt \geq \int_0^p q_y(t) \, dt$, for all $0 \leq p \leq 1$, with strict inequality for some value(s) p.

The tests of FSD and SSD are based on empirical evaluations of conditions (2) or (3). Mounting tests on conditions (3) typically relies on the fact that quantiles are consistently estimated by the corresponding order statistics at a finite number of sample points. Mounting tests on conditions (2) requires empirical cdfs and comparisons at a finite number of observed ordinates. Also, from Shorrocks (1983) or Xu (1995) it is clear that condition (3) of SSD is equivalent to the requirement of GL dominance. FSD implies SSD.

Noting the usual definition of the Lorenz curve of, for instance, X as $L_x(x) = (1/\mu_x) \int_{-\infty}^{x} X \times dF(t)$, and its GL $(x) = \mu_x L_x(x)$, some authors have developed tests for Lorenz and GL dominance on the basis of the sample estimates of conditional interval means and cumulative moments of income distributions; e.g., see Bishop et al. (1989), Bishop et al. (1991), Beach et al. (1995), and Maasoumi (1996a) for a general survey of the same. The asymptotic distributions given by Beach et al. (1995) are particularly relevant for testing for third-order stochastic dominance (TSD). The latter is a useful criterion when Lorenz or GL curves cross at several points and the investigator is willing to adopt "transfer sensitivity" of Shorrocks and Foster (1987), that is a relative preference for progressive transfers to poorer individuals. When either Lorenz or generalized Lorenz curves of two distributions cross unambiguous ranking by FSD and SSD is not possible. Whitmore (1970) introduced the concept of TSD in finance. Shorrocks and Foster (1987) showed that the addition of the "transfer sensitivity" requirement leads to TSD ranking of income distributions. This requirement is stronger than the Pigou-Dalton principle of transfers and is based on the class of welfare functions U_3 which is a subset of U_2 with $u''' \geq 0$. TSD is defined as follows:

Definition (TSD). X third-order stochastic dominates Y, denoted X TSD Y, if and only if any of the following equivalent conditions holds:

(1) $E[u(X)] \geq E[u(Y)]$ for all $u \in U_3$, with strict inequality for some u.

(2) $\int_{-\infty}^{x} \int_{-\infty}^{v} [F(t) - G(t)] \, dt \, dv \leq 0$, for all x in the support, with strict inequality for some x, with the endpoint condition $\int_{-\infty}^{+\infty} [F(t) - G(t)] \, dt \leq 0$.

(3)   When $E[X] = E[Y]$, X TSD Y iff $\lambda_x^2(q_i) \leq \lambda_y^2(q_i)$, for all Lorenz curve crossing points $i = 1, 2, \ldots, (n + 1)$; where $\lambda_x^2(q_i)$ denotes the "cumulative variance" for incomes up to the ith crossing point (Davies and Hoy 1995).

When $n = 1$, Shorrocks and Foster (1987) showed that X TSD Y if (a) the Lorenz curve of X cuts that of Y from above, and (b) $\mathrm{Var}(X) \leq \mathrm{Var}(Y)$. This situation revives the coefficient of variation as a useful statistical index for ranking distributions.

Kaur et al. (1994) assume i.i.d observations for independent prospects X and Y. Their null hypothesis is condition (2) of SSD *for each x* against the alternative of strict violation of the same condition *for all x*. The test of SSD then requires an appeal to union intersection technique which results in a test procedure with maximum asymptotic size of α if the test statistic at each x is compared with the critical value Z_α of the standard normal distribution.

McFadden offers a definition of "maximal" sets, as follows:

Definition (Maximality).   Let $\text{Æ} = \{X_1, X_2, \ldots, X_K\}$ denote a set of K distinct random variables. Let F_k denote the cdf of the kth variable. The set Æ is *first- (second-) order maximal* if not variable in Æ is first- (second-) order weakly dominated by another.

Let $X_{.n} = (x_{1n}, x_{2n}, \ldots, x_{Kn}), n = 1, 2, \ldots, N$, be the observed data. Assume $X_{.n}$ is strictly stationary and α-*mixing*, and assume $F_i(X_i), i = 1, 2, \ldots, K$, and *exchangeable* random variables, so that resampling estimates of the test statistics converge appropriately. This is less demanding than the assumption of independence which is not realistic in many applications (as in mobility analysis, and before- and after-tax scenarios). In general F_k is unknown and estimated by the empirical distribution function $F_{kN}(X_k)$. Finally, if we adopt Klecan et al.'s mathematical regularity conditions pertaining to von Neumann-Morgenstern (VNM) utility functions that generally underlie the expected utility maximization paradigm, the following theorem defines the tests and the hypotheses being tested:

Theorem 6.   Given the mathematical regularity conditions;

(a)   *The variables in Æ are first-order stochastically maximal; i.e.,*

(1) $d = \min_{i \neq j} \max_x [F_i(x) - F_j(x)] > 0$, *if and only if for each i and j. There exists a continuous increasing function u such that $Eu(X_i) > Eu(X_j)$.*

(b) *The variables in Æ are second-order stochastically maximal; i.e.,*

(2) $S = \min_{i \neq j} \max_x \int_{-\infty}^x [F_i(\mu) - F_j(\mu)] \, d\mu > 0$, *if and only if for each i and j, there exists a continuous increasing and strictly concave function u such that* $Eu(X_i) > Eu(X_j)$.

(c) *Assuming the stochastic process* $X_{\cdot n}, n = 1, 2, \ldots$, *to be strictly stationary and* α-mixing with $\alpha(j) = O(j^{-\delta})$, *for some* $\delta > 1$, *we have* $d_{2N} \to d$, *and* $S_{2N} \to S$, *where* d_{2N} *and* S_{2N} *are the empirical test statistics defined as*

(3) $d_{2N} = \min_{i \neq j} \max_x [F_{iN}(x) - F_{jN}(x)]$ *and,*

(4) $S_{2N} = \min_{i \neq j} \max_x \int_0^x [F_{iN}(\mu) - F_{jN}(\mu)] \, d\mu$

Proof. See Theorems 1 and 5 of Klecan et al. (1991).

The null hypotheses tested by these two statistics is that, respectively, Æ is *not* first- (second-) order maximal—i.e., X_i FSD(SSD) X_j, for some i and j. We reject the null when the statistics are positive and large. Since the null hypothesis in each case is composite, power is conventionally determined in the least favorable case of identical marginals $F_i = F_j$. Thus, as is shown in Kaur et al. (1994) and Klecan et al. (1991), tests based on d_{2N} and S_{2N} are consistent. Furthermore, the asymptotic distribution of these statistics are nondegenerate in the least favorable case, being Gaussian (Klecan et al. 1991, Theorems 6–7).

The statistic S_{2N} has, in general, neither a tractable distribution, nor an asymptotic distribution for which there are convenient computational approximations. The situation for d_{2N} is similar except for some special cases; see Durbin (1973, 1985) and McFadden (1989), who assume i.i.d. observations (not crucial), and independent variables in Æ (consequential). Unequal sample sizes may be handled as in Kaur et al.

Klecan et al. (1991) suggest Monte Carlo procedures for computing the significance levels of these tests. This forces a dependence on an assumed parametric distribution for generating MC iterations, but is otherwise quite appealing for very large iterations. Maasoumi et al. (1996) employ bootstrap methods to obtain the empirical distributions of the test statistics and confidence intervals. They report an empirical examination of the U.S. income distribution based on the CPS and PSID data. Their methods are directly applicable to ranking of mobility structures described previously.

V. SOME EMPIRICAL EXAMPLES

Creedy (1985) contains detailed descriptions of empirical studies which implement the transition matrix and other model-based techniques. Shorrocks (1976, 1978a) and Lillard and Willis (1978) also implement the transition matrix method using some of the same U.S. panel data which I will describe below.

The MSZ index method has been implemented by Shorrocks, Maasoumi, Zand-vakili, Trede, and others. Trede's work is based on German panel data and, as mentioned earlier, reports statistical tests of significant change in mobility. The first three authors use the Michigan Panel data. We now exemplify some of these latter studies:

A. Mobility and Gender

The MSZ family of mobility measures was investigated by Maasoumi and Zandvakili (1990) using the Michigan Panel Study of Income Dynamics (PSID). These measures are decomposed in order to learn about components that are due to differences in *gender and income* groups, on the one hand, and within group components which are free of such group characteristics. Several aggregator functions were used to compute the "Permanent income" variable. Their justification and role in robustifying inferences was investigated.

"Household" income data for the period 1969–1981 were taken from the PSID. Household's income (head and spouse, if any) consists of the following: income from wages, salaries, rents, dividends, interest, business, bonuses, commissions, professional practice, aid to dependent children, social security, retirement pay, pension or annuities, unemployment compensation, child support, and other transfer payments. Real total income is obtained using the current consumer price index. Income is adjusted for family size (in 1975) to provide a better measure of family income since family members effectively pool their incomes. We refer to this adjusted income as the per capita family income PCFI).

In computing the permanent incomes three different schemes were used in order to weight income at different times. These α_t weights are (i) equal weights for all years, (ii) the ratio of mean income at time t to the mean income over the entire T periods (MIW in tables), and (iii) the normalized elements of the eigenvector corresponding to the first principal component of the $X'X$ matrix. We did not find any qualitative differences in our results between these three cases, and thus report only the computations based on ratio-of-means weights. The other two cases are reported in Zandvakili (1987).

In our computations the substitution parameter β is restricted by the relations $-\gamma = 1 + \beta$. We computed four different aggregator functions corresponding to four inequality measures with $-\gamma = v = (2, 1, .5, .0)$. $v = 0.0$ and 1.0 correspond to Theil's first and second inequality measures, respectively combined with the linear and the Cobb-Douglas forms of the aggregator function. Tables 1–3 are from Maasoumi and Zandvakili (1990) which provide, respectively, the annual short-run inequalities, the inequalities in the aggregated (long-run) incomes, and the income stability measure R_γ. Decomposition of each based on gender is also given. Note that as one moves toward 1981 the number of periods over which S_i, $I_y(S)$ and R_γ are calculated is increasing from one to 13. The results for every other year are reported to save space.

Table I 1969–1981 per Capita Family Income: Short-Run Inequalities

	Overall	Between	Within	Men	Women
Degree of inequality aversion = 2.0					
1969	1.109	0.014	1.095	0.793	1.676
1971	1.049	0.016	1.033	0.729	1.604
1973	1.002	0.019	0.982	0.636	1.600
1975	1.636	0.023	1.613	0.815	3.015
1977	1.569	0.038	1.530	0.981	2.195
1979	1.895	0.041	1.854	1.211	2.580
1981	2.441	0.047	2.394	1.520	3.292
Degree of inequality aversion = 1.0					
1969	0.430	0.013	0.417	0.340	0.676
1971	0.466	0.014	0.452	0.375	0.709
1973	0.464	0.017	0.446	0.362	0.727
1975	0.531	0.021	0.510	0.422	0.808
1977	0.578	0.033	0.545	0.467	0.808
1979	0.613	0.035	0.579	0.498	0.849
1981	0.706	0.039	0.667	0.578	0.964
Degree of inequality aversion = 0.5					
1969	0.375	0.012	0.363	0.306	0.601
1971	0.407	0.014	0.394	0.341	0.618
1973	0.404	0.017	0.387	0.331	0.635
1975	0.456	0.019	0.436	0.380	0.693
1977	0.494	0.030	0.463	0.419	0.688
1979	0.505	0.032	0.473	0.427	0.707
1981	0.571	0.036	0.535	0.490	0.782
Degree of inequality aversion = 0.0					
1969	0.367	0.012	0.355	0.304	0.606
1971	0.402	0.013	0.388	0.344	0.615
1973	0.395	0.016	0.380	0.332	0.631
1975	0.448	0.018	0.430	0.383	0.693
1977	0.492	0.029	0.463	0.429	0.682
1979	0.483	0.030	0.453	0.417	0.688
1981	0.549	0.034	0.515	0.481	0.753

Table 2 1969–1981 per Capita Family Income (MIW): Long-Run Inequality

	Overall	Between	Within	Men	Women
Degree of inequality aversion = 2.0					
1969	1.109	0.014	1.095	0.793	1.676
1969–71	0.949	0.016	0.933	0.665	1.430
1969–73	0.889	0.018	0.871	0.599	1.365
1969–75	0.951	0.020	0.931	0.588	1.533
1969–77	0.974	0.025	0.950	0.602	1.507
1969–79	1.035	0.030	1.005	0.653	1.509
1969–81	1.100	0.034	1.067	0.690	1.563
Degree of inequality aversion = 1.0					
1969	0.430	0.013	0.417	0.340	0.676
1969–71	0.414	0.014	0.400	0.325	0.651
1969–73	0.408	0.016	0.392	0.316	0.648
1969–75	0.414	0.018	0.396	0.319	0.654
1969–77	0.426	0.021	0.404	0.327	0.662
1969–79	0.434	0.025	0.409	0.336	0.654
1969–81	0.455	0.029	0.425	0.352	0.674
Degree of inequality aversion = 0.5					
1969	0.375	0.012	0.363	0.306	0.601
1969–71	0.364	0.014	0.350	0.296	0.579
1969–73	0.360	0.015	0.345	0.291	0.577
1969–75	0.363	0.017	0.347	0.294	0.579
1969–77	0.373	0.020	0.353	0.302	0.585
1969–79	0.381	0.024	0.357	0.310	0.579
1969–81	0.396	0.027	0.369	0.323	0.591
Degree of inequality aversion = 0.0					
1969	0.367	0.012	0.355	0.304	0.606
1969–71	0.355	0.013	0.342	0.295	0.578
1969–73	0.351	0.014	0.337	0.292	0.572
1969–75	0.354	0.016	0.339	0.295	0.573
1969–77	0.366	0.019	0.347	0.306	0.578
1969–79	0.371	0.023	0.349	0.312	0.566
1969–81	0.386	0.026	0.361	0.326	0.573

Table 3   1969–1981 per Capita Family Income (MIW): Income
Stability

	Overall	Between	Within	Men	Women
Degree of inequality aversion = 2.0					
1969	1.000	0.013	0.987	1.000	1.000
1969–71	0.916	0.016	0.900	0.907	0.918
1969–73	0.885	0.017	0.868	0.873	0.887
1969–75	0.828	0.018	0.811	0.822	0.823
1969–77	0.786	0.020	0.767	0.772	0.786
1969–79	0.744	0.021	0.723	0.723	0.747
1969–81	0.692	0.021	0.671	0.670	0.700
Degree of inequality aversion = 1.0					
1969	1.000	0.030	0.970	1.000	1.000
1969–71	0.928	0.032	0.896	0.912	0.949
1969–73	0.903	0.035	0.869	0.881	0.930
1969–75	0.877	0.038	0.839	0.851	0.907
1969–77	0.855	0.043	0.812	0.824	0.887
1969–79	0.832	0.049	0.783	0.801	0.860
1969–81	0.813	0.052	0.761	0.779	0.841
Degree of inequality aversion = 0.5					
1969	1.000	0.033	0.967	1.000	1.000
1969–71	0.932	0.035	0.896	0.917	0.955
1969–73	0.911	0.038	0.873	0.891	0.942
1969–75	0.885	0.041	0.844	0.862	0.920
1969–77	0.867	0.047	0.820	0.841	0.904
1969–79	0.855	0.054	0.801	0.830	0.885
1969–81	0.840	0.058	0.782	0.813	0.870
Degree of inequality aversion = 0.0					
1969	1.000	0.033	0.967	1.000	1.000
1969–71	0.927	0.034	0.892	0.915	0.950
1969–73	0.904	0.037	0.868	0.889	0.933
1969–75	0.879	0.039	0.840	0.861	0.911
1969–77	0.864	0.045	0.819	0.846	0.895
1969–79	0.852	0.052	0.800	0.835	0.871
1969–81	0.839	0.056	0.783	0.822	0.855

Short-run inequality in Table 1 has generally increased. As expected, inequality is greater with larger degrees of relative inequality aversion (v). There are 1776 male and 529 female headed households in the sample. The "within-group" component of short-run inequalities is dominant. The column heading "within" refers to weighted average of within group inequalities as in the formulae previously discussed in this chapter. The absolute value of the "between-group" component, however, has increased over the 13 years. For both sexes annual inequalities have a rising trend (less uniformly so for $v = 2$). But short-run inequality amongst female headed households is always greater than among men. These annual values, however, contain many transitory components which are partially removed from the aggregated values in Table 2.

Table 2 values exhibit much less volatility. After a decline in the initial years, $I_y(S)$ has increased back to about its original value. Also, long-run inequality is always smaller than the corresponding short-run inequality. Once again, inequality among women is greater than among men, and within-group inequality is several times the between-group component. These relative values are somewhat sensitive to the family size adjustment of incomes. For instance, the between-group component increases to 15–25% of overall inequality for unadjusted incomes; see Zandvakili (1987). This is partly due to a larger proportion of two income earners being among the male headed families.

The corresponding stability measures are presented in Table 3. Again, seven of the 13 possible values are reported without any qualitative loss. $0 < R_y < 1$ in all cases. The following may be concluded from Table 3:

(i)   There is a tendency for the profiles to fall and then level off as the number of aggregated years increases from one to 13.

(ii) The profiles for households headed by men fall faster and further than those of women headed households.

(iii) These patterns are robust with respect to the choice of aggregation function, family size adjustment, and inequality measure.

The fact that the profiles are becoming flatter is an indication that, although there have been some transitory movements in the size distribution of income, there is a lack of any permanent equalization. Further, while some equalization has taken place within each group of households, inequality between men and women headed households has increased in absolute value.

B. Mobility and Income Level

Maasoumi and Zandvakili (1989) give inequality and mobility decompositions by age, education, and race. Similar decompositions by income level can reveal the aggregate impact of all such non-income characteristics (including gender). It is anticipated that if the major causes of variation in incomes are transitory in nature, the length of time spent in any income class will be short. "Permanent" income inequality changes will be very revealing in this context.

The total sample is divided into seven income groups (G1–G7). The assignment to groups is one a one-time basis and according to the simple arithmetic mean income of the individual household over the 13-year period. These real income levels begin with mean incomes of less than $4999, and increasing in increments of $5000. The last group contains mean incomes of $35,000 or more.

Short-run inequalities and their decompositions based on income level are given in Table 4. All the tables and figures in this section are taken from Maasoumi and Zandvakili (1990). There are several recognizable patterns. The between-group inequality has increased steadily over this period. The within-group component of inequality fluctuates around a relatively constant mean value. The observed patterns suggest that the nonincome differences do contribute to the increase in between group inequality. Over 70% of women-headed households earn less than $15,000. Of course, this is confounded by the differential impact of inflation on different income groups (we use real incomes).

In Table 5 long-run inequality levels have risen after an initial decline. Decomposition by income level shows that the between-group component of $I_y(S)$ has increased uniformly. At the same time the within-group inequality has *decreased* steadily. This change has been dramatic so that in the later years the between-group component is larger than the within component. These changes include the well-known life-cycle and human capital effects, and are non inconsistent with the cumulative effects predicted by discrimination theories.

The long-run *within-group* inequalities reveal a falling trend for each of the seven income groups. This is anticipated since transitory components are smoothed out and individual incomes have approached group mean incomes in the long run. These long-run grouping observations are somewhat sensitive to the family size. Within-group aggregate income inequalities are noticeably smaller when income is not adjusted for family size, and there is generally less inequality within the higher-income groups.

Table 6 reports the stability profiles which reveal much higher degrees of permanent equalization *within* income groups than was observed for the gender groups of the last section. Note that as the stability profiles of the whole sample flatten, the corresponding within group profiles continue to fall. In our view some equalization has occurred, but this is mostly confined to within income groups.

On the basis of the approximately 2300 households which remained in the Michigan panel over the period 1969–1981. Maasoumi and Zandvakili (1990) conclude that (i) there is not a great deal of inequality between the men and women-headed households; (ii) the dominant within-group component of inequality is either increasing over this period or, when incomes are smoothed by time aggregation,

Table 4   1969–1981 per Capita Family Income: Short-Run Inequalities

	Overall	Between	Within	$G1$	$G2$	$G3$	$G4$	$G5$	$G6$	$G7$
				Degree of inequality aversion = 2.0						
1969	1.109	0.151	0.958	0.879	1.429	0.668	0.608	0.282	0.293	0.209
1971	1.049	0.224	0.824	0.672	0.560	0.626	0.725	0.281	0.265	0.450
1973	1.002	0.310	0.692	0.524	0.415	0.430	0.479	0.217	0.220	0.212
1975	1.636	0.412	1.225	0.950	0.539	0.459	0.449	0.274	0.236	0.242
1977	1.569	0.524	1.045	0.633	0.487	0.479	0.365	0.290	0.178	0.242
1979	1.895	0.624	1.271	0.639	0.683	0.544	0.460	0.227	0.232	0.211
1981	2.441	0.725	1.716	0.722	0.985	0.679	0.525	0.570	0.310	0.299
				Degree of inequality aversion = 1.0						
1969	0.430	0.114	0.316	0.429	0.413	0.421	0.385	0.236	0.240	0.179
1971	0.466	0.161	0.306	0.413	0.334	0.356	0.411	0.231	0.227	0.224
1973	0.464	0.205	0.259	0.352	0.284	0.302	0.336	0.190	0.199	0.190
1975	0.531	0.259	0.272	0.354	0.324	0.312	0.314	0.224	0.204	0.211
1977	0.578	0.316	0.262	0.364	0.310	0.322	0.273	0.208	0.165	0.218
1979	0.613	0.346	0.267	0.353	0.374	0.335	0.286	0.194	0.193	0.183
1981	0.706	0.384	0.322	0.393	0.490	0.394	0.342	0.252	0.227	0.225
				Degree of inequality aversion = 0.5						
1969	0.375	0.103	0.272	0.396	0.356	0.389	0.353	0.233	0.243	0.179
1971	0.407	0.144	0.264	0.388	0.299	0.325	0.373	0.228	0.230	0.219
1973	0.404	0.180	0.224	0.336	0.263	0.279	0.314	0.188	0.202	0.192
1975	0.456	0.225	0.231	0.326	0.292	0.291	0.294	0.224	0.205	0.212
1977	0.494	0.275	0.219	0.337	0.282	0.303	0.260	0.205	0.167	0.224
1979	0.505	0.293	0.212	0.332	0.331	0.310	0.269	0.191	0.192	0.183
1981	0.571	0.322	0.249	0.364	0.432	0.361	0.315	0.234	0.218	0.221
				Degree of inequality aversion = 0.0						
1969	0.367	0.096	0.270	0.402	0.339	0.391	0.349	0.243	0.263	0.187
1971	0.402	0.134	0.268	0.398	0.290	0.323	0.367	0.235	0.246	0.228
1973	0.395	0.165	0.230	0.348	0.261	0.272	0.313	0.192	0.214	0.202
1975	0.448	0.207	0.242	0.333	0.284	0.290	0.293	0.237	0.215	0.223
1977	0.492	0.254	0.238	0.343	0.274	0.306	0.263	0.214	0.176	0.245
1979	0.483	0.266	0.218	0.345	0.320	0.311	0.271	0.196	0.200	0.191
1981	0.549	0.293	0.256	0.382	0.424	0.362	0.311	0.236	0.222	0.233

Table 5   1969–1981 per Capita Family Income (MIW): Long-Run Inequality

	Overall	Between	Within	G1	G2	G3	G4	G5	G6	G7
					Degree of inequality aversion = 2.0					
1969	1.109	0.151	0.958	0.879	1.429	0.668	0.608	0.282	0.293	0.209
1969–71	0.949	0.208	0.741	0.610	0.708	0.548	0.601	0.230	0.240	0.261
1969–73	0.889	0.255	0.634	0.486	0.504	0.425	0.506	0.208	0.222	0.222
1969–75	0.951	0.311	0.640	0.490	0.404	0.348	0.426	0.188	0.208	0.200
1969–77	0.974	0.365	0.609	0.444	0.341	0.303	0.322	0.166	0.188	0.186
1969–79	1.035	0.425	0.610	0.422	0.313	0.270	0.274	0.142	0.160	0.172
1969–81	1.100	0.481	0.619	0.401	0.308	0.244	0.247	0.147	0.146	0.155
					Degree of inequality aversion = 1.0					
1969	0.430	0.114	0.316	0.429	0.413	0.421	0.385	0.236	0.240	0.179
1969–71	0.414	0.144	0.270	0.354	0.311	0.346	0.374	0.198	0.204	0.176
1969–73	0.408	0.170	0.238	0.309	0.261	0.282	0.328	0.181	0.193	0.166
1969–75	0.414	0.200	0.213	0.275	0.225	0.242	0.290	0.166	0.176	0.159
1969–77	0.426	0.233	0.192	0.259	0.200	0.210	0.233	0.150	0.158	0.154
1969–79	0.434	0.264	0.171	0.241	0.181	0.180	0.192	0.129	0.135	0.143
1969–81	0.455	0.294	0.161	0.232	0.179	0.167	0.180	0.122	0.122	0.130
					Degree of inequality aversion = 0.5					
1969	0.375	0.103	0.272	0.396	0.356	0.389	0.353	0.233	0.243	0.179
1969–71	0.364	0.128	0.236	0.334	0.274	0.320	0.342	0.199	0.208	0.177
1969–73	0.360	0.150	0.210	0.294	0.236	0.262	0.304	0.182	0.198	0.169
1969–75	0.363	0.175	0.188	0.260	0.205	0.225	0.269	0.168	0.179	0.163
1969–77	0.373	0.204	0.169	0.245	0.182	0.196	0.219	0.157	0.161	0.160
1969–79	0.381	0.231	0.150	0.229	0.164	0.172	0.185	0.135	0.139	0.150
1969–81	0.396	0.257	0.139	0.219	0.162	0.163	0.175	0.128	0.127	0.138
					Degree of inequality aversion = 0.0					
1969	0.367	0.096	0.270	0.402	0.339	0.391	0.349	0.243	0.263	0.187
1969–71	0.355	0.117	0.238	0.339	0.261	0.318	0.335	02.07	0.223	0.187
1969–73	0.351	0.137	0.214	0.298	0.226	0.256	0.296	0.188	0.212	0.179
1969–75	0.354	0.159	0.195	0.263	0.197	0.219	0.262	0.176	0.190	0.173
1969–77	0.366	0.186	0.180	0.246	0.174	0.191	0.214	0.168	0.169	0.173
1969–79	0.371	0.210	0.161	0.230	0.153	0.171	0.182	0.144	0.147	0.161
1969–81	0.386	0.235	0.152	0.220	0.153	0.165	0.173	0.136	0.135	0.152

Table 6 1969–1981 per Capita Family Income (MIW): Income Stability

	Overall	Between	Within	$G1$	$G2$	$G3$	$G4$	$G5$	$G6$	$G7$
			Degree of inequality aversion = 2.0							
1969	1.000	0.136	0.864	1.000	1.000	1.000	1.000	1.000	1.000	1.000
1969–71	0.916	0.201	0.715	0.812	0.857	0.872	0.907	0.857	0.878	0.847
1969–73	0.885	0.254	0.631	0.754	0.775	0.766	0.857	0.815	0.853	0.825
1969–75	0.828	0.271	0.557	0.712	0.669	0.680	0.794	0.726	0.793	0.780
1969–77	0.786	0.295	0.492	0.676	0.605	0.592	0.670	0.631	0.756	0.740
1969–79	0.744	0.305	0.439	0.644	0.538	0.507	0.578	0.562	0.675	0.709
1969–81	0.692	0.303	0.390	0.614	0.480	0.437	0.523	0.496	0.577	0.617
			Degree of inequality aversion = 1.0							
1969	1.000	0.265	0.735	1.000	1.000	1.000	1.000	1.000	1.000	1.000
1969–71	0.928	0.323	0.605	0.852	0.861	0.901	0.930	0.880	0.892	0.871
1969–73	0.903	0.376	0.527	0.799	0.785	0.805	0.880	0.840	0.870	0.833
1969–75	0.877	0.425	0.452	0.745	0.689	0.726	0.821	0.768	0.819	0.791
1969–77	0.855	0.468	0.386	0.713	0.627	0.634	0.713	0.693	0.773	0.755
1969–79	0.832	0.505	0.327	0.670	0.554	0.545	0.611	0.617	0.684	0.716
1969–81	0.813	0.526	0.287	0.637	0.510	0.490	0.572	0.567	0.607	0.639
			Degree of inequality aversion = 0.5							
1969	1.000	0.275	0.725	1.000	1.000	1.000	1.000	1.000	1.000	1.000
1969–71	0.932	0.328	0.604	0.864	0.860	0.908	0.936	0.890	0.903	0.885
1969–73	0.911	0.379	0.532	0.812	0.789	0.815	0.887	0.853	0.885	0.853
1969–75	0.885	0.427	0.458	0.758	0.694	0.732	0.825	0.785	0.833	0.812
1969–77	0.867	0.474	0.393	0.722	0.629	0.638	0.721	0.727	0.783	0.781
1969–79	0.855	0.518	0.337	0.685	0.553	0.564	0.629	0.654	0.705	0.750
1969–81	0.840	0.545	0.296	0.647	0.513	0.520	0.595	0.609	0.637	0.675
			Degree of inequality aversion = 0.0							
1969	1.000	0.262	0.738	1.000	1.000	1.000	1.000	1.000	1.000	1.000
1969–71	0.927	0.306	0.621	0.861	0.848	0.904	0.932	0.891	0.906	0.894
1969–73	0.904	0.352	0.553	0.806	0.776	0.806	0.879	0.851	0.886	0.860
1969–75	0.879	0.395	0.484	0.752	0.682	0.718	0.811	0.790	0.834	0.822
1969–77	0.864	0.440	0.424	0.712	0.614	0.627	0.708	0.745	0.780	0.800
1969–79	0.852	0.483	0.369	0.673	0.530	0.563	0.623	0.667	0.705	0.767
1969–81	0.839	0.510	0.330	0.633	0.496	0.528	0.592	0.628	0.649	0.700

relatively stable; (iii) this larger within-group component of inequality is due to high levels of inequality within lower income groups (such as women headed households); (iv) grouping by real income brackets leads predictably to very large between-group inequality values. (v) Some equalization of real incomes has occurred over time *within* most *income* groups, but this is very hard to judge by a comparison of annual inequality measures and most clearly revealed by using our "permanent income" distributions; (vi) modest levels of mobility are recorded as the aggregation interval is expanded, but the corresponding profiles flatten out after about eight or nine years.

We close this subsection with Fig. 1, which summarize the evolution of the income distribution for this sample with the graph of the stability profile P_t.

C. Mobility by Education, Age, and Race

Maasoumi and Zandvakili (1989) is based on the same data as the previous section, but the role of years of schooling of the head of household, his/her age, and race were examined through decomposition of the inequality/mobility measures. Tables 7–15 are from that source.

Table 7 is a summary of short run, long run, and the stability values for all the 13 years. Tables 8–10 provide decompositions by educational attainment which was indicated by the years of completed schooling by the head. The increase in overall short-run inequality is primarily due to increases in within group inequalities. Long-run inequality is quite stable. The R measure declines over longer periods. This indicates that while there is much short-run mobility (change), this does not change permanent income inequality. Note that this phenomenon may be partly due to the anonymity of our measures which are invariant to short-run *switching of positions* by individuals.

It should be noted that education is both a capital good and a provider of a stream of consumption. It has different values for different individuals. This heterogeneity effect is here controlled for leading to conditional inferences. For a discussion of these issues and a multidimensional treatment in which education is regarded as a distinct attribute (with income and wealth) see Maasoumi and Nickelsburg (1988).

Tables 8–10 indicate that the greatest inequality is within the group with fewest schooling years. Indeed, within-group inequality declines steadily with educational attainment: education is an equalizer (some might argue it is a restraint over unusual earnings!). Between groups inequality is rising somewhat over these years but is about one quarter of total inequality, and declining proportionately.

Long-run inequality is much more stable over time and is smaller than short-run inequalities. Looking at these figures a policymaker is less likely drawn to quick reactions to transitory phenomenon, and more likely to focus on stable features, for

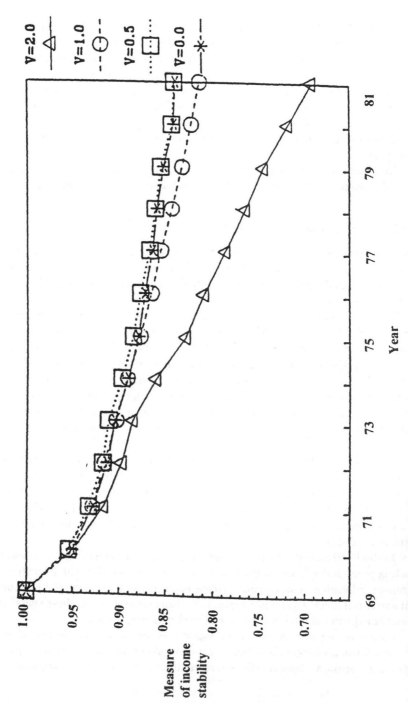

Figure 1   Measure of income stability PCFI 1969–1981, based on mean of income weights.

Table 7 Per Capita Family Income 1969–1981 Based on Mean of Income Weights

	1969	1970	1971	1972	1973	1974	1975	1976	1977	1978	1979	1980	1981
								Short-run inequalities					
$I_{2.0}(Y)$	1.109	0.961	1.049	0.940	1.002	1.190	1.636	1.357	1.569	1.840	1.895	2.072	2.441
$I_{1.0}(Y)$	0.430	0.443	0.466	0.460	0.464	0.488	0.531	0.548	0.573	0.594	0.613	0.669	0.706
$I_{0.5}(Y)$	0.375	0.388	0.407	0.404	0.404	0.423	0.456	0.468	0.494	0.496	0.505	0.551	0.571
$I_{0.0}(Y)$	0.367	0.381	0.402	0.399	0.395	0.415	0.448	0.458	0.492	0.481	0.483	0.541	0.549

V	$I(S_1)$	$I(S_2)$	$I(S_3)$	$I(S_4)$	$I(S_5)$	$I(S_6)$	$I(S_7)$	$I(S_8)$	$I(S_9)$	$I(S_{10})$	$I(S_{11})$	$I(S_{12})$	$I(S_{13})$
							Long-run inequality						
2.0	1.109	0.979	0.949	0.906	0.889	0.896	0.951	0.955	0.974	1.007	1.035	1.060	1.100
1.0	0.430	0.417	0.414	0.412	0.408	0.409	0.414	0.418	0.426	0.429	0.434	0.445	0.455
0.5	0.375	0.364	0.364	0.362	0.360	0.360	0.363	0.368	0.373	0.377	0.381	0.387	0.396
0.0	0.367	0.355	0.355	0.353	0.351	0.350	0.354	0.359	0.366	0.370	0.371	0.378	0.386

V	R_1	R_2	R_3	R_4	R_5	R_6	R_7	R_8	R_9	R_{10}	R_{11}	R_{12}	R_{13}
							Income stability						
2.0	1.000	0.948	0.916	0.897	0.885	0.860	0.828	0.809	0.786	0.764	0.744	0.717	0.692
1.0	1.000	0.955	0.928	0.915	0.903	0.891	0.877	0.865	0.855	0.843	0.832	0.822	0.813
0.5	1.000	0.954	0.932	0.917	0.911	0.897	0.885	0.877	0.867	0.861	0.855	0.843	0.840
0.0	1.000	0.950	0.927	0.912	0.904	0.890	0.879	0.872	0.864	0.859	0.852	0.841	0.839

Table 8 1969–1981 per Capita Family Income, Short-Run Inequalities

V	Overall	Between	Within	0–11th grade	12th grade	College	Advanced degree
2.0	1.112	0.081	1.031	1.124	0.513	0.377	0.224
	0.964	0.096	0.868	0.864	0.506	0.407	0.298
	1.051	0.102	0.949	0.886	0.689	0.469	0.300
	0.942	0.104	0.838	0.760	0.645	0.453	0.252
	1.003	0.101	0.902	0.860	0.603	0.448	0.275
	1.192	0.110	1.083	0.998	0.775	0.513	0.284
	1.641	0.118	1.523	1.478	0.825	0.609	0.283
	1.360	0.117	1.243	1.119	0.909	0.610	0.289
	1.573	0.109	1.464	1.399	0.901	0.706	0.334
	1.844	0.113	1.730	1.529	1.474	0.680	0.406
	1.898	0.114	1.785	1.467	1.831	0.807	0.383
	2.076	0.141	1.935	1.661	1.380	0.888	0.450
	2.447	0.146	2.301	2.018	1.437	0.996	0.572
1.0	0.431	0.079	0.352	0.443	0.294	0.258	0.190
	0.444	0.092	0.352	0.431	0.302	0.263	0.234
	0.467	0.098	0.369	0.435	0.334	0.297	0.223
	0.460	0.099	0.361	0.424	0.323	0.306	0.201
	0.464	0.097	0.367	0.445	0.316	0.292	0.204
	0.488	0.104	0.384	0.457	0.347	0.307	0.213
	0.531	0.112	0.419	0.499	0.370	0.349	0.222
	0.549	0.111	0.438	0.532	0.376	0.353	0.229
	0.579	0.105	0.474	0.598	0.374	0.370	0.261
	0.594	0.109	0.485	0.605	0.402	0.365	0.276
	0.613	0.110	0.503	0.632	0.431	0.360	0.259
	0.669	0.133	0.536	0.668	0.438	0.417	0.301
	0.706	0.138	0.568	0.712	0.473	0.416	0.325
0.5	0.376	0.078	0.297	0.394	0.265	0.243	0.185
	0.388	0.091	0.297	0.381	0.276	0.245	0.227
	0.408	0.098	0.310	0.383	0.300	0.276	0.208
	0.405	0.099	0.306	0.378	0.286	0.287	0.193
	0.404	0.097	0.307	0.393	0.285	0.270	0.194
	0.423	0.104	0.319	0.404	0.303	0.287	0.200
	0.456	0.112	0.345	0.431	0.326	0.321	0.214
	0.468	0.111	0.358	0.464	0.324	0.325	0.221
	0.494	0.105	0.389	0.532	0.321	0.337	0.257
	0.495	0.190	0.386	0.521	0.334	0.330	0.260
	0.504	0.110	0.394	0.540	0.357	0.317	0.240
	0.550	0.133	0.418	0.564	0.358	0.381	0.281
	0.571	0.138	0.433	0.592	0.395	0.361	0.296

(continued)

Table 8   (Continued)

V	Overall	Between	Within	0–11th grade	12th grade	College	Advanced degree
0.0	0.367	0.079	0.288	0.392	0.260	0.245	0.188
	0.381	0.092	0.289	0.373	0.274	0.245	0.233
	0.402	0.099	0.303	0.375	0.300	0.277	0.203
	0.399	0.100	0.299	0.371	0.280	0.290	0.195
	0.396	0.098	0.298	0.383	0.284	0.270	0.193
	0.415	0.105	0.310	0.396	0.295	0.292	0.196
	0.448	0.113	0.336	0.416	0.321	0.324	0.217
	0.458	0.112	0.347	0.456	0.315	0.327	0.223
	0.492	0.107	0.384	0.557	0.311	0.341	0.269
	0.481	0.111	0.370	0.514	0.321	0.328	0.263
	0.483	0.113	0.370	0.526	0.343	0.310	0.238
	0.540	0.135	0.405	0.547	0.334	0.403	0.284
	0.549	0.141	0.408	0.567	0.381	0.352	0.297
	2299			1096	624	452	127

Table 9   1969–1981 per Capita Family Income, (MIW) Long-Run Inequality

V	Overall	Between	Within	0–11th grade	12th grade	College	Advanced degree
2.0	1.112	0.081	1.031	1.124	0.513	0.377	0.224
	0.981	0.090	0.891	0.925	0.481	0.361	0.230
	0.952	0.096	0.856	0.855	0.520	0.363	0.229
	0.908	0.099	0.809	0.789	0.524	0.358	0.217
	0.891	0.101	0.790	0.762	0.519	0.353	0.213
	0.899	0.105	0.794	0.756	0.531	0.354	0.212
	0.053	0.109	0.844	0.805	0.542	0.361	0.211
	0.957	0.113	0.844	0.791	0.556	0.364	0.210
	0.977	0.117	0.860	0.798	0.568	0.374	0.210
	1.010	0.120	0.890	0.811	0.617	0.381	0.214
	1.037	0.123	0.914	0.813	0.684	0.391	0.217
	1.061	0.127	0.935	0.825	0.697	0.399	0.218
	1.103	0.131	0.972	0.855	0.706	0.412	0.227

(continued)

Table 9 (Continued)

V	Overall	Between	Within	0–11th grade	12th grade	College	Advanced degree
1.0	0.431	0.079	0.352	0.443	0.294	0.258	0.190
	0.418	0.086	0.332	0.413	0.283	0.242	0.192
	0.415	0.090	0.325	0.398	0.285	0.240	0.186
	0.413	0.093	0.319	0.388	0.283	0.242	0.178
	0.409	0.095	0.314	0.382	0.279	0.238	0.173
	0.410	0.097	0.312	0.379	0.277	0.237	0.172
	0.414	0.101	0.313	0.380	0.278	0.240	0.172
	0.419	0.104	0.315	0.381	0.279	0.242	0.173
	0.426	0.107	0.319	0.388	0.280	0.246	0.175
	0.430	0.109	0.321	0.392	0.279	0.247	0.178
	0.435	0.110	0.324	0.398	0.282	0.245	0.178
	0.445	0.114	0.331	0.410	0.283	0.246	0.181
	0.455	0.118	0.337	0.422	0.285	0.248	0.184
0.5	0.376	0.078	0.297	0.394	0.265	0.243	0.185
	0.365	0.085	0.279	0.366	0.257	0.226	0.188
	0.364	0.090	0.274	0.353	0.259	0.226	0.181
	0.362	0.093	0.269	0.345	0.255	0.228	0.172
	0.360	0.094	0.266	0.341	0.253	0.225	0.168
	0.360	0.097	0.263	0.339	0.249	0.224	0.166
	0.364	0.100	0.263	0.338	0.250	0.227	0.167
	0.368	0.103	0.265	0.340	0.249	0.231	0.169
	0.373	0.105	0.269	0.347	0.249	0.234	0.174
	0.378	0.107	0.271	0.354	0.248	0.236	0.178
	0.381	0.108	0.273	0.361	0.250	0.234	0.177
	0.387	0.112	0.275	0.369	0.248	0.233	0.180
	0.396	0.116	0.280	0.381	0.249	0.234	0.185
0.0	0.367	0.079	0.288	0.392	0.260	0.245	0.188
	0.356	0.086	0.270	0.359	0.253	0.227	0.193
	0.355	0.091	0.265	0.345	0.257	0.226	0.182
	0.354	0.093	0.260	0.336	0.252	0.230	0.173
	0.351	0.095	0.257	0.331	0.250	0.226	0.169
	0.351	0.097	0.254	0.329	0.245	0.225	0.166
	0.355	0.100	0.254	0.326	0.245	0.230	0.168
	0.359	0.103	0.256	0.329	0.244	0.235	0.170
	0.366	0.104	0.263	0.344	0.244	0.239	0.178
	0.370	0.105	0.265	0.350	0.242	0.240	0.184
	0.372	0.106	0.265	0.356	0.243	0.237	0.182
	0.378	0.111	0.267	0.364	0.239	0.239	0.186
	0.386	0.116	0.270	0.374	0.241	0.238	0.193
	2299			1096	624	452	127

Table 10   1969–1981 per Capita Family Income, (MIW) Income Stability

V	Overall	Between	Within	0–11th grade	12th grade	College	Advanced degree
2.0	1.000	0.073	0.927	1.000	1.000	1.000	1.000
	0.948	0.087	0.861	0.936	0.945	0.920	0.876
	0.916	0.092	0.824	0.899	0.909	0.866	0.833
	0.897	0.098	0.799	0.879	0.883	0.834	0.806
	0.885	0.101	0.785	0.860	0.875	0.817	0.791
	1.860	0.100	0.760	0.832	0.843	0.789	0.780
	1.828	0.095	0.734	0.797	0.815	0.757	0.769
	1.809	0.096	0.713	0.771	0.791	0.731	0.760
	1.786	0.094	0.692	0.737	0.774	0.707	0.737
	1.764	0.091	0.673	0.711	0.740	0.694	0.711
	0.744	0.088	0.656	0.688	0.711	0.673	0.698
	0.717	0.086	0.631	0.663	0.686	0.644	0.664
	0.693	0.082	0.610	0.642	0.663	0.622	0.636
1.0	1.000	0.183	0.817	1.000	1.000	1.000	1.000
	0.955	0.197	0.758	0.946	0.949	0.926	0.901
	0.929	0.202	0.726	0.914	0.919	0.881	0.864
	0.915	0.207	0.708	0.898	0.901	0.858	0.839
	0.903	0.209	0.694	0.881	0.890	0.840	0.826
	0.891	0.212	0.679	0.865	0.866	0.821	0.819
	0.877	0.214	0.663	0.846	0.846	0.802	0.811
	0.865	0.215	0.650	0.826	0.829	0.788	0.805
	0.855	0.214	0.640	0.804	0.820	0.776	0.792
	0.843	0.213	0.630	0.788	0.799	0.765	0.780
	0.832	0.211	0.621	0.775	0.785	0.751	0.769
	0.822	0.211	0.611	0.768	0.768	0.728	0.752
	0.813	0.210	0.603	0.762	0.747	0.715	0.737
0.5	1.000	0.209	0.791	1.000	1.000	1.000	1.000
	0.954	0.223	0.730	0.945	0.949	0.926	0.909
	0.932	0.231	0.701	0.917	0.924	0.886	0.874
	0.917	0.235	0.682	0.899	0.904	0.864	0.847
	0.911	0.238	0.672	0.887	0.897	0.852	0.840
	0.897	0.241	0.656	0.872	0.871	0.831	0.830
	0.885	0.244	0.641	0.853	0.851	0.817	0.826
	0.877	0.246	0.631	0.836	0.837	0.808	0.823
	0.867	0.243	0.624	0.817	0.827	0.799	0.815
	0.861	0.243	0.618	0.809	0.813	0.793	0.814
	0.855	0.242	0.613	0.802	0.803	0.781	0.803
	0.843	0.244	0.599	0.795	0.781	0.751	0.788
	0.840	0.247	0.594	0.796	0.766	0.743	0.782

(*continued*)

Table 10   (Continued)

V	Overall	Between	Within	0–11th grade	12th grade	College	Advanced degree
0.0	1.000	0.215	0.785	1.000	1.000	1.000	1.000
	0.950	0.229	0.721	0.940	0.947	0.925	0.914
	0.927	0.236	0.691	0.910	0.923	0.885	0.875
	0.912	0.241	0.671	0.891	0.902	0.865	0.847
	0.905	0.244	0.661	0.878	0.895	0.851	0.839
	0.890	0.246	0.644	0.863	0.865	0.831	0.828
	0.879	0.249	0.630	0.842	0.847	0.820	0.827
	0.872	0.249	0.623	0.827	0.833	0.816	0.827
	0.864	0.244	0.620	0.815	0.824	0.810	0.827
	0.859	0.244	0.615	0.808	0.810	0.802	0.831
	0.852	0.244	0.608	0.802	0.801	0.789	0.815
	0.841	0.247	0.594	0.796	0.777	0.762	0.807
	0.839	0.252	0.587	0.796	0.765	0.751	0.808
	2299			1096	624	452	127

Table 11   1969–1981 per Capita Family Income, Short-Run Inequalities

V	Overall	Between	Within	18–29	30–39	40–49	50–59	≥60
2.0	1.109	0.004	1.105	1.294	0.667	0.836	1.992	1.059
	0.961	0.006	0.956	0.587	0.641	0.982	1.508	1.083
	1.049	0.007	1.042	0.840	0.834	0.920	1.598	1.081
	0.940	0.009	0.931	0.643	0.777	0.942	1.332	0.854
	1.002	0.013	0.988	0.575	0.751	1.061	1.350	1.019
	1.190	0.017	1.173	0.638	0.854	1.676	1.452	0.852
	1.636	0.022	1.615	0.788	0.999	1.375	3.657	0.937
	1.357	0.027	1.330	0.971	1.298	1.379	1.660	0.916
	1.569	0.031	1.537	0.894	1.332	2.110	1.694	0.973
	1.840	0.051	1.789	1.937	1.572	1.990	1.912	0.965
	1.895	0.061	1.833	0.959	2.309	2.095	1.725	1.012
	2.072	0.083	1.989	1.034	2.050	2.499	1.891	1.036
	2.441	0.098	2.343	1.267	2.292	3.267	1.914	1.135

(continued)

Table 11 (Continued)

V	Overall	Between	Within	18–29	30–39	40–49	50–59	≥60
1.0	0.430	0.004	0.426	0.297	0.373	0.422	0.533	0.510
	0.443	0.006	0.437	0.322	0.366	0.427	0.554	0.537
	0.466	0.007	0.460	0.341	0.406	0.449	0.570	0.542
	0.460	0.009	0.451	0.334	0.416	0.445	0.556	0.496
	0.464	0.012	0.451	0.318	0.391	0.467	0.560	0.508
	0.488	0.015	0.473	0.332	0.419	0.484	0.596	0.518
	0.531	0.019	0.512	0.368	0.445	0.520	0.646	0.572
	0.548	0.023	0.525	0.382	0.478	0.536	0.646	0.565
	0.578	0.027	0.551	0.381	0.495	0.572	0.671	0.619
	0.594	0.042	0.552	0.388	0.500	0.573	0.683	0.591
	0.613	0.051	0.562	0.380	0.488	0.577	0.709	0.648
	0.669	0.067	0.602	0.381	0.553	0.610	0.787	0.644
	0.706	0.080	0.626	0.457	0.539	0.656	0.787	0.664
0.5	0.375	0.004	0.371	0.263	0.331	0.369	0.448	0.446
	0.388	0.006	0.382	0.289	0.326	0.364	0.472	0.481
	0.407	0.007	0.401	0.299	0.362	0.389	0.481	0.492
	0.404	0.009	0.396	0.297	0.373	0.384	0.473	0.467
	0.404	0.012	0.392	0.286	0.345	0.399	0.472	0.474
	0.423	0.014	0.409	0.294	0.371	0.406	0.501	0.497
	0.456	0.018	0.438	0.325	0.387	0.434	0.537	0.562
	0.468	0.021	0.446	0.334	0.400	0.448	0.546	0.557
	0.494	0.025	0.469	0.336	0.418	0.469	0.571	0.639
	0.496	0.038	0.457	0.325	0.417	0.464	0.578	0.584
	0.505	0.047	0.458	0.331	0.395	0.452	0.606	0.652
	0.551	0.061	0.489	0.336	0.466	0.472	0.684	0.632
	0.571	0.073	0.498	0.406	0.423	0.500	0.698	0.637
0.0	0.367	0.004	0.362	0.258	0.323	0.359	0.430	0.437
	0.381	0.006	0.375	0.283	0.318	0.346	0.462	0.488
	0.402	0.007	0.395	0.293	0.361	0.375	0.466	0.506
	0.399	0.008	0.390	0.292	0.374	0.371	0.458	0.493
	0.395	0.012	0.384	0.282	0.388	0.385	0.454	0.501
	0.415	0.014	0.401	0.287	0.369	0.391	0.483	0.539
	0.488	0.017	0.432	0.323	0.382	0.416	0.521	0.635
	0.458	0.020	0.438	0.344	0.384	0.430	0.532	0.634
	0.492	0.024	0.468	0.388	0.411	0.452	0.563	0.802
	0.481	0.036	0.446	0.318	0.408	0.411	0.565	0.669
	0.483	0.044	0.439	0.329	0.376	0.416	0.599	0.768
	0.541	0.057	0.484	0.341	0.489	0.435	0.695	0.718
	0.549	0.068	0.481	0.424	0.394	0.461	0.729	0.707
	2305			359	525	641	471	309

instance, the fact that some mobility is experienced in the early part of this period, but has ceased in the latter part of the sample. Similarly, we note that between-group *long-run* inequality has risen consistently, suggesting that the expected returns to schooling has materialized. In fact the inequality gap between the educational groups here has widened by 50% over time; see John et al. (1993).

Tables 11–13 have the same structure as before but focus on the impact of age. These tables suggest that between-group inequality, both short and long run, has increased dramatically over time. Seniority matters! Within group inequality is larger the older the group. This is also due to accumulation of returns to different investments, opportunities, and attainments. Short-run inequalities increase within groups, while long-run inequalities are stable with a moderate increase toward the end of this period. Maasoumi et al. (1996) find these trends have continued. Finally we note that these figures are based on per capita incomes. Since family size and composition changes over time these figures show greater volatility than the authors found with total family incomes unadjusted for family size (Maasoumi and Zandvakili 1989, Appendix).

Tables 14–15 provide decompositions by race, noting that the "non-white" group includes all heads not classified as "white." This explains the large within-group inequality. The number of households in each group is given in the last column. Inequality among non-whites has increased faster than among whites. Short run inequality has increased within both groups, and somewhat increased in the aggregated incomes. Between-group inequality in the short run distributions declined somewhat in the first half of the period and increased again in the last 4–5 years of the sample. For the long-run incomes, between-group inequalities are rather stable with a slight decline over time. It would appear that within-group characteristics are controlling of the degree of inequality in this sample and for this decomposition. Other grouping criteria that are more race specific than "non-white" are known to indicate greater between group inequality. See John et al. (1991).

Experimentation over the members of the GE family, as well as with different sets of weights given to incomes at different points in time, represent an attempt to robustify summary findings. This is an important element of empirical work in this area since unanimity with respect to weights and degree of aversion to inequality is not likely. Of course, an interpretation of this "robustification" technique is that it is an empirical substitute for unanimous ranking by Lorenz-type comparisons over plausible ranges of parameter values. This is useful when such curves cannot be statistically ordered when they cross only at extreme parametric values.

Several other applications to U.S. and U.K. data are reported in Shorrocks (1978b, 1981). A good deal more is now possible given the dominance testing techniques of Section IV, and the asymptotic distribution theory summarized in Maasoumi (1996a). The bootstrap alternative appears very promising, as demonstrated by Mills and Zandvakili (1996).

Table 12   1969–1981 per Capita Family Income, (MIW) Long-Run Inequalities

V	Overall	Between	Within	18–29	30–39	40–49	50–59	≥60
2.0	1.109	0.004	1.105	1.294	0.667	0.836	1.992	1.059
	0.979	0.005	0.974	0.859	0.618	0.867	1.683	1.016
	0.949	0.005	1.944	0.798	0.648	0.839	1.601	0.968
	0.906	0.006	0.900	0.723	0.647	0.830	1.486	0.883
	0.889	0.007	0.882	0.667	0.645	0.844	1.416	0.859
	0.896	0.008	0.889	0.634	0.651	0.932	1.359	0.812
	0.951	0.009	0.942	0.625	0.667	0.950	1.614	0.784
	0.955	0.010	0.944	0.637	0.707	0.985	1.544	0.764
	0.974	0.012	0.962	0.635	0.741	1.033	1.482	0.740
	1.007	0.014	0.993	0.716	0.782	1.077	1.447	0.735
	1.035	0.016	1.018	0.711	0.866	1.117	1.386	0.733
	1.060	0.020	1.039	0.712	0.909	1.171	1.331	0.729
	1.100	0.025	1.075	0.718	0.963	1.260	1.290	0.725
1.0	0.430	0.004	0.426	0.297	0.373	0.422	0.533	0.510
	0.417	0.005	0.412	0.277	0.353	0.408	0.528	0.505
	0.414	0.005	0.409	0.277	0.352	0.404	0.527	0.492
	0.412	0.006	0.406	0.277	0.355	0.402	0.523	0.475
	0.408	0.007	0.402	0.271	0.351	0.403	0.517	0.461
	0.409	0.007	0.402	0.269	0.352	0.407	0.516	0.455
	0.414	0.009	0.405	0.272	0.355	0.412	0.521	0.456
	0.418	0.010	0.408	0.275	0.360	0.416	0.521	0.455
	0.426	0.012	0.414	0.277	0.369	0.428	0.523	0.454
	0.429	0.014	0.415	0.276	0.372	0.433	0.520	0.455
	0.434	0.017	0.417	0.275	0.374	0.437	0.518	0.464
	0.445	0.021	0.424	0.276	0.380	0.445	0.525	0.471
	0.455	0.026	0.429	0.282	0.386	0.453	0.526	0.473
0.5	0.375	0.004	0.371	0.263	0.331	0.369	0.448	0.446
	0.364	0.005	0.359	0.246	0.313	0.351	0.447	0.448
	0.364	0.005	0.358	0.246	0.314	0.349	0.448	0.442
	0.362	0.006	0.356	0.246	0.318	0.347	0.443	0.433
	0.360	0.006	0.353	0.243	0.316	0.349	0.440	0.426
	0.360	0.007	0.352	0.241	0.315	0.350	0.438	0.425
	0.363	0.008	0.355	0.243	0.318	0.353	0.441	0.433
	0.368	0.010	0.358	0.247	0.320	0.358	0.444	0.437
	0.373	0.011	0.362	0.249	0.325	0.366	0.446	0.442
	0.377	0.014	0.364	0.248	0.329	0.371	0.445	0.450
	0.381	0.017	0.364	0.247	0.328	0.371	0.447	0.466
	0.387	0.021	0.366	0.248	0.330	0.373	0.454	0.472
	0.396	0.026	0.370	0.257	0.334	0.378	0.460	0.476

(continued)

Table 12   (Continued)

V	Overall	Between	Within	18–29	30–39	40–49	50–59	≥60
0.0	0.367	0.004	0.362	0.258	0.323	0.359	0.430	0.437
	0.355	0.005	0.350	0.240	0.304	0.336	0.432	0.444
	0.355	0.005	0.350	0.240	0.308	0.334	0.432	0.443
	0.353	0.006	0.348	0.241	0.314	0.332	0.427	0.440
	0.351	0.006	0.345	0.238	0.311	0.334	0.422	0.436
	0.350	0.007	0.343	0.235	0.310	0.335	0.419	0.439
	0.354	0.008	0.346	0.238	0.313	0.338	0.423	0.455
	0.359	0.010	0.349	0.244	0.313	0.343	0.426	0.466
	0.366	0.011	0.355	0.248	0.319	0.352	0.429	0.487
	0.370	0.013	0.356	0.246	0.323	0.357	0.427	0.499
	0.371	0.016	0.355	0.246	0.321	0.353	0.429	0.522
	0.378	0.021	0.358	0.249	0.329	0.352	0.438	0.529
	0.386	0.026	0.361	0.264	0.328	0.355	0.447	0.533
	2305			359	525	641	471	309

Table 13   1969–1981 per Capita Family Income, (MIW) Income Stability

V	Overall	Between	Within	18–29	30–39	40–49	50–59	≥60
2.0	1.000	0.004	0.996	1.000	1.000	1.000	1.000	1.000
	0.948	0.004	0.944	0.927	0.946	0.951	0.967	0.948
	0.916	0.005	0.911	0.891	0.905	0.919	0.949	0.903
	0.897	0.006	0.891	0.876	0.882	0.901	0.935	0.873
	0.885	0.007	0.878	0.873	0.877	0.887	0.927	0.851
	0.860	0.007	0.853	0.857	0.857	0.850	0.898	0.830
	0.828	0.008	0.821	0.836	0.832	0.829	0.851	0.807
	0.809	0.009	0.801	0.814	0.802	0.810	0.833	0.795
	0.786	0.010	0.777	0.794	0.780	0.782	0.810	0.759
	0.764	0.011	0.753	0.749	0.757	0.763	0.788	0.765
	0.744	0.012	0.732	0.746	0.723	0.746	0.763	0.760
	0.717	0.014	0.703	0.737	0.695	0.721	0.729	0.748
	0.692	0.016	0.677	0.719	0.677	0.693	0.704	0.732

(continued)

Table 13 (Continued)

V	Overall	Between	Within	18–29	30–39	40–49	50–59	≥60
1.0	1.000	0.010	0.990	1.000	1.000	1.000	1.000	1.000
	0.955	0.011	0.944	0.893	0.955	0.961	0.971	0.964
	0.928	0.012	0.917	0.864	0.922	0.934	0.955	0.929
	0.915	0.013	0.902	0.853	0.907	0.922	0.945	0.913
	0.903	0.014	0.889	0.842	0.901	0.912	0.934	0.893
	0.891	0.016	0.875	0.832	0.889	0.903	0.918	0.881
	0.877	0.018	0.859	0.819	0.878	0.890	0.903	0.867
	0.865	0.021	0.844	0.812	0.867	0.878	0.888	0.856
	0.855	0.023	0.831	0.801	0.861	0.876	0.872	0.832
	0.843	0.027	0.815	0.788	0.852	0.868	0.853	0.829
	0.832	0.032	0.800	0.778	0.845	0.860	0.835	0.827
	0.822	0.039	0.783	0.774	0.832	0.854	0.187	0.824
	0.813	0.046	0.767	0.765	0.831	0.846	0.799	0.813
0.5	1.000	0.011	0.989	1.000	1.000	1.000	1.000	1.000
	0.954	0.013	0.941	0.890	0.953	0.959	0.970	0.963
	0.932	0.013	0.918	0.867	0.925	0.936	0.959	0.934
	0.917	0.014	0.903	0.856	0.910	0.922	0.946	0.918
	0.911	0.016	0.894	0.850	0.910	0.918	0.940	0.905
	0.897	0.018	0.878	0.837	0.895	0.908	0.923	0.893
	0.885	0.021	0.865	0.826	0.887	0.896	0.909	0.881
	0.877	0.023	0.853	0.823	0.877	0.889	0.897	0.873
	0.867	0.027	0.840	0.814	0.873	0.888	0.881	0.846
	0.861	0.032	0.830	0.806	0.870	0.887	0.865	0.851
	0.855	0.038	0.817	0.798	0.866	0.881	0.852	0.856
	0.843	0.046	0.797	0.793	0.846	0.871	0.832	0.850
	0.840	0.005	0.785	0.795	0.850	0.868	0.818	0.844
0.0	1.000	0.012	0.988	1.000	1.000	1.000	1.000	1.000
	0.950	0.013	0.937	0.887	0.949	0.954	0.967	0.959
	0.927	0.014	0.913	0.863	0.921	0.930	0.954	0.927
	0.912	0.014	0.897	0.854	0.908	0.915	0.940	0.913
	0.904	0.016	0.888	0.846	0.907	0.911	0.933	0.899
	0.890	0.018	0.871	0.832	0.890	0.901	0.914	0.886
	0.879	0.020	0.859	0.822	0.884	0.890	0.900	0.875
	0.872	0.023	0.849	0.824	0.873	0.885	0.889	0.867
	0.864	0.026	0.838	0.818	0.872	0.887	0.871	0.842
	0.859	0.031	0.828	0.810	0.872	0.887	0.853	0.847
	0.852	0.038	0.814	0.802	0.865	0.877	0.837	0.855
	0.841	0.046	0.795	0.800	0.852	0.865	0.817	0.847
	0.839	0.056	0.784	0.815	0.850	0.862	0.802	0.841
	2305			359	525	641	471	309

Table 14   1969–1981 per Capita Family Income, Short-Run Inequalities

V	Overall	Between	Within	White	Non-white
2.0	1.109	0.087	1.022	0.599	1.099
	0.961	0.085	0.877	0.609	0.870
	1.049	0.088	0.961	0.718	0.901
	0.940	0.085	0.855	0.638	0.810
	1.002	0.076	0.925	0.602	0.980
	1.190	0.079	1.111	0.890	1.022
	1.636	0.074	1.562	0.770	1.891
	1.357	0.070	1.286	0.816	1.414
	1.569	0.069	1.500	0.900	1.704
	1.840	0.070	1.770	1.008	2.048
	1.895	0.066	1.829	1.117	2.085
	2.072	0.079	1.994	1.328	2.068
	2.441	0.078	2.363	1.356	2.653
1.0	0.430	0.073	0.357	0.322	0.428
	0.443	0.071	0.372	0.336	0.444
	0.466	0.074	0.393	0.363	0.451
	0.460	0.071	0.388	0.354	0.456
	0.464	0.065	0.398	0.344	0.507
	0.488	0.067	0.421	0.372	0.519
	0.531	0.063	0.468	0.410	0.582
	0.548	0.060	0.488	0.420	0.623
	0.578	0.059	0.519	0.477	0.661
	0.594	0.060	0.533	0.462	0.676
	0.613	0.057	0.556	0.483	0.703
	0.669	0.067	0.602	0.540	0.727
	0.706	0.066	0.640	0.554	0.810
0.5	0.375	0.068	0.308	0.289	0.388
	0.388	0.066	0.322	0.304	0.398
	0.407	0.068	0.339	0.326	0.403
	0.404	0.066	0.388	0.322	0.410
	0.404	0.061	0.343	0.313	0.453
	0.423	0.062	0.361	0.366	0.457
	0.456	0.059	0.397	0.370	0.500
	0.468	0.056	0.411	0.375	0.542
	0.494	0.056	0.438	0.402	0.568
	0.496	0.057	0.439	0.404	0.566
	0.505	0.053	0.451	0.414	0.583
	0.551	0.062	0.489	0.464	0.591
	0.571	0.062	0.510	0.471	0.656

(continued)

Table 14   (Continued)

V	Overall	Between	Within	White	Non-white
0.0	0.367	0.063	0.303	0.284	0.392
	0.381	0.062	0.319	0.302	0.395
	0.402	0.064	0.338	0.324	0.402
	0.399	0.062	0.337	0.321	0.408
	0.395	0.057	0.338	0.313	0.453
	0.415	0.058	0.357	0.336	0.450
	0.448	0.056	0.393	0.371	0.488
	0.458	0.053	0.405	0.372	0.547
	0.492	0.052	0.439	0.409	0.569
	0.481	0.053	0.428	0.399	0.553
	0.483	0.050	0.433	0.401	0.567
	0.541	0.058	0.482	0.465	0.561
	0.549	0.058	0.491	0.460	0.632
	2305			1534	771

Table 15   1969–1981 per Capita Family Income, (MIW)
Long-Run Inequality

V	Overall	Between	Within	White	Non-white
2.0	1.109	0.087	1.022	0.599	1.099
	0.979	0.087	0.892	0.565	0.923
	0.949	0.089	0.860	0.577	0.858
	0.906	0.089	0.817	0.561	0.804
	0.889	0.088	0.801	0.545	0.795
	0.896	0.089	0.808	0.570	0.784
	0.951	0.088	0.862	0.567	0.872
	0.955	0.087	0.868	0.566	0.885
	0.974	0.086	0.888	0.570	0.916
	1.007	0.088	0.919	0.581	0.950
	1.035	0.087	0.947	0.595	0.984
	1.060	0.087	0.973	0.613	1.012
	1.100	0.087	1.013	0.628	1.062

(continued)

Table 15 (Continued)

V	Overall	Between	Within	White	Non-white
1.0	0.430	0.073	0.357	0.322	0.428
	0.417	0.073	0.345	0.310	0.413
	0.414	0.073	0.341	0.310	0.404
	0.412	0.073	0.339	0.307	0.402
	0.408	0.072	0.337	0.302	0.406
	0.409	0.071	0.338	0.301	0.410
	0.414	0.071	0.343	0.305	0.420
	0.418	0.069	0.349	0.307	0.432
	0.426	0.068	0.357	0.312	0.446
	0.429	0.068	0.361	0.316	0.452
	0.434	0.067	0.367	0.320	0.460
	0.445	0.067	0.377	0.328	0.475
	0.455	0.067	0.387	0.335	0.492
0.5	0.375	0.068	0.308	0.289	0.388
	0.364	0.067	0.297	0.279	0.373
	0.364	0.068	0.296	0.280	0.365
	0.362	0.067	0.294	0.278	0.363
	0.360	0.066	0.294	0.275	0.370
	0.360	0.065	0.294	0.275	0.372
	0.363	0.064	0.299	0.279	0.379
	0.368	0.063	0.304	0.282	0.391
	0.373	0.062	0.311	0.287	0.403
	0.377	0.062	0.316	0.291	0.409
	0.381	0.061	0.320	0.295	0.416
	0.387	0.061	0.326	0.300	0.424
	0.396	0.061	0.335	0.307	0.439
0.0	0.367	0.063	0.303	0.284	0.392
	0.355	0.062	0.293	0.276	0.372
	0.355	0.063	0.292	0.277	0.364
	0.353	0.063	0.291	0.276	0.361
	0.351	0.061	0.290	0.273	0.369
	0.350	0.060	0.290	0.272	0.370
	0.354	0.059	0.295	0.278	0.374
	0.359	0.058	0.301	0.281	0.390
	0.366	0.057	0.309	0.288	0.404
	0.370	0.056	0.313	0.292	0.409
	0.371	0.055	0.316	0.294	0.415
	0.378	0.056	0.322	0.301	0.419
	0.386	0.056	0.330	0.307	0.432
	2305			1534	771

ACKNOWLEDGMENTS

I would like to thank the editors, an anonymous referee, E. Ok, and seminar participants for comments.

REFERENCES

Anderson, G. J. (1994), Nonparametric Tests of Stochastic Dominance in Income Distributions, Working Paper, University of Toronto.

Atkinson, A. (1970), On the Measurement of Inequality, *Journal of Economic Theory*, 11, 244–263.

Atkinson, A. (1983), The Measurement of Economic Mobility, in *Social Justice and Public Policy*, MIT Press, Cambridge, MA.

Atkinson, A. and F. Bourguignon (1982), The Comparison of Multidimensioned Distributions in Economic Status, *Review of Economic Studies*, 49, 183–201.

Atkinson, A. and F. Bourguignon (1987), Income Distribution and Differences in Needs, in G. R. Feiwel (ed.), *Arrow and the Foundations of the Theory of Economic Policy*, Macmillan, London.

Atkinson, A., F. Bourguignon, and C. Morrison (1992), *Empirical Studies of Earnings Mobility*, Fundamentals of Pure and Applied Economics, vol. 52, Distribution Section, Harwood Academic.

Bartholomew, D. J. (1959), A Test of Homogeneity for Ordered Alternatives, *Biometrika*, 46, 36–48.

Bartholomew, D. J. (1982), *Stochastic Models for Social Processes*, 3rd ed., Wiley, New York.

Beach, C. M. and R. Davidson (1983), Distribution-Free Statistical Inference with Lorenz Curves and Income Shares, *Review of Economic Studies*, 50, 723–735.

Beach C. M., R. Davidson, and G. A. Slotve (1995), Distribution-Free Statistical Inference for Inequality Dominance with Crossing Lorenz Curves, Unpublished Paper N95A03, Université d'Aix-Marseille.

Beach, C. M. and J. Richmond (1985), Joint Confidence Intervals for Income Shares and Lorenz Curves, *International Economic Review*, 26, 439–450.

Bishop, J. A., K. V. Chow, and J. P. Formby (1994), Testing for Marginal Changes in Income Distributions with Lorenz and Concentration Curves, *International Economic Review*, 35, 479–488.

Bishop, J. A., J. P. Formby, and P. D. Thistle (1989), Statistical Inference, Income Distributions, and Social Welfare, *Research on Economic Inequality*, 1, 49–82.

Blackorby, C. and D. Donaldson (1978), Measures of Relative Equality and Their Meaning in Terms of Social Welfare, *Journal of Economic Theory*, 18, 59–80.

Bourguignon, F. (1979), Decomposable Income Inequality Measures, *Econometrica*, 47, 901–920.

Chakravarty, S. R. (1995), A Note on the Measurement of Mobility, *Economics Letters*, 48, 33–36.

Conlisk, J. (1989), Ranking Mobility Matrices, *Economics Letters*, 29, 231–235.

Conlisk, J. (1990), Monotone Mobility Matrices, *Journal of Mathematical Sociology*, 15, 173–191.

Creedy, J. (1985), *Dynamics of Income Distribution*, Basil Blackwell, Oxford.

Dardanoni, V. (1992), On Multidimensional Inequality Measurement, Unpublished Paper, University of Perugia.

Davies, J. and M. Hoy (1994), The Normative Significance of Using Third-Degree Stochastic Dominance in Comparing Income Distributions, *Journal of Economic Theory*, 64, 520–530.

Davies, J. and M. Hoy (1995), Making Inequality Comparisons When Lorenz Curves Intersect, *American Economic Review*, 85, 980–986.

Deshpande, J. V. and S. P. Singh (1985), Testing for Second Order Stochastic Dominance, *Communications in Statistics A*, 14, 887–893.

Dardanoni, V. (1993), On Measuring Social Mobility, *Journal of Economic Theory*, 61, 372–394.

Durbin, J. (1973), *Distribution Free Tests Based on the Sample Distribution Function*, SIAM, Philadelphia, PA.

Durbin, J. (1985), The First Passage Density of a Continuous Gaussian Process to a General Boundary, *Journal of Applied Probability*, 22, 99–122.

Dykstra, R. L., S. Kochar and T. Robertson (1995), Inference for Likelihood Ratio Ordering in Two Sample Problems, *Journal of the American Statistical Association*, 90, 1034–1040.

Ebert, U. (1995a), Income Inequality and Differences in Household Size, *Mathematical Social Sciences*, 30, 1034–1040.

Fields, G. and E. A. Ok (1995), The Meaning and the Measurement of Income Mobility, in J. Silber (ed.), *Income Inequality Measurement: From Theory to Practice*, Kluwer, forthcoming.

Geweke, J., R. C. Marshall, and G. A. Zarkin (1986), Mobility Indices in Continuous Time Markov Chains, *Econometrica*, 54, 1407–1423.

Gourieroux, C., A. Holly, and A. Monfort (1982), Likelihood Ratio Test, Wald Test and Kuhn-Tucker Test in Linear Models with Inequality Constraints in the Regression Parameters, *Econometrica*, 50, 1407–1423.

Hadar, J. and W. R. Russell (1969), Rules for Ordering Uncertain Prospects, *American Economic Review*, 59, 25–34.

Hadar, J. and W. R. Russell (1971), Stochastic Dominance and Diversification, *Journal of Economic Theory*, 3, 288–305.

Hanoch, G. and H. Levy (1969), The Efficiency Analysis of Choices Involving Risk, *Review of Economic Studies*, 36, 335–346.

Hart, P. E. (1976a), The Dynamics of Earnings, 1963–1973, *Economic Journal*, 86, 541–565.

Hart, P. E. (1976b), The Comparative Statics and Dynamics of Income Distributions, *Journal of the Royal Statistical Society A*, 139, 108–125.

Hart, P. E. (1981), The Statics and Dynamics of Income Distributions: A Survey, in A. Klevermarken and J. A. Lybeck (eds.), *The Statics and Dynamics of Income*, Tieto, Clevedon.

Hart, P. E. (1983), The Size Mobility of Earnings, *Oxford Bulletin of Economics and Statistics*, 45, 181–193.

Juhn, C., K., M. Murphy, and B. Pierce (1991), Accounting for the Slowdown in Black-White Wage Convergence, in M. H. Kosters (ed.), *Workers and Their Wages: Changing Patterns in the United States*, American Enterprise Institute Press, Washington, DC.

Juhn, C. and K. M. Murphy, and B. Pierce (1993), Wage Inequality and the Rise in Returns to Skill, *Journal of Political Economy*, 101, 410–442.

Kanbur, S. M. R. and J. E. Stiglitz (1986), Intergenerational Mobility and Dynastic Inequality, Woodrow Wilson Discussion paper No. 111, Princeton University.

Kaur, A., B. L. S. Rao, and H. Singh (1994),Testing for Second-Order Stochastic Dominance of Two Distributions, *Econometric Theory*,10, 849–866.

Klecan, L., R. McFadden, and D. McFadden (1991), A Robust Test for Stochastic Dominance, Working Paper, Department of Economics, MIT.

Klevermarken, A. and J. A. Lybeck (eds.), *The Statics and Dynamics of Income*, Tieto, Clevedon.

Kodde, D. A. and F. C. Palm (1986), Wald Criteria for Jointly Testing Equality and Inequality Restrictions, *Econometrica*, 50, 1243–1248.

Kolm, S. C. (1969), The Optimal Production of Social Justice, in J. Margolis and H. Guitton (eds.), *Public Economics*. Also presented at IEA Conference in Public Economics, Biaritz, 1966.

Kolm, S. C. (1976a), Unequal Inequalities: I, *Journal of Economic Theory*, 12, 416–442.

Kolm, S. C. (1976a), Unequal Inequalities: II, *Journal of Economic Theory*, 13, 82–111.

Lehmann, E. L. (1959), *Testing Statistical Hypotheses*, Wiley, New York.

Lillard, L. and R. Willis (1978), Synamic Aspects of Earnings Mobility, *Econometrica*, 46, 985–1012.

Maasoumi, E. (1986), The Measurement and Decomposition of Earnings Mobility, *Econometrica*, 54, 991–997.

Maasoumi, E. (1989a), Composite Indices of Income and Other Developmental Indicators: A General Approach, *Research in Income Inequality*, forthcoming.

Maasoumi, E. (1989b), Continuously Distributed Attributes and Measures of Multivariate Inequality, *Journal of Econometrics*, 42, 131–144.

Maasoumi, E. (1991), ed., *Measurement of Welfare and Inequality*, Annals issue of *Journal of Econometrics*, 50, 1/2.

Maasoumi, E. (1993), A Compendium to Information Theory in Economics and Econometrics, *Econometric Reviews*, 12, 1–49.

Maasoumi, E. (1996a), Empirical Analyses of Welfare and Inequality, in P. Schmidt and M. H. Pesaran (eds.), *Handbook of Applied Microeconometrics*, Blackwell, Oxford, forthcoming.

Maasoumi, E. (1996b), Multidimensioned Approaches to Welfare Analysis, Chapter 15 in J. Silber (ed.), *Income Inequality Measurement: From Theory to Practice*, Kluwer, forthcoming.

Maasoumi, E., J. Mills, and S. Zandvakili (1996), Consensus Ranking of US Income Distributions: A Bootstrap Application of Tests for Stochastic Dominance, mimeo, Department of Economics, Southern Methodist University.

Maasoumi, E. and J. Nickelsburg (1988), Multivariate Measures of Well Being and an Analysis of Inequality in the Michigan Data, *Journal of Business and Economic Statistics*, 6, 327–334.

Maasoumi, E. and M. Trede (1997), Statistical Inference for General Mobility Measure, Working Paper, Department of Economics, Southern Methodist University.

Maasoumi, E. and S. Zandvakili (1986), A Class of Generalized Measures of Mobility with Applications, *Economics Letters*, 22, 97–102.

Maasoumi, E. and S. Zandvakili (1989), Mobility Profiles and Time Aggregates of Individual Incomes, *Research on Income Inequality*, 195–218.

Maasoumi, E. and S. Zandvakili (1990), Generalized Entropy Measures of Mobility for Different Sexes and Income Levels, *Journal of Econometrics*, 43, 121–133.

Markandya, A. (1984), The Welfare Measurement of Changes in Economic Mobility, *Econometrica*, 51, 457–471.

McFadden, D. (1989), Testing for Stochastic Dominance, in T. Formby and T. K. Seo (eds.), *Studies in the Economics of Uncertainty, Part II* (in Honor of J. Hadar), Springer-Verlag, New York.

Mills, J. and S. Zandvakili (1996), Bootstrapping Inequality Measures, Working Paper, Department of Economics, University of Cincinnati.

Mitra, T. and E. A. Ok (1995), The Measurement of Income Mobility: A Partial Ordering Approach, mimeo, Department of Economics, Cornell University.

Perlman, M. D. (1969), One-Sided Testing Problems in Multivariate Analysis, *Annals of Mathematical Statistics*, 40, 549–562.

Prais, S. J. (1955), Measuring Social Mobility, *Journal of the Royal Statistical Society A*, 56–66.

Robertson, T. and F. T. Wright (1981), Likelihood Ratio Tests for and Against Stochastic Ordering Between Multinomial Populations, *Annals of Statistics*, 9, 1248–1257.

Roberston, T. and F. T. Wright, and R. Dykstra (1982), *Order Restricted Statistical Inference*, Wiley, New York.

Sargan, J. D. (1957), The Distribution of Wealth, *Econometrica*, 25, 568–590. Also Reprinted as Chapter 3 in E. Maasoumi (ed.), *Contributions to Econometrics: J. D. Sargan*, Vol. 1, Cambridge University Press, Cambridge.

Sen, A. (1970), *Collective Choice and Social Welfare*, Holden-day, San Francisco, reprinted, North-Holland, Amsterdam, 1979.

Shapiro, A. (1988),Towards a Unified Theory of Inequality Constrained Testing in Multivariate Analysis, *International Statistical Review*, 56, 49–62.

Shorrocks, A. F. (1976), Income Mobility and the Markov Assumption, *Economic Journal*, 86, 566–577.

Shorrocks, A. F. (1978a), The Measurement of Mobility, *Econometrica*, 46, 1013–1024.

Shorrocks, A. F. (1978b), Income Inequality and Income Mobility, *Journal of Economic Theory*, 19, 376–393.

Shorrocks, A. F. (1980), The Class of Adaptively Decomposable Inequality Measures, *Econometrica*, 48, 376–393.

Shorrocks, A. F. (1981), Income Stability in the United States, Chapter 9 in N. A. Klevmarken and J. A. Kybeck (eds.), *The Statics and Dynamics of Income*, Tieto, Clevedon.

Shorrocks, A. F. (1983), Ranking Income Distributions, *Economica*, 50, 3–17.

Shorrocks, A. F. (1984), Inequality Decomposition by Population Subgroups, *Econometrica*, 52, 1369–1385.

Shorrocks, A. F. and J. Foster (1987), Transfer Sensitive Inequality Measures, *Review of Economic Studies*, 54, 485–497.

Shorrocks, A. F. and J. Foster (1992), On the Hart Measure of Income Mobility, in *Industrial Concentration and Income Inequality: Festschrift for Peter Hart*, forthcoming.

Shorrocks, A. F. and J. Foster (1995), Inequality and Welfare Evaluation of Heterogeneous Income Distributions, Unpublished Paper, Department of Economics, University of Essex.

Theil, H. (1972), *Statistical Decomposition Analysis*, North-Holland, Amsterdam.

Trede, M. (1995), Statistical Inference in Mobility Measurement: Sex Differences in Earnings Mobility, Unpublished Paper, University of Koln.

U.S. Bureau of the Census (Various Years), *Current Population Reports: Consumer Income*, Series P-60.

Wolak, F. A. (1988), Duality on Testing Multivariate Hypotheses, *Biometrika*, 75, 611–615.

Wolak, F. A. (1989), Local and Global Testing of Linear and Nonlinear Inequality Constraints in Nonlinear Econometric Models, *Econometric Theory*, 5, 1–35.

Wolak, F. A. (1991), The Local Nature of Hypothesis Tests Involving Inequality Constraints in Nonlinear Models, *Econometrica*, 59, 981–995.

Xu, K. (1995), Asymptotically Distribution-Free Statistical Test for Generalized Lorenz Curves: An Alternative Approach, Department of Economics, Dalhousi University. Paper presented at the 7th World Congress of the Econometric Society, Japan.

Xu, K., G. Fisher, and D. Wilson (1995), New Distribution-Free Tests for Stochastic Dominance, Working Paper No. 95-02, Department of Economics, Dalhousi University.

Zandvakili, S. (1987), *The Measurement of Mobility and Income Stability: An Information Theoretic Approach*, Ph.D. thesis, Department of Economics, Indiana University, Bloomington.

6
Aggregation and Econometric Analysis of Demand and Supply

R. Robert Russell, Robert V. Breunig, and Chia-Hui Chiu
University of California at Riverside, Riverside, California

I. INTRODUCTION AND SUMMARY

A. Overview

The neoclassical development of producer and consumer theory, culminating in the use of duality theory and the introduction of flexible functional forms in the 1970s, focused on the restrictions on demand and supply functions implied by optimizing behavior of producers and consumers. These restrictions are completely characterized by the symmetry and negative semidefiniteness of the (Slutsky) matrix of substitution effects in consumer theory and the symmetry and positive semidefiniteness of the Jacobian of the (net) supply functions in the case of producer theory.* They are important for econometric demand and supply analysis in part because they reduce the number of independent parameters to be estimated. A classic example is development of the linear expenditure system, first estimated by Stone (1954). Beginning with a system of equations in which optimal expenditure on each commodity is a linear function of income and n prices, Klein and Rubin (1947–1948) showed that requiring the system to be generated by income-constrained utility maximization reduces the number of parameters to be estimated from $(n + 1)^2$ to $2n - 1$.

The problem with this nexus between theory and empirical application is that the estimation of demand and supply systems typically uses aggregate, or per capita,

---

*In addition, these matrices must have reduced rank because of homogeneity properties. The conditions are most easily derived in the dual from the (easily proved) concavity and homogeneity of expenditure functions and convexity and homogeneity of profit functions in prices.

data. In the case of aggregate supply functions of producers, this poses no problem so long as all inputs are efficiently allocated, as is the case in competitive equilibrium with no fixed inputs and no production externalities. In this case, profit maximization by all producers on individual technology sets yields the same aggregate net supply as profit maximization on the aggregate technology set, which is obtained by simple summation of the individual technology sets. The equivalence of these two optimization problems is salient in general-equilibrium theory and welfare economics (Debreu 1954, 1959) and has been elegantly illustrated and aptly referred to by Koopmans (1957) as the "interchangeability of set summation and optimization." The essence of Koopman's interchangeability principle is that boundary points of the aggregate technology set are obtained as the sum of boundary points of the individual technology sets where the supports are equal; but this is equivalent to an efficient allocation of net outputs, where the aggregate net output vector and the optimal vectors of each producer are supported by the same price vector. (The vector summation of boundary points of individual technology sets with unequal supports yields interior points of the aggregate technology set.)

From the standpoint of econometric applications (and other applications as well, especially in macroeconomic theory and international trade), the beauty of the aggregation result for profit-maximizing, price-taking producers is that no restrictions (other than those needed for the existence of an optimum) are required. (In particular, convexity of technology sets is not required.) This means that there is no loss of generality in positing the existence of a "representative producer," which generates aggregate net-supply functions by maximizing aggregate profit subject to the constraint that the aggregate net-supply vector be contained in the aggregate technology set. As a result, the Jacobian of the system of aggregate net supply functions has the same properties as those of individual producers.

But aggregation on the consumer side is not so simple; the symmetry and negative semidefiniteness of the substitution matrix does not carry over to aggregate demand systems. In fact, as shown by Debreu (1974), Mantel (1974), and Sonnenschein (1973),* the only restrictions imposed on aggregate demand functions by individual optimization are Walras' law (simply the aggregate budget identity) and homogeneity of degree zero (in prices and income for income-constrained consumers and in prices for endowment-constrained consumers). This, in turn, implies that, without further restrictions, the use of a representative agent in consumer theory is, a fortiori, unjustified. Essentially, the reason for this discouraging result in the aggregation of consumer demand systems is that aggregate demand depends on the arbitrary distribution of incomes or endowments. In fact, if consumer incomes were determined

---

*See also Mas-Colell and Neuefeind (1977). An excellent survey of these results can be found in Shafer and Sonnenschein (1982).

optimally by maximizing a (Bergson-Samuelson) social welfare function, a representative consumer would exist (Samuelson 1956).* (Similarly, if the econometric analysis of producer demand and supply systems entailed an arbitrary distribution of idiosynchratic variables, such as fixed capital stocks, the simple aggregation results would not go through for producers as well.)

There are essentially two ways out of this quandary. One is to impose restrictions on individual preferences that imply certain regularity conditions for aggregate demands. These range from the most restrictive assumption, yielding a representative agent in aggregate demands, to weaker restrictions implying that the aggregate demand systems satisfy the weak axiom of revealed preference (equivalently, that the Jacobian of the aggregate demand system is negative semidefinite).

The first approach to dealing with the consumer aggregation problem predates the Debreu-Sonnenschein-Mantel results and indeed goes all the way back to the classic result of Gorman (1953a). He showed that a representative consumer exists if and only if Engel curves of the consumers are linear and parallel across consumers (for given prices). Gorman's result is derived in the dual. The indirect utility function must be affine in income (for given prices), in which case the (Marshallian) ordinary demand functions, derived by Roy's Identity, are also affine in income. Equivalently, the consumer's expenditure function is affine in utility (or any monotonic transformation of utility), in which case the (Hicksian) income-compensated demand functions are affine in the utility scalar. This structure is commonly referred to as the "Gorman polar form," and has been elucidated in the primal by Blackorby, Boyce, and Russell (1978) and Blackorby and Schworm (1993). Gorman's result is discussed in Section II.A.

Gorman's notion was subsequently generalized by Muellbauer to obtain a weaker type of aggregation and, concomitantly, weaker form of representative consumer, one in which aggregate commodity shares depend on some (generally nonlinear) function of the distribution of income and (possibly) prices. He called the required structure "generalized linear," since it is a generalization of Gorman's "linear" (actual affine) structure. Muellbauer's generalized linear structure can be alternatively characterized by the fact that the Engel curves have "rank 2": they are contained in a two-dimensional subspace of the commodity space. That is, each can be obtained as a linear combination of the others. Requiring that the aggregate income be independent of prices yields Muellbauer's "price-independent generalized linearity" (PIGL) demand system or his "price-independent generalized logarithmic" (PIGLOG) system. Muellbauer's approach, along with the extension by Deaton and Muellbauer (1980), the "almost ideal demand system," is discussed in Section II.B.

---

*See also, Jerison (1984b).

Section II closes with a brief discussion of the tests for the existence of a representative agent, including the parametric tests of Christensen, Jorgenson, and Lau (1975), and others (using flexible functional forms like the translog) and the nonparametric, nonstochastic (mathematical programming) tests of Varian (1982) and Diewert and Parkan (1985). The power of these nonparametric tests has been assessed by Bronars (1987) and Russell and Tengesdal (1996). Lewbel (1991) has provided some evidence in favor of Muellbauer's PIGLOG specification if we exclude incomes in the tails of the distribution, but evidence against it if we include the tails.

The existence of a representative consumer, while necessary for many purposes, is not necessary for the existence of aggregate demand functions that require less-than-complete information about the distribution of incomes. Lau (1977a, 1977b, 1982) and Gorman (1981) spelled out restrictions on individual demand functions that are necessary and sufficient for the existence of an aggregate demand function that depends on prices and summary statistics of the income distribution. This weaker aggregation condition, discussed in Section III.A, is referred to as "exact aggregation" and is a natural generalization of the (rank 2) Muellbauer conditions.

In his remarkable theorem, Gorman (1981) also showed that the requisite individual demand systems are consistent with income-constrained utility maximization (i.e., satisfy the integrability conditions) if and only if the Engel curves can be contained in a three-dimensional subspace (for given prices)—that is, that the demand systems have rank no greater than 3. His theorem, presented in Section III.B, also completely characterizes the class of such functions, which encompasses virtually all demand systems that have been estimated econometrically. Gorman's theorem has been extended and clarified in a series of papers by Heineke (1979, 1993) and Heineke and Shefrin (1982, 1986, 1987, 1988).

Consumer attributes (like household size, geographical region, age of head, etc.) have been incorporated into the exact aggregation framework by Lau (1977a, 1977b, 1982) and implemented using the translog specification, by Jorgenson, Lau, and Stoker (1980). A related issue is the recovery of the parameters of individual demand systems from the estimation of the aggregate demand system. The necessary and sufficient conditions for this identification property have been developed by Heineke and Shefrin (1990). This literature is briefly discussed in Section III.C.

Additional research by, for example, Jorgenson, Lau, and Stoker (1981), Pollak and Wales (1978, 1980), Stoker (1984), Russell (1983), Jorgenson and Slesnick (1984), Stoker (1986a, 1986b), Buse (1992), Blundell, Pashardes, and Weber (1993), and Nicol (1994) have further developed and applied these ideas. A summary of this literature is presented in Section III.D.*

---

*The notion of exact aggregation can also be applied to firms with individual characteristics pertaining to the technology or to fixed inputs or outputs (see Appelbaum 1982, Borooah and Van Der Ploeg 1986, Gouriéroux 1990, Fortin 1991, and Chavas 1993).

Gorman's results on the rank of demand systems have been extended through a series of papers that have characterized well-known—and some lesser-known—specifications. The key contributions have been made by Lewbel (1987, 1989b, 1990, 1991). In a related study, Hausman, Newey, and Powell (1995) provide a cross-sectional test of the Gorman rank-3 condition.

The rank-2 phenomenon shows up in other studies of aggregation (e.g., Jerison's (1984a) results on pairwise aggregation with a fixed income distribution) and as one of the necessary and sufficient conditions for the weak axiom of revealed preference to hold in the aggregate (Freixas and Mas-Collel 1987). Curiously, the rank-2 condition also emerges in a study of a (proportional) budgeting procedure for an organization (Blackorby and Russell 1993). These disparate results, which seem to be crying out for a unifying general theorem, are discussed in Section IV.

The second approach to dealing with the paucity of implications for aggregate consumer demand systems of individual optimization is to restrict the distribution of incomes or preferences of the population. Hildenbrand (1983, 1993), Härdle, Hildenbrand, and Jerison (1991), and Grandmont (1992) seek additional restrictions on aggregate demand implied by reasonable restrictions on the distribution of income or preferences. An early precurser of these results is Becker (1962), who showed that aggregate demand curves will be downward sloping even if individuals are "irrationally" nonoptimizing, in the sense that they distribute their demand "randomly" across the budget hyperplane (in particular, according to a rectangular distribution). The intuition behind this result is fairly obvious: the Giffen paradox will occur only if a sufficient number of individuals is concentrated in the Giffen portions of their demand functions, but this will not happen if the distribution of consumers across the budget plane is rectangular (more generally, "sufficiently" dispersed). This is the motif for the results of Hildenbrand and others, who show that the Jacobian of the aggregate demand system will be negative semidefinite (equivalently, that the weak axiom of revealed preference holds in the aggregate), implying that demand curves are downward sloping if the distribution of income is nonincreasing. Similarly, Grandmont derives the same restriction on aggregate demands by assuming that (neatly parameterized) preferences are sufficiently heterogeneous. These results are surveyed in Section V.

As noted above, there is no aggregation problem on the production side of the economy only if all inputs and outputs are efficiently allocated. If some inputs are not efficiently allocated, aggregation is not so straightforward. Inefficient allocation would occur, for example, if some inputs (e.g., capital) are fixed in the short run (and perfect capital rental markets do not exist). If we take as given the distribution of fixed inputs among firms, aggregation over the variable inputs is straightforward. There has been, however, a persistent interest in the question of the existence of an aggregate amount of fixed inputs, such as an aggregate capital stock that is fixed in the short run. Interest in this area, sparked in part by the "Cambridge controversy," has centered not only on aggregation of fixed inputs across firms, but also across different fixed inputs. At the individual level, the existence of such aggregates is known to be equivalent to certain separability conditions. (See the classic contributions by Gorman 1959, 1968a and the subsequent expositions by Blackorby, Primont, and Russell 1978, 1997.) Requiring the existence of such commodity aggregates at the macro level, however, requires stronger restrictions on individual technologies.

This aggregation problem was, in fact, first posed by Klein (1946a, 1946b), solved by Nataf (1948), and extended by Gorman (1953b). The Klein-Nataf aggregation problem assumes that *no* inputs are efficiently allocated and leads to a very unrealistic (linear) structure for individual technologies. It was pointed out early, however, by May (1946) and Pu (1946), and emphasized later by Solow (1964), that the efficient allocation of *some* inputs could be used to restrict the admissable allocations and hence weaken the aggregation conditions for the fixed inputs. These conjectures turned out to be correct, as rigorously shown by Gorman (1968b).* Blackorby and Schworm (1984, 1988a) provide comprehensive treatments of the problem of obtaining aggregate inputs in aggregate technologies under different assumptions about the existence of efficiently allocated and fixed inputs. The problem of the existence of aggregate commodities also is relevant to the empirical analysis of consumer demand, since most studies employ such aggregates; Blackorby and Schworm (1988b) provide necessary and sufficient conditions for the existence of commodity aggregates in market demand functions.

These results on aggregation across both agents and commodities are surveyed in Section VI. Section VII concludes.

B. Caveats

The aggregation literature surveyed in this chapter overlaps with an extensive literature on the specification of functional form, cross-section (Engel curve) estimation, and other areas of interest to applied econometricians. While we unavoidably touch on these subjects, we limit our discussions to our focus on the issue of aggregation over agents in econometric estimation. For excellent surveys of the literature on econometric demand analysis, specification of functional form, and econometric modeling of producer behavior see, respectively, Deaton (1986), Lau (1986), and Jorgenson (1986).

Another topic that we do not cover is the aggregation of individual preferences to obtain a social welfare function. There is, of course, a huge (social choice) literature on this topic, emanating from the classic impossibility theorem of Arrow (1951). As noted above, aggregate data will be consistent with the existence of a representative consumer if commodities are allocated by maximizing a Bergson-Samuelson

---

*See also the series of papers by Fisher (1965, 1968a, 1968b, 1982, 1983).

social welfare function. This fact, of course, is of little use in the econometric estimation of demand systems in a market economy.

It is important to distinguish the notion of a representative agent that is useful in econometric studies from the notion of a "representative consumer" commonly employed in the macro/finance literature regarding the replication of a competitive equilibrium (Constantinides 1982, Aiyagari 1985, Huang 1987, Eichenbaum, Hansen, and Singleton 1988, and Vilks 1988a, 1988b). This concept requires no restriction on preferences, which is not surprising, since it generates a representative consumer only at *equilibrium points*, but not in even a neighborhood of prices and incomes; as such, it is not a useful construct for econometric applications.*

Finally, despite our efforts to limit the scope of this survey, we discovered that the literature on aggregation across agents is even more extensive than we had expected (and is still burgeoning). Hence, our main focus has been to try to integrate this large body of research (something that has not been done since the publication of Green 1964);[†] we have made no serious attempt to critique it. Even with this limited objective, we have not been as successful as we would have liked; it seems to us that there is a need for a monograph on aggregation over agents—one that would be accessible to econometricians as well as theorists.

II. AGGREGATION AND REPRESENTATIVE AGENTS

A. Gorman Aggregation

The modern theory of consumer aggregation has its genesis in the classic paper by Gorman (1953a). He posed the following question. Suppose we have H optimizing consumers, each maximizing utility subject to a budget constraint with fixed prices and total expenditure. Thus, each consumer has a demand function, d^h, generated by

$$d^h(p, y^h) = \mathrm{argmax}_{x^h}\{U^h(x^h) \mid p \cdot x^h \leq y^h\} \in \mathbf{R}^n_+ \qquad (1)$$

where $p \in \mathbf{R}^n_{++}$ is the (common) price vector and U^h, x^h, and y^h are the utility function, consumption vector, and total expenditure of consumer h. The aggregate demand vector is given straightforwardly by[‡]

$$\sum_h d^h(p, y^h) =: \hat{D}(p, y^1, \dots, y^H) \qquad (2)$$

---

*This characterization is not intended to demean these results: while they cannot be used in econometric modeling, they are relevant to calibration exercises that attempt to simulate a *single* history of an economy (cf., the large body of calibration stimulated by Kydland and Prescott 1982).

[†] See, however, van Daal and Merkies (1984), who cover some of the topics addressed in the survey.

[‡] Notation: $A =: B$ or $B := A$ means B is defined by A.

Simple additive aggregation holds if aggregate demand is independent of the distribution of total expenditures—i.e., if there exists a function, D, such that

$$\hat{D}(p, y^1, \ldots, y^H) = D\left(p, \sum_h y^h\right) =: D(p, y) \tag{3}$$

Theorem.   Assume that $d^h(p, y^h) > 0$ for all h. The necessary and sufficient restriction on the individual demand functions, $d^h, h = 1, \ldots, H$, for additive aggregation, (3), to hold is that they take the form*

$$d^h(p, y^h) = \alpha^h(p) + \beta(p)y^h \quad \forall h \tag{4}$$

Sufficiency of this structure is obvious, since[†]

$$\hat{D}(p, y^1, \ldots, y^H) = \sum_h \alpha^h(p) + \beta(p) \sum_h y^h =: D(p, y) \tag{5}$$

Thus, the individual demand functions must be affine in the (idiosynchratic) total-expenditure variables with a common (price dependent) coefficient on these variables. In fact, this is a general condition for any set of structural equations to aggregate consistently as in (3): the individual functions must be affine, with common coefficients for different households, in all idiosyncratic variables—those that require an individual identifier index h—and the coefficients can depend on the common variables (in this case, prices).[‡] The equations can be made stochastic and still satisfy the aggregation condition as long as the vector of disturbance terms, ϵ^h, enters additively; e.g.,

$$d^h(p, y^h) = \alpha^h(p) + \beta(p)y^h + \epsilon^h \quad \forall h$$

in which case the disturbance term of the macro function is $\sum_h \epsilon^h$:

---

[*]If $d(p, y^h)$ is restricted to be nonnegative, but corner solutions are allowed, it is necessary that $\alpha^h(p) = 0$ for all h in the following restriction.

[†]Sketch of proof of necessity (assuming differentiability): Substitute (2) into (3) and differentiate with respect to each y^h to obtain

$$\frac{\partial d_i^h(y^h)}{\partial y^h} = D_i(p, y) \quad \forall i, \ h = 1, \ldots, H$$

This shows that, for all i, the derivative is identical for all h. Moreover, as the left-hand side (LHS) is independent of $y^{\bar{h}}$ for $\bar{h} \neq h$, the right-hand side (RHS) must be independent of y, which in turn implies that the LHS is, in fact, independent of y^h as well. Integration then yields (4). More generally (eschewing differentiability), (3) is a system of Pexider equations, whose solution is (4); see Corollary 10 on page 43 of Aczel and Dhombres 1989.

[‡]If one does not require that demands aggregate exactly, as in (3), but only that the expected value of x^h be independent of the distribution of y^h, then it is sufficient that the β coefficients be distributed independently of x^h.

$$x = \sum_h \alpha^h(p) + \sum_h \beta(p)y^h + \sum_h \epsilon^h$$

$$=: \alpha(p) + \beta(p)y + \epsilon \tag{6}$$

The result in the above theorem has been known for a long time; see, for example, the papers culminating in Theil (1954) and reviewed by Green (1964). Gorman's (1953a) contribution is to characterize this aggregation problem in terms of the restrictions on consumer preferences.* Formally, the problem is posed as follows. Roy's identity, combined with (4), yields the following system of (nonlinear) partial differential equations for consumer h.

$$-\frac{\partial V^h(p, y^h)/\partial p_i}{\partial V^h(p, y^h)/\partial y^h} = \alpha^h(p) + \beta(p)y^h, \qquad i = 1, \ldots, n \tag{7}$$

where V^h is the indirect utility function. Integration of this system yields the required structure of preferences:[†]

Theorem (Gorman 1953a). Assume interior solutions to the individual optimization problems.[‡] The aggregation condition (3) holds if and only if consumer preferences can be represented by expenditure functions with structure

$$E^h(u, p) = \Pi(p)u^h + \Lambda^h(p) \quad \forall h \tag{8}$$

or, equivalently, by indirect utility functions with the structure[§]

$$V^h(p, y^h) = \frac{y^h}{\Pi(p)} - \frac{\Lambda^h(p)}{\Pi(p)}$$

$$=: \frac{y^h}{\Pi(p)} - \Gamma^h(p) \quad \forall h \tag{9}$$

---

*Antonelli (1886) seems to have been the first to notice that homothetic and identical preferences are necessary and sufficient for consistent aggregation if we require that the conditions hold globally. Requiring that they hold only in a neighborhood, as in Gorman (1953a, 1961), yields a richer structure of preferences, allowing preference heterogeneity.

[†]This integration problem can be simplified, since affinity of the ordinary (Marshallian) demand functions in income implies affinity of the constant-utility (income compensated, or Hicksian) demand functions in (a particular normalization of) the utility variable:

$$\delta^h(u, p) = \hat{\alpha}^h(p) + \hat{\beta}(p)u^h \quad \forall h$$

Integration of this system of differential equations yields the consumer expenditure functions given by (8) below.

[‡]This restriction is inconsequential in most empirical studies. If the restriction is eschewed, the necessary and sufficient conditions for aggregation is homotheticity of preferences (see Samuelson 1956), in which case $\Lambda^h(p) = 0$ in (8) and (9) and, as noted earlier, $\alpha^h(p) = 0$ for all h in (4) or (6).

[§]This structure presumes a particular normalization of the utility representation; see the discussion to come.

where Π^h *and* Λ^h *are concave and homogeneous of degree 1 (hence,* Γ^h *is homogeneous of degree 0).*

Proof of sufficiency is trivial. By Roy's identity:

$$d^h(p, y) = \Pi(p)\nabla\Gamma^h(p) + \frac{\nabla\Pi(p)}{\Pi(p)}y^h$$

$$=: \alpha^h(p) + \beta(p)y^h \quad \forall h \tag{10}$$

so that aggregate demand is given by

$$\sum_h d^h(p, y) = \Pi(p)\sum_h \nabla\Gamma^h(p) + \frac{\nabla\Pi(p)}{\Pi(p)}\sum_h y^h$$

$$=: \alpha(p) + \beta(p)y \tag{11}$$

The salient feature of (8) or (9) is that they completely characterize, in the dual, the preferences that generate demands with the affine structure in the one idiosyncratic variable y^h. The marginal propensity to consume the ith commodity is

$$\frac{\partial d_i^h(p, y^h)}{\partial y^h} = \frac{\partial\Pi(p)}{\partial p_i} \bigg/ \Pi(p) =: \Pi_i(p)/\Pi(p) \quad \forall h \tag{12}$$

which is independent of h as well as y^h, indicating that Engel curves, and income consumption curves, are linear and parallel across consumers. Hence, transferring a dollar of total expenditure from one consumer to another will leave total demand unchanged.

The direct preferences corresponding to (8) and (9) can be further explicated. The constant-utility (vector valued) demand function is given by applying Shephard's lemma to (8):

$$\delta^h(u, p) = \nabla_p E(u, p) = \nabla\Lambda^h(p) + \nabla\Pi(p)u^h \tag{13}$$

Thus, the ICCs are linear but not necessarily parallel for different prices (although, or course, they cannot intersect). Consumption bundles on the base (zero utility) indifference surface are given by

$$\delta^h(0, p) = \nabla\Lambda^h(p) \tag{14}$$

Note that preferences are well defined for all consumption bundles on or above the base indifference surface, but may not be well defined below this surface. This is so because (8) represents consumer preferences only if it is concave in prices. But when $u < 0$, the first term in (8) is convex in prices and for sufficiently small values of u the convexity of this term will dominate the concavity of the second term, violating the fundamental regularity condition for consumer expenditure functions. Of course, if the base indifference surface does not intersect the positive orthant, preferences are well defined globally.

The structure (8) has an evocative interpretation. The "intercept" term $\Gamma^h(p)$ can be interpreted as the "fixed cost" of obtaining the base indifference surface—or the base utility level (normalized to be zero)—and $\Pi(p)$ is the marginal price of the composite commodity "utility."

The individual structure of consumer preferences that are necessary and sufficient for Gorman aggregation is often referred to as the "Gorman polar form" (GPF), following Blackorby, Primont, and Russell (1978), but Gorman refers to it as "quasi-homotheticity," since it is a generalization of homotheticity. By Shephard's (1970) decomposition theorem, homotheticity is characterized in the dual by

$$E^h(u, p) = \Pi(p)u^h \tag{15}$$

in which case the base indifference surface degenerates to a single point—the origin—and all income consumption curves are rays. An intermediate special case is "affine homotheticity," generated by

$$\Lambda^h(p) = \sum_h \gamma_i^h p_i \tag{16}$$

in which case the base indifference surface degenerates to a point $\gamma^h = \langle \gamma_1^h, \ldots, \gamma_n^h \rangle$, not necessarily the origin. A prominent example of affine homotheticity is the Stone-Geary structure, generated by (16) and

$$\Pi(p) = \prod_i p_i^{\beta_i} \tag{17}$$

The direct utility function dual to this structure is a Cobb-Douglas function in affine transformations of the consumption quantities,

$$U^h(x^h) = \prod_i (x_i^h - \gamma_i^h)^{\beta_i} \tag{18}$$

and the demand system is*

---

*The Stone-Geary structure evolved as the solution to a classic demand-system integrability problem. Klein and Rubin (1947–48) showed that the unconstrained linear expenditure system

$$p_i d_i(y, p) = \beta_i y + \sum_j \alpha_{ij} p_j \quad \forall i$$

is integrable if and only if the following parameter restrictions hold: $\beta_i > 0$ $\forall i$; $\sum_i \beta_i = 1$; $\alpha_{ij} = \gamma_j \beta_i$ for $j \neq i$; and $\alpha_{ii} = \gamma_i(1 - \beta_i)$ $\forall i$, in which case the demand system simplifies to (19). Later, Geary (1950), using Roy's identity, solved the partial integration problem to obtain the form of the utility function and Stone (1954) implemented the system by estimating the parameters using British consumption data.

$$d_i^h(p, y) = \gamma_i^h + \frac{\beta_i}{p_i}\left(y^h - \sum_j p_j \gamma_j^h\right), \qquad \forall i \tag{19}$$

The Stone-Geary specification was generalized by Brown and Heien (1972), who replaced (17) with a CES function of prices, in which case (18) becomes a CES function in affine translations of the consumption quantities. A further generalization was estimated by Blackorby, Boyce, and Russell (1978), who specified flexible functional forms (in particular, Diewert's (1971) generalized Leontief) for Λ^h and Π^h. But the Stone-Geary specification was the real workhorse of empirical demand analysis for two decades, until its replacement by flexible functional forms in the 1970s.*

The principal thrust of Gorman's aggregation theorem is that aggregate demand systems are consistent with individual utility maximization (individual "rationality") and simple aggregation if and only if individual preferences have the structure (8) with interpersonal commonality of the function Π. but more can be said. The total expenditure of the H optimizing consumers is

$$\sum_h E^h(u^h, p) = \Pi(p)\sum_h u^h + \sum_h \Lambda^h(p) \tag{20}$$

If we interpret $u = \sum_h u^h$ as aggregate utility and set $\Lambda(p) = \sum_h \Lambda^h(p)$,

$$E(u, p) := \sum_h E^h(u^h, p) = \Pi(p)u + \Lambda(p) \tag{21}$$

has the structure of an expenditure function in aggregate utility (since Π and Λ have the requisite homogeneity, monotonicity, and curvature properties); indeed, (21) has the same (GPF) structure as the individual expenditure functions in (8). Corresponding to (21) is the aggregate indirect utility function

$$V(p, y) = \frac{y}{\Pi(p)} - \frac{\Lambda(p)}{\Pi(p)} =: \frac{y}{\Pi(p)} - \Gamma(p) \tag{22}$$

which, of course, has the same structure as the individual utility functions (9). Finally, the application of Roy's Identity to (22) yields the aggregate demand system

$$D(p, y) = \Pi(p)\nabla\Gamma(p) + \frac{\nabla\Pi(p)}{\Pi(p)}y \tag{23}$$

---

*The Stone-Geary system has been generalized in other directions. Howe, Pollak, and Wales (1979), with some amendments by van Daal and Merkies (1989), constrained a system that is *quadratic* in income to satisfy integrability conditions; their structure is a generalization of the Gorman polar form (and hence does not satisfy his aggregation condition).

which, with $y = \sum_h y^h$, is identical to the aggregate demand system (11), obtained by direct aggregation of the individual demand systems. Thus, the aggregate demand system is rationalized by a utility function defined on aggregate commodity quantities, $x = \sum_h x^h$. In other words, the aggregate demand system is generated by an optimizing "representative agent," with utility $u = \sum_h u^h$ and aggregate income $y = \sum_h y^h$. Thus, the econometric estimation of any aggregate demand system with the structure (23) is consistent with individual optimization and aggregation of the individual demand systems.

The insight in Gorman's theorem can be further elucidated by a comparison of the representative-agent phenomenon in production theory and in consumer theory.* Producers, labeled $f = 1, \ldots, F$, are assumed to maximize profit, $p \cdot z^f$, on a technology set, $T^f \subseteq \mathbf{R}^n$, defined as the set of technologically feasible net-output vectors. The aggregate technology set is $T := \sum_f T^f$.† The net-supply functions, $\phi^f, f = 1, \ldots, F$, are defined by

$$\phi^f(p) = \text{argmax}_{z^f}\{p \cdot z^f \mid z^f \in T^f\} \quad \forall f \tag{24}$$

These net-supply functions can be trivially aggregated to a macro net-supply function

$$\Phi(p) := \sum_f \phi^f(p) \tag{25}$$

But much more can be said, since

$$\Phi(p) = \text{argmax}_z\{p \cdot z \mid z \in T\} \tag{26}$$

Thus, the aggregate net-supply function can be generated by maximizing aggregate profit on the aggregate technology set; i.e., the aggregate net-supply functions are generated by the optimizing behavior of a "representative producer." Thus, *there is no aggregation problem on the production side of the economy*, so long as all inputs are efficiently allocated.‡

As indicated by Gorman's theorem, the existence of a representative consumer is more problematical. To illustrate the problem, recall that expenditure-constrained utility maximization and utility-constrained expenditure minimization are solved

---

*Although the discussion is couched in the simplest possible terms, the producer case could be elaborated upon to encompass intertemporal production technologies (see, however, Blackorby and Schworm 1982), and the consumer case can be adapted to the study of labor supply, asset demand, and other problems characterized by optimization subject to a preference ordering and a budget constraint (with fixed endowments).

†This construction, of course, assumes that there are no externalities in production.

‡See Section VI.

by the same consumption vector, so long as preferences satisfy local nonsatiation.[*] Thus, the optimal consumption vectors are given by

$$\overset{*}{x}^h = \mathrm{argmin}_{x^h}\{p \cdot x^h \mid x^h \in N^h(\overset{*}{x}^h)\} \ \forall h \tag{27}$$

where

$$N^h(\overset{*}{x}^h) = \{x^h \in \mathbf{R}^n_+ \mid U^h(x^h) \geq U^h(\overset{*}{x}^h)\} \tag{28}$$

is the no-worse-than-$\overset{*}{x}^h$ set of consumer h. The community level set (the Scitovsky 1942 set), the set of aggregate consumption bundles that can be distributed to the H consumers in ways that place each consumer in the individual no-worse-than-$\overset{*}{x}^h$ set, is given by[†]

$$N^S(\overset{*}{x}^1, \ldots, \overset{*}{x}^H) = \sum_h N^h(\overset{*}{x}^h) \tag{29}$$

(The boundary of $N^S(\overset{*}{x}^1, \ldots, \overset{*}{x}^h)$ is a community (Scitovsky) indifference surface.) Note that, by the interchangeability of set summation and optimization,

$$\overset{*}{x} = \mathrm{argmin}_x\{p \cdot x \mid x \in N^S(\overset{*}{x}^1, \ldots, \overset{*}{x}^H)\} \tag{30}$$

giving the appearance of a representative consumer.[‡] Consider, however, a reallocation of income, adding Δ to y^h and subtracting Δ from $y^{\bar{h}}$. The new aggregate demand vector, say $\hat{x}$, is not, in general, equal to $\overset{*}{x}$, even though prices and total income are unchanged. There is, of course, a new Scitovsky set, $N^S(\hat{x}^1, \ldots, \hat{x}^H)$ and

$$\hat{x} = \mathrm{argmin}_x\{p \cdot x \mid x \in N^S(\hat{x}^1, \ldots, \hat{x}^H)\} \tag{31}$$

but, in general, the two Scitovsky indifference surfaces intersect. Thus, there does not exist a preference ordering defined on aggregate consumption vectors that rationalizes aggregate demands. If, on the other hand, the income consumption curves of h and $\bar{h}$ were both linear, with identical slopes, in the neighborhoods of $\overset{*}{x}^h$ and $\overset{*}{x}^{\bar{h}}$, the aggregate demand would be unchanged by this redistribution of income; moreover, if this condition held globally (or at least in a neighborhood) the Scitovsky sets would also be unchanged.

As pointed out in Section I, if incomes are optimally allocated by the maximization of a Bergon-Samuelson (utilitarian) social welfare function, the aggregate data are rationalized by an aggregate utility function. In this case the social planner's optimization problem is

---

[*]See, e.g., Debreu (1959) or Varian (1992).

[†]This construction assumes there are no consumption externalities.

[‡]In fact, this is an example of a "representative agent at a point," a concept commonly employed in finance and macro.

$$\max_{x^1,\ldots,x^H} W(U^1(x^1),\ldots,U^H(x^H)) \quad \text{s.t.} \quad \sum_h p \cdot x^h \leq y \tag{32}$$

or, equivalently,

$$\max_x \hat{W}(x_1,\ldots,x_2) := W(U^1(x^1),\ldots,U^H(x^H)) \quad \text{s.t.} \quad p \cdot x \leq y \tag{33}$$

Thus, $\hat{W}$ induces an ordering on the aggregate consumption space that rationalizes the aggregate choices. This construct, however, is not useful for econometric analysis of demand in a market economy.

To summarize, the problem of aggregation consistency is fundamental to the theory and econometric estimation of aggregate consumer demand functions, even under the assumption of efficient allocation of consumption goods, but the problem does not exist for producers operating in an economy with an efficient allocation of inputs and outputs.

B. Muellbauer Aggregation

The necessary conditions for Gorman aggregation are quite stringent. This fact inspired Muellbauer's (1975, 1976) formulation of a weaker form of aggregation—and, concomitantly, a weaker representative-agent concept. The two fundamental differences between Gorman aggregation and Muellbauer aggregation are that (i) Muellbauer requires consistent aggregation of *expenditure shares* rather than commodity demands and (ii) the aggregate income of Muellbauer's representative agent is a possibly *nonlinear* function of individual incomes, rather than a simple sum, and can depend on prices as well as the distribution of income.

Individual commodity shares are defined by

$$s_i^h(p, y^h) =: \frac{p_i d_i^h(p, y^h)}{y^h} \quad \forall i \, \forall h \tag{34}$$

Aggregate commodity shares are then given by

$$\frac{\sum_h p_i d_i^h(p, y^h)}{\sum_h y^h} = \sum_h \frac{y^h s_i^h(p, y^h)}{\sum_h y^h} =: S_i(y^1,\ldots,y^H, p) \quad \forall i \tag{35}$$

Thus, aggregate shares, like the aggregate demands in (2), depend in general on the distribution of income and prices. Muellbauer aggregation holds if there exists a function Y such that

$$S_i(p, y^1,\ldots,y^H) = \hat{S}_i(p, Y(p, y^1,\ldots,y^H)) \, \forall i \tag{36}$$

This condition is nontrivial because the function Y is identical for all n aggregate demand shares. Thus, if Muellbauer aggregation holds, aggregate commodity shares depend on a common scalar $Y(y^1, \ldots, y^H, p)$, interpreted as "aggregate income," and the commodity prices.

It is easy to see that Muellbauer aggregation subsumes Gorman aggregation as a special case. The (GPF) demand system (10) yields the following system of share equations:

$$s_i^h(p, y^h) = \frac{p_i \Pi_i(p)}{\Pi(p)} + \frac{p_i \Gamma_i^h(p) \Pi(p)}{y^h} \quad \forall i \tag{37}$$

Aggregation over consumers yields

$$S_i(p, y^1, \ldots, y^H) = \frac{p_i \Pi_i(p)}{\Pi(p)} + \frac{p_i \Pi(p) \sum_h \Gamma_i^h(p)}{\sum_h y^h}$$

$$=: \hat{S}_i \left(p, \sum_h y^h \right) \quad \forall i \tag{38}$$

which is the special case of Muellbauer aggregation where

$$Y(p, y^1, \ldots, y^H) = \sum_h y^h \tag{39}$$

Theorem (Muellbauer 1975, 1976).   Assume interior solutions to the individual optimization problems. Muellbauer aggregation holds if the individual expenditure functions have the structure

$$E^h(u^h, p) = \Psi^h(u^h, \Gamma(p)) \Pi(p) \quad \forall h \tag{40}$$

in which case the indirect utility functions are

$$V^h \left(\frac{p}{y^h} \right) = \Theta^h \left(\frac{y^h}{\Pi(p)}, \Gamma(p) \right) \quad \forall h \tag{41}$$

where each Ψ^h and each Θ^h is increasing in its first argument.

---

*This structure is necessary as well as sufficient if we add to (40) a function of prices, $\Lambda^h(p)$, that vanishes when we sum over households: $\sum_h \Lambda^h(p) = 0$. While this possibility is formally required for the structure to be necessary as well as sufficient, it is uninteresting—a nuisance term. It does not make a lot of sense for preferences to satisfy a condition like this. For one thing, it means that individual preferences would have to depend on the number of households H; otherwise, the sum of these nuisance terms would not vanish when we changed H. For these reasons, we carry out the analysis of Muellbauer aggregation ignoring this term (as has been the case in subsequent studies building on Muellbauer's ideas). (See, however, Blackorby, Davidson, and Schworm 1993 for a different approach.) Note also that the functions in these representations must satisfy certain homogeneity conditions, but because of the ways in which they enter the expenditure and indirect utility functions, there are degrees of freedom in choosing the degrees of homogeneity.

To see that this structure is sufficient, apply Roy's identity to obtain

$$S_i^h(p, y^h) = v^h(y^h, p)A^i(p) + B^i(p) \tag{42}$$

where

$$v^h(p, y^h) = \frac{\Theta_2^h(y^h/\Pi(p), \Gamma(p))}{y^h \Theta_1^h(y^h/\Pi(p), \Gamma(p))} \tag{43}$$

$$A^i(p) = p_i \Gamma_i(p) \Pi(p) \tag{44}$$

$$B^i(p) = \frac{p_i \Pi_i(p)}{\Pi(p)} \tag{45}$$

The subscripts on Θ indicate differentiation with respect to the first or second term, and the subscripts on Π and Λ indicate differentiation with respect to the indicated price variable. Summation yields

$$\begin{aligned} S_i(p, y^1, \ldots, y^H) &= A^i(p) \frac{\sum_h y^h v^h(y^h, p)}{\sum_h y^h} + B^i(p) \\ &=: A^i(p)v(Y(y^1, \ldots, y^H, p), p) + B^i(p) \\ &= \hat{S}_i(Y(y^1, \ldots, y^H, p), p) \end{aligned} \tag{46}$$

where $Y(p, y^1, \ldots, y^H)$ is implicitly defined by

$$v(Y(y^1, \ldots, y^H, p), p) = \frac{\sum_h y^h v^h(y^h, p)}{\sum_h y^h} \tag{47}$$

(assuming a solution exists*).

Note that setting

$$\Psi^h(u^h, \Gamma(p)) = \psi^h(u^h) + \Gamma(p) \tag{48}$$

in (40) yields the Gorman polar form. Because Muellbauer's structure is a straightforward generalization of Gorman's "linear" (actually affine) structure, he refers to (40) or (41) as "generalized linearity."

The system of aggregate share equations, along with the implicit definition of aggregate income in (47), is rationalized by an aggregate utility function with the structure

$$V\left(\frac{p}{y}\right) = \Phi\left(\frac{Y(y^1, \ldots, y^H, p)}{\Pi(p)}, \Gamma(p)\right) \tag{49}$$

Application of Roy's identity yields the $\hat{S}_i$ in (46). Inversion in $Y(y_1, \ldots, y^H, p)$ yields

$$E(u, p) = \Psi(u, \Gamma(p)) \Pi(p) \tag{50}$$

---

*Muellbauer shows that it will exist under certain conditions.

where u is the "utility" of the representative consumer. Thus, as in the case of Gorman aggregation, the Muellbauer representative consumer has the same preference structure as that of the individual consumers. Similarly, the preference structure of Gorman's representative agent in (8) is obtained by setting $\Psi(u, \Gamma(p)) = \Psi(u) + \Gamma(p)$. Moreover, just as Gorman's preference structure is equivalent to the differential condition (12), Muellbauer's generalized linearity structure is equivalent to the following differential conditions:*

$$
\begin{aligned}
\frac{\partial s_i(u, p)/\partial y}{\partial s_j(u, p)/\partial y} &= \frac{\partial v(y, p)/\partial y}{\partial v(y, p)/\partial y} \frac{A^i(p)}{A^j(p)} \\
&= \frac{\partial v^h(y^h, p)/\partial y^h}{\partial v^h(y^h, p)/\partial y^h} \frac{A^i(p)}{A^j(p)} = \frac{\partial s_i^h(y^h, p)/\partial y^h}{\partial s_j^h(y^h, p)/\partial y^h} \quad \forall i, j \ \forall h
\end{aligned}
$$

(51)

That is, ratios of marginal effects of income changes on shares are independent of income levels for both individuals and the representative consumer.

Muellbauer's generalized linearity is further explicated by rewriting the share equations in matrix notation:

$$
\begin{bmatrix} s_1^h(y^h, p) \\ \vdots \\ s_n^h(y, p) \end{bmatrix} = \begin{bmatrix} A^1(p) & B^1(p) \\ \vdots & \vdots \\ A^n(p) & B^n(p) \end{bmatrix} \begin{bmatrix} v^h(y^h, p) \\ 1 \end{bmatrix} \quad \forall h
$$

(52)

The $n \times 2$ matrix on the RHS has rank 2 and $v^h(y^h, p)$ is independent of i. (A similar identity holds for the representative consumer.) Hence, any share equation can be obtained as a linear combination of any other, and for given prices, the share equations lie in a two-dimensional linear subspace. This "rank-2 condition," which has been explored in some depth in recent years, is examined further in Sections III and IV.

Another interpretation of Muellbauer's representative consumer in commodity shares is attributable to Gorman (1976). Note that the existence of a Muellbauer representative agent can be characterized, using Shephard's lemma, as the existence of an aggregate expenditure function, E, such that

$$
\frac{\sum_h E_i^h(u^h, p)}{\sum_h E^h(u^h, p)} = \frac{E_i(u, p)}{E(u, p)} \quad \forall i
$$

(53)

and a function U such that

$$
u = U(u^1, \ldots, u^h, p)
$$

(54)

---

*See Muellbauer (1975, 1976) for proof of sufficiency of this condition.

Gorman first noted that Muellbauer aggregation is trivially satisfied if there are only two commodities, since determining one aggregate share determines the other and U therefore is implicitly defined by, e.g.,

$$\frac{\sum_h E_1^h(u^h, p)}{\sum_h E^h(u^h, p)} = \frac{E_1(u, p)}{E(u, p)} \tag{55}$$

This fact is not surprising, since in the two-commodity case there is no integrability problem. More importantly, it suggests that the solution to Muellbauer's aggregation problem might be characterized by the existence of two aggregate pseudocommodities, and in fact this is an evocative interpretation of the solution given by (40), in which $\Gamma(p)$ and $\Pi(p)$ are interpreted as the "prices," or "unit costs," of two "intermediate commodities" in the production of utility. A complete explication of this interpretation requires quite a bit of duality theory, so we refer the reader to Gorman's paper for further study. (Compare this interpretation, however, to the interpretation of the Gorman polar form expenditure function (8).)

Gorman also pointed out a problem with the Muellbauer representative agent: there is no requirement that Y in (36) or (47) be increasing (or even nondecreasing) in individual incomes—equivalently, that U defined by (53) be increasing in individual utility levels. Thus, the income or utility of the representative agent could be increasing when the incomes or utilities of the consumers it represents are declining. This seems to be a consequence of requiring only that the representative consumer replicate the aggregate shares, but not aggregate demands.

Muellbauer—as well as most follow-up studies—focused on the special case where $Y(y^1, \ldots, y^H, p)$ is independent of p. In this case, $v^h(y^h, p)$ in (47) must be independent of p for all h (for arbitrary distributions of incomes, $y^1, \ldots, y^h$):

$$v^h(y^h, p) = \xi^h(y^h) \quad \forall h \tag{56}$$

Muellbauer shows that this, in turn, implies that*

$$\xi^h(y^h) = y_h^\rho \quad \text{or} \quad \rho \ln y^h \quad \forall h \tag{57}$$

Thus,

$$s_i^h(p, y^h) = A^i(p)y_h^\rho + B^i(p) \quad \text{or} \quad A^i(p) \ln y^h + B^i(p) \tag{58}$$

(where ρ has been incorporated into $A^i(p)$) and

$$S_i(p, y^1, \ldots, y^H) = \frac{A^i(p) \sum_h y_h^{\rho+1}}{\sum_h y^h} + B^i(p) \tag{59}$$

---

*To avoid confusion, where the income variable is raised to a power, we move the household indicator index h to the subscript position.

or

$$S_i(p, y^1, \ldots, y^H) = \frac{A^i(p) \sum_h y^h \ln y^h}{\sum_h y^h} + B^i(p) \tag{60}$$

The first of these commodity-share equations is called *price-independent generalized linearity* (PIGL), and the second is *price-independent generalized logarithmic* (PIGLOG).

Empirical application requires that specific forms of the functions Γ and Π be specified. Deaton and Muellbauer (1980) adopted a translog specification for Γ and the sum of a translog and a Cobb-Douglas for Π. They called this specification the *almost ideal demand system*, with the (unfortunate) acronym AIDS. This specification has been widely applied in empirical demand analysis, and Lewbel (1989a) has estimated a wider class of functions that nests both the translog and the AIDS specification. Another generalization of the AIDS specification by Banks, Blundell, and Lewbel (1993), the quadratic AIDS (QUAIDS), has rank-3 Engel curves and hence does not satisfy aggregation conditions.

C.   Testing the Representative-Consumer Hypotheses

Most tests of the representative-consumer hypotheses have not been designated as such. Rather they have purported simply to test consumer "rationality"; but since they have used aggregated data (typically the only data for which consumption-good prices are available), they have in fact been testing for the existence of a representative consumer that rationalizes the aggregate data.* As such, these tests have anticipated the widespread use of representative consumers in recent macroeconomic research.

Two types of tests have been employed, and there is a sharp difference in the results using these two approaches. Most commonly, researchers have specified a functional form representing preferences and then tested for parametric restrictions consistent with the symmetry of the estimated Slutsky matrix of substitution effects of price changes. Other tests have eschewed parametric forms, using revealed preference methods to test for rationalization of the aggregate data by expenditure-constrained utility maximization.

The first approach began with the Christensen-Jorgenson-Lau (1975) tests for symmetry of the Slutsky matrix generated by the demand functions derived from a translog indirect utility function. These tests turn out to be identical to testing for symmetry of the second-order coefficients in the translog specification. That is,

---

*See, however, Pollak and Wales (1979) and Shapiro and Braithwait (1979) for direct tests of the GPF structure, which doesn't fare at all well.

the approach is to estimate the derived demand system with and without symmetry of these coefficients and use a maximum likelihood ratio or Wald test to test for symmetry.* There is by now a large literature on this approach, inspired in part by the competition between proponents of alternative flexible-form specifications, and we make no attempt to survey it. Suffice it to say that the symmetry conditions are commonly rejected.

Of course, the weakness of these parametric tests is that rejection can be attributable not to the failure of consistent aggregation to hold, but rather because the preferences of the representative agent are misspecified. Although, by definition and design, flexible functional forms can provide a second-order approximation to an arbitrary "true" specification at a point, they are in fact employed as global specifications in tests of symmetry (and other properties). The second-order parameters are estimated from data over the entire sample space.[†]

The alternative approach, which eschews specification of functional form, is based on the revealed preference approach to testing the consumer-optimization hypothesis, formulated by Samuelson (1938, 1946–47) and Houthakker (1950) and first implemented empirically by Houthakker (1963) and Koo (1963). But the modern formulation has its roots in a remarkable theorem of Afriat (1967, 1972) (which was nicely elucidated by Diewert 1973). Varian (1982) applied Afriat's method to annual U.S. data on nine consumption categories from 1947 to 1978. Remarkably, he found no violations of the revealed preference axioms. As the data are *aggregate* U.S. consumption quantities and prices, this result apparently provides strong support for the existence of a Gorman representative consumer.

Varian raised the possibility, however, that these tests may be plagued by low power, given the data with which economists have to work. In particular, they will have low power if the variation in total expenditure over time is large relative to the variation in relative prices. (If, for example, the budget hyperplanes do not intersect, the tests have zero power.)

Varian's concerns were apparently allayed by Bronars (1987): in a Monte Carlo assessment of the power of Varian's test against the alternative hypothesis of Becker-type irrational behavior, he found that these tests have a considerable amount of power, especially when per capita data are used. Roughly speaking, Becker's notion of irrational behavior entails a (uniformly) random distribution of consumption quantities across the budget hyperplane. To implement this notion as an alternative hypothesis, Bronars took Varian's price and total expenditure data as given, replaced the consumption quantities with randomly constructed values (using three different

---

*Asymmetry of these second-order coefficients is equivalent to violation of Young's Theorem on the symmetry of second-order cross derivatives.

[†] See White (1980) for a penetrating analysis of the problems with interpreting tests using flexible forms as tests at "the" point of approximation.

algorithms), and applied Varian's GARP test, alternatively using per capita and aggregate data. He repeated this exercise a large number of times to obtain a measure of the power of Varian's test against the Becker alternative hypothesis. The upshot of his results is that Varian's test has a substantial amount of power against the Becker irrationality hypothesis.

Bronars noted that his alternative hypothesis is "rather naive" (page 697). It is not clear what type of individual behavior, if any, together with aggregation across consumers, would yield a (uniformly) random allocation of aggregate consumption quantities across the budget plane. Recently, Russell and Tengesdal (1996) raise the question of whether Varian's tests have substantial power against a less naive alternative hypothesis, predicated on the fact that the null hypothesis of Varian's test, given the aggregate data employed, is a compound one, entailing both individual rationality and aggregation consistency.

Using Monte Carlo methods analogous to Bronars', Russell and Tengesdal use actual price and aggregate total expenditure data, but generate aggregate consumption quantity data by (1) specifying heterogeneous individual utility functions (that do not satisfy Gorman aggregation)* and a distribution of total expenditures among individuals, (2) generating optimal individual demands, given the prices and the preferences and total expenditures of individuals, and (3) aggregating demands across individuals. Using Varian's algorithm to calculate power indices in terms of the percentage of the simulations with GARP violations, they find few violations, despite a large number of sensitivity tests. These results suggest that Varian's tests using aggregate U.S. consumption data lack power and hence provide little support for the existence of a Gorman representative consumer.

A recent paper by Fleissig, Hall, and Seater (1994) reinforces this point. They find that Varian's GARP tests are violated using both monthly and quarterly U.S. consumption data. Another suggestion that these nonparametric (nonstochastic) tests lack power is the surprising result of Manser and McDonald (1988), in which it is shown that the Afriat-Varian tests fail to reject the hypothesis of a *homothetic* utility function rationalizing aggregate U.S. consumption data. But the concomitant implication of constant budget shares—assuming strictly convex indifference surfaces—is easily rejected statistically. As Manser and McDonald point out, these two outcomes are explained by the fact that the rationalizing utility function in Afriat's theorem is piecewise linear, as are the indifference surfaces, so that substantial changes in budget shares can be consistent with homotheticity.

---

*In particular, they employ the Stone-Geary specification described in Section II.A, but with heterogeneous β_i parameters. Gorman aggregation holds for this specification if and only if these parameters are identical for all consumers. This specification allows a simple parameterization of consumer heterogeneity and hence the "degree" of aggregation inconsistency, ranging all the way to "maximal" heterogeneity (a uniform distribution over the [0, 1] interval or, alternatively, preferences in which each consumer has just one positive β_i—equal of course to 1).

The above studies pertain to the existence of a Gorman representative consumer, defined in terms of the simple sums (or means) or expenditures. Of course, confirmation of the existence of a Gorman representative consumer implies confirmation of the existence of the weaker representative-consumer notion of Muellbauer. A (statistical) nonparametric study of aggregate consumption by Lewbel (1991) suggests that data in the middle 90% or so of the income distribution are consistent with Muellbauer's PIGLOG specification and concomitant representative consumer in budget shares. The data at both the low and the high end of the income distribution are inconsistent with Muellbauer's rank-2 condition. As will be shown in the next section, this implies that the Engel curves of optimizing consumers have rank 3 (and hence are incompatible with both forms of representative consumer).

III. "EXACT" AGGREGATION AND HOUSEHOLD ATTRIBUTES

A. The Exact Aggregation Theorem

As noted at the outset of the previous section, aggregate demand generally depends on the entire distribution of income. When the necessary and sufficient conditions for a Gorman representative consumer are satisfied, however, a single statistic—namely the first moment of the income distribution—suffices to determine aggregate demand (see (11)). The existence of a Muellbauer representative agent (in budget shares) is weaker, and the aggregate demand functions accordingly depend on *two* income distribution statistics. To see this, note that the aggregate demand functions implied by (36) are

$$
\begin{aligned}
D_i(p, y^1, \ldots, y^H) &= p_i^{-1} \hat{S}_i(Y(y^1, \ldots, y^H), p) \sum_h y^h \\
&=: \tilde{D}_i(p, \theta_1(y^1, \ldots, y^H), \theta_2(y^1, \ldots, y^H)) \quad \forall i
\end{aligned}
\tag{61}
$$

where $\theta_1(y^1, \ldots, y^H) = \sum_h y^h$ and $\theta_2(y^1, \ldots, y^H) = Y(y^1, \ldots, y^H)$. (Note that the second of the identities in (61), being more general than the first, imposes weaker restrictions on preferences than those needed for a Muellbauer representative agent.) More specifically, with price-independent aggregate income (58), Muellbauer aggregation requires that the individual demand functions take the following form:

$$
d_i^h(p, y^h) = \beta_{i1}(p)\xi_1(y^h) + \beta_{i2}(p)\xi_2(y^h)
\tag{62}
$$

where $\beta_{i1}(p) = A_i(p)/p_i$ and $\beta_{i2}(p) = B_i(p)/p_i$, for all i,

$$
\xi_2(y^h) = y^h
\tag{63}
$$

and

$$
\xi_1(y^h) = y_h^{\rho+1} \quad \text{or} \quad y^h \ln y^h
\tag{64}
$$

for all h. The two cases in (64) generate the PIGL and PIGLOG systems, in which case individual demands as well as aggregate demands have rank 2 and aggregate demands are given by

$$
\begin{aligned}
\hat{D}_i(p, y^1, \ldots, y^H) &= \beta_{i2}(p) \sum_h y^h + \beta_{i1}(p) \sum_h y_h^{\rho+1} \\
&=: \beta_{i1}(p)\theta_1(y^1, \ldots, y^H) + \beta_{i2}(p)\theta_2(y^1, \ldots, y^H) \\
&= D_i(p, \theta_1(y^1, \ldots, y^H), \theta_2(y^1, \ldots, y^H)) \quad \forall i
\end{aligned}
\tag{65}
$$

or

$$
\begin{aligned}
\hat{D}_i(p, y^1, \ldots, y^H) &= \beta_{i2}(p) \sum_h y^h + \beta_{i1}(p) \sum_h y^h \ln y^h \\
&=: D_i(p, \theta_1(y^1, \ldots, y^H), \theta_2(y^1, \ldots, y^H)) \quad \forall i
\end{aligned}
\tag{66}
$$

A specific case of (65), with $\rho = 1$ is that where individual demand functions are quadratic in income:

$$
d_i^h(p, y^h) = \beta_{i1}(p) \cdot y^h + \beta_{i2}(p) \cdot y_h^2, \quad \forall i, \forall i \, \forall h
\tag{67}
$$

in which case aggregate demands are

$$
\begin{aligned}
\hat{D}_i(p, y^1, \ldots, y^H) &= \beta_{i1}(p) \cdot \sum_h y^h + \beta_{i2}(p) \cdot \sum_h y_h^2 \\
&=: D_i(p, \theta_1(y^1, \ldots, y^H), \theta_2(y^1, \ldots, y^H)) \quad \forall i
\end{aligned}
\tag{68}
$$

Thus, in this special case of Muellbauer aggregation, we need information about the second as well as the first moment of the distribution to determine aggregate demand. But knowledge of these two statistics allows us to determine aggregate demand *exactly*, no matter how many individuals are in the economy.

This characterization of Muellbauer's results raises the question of whether it generalizes to cases where there are more than two aggregator (θ_r) functions and where the forms of these functions are arbitrary. This is the motivation for the notion of *exact aggregation*, defined by the existence of T symmetric* functions of individual incomes, say $\theta_t(y_1, \ldots, y^H)$, $t = 1, \ldots, T$, where $T < H$, such that[†]

$$
\hat{D}_i(p, y^1, \ldots, y^h) = \tilde{D}_i(p, \theta_1(y^1, \ldots, y^h), \ldots, \theta_T(y^1, \ldots, y^h)) \quad \forall i
\tag{69}
$$

---

*By symmetry, we mean anonymity—that is, that the values of the aggregator functions are unaffected by an arbitrary permutation of the y^h variables. Without this condition, calculations would be intractible for large H.

[†]Obviously, if $T \geq H$, there is no economy of information and this structure imposes no restrictions on the individual demand functions.

This aggregation notion appears to have been formulated independently by Gorman (1981) and Lau (1977a, 1977b, 1982);* the necessary and sufficient conditions for exact aggregation are as follows.

Theorem (Lau 1982, Gorman 1981).   Given certain regularity conditions,[†] the aggregate demand functions satisfy (69) if and only if the individual demand functions have the following form:

$$d_i^h(p, y^h) = \alpha_i^h(p) + \sum_{t=1}^{T} \beta_{it}(p) \cdot \xi_t(y^h) \quad \forall h \ \forall i \tag{70}$$

Accordingly, aggregate demands are given by

$$\sum_h d_i^h(p, y^h) = \sum_h \alpha_i^h(p) + \sum_{t=1}^{R} \beta_{it}(p) \cdot \left(\sum_h \xi_t(y^h) \right)$$

$$=: \alpha_i^h(p) + \sum_{t=1}^{T} \beta_{it}(p) \cdot \theta_t(y^1, \ldots, y^H) \tag{71}$$

$$= \bar{D}_i(p, \theta_1(y^1, \ldots, y^H), \ldots, \theta_T(y^1, \ldots, y^H)) \quad \forall i$$

Thus, for exact aggregation to hold, individual demand functions must be affine in functions of income and identical across consumers up to the addition of a function of prices that is independent of income. Moreover, the identical part of the function must be multiplicatively separable into a function of prices and a function of income. Finally, the symmetric functions, θ_t, $t = 1, \ldots, T$, in the aggregate demand system must be additive functions of (identical) transformations, ξ_t, of individual incomes.

Note t at exact aggregation, unlike Muellbauer (hence Gorman) aggregation, does not im ıly the existence of a representative consumer; that is, without additional restriction.. the aggregate demand functions (71) cannot be generated by the max-imizatior. ɔf an aggregate utility function subject to an aggregate budget constraint. Thus, exact aggregation is weaker than the existence of a representative consumer.

B.   Gorman's Theorem on the Maximal Rank of Demand Systems

In matrix notation, the necessary and sufficient condition for exact aggregation is

$$d^h(p, y^h) = \alpha^h(p) + \hat{B}(p)\xi(y^h) =: B^h(p)\xi(y^h) \quad \forall h \tag{72}$$

---

*Gorman uses this aggregation condition as motivation for his theorem on the maximal rank of demand systems, discussed below. He first presented the paper at a London School of Economics workshop in January 1977.

[†]See Lau (1982) and Heineke and Shefrin (1988) for the particulars.

where $B^h(p) = [\alpha^h(p) \; \hat{B}(p)]$ is an $n \times R$ matrix, with $R = T + 1$, and $\xi(y^h)$ is an R-dimensional vector. This structure, however, is not necessarily consistent with individual optimization without further restrictions. In a remarkable paper, Gorman (1981) examined the restrictions on demand functions of type (72) implied by individual optimization. Gorman's theorem is as follows.

Theorem (Gorman 1981). *If the complete system of demand equations (72) reflects well-behaved preferences, then the rank of its coefficient matrix $B(p)$ is at most 3. Moreover, one of the following must hold (where $\mathbf{I}^+$ is the set of nonnegative integers):*[*]

$$
\begin{aligned}
&\text{(i)} \quad \xi_r(y^h) = y^h (\ln y^h)^{\kappa_r}, \qquad \kappa_r \in \mathbf{I}^+ \; \forall r, \quad \kappa_r = 0 \quad \text{for some } r \\
&\text{(ii)} \quad \xi_r(y_h) = y_h^{\rho_r + 1} \qquad \forall r, \quad \rho_r = 0 \quad \text{for some } r \\
&\text{(iii)} \quad \xi_r(y^h) = y^h \sin(\rho_r \ln y^h) \quad \text{or} \quad y^h \cos(\rho_r \ln y^h) \\
&\qquad\qquad \rho_r \geq 0 \; \forall r, \quad \rho_r = 0 \quad \text{for some } r
\end{aligned}
\tag{73}
$$

Note that the Muellbauer's PIGLOG and PIGL systems are generated as special cases of (i) or (ii), with $R = 2$ (hence rank 2). PIGLOG is a special case of (i) with $\kappa_1 = 1$ and $\kappa_2 = 0$. PIGL is a special case of (ii) with $\rho_2 = 0$. Similarly, the Gorman polar form (quasi-homotheticity) is a special case of (ii) with $R = 2$ (rank 2), $\rho_1 = -1$, and $\rho_2 = 0$. The rank-3 quadratic expenditure system is obtained by setting $\rho_1 = -1, \rho_2 = 0, \rho_3 = 1$ in (ii). Homotheticity is rank 1 with $\rho_1 = -1$ in (ii). In fact, virtually all consumer demand systems that have been estimated belong to the class of rank-2 demand systems satisfying the conditions of Gorman's theorem. Section IV contains more on rank-2 (and rank-3) demand systems. The key point here, however, is that exact aggregation is considerably more general than the existence of a representative agent; the latter is but one way of specifying demand systems that can be consistently aggregated.

C. Household Attributes

In the Lau-Gorman exact aggregation theorem, heterogeneity of preferences enters only through the term $\alpha^h(p)$ in (70), which is independent of income. Additional heterogeneity can be introduced into the exact aggregation framework by incorporating an attribute vector a^h into $\beta_{ir}(p)$ and $\xi_r(y^h)$, for each r. (Commonly used attributes include household size, age of household head, race, and geographical

---

[*]As noted by Gorman, the consumption space can divide into subsets with different forms among (i)–(iii) holding over different regions. Heineke and Shefrin (1987) have shown that, if we require the following specifications to hold globally, boundedness of budget shares implies that the only integrable system is the homothetic (rank 1) specification. See Heineke (1979, 1993) and Heineke and Shefrin (1982, 1986, 1987, 1988, 1990) for extensions and clarifications of Gorman's theorem.

region.) For the sake of symmetry, assume that the heterogeneity of α^h is also captured by these attributes, so that*

$$d_i(p, y^h, a^h) = \sum_{r=1}^{R} \beta_{ir}(p, a^h)\xi_r(y^h, a^h) \quad \forall i \ \forall h \tag{74}$$

where $\beta_{i1}(p, a^h) = \alpha_i^h(p)$.

Integrability implies that the ρ_r parameters in (73) depend on the attribute vector a^h and exact aggregation implies that the β_{ir} functions separate multiplicatively into a function of p and a function of a^h, so that†

$$
\begin{aligned}
d_i(p, y^h, a^h) &= \sum_{r=1}^{R} \gamma_{ir}(p)\zeta_{ir}(a^h)\xi_r(y^h, a^h) \\
&=: \sum_{r=1}^{R} \gamma_{ir}(p)\tilde{\xi}_{ir}(y^h, a^h) \quad \forall i \ \forall h
\end{aligned}
\tag{75}
$$

where one of the following holds:

$$
\begin{aligned}
&\text{(i)} \quad \tilde{\xi}_{ir}(y^h, a^h) = \zeta_{ir}(a^h)y^h(\ln y^h)^{k_r}, \qquad k_r \in I^+ \forall r, \\
&\qquad\qquad \kappa_r = 0 \text{ for some } r \\
&\text{(ii)} \quad \tilde{\xi}_{ir}(y^h, a^h) = \zeta_{ir}(a^h)y_h^{\rho_r(a^h)+1}, \qquad \rho_r(a^h) = 0 \text{ for some } r \\
&\text{(iii)} \quad \tilde{\xi}_{ir}(y^h, a^h) = \zeta_{ir}(a^h)y^h \sin(\rho_r(a^h)\ln y^h) \quad \text{or} \\
&\qquad\qquad\qquad \zeta_{ir}(a^h)y^h \cos(\rho_r(a^h)\ln y^h) \\
&\qquad\qquad \rho_r(a^h) \geq 0 \ \forall r, \quad \rho_r(a^h) = 0 \text{ for some } r
\end{aligned}
\tag{76}
$$

The aggregate demand system then becomes

$$
\begin{aligned}
\sum_h d_i(p, y^h, a^h) &= \sum_{r=1} \gamma_{ir}(p) \sum_h \tilde{\xi}_{ir}(y^h, a^h) \\
&=: \sum_r \gamma_{ir}(p)\theta_{ir}(y^1, y^2, \ldots, y^H, a^1, a^2, \ldots, a^H) \\
&= D_i(p, \theta_{i1}(y^1, y^2, \ldots, y^H, a^1, a^2, \ldots, a^H), \\
&\qquad \ldots, \theta_{iR}(y^1, y^2, \ldots, y^H, a^1, a^2, \ldots, a^H)) \quad \forall i
\end{aligned}
\tag{77}
$$

---

*As heterogeneity of preferences is incorporated entirely through the heterogenous attribute vectors, the demand functions do not require an h superscript.

† More precisely, these functions separate into a finite sum of multiplicatively separable terms, so that the number of terms in the following structure may be larger than R; see Heineke and Shefrin (1988), who refer to this structure as the "finite basis property," for details.

Thus, specifying functional forms for $\gamma_{ir}(p)$ and $\tilde{\xi}_{ir}(y^h, a^h)$ in (77) and then summing across consumers of different types, defined by a^h, results in an aggregate demand system that is consistent with individual rationality and exact aggregation.

The question that immediately follows is: under what conditions do the parameters estimated from (77) allow us to recover (uniquely) the parameters of the microfunctions in (75)? Given a joint distribution of income and attributes, expected per capita demand is given by

$$E(x_i) = \sum_{r=1}^{R} \gamma_{ir}(p)E(\zeta_{ir}(a^h)\tilde{\xi}_{ir}(y^h, a^h)) \tag{78}$$

Since the $\gamma_{ir}(p)$ terms enter individual and aggregate demand functions in the same way, the parameters in $\gamma_{ir}(p)$ estimated from (77) will be the parameters in $\gamma_{ir}(p)$ of (75). On the other hand, the term $E(\zeta_{ir}(a^h)\tilde{\xi}_{ir}(y^h, a^h))$ in the aggregate demand function may not include the parameters in the microeconomic demand function. Heineke and Shefrin (1990) show that (75) is recoverable from the estimation of (77) if

(i)   The $\rho_r(a^h)$ functions in (76) do not depend on any unknown parameters.
(ii) The functions $\zeta_{ir}(a^h)$ in (76) are linear in the parameters.

Stoker (1984) has shown, for the general problem of microequation recovery from macroequation estimation, that linearity of the microfunctions in parameters is a sufficient condition for identification when there are no distributional restrictions (in this case, on the distribution of incomes and attributes).

As an example, consider Jorgenson, Lau, and Stoker's (1982) paper on estimating demand systems derived from a translog indirect utility function. The vector of expenditure shares for household h take the form*

$$s\left(\frac{p}{y^h}, a^h\right) = \frac{1}{Q(p/y^h, a^h)}\left(\alpha + B_{pp}\ln\frac{p}{y^h} + B_{pa}a^h\right) \quad \forall h \tag{79}$$

where α is an n-vector of parameters, B_{pp} and B_{ba} are (appropriately dimensioned) matrices of parameters, and

$$Q\left(\frac{p}{y^h}, a^h\right) = \iota'\alpha + \iota'B_{pp}\ln\frac{p}{y^h} + \iota'B_{pa}a^h \tag{80}$$

Note that the parameters are the same for all households, so that preference heterogeneity is entirely captured by the different attributes, $a^h, h = 1, \ldots, H$. It follows from the conditions for exact aggregation that the individual demand functions are linear in functions of y^h and a^h. These conditions will be satisfied if $Q(p/y^h, a^h)$, the denominator of (79), does not contain y^h and a^h. Therefore, the following two restrictions must be imposed:

---

*$\ln(p/y^h)$ is the n-vector of logs of normalized prices and ι is an n-vector of ones.

$$\iota' B_{pp} \iota = 0 \quad \text{and} \quad \iota' B_{pa} = 0 \tag{81}$$

Thus, $Q(p/y^h, a^h)$ is reduced to $Q(p) = \iota'\alpha + \iota' B_{pp} \ln p$, which is independent of h, and the household shares can be rewritten as

$$s\left(\frac{p}{y^h}, a^h\right) = \frac{1}{Q(p)}\left(\alpha + B_{pp} \ln \frac{p}{y^h} + B_{pa} a^h\right) \quad \forall h \tag{82}$$

Accordingly, the aggregate expenditure shares are

$$S(p, y^1, \ldots, y^h, a^1, \ldots, a^H)$$

$$= \frac{\sum_h y^h s(p/y^h, a^h)}{\sum y^h} \tag{83}$$

$$= \frac{1}{Q(p)}\left(\alpha + B_{pp} \ln p - B_{pp} e \frac{\sum_h y^h \ln y_h}{\sum_h y^h} + B_{pa} \frac{\sum_h y^h a^h}{\sum_h y^h}\right)$$

Note that (82) satisfies the conditions of exact aggregation; specifically, the demand equations (given here in expenditure share form) are linear in the parameters and the microequations (82) contain only parameters that appear in the aggregate equation (83). The parameters in (83) can be estimated by using both cross-sectional and time-series data. Thus, the individual expenditure shares can also be obtained by replacing the unknown parameters in (82) with those estimated from (83).

As shown by Heineke and Shefrin (1990), imposing (81) is equivalent to imposing identification conditions; that is, the translog expenditure share system automatically possesses their restricted finite basis property. The JLS translog system is in fact a restricted version of (76)(i). As Heineke and Shefrin (1990) have pointed out, it restricts (76)(i) to being rank 2, when in fact the Gorman theorem allows the system to be up to rank 3. Also, JLS conflate the rank-2 specification and the requirement that there be only two functions of income and attributes. The restrictions on rank do not imply that the number of income/attribute functions must be only as great as the rank of the coefficient matrix $B^h(p)$ in (72).

The JLS translog specification can also be shown to incorporate the following restrictions on (76)(i):

(i) $\zeta_{i1}(a^h)$ must be linear in the attributes and must have an intercept term that is constant for all h.*
(ii) $\zeta_{i2}(a^h)$ must be constant for all h. There are only two ζ_{ir} functions, since the rank of the system is 2 and only two income/attribute functions are posited.

---

*In estimation, the attributes considered by JLS can be divided into discrete categories, allowing for the use of dummy variables. This implies further restriction on the ζ_{i1} functions.

The translog system is thus a restricted version of (76)(i) and a much wider class of similar demand functions is in fact compatible with exact aggregation.

D. Empirical Applications

Empirical applications of the theory of exact aggregation are based primarily on the work of Jorgenson, Lau, and Stoker (1980, 1982).* Their econometric model incorporates household attributes into the translog indirect utility function with a set of 18 dummy variables for attributes like family size, age of head, region of residence, race, and type of residence. They impose restrictions on the parameters of the share equations such that conditions for exact aggregation hold and aggregate across individuals to obtain the aggregate share demand functions. They estimate this aggregate demand system by pooling cross-sectional data on expenditures by individual households and aggregate annual time-series data on expenditure, price levels, and statistics of the distribution of family income and demographic variables. In addition to estimating the model, they analyze the welfare changes from oil price shocks.

Jorgenson, Lau, and Stoker use dummy variables and simple linear equations for the incorporation of demographic attributes. This method remains the most commonly used one in applications. On the other hand, Pollak and Wales (1978, 1980) propose two alternative methods for inclusion of demographic variables: demographic translating and demographic scaling. In both demographic translating and demographic scaling, new parameters are introduced to capture the effects of demographic heterogeneity. For the demographic translating procedure, n "translating parameters," $\tau_1, \tau_2, \ldots, \tau_n$, are added to the model and the original demand function d_i^h is modified as follows:[†]

$$\tilde{d}_i(p, y^h, a^h) = \tau_i + d_i^h \left(p, y^h - \sum_{j=1}^{n} p_j \tau_j \right) \quad \forall i \; \forall h \tag{84}$$

where

$$\tau_i = T_i(a^h) = \alpha_i + \sum_j \beta_{ij} a_j^h \quad \forall i \; \forall h \tag{85}$$

---

*See also Jorgenson, Lau, and Stoker (1981), Stoker (1986a), Jorgenson and Slesnick (1984), and Nicol (1994).

[†] This demand is assumed to be derived by utility maximization. The authors investigated the LES, QES, basic translog (i.e., the translog of Christensen, Jorgenson, and Lau 1975), and generalized translog demand systems. The primary empirical result of their work is that both the number and age of children in a household have significant effects on consumption patterns.

Thus, we can write the demand as

$$\bar{d}(p, y^h, a^h) = \alpha_i + \sum_j \beta_{ij} a_j^h + d_i^h \left(p, y^h - \sum_{j=1}^n p_j \tau_j \right) \quad \forall i \; \forall h \qquad (86)$$

Similarly, demographic scaling introduces n "scaling parameters," $\sigma_1, \sigma_2, \ldots,$ σ_n, into the original demand system and the new demand system is

$$d_i^h(p, y^h, a^h) = \sigma_i \cdot d_i^h(p_1 \sigma_1, p_2 \sigma_2, \ldots, p_n \sigma_n) \quad \forall i \; \forall h \qquad (87)$$

where

$$\sigma_i = \sum_i (a^h) = 1 + \sum_j c_{ij} a_j^h \qquad (88)$$

Note that, in (84) and (87), τ_i and σ_i are the only parameters in the demand system that depend on demographic variables. Unfortunately, as Pollak and Wales point out, their aggregation procedure, unlike that of Jorgenson, Lau, and Stoker, is inappropriate. Moreover, there do not appear to be any applications of demographic translating and demographic scaling to appropriately aggregable demand systems.

In a related development that accounts for serial correlation, Blundell, Pashardes, and Weber (1993) incorporate household demographic attributes and time-dependent variables (time-trend and seasonal dummy components) into demand systems. The aggregate models they derive from this specification seem to perform quite well compared to models using microlevel data in both forecasting efficacy and evaluation of aggregate consequences of public policy changes.

In demand system estimation, most authors include some type of dynamic specification (commonly an AR(1) process for the error terms) to account for habit formation. Few, however, include forms that adjust for the distribution of incomes. In an interesting paper, Stoker (1986b) compares an LES model with the habit formation structure with one that does not have any dynamics, but instead allows for distributional effects on demand. He shows that the distributional effect is statistically significant and that it can displace AR(1) dynamics in the widely used models. His results were confirmed and extended by Buse (1992), who used Canadian rather than U.S. data and estimated a quadratic expenditure systems (QES) model as well as an LES model. He concluded that Stoker's results are widely applicable. Stoker (1993) provides further extension of this work.

To summarize, "exact" aggregation allows summary statistics of income and attributes to be used to create aggregate functions that are consistent with aggregation over individual demand functions generated by income-constrained utility maximization. The conditions for existence of such functions and for their identifiability have been briefly discussed here. Even if the conditions for the existence of a Gorman representative agent do not hold, "exact" aggregation demonstrates the conditions under which aggregate data may still be used for econometric estimation when appropriate distributional information is available.

IV. ADDITIONAL RESULTS ON AGGREGATION AND THE RANK OF DEMAND SYSTEMS

A. Tests and Extensions of Gorman Rank Conditions

Several papers have extended Gorman's (1981) results on the rank of Engel curves and some testing of the rank conditions has begun. Lewbel (1987) characterized demand systems that are affine in income and an arbitrary function of income:

$$d_i(p, y) = a_i(p) + b_i(p)y + c_i(p)\psi(y) \quad \forall i \tag{89}$$

(For notational convenience, we have temporarily suspended use of the household index.) This subclass of demand systems encompasses most of those that have been used to estimate demand taking aggregation into explicit account. Here, demands are linear in functions of income, but unlike Gorman's formulation, only one of those functions is of unspecified form. Lewbel shows that eight different possibilities fully characterize the demand systems of the form (89): homothetic, quasi-homothetic (GPF), quadratic (QES), PIGL, PIGLOG, extended PIGL, extended PIGLOG, and LINLOG.*

Only the QES among all these systems has rank 3. Indeed, it can be seen that the QES is case (ii) in (73) characterized by Gorman (1981). The rest have rank 2, except for homothetic demands, which have rank 1. All of the rank-2 systems have indirect utility functions (or cost functions) that can be written as the generalized Gorman polar form necessary for two-stage budgeting; see Gorman (1959), Blackorby, Primont and Russell (1978), and Blackorby and Russell (1997). Lewbel (1990) characterized full-rank demand systems—those in which the $n \times R$ matrix, $B^h(p)$, of (72) has rank R. It turns out that PIGL and PIGLOG are the only rank-2 demand systems with full rank. The rank-3 systems had previously been characterized by Gorman.

Lewbel (1989b) extends the rank result to functions of real income instead of nominal income. As Gorman himself pointed out, the result on the rank of Engel curves is rendered less interesting by the fact that the Engel curves are functions of nominal (money) income and not real income. If we write the income shares for each consumer as

$$s_i(q, z) = \sum_{r=1}^{R} b_{ri}(q)\psi_r(z) \quad \forall i \tag{90}$$

---

*The GPF is described in (10), the PIGL and PIGLOG were characterized by Muellbauer (58), and the QES was characterized by Pollak and Wales (64) with the intercept term $\alpha_i(p)$ appended). Lewbel introduces three new demand systems: extended PIGL, extended PIGLOG, and LINLOG. The extensions to PIGL and PIGLOG demands involve the addition of an additive constant term and, in the extended PIGL case, the $A_i(p)$ term is multiplied by a differentiable function of prices. None of these has been used in empirical application.

where q is the vector of log prices and z is log income, we know that the rank of the matrix of coefficients $[b_{ri}]$ is at most 3. Lewbel shows that if we replace $\psi_r(z)$ with $\tilde{\psi}_r(\tilde{z})$, where $\tilde{z}$ is the log of *deflated* income, then the system

$$s_i(q, z) = \sum_{r=1}^{R} c_{ri}(q)\tilde{\psi}_r(\tilde{z}) \quad \forall i \tag{91}$$

has at most rank 4. In general, all the other properties that hold for Gorman demand systems of the form (72) hold for the deflated demand systems of (91). If $V(q, z)$ is the indirect utility function corresponding to a Gorman system, then $V(q, \phi(\tilde{z}))$, where $\phi(\tilde{z})$ is a nonzero, bounded, continuous differentiable function, is the indirect utility function for the deflated demand system.

Lewbel (1991) constructed nonparametric tests for the rank of a demand system and applied these tests to family expenditure survey data from the U.K. and consumer expenditure surveys from the United States. To minimize income-correlated demographic variation, Lewbel selected only a subset of households that were fairly homogeneous in terms of household attributes. (Specifically, only married couples with two children, where the head of household was employed full-time, were chosen. Note that another approach would be to define rank as the space spanned by function of "attributes" as well as income. This is the approach explored in Section III.C. Here, Lewbel only considered households in one cross section of attributes.) Using this reduced data set, he found that, in the middle part of the distribution the data are consistent with rank 2 and that the PIGLOG system gives a fairly good fit. In the tails of the distribution (the lower and upper 5%), however, he finds that the data appear to be of higher rank and that rank-3 systems would be a better fit.

Hausman, Newey, and Powell (1995) (HNP) also find that rank-3 demand systems fit the data better than rank-2 systems. They are primarily concerned with the problem of estimating systems of demand when there are errors in variables. This is a common problem with survey data dealing with income and expenditure. Using the specification for shares,

$$s_i(y) = \beta_{i0} + \beta_{i1} \ln y + \beta_{i2}(\ln y)^2 \quad \forall i \tag{92}$$

where the β_i's are constants since the estimation is done at only one price situation, HNP estimate demands for five categories of goods. They compare the results from OLS and instrumental variable techniques using a repeated observation as an instrument. They find that the inclusion of the quadratic term gives better estimates of the demand system than a rank-2 specification. To test whether or not a rank-4 specification gives any additional information, HNP add another term and reestimate the system. Thus, the specification becomes

$$s_i(y) = \beta_{0i} + \beta_{1i} \ln y + \beta_{i2}(\ln y)^2 + \beta_{i3}(\ln y)^3 \quad \forall i \tag{93}$$

If the demand system is rank 3, the ratio β_3/β_2 should be constant. (In other words, the addition of the last term provides no new information.) HNP find that, in fact, this ratio is almost perfectly constant. Thus, Gorman's rank condition seems to be confirmed in the data for this specification.

B.   Other Rank-Two Results

In Sections II and III, the question of aggregation was considered without placing any restrictions on the economy except the usual regularity conditions on individual consumer demand functions. Several authors have considered weaker forms of aggregation. In particular, they have asked whether the class of individual demand functions consistent with representative-agent-type aggregation might be broadened by placing other restrictions on the economy.

One approach is to restrict the distribution of income. Jerison (1984a) showed that, when the income distribution is restricted to a set of allowable distributions (distributions that are relatively open on the unit simplex), a wider class of consumer demand functions aggregates. Jerison was concerned with a representative consumer of the Gorman type—where the demand of a group of consumers can be derived from utility maximization based upon the mean income of the group. Jerison introduced the concept of pairwise aggregation—a representative consumer existing for each pair of demand functions. In other words, pairwise symmetry is simply the condition that mean demand functions for each pair have a symmetric Slutsky matrix. Define the vector of share differences as

$$G^{hh'}(p, y^h, y^{h'}) = s^h(p, y^h) - s^{h'}(p, y^{h'}) \tag{94}$$

Jerison defined the demand system as exhibiting "nonconvergence" if $y|G^{hh'}(p, y^h, y^{h'})|$ is nondecreasing as y changes. If $G^{hh'}(p, y^h, y^{h'})$ does not change direction (i.e., does not change from positive to negative, or vice versa) for small changes in mean income, the demand pair is considered to exhibit "local invariance." If $G^{hh'}(p, y^h, y^{h'})$ does not change direction for any changes in mean income, the demand pair exhibits "invariance."

Jerison derived the relationship between aggregation, pairwise aggregation, local invariance, nonconvergence, and pairwise symmetry. When the set of allowable fixed income distributions is limited to a singleton (where all consumers have a strictly positive share of total income), pairwise symmetry and nonconvergence imply consistent aggregation for all pairs of consumers and the existence of a Gorman-type representative agent for the whole economy. Jerison provided a characterization of the broadest class of pairwise aggregable demands for the case of a single, fixed income distribution.

Perhaps of wider interest, however, is the characterization of demand systems when the income distribution is less restricted:

Theorem (Jerison 1984a). If the number of consumers H is greater than 2, the income distribution is fixed, and certain regularity conditions hold,* then the demand system is pairwise symmetric iff the consumer budget shares have one of the following forms:

(i) $w^h(p, y^h) = (\ln y^h)B(p) + C^h(p),$

(ii) $w^h(p, y^h) = (y^h)^\rho B^h(p) + C(p),$ (95)

(iii) $w^h(p, y^h) = v^h(p, y^h)B(p) + C(p),$

where B, B^h, C, and C^h satisfy certain regularity conditions.[†] Furthermore, if the shares of all pairs of consumers exhibit nonconvergence and if all consumers have budget shares of the form (i), (ii), or (iii), the demand system exhibits pairwise aggregation.

All of these are rank-2 demand systems. Thus, even when the weaker condition of pairwise aggregation is considered, the rank-2 condition continues to hold. Case (iii) is Muellbauer's generalized linear demand structure.[‡] Jerison's aggregation question, however, differs from Muellbauer's in that (i) and (ii) are neither stronger nor weaker than Muellbauer's GL demands restricted to satisfy his aggregation condition.

Freixas and Mas-Colell (1987) likewise find that the generalized linear category of demands provides an important condition for the weak axiom of revealed preference (WARP) to hold in the aggregate. Rather than requiring aggregate demand to be rationalizable by a utility maximizing representative consumer, they ask what conditions must be placed on Engel curves such that aggregate data satisfy WARP. They assume that preferences are identical across consumers and they put no restrictions on the income distribution.

Freixas and Mas-Colell introduce two conditions on Engel curves. The first they call uniform curvature (UC). Uniform curvature essentially requires that goods are either luxuries or necessities in all ranges of allowable income. No torsion (NT), the other condition, requires that Engel curves lie in a plane through the origin; in other words, Engel curves must have rank 2, or the generalized linear structure of (42). They then prove the following theorem.

---

[*]Specifically, demands are analytic (i.e., demands can be represented by a power series arbitrarily closely at any point) and the income distribution belongs to a convex, open set of income distributions confined to the $H - 1$ unit simplex.

[†]Specifically, they are analytic with $\iota'B = 0 = \iota'B^h$ and $\iota'C = 0 = \iota'C^h$ (where ι is the appropriately dimensioned unit vector) for all consumers $h = 1, \ldots, H$.

[‡]Note, however, that Jerison's conditions rule out the additive terms that sum to zero in the Muellbauer necessity conditions (see the footnote to Muellbauer's theorem on p. 192).

Theorem (Freixas and Mas-Colell 1987). *If agents have identical preferences, the distribution of income has support (a, b), and the Engel functions satisfy NT and UC on the interval (a, b), then the weak axiom of revealed preference holds in the aggregate.*

Again, rank-2 demands play an important role in this type of restriction of preferences.

Clearly, the results of Freixas and Mas-Colell are important. It is quite desirable that the weak axiom of revealed preference be satisfied in the aggregate. In a sense, this is demanding some minimal rationality from aggregate demand. It also provides testable implications—though since aggregate income tends to grow over time while relative prices change only very slowly one would not expect to uncover violations of WARP in the aggregate very frequently.

Jerison's results are perhaps less empirically useful. Conceivably, one could create a test for the presence of pairwise aggregation. However, it is unclear why that would be a desirable attribute for an economy to have. The restrictions necessary for pairwise aggregation to imply aggregation are likely to be violated in any actual economy over time. Thus, pairwise aggregation lends no practical help to solving the aggregation question, though the results are fascinating from a technical point of view.

The rank-2 restriction also arises in the context of proportional budgeting. Blackorby and Russell (1993) characterize the optimality of the two-stage budgeting procedure, where a society allocates shares of total income to individual members independent of the level of aggregate income. (This could also be the two-stage budgeting process of a firm.) They use a linear, additive social welfare function, where individual shares are allowed to depend on social welfare weights and prices. Shares that do not depend on prices give rise to price-independent proportional budgeting (PIPB).

For both proportional budgeting and price-independent proportional budgeting, it turns out that the individual indirect utility functions must have the generalized linear structure. Here we give only the result for the PIPB case.

Theorem (Blackorby and Russell 1993). *Price-independent proportional budgeting is optimal iff individual indirect utility functions have the form in (i) or (ii):*

$$
\text{(i)} \quad V^h(p, y^h) = \left[\frac{y^h}{\alpha_h \Pi(p)} \right]^{\rho} + \Theta^h(p), \qquad \rho > 0
$$

$$
\text{(ii)} \quad V^h(p, y^h) = \alpha_h \Gamma(p) \ln \left[\frac{y^h}{\Pi^h(p)} \right] + \Theta^h(p)
$$

(96)

It is straightforward to show that case (i) corresponds to Jerison's (ii) in (95). Case (ii) here corresponds to Jerison's case (i). Jerison's third possibility for pairwise aggregation is not sufficient for PIPB. Blackorby and Russell also discuss the relationship of their results to those of Muellbauer for the more general proportional budgeting (PB). They show that aggregate share functions have the Muellbauer PIGL and PIGLOG structure in aggregate expenditure. However, these could not have been derived by aggregating over *individual* PIGL or PIGLOG demand structures. The aggregation rule that emerges from an economy practicing proportional budgeting is linear. This implies a greater restriction than the results of Muellbauer (see Section II).

V. MARKET DEMAND AND HETEROGENEITY

While the exact aggregation conditions surveyed in Section III enable an investigator to infer values of parameters of individual demand systems, they do not yield useful restrictions on aggregate demand systems. Moreover, the well-known results of Sonnenschein (1973, 1974), Mantel (1974), and Debreu (1974) indicate that almost no aggregate regularities follow from individual optimization. In fact, they proved that any continuous, aggregate excess demand system that satisfies Walras' law could have been generated by a pure exchange economy of consumers with continuous, strictly convex, and monotone orderings.

Mas-Colell and Neuefeind (1977) extend Debreu's result, allowing individual demands to be correspondences instead of functions. Mantel (1979) considers the case where the excess demand functions are required to be differentiable, with similar results. Shafer and Sonnenschein (1982) review this literature and its extensions. Kirman and Koch (1986) consider the Debreu-Sonnenschein-Mantel (DSM) result under the restriction that all individuals have the same preferences, and show that even under this strict assumption, the primary negative result—no restrictions on aggregate demand—still holds. Kirman (1992) discusses how results from estimation of representative agent models that ignore Debreu-Sonnenschein-Mantel can be misleading.

Individual optimization alone simply provides too little structure to explain anything about macro behavior except simple adding-up conditions. Properties that we might wish to find in the aggregate, such as aggregate revealed preference or downward-sloping aggregate demand curves must depend on something other than individual optimization. It is also important to be able to explain why such properties seem to hold in the aggregate despite the results of DSM.

There appear to be three avenues to dealing with the results of DSM. The first is to take the approach of the literature on the representative consumer in Section III and find restrictions on demand functions that imply and/or are implied by the desired macro property. The second is to restrict the distribution of income. The third is to restrict the distribution of preferences. The second and third approaches, while imposing restrictions on the degree of heterogeneity of incomes or preferences can be contrasted with the first approach in that they allow preferences to take any form. Also, such approaches need not be based on any assumptions about individual rationality or optimization.

A.   The Irrational Consumer and Market Demand

The importance of heterogeneity to demand regularity has long been recognized, but Becker (1962) was one of the first to formalize the concept. Becker hoped to demonstrate that individual rationality and the existence of a well-defined preference relationship were not fundamental to aggregate regularity of demand.[*] Becker imposed minimumal rationality—that agents consume their entire income, but allowed agents to be "irrational" in a particular way. Two formulations of agent irrationality were proposed: (1) agents randomly choose their consumption bundles along the budget hyperplane; and (2) agents have completely inflexible demands. In the second case, they demand exactly what they consumed in the last period; if that bundle is no longer available, they move along the old budget hyperplane to the nearest point that they can now afford.

Becker showed that, provided the distribution of choices is spread across the budget hyperplane, aggregate demand will be downward sloping. Thus, regularity conditions in aggregate demand, which are often modeled as corresponding to the choice of a "representative consumer," can be modeled equally well as a collection of "irrational" agents. In addition to pointing the way toward agent heterogeneity as a promising line of explanation of aggregate demand regularity, this paper also provided an interesting rejoinder to critics of the rationality hypothesis in economics.

B.   Income and Preference Heterogeneity

The early result by Becker regarding the spread of agents across the budget hyperplane leads to the question of what restrictions on aggregate demand might result from restricting the distribution of income while allowing preference heterogeneity among agents. For the case where each consumer has a fixed share of total income, Eisenberg (1961) showed that market demand could be generated by a multiplicative social welfare function with exponential weights equal to each agent's share of aggregate income. Though preferences are allowed to differ between agents in this formulation, they must be homothetic. Of course, any shift in the relative income distribution would demand a new set of weights to rationalize the allocation.[†]

---

[*] As it turns out, Becker was incorrect in assuming that individual rationality gave downward-sloping aggregate demand curves. The DSM result 20 years after his work demonstrated this.

[†] Chipman (1974) clarified this result and proved it in a slightly different manner. See also Shapiro (1977).

Hildenbrand (1983) seeks conditions that do not involve restrictions on individual preferences under which the law of demand will be fulfilled in the aggregate—in other words, conditions sufficient for aggregate demand curves to be downward-sloping. Hildenbrand argues that individual demand functions are in fact a counterfactual construct and that assumptions about individual preferences or demand functions are inherently untestable.[*] He proposes to replace such untestable assumptions with testable assumptions about distributions of agents and income.[†]

Consider the vector of individual demands, $d^h(p, y^h) = d(p, y)$, which satisfy WARP (or which have been derived from a preference relation) and are identical for all consumers. Assuming a continuum of consumers, we can write (mean) market demand as

$$D(p) = \int d(p, y)\rho(y) \, dy \tag{97}$$

where ρ is the income distribution. A sufficient condition for the aggregate law of demand is that the Jacobian matrix

$$\left[\frac{\partial D(p)}{\partial p} \right] \tag{98}$$

be negative definite for all p. By the definition of (mean) market demand,

$$\begin{aligned}
\frac{\partial D(p)}{\partial p} &= \frac{\int \partial d(p, y)\rho(y) \, dy}{\partial p} \\
&= \frac{\int \partial \delta(p, u)\rho(y) \, dy}{\partial p} - \frac{\int \partial d(p, y)}{\partial y} \cdot d(p, y)^T \rho(y) \, dy \\
&= \overline{S} - \overline{M}
\end{aligned} \tag{99}$$

where δ is the compensated demand function (identical for all consumers here), $\overline{S}$ is the mean Slutsky substitution matrix, and $\overline{M}$ is the mean Slutsky income-effect matrix. The T superscript denotes transpose. $\overline{S}$ is negative semidefinite, since all the individual substitution matrices must be negative semidefinite. So the law of demand reduces to the condition that $\overline{M}$ be positive definite. Hildenbrand shows that if we restrict the distribution of income of agents such that the density of incomes is monotone and nonincreasing, then all aggregate partial demand curves will be downward-sloping:

---

[*]See Hildenbrand (1989) and, in particular, Hildenbrand's (1993) *Market Demand*.
[†]Lewbel (1994) critically reviews Hildenbrand's approach.

Theorem (Hildenbrand 1983).   For every individual demand function d^h, where each d^h satisfies the weak axiom of revealed preference and for every decreasing density ρ on $\mathbf{R}_+$, the mean demand function $D(p)$ is monotone, i.e., for every p and q in the set of all prices, we have

$$(p - q)(D(p) - D(q)) \leq 0 \tag{100}$$

Notice that restrictions are placed on the distribution of income of agents, while no restrictions are placed on the form of the preference function. Preferences here are assumed to satisfy the weak axiom of revealed preference and to be identical across consumers, but no assumptions are placed on the form of Engel curves.

The restriction that all consumers have identical preferences is certainly an undesirable one. Let us assume that preferences belong to some allowable set, A. And let demand vary between consumers, where individual demands are now

$$d^\alpha(p, y) \tag{101}$$

where $\alpha \in A$ indexes consumers. Market demand is

$$D(p) = \int_{A \times \mathbf{R}_+} d^\alpha(p, y) \, d\mu \tag{102}$$

where μ is the joint distribution of incomes and attributes. The mean Slutsky matrix is now

$$\frac{\int \partial \delta^\alpha(p, u) \, d\mu}{\partial p} - \frac{\int \partial d^\alpha(p, y)}{\partial y} \cdot d^\alpha(p, y)^T \, d\mu = \overline{S}^* - \overline{M}^* \tag{103}$$

Hildenbrand (1989) considers the conditions that ensure the positive definiteness of $\overline{M}^*$ when preferences are allowed to vary among agents. He introduces the assumption of metonymy—that the distribution of preferences, α, conditional on income, does not depend on the level of income.[*] Without placing any restrictions on individual preferences, Hildenbrand is able to show that a decreasing density, ρ, and metonymy are sufficient to imply the aggregate law of demand. Here individual preferences are not assumed to satisfy WARP. In fact, this result holds even if all individual income effects are negative semidefinite! This is a pretty astounding result and well demonstrates one of Hildenbrand's main points: that there is something qualitatively different about aggregate demand that cannot be captured in a mere summing of individual demands.

Caution is nonetheless encouraged. In practice, the density function of household incomes will not usually be decreasing. Thus, this result is not useful in explaining why aggregate demand data seem to obey general regularity conditions. But

---

[*]See Hildenbrand (1989) and Härdle, Hildenbrand, and Jerison (1991) for technical details.

Hildenbrand's result does show that restricting the income distribution is, at some level, analogous to restricting the shape of individual Engel curves. It leaves open the possibility that some combination of restrictions can be shown to hold empirically in the economy.

Jean-Michel Grandmont (1987) extends Hildenbrand's result to a slightly broader class of densities that allow for discontinuities and also allow for unbounded densities, provided certain convergence conditions are met. Grandmont also shows that a result similar to Hildenbrand's can be derived by restricting the distribution of preferences.

Härdle, Hildenbrand, and Jerison (1991) (HHJ) consider the aggregate data and the question of whether market demand adheres to the law of demand. They consider the mean demand function of (102). A sufficient condition of monotonicity of the mean demand function is that $\overline{S}^*$ be negative semidefinite and $\overline{M}^*$ be positive semidefinite. We know that $\overline{S}^*$ will be negative semidefinite, since all the individual substitution effects are negative. Thus, positive semidefiniteness of $\overline{M}^*$ is sufficient for the law of demand to hold in the aggregate.

Using the assumption of metonymy, HHJ are able to estimate the mean Slutsky income-effect matrix. Utilizing the symmetricized matrix M,

$$M = \overline{M}^* + \overline{M}^{*T} \tag{104}$$

(which will only be positive definite when $\overline{M}^*$ is), they estimate a related matrix, $\hat{M}^*$, which is identical to $\overline{M}^*$ under the metonymy assumption. They consider nine commodity aggregates in U.K. expenditure data from 1969 through 1983 and show that in all cases (for all years and for all commodities) the matrix $\hat{M}^*$ is positive definite. Unfortunately, it is hard to judge the robustness of this result since their tests rely on the eigenvalues of the estimated matrix, and there is no available distribution theory for these estimated eigenvalues. Using bootstrapping techniques, they give some idea that their results may be robust.

From nonparametric density estimates presented in their paper, HHJ clearly show that the distribution of incomes is not decreasing. So decreasing density of income is not driving the adherance to the law of demand observed in the data. HHJ identify two possible explanations. One is that, at higher income levels, preferences seem to be more heterogenous. In other words, Engel curves tend to spread out at higher income levels. This reduces the possibility of pathological outcomes. It would also seem to provide some corroboration of Lewbel's results that at higher income levels, rank-2 Engel curves do not describe demand adequately (see Section IV.A). Also, the slopes of cross-product curves for the estimated $\hat{M}^*$ matrix are fairly small compared to the slopes of the own-product curves. Thus own-price effects seem to be dominating.

Härdle and Hart (1992) develop the asymptotic theory necessary for testing the positive definiteness of income-effect matrices using a bootstrapping approach. They create a bootstrapped asymptotic distribution for the smallest estimated eigenvalue, λ_1, using the assumption that the income-effect matrix is symmetric. This allows for testing of a null hypothesis of positive semidefiniteness versus the one-sided alternative of $\hat{\lambda}_1 > 0$. They follow the average derivative techniques developed by Härdle and Stoker (1989).

Grandmont (1992) proposes restricting agent preferences instead of the income distribution. Following Becker's formulation of the idea, one would like to know if agent preferences are such that individual choices will be spread (enough) along the budget hyperplane to generate downward-sloping aggregate demand curves. The problem is the lack of a metric for preferences. How do we measure how far apart two preference orderings are? How can we measure the dispersion of a distribution of preferences without such a measure?

Grandmont proposes an algebraic structure on preferences using affine transformations of the commodity space. Given a vector of commodities, $x \in \mathbf{R}^n$, and any collection of real numbers $\alpha = (\alpha_1, \ldots, \alpha_n)$, a new vector can be generated by multiplying each commodity x_i by the factor e^{α_i}:

$$x_\alpha = e^\alpha \otimes x = (e^{\alpha_1} x_1, \ldots, e^{\alpha_n} x_n) \tag{105}$$

These transforms can then be used to generate a linear structure on preference relations and/or demand functions. They are affine, since each point in the commodity space is transformed linearly to a new point based on the values of the α-vector. If $\alpha_i = \alpha$ for every i, then the transformation is said to be homothetic, as each point in the commodity space is transformed to a point on a ray passing through the original point.

Grandmont considers one such equivalence class (i.e., one class of demand functions that are all related to each other as α-transforms). Within that class, any preference can be represented as an n-dimensional vector, α. If demands in the economy are all members of one such class, and if the distribution of α is flat enough, then aggregate demand will be monotone and individual demand curves are downward-sloping. Aggregate demand also obeys the weak axiom of revealed preference.

What is the meaning of this? The distribution of α represents the degree of heterogeneity of preferences within any equivalence class. If the distribution of α is not flat, then preferences are concentrated around some point and aggregate demand may behave strangely depending on those individual preferences. If the distribution of α is flat, then preferences are sufficiently heterogenous and aggregate demand is well behaved.

These results do not depend on any restrictions on individual demands except that they obey homogeneity and Walras' law. Demands are not even required to be preference-driven. Thus, individual rationality, as in Becker's formulation, is simply unnecessary for aggregate demand regularity.

Grandmont does not consider heterogeneity across equivalence classes. Nor does he consider heterogeneity in both preference and income dimensions. These ar-

eas seem promising for future research. Given the results of Hildenbrand and Grandmont, it seems that some restrictions on heterogeneity across multiple dimensions
should also yield regularity of aggregate demand. Perhaps the unrealistic restrictions on the distribution of income (that the density must be nonincreasing) may be
weakened and combined with some restriction on the distribution of preferences to
yield the law of demand.

VI. JOINT AGGREGATION OVER AGENTS AND COMMODITIES

A.   Fixed Inputs or Outputs

In Section I.A, we noted that, so long as all commodities (net outputs in that exposition) are efficiently allocated, there is no aggregation problem on the production side
of the economy: a "representative producer" exists under the most basic regularity
conditions. But there is a long history of research (and not a little controversy) over
the issue of aggregation in production when inputs are *not* efficiently allocated. Thus,
some inputs—notably capital (in the absence of perfect capital rental markets)—are
fixed in the proverbial short run. In addition, outputs of, say, regulated firms might
be fixed, in which case they are modeled as minimizing cost subject to an output
constraint. This section surveys the literature on aggregation of fixed and variable
inputs.*

Let v^f and z^f, $f = 1, \ldots, F$, be the vectors of variable and fixed commodities, respectively, in the technology set of firm f. The set of feasible variable quantities is dependent on the quantities of fixed variables and is denoted $T^f(z^f)$ for firm
f. Also, define the production functions of the firms, G^f, $f = 1, \ldots, F$, by

$$v^f \in T^f(z^f) \iff G^f(v^f, z^f) \leq 0 \tag{106}$$

Conditional on the quantities of fixed inputs or outputs of all firms, the aggregate
technology set is then defined (assuming no externalities exist) by

$$T(z^1, \ldots, z^F) = \sum_f T^f(z^f) \tag{107}$$

so that the aggregate production function G is defined by

$$v \in T(z^1, \ldots, z^F) \iff G(v, z^1, \ldots, z^F) \leq 0 \tag{108}$$

---

*Parts of this section borrow shamelessly from the excellent editorial summaries of Gorman (1953b) (writ-
ten by Bill Schworm) and Gorman (1968b) (written by the editors) in Blackorby and Shorrocks (1995).

Thus, the aggregate technology set and the aggregate production function depend on the quantity of each fixed input held by each firm (or each fixed output). From an econometric point of view, this is a nightmare. Thus, it is standard practice (explicitly or implicitly) to aggregate over these fixed inputs or outputs, in order to reduce the dimensions of estimation problems to manageable proportions.

The existence of an aggregate fixed input or output for an individual firm is represented formally by the existence of functions, $\hat{G}^f$ and Z^f such that

$$G^f(v^f, z^f) = \hat{G}^f(v^f, Z^f(z^f)) \tag{109}$$

Thus, the vector z^f is reduced to a scalar $Z^f(z^f)$ on the right-hand side. The conditions required for this structure are well known to be separability of the fixed inputs/outputs from the variable inputs/outputs. That is, technical rates of substitution between fixed inputs/outputs must be independent of the quantities of variable inputs/outputs.* The existence of an aggregator function over fixed inputs in the economy-wide production function

$$G(v, z^1, \ldots, z^F) = \hat{G}(v, Z(z^1, \ldots, z^F)) \tag{110}$$

however, is more problematical.

B. Klein Aggregation

Several ways of simplifying the structure in (108), taking full account of the aggregation problem, have been proposed. The first suggestion was that of Klein (1946a, 1946b), who proposed an aggregation procedure predicated on the assumption that *none* of the inputs or outputs is efficiently allocated. He posited multiple types of labor and capital inputs producing a set of outputs in accordance with the following production function:

$$G^f(x^f, n^f, z^f) = 0 \quad \forall f \tag{111}$$

where x^f, n^f, and z^f are the vectors of output, labor, and capital quantities, respectively, for firm f. To simplify the analysis of economy-wide production functions, Klein proposed the existence of functions, $\hat{G}$, X, N, and Z such that

$$\hat{G}(X(x^1, \ldots, x^F), N(n^1, \ldots, n^F), Z(z^1, \ldots, z^F)) = 0$$
$$\Longleftrightarrow G^f(x^f, n^f, z^f) = 0 \quad \forall f \tag{112}$$

Thus, a problem of gargantuan proportions is reduced to one in which there are just three variables in the economy's production function: aggregate output, aggregate labor input, and aggregate capital input.

---

*The equivalence between separability and the functional structure in (109) was developed independently by Leontief (1947a, 1947b) and Sono (1961) and subjected to a penetrating analysis by Gorman (1968a). For a comprehensive treatment and a recent survey, see Blackorby, Primont, and Russell (1978, 1997).

The question posed by Klein was answered by Nataf (1948), who showed that (112) holds if and only if each firm's production function can be written as

$$G^f(x^f, n^f, z^f) = X^f(x^f) + N^f(n^f) + Z^f(z^f) = 0 \tag{113}$$

That is, there exist aggregators over each of the three types of variables in each production function—hence, each set of variables is separable from its complement—and the technology is linear in these aggregates.* It is easy to see that this structure is sufficient for the Klein condition (112). Define

$$X(x^1, \ldots, x^F) := \sum_f X^f(x^1, \ldots, x^F)$$

$$N(n^1, \ldots, n^F) := \sum_f N^f(n^1, \ldots, n^F) \tag{114}$$

$$Z(z^1, \ldots, z^F) := \sum_f Z^f(z^1, \ldots, z^F)$$

and

$$\hat{G}(X(x^1, \ldots, x^F), N(n^1, \ldots, n^F), Z(z^1, \ldots, z^F))$$
$$:= X(x^1, \ldots, x^F) + N(n^1, \ldots, n^F) + Z(z^1, \ldots, z^F) \tag{115}$$

It is now clear that

$$X(x^1, \ldots, x^F) + N(n^1, \ldots, n^F) + Z(z^1, \ldots, z^F) = 0$$
$$\Longleftrightarrow X^f(x^f) + N^f(n^f) + Z^f(z^f) = 0 \quad \forall f \tag{116}$$

Gorman (1953b) noted that the Klein-Nataf structure is not very useful for empirical implementation because the aggregator functions for output, labor, and capital in (112) and (116) depend on the entire distribution of these variables among firms. He suggested that, to be useful, these aggregate commodities should depend only on the sum across firms of the component variables: i.e., that the aggregate production function take the form

$$\tilde{G}\left(X\left(\sum_f x^f\right), N\left(\sum_f n^f\right), Z\left(\sum_f z^f\right)\right) = 0 \tag{117}$$

He then showed that the necessary and sufficient conditions for this aggregate production structure is that the individual production equation have the affine form

---

*Hence, the variables are, in Gorman's (1968a) terminology, *completely* separable in G^f: in addition to each group being separable, arbitrary unions of groups are separable (so that, e.g., technical rates of substitution between labor and capital inputs are independent of output quantities).

$$\alpha \cdot x^f + \beta \cdot n^f + \gamma \cdot z^f + \theta^f = 0 \tag{118}$$

where α, β, and γ are appropriately dimensioned vectors of parameters that are identical for all firms and θ^f is an arbitrary, idiosynchratic scalar for firm f. In this case, the economy production equation is

$$\alpha \cdot \sum_f x^f + \beta \cdot \sum_f n^f + \gamma \cdot \sum_f z^F + \sum_f \theta^f = 0 \tag{119}$$

so that the Klein aggregates have the linear structures

$$X\left(\sum_f x^f\right) = \alpha \cdot \sum_f x^f$$

$$N\left(\sum_f n^f\right) = \beta \cdot \sum_f n^f \tag{120}$$

and

$$Z\left(\sum_f z^f\right) = \gamma \cdot \sum_f z^f$$

C. Aggregation with Both Efficiently Allocated ("Labor") Inputs and Fixed ("Capital") Inputs or Outputs

The restrictions on individual technologies that are necessary and sufficient for the existence of Klein aggregates are too demanding to be useful in empirical analysis. Even if we allow the aggregates to depend on the entire distribution of component variables, as in (112), the individual technologies are linear in the firm-specific aggregates, and if we more realistically require that the economy aggregates depend only on the sum of component variables, as in (117), the individual technologies are linear in each output and input variable (implying, e.g., linear isoquants and linear production possibility surfaces).

Subsequent literature on "capital" aggregation has focused on the unreasonableness of Klein's requirement that the allocation of outputs and inputs be entirely arbitrary. In fact, May (1946) and Pu (1946) immediately pointed out that competitive-equilibrium conditions—in particular, profit maximization by firms—might be exploited to weaken the requirements for the existence of composite commodities in aggregate production technologies. In particular, Solow (1964) showed how the efficient allocation of labor inputs might facilitate the aggregation of (fixed) capital inputs. His ingenious insight was to note that, under certain conditions, the efficient allocation of inputs not being aggregated can render efficient an arbitrary allocation of fixed inputs as well.

The ultimate—and naturally elegant—solution to the problem of aggregating over fixed inputs in the aggregate technology and in the presence of efficiently allocated inputs was presented by Gorman (1965, 1968b). The existence of efficiently allocated inputs and outputs allowed him to exploit duality theory to solve the problem (as in the case of the representative consumer in Section II). Given the technology represented in (106), the variable profit function* of firm f, denoted P^f, is defined by

$$P^f(p, z^f) := \max_{v^f}\{p \cdot v^f \mid G^f(v^f, z^f) = 0\} \tag{121}$$

where p is the (competitive equilibrium) price vector for variable inputs and outputs. For the economy, the variable profit function P is defined by

$$P(p, z^1, \ldots, z^F) = \max_{v}\{p \cdot v \mid G(v, z^1, \ldots, z^F) = 0\}$$
$$= \sum_f P^f(p, z^f) \tag{122}$$

where the second identity follows from the existence of a "representative firm" conditional on the allocation of fixed inputs (i.e., the interchangeability of set summation and optimization), discussed in Section II.A. Note that, if outputs are fixed and inputs are variable, P is the cost function of the firm, showing minimal cost as a function of input prices and the fixed output quantities.

Theorem (Gorman 1968b). *The aggregator function in (110) exists if and only if the individual variable profit functions (or cost functions) have the structure*

$$P^f(p, z^f) = \Pi(p)Z^f(z^f) + \Lambda^f(p) \quad \forall f \tag{123}$$

in which case the variable profit function for the economy is

$$P(p, z^1, \ldots, z^F) = \Pi(p) \sum_f Z^f(z^f) + \sum_f \Lambda^f(p)$$
$$=: \Pi(p)Z(z^1, \ldots, z^F) + \Lambda(p) \tag{124}$$

The similarity between this structure and the structure of the expenditure function that is necessary and sufficient for the existence of a representative consumer in Section II is palpable. In fact, as nicely elucidated in the editorial summary

---

*See, e.g., Diewert (1974) for a discussion of variable profit functions.

of Gorman (1968b) in Blackorby and Shorrocks (1995), Gorman's proof of the existence of an aggregate fixed input for the economy constitutes, with little more than a change in notation, an alternative proof of the existence of a community indifference map and hence the existence of a Gorman-type representative consumer.

Recall from Section II.A that the community (Scitovsky) preference set (the set of aggregate commodity bundles that makes possible a particular profile of utility levels of the H consumers) is the sum of the individual level sets:

$$N(u_1, \ldots, u^h) = \sum_h N^h(u^h) \tag{125}$$

The individual and community expenditure functions are given by

$$E^h(u^h, p) = \min_{x^h}\{p \cdot x^h \mid x^h \in N^h(u^h)\} \tag{126}$$

and

$$E(u^1, \ldots, u^h, p) = \min_x\{p \cdot x \mid x \in N(u_1, \ldots, u^h)\}$$
$$= \sum_h E^h(u^h, p) \tag{127}$$

A community preference map, and hence a representative consumer, exists if and only if the community preference set depends only on a scalar utility (of the representative consumer),

$$N(u^1, \ldots, u^h) = \hat{N}(U(u^1, \ldots, u^h)) \tag{128}$$

in which case the community expenditure function is

$$E(u^1, \ldots, u^h, p) = \hat{E}(U(u^1, \ldots, u^h), p) \tag{129}$$

Apart from a change in notation (substituting consumer utilities for fixed inputs and expenditure minimization for profit maximization), this is the same structure as the above results on the aggregation of fixed inputs. Thus, the individual expenditure functions must have the form

$$E^h(u^h, p) = \Pi(p)\psi^h(u^h) + \Lambda^h(p) \tag{130}$$

which implies that

$$E(u^1, \ldots, u^h, p) = G(p) \sum_h \psi^h(u^h) + \sum_h \Lambda^h(p)$$
$$=: \Pi(p)U(u^1, \ldots, u^h) + \Lambda(p) \tag{131}$$

which is (apart from the harmless monotonic transformation of the utility levels) identical to the structure in the Gorman theorem on representative consumers in Section II. The parallel nature of these (affine) structures underscores the common thread running through apparently diverse aggregation problems.

D. Extension to Multiple Fixed Inputs

The Gorman theorem in Section VI.C on the aggregation of fixed inputs in the presence of some efficiently allocated inputs was essentially worked out in his analysis of capital aggregation in vintage models for a single firm (Gorman 1965). If we interpret (106) as representing the technologies for different capital vintages, then the structure (123) for the vintage-specific variable profit functions is necessary and sufficient for the existence of an aggregate capital stock, obtained by aggregating over the different capital vintages, as in (110). This analysis was extended, and the proof made much more elegant, in Gorman (1968b). The problem solved in this paper generalizes the structure in (110) to include multiple aggregate fixed inputs. Thus, rewrite (106) as

$$v^f \in T^f(z_1^f, \ldots, z_R^f) \iff G^f(v^f, z_1^f, \ldots, z_R^f) = 0 \quad \forall f \tag{132}$$

where z_r^f is the vector of fixed inputs of type $r, r = 1, \ldots, R$, held by firm f. The aggregate technology set then depends on the distribution of quantities of each type of fixed input among firms and the question is when can the aggregate technology by represented by the production equation

$$\tilde{G}(v, Z^1(z_1^1, \ldots, z_1^F), \ldots, Z^R(z_R^1, \ldots, z_R^F)) = 0 \tag{133}$$

The answer, as elegantly shown by Gorman (1968b), is a straightforward generalization of the earlier result: the variable profit functions of the individual firms must be

$$P^f(p, z_1^f, \ldots, z_R^f) = \sum_r \Pi_r(p) Z_r^f(z^r) + \Lambda^f(p) \quad \forall f \tag{134}$$

and the variable profit function of the economy therefore has the structure

$$P(p, Z^1(z_1^1, \ldots, z_1^F), \ldots, Z^R(z_R^1, \ldots, z_R^F))$$

$$= \sum_r \Pi_r(p) \sum_f \sum_r Z_r^f(z_r^f) + \sum_f \Lambda^f(p) \tag{135}$$

$$=: \sum_r \Pi_r(p) Z^r(z_r^1, \ldots, z_r^f) + \Lambda(p)$$

where

$$Z^r(z_r^1, \ldots, z_r^F) = \sum_f Z_r^f(z_r^f) \quad \forall r \tag{136}$$

Again, the similarity of this structure, which rationalizes the existence of multiple aggregate inputs in an economy-wide production function, with the structure of exact aggregation of consumer demand functions with multiple income distribution statistics, surveyed in Section III, is palpable. In fact, using the envelope theorem to derive the net-supply functions of variable netput vectors for firm f, we obtain

$$\phi^f(p, z^f) = \sum_r \nabla \Pi_r(p) Z_r^f(z_r^f) + \nabla \Lambda^f(p) \tag{137}$$

which has a structure that is very similar to that analyzed by Gorman (1981). Aggregate (vector valued) net-supply functions are then given by

$$\Phi(p, z^1, \ldots, z^F) = \sum_f \sum_r \nabla \Pi_r(p) Z^f(z_r^f) + \sum_f \nabla \Lambda^f(p)$$

$$=: \sum_r \nabla \Pi_r(p) \Theta_r(z_r^1, \ldots, z_r^F) + \Lambda(p) \tag{138}$$

$$=: \hat{\Phi}(p, \Theta_1(z_1^1, \ldots, z_1^F) \ldots, \Theta_R(z_R^1, \ldots, z_R^F))$$

This structure of net-demand systems is analogous to that of exact aggregation in consumer theory, the principal difference being that the θ functions in the aggregate consumer demand functions in exact aggregation depend on the H incomes of the consumers, whereas in the aggregation of fixed inputs, the Θ functions depend on vectors of fixed inputs of firms.

In fact, in econometric applications, it is standard to represent heterogeneous technologies of firms through the use of technological "attributes," analogous to those representing heterogeneity of consumer preferences in Section III. This raises problems that are identical to those posed by the existence of fixed inputs; indeed, one can think of fixed inputs as one type of technological attribute. These attributes can be modeled exactly as fixed inputs are modeled in constructing aggregate technologies. Indeed, one can interpret some of the z_r^f variables in (132) as attributes and the Z^r functions in (133) as aggregate attributes. The approach of Heineke and Shefrin (1990) can then be used to identify attribute parameters. See Appelbaum (1982), Borooah and Van Der Ploeg (1986), Chavas (1993), Gouriéroux (1990), and Fortin (1991) for applications of these aggregation notions in producer theory. In particular, Fortin and Gouriéroux find that significant bias results from ignoring aggregation issues in econometric analysis of production.

Owing to the difference between the optimization problems of consumers and producers, two caveats need to be emphasized. First, while the structures of exact aggregation, on the one hand, and aggregation over fixed inputs and technological attributes are analogous, the proofs of these aggregation results are quite different. Second, Gorman's theorem on exact aggregation and the rank of demand systems does not go through for producers.

E. Other Extensions

Econometric studies using aggregate data also commonly aggregate over variable inputs (e.g., different types of labor). Again, at the level of the individual firm the necessary and sufficient condition for the existence of such aggregates is separability, but at the aggregate level more is required. The necessary and sufficient conditions for aggregation of efficiently allocated inputs are different from, but similar in many respects to, the conditions for aggregation of fixed inputs. Moreover, as in the case of fixed inputs, the conditions are less demanding if aggregation is required only over a subset of the efficiently allocated inputs. In addition, aggregation of variable inputs is made more difficult if there exist fixed inputs in the technology. These issues were analyzed in Gorman (1967, 1982a), and Blackorby and Schworm (1988a) provide an excellent survey and analysis of the various results obtained under different assumptions about which subsets of efficiently allocated and fixed inputs are being aggregated in the economy-wide technology.* To a large extent, the common theme of affine structures runs through these results, but there are some surprises (which, in the interest of space, are left to the reader to discover).

The use of common duality concepts in each of these aggregation problems entails the use of strong convexity assumptions regarding the technology. The possible inadequacy of this approach for long-run analysis led Gorman (1982b) to examine aggregation conditions under constant returns to scale.

Joint aggregation over commodities and agents is also common in studies of consumer behavior. As one would expect, the necessary and sufficient conditions, as shown by Blackorby and Schworm (1988b), are stronger than those required for either type of aggregation taken separately.

Finally, Gorman (1990) has more recently examined aggregation problems in which the aggregate inputs for the economy depend on the quantities of *all* inputs, perhaps because of externalities. These studies lead to structures that are quite similar to the affine structures above and, interestingly, reminiscent of the rank conditions for Engel curves in Sections III and IV.

VII. CONCLUDING REMARKS

As Section I.A summarizes the literature, our concluding remarks will be limited to two observations. First, when we began research on this survey, we expected the chapter to be about half theory and half econometrics; the reader has undoubtedly noted, however, that the *theory* of aggregation dominates. Thus, the main lesson we take from this endeavor is that, while the theory of aggregation is fairly well developed, econometric application is in its infancy. The principal empirical literature— most notably Jorgenson, Lau, and Stoker (1980, 1981, 1982)—*maintains* exact aggregation conditions, generates results on demographics and demand, and studies various issues of welfare economics. Less well developed is the testing of restrictions on preferences and on the distribution of incomes or preferences that are nec-

---

*Also see Fisher (1965, 1968a, 1968b, 1969, 1982, 1983) and Blackorby and Schworm (1984).

essary or sufficient for various types of aggregation consistency. The most promising avenues for research in this direction have been forged by Lewbel (1991), Härdle, Hildenbrand, and Jerison (1991), and Hausman, Newey, and Powell (1995). But it seems that the potential for testing aggregation conditions has barely been tapped.

Second, just as the theory of aggregation has dominated this literature, it should be obvious to the reader that one person, Terence Gorman, has in turn dominated the search for theoretical results. His towering achievements in this area are testimony to his penetrating intellect. Moreover, while many of his papers are challenging reading, he has always kept his eye on the potential for empirical application. One would hope that applied econometricians would focus more on testing the aggregation theories promulgated primarily by Gorman.

ACKNOWLEDGMENTS

We are deeply indebted to Chuck Blackorby and Tony Shorrocks for their masterful editorial effort in making Terence Gorman's works, many previously unpublished, available in Volume I of the *Collected Works of W. M. Gorman*; without this towering volume, including the illuminating editorial introductions to Gorman's papers, our task would have been impossible. Nor would the chapter have been possible without the investment in human capital accruing to the long collaboration between Blackorby and Russell. We also thank Kusum Mundra for comments, and we are especially grateful to Chuck Blackorby and Bill Schworm for reading the manuscript and for many valuable suggestions. Needless to say, they are in no way responsible for any remaining errors.

REFERENCES

Aczel, J. and J. Dhombres (1989), *Functional Equations in Several Variables with Applications to Mathematics, Information Theory and to the Natural and Social Sciences*, Cambridge University Press, Cambridge.

Afriat, S. N. (1967), The Construction of a Utility Function From Expenditure Data, *International Economic Review*, 8, 460–472.

Afriat, S. N. (1972), Efficiency Estimation of Production Functions, *International Economic Review* 13, 568–598.

Aiyagari, S. (1985), Observational Equivalence of the Overlapping Generations and Discounted Dynamic Programming Frameworks for One-Sector Growth, *Journal of Economic Theory*, 35, 201–221.

Antonelli, G. B. (1886), *Sulla Teoria Matematica della Economia Politica*, Pisa: nella tipografia del Folchetto; English translation in *Preferences, Utility, and Demand* (J. S. Chipman, L. Hurwicz, M. K. Richter, and H. F. Sonnenschein, eds.), Harcourt Brace Jovanovich, New York, 1971).

Appelbaum, E. (1982), The Estimation of the Degree of Oligopoly Power, *Journal of Econometrics*, 19, 287–299.

Arrow, K. J. (1951), *Social Choice and Individual Values*, Wiley, New York.

Banks, J., R. Blundell, and A. Lewbel (1993), Quadratic Engel Curves and Welfare Measurement, Working Paper, Institute of Fiscal Studies, London.

Becker, G. S. (1962), Irrational Behavior and Economic Theory, *Journal of Political Economy*, 70, 1–13.

Blackorby, C., R. Boyce, and R. R. Russell (1978), Estimation of Demand Systems Generated by the Gorman Polar Form: A Generalization of the S-Branch Utility Tree, *Econometrica*, 46, 345–364.

Blackorby, C., R. Davidson, and W. Schworm (1993), Economies with a Two-Sector Representation, *Economic Theory*, 3, 717–734.

Blackorby, C., D. Primont, and R. R. Russell (1978), *Duality, Separability, and Functional Structure: Theory and Economic Applications*, North-Holland/American Elsevier, New York.

Blackorby, C., D. Primont, and R. R. Russell (1997), Separability, in *Handbook of Utility Theory* (S. Barbara, P. J. Hammond, and C. Seidl, eds.), Kluwer, Boston.

Blackorby, C. and R. R. Russell (1993), Samuelson's "Shibboleth" Revisited: Proportional Budgeting among Agents and Rank-Two Demand Systems, in *Mathematical Modelling in Economics* (W. E. Diewert, K. Spremann, and F. Stehling, eds.), Springer-Verlag, Heidelberg, 35–46.

Blackorby, C. and R. R. Russell (1997), Two-Stage Budgeting: An Extension of Gorman's Theorem, *Economic Theory*, 9, 185–193.

Blackorby, C. and W. Schworm (1982), Aggregate Investment and Consistent Intertemporal Technologies, *Review of Economic Studies*, 49, 595–614.

Blackorby, C. and W. Schworm (1984), The Structure of Economies with Aggregate Measures of Capital: A Complete Characterisation, *Review of Economic Studies*, 51, 633–650.

Blackorby, C. and W. Schworm (1988a), The Existence of Input and Output Aggregates in Aggregate Production Functions, *Econometrica*, 56, 613–643.

Blackorby, C. and W. Schworm (1988b), Consistent Commodity Aggregates in Market Demand Equations, in *Measurement in Economics* (W. Eichhorn, ed.), Physica-Verlag, Heidelberg, 577-606.

Blackorby, C. and W. Schworm (1993), The Implications of Additive Consumer Preferences in a Multi-Consumer Economy, *Review of Economic Studies*, 60, 209–228.

Blackorby, C. and A. F. Shorrocks (1995), *Separability and Aggregation: Collected Works of W. M. Gorman*, Vol. I, Clarendon Press, Oxford.

Blundell, R., P. Pashardes, and G. Weber (1993), What do We Learn About Consumer Demand Patterns from Micro Data? *American Economic Review*, 83, 570–597.

Borooah, V. K. and F. Van Der Ploeg (1986), Oligopoly Power in British Industry, *Applied Economics*, 18, 583–598.

Bronars, S.G. (1987), The Power of Nonparametric Tests of Preference Maximization, *Econometrica*, 55, 693–698.

Brown, M. and D. Heien (1972), The S-Branch Utility Tree: A Generalization of the Linear Expenditure System, *Econometrica*, 40, 737–747.

Buse, A. (1992), Aggregation, Distribution and Dynamics in the Linear and Quadratic Expenditure Systems, *Review of Economics and Statistics*, 74, 45–53.

Chavas, J.-P. (1993), On Aggregation and Its Implications for Aggregate Behavior, *Ricerche Economiche*, 47, 201–214.

Chipman, J. (1974), Homothetic Preferences and Aggregation, *Journal of Economic Theory*, 8, 26–38.

Christensen, L., D. C. Jorgenson, and L. J. Lau (1975), Transcendental Logarithmic Utility Functions, *American Economic Review*, 65, 367–383.

Constantinides, G. (1982), Intertemporal Asset Pricing with Heterogeneous Consumers and without Demand Aggregation, *Journal of Business*, 55, 253–267.

Deaton, A. (1986), Demand Analysis, in *Handbook of Mathematical Economics*, Vol. 3 (K. J. Arrow and M. D. Intriligator, eds.), North-Holland, Amsterdam, 1767–1839.

Deaton, A. S. and J. Muellbauer (1980), An Almost Ideal Demand System, *American Economic Review*, 70, 312–326.

Debreu, G. (1954), Valuation Equilibrium and Pareto Optimum, in *Proceedings of the National Academy of Sciences of the USA*, 46, 588–592.

Debreu, G. (1959), *Theory of Value*, Wiley, New York.

Debreu, G. (1974), Excess Demand Functions, *Journal of Mathematical Economics*, 1, 15–21.

Diewert, W. E. (1971), An Application of the Shephard Duality Theorem: A Generalized Leontief Production Function, *Journal of Political Economy*, 79, 481–507.

Diewert, W. E. (1973), Afriat and Revealed Preference Theory, *Review of Economic Studies*, 40, 419–426.

Diewert, W. E. (1974), Applications of Duality Theory, in *Frontiers of Quantitative Economics*, Vol. 2 (M. Intriligator and D. Kendrick, eds.), North-Holland, New York, 106–171.

Diewert, W. E. and C. Parkan (1985), Tests for the Consistency of Consumer Data, *Journal of Econometrics*, 30, 127–147.

Eichenbaum, M., L. Hansen, and K. Singleton (1988), A Time Series Analysis of Representative Agent Models of Consumption and Leisure Choice Under Uncertainty, *Quarterly Journal of Economics*, 103, 51–78.

Eisenberg, B. (1961), Aggregation of Utility Functions, *Management Science*, 7, 337–350.

Fisher, F. (1965), Embodied Technical Change and the Existence of an Aggregate Capital Stock, *Review of Economic Studies*, 32, 263–288.

Fisher, F. (1968a), Embodied Technology and the Existence of Labour and Output Aggregates, *Review of Economic Studies*, 35, 391–412.

Fisher, F. (1968b), Embodied Technology and Aggregation of Fixed and Movable Capital Goods, *Review of Economic Studies*, 35, 417–428.

Fisher, F. (1969), The Existence of Aggregate Production Functions, *Econometrica*, 37, 553–577.

Fisher, F. (1982), Aggregate Production Functions Revisited: The Mobility of Capital and the Rigidity of Thought, *Review of Economic Studies*, 49, 615–626.

Fisher, F. (1983), On the Simultaneous Existence of Full and Partial Capital Aggregates, *Review of Economic Studies*, 50, 197–208.

Fleissig, A. R., A. R. Hall, and J. J. Seater (1994), GARP, Separability, and the Representative Consumer, Working Paper, Department of Economics, North Carolina State University.

Fortin, N. M. (1991), Fonctions de Production et Biais d'Agrégation, *Annales d'Économie et de Statistique*, 20/21, 41–68.

Freixas, X. and A. Mas-Colell (1987), Engel Curves Leading to the Weak Axiom in the Aggregate, *Econometrica*, 55, 515–531.

Geary, R. C. (1950), A Note on "A Constant-Utility Index of the Cost of Living," *Review of Economic Studies*, 18, 65–66.

Gorman, W. M. (1953a), Community Preference Fields, *Econometrica*, 21, 63–80 (reprinted as Chapter 15 in Blackorby and Shorrocks 1995).

Gorman, W. M. (1953b), Klein Aggregates and Conventional Index Numbers, presented to the Innsbruck Meeting of the Econometric Society (published as Chapter 16 in Blackorby and Shorrocks 1995).

Gorman, W. M. (1959), Separable Utility and Aggregation, *Econometrica*, 27, 469–481 (reprinted as Chapter 3 in Blackorby and Shorrocks 1995).

Gorman, W. M. (1961), On a Class of Preference Fields, *Metroeconomica*, 13, 53–56 (reprinted as Chapter 17 in Blackorby and Shorrocks 1995).

Gorman, W. M. (1965), Capital Aggregation in Vintage Model, presented to the First World Congress of the Econometric Society (published as Chapter 18 in Blackorby and Shorrocks 1995).

Gorman, W. M. (1967), Aggregates for Variable Goods, typescript, London School of Economics (published as Chapter 18 in Blackorby and Shorrocks 1995).

Gorman, W. M. (1968a), Measuring the Quantities of Fixed Factors, in *Value, Capital, and Growth: Essays in Honor of Sir John Hicks* (J. N. Wolfe, ed.), Edinburgh University Press, Edinburgh, 141–172 (reprinted as Chapter 19 in Blackorby and Shorrocks 1995).

Gorman, W. M. (1968b), The Structure of Utility Functions, *Review of Economic Studies* 5, 369–390 (reprinted as Chapter 12 in Blackorby and Shorrocks 1995).

Gorman, W. M. (1976), Muellbauer's Representative Agent, typescript (published as Chapter in Blackorby and Shorrocks 1995).

Go·   , W. M. (1981), Some Engel Curves, in *Essays in the Theory and Measurement of Consumer Behaviou·* (A. S. Deaton, ed.), Cambridge University Press, Cambridge, 7–29 (reprinted as Chapter 20 in Blackorby and Shorrocks 1995).

Gorman, W. M. (1982a), Aggregation in the Short and Long Run, typescript, Oxford (published as Chapter 25 in Blackorby and Shorrocks 1995).

Gorman, W. M. (1982b), Long-Run Aggregates under Constant Returns, typescript, Oxford (published as Chapter 26 in Blackorby and Shorrocks 1995).

Gorman, W. M. (1990), More Measures for Fixed Factors, in *Measurement and Modeling in Economics* (G. D. Myles, ed.), North-Holland, Amsterdam (reprinted as Chapter 21 in Blackorby and Shorrocks 1995).

Gour.éroux, C. (1990), Hétérogénéité II. Étude des Biais de Représentativité, *Annales d'Économie e, de Statistique*, 17, 185–204.

Grandmont J. M. (1987), Distributions of Preferences and the Law of Demand, *Econometrica*, 5⸳, 155–161.

Grandmont, J. M. (1992), Transformations of the Commodity Space, Behavioral Heterogeneity, and the Aggregation Problem, *Journal of Economic Theory*, 57, 1–35.

Green, H. A. J. (1964), *Aggregation in Economic Analysis: An Introductory Survey*, Princeton University Press, Princeton.

Härdle, W. and J. Hart (1992), A Bootstrap Test for Positive Definiteness of Income Effect Matrices, *Econometric Theory*, 8, 276–290.

Härdle, W., W. Hildenbrand, and M. Jerison (1991), Empirical Evidence on the Law of Demand, *Econometrica*, 59, 1525–1549.

Härdle, W. and T. M. Stoker (1989), Investigating Smooth Multiple Regression by the Method of Average Derivatives, *Journal of the American Statistical Association*, 84, 986–995.

Hausman, J. A., W. K. Newey, and J. L. Powell (1995), Nonlinear Errors in Variables Estimation of Some Engel Curves, *Journal of Econometrics*, 65, 205–233.

Heineke, J. M. (1979), Exact Aggregation and Estimation, *Economics Letters*, 4, 157–162.

Heineke, J. M. (1993), Exact Aggregation and Consumer Demand Systems, *Richerche Economiche*, 47, 215–232.

Heineke, J. M. and H. M. Shefrin (1982), The Finite Basis Property and Exact Aggregation, *Economics Letters*, 9, 209–213.

Heineke, J. M. and H. M. Shefrin (1986), On an Implication of a Theorem Due to Gorman, *Economics Letters*, 21, 321–323.

Heineke, J. M. and H. M. Shefrin (1987), On Some Global Properties of Gorman Class Demand Systems, *Economics Letters*, 25, 155–160.

Heineke, J. M. and H. M. Shefrin (1988), Exact Aggregation and the Finite Basis Property, *International Economic Review*, 29, 525–538.

Heineke, J. M. and H. M. Shefrin (1990), Aggregation and Identification in Consumer Demand Systems, *Journal of Econometrics*, 44, 377–390.

Hildenbrand, W. (1983), On the Law of Demand, *Econometrica*, 51, 997–1019.

Hildenbrand, W. (1989), Facts and Ideas in Microeconomic Theory, *European Economic Review*, 33, 251–276.

Hildenbrand, W. (1993), *Market Demand: Theory and Empirical Evidence*, Princeton University Press, Princeton.

Houthakker, H. S. (1950), Revealed Preference and the Utility Function, *Economica*, 17, 159–174.

Houthhakker, H. S. (1963), Some Problems in the International Comparison of Consumption Patterns, in *Les Besoins de Biens de Consommation* (R. Mosse, ed.), Centre National de la Recherche Scientifique, Grenoble.

Howe, H., R. A. Pollak, and T. J. Wales (1979), Theory and Time Series Estimation of the Quadratic Expenditure System, *Econometrica*, 47, 1231–1247.

Huang, C. (1987), An Intertemporal General Equilibrium Asset Pricing Model: The Case of Diffusion Information, *Econometrica*, 55, 117–142.

Jerison, M. (1984a), Aggregation and Pairwise Aggregation of Demand When the Distribution of Income Is Fixed, *Journal of Economic Theory*, 33, 1–31.

Jerison, M. (1984b), Social Welfare and the Unrepresentative Representative Consumer, manuscript, SUNY Albany.

Jorgenson, D. W. (1986), Econometric Methods for Modeling Producer Behavior, in *Handbook of Mathematical Economics*, Vol. 3 (K. J. Arrow and M. D. Intriligator, eds.), North-Holland, Amsterdam, 1841–1915.

Jorgenson, D. W., L. J. Lau, and T. M. Stoker (1980), Welfare Comparison under Exact Aggregation, *American Economic Review*, 70, 268–272.

Jorgenson, D. W., L. J. Lau, and T. M. Stoker (1981), Aggregate Consumer Behavior and Individual Welfare, in *Macroeconomic Analysis* (D. Currie, R. Nobay, and D. Peel, eds.), Croom-Helm, London, 35–61.

Jorgenson, D. W., L. J. Lau, and T. M. Stoker (1982), The Transcendental Logarithmic Model of Aggregate Consumer Behavior, in *Advances in Econometrics*, Vol. 1 (R. Baseman and G. Rhodes, eds.), JAI Press, Greenwich, 97–238.

Jorgenson, D. W. and D. T. Slesnick (1984), Aggregate Consumer Behaviour and the Measurement of Inequality, *Review of Economic Studies*, 51, 369–392.

Kirman, A. (1992), Whom or What Does the Representative Individual Represent? *Journal of Economic Perspectives*, 6, 117–136.

Kirman, A. and K. Koch (1986), Market Excess Demand Functions in Exchange Economies: Identical Preferences and Collinear Endowments, *Review of Economic Studies*, 53, 457–463.

Klein, L. (1946a), Macroeconomics and the Theory of Rational Behaviour, *Econometrica*, 14, 93–108.

Klein, L. R. (1946b), Remarks on the Theory of Aggregation, *Econometrica*, 14, 303–312.

Klein, L. R. and H. Rubin (1947–1948), A Constant Utility Index of the Cost of Living, *Review of Economic Studies*, 15, 84–87.

Koo, Y. C. (1963), An Empirical Test of Revealed Preference Theory, *Econometrica*, 31, 646–664.

Koopmans, T. C. (1957), *Three Essays on the State of Economic Science*, McGraw-Hill, New York.

Kydland, F. E. and E. C. Prescott (1982), Time to Build and Aggregate Fluctuations, *Econometrica*, 50, 1345–1370.

Lau, L. J. (1977a), Existence Conditions for Aggregate Demand Functions: The Case of a Single Index, IMSSS Technical Report 248, Stanford University.

Lau, L. J. (1977b), Existence Conditions for Aggregate Demand Functions: The Case of Multiple Indexes, IMSSS Technical Report 249, Stanford University.

Lau, L. J. (1982), A Note on the Fundamental Theorem of Exact Aggregation, *Economics Letters*, 9, 119–126.

Lau, L. J. (1986), Functional Forms in Econometric Model Building, in *Handbook of Econometrics*, Vol. III (Z. Griliches and M. D. Intriligator, eds.), North-Holland, Amsterdam.

Leontief, W. W. (1947a), Introduction to a Theory of the Internal Structure of Functional Relationships, *Econometrica*, 15, 361–373.

Leontief, W. W. (1947b), A Note on the Interrelation of Subsets of Independent Variables of a Continuous function with Continuous First Derivatives, *Bulletin of the American Mathematical Society*, 53, 343–350.

Lewbel, A. (1987), Characterizing Some Gorman Engel Curves, *Econometrica*, 55, 1451–1459.

Lewbel, A. (1989a), Nesting the AIDS and Translog Demand Systems, *International Economic Review*, 30, 349–356.

Lewbel, A. (1989b), A Demand System Rank Theorem, *Econometrica*, 57, 701–705.

Lewbel, A. (1990), Full Rank Demand Systems, *International Economic Review*, 31, 289–300.

Lewbel, A. (1991), The Rank of Demand Systems: Theory and Nonparametric Estimation, *Econometrica*, 59, 711–730.

Lewbel, A. (1994), An Examination of Werner Hildenbrand's Market Demand, *Journal of Economic Literature*, 32, 1832–1841.

Manser, M. E. and R. J. McDonald (1988), An Analysis of Substitution Bias in Measuring Inflation, 1959–85, *Econometrica*, 56, 909–930.

Mantel, R. (1974), On the Characterization of Aggregate Excess Demand, *Journal of Economic Theory*, 7, 348–353.

Mantel, R. (1979), Homothetic Preferences and Community Excess Demand Functions, *Journal of Economic Theory*, 12, 197–201.

Mas-Colell, A. and W. Neuefeind (1977), Some Generic Properties of Aggregate Excess Demand and an Application, *Econometric*, 45, 591–599.

May, K. (1946), The Aggregation Problem for a One-Industry Model, *Econometrica*, 14, 285–298.

Muellbauer, J. (1975), Aggregation, Income Distribution, and Consumer Demand, *Review of Economic Studies*, 62, 269–283.

Muellbauer, J. (1976), Community Preferences and the Representative Consumer, *Econometrica*, 44, 525–543.

Nataf, A. (1948), Sur la Possibilité de la Construction de Certains Macromodeles, *Econometrica*, 16, 232–244.

Nicol, C. J. (1994), Identifiability of Household Equivalence Scales through Exact Aggregation: Some Empirical Results, *Canadian Journal of Economics*, 27, 307–328.

Pollak, R. A. and T. J. Wales (1978), Estimation of Complete Demand Systems from Household Budget Data: The Linear and Quadratic Expenditure Systems, *American Economic Review*, 68, 348–359.

Pollak, R. A. and T. J. Wales (1979), Welfare Comparisons and Equivalent Scales, *American Economic Review*, 69, 216–221.

Pollak, R. A. and T. J. Wales (1980), Comparison of the Quadratic Expenditure System and Translog Demand Systems with Alternative Specifications of Demographic Effects, *Econometrica*, 48, 595–612.

Pu, S. (1946), A Note on Macroeconomics, *Econometrica*, 14, 299–302.

Russell, R. R. and M. O. Tengesdal (1996), On the Power of Non-Parametric Tests of the Representative-Consumer Hypothesis, Discussion Paper, University of California, Riverside.

Russell, T. (1983), On a Theorem of Gorman, *Economics Letters*, 11, 233-224.

Samuelson, P. A. (1938), A Note on the Pure Theory of Consumer's Behavior, *Economica*, 5, 61–71.

Samuelson, P. A. (1946–47), Consumption Theory in Terms of Revealed Preference. *Economica*, 15, 243–253.

Samuelson, P. A. (1956), Social Indifference Curves, *Quarterly Journal of Economics*, 70, 1–22.

Scitovsky, T. (1942), A Reconsideration of the Theory of Tariffs, *Review of Economic Studies*, 9, 89–110.

Shafer, W. and H. Sonnenschein (1982), Market Demand and Excess Demand Functions, in *Handbook of Mathematical Economics* (K. Arrow and M. D. Intrilligator, eds.), North-Holland, Amsterdam, 671–692.

Shapiro, P. (1977), Aggregation and the Existence of a Social Utility Function, *Journal of Economic Theory*, 16, 475–480.

Shapiro, P. and S. Braithwait (1979), Empirical Tests for the Existence of Group Utility Functions, *Review of Economic Studies*, 46, 653–665.

Shephard, R. W. (1970), *Theory of Cost and Production Functions*, Princeton University Press, Princeton.

Solow, R. (1964), Capital, Labor, and Income in Manufacturing, in *The Behavior of Income Shares*, Studies in Income and Wealth, National Bureau of Economic Research 27, Princeton University Press, Princeton, 101–128.

Sonnenschein, H. (1973), Do Walras' Identity and Continuity Characterize the Class of Community Excess Demand Functions? *Journal of Economic Theory*, 6, 345–354.

Sonnenschein, H. (1974), Market Excess Demand Functions, *Econometrica*, 40, 549–563.

Sono, M. (1961), The Effect of Price Changes on the Demand and Supply of Separable Goods, *International Economic Review*, 2, 239–271.

Stoker, T. M. (1984), Completeness, Distribution Restrictions, and the Form of Aggregate Functions, *Econometrica*, 52, 887–907.

Stoker, T. M. (1986a), The Distributional Welfare Effects of Rising Prices in the United States: The 1970's Experience, *American Economic Review*, 76, 335-349.

Stoker, T. M. (1986b), Simple Tests of Distributional Effects on Macroeconomic Equations, *Journal of Political Economy*, 94, 763–795.

Stoker, T. M. (1993), Empirical Approaches to the Problem of Aggregation over Individuals, *Journal of Economic Literature*, 31, 1827–1874.

Stone, R. (1954), Linear Expenditure Systems and Demand Analysis: An Application to the Pattern of British Demand, *The Economic Journal*, 64, 511–527.

Theil, H. (1954), *Linear Aggregation of Economic Relations*, North-Holland, Amsterdam.

Van Daal, J. and A. H. Q. M. Merkies (1984), *Aggregation in Economic Research*, D. Reidel, Dordrecht.

Van Daal, J. and A. H. Q. M. Merkies (1989), A Note on the Quadratic Expenditure Model, *Econometrica*, 57, 1439–1443.

Varian, H. R. (1982), The Nonparametric Approach to Demand Analysis, *Econometrica*, 50, 945–973.

Varian, H. R. (1992), *Microeconomic Analysis*, W. W. Norton, New York.

Vilks, A. (1988a), Approximate Aggregation of Excess Demand Functions, *Journal of Economic Theory*, 45, 417–424.

Vilks, A. (1988b), Consistent Aggregation of a General Equilibrium Model, in *Measurement in Economics* (Wolfgang Eichhorn, ed.), Physica-Verlag, Heidelberg, 691–703.

White, H. (1980), Using Least Squares to Approximate Unknown Regression Functions, *International Economic Review*, 21, 149–170.

7
Spatial Dependence in Linear Regression Models with an Introduction to Spatial Econometrics

Luc Anselin
West Virginia University, Morgantown, West Virginia

Anil K. Bera
University of Illinois, Champaign, Illinois

I. INTRODUCTION

Econometric theory and practice have been dominated by a focus on the time dimension. In stark contrast to the voluminous literature on serial dependence over time (e.g., the extensive review in King 1987), there is scant attention paid to its counterpart in cross-sectional data, spatial autocorrelation. For example, there is no reference to the concept nor to its relevance in estimation or specification testing in any of the commonly cited econometrics texts, such as Judge et al. (1982), Greene (1993), or Poirier (1995), or even in more advanced ones, such as Fomby et al. (1984), Amemiya (1985), Judge et al. (1995), and Davidson and MacKinnon (1993) (a rare exception is Johnston 1984). In contrast, spatial autocorrelation and spatial statistics in general are widely accepted as highly relevant in the analysis of cross-sectional data in the physical sciences, such as in statistical mechanics, ecology, forestry, geology, soil science, medical imaging, and epidemiology (for a recent review, see National Research Council 1991).

In spite of this lack of recognition in "mainstream" econometrics, applied workers saw the need to explicitly deal with problems caused by spatial autocorrelation in cross-sectional data used in the implementation of regional and multiregional econometric models. In the early 1970s, the Belgian economist Jean Paelinck coined the term "spatial econometrics" to designate a field of applied econometrics dealing with estimation and specification problems that arose from this. In their classic book *Spatial Econometrics*, Paelinck and Klaassen (1979) outlined five characteristics of the field: (1) the role of spatial interdependence in spatial models; (2) the asymmetry in spatial relations; (3) the importance of explanatory factors located in other spaces; (4) differentiation between ex post and ex ante interaction; and (5) explicit modeling of space (Paelinck and Klaassen 1979, pp. 5–11; see also Hordijk and Paelinck 1976, Paelinck 1982). In Anselin (1988a, p. 7), spatial econometrics is defined more broadly as "the collection of techniques that deal with the peculiarities caused by space in the statistical analysis of regional science models." The latter incorporate regions, location and spatial interaction explicitly and form the basis of most recent empirical work in urban and regional economics, real estate economics, transportation economics, and economic geography. The emphasis on the model as the starting point differentiates spatial econometrics from the broader field of spatial statistics, although they share a common methodological framework. Much of the contributions to spatial econometrics have appeared in specialized journals in regional science and analytical geography, such as the *Journal of Regional Science, Regional Science and Urban Economics, Papers in Regional Science, International Regional Science Review, Geographical Analysis*, and *Environment and Planning A*. Early reviews of the relevant methodological issues are given in Hordijk (1974, 1979), Bartels and Hordijk (1977), Arora and Brown (1977), Paelinck and Klaassen (1979), Bartels and Ketellapper (1979), Cliff and Ord (1981), Blommestein (1983), and Anselin (1980, 1988a, 1988b). More recent collections of papers dealing with spatial econometric issues are contained in Anselin (1992a), Anselin and Florax (1995a), and Anselin and Rey (1997).

Recently, an attention to the spatial econometric perspective has started to appear in mainstream empirical economics as well. This focus on spatial dependence has occurred in a range of fields in economics, not only in urban, real estate, and regional economics, where the importance of location and spatial interaction is fundamental, but also in public economics, agricultural and environmental economics, and industrial organization. Recent examples of empirical studies in mainstream economics that explicitly incorporated spatial dependence are, among others, the analysis of U.S. state expenditure patterns in Case et al. (1993), an examination of recreation expenditures by municipalities in the Los Angeles region in Murdoch et al. (1993), pricing in agricultural markets in LeSage (1993), potential spillovers from public infrastructure investments in Holtz-Eakin (1994), the determination of agricultural land values in Benirschka and Binkley (1994), the choice of retail sales contracts by integrated oil companies in Pinkse and Slade (1995), strategic interaction among local governments in Brueckner (1996), and models of nations' decisions to ratify environmental controls in Beron et al. (1996) and Murdoch et al. (1996). Substantively, this follows from a renewed focus on Marshallian externalities, spatial spillovers, copy-catting, and other forms of behavior where an economic actor mimics or reacts to the actions of other actors, for example in the new economic geography of Krugman (1991), in theories of endogenous growth (Romer 1986), and in analyses of local political economy (Besley and Case 1995). Second, a number of important policy issues have received an explicit spatial dimension, such as the designation of target areas or enterprise zones in development economics and the identification of underserved mortgage markets in urban areas. A more practical reason is the increased availability of large socioeconomic data sets with detailed spatial information, such as county-level economic information in the REIS CD-ROM (Regional Economic Information System) of the U.S. Department of Commerce, and tract-level data on mortgage transactions collected under the Housing Mortgage Disclosure Act (HMDA) of 1975.

From a methodological viewpoint, spatial dependence is not only important when it is part of the model, be it in a theoretical or policy framework, but it can also arise due to certain misspecifications. For instance, often the cross-sectional data used in model estimation and specification testing are imperfect, which may cause spatial dependence as a side effect. For example, census tracts are not housing markets and counties are not labor markets, but they are used as proxies to record transactions in these markets. Specifically, a mismatch between the spatial unit of observation and the spatial extent of the economic phenomena under consideration will result in spatial measurement errors and spatial autocorrelation between these errors in adjoining locations (Anselin 1988a).

In this chapter, we review the methodological issues related to the explicit treatment of spatial dependence in linear regression models. Specifically, we focus on the specification of the structure of spatial dependence (or spatial autocorrelation), on the estimation of models with spatial dependence and on specification tests to detect spatial dependence in regression models. Our review is organized accordingly into three main sections. We have limited the review to cross-sectional settings for linear regression models and do not consider dependence in space-time nor models for limited dependent variables. Whereas there is an established body of theory and methodology to deal with the standard regression case, this is not (yet) the case for techniques to analyze the other types of models. Both areas are currently the subject of active ongoing research (see, e.g., some of the papers in Anselin and Florax 1995a). Also, we have chosen to focus on a classical framework and do not consider Bayesian approaches to spatial econometrics (e.g., Hepple 1995a, 1995b, LeSage 1997).

In our review, we attempt to outline the extent to which general econometric principles can be applied to deal with spatial dependence. Spatial econometrics is often erroneously considered to consist of a straightforward extension of techniques to handle dependence in the time domain to two dimensions. In this chapter, we emphasize the limitations of such a perspective and stress the need to explicitly tackle the spatial aspects of model specification, estimation, and diagnostic testing.

II. THE PROBLEM OF SPATIAL AUTOCORRELATION

We begin this review with a closer look at the concept of spatial dependence, or its weaker expression, spatial autocorrelation, and how it differs from the more familiar serial correlation in the time domain. While, in a strict sense, spatial autocorrelation and spatial dependence clearly are not synonymous, we will use the terms interchangeably. In most applications, the weaker term autocorrelation (as a moment of the joint distribution) is used and only seldom has the focus been on the joint density as such (a recent exception is the semiparametric framework suggested in Brett and Pinkse 1997).

In econometrics, an attention to serial correlation has been the domain of time-series analysis and the typical focus of interest in the specification and estimation of models for cross-sectional data is heteroskedasticity. Until recently, spatial auto-correlation was largely ignored in this context, or treated in the form of groupwise equicorrelation, e.g., as the result of certain survey designs (King and Evans 1986). In other disciplines, primarily in physical sciences, such as geology (Isaaks and Srivastava 1989, Cressie 1991) and ecology (Legendre 1993), but also in geography (Griffith 1987, Haining 1990) and in social network analysis in sociology and psychology (Dow et al. 1982, Doreian et al. 1984, Leenders 1995), the dependence across "space" (in its most general sense) has been much more central. For example, Tobler's (1979) "first law of geography" states that "everything is related to everything else, but closer things more so," suggesting spatial dependence to be the rule rather than exception. A large body of spatial statistical techniques has been developed to deal with such dependencies (for a recent comprehensive review, see Cressie 1993; other classic references are Cliff and Ord 1973, 1981, Ripley 1981, 1988, Upton and Fingleton 1985, 1989). Useful in this respect is Cressie's (1993) taxonomy of spatial data strucures differentiating between point patterns, geostatistical data, and lattice data. In the physical sciences, the dominant underlying assumption tends to be that of a continuous spatial surface, necessitating the so-called geostatistical perspective rather than discrete observation points (or regions) in space, for which the so-called lattice perspective is relevant. The latter is more appropriate for economic data, since it is to some extent an extension of the ordering of observations on a one-dimensional time axis to an ordering in a two-dimensional space. It will be the almost exclusive focus of our review.

The traditional emphasis in econometrics on heterogeneity in cross-sectional data is not necessarily misplaced, since the distinction between spatial heterogeneity and spatial autocorrelation is not always obvious. More specifically, in a single cross section the two may be observationally equivalent. For example, when a spatial cluster of exceptionally large residuals is observed for a regression model, it cannot be ascertained without further structure whether this is an instance of heteroskedasticity (i.e., clustering of outliers) or spatial autocorrelation (a spatial stochastic process yielding clustered outliers). This problem is known in the literature as "true" contagion versus "apparent" contagion and is a major methodological issue in fields such as epidemiology (see, e.g., Johnson and Kotz 1969, Chapter 9, for a formal distinction between different forms of contagious distributions). The approach taken in spatial econometrics is to impose structure on the problem through the specification of a model, coupled with extensive specification testing for potential departures from the null model. This emphasis on the "model" distinguishes (albeit rather subtly) spatial econometrics from the broader field of spatial statistics (see also Anselin 1988a, p. 10, for further discussion of the distinction between the two). In our review, we deal almost exclusively with spatial autocorrelation. Once this aspect of the model is specified, the heterogeneity may be added in a standard manner (see Anselin 1988a, Chap. 9, and Anselin 1990a).

In this section, we first focus on a formal definition of spatial autocorrelation. This is followed by a consideration of how it may be operationalized in tests and econometric specifications by means of spatial weights and spatial lag operators. We close with a review of different ways in which spatial autocorrelation may be incorporated in the specification of econometric models in the form of spatial lag dependence, spatial error dependence, or higher-order spatial processes.

A. Defining Spatial Autocorrelation

Spatial autocorrelation can be loosely defined as the coincidence of value similarity with locational similarity. In other words, high or low values for a random variable tend to cluster in space (positive spatial autocorrelation), or locations tend to be surrounded by neighbors with very dissimilar values (negative spatial autocorrelation). Of the two types of spatial autocorrelation, positive autocorrelation is by far the more intuitive. Negative spatial autocorrelation implies a checkerboard pattern of values and does not always have a meaningful substantive interpretation (for a formal discussion, see Whittle 1954). The existence of positive spatial autocorrelation implies that a sample contains less information than an uncorrelated counterpart. In order to properly carry out statistical inference, this loss of information must be explicitly acknowledged in estimation and diagnostics tests. This is the essence of the problem of spatial autocorrelation in applied econometrics.

A crucial issue in the definition of spatial autocorrelation is the notion of "locational similarity," or the determination of those locations for which the values of the random variable are correlated. Such locations are referred to as "neighbors," though strictly speaking this does not necessarily mean that they need to be collocated (for a more formal definition of neighbors in terms of the conditional density function, see Anselin 1988a, pp. 16–17; Cressie 1993, p. 414).

More formally, the existence of spatial autocorrelation may be expressed by the following moment condition:

$$\text{Cov}(y_i, y_j) = E(y_i y_j) - E(y_i) \cdot E(y_j) \neq 0 \qquad \text{for } i \neq j \tag{1}$$

where y_i and y_j are observations on a random variable at locations i and j in space, and i, j can be points (e.g., locations of stores, metropolitan areas, measured as latitude and longitude) or areal units (e.g., states, counties or census tracts). Of course, there is nothing *spatial* per se to the nonzero covariance in (1). It only becomes spatial when the pairs of i, j locations for which the correlation is nonzero have a meaningful interpretation in terms of spatial structure, spatial interaction or spatial arrangement of observations.

For a set of N observations on cross-sectional data, it is impossible to estimate the potentially N by N covariance terms or correlations directly from the data. This is a fundamental problem in dealing with spatial autocorrelation and necessitates the imposition of structure. More specifically, in order for the problem to become tractable, it is necessary to impose sufficient constraints on the N by N spatial interaction (covariance) matrix such that a finite number of parameters characterizing the correlation can be efficiently estimated. Note how this contrasts with the situation where repeated observations are available, e.g., in panel data sets. In such instances, under the proper conditions, the elements of the covariance matrix may be estimated explicitly, in a vector autoregressive approach (for a review, see Lütkepohl 1991) or by means of so-called generalized estimating equations (Liang and Zeger 1986, Zeger and Liang 1986, Zeger et al. 1988, Albert and McShane 1995).

In contrast, when the N observations are considered as fixed effects in space, there is insufficient information in the data to estimate the N by N interactions. Increasing the sample size does not help, since the number of interactions increases with N^2, or, in other words, there is an incidental parameter problem. Alternatively, when the locations are conceptualized in a random-effects framework, sufficient constraints must be imposed to preclude that the range of interaction implied by a particular spatial stochastic process increases faster than the sample size as asymptotics are invoked to obtain the properties of estimators and test statistics.

Two main approaches exist in the literature to impose constraints on the interaction. In geostatistics, all *pairs* of locations are sorted according to the distance that separates them, and the strength of covariance (correlation) between them is expressed as a continuous function of this distance, in a so-called variogram or semivariogram (Cressie 1993, Chap. 2). As pointed out, the geostatistical perspective is seldom taken in empirical economics, since it necessitates an underlying process that is continuous over space. In such an approach, observations (points) are considered to form a sample from an underlying continuous spatial process, which is hard to maintain when the data consist of counties or census tracts. A possible exception may be the study of real estate data, where the locations of transactions may be conceptualized as points and analyzed using a geostatistical framework, as in Dubin (1988, 1992). Such an approach is termed "direct representation" in the literature, since the elements of the covariance (or correlation) matrix are modeled *directly* as functions of distances.

Our main focus in this review will be on the second approach, the so-called lattice perspective. For each data point, a relevant "neighborhood set" must be defined, consisting of those other locations that (potentially) interact with it. For each observation i, this yields a spatial ordering of locations $j \in S_i$ (where S_i is the neighborhood set), which can then be exploited to specify a spatial stochastic process. The covariance structure between observations is thus not modeled directly, but follows from the particular form of the stochastic process. We return to this issue below. First, we review the operational specification of the neighborhood set for each observation by means of a so-called spatial weights matrix.

B.   Spatial Weights

A spatial weights matrix is a N by N positive and symmetric matrix W which expresses for each observation (row) those locations (columns) that belong to its neighborhood set as nonzero elements. More formally, $w_{ij} = 1$ when i and j are neighbors, and $w_{ij} = 0$ otherwise. By convention, the diagonal elements of the weights matrix are set to zero. For ease of interpretation, the weights matrix is often standardized such that the elements of a row sum to one. The elements of a row-standardized weights matrix thus equal $w_{ij}^s = w_{ij}/\sum_j w_{ij}$. This ensures that all weights are between 0 and 1 and facilitates the interpretation of operations with the weights matrix as an averaging of neighboring values (see Section II.C). It also ensures that the spatial parameters in many spatial stochastic processes are comparable between models. This is not intuitively obvious, but relates to constraints imposed in a maximum likelihood estimation framework. For the latter to be valid, spatial autoregressive parameters must be constrained to lie in the interval $1/\omega_{\min}$ to $1/\omega_{\max}$, where $\omega_{\min}$ and $\omega_{\max}$ are respectively the smallest (on the real line) and largest eigenvalues of the matrix W (Anselin 1982). For a row-standardized weights matrix, the largest eigenvalue is always $+1$ (Ord 1975), which facilitates the interpretation of the autoregressive coefficient as a "correlation" (for an alternative view, see Kelejian and Robinson 1995). A side effect of row standardization is that the resulting matrix is likely to become asymmetric (since $\sum_j w_{ij} \neq \sum_i w_{ji}$), even though the original matrix may have been symmetric. In the calculation of several estimators and test statistics, this complicates computational matters considerably.

The specification of which elements are nonzero in the spatial weights matrix is a matter of considerable arbitrariness and a wide range of suggestions have been offered in the literature. The "traditional" approach relies on the geography or spatial arrangement of the observations, designating areal units as "neighbors" when they have a border in common (first-order contiguity) or are within a given distance of each other; i.e., $w_{ij} = 1$ for $d_{ij} \leq \delta$, where d_{ij} is the distance between units i and j, and δ is a distance cutoff value (distance-based contiguity). This geographic approach has been generalized to so-called Cliff-Ord weights that consist of a function of the relative length of the common border, adjusted by the inverse distance between two observations (Cliff and Ord 1973, 1981). Formally, Cliff-Ord weights may be expressed as:

$$w_{ij} = \frac{b_{ij}^{\beta}}{d_{ij}^{\alpha}} \qquad (2)$$

where b_{ij} is the share of the common border between units i and j in the perimeter of i (and hence b_{ij} does not necessarily equal b_{ji}), and α and β are parameters. More generally, the weights may be specified to express any measure of "potential interaction" between units i and j (Anselin 1988a, Chap. 3). For example, this may be related directly to spatial interaction theory and the notion of potential, with $w_{ij} = 1/d_{ij}^{\alpha}$ or $w_{ij} = e^{-\beta d_{ij}}$, or more complex distance metrics may be implemented (Anselin 1980, Murdoch et al. 1993). Typically, the parameters of the distance function are set a priori (e.g., $\alpha = 2$, to reflect a gravity function) and not estimated jointly with the other coefficients in the model. Clearly, when they are estimated jointly, the resulting specification will be highly nonlinear (Anselin 1980, Chap. 8, Ancot et al. 1986, Bolduc et al. 1989, 1992, 1995).

Other specifications of spatial weights are possible as well. In sociometrics, the weights reflect whether or not two individuals belong to the same social network (Doreian 1980). In economic applications, the use of weights based on "economic" distance has been suggested, among others, in Case et al. (1993). Specifically, they suggest to use weights (before row standardization) of the form $w_{ij} = 1/|x_i - x_j|$, where x_i and x_j are observations on "meaningful" socioeconomic characteristics, such as per capita income or percentage of the population in a given racial or ethnic group.

It is important to keep in mind that, irrespective of how the spatial weights are specified, the resulting spatial process must satisfy the necessary regularity conditions such that asymptotics may be invoked to obtain the properties of estimators and test statistics. For example, this requires constraints on the extent of the range of interaction and/or the degree of heterogeneity implied by the weights matrices (the so-called mixing conditions; Anselin 1988a, Chap. 5). Specifically, this means that weights must be nonnegative and remain finite, and that they correspond to a proper metric (Anselin 1980). Clearly, this may pose a problem with socioeconomic weights when $x_i = x_j$ for some observation pairs, which may be the case for poorly chosen economic determinants (e.g., when two states have the same percentage in a given racial group). Similarly, when multiple observations belong to the same areal unit (e.g., different banks located in the same county) the distance between them must be set to something other than zero (or $1/d_{ij} \to \infty$). Finally, in the standard estimation and testing approaches, the weights matrix is taken to be *exogenous*. Therefore, indicators for the socioeconomic weights should be chosen with great care to ensure their exogeneity, unless their endogeneity is considered explicitly in the model specification.

Operationally, the derivation of spatial weights from the location and spatial arrangement of observations must be carried out by means of a geographic information system, since for all but the smallest data sets a visual inspection of a map is impractical (for implementation details, see Anselin et al. 1993a, 1993b, Anselin 1995, Can 1996). A mechanical construction of spatial weights, particularly when based on a distance criterion, may easily result in observations to become "unconnected" or isolated islands. Consequently, the row in the weights matrix that corresponds to these observations will consist of zero values. While not inherently invalidating estimation or testing procedures, the unconnected observations imply a loss of degrees of freedom, since, for all practical purposes, they are eliminated from consideration in any "spatial" model. This must be explicitly accounted for.

C. Spatial Lag Operator

In time-series analysis, values for "neighboring" observations can be easily expressed by means of a backward- or forward-shift operator on the one-dimensional time axis, yielding lagged variables y_{t-k} or y_{t+k}, where k is the desired shift (or lag). By contrast, there is no equivalent and unambiguous *spatial* shift operator. Only on a regular grid structure is there a potential solution, although not as straightforward as in the time domain. Following the so-called rook criterion for contiguity, each grid cell or vertex on a regular lattice, (i, j), has four neighbors: $(i+1, j)$ (east), $(i-1, j)$ (west), $(i, j + 1)$ (north), and $(i, j - 1)$ (south). Corresponding to this are four spatially shifted variables: $y_{i+1,j}$, $y_{i-1,j}$, $y_{i,j+1}$, and $y_{i,j-1}$, each of which may require its own parameter in a spatial process model. However, the rook criterion is not the only way spatial neighbors may be defined on a regular lattice, nor does the number of neighbors necessarily equal 4. For example, following the queen criterion, each observation has eight neighbors, yielding eight spatially shifted variables; the four for the rook criterion, as well as $y_{i-1,j+1}$, $y_{i-1,j-1}$, $y_{i+1,j+1}$ and $y_{i+1,j-1}$, again each possibly with its own parameter. This notion of a spatial shift operator on a regular lattice has received only limited attention in the literature, mostly with a theoretical focus and primarily in statistical mechanics, in so-called Ising models (for details, see Cressie 1993, pp. 425–426).

On an irregular spatial structure, which characterizes most economic applications, this formal notion of spatial shift is impractical, since the number of shifts would differ by observation, thereby making any statistical analysis extremely unwieldy. Instead, the concept of a *spatial lag operator* is used, which consists of a weighted average of the values at neighboring locations. The weights are fixed and exogenous, similar to a distributed lag in time series. Formally, a spatial lag operator is obtained as the product of a spatial weights matrix W with the vector of observations on a random variable y, or Wy. Each element of the resulting *spatially lagged variable* equals $\sum_j w_{ij} y_j$, i.e., a weighted average of the y values in the neighbor set S_i, since $w_{ij} = 0$ for $j \notin S_i$. Row standardization of the spatial weights matrix ensures that a spatial lag operation yields a smoothing of the neighboring values, since the positive weights sum to one.

Higher-order spatial lag operators are defined in a recursive manner, by applying the spatial weights matrix to a lower-order lagged variable. For example, a second-order spatial lag is obtained as $W(Wy)$, or W^2y. However, in contrast to time series, where such an operation is unambiguous, higher-order spatial operators yield redundant and circular neighbor relations, which must be eliminated to ensure proper estimation and inference (Blommestein 1985, Blommestein and Koper 1992, Anselin and Smirnov 1996).

In spatial econometrics, spatial autocorrelation is modeled by means of a functional relationship between a variable. y, or error term, ε, and its associated spatial lag, respectively Wy for a spatially lagged dependent variable and $W\varepsilon$ for a spatially lagged error term. The resulting specifications are referred to as *spatial lag* and *spatial error* models, the general properties of which we consider next.

D. Spatial Lag Dependence

Spatial lag dependence in a regression model is similar to the inclusion of a serially autoregressive term for the dependent variable (y_{t-1}) in a time-series context. In spatial econometrics, this is referred to as a mixed regressive, spatial autoregressive model (Anselin 1988a, p. 35). Formally,

$$y = \rho Wy + X\beta + \varepsilon \tag{3}$$

where y is a N by 1 vector of observations on the dependent variable, Wy is the corresponding spatially lagged dependent variable for weights matrix W, X is a N by K matrix of observations on the explanatory (exogenous) variables, ε is a N by 1 vector of error terms, ρ is the spatial autoregressive parameter, and β is a K by 1 vector of regression coefficients. The presence of the spatial lag term Wy on the right side of (3) will induce a nonzero correlation with the error term, similar to the presence of an endogenous variable, but different from a serially lagged dependent variable in the time-series case. In the latter model, y_{t-1} is uncorrelated with ε_t, in the absence of serial correlation in the errors. In contrast, $(Wy)_i$ is always correlated with ε_i, irrespective of the correlation structure of the errors. Moreover, the spatial lag for a given observation i is not only correlated with the error term at i, but also with the error terms at all other locations. Therefore, unlike what holds in the time-series case, an ordinary least-squares estimator will not be consistent for this specification (Anselin 1988a, Chap. 6). This can be seen from a slight reformulation of the model:

$$y = (\mathbf{I} - \rho W)^{-1}X\beta + (\mathbf{I} - \rho W)^{-1}\varepsilon \tag{4}$$

The matrix inverse $(\mathbf{I} - \rho W)^{-1}$ is a full matrix, and not triangular as in the time-series case (where dependence is only one-directional), yielding an infinite series that involves error terms at all locations, $(\mathbf{I}+\rho W+\rho^2 W^2+\rho^3 W^3+\cdots)\varepsilon$. It therefore readily follows that $(Wy)_i$ contains the element ε_i, as well as other ε_j, $j \neq i$. Thus,

$$E[(Wy)_i\varepsilon_i] = E[\{W(\mathbf{I}-\rho W)^{-1}\varepsilon\}_i\varepsilon_i] \neq 0 \tag{5}$$

The spatial dynamics embedded in the structure of the spatial process model (3) determine the form of the covariance between the observations at different locations (i.e., the spatial autocorrelation). For the mixed regressive, spatial autoregressive model this can easily be seen to equal $(\mathbf{I} - \rho W)^{-1}\Omega(\mathbf{I} - \rho W')^{-1}$, where Ω is the variance matrix for the error term ε (note that for a row-standardized spatial weights matrix, $W \neq W'$). Without loss of generality, the latter can be assumed to be diagonal and homoskedastic, or, $\Omega = \sigma^2\mathbf{I}$, and hence, $\mathrm{Var}[y] = \sigma^2(\mathbf{I} - \rho W)^{-1}(\mathbf{I} - \rho W')^{-1}$ The resulting variance matrix is full, implying that each location is correlated with every other location, but in a fashion that decays with the order of contiguity (the powers of W in the series expansion of $(\mathbf{I} - \rho W)^{-1}$).

The implication of this particular variance structure is that the simultaneity embedded in the Wy term must be explicitly accounted for, either in a maximum likelihood estimation framework, or by using a proper set of instrumental variables. We turn to this issue in Section III. When a spatially lagged dependent variable is ignored in a model specification, but present in the underlying data generating process, the resulting specification error is of the omitted variable type. This implies that OLS estimates in the nonspatial model (i.e., the "standard" approach) will be biased and inconsistent.

The interpretation of a significant spatial autoregressive coefficient ρ is not always straightforward. Two situations can be distinguished. In one, the significant spatial lag term indicates true contagion or substantive spatial dependence, i.e., it measures the extent of spatial spillovers, copy-catting or diffusion. This interpretation is valid when the actors under consideration match the spatial unit of observation and the spillover is the result of a theoretical model. For example, this holds for the models of farmers' innovation adoption in Case (1992), state expenditures and tax setting behavior in Case et al. (1993) and Besley and Case (1995), strategic interaction among California cities in the choice of growth controls in Brueckner (1996), and in the median voter model for recreation expenditures of Murdoch et al. (1993). Alternatively, the spatial lag model may be used to deal with spatial autocorrelation that results from a mismatch between the spatial scale of the phenomenon under study and the spatial scale at which it is measured. Clearly, when data are based on administratively determined units such as census tracts or blocks, there is no good reason to expect economic behavior to conform to these units. For example, this interpretation is useful for the spatial autoregressive models of urban housing and mortgage markets in Can (1992), Can and Megbolugbe (1997), and Anselin and Can (1996). Since urban housing and mortgage markets operate at a different spatial scale than census tracts, positive spatial autocorrelation may be expected and will in fact result in the sample containing less information than a truly "independent" sample of observations. The inclusion of a spatially lagged dependent variable in the model specification is a way to correct for this loss of information. In other words, it allows for the proper interpretation of the significance of the exogenous variables in the model (the X), after the spatial effects have been corrected for, or filtered out (see also Getis 1995 for a discussion of alternative approaches to spatial filtering). More formally, the spatial lag model may be reexpressed as

$$(\mathbf{I} - \rho W)y = X\beta + \varepsilon \tag{6}$$

where $(\mathbf{I} - \rho W)y$ is a spatially filtered dependent variable, i.e., with the effect of spatial autocorrelation taken out. This is roughly similar to first differencing of the dependent variable in time series, except that a value of $\rho = 1$ is not in the allowable parameter space for (3) and thus ρ must be estimated explicitly (Section III).

E. Spatial Error Dependence

A second way to incorporate spatial autocorrelation in a regression model is to specify a spatial process for the disturbance terms. The resulting error covariance will be nonspherical, and thus OLS estimates, while still unbiased, will be inefficient. More efficient estimators are obtained by taking advantage of the particular structure of the error covariance implied by the spatial process. Different spatial processes lead to different error covariances, with varying implications about the range and extent of spatial interaction in the model. The most common specification is a spatial autoregressive process in the error terms:

$$y = X\beta + \varepsilon \tag{7}$$

i.e., a linear regression with error vector ε, and

$$\varepsilon = \lambda W\varepsilon + \xi \tag{8}$$

where λ is the spatial autoregressive coefficient for the error lag $W\varepsilon$ (to distinguish the notation from the spatial autoregressive coefficient ρ in a spatial lag model), and ξ is an uncorrelated and (without loss of generality) homoskedastic error term. Alternatively, this may be expressed as

$$y = X\beta + (\mathbf{I} - \lambda W)^{-1}\xi \tag{9}$$

From this follows the error covariance as

$$E[\varepsilon\varepsilon'] = \sigma^2(\mathbf{I} - \lambda W)^{-1}(\mathbf{I} - \lambda W')^{-1} = \sigma^2[(\mathbf{I} - \lambda W)'(\mathbf{I} - \lambda W)]^{-1} \tag{10}$$

a structure identical to that for the dependent variable in the spatial lag model. Therefore, a spatial autoregressive error process leads to a nonzero error covariance between every pair of observations, but decreasing in magnitude with the order of contiguity. Moreover, the complex structure in the inverse matrices in (10) yields nonconstant diagonal elements in the error covariance matrix, thus inducing heteroskedasticity in ε, irrespective of the heteroskedasticity of ξ (an illuminating numerical illustration of this feature is given in McMillen 1992). We have a much simpler situation for the case of autocorrelation in the time-series context where the model is written as $\varepsilon_t = \lambda\varepsilon_{t-1} + \xi_t$. Therefore, this is a special case of (8) with

$$
W = W^T = \begin{bmatrix}
0 & 0 & 0 & \cdot & \cdot & \cdot & 0 & 0 \\
1 & 0 & 0 & \cdot & \cdot & \cdot & 0 & 0 \\
0 & 1 & 0 & \cdot & \cdot & \cdot & 0 & 0 \\
\cdot & & & \cdot & \cdot & \cdot & & \cdot \\
0 & 0 & 0 & \cdot & \cdot & \cdot & 1 & 0
\end{bmatrix}
$$

where each observation is connected to only its immediate past value. As we know, for this case, $\mathrm{Var}(\varepsilon_t) = \sigma^2/(1-\lambda^2)$ for all t. That is, autocorrelation does not induce heteroskedasticity. In a time-series model, heteroskedasticity can come only through ξ_t given the above AR(1) model.

A second complicating factor in specification testing is the great degree of similarity between a spatial lag and a spatial error model, as suggested by the error covariance structure. In fact, after premultiplying both sides of (9) by $(\mathbf{I} - \lambda W)$ and moving the spatial lag term to the right side, a *spatial Durbin model* results (Anselin 1980):

$$y = \lambda W y + X\beta - \lambda W X\beta + \xi \tag{11}$$

This model has a spatial lag structure (but with the spatial autoregressive parameter λ from (8)) with a well-behaved error term ξ. However, the equivalence between (7)–(8) and (11) imposes a set of nonlinear common factor constraints on the coefficients. Indeed, for (11) to be a proper spatial error model, the coefficients of the lagged exogenous variables WX must equal minus the product of the spatial autoregressive coefficient λ and the coefficients of X, for a total of K constraints (for technical details, see Anselin 1988a, pp. 226–229).

Spatial error dependence may be interpreted as a nuisance (and the parameter λ as a nuisance parameter) in the sense that it reflects spatial autocorrelation in measurement errors or in variables that are otherwise not crucial to the model (i.e., the "ignored" variables spillover across the spatial units of observation). It primarily causes a problem of inefficiency in the regression estimates, which may be remedied by increasing the sample size or by exploiting consistent estimates of the nuisance parameter. For example, this is the interpretation offered in the model of agricultural land values in Benirschka and Binkley (1994).

The spatial autoregressive error model can also be expressed in terms of spatially filtered variables, but slightly different from (6). After moving the spatial lag variable in (11) to the left hand side, the following expression results:

$$(\mathbf{I} - \lambda W)y = (\mathbf{I} - \lambda W)X\beta + \xi \tag{12}$$

This is a regression model with spatially filtered dependent and explanatory variables and with an uncorrelated error term ξ, similar to first differencing of both y and X in time-series models. As in the spatial lag model, $\lambda = 1$ is outside the parameter space and thus λ must be estimated jointly with the other coefficients of the model (see Section III).

Several alternatives to the spatial autoregressive error process (8) have been suggested in the literature, though none of them have been implemented much in practice. A spatial moving average error process is specified as (Cliff and Ord 1981, Haining 1988, 1990):

$$\varepsilon = \gamma W \xi + \xi \tag{13}$$

where γ is the spatial moving average coefficient and ξ is an uncorrelated error term. This process thus specifies the error term at each location to consist of a location-specific part, ξ_i ("innovation"), as well as a weighted average (smoothing) of the errors at neighboring locations, $W\xi$. The resulting error covariance matrix is

$$E[\varepsilon\varepsilon'] = \sigma^2(\mathbf{I} + \gamma W)(\mathbf{I} + \gamma W') = \sigma^2[\mathbf{I} + \gamma(W + W') + \gamma^2 WW'] \tag{14}$$

Note that in contrast to (10), the structure in (14) does not yield a full covariance matrix. Nonzero covariances are only found for first-order $(W + W')$ and second-order (WW') neighbors, thus implying much less overall interaction than the autoregressive process. Again, unless all observations have the same number of neighbors and identical weights, the diagonal elements of (14) will not be constant, inducing heteroskedasticity in ε, irrespective of the nature of ξ.

A very similar structure to (13) is the spatial error components model of Kelejian and Robinson (1993, 1995), in which the disturbance is a sum of two independent error terms, one associated with the "region" (a smoothing of neighboring errors) and one which is location-specific:

$$\varepsilon = W\xi + \psi \tag{15}$$

with ξ and ψ as independent error components. The resulting error covariance is

$$E[\varepsilon\varepsilon'] = \sigma_\psi^2 \mathbf{I} + \sigma_\xi^2 WW' \tag{16}$$

where σ_ψ^2 and σ_ξ^2 are the variance components associated with respectively the location-specific and regional error parts. The spatial interaction implied by (16) is even more limited than for (14), pertaining only to the first- and second-order neighbors contained in the nonzero elements of WW'. Heteroskedasticity is implied unless all locations have the same number of neighbors and identical weights, a situation excluded by the assumptions needed for the proper asymptotics in the model (Kelejian and Robinson 1993, p. 301).

In sum, every type of spatially dependent error process induces heteroskedasticity as well as spatially autocorrelated errors, which will greatly complicate specification testing in practice. Note that the "direct representation" approach based on geostatistical principles does not suffer from this problem. For example, in Dubin (1988, 1992), the elements of the error covariance matrix are expressed directly as functions of the distance d_{ij} between the corresponding observations, e.g., $E[\varepsilon_i \varepsilon_j] = \gamma_1 e^{(-d_{ij}/\gamma_2)}$, with γ_1 and γ_2 as parameters. Since $e^{-d_{ii}/\gamma_2} = 1$, irrespective of the value of γ_2, the errors ε will be homoskedastic unless explicitly modeled otherwise.

F. Higher-Order Spatial Processes

Several authors have suggested processes that combine spatial lag with spatial error dependence, though such specifications have seen only limited applications. The most general form is the spatial autoregressive, moving-average (SARMA) process outlined by Huang (1984). Formally, a SARMA(p, q) process can be expressed as

$$y = \rho_1 W_1 y + \rho_2 W_2 y + \cdots + \rho_p W_p y + \varepsilon \tag{17}$$

for the spatial autoregressive part, and

$$\varepsilon = \gamma_1 W_1 \xi + \gamma_2 W_2 \xi + \cdots + \gamma_q W_q \xi + \xi \tag{18}$$

for the moving-average part, in the same notation as above. For greater generality, a regressive component $X\beta$ can be added to (17) as well. The spatial autocorrelation pattern resulting from this general formulation is highly complex. Models that implement aspects of this form are the second-order SAR specification in Brandsma and Ketellapper (1979a) and higher-order SAR models in Blommestein (1983, 1985).

A slightly different specification combines a first-order spatial autoregressive lag with a first-order spatial autoregressive error (Anselin 1980, Chap. 6; Anselin 1988a, pp. 60–65). It has been applied in a number of empirical studies, most notably in the work of Case, such as the analysis of household demand (Case 1987, 1991), of innovation diffusion (Case 1992), and local public finance (Case et al. 1993, Besley and Case 1995). Formally, the model can be expressed as a combination of (3) with (8), although care must be taken to differentiate the weights matrix used in the spatial lag process from that in the spatial error process:

$$y = \rho W_1 y + X\beta + \varepsilon \tag{19}$$

$$\varepsilon = \lambda W_2 \varepsilon + \xi \tag{20}$$

After some algebra, combining (20) and (19) yields the following reduced form:

$$y = \rho W_1 y + \lambda W_2 y - \lambda \rho W_2 W_1 y + X\beta - \lambda W_2 X\eta + \xi \tag{21}$$

i.e., an extended form of the spatial Durbin specification but with an additional set of nonlinear constraints on the parameters. Note that when W_1 and W_2 do not overlap, for example when they pertain to different orders of contiguity, the product $W_2 W_1 = 0$

and (21) reduces to a biparametric spatial lag formulation, albeit with additional constraints on the parameters. On the other hand, when W_1 and W_2 are the same, the parameters ρ and λ are only identified when at least one exogenous variable is included in X (in addition to the constant term) and when the nonlinear constraints are enforced (Anselin 1980, p. 176). When $W_1 = W_2 = W$, the model becomes

$$y = (\rho + \lambda)Wy - \lambda\rho W^2 y + X\beta - \lambda WX\beta + \xi \qquad (22)$$

Clearly, the coefficients of Wy and $W^2 y$ alone do not allow for a separate identification of ρ and λ. Using the nonlinear constraints between the β and $-\lambda\beta$ (the coefficients of X and WX) yields an estimate of λ, but this will only be unique when the constraints are strictly enforced. Similarly, an estimate of λ may result in two possible estimates for ρ (one using the coefficient of Wy, the other of $W^2 y$) unless the nonlinear constraints are strictly enforced. This considerably complicates estimation strategies for this model. In contrast, a SARMA(1, 1) model does not suffer from this problem.

In empirical practice, an alternative perspective on the need for higher-order processes is to consider them to be a result of a poorly specified weights matrix rather than as a realistic data generating process. For example, if the weights matrix in a spatial lag model underbounds the true spatial interaction in the data, there will be remaining spatial error autocorrelation. This may lead one to implement a higher-order process, while for a properly specified weights matrix no such process is needed (see Florax and Rey 1995 for a discussion of the effects of misspecified weights). In practice, this will require a careful specification search for the proper form of the spatial dependence in the model, an issue to which we return in Section IV. First, we consider the estimation of regression models that incorporate spatial autocorrelation of a spatial lag or error form.

III.   ESTIMATING SPATIAL PROCESS MODELS

Similar to when serial dependence is present in the time domain, classical sampling theory no longer holds for spatially autocorrelated data, and estimation and inference must rely on the asymptotic properties of stochastic processes. In essence, rather than considering N observations as independent pieces of information, they are conceptualized as a single realization of a process. In order to carry out meaningful inference on the parameters of such a process, constraints must be imposed on both heterogeneity and the range of interaction. While many properties of estimators for spatial process models may be based on the same principles as developed for dependent (and heterogeneous) processes in the time domain (e.g., the formal properties outlined in White 1984, 1994), there are some important differences as well. Before covering specific estimation procedures, we discuss these differences in some detail, focusing in particular on the notion of stationarity in space and the distinction between simultaneous and conditional spatial processes. Next, we turn to a review of maximum likelihood and instrumental variables estimators for spatial regression models. We close with a brief discussion of operational implementation and software issues.

A. Spatial Stochastic Processes

As in the time domain, in order to carry out meaningful inference for a spatial process, some degree of equilibrium must be assumed in the sense that the stochastic generating mechanism is taken to work uniformly over space. In a strict sense, a notion of "spatial stationarity" accomplishes this objective since it imposes the condition that any joint distribution of the random variable under consideration over a subset of the locations depends only on the relative position of these observations in terms of their relative orientation (angle) and distance. Even stricter is a notion of isotropy, for which only distance matters and orientation is irrelevant. For practical purposes, the notions of stationarity and isotropy are too demanding and not verifiable. Hence, weaker conditions are typically imposed in the form of stationarity of the first (mean) and second moments (variance, covariance, or spatial autocorrelation). Even weaker requirements follow from the so-called intrinsic hypothesis in geostatistics, which requires only stationarity of the variance of the increments, leading to the notion of a variogram (for technical details, see Ripley 1988, pp. 6–7; Cressie 1993, pp. 52–68).

For stationary processes in the time domain, the careful inspection of autocovariance and autocorrelation functions is a powerful aid in the identification of the model, e.g., following the familiar Box-Jenkins approach (Box et al. 1994). One could transpose this notion to spatial processes and consider spatial autocorrelation functions indexed by order of contiguity as the basis for model identification. However, as Hooper and Hewings (1981) have shown, this is only appropriate for a very restrictive class of spatial processes on regular lattice structures. For applied work in empirical economics, such restrictions are impractical and the spatial dependence in the model must be specified explicitly by means of the spatial lag and spatial error structures reviewed in the previous section. Inference may be based on the asymptotic properties (central limit theorems and laws of large numbers) of so-called dependent and heterogeneous processes, as developed in White and Domowitz (1984) and White (1984, 1994). Central to these notions is the concept of mixing sequences, allowing for a trade-off between the range of dependence and the extent of heterogeneity (see Anselin 1988a, pp. 45–46 for an intuitive extension of this to spatial econometric models). While rigorous proofs of these properties have not been derived for the explicit spatial case, the notion of a spatial weights matrix based on a proper metric is general enough to meet the criteria imposed by mixing conditions. In a spatial econometric approach then, a spatial lag model is considered to be a special case of simultaneity or endogeneity with dependence, and a spatial error model is a special case of a nonspherical error term, both of which can be tackled by means of generally established econometric theory, though not as direct extensions of the time-series analog.

The emphasis on "simultaneity" in spatial econometrics differs somewhat from the approach taken in spatial statistics, where conditional models are often considered to be more natural (Cressie 1993, p. 410). Again, the spatial case differs substantially from the time-series one since in space a conditional and simultaneous approach are no longer equivalent (Brook 1964, Besag 1974, Cressie 1993, pp. 402–410). More specifically, in the time domain a Markov chain stochastic process can be expressed in terms of the joint density (ignoring a starting point to ease notation) as

$$\text{Prob}[z] = \prod_{t=1}^{N} Q_t[z_t, z_{t-1}] \tag{23}$$

where z refers to the vector of observations for all time points, and Q_t is a function that only contains the observation at t and at $t - 1$ (hence, a Markov chain). The conditional density for this process is

$$\text{Prob}[z_t|z_1, z_2, \ldots, z_{t-1}] = \text{Prob}[z_t|z_{t-1}] \tag{24}$$

illustrating the lack of memory of the process (i.e., the conditional density depends only on the first-order lag). Due to the one-directional nature of dependence in time, (23) and (24) are equivalent (Cressie 1993, p. 403). An extension of (23) to the spatial domain may be formulated as

$$\text{Prob}[z] = \prod_{i=1}^{N} Q_i[z_j, z_j; j \in S_i] \tag{25}$$

where the z_j only refer to those locations that are part of the neighborhood set S_i of i. A conditional specification would be

$$\text{Prob}[z_i|z_j, j \neq i] = \text{Prob}[z_i|z_j; j \in S_i] \tag{26}$$

i.e., the conditional density of z_i, given observations at all other locations only depends on those locations in the neighborhood set of i. The fundamental result in this respect goes back to Besag (1974), who showed that the conditional specification only yields a proper joint distribution when the so-called Hammersley-Clifford theorem is satisfied, which imposes constraints on the type and range of dependencies in (26). Also, while a joint density specification always yields a proper conditional specification, in range of spatial interaction implied is not necessarily the same. For example, Cressie (1993, p. 409) illustrates how a first-order symmetric spatial autoregressive process corresponds with a conditional specification that includes third-order neighbors (Haining 1990, pp. 89–90). Consequently, it does make a difference whether one approaches a spatially autocorrelated phenomenon by means of

(26) versus (25). This also has implications for the substantive interpretation of the model results, as illustrated for an analysis of retail pricing of gasoline in Haining (1984).

In practice, it is often easier to estimate a conditional model, especially for nonnormal distributions (e.g., auto-Poisson, autologistic). Also, a conditional specification is more appropriate when the focus is on spatial prediction or interpolation. For general estimation and inference, however, the constraints imposed on the type and range of spatial interaction in order for the conditional density to be proper are often highly impractical in empirical work. For example, an auto-Poisson model (conditional model for spatially autocorrelated counts) only allows negative autocorrelation and hence is inappropriate for any analysis of clustering in space.

In the remainder, our focus will be exclusively on simultaneously specified models, which is a more natural approach from a spatial econometric perspective (Anselin 1988a, Cressie 1993, p. 410).

B. Maximum Likelihood Estimation

The first comprehensive treatment of maximum likelihood estimation of regression models that incorporate spatial autocorrelation in the form of a spatial lag or a spatial error term was given by Ord (1975). The point of departure is a joint normal density for the errors in the model, from which the likelihood function is derived. An important aspect of this likelihood function is the Jacobian of the transformation, which takes the form $|\mathbf{I} - \rho W|$ and $|\mathbf{I} - \lambda W|$ in respectively the spatial lag and spatial autoregressive error models, with ρ and λ as the autoregressive coefficient and W as the spatial weights matrix. The need for this Jacobian can be seen from expression (4) for the spatial lag model and (12) for the spatial autoregressive error model (for a more extensive treatment, see Anselin 1988a, Chap. 6). In contrast to the time-series case, the spatial Jacobian is not the determinant of a triangular matrix, but of a full matrix. This would complicate computational matters considerably, were it not that Ord (1975) showed how it can be expressed in function of the eigenvalues ω_i of the spatial weights matrix as

$$|\mathbf{I} - \rho W| = \prod_{i=1}^{N}(1 - \rho\omega_i) \tag{27}$$

Using this simplification, under the normality assumption, the log-likelihood function for the spatial lag model (3) follows in a straightforward manner as

$$L = \sum_i \ln(1 - \rho\omega_i) - \frac{N}{2}\ln(2\pi) - \frac{N}{2}\ln(\sigma^2) \\ - \frac{(y - \rho Wy - X\beta)'(y - \rho Wy - X\beta)}{2\sigma^2} \tag{28}$$

in the same notation as used in Section II. This expression clearly illustrates why, in contrast to the time-series case, ordinary least squares (i.e., the minimization of the last term in (28)) is not maximum likelihood, since it ignores the Jacobian term. From the usual first-order conditions, the ML estimates for β and σ^2 in a spatial lag model are obtained as (for details, see Ord 1975, Anselin 1980, Chap. 4: Anselin 1988a, Chap. 6):

$$\beta_{ML} = (X'X)^{-1}X'(\mathbf{I} - \rho W)y \tag{29}$$

and

$$\sigma^2_{ML} = \frac{(y - \rho Wy - X\beta_{ML})'(y - \rho Wy - X\beta_{ML})}{N} \tag{30}$$

Conditional upon ρ, these estimates are simply OLS applied to the spatially filtered dependent variable and the explanatory variables in (6). Substitution of (29) and (30) in the log-likelihood function yields a concentrated log-likelihood as a nonlinear function of a single parameter ρ:

$$L_c = -\frac{N}{2}\ln\left[\frac{(e_0 - \rho e_L)'(e_0 - \rho e_L)}{N}\right] + \sum_i \ln(1 - \rho\omega_i) \tag{31}$$

where e_0 and e_L are residuals in a regression of y on X and Wy on X, respectively (for technical details, see Anselin 1980, Chap. 4). A maximum likelihood estimate for ρ is obtained from a numerical optimization of the concentrated log-likelihood function (31). Based on the framework outlined in Heijmans and Magnus (1986a, 1986b), it can be shown that the resulting estimates have the usual asymptotic properties, including consistency, normality, and asymptotic efficiency. The asymptotic variance matrix follows as the inverse of the information matrix

AsyVar$[\rho, \beta, \sigma^2]$

$$= \begin{bmatrix} \text{tr}[W_A]^2 + \text{tr}[W_A'W_A] + \dfrac{[W_A X\beta]'[W_A X\beta]}{\sigma^2} & \dfrac{(X'W_A X\beta)'}{\sigma^2} & \dfrac{\text{tr}(W_A)}{\sigma^2} \\ \dfrac{X'W_A X\beta}{\sigma^2} & \dfrac{X'X}{\sigma^2} & 0 \\ \dfrac{\text{tr}(W_A)}{\sigma^2} & 0 & \dfrac{N}{2\sigma^4} \end{bmatrix}^{-1}$$

$$\tag{32}$$

where $W_A = W(\mathbf{I} - \rho W)^{-1}$ to simplify notation. Note that while the covariance between β and the error variance is zero, as in the standard regression model, this is not the case for ρ and the error variance. This lack of block diagonality in the information matrix for the spatial lag model will lead to some interesting results on the structure of specification tests, to which we turn in Section IV. It is yet another distinguishing characteristic between the spatial case and its analog in time series.

Maximum likelihood estimation of the models with spatial error autocorrelation that were covered in Section II.E can be approached by considering them as special cases of general parametrized nonspherical error terms, for which $E[\varepsilon\varepsilon'] = \sigma^2\Omega(\theta)$, with θ as a vector of parameters. For example, from (32) for a spatial autoregressive error term, it follows that

$$\Omega(\lambda) = [(\mathbf{I} - \lambda W)'(\mathbf{I} - \lambda W)]^{-1} \tag{33}$$

As shown in Anselin (1980, Chap. 5), maximum likelihood estimation of such specifications can be carried out as an application of the general framework outlined in Magnus (1978). Most spatial processes satisfy the necessary regularity conditions, although this is not necessarily the case for direct representation models (Mardia and Marshall 1984, Warnes and Ripley 1987, Mardia and Watkins 1989). Under the assumption of normality, the log-likelihood function takes on the usual form:

$$L = -\frac{1}{2}\ln|\Omega(\lambda)| - \frac{N}{2}\ln(2\pi) - \frac{N}{2}\ln(\sigma^2)$$
$$- \frac{(y - X\beta)'\Omega(\lambda)^{-1}(y - X\beta)}{2\sigma^2} \tag{34}$$

for example, with $\Omega(\lambda)$ as in (33). First-order conditions yield the familiar generalized least-squares estimates for β, conditional upon λ:

$$\beta_{\text{ML}} = [X'\Omega(\lambda)^{-1}X]^{-1}X'\Omega(\lambda)^{-1}y \tag{35}$$

For a spatial autoregressive error process, $\Omega(\lambda)^{-1} = (\mathbf{I} - \lambda W)'(\mathbf{I} - \lambda W)$, so that for known λ, the maximum likelihood estimates are equivalent to OLS applied to the spatially filtered variables in (12). Note that for other forms of error dependence, the GLS expression (35) will involve the inverse of an N by N error covariance matrix. For example, for the spatial moving average errors, as in (13), $\Omega(\gamma)^{-1} = [\mathbf{I}+\gamma(W + W') + \gamma^2 WW']^{-1}$, which does not yield a direct expression in terms of spatially transformed y and X.

Obtaining a consistent estimate for λ is not as straightforward as in the time-series case. As pointed out, OLS does not yield a consistent estimate in a spatial lag model. It therefore cannot be used to obtain an estimate for λ from a regression of residuals e on We, as in the familiar Cochrane-Orcutt procedure for serially autoregressive errors in the time domain. Instead, an explicit optimization of the likelihood function must be carried out. One approach is to use the iterative solution of the first-order conditions in Magnus (1978, p. 283):

$$\text{tr}\left[\left(\frac{\partial\Omega^{-1}}{\partial\lambda}\right)\Omega\right] = e'\left(\frac{\partial\Omega^{-1}}{\partial\lambda}\right)e \tag{36}$$

where $e = y - X\beta$ are GLS residuals. For a spatial autoregressive error process, $\partial \Omega^{-1}/\partial \lambda = -W - W' + \lambda W'W$. Solution of condition (36) can be obtained by numerical means. Alternatively, the GLS expression for β and similar solution of the first-order conditions for σ^2 can be substituted into the log-likelihood function to yield a concentrated log-likelihood as a nonlinear function of the autoregressive parameter λ (for technical details, see Anselin 1980, Chap. 5):

$$L_C = -\frac{N}{2} \ln \left(\frac{u'u}{N} \right) + \sum_i \ln(1 - \lambda \omega_i) \tag{37}$$

with $u'u = y'_L y_L - y'_L X_L [X'_L X_L]^{-1} X'_L y_L$, and y_L and X_L as spatially filtered variables, respectively $y - \lambda Wy$ and $X - \lambda WX$. The Jacobian term follows from $\ln |\Omega(\lambda)| = 2 \ln |\mathbf{I} - \lambda W|$ and the Ord simplification in terms of eigenvalues of W.

The asymptotic variance for the ML estimates conforms to the Magnus (1978) and Breusch (1980) general form and is block diagonal between the regression (β) and error variance parameters σ^2 and θ. For example, for a spatial autoregressive error, the asymptotic variance for the regression coefficients is $\mathrm{AsyVar}[\beta] = \sigma^2 [X'_L X_L]^{-1}$. The variance block for the error parameters is

$$\mathrm{AsyVar}[\sigma^2, \lambda] = \begin{bmatrix} N/2\sigma^4 & \dfrac{\mathrm{tr}(W_B)}{\sigma^2} \\ \dfrac{\mathrm{tr}(W_B)}{\sigma^2} & \mathrm{tr}(W_B)^2 + \mathrm{tr}(W'_B W_B) \end{bmatrix}^{-1} \tag{38}$$

where, for ease of notation, $W_B = W(\mathbf{I} - \lambda W)^{-1}$. Due to the block-diagonal form of the asymptotic variance matrix, knowledge of the precision of λ does not affect the precision of the β estimates. Consequently, if the latter is the primary interest, the complex inverse and trace expressions in (38) need not be computed, as in Benirschka and Binkley (1994). A significance test for the spatial error parameter can be based on a likelihood ratio test, in a straightforward way (Anselin 1988a, Chap. 8).

Higher-order spatial processes can be estimated using the same general principles, although the resulting log-likelihood function will be highly nonlinear and the use of a concentrated log-likelihood becomes less useful (Anselin 1980, Chap. 6).

The fit of spatial process models estimated by means of maximum likelihood procedures should not be based on the traditional R^2, which will be misleading in the presence of spatial autocorrelation. Instead, the fit of the model may be assessed by comparing the maximized log-likelihood or an adjusted form to take into account the number of parameters in the models, such as the familiar AIC (Anselin 1988b).

C. GMM/IV Estimation

The view of a spatially lagged dependent variable Wy in the spatial lag model as a form of endogeneity or simultaneity suggests an instrumental variable (IV) approach to estimation (Anselin 1980, 1988a, Chap. 7; 1990b). Since the main problem is the correlation between Wy and the error term in (3), the choice of proper instruments for Wy will yield consistent estimates. However, as usual, the efficiency of these estimates depends crucially on the choice of the instruments and may be poor in small samples. On the other hand, in contrast to the maximum likelihood approach just outlined, IV estimation does not require an assumption of normality.

Using the standard econometric results (for a review, see Bowden and Turkington 1984), and with Q as a P by N matrix ($P \geq K + 1$) of instruments (including K "exogenous" variables from X), the IV or 2SLS estimate follows as

$$\beta_{IV} = [Z'Q(Q'Q)^{-1}Q'Z]^{-1}Z'Q(Q'Q)^{-1}Q'y \tag{39}$$

with $Z = [Wy\ X]$, $\text{AsyVar}(\beta_{IV}) = \sigma^2[Z'Q(Q'Q)^{-1}Q'Z]^{-1}$, and $\sigma^2 = (y - Z\beta_{IV})'(y - Z\beta_{IV})/N$.

Clearly, this approach can also be applied to models where other endogenous variables appear in addition to the spatially lagged dependent variable, as in a simultaneous equation context, provided that the instrument set is augmented to deal with this additional endogeneity. It also forms the basis for a bootstrap approach to the estimation of spatial lag models (Anselin 1990b). Moreover, it is easily extended to deal with more complex error structures, e.g., reflecting forms of heteroskedasticity or spatial error dependence (Anselin 1988a, pp. 86–88). The formal properties of such an approach are derived in Kelejian and Robinson (1993) for a general methods of moments estimator (GMM) in the model $y = \rho Wy + X\beta + \varepsilon$ with spatial error components, $\varepsilon = W\xi + \psi$. The GMM estimator takes the form

$$\beta_{GMM} = [Z'Q(Q'\hat{\Omega}Q)^{-1}Q'Z]^{-1}Z'Q(Q'\hat{\Omega}Q)^{-1}Q'y \tag{40}$$

where $\hat{\Omega}$ is a consistent estimate for the error covariance matrix. The asymptotic variance for β_{GMM} is $[Z'Q(Q'\hat{\Omega}Q)^{-1}Q'Z]^{-1}$. For the spatial error components model, Kelejian and Robinson (1993, pp. 302–304) suggest an estimate for $\hat{\Omega} = \hat{\varphi}_1 I + \hat{\varphi}_2 WW'$, with $\hat{\varphi}_1$ and $\hat{\varphi}_2$ as the least-squares estimates in an auxilliary regression of the squared IV residuals $(y - Z\beta_{IV})$ on a constant and the diagonal elements of WW'.

A particularly attractive application of GLS-IV estimation in spatial lag models is a special case of the familiar White heteroskedasticity-consistent covariance estimator (White 1984, Bowden and Turkington 1984, p. 91). The estimator is as in (40), but $Q'\hat{\Omega}Q$ is estimated by $Q'\tilde{\Omega}Q$, where $\tilde{\Omega}$ is a diagonal matrix of squared IV residuals, in the usual fashion. This provides a way to obtain consistent estimates for the spatial autoregressive parameter ρ in the presence of heteroskedasticity of unknown form, often a needed feature in applied empirical work.

A crucial issue in instrumental variables estimation is the choice of the instruments. In spatial econometrics, several suggestions have been made to guide the selection of instruments for Wy (for a review, see Anselin 1988a, pp. 84–86; Land and Deane 1992). Recently, Kelejian and Robinson (1993 p. 302) formally demonstrated the consistency of β_{GMM} in the spatial lag model with instruments consisting of first-order and higher-order spatially lagged explanatory variables (WX, W^2X, etc.).

An important feature of the instrumental variables approach is that estimation can easily be carried out by means of standard econometric software, provided that the spatial lags can be computed as the result of common matrix manipulations (Anselin and Hudak 1992). In contrast, the maximum likelihood approach requires specialized routines to implement the nonlinear optimization of the log-likelihood (or concentrated log-likelihood). We next turn to some operational issues related to this.

D. Operational Implementation and Illustration

To date, none of the widely available econometric software packages contain specific routines to implement maximum likelihood estimation of spatial process models or to carry out specification tests for spatial autocorrelation in regression models. This lack of attention to the analysis of the lattice data structures that are most relevant in empirical economics contrasts with a relatively large range of software for spatial data analysis in the physical sciences, geared to point patterns and geostatistical data. Examples of these are the GSLIB library (Deutsch and Journel 1992) and the recent S+Spatialstats add-on to the S-PLUS statistical software (MathSoft 1996). While the latter does include some analyses for lattice data, estimation is limited to maximum likelihood of spatial error models with autoregressive or moving-average structures. However, the spatial lag model is not covered and specification diagnostics are totally absent.

The only self-contained software package specifically geared to spatial econometric analysis in SpaceStat (Anselin 1992b, 1995). It contains both maximum likelihood and instrumental variables estimators for spatial lag and error models, as well as ways to estimate heteroskedastic specifications and a wide range of diagnostics for spatial effects. In addition, SpaceStat also includes extensive features to carry out exploratory spatial data analysis as well as utilities to create and manipulate spatial weights matrices and interface with geographic information systems.

There are two major practical issues that must be resolved to implement the estimation of spatial lag and spatial error models. The first is the need to construct spatially lagged variables from observations on the dependent variable or residual term. This is relevant for both instrumental variables (IV, 2SLS, GMM) as well as maximum likelihood estimation. In principle, the lag can be computed as a simple matrix multiplication of the spatial weights matrix W with the vector of observations, say Wy. This is straightforward to implement in most econometric software packages that contain matrix algebra routines (specific examples for Gauss, Splus, Limdep, Rats and Shazam are given in Anselin and Hudak 1992, Table 2, p. 514). In practice, however, the size of the matrix that can be manipulated by econometric software is severely limited and insufficient for most empirical applications, unless sparse matrix routines can be exploited (avoiding the need to store a full N by N matrix). This is increasingly the case in state-of-the-art matrix algebra packages (e.g., Matlab, Gauss), but still fairly uncommon in application-oriented econometric software; hence, the computation of spatial lags will typically necessitate some programming effort on the part of the user (the construction of spatial lags based on sparse spatial weights formats in SpaceStat is discussed in Anselin 1995). Once the spatial lagged dependent variables are computed, IV estimation of the spatial lag model can be carried out with any standard econometric package.

The other major operational issue pertains only to maximum likelihood estimation. It is the need to manipulate large matrices of dimension equal to the number of observations in the asymptotic variance matrices (32) and (38) and in the Jacobian term (27) of the log-likelihoods (31) and (37). In contrast to the time-series case, the matrix W is not triangular and hence a host of computational simplifications are not applicable. The problem is most serious in the computation of the asymptotic variance matrix of the maximum likelihood estimates. The inverse matrices in both $W_A = W(I - \rho W)^{-1}$ of (32) and $W_B = W(I - \lambda W)^{-1}$ of (38) are full matrices which do not lend themselves to the application of sparse matrix algorithms. For low values of the autoregressive parameters, a power expansion of $(I - \rho W)^{-1}$ or $(I - \lambda W)^{-1}$ may be a reasonable approximation to the inverse, e.g., $(I - \rho W)^{-1} = \sum_k \rho^k W^k +$ error, with $k = 0, 1, \ldots, K$, such that $\rho^K < \delta$, where δ is a sufficiently small value. However, this will involve some computing effort in the construction of the powers of the weights matrices and is increasingly burdensome for higher values of the autoregressive parameter. In general, for all practical purposes, the size of the problem for which an asymptotic variance matrix can be computed is constrained by the largest matrix inverse that can be carried out with acceptable numerical precision in a given software/hardware environment. In current desktop settings, this typically ranges from a few hundred to a few thousand observations. While this makes it impossible to compute asymptotic t-tests for all the parameters in spatial models with very large numbers of observations, it does not preclude asymptotic inference. In fact, as we argued in Section III.B, due to the block diagonality of the asymptotic variance matrix in the spatial error case, asymptotic t-statistics can be constructed for the estimated β coefficients without knowledge of the precision of the autoregressive parameter λ (see also Benirschka and Binkley 1994, Pace and Barry 1996). Inference on the autoregressive parameter can be based on a likelihood ratio test (Anselin 1988a, Chap. 6). A similar approach can be taken in the spatial lag model. However, in contrast to the error case, asymptotic t-tests can no longer be constructed for the estimated β coefficients, since the asymptotic variance matrix (32) is not block diagonal. Instead, likelihood ratio tests must be considered explicitly for any subset of coefficients of interest (requiring a separate optimization for each specification; see Pace and Barry 1997).

With the primary objective of obtaining consistent estimates for the parameters in spatial regression models, a number of authors have suggested ways to manipulate popular statistical and econometric software packages in order to maximize the log-likelihoods (28) and (37). Examples of such efforts are routines for ML estimation of the spatial lag and spatial autoregressive error model in Systat, SAS, Gauss, Limdep, Shazam, Rats and S-PLUS (Bivand 1992, Griffith 1993, Anselin and Hudak 1992, Anselin et al. 1993b). The common theme among these approaches is to find a way to convert the log-likelihoods for the spatial models to a form amenable for use with standard nonlinear optimization routines. Such routines proceed incrementally, in the sense that the likelihood is built up from a sum of elements that correspond to individual observations. At first sight, the Jacobian term in the spatial models would preclude this. However, taking advantage of the Ord decomposition in terms of eigenvalues, pseudo-observations can be constructed for the elements of the Jacobian. Specifically, each term $1 - \rho\omega_i$ is considered to correspond to a pseudo-variable ω_i, and is summed over all "observations." For example, for the spatial lag model, the log-likelihood (ignoring constant terms) can be expressed as

$$L = \sum_i \left[\ln(1 - \rho\omega_i) - \frac{\ln(\sigma^2)}{2} - \frac{(y_i - \rho\{Wy\}_i - x_i\beta)^2}{2\sigma^2} \right] \qquad (41)$$

which fits the format expected by most nonlinear optimization routines. Examples of practical implementations are listed in Anselin and Hudak (1992, Table 10, p. 533) and extensive source code for various econometric software packages is given in Anselin et al. (1993b).

One problem with this approach is that the asymptotic variance matrices computed by the routines tend to be based on a numerical approximation and do not necessarily correspond to the analytical expressions in (32) and (38). This may lead to slight differences in inference depending on the software package that is used (Anselin and Hudak 1992, Table 10, p. 533). An alternative approach that does not require the computation of eigenvalues is based on sparse matrix algorithms to efficiently compute the determinant of the Jacobian at each iteration of the optimization routine. While this allows the estimation of models for very large data sets (tens of thousands of observations), for example, by using the specialized routines in the Matlab software, this does not solve the asymptotic variance matrix problem. Inference therefore must be based on likelihood ratio statistics (for details and implementation, see Pace and Barry 1996, 1997).

To illustrate the various spatial models and their estimation, the results for the parameters in a simple spatial model of crime estimated for 49 neighborhoods in Columbus, Ohio, are presented in Table 1. The model and results are based on Anselin (1988a, pp. 187–196) and have been used in a number of papers to benchmark different estimators and specification tests (e.g., McMillen 1992, Getis 1995, Anselin et al. 1996, LeSage 1997). The data are also available for downloading via the internet from http://www.rri.wvu.edu/spacestat.htm. The estimates reported in Table 1 include OLS in the standard regression model, OLS (inconsistent), ML, IV, and heteroskedastic-robust IV for the spatial lag model, and ML for the spatial error

Table 1   Estimates in a Spatial Model of Crime[a]

	OLS	Lag-OLS	Lag-ML	Lag-IV	Lag-GIVE	Err-ML
Constant	68.629	38.783	45.079	43.963	46.667	59.893
	(4.73)	(9.32)	(7.18)	(11.23)	(7.61)	(5.37)
ρ		0.549	0.431	0.453	0.419	
		(0.153)	(0.118)	(0.191)	(0.139)	
Income	−1.597	−0.886	−1.032	−1.010	−1.185	−0.941
	(0.334)	(0.358)	(0.305)	(0.389)	(0.434)	(0.331)
Housing	−0.274	−0.264	−0.266	−0.266	−0.234	−0.302
value	(0.103)	(0.092)	(0.088)	(0.092)	(0.173)	(0.090)
λ						0.562
						(0.134)
R^2	0.552	0.652		0.620	0.633	
Log-lik	−187.38		−182.39			−183.38

[a]Data are for 49 neighborhoods in Columbus, Ohio, 1980. Dependent variable is per capita residential burglaries and vehicle thefts. Income and housing values are in thousand dollars. A first-order contiguity spatial weights matrix was used to construct the spatial lags.

model. The spatial lags for the exogenous variables (WX) were used as instruments in the IV estimation. In addition to the estimates and their standard errors, the fit of the different specifications estimated by ML is compared by means of the maximized log-likelihood. For OLS and the IV estimates, the R^2 is listed. However, this should be interpreted with caution, since R^2 is inappropriate as a measure of fit when spatial dependence is present. All estimates were obtained by means of the SpaceStat software.

A detailed interpretation of the results in Table 1 is beyond the scope of this chapter, but a few noteworthy features may be pointed out. The two spatial models provide a superior fit relative to OLS, strongly suggesting the presence of spatial dependence. Of the two, the spatial lag model fits better, indicating it is the preferred alternative. Given the lack of an underlying behavioral model (unless one is willing to make heroic assumptions to avoid the ecological fallacy problem), the results should be interpreted as providing consistent estimates for the coefficients of income and housing value after the spatial dependence in the crime variable is filtered out. The most affected coefficient (besides the constant term) pertains to the income variable, and is lowered by about a third while remaining highly significant. The estimates for the autoregressive coefficient vary substantially between the inconsistent and biased OLS and the consistent estimates, but the Lag-IV coefficient has a considerably higher standard error. In some instances, OLS can thus yield "better" estimates in an MSE sense relative to IV. Diagnostics in the Lag-ML model indicate strong remaining presence of heteroskedasticity (the spatial Breusch-Pagan test from Anselin

1988a, p. 123, yields a highly significant value of 25.35, $p < 0.00001$). The robust Lag-GIVE estimates support the importance of this effect: the estimate for the autoregressive parameter is quite close to the ML value while obtaining a significantly smaller standard error relative to both OLS and the nonrobust IV. Moreover, the estimate for the Housing variable is no longer significant. This again illustrates the complex interaction between heterogeneity and spatial dependence.

IV. TESTS FOR SPATIAL DEPENDENCE

As it happened in the mainstream econometrics literature, the initial stages of development in spatial econometrics were characterized by an emphasis on estimation. As discussed in the last section, Cliff and Ord (1973) and others formulated the maximum likelihood approach which goes to back to work of Whittle (1954). In mainstream econometrics, the test for serial correlation developed by Durbin and Watson (1950, 1951) was the first explicit specification test for the regression model. It has gained widespread acceptance since its inception. However, routine testing for other specifications (such as homoskedasticity, normality, exogeneity, and functional form) did not take prominence until the early eighties. A major breakthrough was the rediscovery of the Rao (1947) score (RS) test (known as the Lagrange multiplier test in econometrics). The RS test became very popular due to its computational ease compared to the other two asymptotically "equivalent" test procedures, namely the likelihood ratio (LR) and Wald (W) tests (see Godfrey 1988 and Bera and Ullah 1991).

In a similar fashion, the origins of specification testing in spatial econometrics can be traced back to Moran's (1950a, 1950b) test for autocorrelation. This test laid in obscurity until it was revived by Cliff and Ord (1972). It received further impetus by Burridge (1980) as an RS test. However, the early spatial econometrics literature on testing was dominated by the Wald and LR tests (for example, see Brandsma and Ketellapper 1979a, 1979b; Anselin 1980). Since the latter require the estimation of the alternative model by means of nonlinear optimization (as discussed in Section III), the advantages of basing a test on the least-squares regression of the null model, offered by the RS test, were quickly realized. During the last 15 years, a number of such tests were developed (see Anselin 1988a, 1988c).

Although mainstream econometrics and spatial econometrics literature went through similar developments in terms of specification testing, the implementation of the tests in spatial models turns out to be quite different from the standard case. For example, most of the RS specification tests cannot be written in the familiar "NR^2" form (where R^2 is a coefficient of determination) nor they can be computed by running any artificial regression. In addition, the interaction between spatial lag dependence and spatial error dependence in terms of specification testing is stronger and more complex than its standard counterpart. There are, however, some common threads. As in the standard case, most of the tests for dependence in the spatial model can be constructed based on the OLS residuals. In our discussion we will emphasize the similarities and the differences between specification testing in spatial econometric models and the standard case.

We start the remainder of the section with a discussion of Moran's I statistic and stress its close connection to the familiar Durbin-Watson test. Moran's I was not developed with any specific kind of dependence as the alternative hypothesis, although it has been found to have power against a wide range of forms of spatial dependence. We next consider a test developed in the same spirit by Kelejian and Robinson (1992). This is followed by a focus on tests for specific alternative hypothesis in the form of either spatial lag or spatial error dependence. Tests for these two kinds of autocorrelations are not independent even asymptotically, and their separate applications when other or both kinds of autocorrelations are present will lead to unreliable inference. Therefore, it is natural to discuss a test for joint lag and error autocorrelations. However, the problem with such a test is that we cannot make any specific inference regarding the exact nature of dependence when the joint null hypothesis is rejected. One approach to deal with this problem is to test for spatial error autocorrelation after estimating a spatial lag model, and vice versa. This, however, requires ML estimation, and the simplicity of tests based on OLS residuals is lost. We therefore consider a recently developed set of diagnostics in which the OLS-based RS test for error (lag) dependence is adjusted to take into account the local presence of lag (error) dependence (Anselin et al. 1996). We then provide a brief review of the small-sample properties of the various tests. Finally, the section is closed into a discussion of implementation issues and our illustrative example of the spatial model of crime.

A. Moran's I Test

Moran's (1950a, 1950b) I test was originally developed as a two-dimensional analog of the test of significance of the serial correlation coefficient in univariate time series. Cliff and Ord (1972, 1973) formally presented Moran's I statistics as

$$I = \frac{N}{S_o} \left(\frac{e'We}{e'e} \right) \tag{42}$$

where $e = y - X\tilde{\beta}$ is a vector of OLS residuals, $\tilde{\beta} = (X'X)^{-1}X'y$, W is the spatial weights matrix, N is the number of observations, and S_o is a standardization factor equal to the sum of the spatial weights, $\sum_i \sum_j w_{ij}$. For a row-standardized weights matrix W, S_o simplifies to N (since each row sum equals 1) and the statistic becomes

$$I = \frac{e'We}{e'e} \tag{43}$$

Moran did not derive the statistic from any basic principle; instead, it was suggested as a simple test for correlation between nearest neighbors which generalized one of his earlier tests in Moran (1948). Consequently, the test could be given different interpretations. The first striking characteristic is the similarity between Moran's I and the familiar Durbin-Watson (DW) statistic

$$DW = \frac{e'Ae}{e'e} \tag{44}$$

where

$$A = \begin{bmatrix} 1 & -1 & 0 & 0 & \cdots & 0 & 0 & 0 \\ -1 & 2 & -1 & 0 & \cdots & 0 & 0 & 0 \\ 0 & -1 & 2 & -1 & \cdots & 0 & 0 & 0 \\ \cdot & \cdot & \cdot & \cdot & \cdots & \cdot & \cdot & \cdot \\ \cdot & \cdot & \cdot & \cdot & \cdots & \cdot & \cdot & \cdot \\ 0 & 0 & 0 & 0 & \cdots & -1 & 2 & -1 \\ 0 & 0 & 0 & 0 & \cdots & 0 & -1 & 1 \end{bmatrix}$$

Therefore, both statistics equal the ratio of quadratic forms in OLS residuals and they differ only in the specification of the interconnectedness between the observations (neighboring locations). It is well known that the DW test is a uniformly most powerful (UMP) test for one sided alternatives with error distribution $\varepsilon_t = \lambda\varepsilon_{t-1} + \xi_t$ (see, e.g., King 1987). Similarly Moran's I possesses some optimality properties. More precisely, Cliff and Ord (1972) established a link between the LR and I tests. If we take the alternative model as (8), i.e.,

$$\varepsilon = \lambda W \varepsilon + \xi$$

then the LR statistic for testing H_0: $\lambda = 0$ against the alternative H_a: $\lambda = \lambda_1$, when ε and σ^2 are known, is proportional to

$$\frac{\varepsilon'W\varepsilon}{\varepsilon'(I + \lambda_1^2 G)\varepsilon} \tag{45}$$

where G is a function of W. Therefore, I approaches the LR statistic as $\lambda_1 \to 0$, and it can be shown to be consistent for H_0: $\lambda = 0$ against H_a: $\lambda \neq 0$. As we discuss later, Burridge (1980) also showed that I is equivalent to the RS test for $\lambda = 0$ in (8) (or $\gamma = 0$ in the spatial moving average process (13)) with an unscaled denominator. Since we know that the LR and RS tests are asymptotically equivalent under the null and local alternatives, Cliff and Ord's result regarding asymptotic equivalence of I and LR becomes very apparent. King and Hillier (1985) derived the locally best invariant (LBI) test for the wider problem of testing H_0: $\lambda = 0$ against H_a: $\lambda > 0$ when the covariance matrix of the regression disturbance is of the known form $\sigma^2\Omega(\lambda)$ (as in our (10)), and showed the test to be identical to the one-sided version of the RS test. Combining this result with that of Burridge (1980), we can conclude that Moran's I must be an LBI test, which was demonstrated by King (1981).

In practice the test is implemented on the basis of an asymptotically normal standardized z-value, obtained by subtracting the expected value and dividing by the standard deviation. One advantage of statistic like I is that under $H_0 : \lambda = 0$ and normality of ε, $e'e$ is distributed as central χ^2. Cliff and Ord (1972) exploited this to derive the first two moments as

$$E(I) = \frac{\text{tr}(MW)}{N - K} \tag{46}$$

and

$$V(I) = \frac{\text{tr}(MWMW') + \text{tr}(MW)^2 + \{\text{tr}(MW)\}^2}{(N - K)(N - K + 2)} - [E(I)]^2 \tag{47}$$

where $M = I - X(X'X)^{-1}X'$, and W is a row-standardized weights matrix.

It is possible to develop a finite-sample-bound test for I following Durbin and Watson (1950, 1951). However, for I, we need to make the bounds independent of not only X but also of the weight matrix W. This poses some difficulties. Tiefelsdorf and Boots (1995), using the results of Imhof (1961) and Koerts and Abrahamse (1968), showed that exact critical values of I can be computed by numerical integration. They first expressed I in terms of the eigenvalues $\gamma_1, \gamma_2, \ldots, \gamma_{N-K}$ of MW, other than the K zeros, and $N - K$ independent $N(0, 1)$ variables $\delta_1, \delta_2, \ldots, \delta_{N-K}$; more specifically,

$$I = \sum_{i=1}^{N-K} \gamma_i \delta_i^2 \Big/ \sum_{i=1}^{N-K} \delta_i^2 \tag{48}$$

Then

$$\Pr(I \leq I_0 | H_0) = \Pr \left(\sum_{i=1}^{N-K} (\gamma_i - I_0)\delta_i^2 \leq 0 | H_0 \right) \tag{49}$$

Note that $\sum_{i=1}^{N-K}(\gamma_i - I_0)\delta_i^2$ is a weighted sum of $(N - K)$ χ_1^2 variables. Imhof's method simplifies the probability in (49) to

$$\Pr(I \leq I_0 | H_0) = \frac{1}{2} - \frac{1}{\pi} \int_0^\infty \frac{\sin\{a(u)\}}{ub(u)} \, du \tag{50}$$

where

$$a(u) = \frac{1}{2} \sum_{i=1}^{N-K} \arctan\{(\gamma_i - I_0)u\}$$

$$b(u) = \prod_{i=1}^{N-K} [1 + (\gamma_i - I_0)^2 u^2]^{1/4}$$

The integral in (50) can be evaluated by numerical integration (for more on this, see Tiefelsdorf and Boots 1995).

It is instructive to note that the computation of exact critical values of the DW statistic involves the same calculations as for Moran's I except that the γ_i is the eigenvalues of MA, where A is the fixed matrix given by in (44). Even with the recent dramatic advances in computer technology, it will take some time for practitioners to use the above numerical integration technique to implement Moran's I test.

B. Kelejian-Robinson Test

The test developed by Kelejian and Robinson (1992) is in the same spirit of Moran's I in the sense that it is not based on an explicit specification of the generating process of the disturbance term. At the same time the test does not require the model to be linear or the disturbance term to be normally distributed. Although the test does not attempt to identify the cause of spatial dependence, Kelejian and Robinson (1992) made the following assumption about spatial autocorrelation:

$$\text{Cov}(\varepsilon_i, \varepsilon_j) = \sigma_{ij} = Z_{ij}\alpha \tag{51}$$

where Z_{ij} is 1 by q vector which can be constructed from the independent variables X, α is q by 1 vector of parameters, and i, j are contiguous in the sense that they are neighbors in a general spatial "ordering" of the observations. The null hypothesis of no spatial correlation can be tested by $H_0 : \alpha = 0$ in (51).

For a given sample of size N, let C denote h_N by 1 vector σ_{ij}'s which are not zero for $i < j$. Therefore, a test for $\alpha = 0$ can be achieved by running a regression of C on the observation matrix Z which is of dimension h_N by q consisting of Z_{ij} values. Since we do not observe the elements of C, they are replaced by the cross product of OLS residuals, $e_i e_j$. The resulting h_N by 1 vector is denoted by $\hat{C}$. The test is based on $\hat{\gamma} = (Z'Z)^{-1}Z'\hat{C}$ and is given by

$$KR = \frac{\hat{\gamma}'Z'Z\hat{\gamma}}{\tilde{\sigma}^4} \tag{52}$$

where $\tilde{\sigma}^4$ is a consistent estimator of σ^4. For example, we can use $[e'e/N]^2$ or $(\hat{C} - Z\hat{\gamma})'(\hat{C} - Z\hat{\gamma})/h_N$ for $\tilde{\sigma}^4$. Under $H_0 : \alpha = 0$, $KR \xrightarrow{D} \chi_q^2$ (central chi-square with q degrees of freedom), where $\xrightarrow{D}$ denotes convergence in distribution. Putting $\hat{\gamma} = (Z'Z)^{-1}Z'\hat{C}$, KR can be expressed as

$$KR = \frac{\hat{C}'Z(Z'Z)^{-1}Z'\hat{C}}{\tilde{\sigma}^4} \tag{53}$$

Since for the implementation of the test we need the distribution only under the null hypothesis, it is legitimate to replace σ^4 by a consistent estimate under $\alpha = 0$.

Note that under H_0, $\hat{C}'\hat{C}/h_N \xrightarrow{P} \sigma^4$, where $\xrightarrow{P}$ means convergence in probability. Therefore, an asymptotically equivalent form of the test is

$$h_N \cdot \frac{\hat{C}'Z(Z'Z)^{-1}Z'\hat{C}}{\hat{C}'\hat{C}} \tag{54}$$

which has the familiar NR^2 form. Here R^2 is the uncentered coefficient of determination of $\hat{C}$ on Z and h_N is the sample size of this regression.

It is also not difficult to see an algebraic connection between KR and Moran's I. From (43)

$$I^2 = \frac{(e'We)^2}{(e'e)^2} = \frac{1}{N^2\tilde{\sigma}^4}\left(\sum_{i=1}^{N}\sum_{j=1}^{N} W_{ij}e_ie_j\right)^2$$

$$= \sum_{k=1}^{N}\sum_{l=1}^{N}\sum_{m=1}^{N}\sum_{n=1}^{N} \frac{W_{kl}W_{mn}(e_ke_l)(e_me_n)}{N^2\tilde{\sigma}^4} \tag{55}$$

Using (53), we can write

$$KR = \sum_{i=1}^{h_N}\sum_{j=1}^{h_N} \frac{p_{ij}\hat{C}_i\hat{C}_j}{\tilde{\sigma}^4} \tag{56}$$

where p_{ij} are the elements of $Z(Z'Z)^{-1}Z'$. Given that $\hat{C}_i$'s contain terms like e_ke_l, $k < l$, it appears that the I^2 and KR statistics have similar algebraic structure.

C. Tests for Spatial Error Autocorrelation

In contrast to the earlier two tests, the alternative hypothesis is now stated explicitly through the data generating process of ε as in (8), i.e.,

$$\varepsilon = \lambda W\varepsilon + \xi$$

and we test $\lambda = 0$. All three general principles of testing, namely LR, W, and RS can be applied. Out of the three, the RS test as described in Rao (1947) is the most convenient one to use since it requires estimation only under the null hypothesis. That is, the RS test can be based on the OLS estimation of the regression model (7). Silvey (1959) derived the RS test using the Lagrange multiplier(s) of a constrained optimization problem.

Burridge (1980) used Silvey's form to test $\lambda = 0$, although the Rao's score form, namely

$$RS = d'(\tilde{\theta})\mathcal{I}(\tilde{\theta})^{-1}d(\tilde{\theta}) \tag{57}$$

is more popular and much easier to use. In (57), $d(\theta) = \partial L(\theta)/\partial \theta$ is the score vector, $\mathcal{I}(\theta) = -E[\partial^2 L(\theta)/\partial(\theta)\partial(\theta)']$ is the information matrix, $L(\theta)$ is the log-likelihood function, and $\tilde{\theta}$ is the *restricted* (under the tested hypothesis) maximum likelihood estimator of the parameter vector θ. For the spatial error autocorrelation model $\theta = (\beta', \sigma^2, \lambda)'$ and the log-likelihood function is given in (34). The test is essentially based on the score with respect to λ, i.e., on

$$d_\lambda = \left. \frac{\partial L}{\partial \lambda} \right|_{\lambda=0} = \frac{\varepsilon' W \varepsilon}{\sigma^2} \tag{58}$$

We can immediately see the connection of this to Moran's I statistic. After computing $\mathcal{I}(\theta)$ under H_0, from (36), we have the test statistic

$$RS_\lambda = \frac{\tilde{d}_\lambda^2}{T} = \frac{[e'We/\tilde{\sigma}^2]^2}{T} \tag{59}$$

where $T = \text{tr}[(W' + W)W]$. Therefore, the test requires only OLS estimates, and under H_0, $RS_\lambda \overset{D}{\to} \chi_1^2$. It is interesting to put $W = W^T$ (Section II.E) and obtain $T = N - 1$ and $RS_\lambda = (N-1)\tilde{\lambda}^2$ where $\tilde{\lambda} = \sum_t e_t e_{t-1} / \sum_t e_{t-1}^2$ in the time-series context. Burridge (1980) derived the RS test (59) using the estimates of the Lagrange multiplier following Silvey (1959). The Lagrangian function for this problem is

$$L^R(\theta, \mu) = L(\theta) - \mu\lambda \tag{60}$$

where μ is the associated Lagrange multiplier. From the first-order conditions, we have

$$\left. \frac{\partial L^R(\theta, \mu)}{\partial \lambda} \right|_{\tilde{\theta}, \tilde{\mu}} = 0$$

i.e.,

$$\tilde{\mu} = \left. \frac{\partial L(\theta)}{d\lambda} \right|_{\tilde{\theta}} = \tilde{d}_\lambda \tag{61}$$

and this results in the same statistic RS_λ.

A striking feature of the RS test is its invariance to different alternatives (for details, see Bera and McKenzie 1986). The RS test uses the slope $\partial L/\partial \theta$ at $\theta = \tilde{\theta}$, and there may be many likelihood functions (models) which have the same slope at $\tilde{\theta}$. If we specify the alternative hypothesis as a spatial moving-average process (13) and test $H_0 : \gamma = 0$, we obtain the same Rao's score statistic RS_λ. Therefore, RS_λ will be locally optimal for both autoregressive and moving-average alternatives. But this also means that when the null hypothesis is rejected, the test does not provide any guidance regarding the nature of the disturbance process, even when other aspects of the spatial model are resolved. This also raises the question whether RS_λ will be inferior to other asymptotically equivalent tests such as LR and W, with respect to power, since it does not use the precise information contained in the alternative hypothesis. In the context of the standard regression model, Monte Carlo results of Godfrey (1981) and Bera and McKenzie (1986) suggest that there is no setback in the performance of RS test compared to the LR test. In Section IV.G, we discuss the finite sample performance of RS_λ and other tests.

Computationally, the W and LR tests are more demanding since they require ML estimation under the alternative, and the explicit forms of the tests are more complicated. For instance, let us consider the W test which can be computed using the ML estimate $\hat{\lambda}$ by maximizing (34) with respect to β, σ^2, and λ. We can write the W statistic as (Anselin 1988a, p. 104)

$$WS_\lambda = \frac{\hat{\lambda}^2}{\widehat{\text{AsyVar}}(\hat{\lambda})} \tag{62}$$

where AsyVar $(\hat{\lambda})$ can be obtained from (38) as

$$\text{AsyVar}(\hat{\lambda}) = \left[\text{tr}(W_B^2) + \text{tr}(W_B' W_B) - \frac{\{\text{tr}(W_B)\}^2}{N} \right]^{-1} \tag{63}$$

For implementation we need to replace λ by $\hat{\lambda}$ in the above expression. In the standard time-series regression case the results are much simpler. For example, AsyVar $[\sigma^2, \lambda]$ is a diagonal matrix and AsyVar$(\hat{\lambda})$ is simply $(1 - \lambda^2)/(N - 1)$. Therefore the Wald test statistic can be simply written as

$$WS_\lambda^T = \frac{(N - 1)\hat{\lambda}^2}{1 - \hat{\lambda}^2} \tag{64}$$

Note that under $\lambda = 0$, the asymptotic variance $(1 - \lambda^2)/(N - 1)$ reduces to $1/(N - 1)$, the expression for AsyVar$(\hat{\lambda})$ used in the time series case to test the significance of λ.

The LR statistic can be easily obtained using the concentrated log-likelihood function L_C in (37). We can write

$$LR_\lambda = 2[\hat{L}_C - \tilde{L}_C] \tag{65}$$

where the "hat" denotes that the quantities are evaluated at the unrestricted ML estimates $\hat{\beta}, \hat{\sigma}^2$, and $\hat{\lambda}$. It is easy to see that LR_λ reduces to (Anselin 1988a, p. 104)

$$LR_\lambda = N[\ln \tilde{\sigma}^2 - \ln \hat{\sigma}^2] + 2 \sum_{i=1}^{N} \ln(1 - \hat{\lambda}\omega_i) \tag{66}$$

The appearance of the last term in (66) differentiate the spatial dependence situation from the serial correlation case for time-series data.

Finally, for higher-order spatial processes, it is easy to generalize the RS statistic (59). For example, if we consider a qth-order spatial autoregressive model

$$\varepsilon = \lambda_1 W_1 \varepsilon + \lambda_1 W_2 \varepsilon + \cdots + \lambda_q W_q \varepsilon + \xi \tag{67}$$

and test $H_0 : \lambda_1 = \lambda_2 = \cdots = \lambda_q = 0$, the RS statistic will be given by

$$RS_{\lambda_1 \ldots \lambda_q} = \sum_{l=1}^{q} \frac{[e' W_l e / \tilde{\sigma}^2]^2}{T_l} \tag{68}$$

where $T_l = \text{tr}[W_l' W_l + W_l^2], l = 1, 2, \ldots, q$. Under the null of no spatial dependence, $RS_{\lambda_1 \ldots \lambda_q} \overset{D}{\to} \chi_q^2$. Therefore, the test statistic for higher-order dependence is simply the sum of corresponding individual tests. The same test statistic will result when a moving average model as in (18) is taken as the alternative instead of (67). As expected, the Wald and LR tests in this context will be more complicated as they require ML estimation of $\lambda_1, \lambda_2, \ldots, \lambda_q$.

D.   Tests for Spatial Lag Dependence

In this section, we consider tests on the null hypothesis $H_0 : \rho = 0$ in (3) using the log-likelihood function (26). Once again the RS test is the easiest one to use, and Anselin (1988c) derived it explicitly (his equation (32)). The score with respect to ρ is

$$d\rho = \left. \frac{\partial L}{\partial \rho} \right|_{\rho=0} = \frac{\varepsilon' W y}{\sigma^2} \tag{69}$$

The inverse of the information matrix is given in (30). The complicating feature of this matrix is that even under $\rho = 0$, it is not block diagonal; the (ρ, β) term is equal to $(X'WX\beta)/\sigma^2$, obtained by putting $\rho = 0$; i.e., $W_A = W$. This absence of block diagonality causes two problems. First, as we mentioned in Section II, the presence of spatial dependence implies that a sample contains less information than an independent counterpart. This can now be easily demonstrated using (30). In the absence of dependence ($\rho = 0$ in (3)), the ML estimate of β will have variance $\sigma^2 (X'X)^{-1}$ which is the inverse of the information. But when $\rho \neq 0$, to compute the variance of the ML estimate of β we need to add a positive-definite part to $\sigma^2 (X'X)^{-1}$ due to absence of block diagonality. Second, to obtain the asymptotic variance of d_ρ, even under $\rho = 0$ from (30), we cannot ignore one of the off-diagonal terms. This was not the case for d_λ in Section IV.C. Asymptotic variance of d_λ was obtained just using the $(2, 2)$ element of (36) (see (59)). For the spatial lag model, asymptotic variance of d_ρ is obtained from the reciprocal of the $(1,1)$ element of

$$\begin{bmatrix} \text{tr}[W^2 + W'W] + [WX\beta]'[WX\beta]/\sigma^2 & (X'WX\beta)'/\sigma^2 \\ (X'WX\beta)/\sigma^2 & (X'X)/\sigma^2 \end{bmatrix}^{-1} \tag{70}$$

Since under $\rho = 0$, $W_A = W$ and $\text{tr}(W) = 0$, the expression is $T_1 = [(WX\beta)'$ $M(WX\beta) + T\sigma^2]/\sigma^2$, where T is given in (59). Therefore, the RS statistic is given by

$$RS_\rho = \frac{\tilde{d}_\rho^2}{\tilde{T}_1} = \frac{[e'Wy/\tilde{\sigma}^2]^2}{\tilde{T}_1} \tag{71}$$

where in $\tilde{T}_1$, β, and σ^2 are replaced by $\tilde{\beta}$ and $\tilde{\sigma}^2$, respectively. Under $H_0 : \rho = 0$, $RS_\rho \xrightarrow{D} \chi_1^2$, the Wald and LR tests will require maximization of the log-likelihood function (26) or (29). Let $\hat{\rho}$ be the ML estimate of ρ. To get the asymptotic variance of $\hat{\rho}$, we need the $(1, 1)$ element of (30). Since the Wald test requires estimation under the alternative hypothesis (i.e., $\rho \neq 0$), the $(1, 3)$ element $\text{tr}(W_A)/\sigma^2$ will also be nonzero and the resulting expression will more complicated than T_1 given above (Anselin 1988a, p. 104). The LR statistic will have the same form as in (66) except for the last term:

$$LR_\rho = N[\ln \tilde{\sigma}^2 - \ln \hat{\sigma}^2] + 2 \sum_{i=1}^{N} \ln(1 - \hat{\rho}\omega_i) \tag{72}$$

If ML estimation is already performed, LR_ρ is much easier to compute than its Wald counterpart. Under $\rho = 0$ both Wald and LR statistics will be asymptotically distributed as χ_1^2.

E. Testing in the Possible Presence of Both Spatial Error and Lag Autocorrelation

The test described in the Sections IV.C and IV.D can be termed as *one-directional* tests in the sense that they are designed to test a single specification assuming correct specification for the rest of the model. For example, we discussed RS_λ, WS_λ, and LR_λ statistics for the null hypothesis $H_0 : \lambda = 0$ assuming that $\rho = 0$. Because of the nature of the information matrix, these tests will not be valid even asymptotically, when $\rho \neq 0$. For instance, we noted that under the null, $H_0 : \lambda = 0$ all the three statistics are asymptotically distributed as *central* χ^2 with one degree of freedom. This result is valid only when $\rho = 0$. To evaluate the effects of nonzero ρ on RS_λ, WS_λ, and LR_λ, let us write the model when both the spatial error and lag autocorrelation are present:

$$y = \rho W_1 y + X\beta + \varepsilon$$
$$\varepsilon = \lambda W_2 \varepsilon + \xi \qquad \xi \sim N(0, \sigma^2 I) \tag{73}$$

where W_1 and W_2 are spatial weights matrices associated with the spatially lagged dependent variable and the spatial autoregressive disturbances, respectively. Recall from Section II.F that for model (73) to be identified, it is necessary that $W_1 \neq W_2$ or that the matrix X contain at least one exogenous variable in addition to the constant term. An alternative specification of spatial moving-average error process for ε as in (13),

$$\varepsilon = \lambda W_2 \xi + \xi \tag{74}$$

has no such problems and it also leads to identical results in terms of test statistics discussed here. Using the results of Davidson and MacKinnon (1987) and Saikkonen (1989), we evaluate the impact of local presence of ρ on the asymptotic null distribution of RS_λ, LR_λ, and WS_λ. Let $\rho = \delta/\sqrt{N}$, $\delta < \infty$, then it can be shown that under $H_0 : \lambda = 0$, all three tests asymptotically converge to a *noncentral* χ_1^2, with noncentrality parameter

$$R_\rho = \frac{\delta^2 T_{12}^2}{N T_{22}} \tag{75}$$

where $T_{ij} = \text{tr}[W_i W_j + W_i' W_j]$, $j = 1, 2$ (note that $T_{12} = T_{21}$). Therefore, the tests will reject the null of error autocorrelation even when $\lambda = 0$ due to the local presence of the lag dependence. In a similar way we can express the asymptotic distributions of RS_ρ, LR_ρ, and WS_ρ. Under $\rho = 0$ and local presence of error dependence, say, $\lambda = \tau/\sqrt{N}$, $\tau < \infty$. In this case the distributions remain χ_1^2, but with a noncentrality parameter

$$R_\lambda = \frac{\tau^2 T_{12}^2 \sigma^2}{N D} \tag{76}$$

where $D = (W_1 X \beta)' M (W_1 X \beta) + T_{11} \sigma^2$. Therefore, again we will have unwanted "power" due to the presence of local error dependence. In the noncentrality parameters R_ρ and R_λ, the crucial quantity is $T_{12}/\sqrt{N}$, which can be interpreted as the covariance between the scores d_λ and d_ρ. Note that if $T_{12} = 0$, then both R_ρ and R_λ vanish, and local presence of one kind of dependence cannot affect the test for the other one. The trace term $T_{12} = \text{tr}[W_1 W_2 + W_1' W_2]$, which will only be zero when the nonzero elements in each row/column of the weights matrices W_1 and W_2 do not overlap. In other words, this will be the case when the pattern of spatial dependence in the lag term and in the error term pertain to a completely different set of neighbors for each observation. However, in the typical case where $W_1 = W_2$ (or overlap to any extent) then the noncentrality parameter will not vanish.

For valid statistical inference there is a need to take account of possible lag dependence while we test for error dependence, and vice versa. In Anselin (1988c) two different approaches are suggested. One is to test jointly for $H_0 : \lambda = \rho = 0$ in (73) using the RS principle so that the test can be implemented with OLS residuals (see Anselin 1988c). The resulting joint test statistic is given by

$$RS_{\lambda\rho} = \tilde{E}^{-1} \left[(\tilde{d}_\lambda)^2 \frac{\tilde{D}}{\tilde{\sigma}^2} + (\tilde{d}_\rho)^2 T_{22} - 2\tilde{d}_\lambda \tilde{d}_\rho T_{12} \right] \tag{77}$$

where $E = (D/\sigma^2)T_{22} - (T_{12})^2$. Note that this joint test not only depends on d_λ and d_ρ but also on their interaction factor with a coefficient T_{12}. Expression (77) appears to be somewhat complicated but can be computed quite easily using only OLS residuals. Also the expression simplifies greatly when the spatial weights matrices W_1 and W_2 are assumed to be the same which is the case in most applications. Under $W_1 = W_2 = W$, $T_{11} = T_{21} = T_{22} = T = \mathrm{tr}[(W' + W)W]$, and (77) reduces to

$$RS_{\lambda\rho} = \frac{\tilde{d}_\lambda^2}{T} + \frac{(\tilde{d}_\lambda - \tilde{d}_\rho)^2}{\tilde{\sigma}^{-2}(\tilde{D} - T\tilde{\sigma}^2)} \tag{78}$$

Under $H_0 : \lambda = \rho = 0$, $RS_{\lambda\rho}$ will converge to a central χ^2 with *two* degrees of freedom. Because of this two degrees of freedom, the statistic will result in loss of power compared to the proper one-directional test when only one of the two forms of misspecification is present. To see this consider the presence of only $\lambda = \tau/\sqrt{N}$, with $\rho = 0$. In this case the noncentrality parameter for both RS_λ and $RS_{\lambda\rho}$ is the same $\tau^2 NT$. Due to the higher degrees of freedom of the joint test $RS_{\lambda\rho}$, we can expect some loss of power (Dasgupta and Perlman 1974). Another problem with $RS_{\lambda\rho}$ is that since it is an omnibus test, if the null hypothesis is rejected, it is not possible to infer whether the misspecification is due to lag or error dependence.

A second approach is to carry out an RS test for one form of misspecification in a model where the other form is unconstrained. For example, this consists of testing the null hypothesis $H_0 : \lambda = 0$ in the presence of ρ, i.e., based on the residuals of a maximum likelihood estimation of the spatial lag model. The resulting statistic $RS_{\lambda|\rho}$ is given as

$$RS_{\lambda|\rho} = \frac{\hat{d}_\rho^2}{T_{22} - (T_{21A})^2 \widehat{\mathrm{Var}}(\hat{\rho})} \tag{79}$$

where the "hat" denotes quantities are evaluated at the maximum likelihood estimates of the parameters of the model $Y = \rho W_1 y + X\beta + \xi$ obtained by means of nonlinear optimization. In (79) T_{21A} stands for $\mathrm{tr}[W_2 W_1 A^{-1} + W_2' W_1 A^{-1}]$, with $A = I - \hat{\rho} W_1$. Under $H_0 : \lambda = 0$, $RS_{\lambda/\rho}$ will converge to a central χ^2 with one degree of freedom. Similarly, an RS test can be developed for $H_0 : \rho = 0$ in the presence of error dependence (Anselin et al. 1996). This test statistic can be written as

$$RS_{\rho|\lambda} = \frac{[\hat{\varepsilon}' B' B W_1 y]^2}{H_\rho - H_{\theta\rho} \widehat{\mathrm{Var}}(\hat{\theta}) H_{\theta\rho}'} \tag{80}$$

where $\hat{\varepsilon}$ is a vector of residuals in the ML estimation of the null model with spatial AR errors, $y = X\beta + (I - \lambda W_2)^{-1}\xi$ with $\theta = (\beta', \lambda, \sigma^2)'$, and $B = I - \hat{\lambda} W_2$. The terms in the denominator of (80) are

$$H_\rho = \mathrm{tr}\, W_1^2 + \mathrm{tr}(BW_1B^{-1})'(BW_1B^{-1}) + \frac{(BW_1X\beta)'(BW_1X\beta)}{\sigma^2}$$

$$H'_{\theta\rho} = \begin{bmatrix} \dfrac{(BX)'BW_1X\beta}{\sigma^2} \\ \mathrm{tr}(W_2B^{-1})'BW_1B^{-1} + \mathrm{tr}\, W_2W_1B^{-1} \\ 0 \end{bmatrix}$$

and $\widehat{\mathrm{Var}}(\hat\theta$ is the estimated variance-covariance matrix for the parameter vector θ.

It is also possible to obtain the W and LR statistics in the above three cases, though these will involve the estimation of a spatial model with two parameters, requiring considerably more complex nonlinear optimization. In contrast, $RS_{\lambda|\rho}$ and $RS_{\rho|\lambda}$ are theoretically valid statistics that have the potential to identify the possible source(s) of misspecification and can be derived from the results of the maximization of the log-likelihood functions (32) and (26). However, this is clearly more computationally demanding than tests based on OLS residuals. We now turn to an approach that accomplishes carrying out the tests without maximum likelihood estimation of λ and ρ.

F.   Robust Test in the Presence of Local Misspecification

It is not possible to robustify tests in the presence of *global* misspecification (i.e., λ and ρ taking values far away from zero). However, using the general approach of Bera and Yoon (1993), Anselin et al. (1996) suggested tests which are robust to *local* misspecifications, as defined in the previous subsection. The idea is to adjust the one-directional score tests RS_λ and RS_ρ by taking account of the noncentrality parameters R_ρ and R_λ, given in (75) and (76), so that under the null the resulting test statistics have *central* χ_1^2 distributions.

The modified test for $H_0 : \lambda = 0$ in the local presence of ρ is given by

$$RS_\lambda^* = \frac{[\tilde{d}_\lambda - T_{12}\tilde\sigma^2\tilde{D}^{-1}\tilde{d}_\rho]^2}{T_{22} - (T_{12})^2\tilde\sigma^2\tilde{D}} \tag{81}$$

When $W_1 = W_2 = W$, RS_λ^* becomes

$$RS_\lambda^* = \frac{[\tilde{d}_\lambda - T\tilde\sigma^2\tilde{D}^{-1}\tilde{d}_\rho]^2}{T(1 - T\tilde\sigma^2\tilde{D})} \tag{82}$$

Comparing RS_λ^* in (81) and RS_λ in (59), it is clear that the adjusted test modifies RS_λ by correcting for the presence of ρ through $\tilde{d}_\rho$ and T_{12}, where the latter quantity represents the covariance between d_λ and d_ρ. Under H_0: $\lambda = 0$ (and $\rho = \delta/\sqrt{N}$), RS_λ^* converges to a *central* χ_1^2 distribution; i.e., RS_λ^* has the same asymptotic distribution as RS_λ based on the correct specification. This therefore produces asymptotically the correct size in the presence of local lag dependence. Also as noted for RS_λ^*, we only need OLS estimation thus circumventing direct estimation of the nuisance parameter ρ. However, there is a price to be paid for robustification and simplicity in estimation. Consider the case when there is no lag dependence ($\rho = 0$), but only spatial error dependence through $\lambda = \tau/\sqrt{N}$. Under this setup, the noncentrality parameters of RS_λ and RS_λ^* are respectively $\tau^2 T_{22}/N$ and $\tau^2 (T_{22} - T_{12}^2 \sigma^2 D^{-1})/N$. Since $\tau^2 T_{12}^2 \sigma^2 D^{-1}/N \geq 0$, the asymptotic power of RS_λ^* will be less than that of RS_λ when there is no lag dependence. The above quantity can be regarded as a cost of robustification. Once again, note its dependence on T_{12}. It is also instructive to compare RS_λ^* with Anselin's $RS_{\lambda|\rho}$ in (79). It is readly seen that $RS_{\lambda|\rho}$ does not have the mean correction factor. $RS_{\lambda|\rho}$ uses the restricted ML estimator of ρ (under $\lambda = 0$) for which $\hat{d}_\rho = 0$. We may view $RS_{\lambda|\rho}$ as the spatial version of Durbin's h statistic, which can also be derived from the general RS principle. Unlike Durbin's h, however, $RS_{\lambda|\rho}$ cannot be computed using the OLS residuals.

In a similar way, the adjusted score test for $H_0 : \rho = 0$, in the presence of local misspecification involving spatial-dependent error process can be expressed as

$$RS_\rho^* = \frac{[\tilde{d}_\rho - T_{12}T_{22}^{-1}\tilde{d}_\lambda]^2}{\tilde{\sigma}^{-2}\tilde{D} - (T_{12})^2 T_{22}^{-1}} \tag{83}$$

Under $W_1 = W_2 = W$, the above expression simplifies to

$$RS_\rho^* = \frac{[\tilde{d}_\rho - \tilde{d}_\lambda]^2}{\tilde{\sigma}^{-2}\tilde{D} - T} \tag{84}$$

All our earlier discussion of RS_λ^* also applies to RS_ρ^*.

Finally, consider the relationship among the five statistics RS_λ, RS_ρ, RS_λ^*, RS_ρ^*, and $RS_{\lambda\rho}$ given in (59), (71), (82), (84), and (78) respectively. $RS_{\lambda\rho}$ is not the sum of RS_λ and RS_ρ; i.e., there is no additivity of the score tests along the lines discussed in Bera and Jarque (1982) and Bera and McKenzie (1987). From (77), it is clear that additivity follows only if $T_{12} = 0$ or $T = 0$ for the case of $W_1 = W_2$, i.e., when d_λ and d_ρ are asymptotically uncorrelated. In that case also $RS_\lambda^* = RS_\lambda$ and $RS_\rho^* = RS_\rho$ (see (81), (59), (83), and (71)). Hence, for $T = 0$, the conventional one-directional tests RS_λ and RS_ρ are asymptotically valid in the presence of local misspecification. However, as noted earlier $T > 0$ and $T_{12} > 0$ when W_1 and W_2 have some overlap in the neighbor structure. Under these circumstances (which are the most common situation encountered in practice), the following very intriguing result is obtained:

$$RS_{\lambda\rho} = RS_\lambda^* + RS_\rho = RS_\lambda + RS_\rho^* \tag{85}$$

i.e., the two-directional test for λ and ρ can be decomposed into the sum of the adjusted one-directional test of one type of alternative and the unadjusted form for the other. By construction, under $\lambda = \rho = 0$, RS_λ^* and RS_ρ are asymptotically independently distributed, which cannot be said about RS_λ and RS_ρ. By applying all the unadjusted and adjusted tests and exploiting the result (85), it is possible to identify the exact nature of dependence in practice (Anselin et al. 1996). Finally, we should mention that because of the complexity of the Wald and LR tests, it is not possible to derive their adjusted versions that would be valid under local misspecification. Of course, it is not computationally prohibitive to obtain these tests after the joint estimation of both λ and ρ.

G. Small Sample Properties

We have covered a number of procedures for testing spatial dependence. For ease of implementation, we have emphasized Rao's score test which in many cases can be computed based on the OLS residuals. As we indicated, all these tests are of asymptotic nature; i.e., their justification derives from the presence of very large samples. That is, however, not the case in most applications. The small sample performance of the above tests both in terms of size and power is of major concern to practitioners.

There are only a few papers on the finite sample properties of tests on spatial dependence compared to the vast literature on those for testing for serial correlation for time-series data as summarized in King (1987). Bartels and Hordijk (1977) studied the behavior of Moran's I. However, their focus was on the performance of different residuals, and they found that OLS residuals give the best results. Brandsma and Ketellapper (1979b) included the LR test (LR_λ) in their study, but it performed poorly compared to I. Both these studies were quite limited in terms of a small number of replications, few sample sizes, the use of only one type (irregular) weights matrix and the narrow range of alternative values for the autocorrelation coefficient. A first extensive set of Monte Carlo simulations was carried out by Anselin and Rey (1991), who compared Moran's I to RS_λ and RS_ρ for different weights matrices and error distributions. In terms of size, the small sample distributions of the statistics corresponded close to their theoretical counterparts, except for the smallest size ($N = 25$). In terms of power, Moran's I had power against both kinds of dependence, spatial lag and error autocorrelations. RS_λ and RS_ρ had highest power for their respective designated alternatives. These tests were found to possess superior performance, but they fall short of providing a good strategy for identifying the exact nature of dependence.

Anselin and Florax (1995b) provide the most comprehensive set of simulation results to date. They carried out experiments for both regular (rook and queen) and nonregular weight matrices, single- and multidirectional alternatives, and for different error distributions, and included all the tests discussed earlier except the Wald and LR tests. The results are too extensive to discuss in detail, and here we provide only a brief summary of the main findings. First, the earlier results of Anselin and Rey (1991) were confirmed on the power of I against any form of dependence and the optimality of the RS tests against the alternatives for which they were designed. Second, the specification of the spatial weights matrix impacted the performance of all tests, with a higher power obtained in the rook case. Third, as in Anselin and Rey (1991), higher powers were achieved by lag tests relative to tests against error dependence. This is important, since the consequences of ignoring lag dependence are more serious. Fourth, the KR statistic did not perform well. For example, when the errors were generated as lognormal, it significantly over rejected the true null hypothesis in all configurations. There are two possible explanations. One is its higher degrees of freedom. Another is that its power depends on the degree of autocorrelation in the explanatory variables which substitute for the weights matrix (compare (55) and (56)). It is interesting to note that White's (1980) test for heteroskedasticity which is very similar to KR encounters problems of the same type. Fifth, the most striking result is that the adjusted tests RS_λ^* and $RS\rho^*$ performed remarkably well. They had reasonable empirical sizes, remaining within the confidence interval in all cases. In terms of power they performed exactly the way they were supposed to. For instance, when the data were generated under $\rho > 0$, $\lambda = 0$, although RS_ρ had the most power, the powers of RS_ρ^* was very close to that of RS_ρ. That is, the price paid for adjustments that were not needed turned out to be small. The real superiority of RS_ρ^* was revealed when $\lambda > 0$ and $\rho = 0$. It yielded low rejection frequencies even for $\lambda = 0.9$. The correction for error dependence in RS_ρ^* worked in the right direction when no lagged dependence was present for all configurations. When $\rho > 0$, the power function of RS_ρ^* was seen to be almost unaffected by the values of λ, even for those far away from zero (global misspecification). For these alternatives $RS_{\lambda\rho}$ also had good power, but could not point to the correct alternative when only one kind of dependence is present. RS_λ^* also performed well though not as spectacularly as RS_ρ^*. The adjusted tests thus seem more appropriate to test for lag dependence in the presence of error correlation than for the reverse case. Again, this is important since ignoring lag dependence has more severe consequences. Based on these results Anselin and Florax (1995b) suggested a simple decision rule. When RS_ρ is more significant than RS_λ, and RS_ρ^* is significant while RS_λ^* is not, a lag dependence is the likely alternative. In a similar way presence of error dependence can be identified through RS_λ^*. Finally, the finite-sample performance of tests against higher-order dependence $RS_{\lambda_1\lambda_2}$ (see (68)) and the joint test $RS_{\lambda\rho}$ were satisfactory, although these type of tests provide less insightful guidance for an effective specification search. For joint and higher-order alternatives, these tests are optimal, and in practice they should be used along with the unadjusted and adjusted one-directional tests.

H. Operational Implementation and Illustration

As is the case for the estimation of spatial regression models, specification tests for spatial dependence are notably absent from econometric software, with the exception of SpaceStat (Anselin 1992b, 1995). Moreover, as pointed out, these tests cannot be obtained in the usual NR^2 format, which lends itself to straightforward implementation by means of auxiliary or augmented regressions. The closest to this situation is the Kelejian-Robinson test (54), provided one has an easy way to select the pairs of neighboring data points from the data. Typically, specification tests for spatial dependence must be implemented explicitly either by writing special-purpose software or by taking advantage of macros in econometric and statistical software. As in maximum likelihood estimation, the size of the weights matrix may be a constraint when the number of observations is large. This is particularly the case for Moran's I, where several operations are involved in the computation of the expected value and variance (46) and (47). Examples of the implementation of this test for small data sets in standard econometric software are given in Anselin and Hudak (1992) and Anselin et al. (1993b), for Shazam, Rats and Limdep, among others.

Given their importance for applied work, we now briefly describe implementation strategies for the RS tests for spatial error and spatial lag autocorrelation, RS_λ (59) and RS_ρ (71). First, note that the squared expression in the numerator equals N times a regression coefficient of an auxiliary regression of respectively We on e (in (59)) and Wy on e (in (71)). Once the lags are constructed, these coefficients can be obtained using standard software. The denominator in the expressions is slightly more complex. The trace elements $T = \text{tr}(WW) + \text{tr}(W'W)$ can easily be seen to equal, respectively, $\sum_i \sum_j w_{ij}.w_{ji}$ and $\sum_i \sum_j (w_{ij})^2$. When the spatial weights matrix consists of simple row-standardized contiguity weights, each element w_{ij} for a given i equals $1/k_i$, where k_i is the number of neighbors for observation i. Hence, $\sum_i \sum_j (w_{ij})^2 = \sum_i (1/k_i)$, which can easily be computed. The other trace term is $\sum_i \sum_j w_{ij}.w_{ji} = \sum_i (1/k_i)[\sum_j \delta_{ij}/k_j]$, where δ_{ij} is a binary variable indicating whether or not $w_{ij} \neq 0$. This requires only slighly more work to compute, similar to the sorting needed to establish the neighbor pairs in the Kelejian-Robinson test. Most importantly, the trace operations can be carried out without having to store a full matrix in memory, taking advantage of the sparse nature of spatial weights (for technical details, see Anselin 1995). Of course, for symmetric weights, the two traces are equal. In practice, this may occur when all observations are considered to have an equal number of neighbors, as in Pace and Barry (1996). The other term in the denominator of (71) is the residual sum of squares in a regression with WXb (i.e., the spatial lags for the predicted values from the OLS regression) on X, which can be obtained in a straightforward way.

To illustrate the various specification tests, we list the results of Moran's I, KR, and the RS and LR tests for the spatial model of crime in Table 2 (using a slightly different notation, most of these results are reported in Table 2, p. 87 of Anselin et al. 1996). All results are part of the standard SpaceStat regression diagnostic output. They reflect a situation that is often encountered in practice: strong significance of Moran's I and KR, as well as of both one-directional RS and LR tests. Clearly, spatial dependence is a problem, although without further investigation it is not obvious which form of spatial dependence is the proper alternative. Convincing evidence is provided by the robust tests RS_λ^* and RS_ρ^*. While the former is not at all significant,

Table 2 Specification Tests against Spatial Dependence[a]

Estimates	Test (equation number)	Value	p-value
OLS	Moran's I (z-value) (43)	2.95	0.003
OLS	Kelejian-Robinson (54)	11.55	0.009
OLS	$RS_{\lambda\rho}$ (78)	9.44	0.009
OLS	RS_λ (59)	5.72	0.02
OLS	RS_λ^* (82)	0.08	0.78
OLS	RS_ρ (71)	9.36	0.002
OLS	RS_ρ^* (84)	3.72	0.05
Lag-ML	LR_ρ (72)	9.97	0.002
Lag-ML	$RS_{\lambda|\rho}$ (79)	0.32	0.57
Err-ML	LR_λ (66)	7.99	0.005
Err-ML	$RS_{\rho|\lambda}$ (80)	1.76	0.18

[a]*Source*: From Anselin (1988a, Chap. 12; 1992a, Chap. 26; 1995) and Anselin et al. (1996).

the latter is significant at ρ slightly higher than 0.05. In other words, the impression of spatial error autocorrelation that may be given by an uncritical interpretation of Moran's I is spurious, since no evidence of such autocorrelation remains after robustifying for spatial lag dependence. Instead, a spatial lag model is the suggested alternative, consistent with the estimation results in Table 1.

V. CONCLUSIONS

In our review of methods to deal with spatial dependence in regression analysis, we have emphasized the distinguishing characteristics of spatial econometrics relative to time-series analysis. We highlighted the concept of spatial weights and the associated spatial lag operator which allow for the formal specification of neighbors in space, a much more general concept than its counterpart in time. In the estimation of spatial regression models, the maximum likelihood approach was shown to be prevalent and requiring nonlinear optimization of the likelihood function. The simplifying results from serial correlation in time series do not hold and estimation necessitates the explicit manipulation of matrices of dimension equal to the number of observations. Diagnostics for spatial effects in regression models may be based on the powerful score principle, but they do not boil down to simple significance tests of the coefficients in an auxiliary regression, as they do for time series.

The differences between the time domain and space are both puzzling and challenging, in terms of theory as well as from an applied perspective. They are the subject of active research efforts to develop diagnostics for multiple sources of misspecification, to discriminate between heterogeneity and spatial dependence, and to estimate models for complex forms of interaction in realistic data settings. Extensions to the space-time domain and to models for limited dependent variables are particularly challenging. We hope that our review of the fundamental concepts and basic methods will stimulate others to both apply these techniques as well as to pursue solutions for the remaining research questions.

ACKNOWLEDGMENTS

We would like to thank Aman Ullah and an anonymous referee for helpful suggestions, and Robert Rozovsky for very able research assistance. We also would like to thank Naoko Miki for her help in preparing the manuscript. However, we retain the responsibility for any remaining errors. The first author acknowledges ongoing support for the development of spatial econometric methods by the U.S. National Science Foundation, notably through grants SES 87-21875, SES 89-21385, and SBR 94-10612 as well as grant SES 88-10917 to the National Center for Geographic Information and Analysis (NCGIA). The second author acknowledges financial support by the Bureau of Economic and Business Research of the University of Illinois.

REFERENCES

Albert, P. and L. M. McShane (1995), A Generalized Estimating Equations Approach for Spatially Correlated Binary Data: Applications to the Analysis of Neuroimaging Data, *Biometrics*, 51, 627–638.

Amemiya, T. (1985), *Advanced Econometrics*, Harvard University Press, Cambridge, MA.

Ancot, J.-P., J. Paelinck, and J. Prins (1986), Some New Estimators in Spatial Econometrics, *Economics Letters*, 21, 245–249.

Anselin, L. (1980), Estimation Methods for Spatial Autoregressive Structures, *Regional Science Dissertation and Monograph Series 8*, Cornell University, Ithaca, NY.

Anselin, L. (1982), A Note on Small Sample Properties of Estimators in a First-Order Spatial Autoregressive Model, *Environment and Planning A*, 14, 1023–1030.

Anselin, L. (1988a), *Spatial Econometrics: Methods and Models*, Kluwer, Dordrecht.

Anselin, L. (1988b), Model Validation in Spatial Econometrics: A Review and Evaluation of Alternative Approaches, *International Regional Science Review*, 11, 279–316.

Anselin, L. (1988c), Lagrange Multiplier Test Diagnostics for Spatial Dependence and Spatial Heterogeneity, *Geographical Analysis*, 20, 1–17.

Anselin, L. (1990a), Spatial Dependence and Spatial Structural Instability in Applied Regression Analysis, *Journal of Regional Science*, 30, 185–207.

Anselin, L. (1990b), Some Robust Approaches to Testing and Estimation in Spatial Econometrics, *Regional Science and Urban Economics*, 20, 141–163.

Anselin, L. (ed.) (1992a), *Space and Applied Econometrics*. Special Issue, *Regional Science and Urban Economics*, 22.

Anselin, L. (1992b), *SpaceStat: A Program for the Analysis of Spatial Data*, National Center for Geographic Information and Analysis, University of California, Santa Barbara, CA.

Anselin, L. (1995), *SpaceStat Version 1.80 User's Guide*, Regional Research Institute, West Virginia University, Morgantown, WV.

Anselin, L., A. K. Bera, R. Florax, and M. J. Yoon (1996), Simple Diagnostic Tests for Spatial Dependence, *Regional Science and Urban Economics*, 26, 77–104.

Anselin, L. and A. Can (1996), Spatial Effects in Models of Mortgage Origination, Paper presented at the Mid Year AREUEA Conference, Washington, DC, May 28–29.

Anselin, L., R. Dodson, and S. Hudak (1993a), Linking GIS and Spatial Data Analysis in Practice, *Geographical Systems*, 1, 3–23.

Anselin, L., R. Dodson, and S. Hudak (1993b), *Spatial Data Analysis and GIS: Interfacing GIS and Econometric Software*, Technical Report 93-7, National Center for Geographic Information and Analysis, University of California, Santa Barbara.

Anselin, L. and R. Florax (eds.) (1995a), *New Directions in Spatial Econometrics*, Springer-Verlag, Berlin.

Anselin, L. and R. Florax (1995b), Small Sample Properties of Tests for Spatial Dependence in Regression Models: Some Further Results, in L. Anselin and R. Florax (eds.), *New Directions in Spatial Econometrics*, Springer-Verlag, Berlin, 21–74.

Anselin, L. and S. Hudak (1992), Spatial Econometrics in Practice, a Review of Software Options, *Regional Science and Urban Economics*, 22, 509–536.

Anselin, L. and S. Rey (1991), Properties of Tests for Spatial Dependence in Linear Regression Models, *Geographic Analysis*, 23, 112–131.

Anselin, L. and S. Rey (eds.) (1997), *Spatial Econometrics*. Special Issue, *International Regional Science Review*, 20.

Anselin, L. and O. Smirnov (1996), Efficient Algorithms for Constructing Proper Higher Order Spatial Lag Operators, *Journal of Regional Science*, 36, 67–89.

Arora, S. and M. Brown (1977), Alternative Approaches to Spatial Autocorrelation: An Improvement over Current Practice, *International Regional Science Review*, 2, 67–78.

Bartels, C. P. A. and L. Hordijk (1977), On the Power of the Generalized Moran Contiguity Coefficient in Testing for Spatial Autocorrelation among Regression Disturbances, *Regional Science and Urban Economics*, 7, 83–101.

Bartels, C. and R. Ketellapper (eds.) (1979), *Exploratory and Explanatory Analysis of Spatial Data*, Martinus Nijhoff, Boston.

Benirschka, M. and J. K. Binkley (1994), Land Price Volatility in a Geographically Dispersed Market, *American Journal of Agricultural Economics*, 76, 185–195.

Bera, A. K. and C. M. Jarque (1982), Model Specification Tests: A Simultaneous Approach, *Journal of Econometrics*, 20, 59–82.

Bera, A. K. and C. R. McKenzie (1986), Alternative Forms and Properties of the Score Test, *Journal of Applied Statistics*, 13, 13–25.

Bera, A. K. and C. R. McKenzie (1987), Additivity and Separability of the Lagrange Multiplier, Likelihood Ratio and Wald Tests, *Journal of Quantitative Economics*, 3, 53–63.

Beron, K. J., J. C. Murdoch, and W. P. M. Vijverberg (1996), Why Cooperate? An Interdependent Probit Model of Network Correlations, Working Paper, School of Social Sciences, University of Texas at Dallas, Richardson, TX.

Bera, A. K. and A. Ullah (1991), Rao's Score Test in Econometrics, *Journal of Quantitative Economics*, 7, 189–220.

Bera, A. K. and M. J. Yoon (1993), Specification Testing with Misspecified Alternatives, *Econometric Theory*, 9, 649–658.

Besag, J. (1974), Spatial Interaction and the Statistical Analysis of Lattice Systems, *Journal of the Royal Statistical Society*, B, 36, 192–225.

Besley, T. and A. Case (1995), Incumbent Behavior: Vote-Seeking, Tax-Setting, and Yardstick Competition, *American Economic Review*, 85, 25–45.

Bivand, R. (1992), Systat Compatible Software for Modeling Spatial Dependence among Observations, *Computers and Geosciences*, 18, 951–963.

Blommestein, H. (1983), Specification and Estimation of Spatial Econometric Models: A Discussion of Alternative Strategies for Spatial Economic Modelling, *Regional Science and Urban Economics*, 13, 250–271.

Blommestein, H. (1985), Elimination of Circular Routes in Spatial Dynamic Regression Equations, *Regional Science and Urban Economics*, 15, 121–130.

Blommestein, H. J. and N. A. Koper (1992), Recursive Algorithms for the Elimination of Redundant Paths in Spatial Lag Operators, *Journal of Regional Science*, 32, 91–111.

Bolduc, D., M. G. Dagenais, and M. J. Gaudry (1989), Spatially Autocorrelated Errors in Origin-Destination Models: A New Specification Applied to Urban Travel Demand in Winnipeg, *Transportation Research*, B 23, 361–372.

Bolduc, D., R. Laferrière, and G. Santarossa (1992), Spatial Autoregressive Error Components in Travel Flow Models, *Regional Science and Urban Economics*, 22, 371–385.

Bolduc, D., R. Laferrière, and G. Santarossa (1995), Spatial Autoregressive Error Components in Travel Flow Models, an Application to Aggregate Mode Choice, in L. Anselin and R. Florax (eds.), *New Directions in Spatial Econometrics*, Springer-Verlag, Berlin, 96–108.

Bowden, R. J. and D. A. Turkington (1984), *Instrumental Variables*, Cambridge University Press, Cambridge.

Box, G. E. P., G. M. Jenkins, and G. C. Reinsel (1994), *Time Series Analysis, Forecasting and Control*, 3rd ed., Prentice Hall, Englewod Cliffs, NJ.

Brandsma, A. S. and R. H. Ketellapper (1979a), A Biparametric Approach to Spatial Autocorrelation, *Environment and Planning A*, 11, 51–58.

Brandsma, A. S. and R. H. Ketellapper (1979b), Further Evidence on Alternative Procedures for Testing of Spatial Autocorrelation among Regression Disturbances, in C. Bartels and R. Ketellapper (eds.), *Exploratory and Explanatory Analysis in Spatial Data*, Martin Nijhoff, Boston, 111–136.

Brett, C. and C. A. P. Pinkse (1997), Those Taxes Are All over the Map! A Test for Spatial Independence of Municipal Tax Rates in British Columbia, *International Regional Science Review*, 20, 131–151.

Breusch, T. (1980), useful Invariance Results for Generalized Regression Models, *Journal of Econometrics*, 13, 327–340.

Brook, D. (1964), On the Distinction between the Conditional Probability and Joint Probability Approaches in the Specification of Nearest Neighbor Systems, *Biometrika*, 51, 481–483.

Brueckner, J. K. (1996), Testing for Strategic Interaction among Local Governments: The Case of Growth Controls, Discussion Paper, Department of Economics, University of Illinois, Champaign.

Burridge, P. (1980), On the Cliff-Ord Test for Spatial Autocorrelation, *Journal of the Royal Statistical Society B*, 42, 107–108.

Can, A. (1992), Specification and Estimation of Hedonic Housing Price Models, *Regional Science and Urban Economics*, 22, 453–474.

Can, A. (1996), Weight Matrices and Spatial Autocorrelation Statistics Using a Topological Vector Data Model, *International Journal of Geographical Information Systems*, 10, 1009–1017.

Can, A. and I. F. Megbolugbe (1997), Spatial Dependence and House Price Index Construction, *Journal of Real Estate Finance and Economics*, 14, 203–222.

Case, A. (1987), On the Use of Spatial Autoregressive Models in Demand Analysis, Discussion Paper 135, Research Program in Development Studies, Woodrow Wilson School, Princeton University.

Case, A. (1991), Spatial Patterns in Household Demand, *Econometrica*, 59, 953–965.

Case, A. (1992). Neighborhood Influence and Technological Change, *Regional Science and Urban Economics*, 22, 491–508.

Case, A. C., H. S. Rosen, and J. R. Hines (1993), Budget Spillovers and Fiscal Policy Interdependence: Evidence from the States, *Journal of Public Economics*, 52, 285–307.

Cliff, A. and J. K. Ord (1972), Testing for Spatial Autocorrelation among Regression Residuals, *Geographic Analysis*, 4, 267–284.

Cliff, A. and J. K. Ord (1973), *Spatial Autocorrelation*, Pion, London.

Cliff, A. and J. K. Ord (1981), *Spatial Processes: Models and Applications*, Pion, London.

Cressie, N. (1991), Geostatistical Analysis of Spatial Data, in National Research Council, *Spatial Statistics and Digital Image Analysis*, National Academy Press, Washington, DC, 87–108.

Cressie, N (1993), *Statistics for Spatial Data*, Wiley, New York.

Dasgupta, S. and M. D. Perlman (1974), Power of the Noncentral F-test: Effect of Additional Variate on Hotelling's T^2-test, *Journal of the American Statistical Association*, 69, 174–180.

Davidson, R. and J. G. MacKinnon (1987), Implicit Alternatives and Local Power of Test Statistics, *Econometrica*, 55, 1305–1329.

Davidson, R. and J. G. MacKinnon (1993), *Estimation and Inference in Econometrics*, Oxford University Press, New York.

Deutsch, C. V. and A. G. Journel (1992), *GSLIB: Geostatistical Software Library and User's Guide*, Oxford University Press, Oxford.

Doreian, P. (1980), Linear Models with Spatially Distributed Data, Spatial Disturbances or Spatial Effects, *Sociological Methods and Research*, 9, 29–60.

Doreian, P., K. Teuter, and C-H. Wang (1984), Network Autocorrelation Models, *Sociological Methods and Research*, 13, 155–200.

Dow, M. M., M. L. Burton, and D. R. White (1982), Network Autocorrelation: A Simulation Study of a Foundational Problem in Regression and Survey Study Research, *Social Networks*, 4, 169–200.

Dubin, R. (1988), Estimation of Regression Coefficients in the Presence of Spatially Autocorrelated Error Terms, *Review of Economics and Statistics*, 70, 466–474.

Dubin, R. (1992), Spatial Autocorrelation and Neighborhood Quality, *Regional Science and Urban Economics*, 22, 433–452.

Durbin, J. and G. S. Watson (1950), Testing for Serial Correlation in Least Squares Regression I, *Biometrika*, 37, 409–428.

Durbin, J. and G. S. Watson (1951), Testing for Serial Correlation in Least Squares Regression II, *Biometrika*, 38, 159–179.

Florax, R. and S. Rey (1995), The Impacts of Misspecified Spatial Interaction in Linear Regression Models, in L. Anselin and R. Florax (eds.), *New Directions in Spatial Econometrics*, Springer-Verlag, Berlin, 111–135.

Fomby, T. B., R. C. Hill, and S. R. Johnson (1984), *Advanced Econometric Methods*, Springer-Verlag, New York.

Getis, A. (1995), Spatial Filtering in a Regression Framework: Examples Using Data on Urban Crime, Regional Inequality, and Government Expenditures, in L. Anselin and R. Florax (eds.), *New Directions in Spatial Econometrics*, Springer-Verlag, Berlin, 172–185.

Godfrey, L. (1981), On the Invariance of the Lagrange Multiplier Test with Respect to Certain Changes in the Alternative Hypothesis, *Econometrica*, 49, 1443–1455.

Godfrey, L. (1988), *Misspecification Tests in Econometrics*, Cambridge University Press, New York.

Greene, W. H. (1993), *Econometric Analysis*, 2nd ed., Macmillan, New York.

Griffith, D. A. (1987), *Spatial Autocorrelation, A Primer*, Association of American Geographers, Washington, DC.

Griffith, D. A. (1993), *Spatial Regression Analysis on the PC: Spatial Statistics Using SAS*, Association of American Geographers, Washington, DC.

Haining, R. (1984), Testing a Spatial Interacting-Markets Hypothesis, *The Review of Economics and Statistics*, 66, 576–583.

Haining, R. (1988), Estimating Spatial Means with an Application to Remotely Sensed Data, *Communications in Statistics: Theory and Methods*, 17, 573–597.

Haining, R. (1990), *Spatial Data Analysis in the Social and Environmental Sciences*, Cambridge University Press, Cambridge.

Heijmans, R. and J. Magnus (1986a), On the First-Order Efficiency and Asymptotic Normality of Maximum Likelihood Estimators Obtained from Dependent Observations, *Statistica Neerlandica*, 40, 169–188.

Heijmans, R. and J. Magnus (1986b), Asymptotic Normality of Maximum Likelihood Estimators Obtained from Normally Distributed but Dependent Observations, *Econometric Theory*, 2, 374–412.

Hepple, L. W. (1995a), Bayesian Techniques in Spatial and Network Econometrics. 1: Model Comparison and Posterior Odds, *Environment and Planning A*, 27, 447–469.

Hepple, L. W. (1995b), Bayesian Techniques in Spatial and Network Econometrics. 2: Computational Methods and Algorithms, *Environment and Planning A*, 27, 615–644.

Holtz-Eakin, D. (1994), Public-Sector Capital and the Productivity Puzzle, *Review of Economics and Statistics*, 76, 12–21.

Hooper, P. and G. Hewings (1981), Some Properties of Space-Time Processes, *Geographical Analysis*, 13, 203–223.

Hordijk, L. (1974), Spatial Correlation in the Disturbances of a Linear Interregional Model, *Regional and Urban Economics*, 4, 117–140.

Hordijk, L. (1979), Problems in Estimating Econometric Relations in Space, *Papers, Regional Science Association*, 42, 99–115.

Hordijk, L. and J. Paelinck (1976), Some Principles and Results in Spatial Econometrics, *Recherches Economiques de Louvain*, 42, 175–197.

Huang, J. S. (1984). The Autoregressive Moving Average Model for Spatial Analysis, *Australian Journal of Statistics*, 26, 169–178.

Imhof, J. P. (1961), Computing the Distribution of Quadratic Forms in Normal Variables, *Biometrika*, 48, 419–426.

Isaaks, E. H. and R. M. Srivastava (1989), *An Introduction to Applied Geostatistics*, Oxford University Press, Oxford.

Johnson, N. L., and S. Kotz (1969), *Distributions in Statistics: Discrete Distributions*, Houghton Mifflin, Boston.

Johnston, J. (1984), *Econometric Models*, McGraw-Hill, New York.

Judge, G., R. C. Hill, W. E. Griffiths, H. Lütkepohl, and T-C. Lee (1982), *Introduction to the Theory and Practice of Econometrics*, Wiley, New York.

Judge, G., W. E. Griffiths, R. C. Hill, H. Lütkepohl, and T.-C. Lee (1985), *The Theory and Practice of Econometrics*, 2nd ed., Wiley, New York.

Kelejian, H. and D. Robinson (1992), Spatial Autocorrelation: A New Computationally Simple Test with an Application to Per Capita Country Police Expenditures, *Regional Science and Urban Economics*, 22, 317–331.

Kelejian, H. H. and D. P. Robinson (1993), A Suggested Method of Estimation for Spatial Interdependent Models with Autocorrelated Errors, and an Application to a County Expenditure Model, *Papers in Regional Science*, 72, 297–312.

Kelejian, H. H. and D. P. Robinson (1995), Spatial Correlation: A Suggested Alternative to the Autoregressive Model, in L. Anselin and R. Florax (eds.), *New Directions in Spatial Econometrics*, Springer-Verlag, Berlin, 75–95.

King, M. L. (1981), A Small Sample Property of the Cliff-Ord Test for Spatial Correlation, *Journal of the Royal Statistical Society B*, 43, 263–264.

King, M. L. (1987), Testing for Autocorrelation in Linear Regression Models: A Survey, in M. L. King and D. E. A. Giles (eds.), *Specification Analysis in the Linear Model*, Routledge and Kegan Paul, London, 19–73.

King, M. L. and M. A. Evans (1986), Testing for Block Effects in Regression Models Based on Survey Data, *Journal of the American Statistical Association*, 81, 677–679.

King, M. L. and G. H. Hillier (1985), Locally Best Invariant Tests of the Error Covariance Matrix of the Linear Regression Model, *Journal of the Royal Statistical Society B*, 47, 98–102.

Koerts, J. and A. P. I. Abrahamse (1968), On the Power of the BLUS Procedure, *Journal of the American Statistical Association*, 63, 1227–1236.

Krugman, P. (1991), Increasing Returns and Economic Geography, *Journal of Political Economy*, 99, 483–499.

Land, K. and G. Deane (1992), On the Large-Sample Estimation of Regression Models with Spatial- or Network-Effects Terms: A Two Stage Least Squares Approach, in P. Marsden (ed.), *Sociological Methodology*, Jossey Bass, San Francisco, 221–248.

Leenders, R. T. (1995), *Structure and Influence. Statistical Models for the Dynamics of Actor Attributes, Network Structure and Their Interdependence*, Thesis Publishers, Amsterdam.

Legendre, P. (1993), Spatial Autocorrelation: Trouble or New Paradigm, *Ecology*, 74, 1659–1673.

LeSage, J. P. (1993), Spatial Modeling of Agricultural Markets, *American Journal of Agricultural Economics*, 75, 1211–1216.

LeSage, J. P. (1997), Bayesian Estimation of Spatial Autoregressive Models, *International Regional Science Review*, 20, 113–129.

Liang, K. Y. and S. L. Zeger (1986), Longitudinal Data Analysis Using Generalized Linear Models, *Biometrika*, 73, 13–22.

Lütkepohl, H. (1991), *Introduction to Multiple Time Series Analysis*, Springer-Verlag, Berlin.

Magnus, J. (1978), Maximum Likelihood Estimation of the GLS Model with Unknown Parameters in the Disturbance Covariance Matrix, *Journal of Econometrics*, 7, 281–312; Corrigenda, *Journal of Econometrics*, 10, 261.

Mardia, K. V. and R. J. Marshall (1984), Maximum Likelihood Estimation of Models for Residual Covariance in Spatial Regression, *Biometrika*, 71, 135–146.

Mardia, K. V. and A. J. Watkins (1989), On Multimodality of the Likelihood for the Spatial Linear Model, *Biometrika*, 76, 289–295.

MathSoft (1996), *S+Spatialstats User's Manual, Version 1.0*, MathSoft, Inc., Seattle.

McMillen, D. P. (1992), Probit with Spatial Autocorrelation, *Journal of Regional Science*, 32 335–348.

Moran, P. A. P. (1948), The Interpretation of Statistical Maps, *Biometrika*, 35, 255–260.

Moran, P. A. P. (1950a), Notes on Continuous Stochastic Phenomena, *Biometrika*, 37, 17–23.

Moran, P. A. P. (1950b), A Test for the Serial Independence of Residuals, *Biometrika*, 37, 178–181.

Murdoch, J. C., M. Rahmatian, and M. A. Thayer (1993), A Spatially Autoregressive Median Voter Model of Recreation Expenditures, *Public Finance Quarterly*, 21, 334–350.

Murdoch, J. C., T. Sandler, and K. Sargent (1996), A Tale of Two Collectives: Sulfur versus Nitrogen Oxides Emission Reduction in Europe, Working Paper, Department of Economics, Iowa State University, Ames, IA.

National Research Council (1991), *Spatial Statistics and Digital Image Analysis*, National Academy Press, Washington, DC.

Ord, J. K. (1975), Estimation Methods for Models of Spatial Interaction, *Journal of the American Statistical Association*, 70, 120–126.

Pace, R. K. and R. Barry (1996), Sparse Spatial Autoregressions, *Statistics and Probability Letters*, 2158, 1–7.

Pace, R. K. and R. Barry (1997), Quick Computation of Spatial Autoregressive Estimators, *Geographical Analysis*, 29 (forthcoming).

Paelinck, J. (1982), Operational Spatial Analysis, *Papers, Regional Science Association*, 50, 1–7.

Paelinck, J. and L. Klaassen (1979), Spatial Econometrics, Saxon House, Farnborough.

Pinkse, J. and M. E. Slade (1995), Contracting in Space, an Application of Spatial Statistics to Discrete-Choice Models, Working Paper, Department of Economics, University of British Columbia, Vancouver, BC.

Poirier, D. J. (1995), *Intermediate Statistics and Econometrics. A Comparative Approach*, The MIT Press, Cambridge, MA.

Rao, C. R. (1947), Large Sample Tests of Statistical Hypotheses Concerning Several Parameters with Applications to Problems of Estimation, *Proceedings of the Cambridge Philosophical Society*, 44, 50–57.

Ripley, B. D. (1981), *Spatial Statistics*, Wiley, New York.

Ripley, B. D. (1988), *Statistical Inference for Spatial Processes*, Cambridge University Press, Cambridge.

Romer, P. M. (1986), Increasing Returns and Long-Run Growth, *Journal of Political Economy*, 94, 1002–1037.

Saikkonen, P. (1989), Asymptotic Relative Efficiency of the Classical Test Statistics Under Misspecification, *Journal of Econometrics*, 42, 351–369.

Silvey, S. D. (1959), The Lagrange Multiplier Test, *Annals of Mathematical Statistics*, 30, 389–407.

Tiefelsdorf, M. and B. Boots (1995), The Exact Distribution of Moran's *I*, *Environment and Planning A*, 27, 985–999.

Tobler, W. (1979), Cellular Geography, in S. Gale and G. Olsson (eds.), *Philosophy in Geography*, Reidel, Dordrecht, 379–386.

Upton, G. J. and B. Fingleton (1985), *Spatial Data Analysis by Example. Volume 1: Point Pattern and Quantitative Data*, Wiley, New York.

Upton, G. J. and B. Fingleton (1989), *Spatial Data Analysis by Example. Volume 2: Categorical and Directional Data*, Wiley, New York.

Warnes, J. J. and B. D. Ripley (1987), Problems with Likelihood Estimation of Covariance Functions of Spatial Gaussian Processes, *Biometrika*, 74, 640–642.

White, H. (1980), A Heteroskedastic-Consistent Covariance Matrix Estimator and a Direct Test for Heteroskedasticity, *Econometrica*, 48, 817–838.

White, H. (1984), *Asymptotic Theory for Econometricians*, Academic Press, Orlando.

White, H. (1994), *Estimation, Inference and Specification Analysis*, Cambridge University Press, Cambridge.

White, H. and I. Domowitz (1984), Nonlinear Regression with Dependent Observations, *Econometrica*, 52, 143–161.

Whittle, P. (1954), On Stationary Processes in the Plane, *Biometrika*, 41, 434–449.

Zeger, S. L. and K. Y. Liang (1986), Longitudinal Data Analysis for Discrete and Continuous Outcomes, *Biometrics*, 42, 121–130.

Zeger, S. L., K. Y. Liang, and P. S. Albert (1988), Models for Longitudinal Data: A Generalized Estimating Equations Approach, *Biometrics*, 44, 1049–1060.

8
Panel Data Methods

Badi H. Baltagi
Texas A&M University, College Station, Texas

I. INTRODUCTION

Panel data refers to data sets consisting of multiple observations on each sampling unit. This could be generated by pooling time-series observations across a variety of cross-sectional units including countries, states, regions, firms, or randomly sampled individuals or households. Two well-known examples in the United States are the Panel Study of Income Dynamics (PSID) and the National Longitudinal Survey (NLS). The PSID began in 1968 with 4802 families, including an oversampling of poor households. Annual interviews were conducted and socioeconomic characteristics of each of the families and of roughly 31000 individuals who have been in these or derivative families were recorded. The list of variables collected is over 5000. The NLS followed five distinct segments of the labor force. The original samples include 5020 older men, 5225 young men, 5083 mature women, 5159 young women, and 12686 youths. There was an oversampling of blacks, hispanics, poor whites, and military in the youths survey. The list of variables collected runs into the thousands. Panel data sets have also been constructed from the U.S. Current Population Survey (CPS), which is a monthly national household survey conducted by the Census Bureau. The CPS generates the unemployment rate and other labor force statistics. Compared with the NLS and PSID data sets, the CPS contains fewer variables, spans a shorter period, and does not follow movers. However, it covers a much larger sample and is representative of all demographic groups. European panel data sets include the German Social Economic Panel, the Swedish study of household market and nonmarket activities, and the Intomart Dutch panel of households.

Some of the benefits and limitations of using panel data sets are listed in Hsiao (1986). Obvious benefits are a much larger data set with more variability and less collinearity among the variables than is typical of cross-sectional or time-series data. With additional, more informative data, one can get more reliable estimates and test more sophisticated behavioral models with less restrictive assumptions. Another advantage of panel data sets are their ability to control for individual heterogeneity. Not controlling for these unobserved individual specific effects leads to bias in the resulting estimates. Panel data sets are also better able to identify and estimate effects that are simply not detectable in pure cross sections or pure time-series data. In particular, panel data sets are better able to study complex issues of dynamic behavior. For example, with cross-sectional data set one can estimate the rate of unemployment at a particular point in time. Repeated cross sections can show how this proportion changes over time. Only panel data sets can estimate what proportion of those who are unemployed in one period remain unemployed in another period.

Limitations of panel data sets include problems in the design, data collection, and data management of panel surveys (Kasprzyk et al. 1989). These include the problems of coverage (incomplete account of the population of interest), nonresponse (due to lack of cooperation of the respondent or because of interviewer error), recall (respondent not remembering correctly), frequency of interviewing, interview spacing, reference period, the use of bounding to prevent the shifting of events from outside the recall period into the recall period and time in sample bias. Another limitation of panel data sets is the distortions due to measurement errors. Measurement errors may arise because of faulty response due to unclear questions, memory errors, deliberate distortion of responses (e.g., prestige bias), inappropriate informants, misrecording of responses, and interviewer effects. Although these problems can occur in cross-sectional studies, they are aggravated in panel data studies. Duncan and Hill (1985) in a validation study of the PSID data set compare the records of a large firm with the response of its employees and find the ratio of measurement error variance to true variance to be of the order of 184% for average hourly earnings. These figures are for a one-year recall (i.e., 1983 for 1982) and are more than doubled with two years' recall. Panel data sets may also exhibit bias due to sample selection problems. For the initial wave of the panel, respondents may refuse to participate or the interviewer may not find anybody at home. This may cause some bias in the inference drawn from this sample. While this nonresponse can also occur in cross-sectional data sets, it is more serious with panels because subsequent waves of the panel are still subject to nonresponse. Respondents may die, or move, or find that the cost of responding is high. The rate of attrition differs across panels and usually increases from one wave to the next, but the rate of increase declines over time. Becketti et al. (1988) studied the representativeness of the PSID, 14 years after it started. They find that only 40% of those originally in the sample in 1968 remained in the sample in 1981. Typical panels involve annual data covering a short span of time for each individual. This means that asymptotic arguments rely crucially on the number of individuals in the panel tending to infinity. Increasing the time span of the panel is not without cost either. In fact, this increases the chances of attrition with every new wave and increases the degree of computational difficulty in the estimation of qualitative limited dependent variable panel data models (Baltagi 1995b).

II. THE ERROR COMPONENTS REGRESSION MODEL

Although, random coefficient regressions can be used in the estimation and specification of panel data models (Swamy 1971, Hsiao 1986, Dielman 1989), most panel data applications have been limited to a simple regression with error components disturbances

$$y_{it} = x'_{it}\beta + \mu_i + \lambda_t + \nu_{it}, \qquad i = 1, \ldots, N; \quad t = 1, \ldots, T \qquad (1)$$

where i denotes individuals and t denotes time, x_{it} is a vector of observations of k explanatory variables, β is a k-vector of unknown coefficients, μ_i is an unobserved individual specific effect, λ_t is an unobserved time specific effect and ν_{it} is a zero-mean random disturbance with variance σ_ν^2. The error components disturbances follow a two-way analysis of variance (ANOVA). If μ_i and λ_t denote fixed parameters to be estimated, this model is known as the fixed-effects (FE) model. The x_{it}'s are assumed independent of the ν_{it}'s for all i and t. Inference in this case is conditional on the particular N individuals and over specific time-periods observed. Estimation in this case amounts to including $N - 1$ individual dummies and $T - 1$ time dummies to estimate these time invariant and individual invariant effects. This leads to an enormous loss in degrees of freedom. In addition, this attenuates the problem of multicollinearity among the regressors. Furthermore, this may not be computationally feasible for large N and/or T. In this case, one can eliminate the μ_i's and λ_t's and estimate β by running least squares of $\tilde{y}_{it} = y_{it} - \bar{y}_{i.} - \bar{y}_{.t} + \bar{y}_{..}$ on the $\tilde{x}_{it}$'s similarly defined, where the dot indicates summation over that index and the bar denotes averaging. This transformation is known as the within transformation and the corresponding estimator of β is called the within estimator or the FE estimator. Note that the FE estimator cannot estimate the effect of any time-invariant variable like sex, race, religion, or union participation. Nor can it estimate the effect of any individual invariant like price, interest rate, etc., that vary only with time. These variables are wiped out by the within transformation.

If μ_i and λ_t are random variables with zero means and constant variances σ_μ^2 and σ_λ^2, this model is known as the random-effects model. The preceding moments are conditional on the x_{it}'s. In addition, μ_i, λ_t, and ν_{it} are assumed to be conditionally independent. The random-effects (RE) model can be estimated by GLS which can be obtained using a least-squares regression of $y_{it}^* = y_{it} = \theta_1\bar{y}_{i.} - \theta_2\bar{y}_{.t} + \theta_3\bar{y}_{..}$ on x_{it}^* similarly defined, where θ_1, θ_2 and θ_3 are simple functions of the variance components σ_μ^2, σ_λ^2, and σ_ν^2 (Fuller and Battese 1974). The corresponding GLS estimator of β is known as the RE estimator. Note that for this RE model one can estimate the effects of time-invariant and individual-invariant variables. The best quadratic unbiased (BQU) estimators of the variance components are ANOVA type estimators based on the true disturbances and these are minimum variance unbiased (MVU) under normality of the disturbances. One can obtain feasible estimates of the variance components by replacing the true disturbances by OLS residuals (Wallace and Hussain 1969). Alternatively, one could substitute the fixed-effects residuals as proposed by Amemiya (1971). In fact, Amemiya (1971) shows that the Wallace and Hussain (1969) estimates of the variance components have a different asymptotic distribution from that knowing the true disturbances, while the Amemiya (1971) estimates of the variance components have the same asymptotic distribution as that knowing the true disturbances. Other estimators of the variance components were proposed by Swamy and Arora (1972) and Fuller and Battese (1974). Maximum likelihood estimation (MLE) under the normality of the disturbances is derived by Amemiya (1971). The first-order conditions are nonlinear, but can be solved using an iterative GLS scheme (Breusch 1987). Finally one can apply Rao's (1972) minimum norm quadratic unbiased estimation (MINQUE) methods. These methods are surveyed in Baltagi (1995b).Wallace and Hussain (1969) compare the RE and FE estimators of β in the case of nonstochastic (repetitive) x_{it}'s and find that both are (i) asymptotically normal, (ii) consistent and unbiased, and that (iii) $\hat{\beta}_{RE}$ has a smaller generalized variance (i.e., more efficient) in finite samples. In the case of nonstochastic (nonrepetitive) x_{it}'s they find that both $\hat{\beta}_{RE}$ and $\tilde{\beta}_{FE}$ are consistent, asymptotically unbiased and have equivalent asymptotic variance-covariance matrices, as both N and $T \rightarrow \infty$. Under the random effects model, GLS based on the true variance components is BLUE, and all the feasible GLS estimators considered are asymptotically efficient as N and $T \rightarrow \infty$. Maddala and Mount (1973) compared OLS, FE, RE, and MLE methods using Monte Carlo experiments. They found little to choose among the various feasible GLS estimators in small samples and argued in favor of methods that were easier to compute. MINQUE was dismissed as more difficult to compute, and the applied researcher given one shot at the data was warned to compute at least two methods of estimation. If these methods give different results, the authors diagnose misspecification. Taylor (1980) derived exact finite sample results for the one-way error component model ignoring the time effects. He found the following important results. (1) Feasible GLS is more efficient that the FE estimator for all but the fewest degrees of freedom. (2) The variance of feasible GLS is never more than 17% above the Cramer-Rao lower bound. (3) More efficient estimators of the variance components do not necessarily yield more efficient feasible GLS estimators. These finite-sample results are confirmed by the Monte Carlo experiments carried out by Baltagi (1981a).

One test for the usefulness of panel data models is their ability to predict. For the FE model, the best linear unbiased predictor (BLUP) was derived by Wansbeek and Kapteyn (1978) and Taub (1979). This derivation was generalized by Baltagi and Li (1992) to the RE model with serially correlated remainder disturbances. More recently, Baillie and Baltagi (1995) derived the asymptotic mean square prediction error for the FE and RE predictors as well as two other misspecified predictors and compared their performance using Monte Carlo experiments.

III. TEST OF HYPOTHESES

Fixed versus random effects has generated a lively debate in the biometrics literature. In econometrics, see Mundlak (1978). The random and fixed effects models yield different estimation results, especially if T is small and N is large. A specification test based on the difference between these estimates is given by Hausman (1978). The null hypothesis is that the individual and time effects are not correlated with the x_{it}'s. The basic idea behind this test is that the fixed effects estimator $\tilde{\beta}_{FE}$ is consistent whether the effects are or are not correlated with the x_{it}'s. This is true because the fixed effects transformation described by $\tilde{y}_{it}$ wipes out the μ_i and λ_t effects from the model. However, if the null hypothesis is true, the fixed effects estimator is not efficient under the RE specification, because it relies only on the within variation in the data. On the other hand, the RE estimator $\hat{\beta}_{RE}$ is efficient under the null hypothesis but is biased and inconsistent when the effects are correlated with the x_{it}'s. The difference between these estimators $\hat{q} = \tilde{\beta}_{FE} - \hat{\beta}_{RE}$ tend to zero in probability limits under the null hypothesis and is nonzero under the alternative. The variance of this difference is equal to the difference in variances, $\text{var}(\hat{q}) = \text{var}(\tilde{\beta}_{FE}) - \text{var}(\hat{\beta}_{RE})$, since $\text{cov}(\hat{q}, \hat{\beta}_{RE}) = 0$ under the null hypothesis. Hausman's test statistic is based upon $m = \hat{q}'[\text{var}(\hat{q})]^{-1}\hat{q}$ and is asymptotically distributed as χ^2 with k degrees of freedom under the null hypothesis.[*] The Hausman test can also be computed as a variable addition test by running y^* on the regressor matrices X^* and $\bar{X}$ testing that the coefficients of $\bar{X}$ are zero using the usual F-test. This test was generalized by Arellano (1993) to make it robust to heteroskedasticity and autocorrelation of arbitrary forms. In fact, if either heteroskedasticity or serial correlation is present, the variances of the FE and RE estimators are not valid and the corresponding Hausman test statistic is inappropriate. Ahn and Low (1996) show that the Hausman test statistic can be obtained as $(NT)R^2$ from the regression of GLS residuals on $\bar{X}$ and $\bar{X}$ where the latter denotes the matrix of regressors averaged over time. Also, an alternative generalized method of moments (GMM) test is recommended for testing the joint null hypothesis of exogeneity of the regressors *and* the stability of regression parameters over time. If the regression parameters are nonstationary over time, then both $\hat{\beta}_{RE}$ and $\tilde{\beta}_{FE}$ are inconsistent even though the regressors may be exogenous. Ahn and Low perform Monte Carlo experiments which show that both the Haus-

---

[*]For the one-way error components model with individual effects only, Hausman and Taylor (1981) show that Hausman's specification test can also be based on two other contrasts that yield numerically identical results. Kang (1985) extends this analysis to the two-way error components model.

man and the alternative GMM test have good power in detecting endogeneity of the regressors. However, the alternative GMM test dominates if the coefficients of the regressors are nonstationary. Li and Stengos (1992) propose a Hausman specification test based on $\sqrt{N}$-consistent semiparametric estimators. They apply it in the context of a dynamic panel data model of the form

$$y_{it} = \delta y_{i,t-1} + g(x_{it}) + u_{it}, \qquad i = 1, \ldots, N; \quad t = 1, \ldots, T \tag{2}$$

where the function $g(\cdot)$ is unknown, but satisfies certain moment and differentiability conditions. The x_{it} observations are IID with finite fourth moments and the disturbances u_{it} are IID$(0, \sigma^2)$ under the null hypothesis. Under the alternative, the disturbances u_{it} are IID in the i subscript but are serially correlated in the t subscript. Li and Stengos base the Hausman test for $H_0: E(u_{it}|y_{i,t-1}) = 0$ on the difference between two $\sqrt{N}$-consistent instrumental variables estimators for δ, under the null and the alternative respectively.

For panels with large N and small T, testing for poolability of the data amounts to testing the stability of the cross-sectional regression across time. In practice, the Chow (1960) test for the equality of the regression coefficients is popular. This is proper only under the spherical disturbances assumption. It leads to improper inference under the random-effects specification. In fact, Baltagi (1981a) shows that in this case the Chow test leads to rejection of the null too often when in fact it is true. However, if one accounts for the random-effects variance-covariance matrix, the F-test for the equality of slopes performs well in Monte Carlo experiments. Recently, Baltagi, Hidalgo, and Li (1996) derive a nonparametric test for poolability which is robust to functional form misspecification. In particular, they consider the following nonparametric panel data model:

$$y_{it} = g_t(x_{it}) + \epsilon_{it} \qquad (i = 1, \ldots, N; t = 1, \ldots, T) \tag{3}$$

where $g_t(\cdot)$ is an unspecified functional form that may vary over time, and x_{it} is a $k \times 1$ column vector of predetermined explanatory variables with ($p \geq 1$) variables being continuous and $k - p$ (≥ 0). Poolability of the data over time is equivalent to testing that $g_t(x) = g_s(x)$ almost everywhere for all t and $s = 1, 2, \ldots, T$ versus $g_t(x) \neq g_s(x)$ for some $t \neq s$ with probability greater than zero. The test statistic is shown to be consistent and asymptotically normal and is applied to a panel data set on earnings.

In choosing between pooled homogeneous parameter estimators versus non-pooled heterogeneous parameter estimators, some mean-square error (MSE) criteria can be used as described in Wallace (1972) to capture the trade-off between bias and variance. Bias is introduced when the poolability restriction is not true. However, the variance is reduced by imposing the poolability restriction. Hence, the MSE criteria may choose the pooled estimator despite the fact that the poolability restriction is not true. Ziemer and Wetzstein (1983) suggest comparing pooled and nonpooled estimators according to their forecast risk performance. They show for a wilderness recreation demand model that a Stein-rule estimator give better forecast risk performance than the pooled or individual cross-sectional estimates. More recently, the fundamental assumption underlying pooled homogeneous parameters models has been called into question. For example, Robertson and Symons (1992) warned about the bias from pooled estimators when the estimated model is dynamic and homogeneous when in fact the true model is static and heterogeneous. Pesaran and Smith (1995) argued in favor of heterogeneous estimators rather than pooled estimators for panels with large N and T. They showed that when the true model is dynamic and heterogeneous, the pooled estimators are inconsistent, whereas an average estimator of heterogeneous parameters can lead to consistent estimates as long as both N and T tend to infinity. Using a different approach, Maddala, Srivastava, and Li (1994) argued that shrinkage estimators are superior to either heterogeneous or homogeneous parameter estimates especially for prediction purposes. In this case, one shrinks the individual heterogeneous estimates toward the pooled estimate using weights depending on their corresponding variance-covariance matrices. Baltagi and Griffin (1997) compare the short-run and long-run forecast performance of the pooled homogeneous, individual heterogeneous, and shrinkage estimators for a dynamic demand for the gasoline across 18 OECD countries. Based on 1-, 5-, and 10-year forecasts, the results support the case for pooling. Alternative tests for structural change in panel data include Han and Park (1989), who used the cumulative sum and cusum of squares to test for structural change based on recursive residuals. They find no structural break over the period 1958–1976 in U.S. foreign trade of manufacturing goods.

Testing for random individual effects is of utmost importance in panel data applications. Ignoring these effects lead to huge bias in estimation (Moulton 1986). A popular Lagrange multiplier (LM) test for the significance of the random effects $H_0^a; \sigma_\mu^2 = 0$ was derived by Breusch and Pagan (1980). This test statistic can be easily computed using least-squares residuals. This assumes that the alternative hypothesis is two-sided when we know that the variance components are nonnegative. A one-sided version of this test is given by Honda (1985). This is shown to be *uniformly most powerful* and robust to nonnormality. However, Moulton and Randolph (1989) showed that the asymptotic $N(0, 1)$ approximation for this one-sided LM statistic can be poor even in large samples. They suggest an alternative standardized Lagrange multiplier (SLM) test whose asymptotic critical values are generally closer to the exact critical values than those of the LM test. This SLM test statistic centers and scales the one-sided LM statistic so that its mean is zero and its variance is one.

For $H_0^b; \sigma_\mu^2 = \sigma_\lambda^2 = 0$, the two-sided LM test is given by Breusch and Pagan (1980) and is distributed as χ_2^2 under the null. Honda (1985) does not derive a uniformly most powerful one-sided test for H_0^b, but he suggests a "handy" one-sided test which is distributed as $N(0, 1)$ under H_0^b. Later Honda (1991) derives the SLM version of this one-sided test. Baltagi, Chang, and Li (1992) derive a locally mean most powerful (LMMP) one-sided test for H_0^b and its SLM version is given by Baltagi (1995b). Under H_0^b, $\sigma_\mu^2 = \sigma_\lambda^2 = 0$, these standardized Lagrange multiplier statistics are asymptotically $N(0, 1)$ and their asymptotic critical values should be closer to the exact critical values than those of the corresponding unstandardized tests. Alternatively, one can perform a likelihood ratio test or an ANOVA-type F-test. Both tests have the same asymptotic distribution as their LM counterparts. Moulton and Randolph (1989) find that although the F-test is not locally most powerful, its power function is close to the power function of the exact LM test and is therefore recommended. A comparison of these various testing procedures using Monte Carlo experiments is given by Baltagi, Chang, and Li (1992). Recent developments include a generalization by Li and Stengos (1994) of the Breusch-Pagan test to the case where the remainder error is heteroskedastic of unknown form. Also, Baltagi and Chang (1996) propose a simple ANOVA F-statistic based on recursive residuals to test for random individual effects.

For incomplete (or unbalanced) panels, the Breusch-Pagan test can be easily extended; see Moulton and Randolph (1989) for the one-way error components model and Baltagi and Li (1990) for the two-way error components model. For nonlinear models, Baltagi (1996) suggests a simple method for testing for zero random individual and time effects using a Gauss-Newton regression. In case the regression model is linear, this test amounts to a variable addition test, i.e., running the original regression with two additional regressors. The first is the average of the least-squares residuals over time, while the second is the average of the least-squares residuals over individuals. The test statistic becomes the F-statistic for the significance of the two additional regressors.

Baltagi and Li (1995) derive three LM test statistics that jointly test for serial correlation and individual effects. The first LM statistic jointly tests for zero first-order serial correlation *and* random individual effects, the second LM statistic tests for zero first-order serial correlation assuming fixed individual effects, and the third LM statistic tests for zero first-order serial correlation assuming random individual effects. In all three cases, Baltagi and Li (1995) showed that the corresponding LM statistic is the *same* whether the alternative is AR(1) or MA(1). In addition, Baltagi and Li (1995) derive two simple tests for distinguishing between AR(1) and MA(1) remainder disturbances in error components regressions and perform Monte Carlo experiments to study the performance of these tests. For the fixed-effects model, Bhargava, Franzini, and Narendranathan (1982) derived a modified Durbin-Watson test statistic based on FE residuals to test for first-order serial correlation and a test for random walk based on differenced OLS residuals. Chesher (1984) derived a score test for neglected heterogeneity, which is viewed as causing parameter variation. Also, Hamerle (1990) and Orme (1993) suggest a score test for neglected heterogeneity for qualitative limited dependent-variable panel data models.

Holtz-Eakin (1988) derives a simple test for the presence of individual effects in dynamic (autoregressive) panel data models, while Holtz-Eakin, Newey, and

Rosen (1988) formulate a coherent set of procedures for estimating and testing VAR (vector autoregression) with panel data. Arellano and Bond (1991) consider tests for serial correlation and overidentification restrictions in a dynamic random-effects model, while Arellano (1990) considers testing covariance restrictions for error components or first-difference structures with white noise, MA, or AR schemes.

Chamberlain (1982, 1984) finds that the fixed effects specification imposes testable restrictions on coefficients from regressions of all leads and lags of the dependent variable on all leads and lags of independent variables. These overidentification restrictions are testable using minimum chi-squared statistics. Angrist and Newey (1991) show that, in the standard fixed effects model, this overidentification test statistic is simply the degrees of freedom times the R^2 from a regression of *within residuals* on all leads and lags of the independent variables. They apply this test to models of the union-wage effect using five years of data from the National Longitudinal Survey of Youth and to a conventional human capital earnings function estimating the return to schooling. They do not reject a fixed effect specification in the union-wage example, but they do reject it in the return to schooling example.

Testing for unit roots using panel data has been recently reconsidered by Quah (1994), Levin and Lin (1996), and Im, Pesaran, and Shin (1996). This has been applied by MacDonald (1996) to real exchange rates for 17 OECD countries based on a wholesale price index, and 23 OECD countries based on a consumer price index, all over the period 1973–1992. The null hypothesis that real exchange rates contain a unit root is rejected. Earlier applications include Bourmahdi and Thomas (1991), who apply a likelihood ratio unit root panel data test to assess efficiency of the French capital market. Using 140 French stock prices observed weekly from January 1973 to February 1986 ($T = 671$) on the Paris Stock Exchange, Boumahdi and Thomas (1991) do not reject the null hypothesis of a unit root. Also, Breitung and Meyer (1994) apply panel data unit roots test to contract wages negotiated on firm and industry level in western Germany over the period 1972–1987. They find that both firm and industry wages possess a unit root in the autoregressive representation. However, there is weak evidence for a cointegration relationship.

IV. GENERALIZATIONS OF THE ERROR COMPONENTS MODEL

The error components disturbances are homoskedastic across individuals. This may be an unrealistic assumption and has been relaxed by Mazodier and Trognon (1978) and Baltagi and Griffin (1988). A more general heteroskedastic model is given by Randolph (1988) in the context of unbalanced panels. Also, Li and Stengos (1994) proposed estimating a one-way error component model with heteroskedasticity of unknown form using adaptive estimation techniques.

300   BALTAGI

The error components regression model has been also generalized to allow for serial correlation in the remainder disturbances by Lillard and Willis (1978), Revankar (1979), MaCurdy (1982), Baltagi and Li (1991, 1995), and Galbraith and Zinde-Walsh (1995). Chamberlain (1982, 1984) allows for arbitrary serial correlation and heteroskedastic patterns by viewing each time period as an equation and treating the panel as a multivariate setup. Also, Kiefer (1980), Schmidt (1983), Arellano (1987), and Chowdhury (1994) extend the fixed-effects model to cover cases with an arbitrary intertemporal covariance matrix.

The normality assumption on the error components disturbances may be untenable. Horowitz and Markatou (1996) show how to carry out nonparametric estimation of the densities of the error components. Using data from the Current Population Survey, they estimate an earnings model and show that the probability that individuals with low earnings will become high earners in the future are much lower than that obtained under the assumption of normality. One drawback of this nonparametric estimator is its slow convergence at a rate of $1/(\log N)$, where N is the number of individuals. Monte Carlo results suggest that this estimator should be used for N larger than 1000.

Micro panel data on households, individuals, and firms are highly likely to exhibit measurement error; see Duncan and Hill (1985) who found serious measurement error in average hourly earnings in the Panel of Income Dynamics. Using panel data, Griliches and Hausman (1986) showed that one can identify and estimate a variety of errors in variables models *without* the use of external instruments. Griliches and Hausman suggest differencing the data j periods apart $(y_{it} - y_{i,t-j})$, thus generating "different-lengths" difference estimators. These transformations wipe out the individual effect, but they may aggravate the measurement error bias. One can calculate the bias of the different-lengths differenced estimators and use this information to obtain consistent estimators of the regression coefficients. Extensions of this model include Kao and Schnell (1987a, 1987b), Wansbeek and Koning (1989), Hsiao (1991), Wansbeek and Kapteyn (1992), and Biorn (1992). See also Baltagi and Pinnoi (1995) for an application to the productivity of the public capital stock.

The error components model has been extended to the seemingly unrelated regressions case by Avery (1977), Baltagi (1980), Magnus (1982), Prucha (1984), and Kinal and Lahiri (1990). Some applications include Howrey and Varian (1984) on the estimation of a system of demand equations for electricity by time of day, and Sickles (1985) on the analysis of productivity growth in the U.S. airlines industry.

For the simultaneous equation with error components. Baltagi (1981b) derives the error component two-stage (EC2SLS) and three-stage (EC3SLS) least-squares estimators, while Prucha (1985) derives the full-information MLE under the normality assumption. These estimators are surveyed in Krishnakumar (1988). Monte Carlo experiments are given by Baltagi (1984) and Mátyás and Lovrics (1990). Recent applications of EC2SLS and EC3SLS include (i) an econometric rational-expectations macroeconomic model for developing countries with capital controls (Haque, Lahiri, and Montiel 1993), and (ii) an econometric model measuring income and price elasticities of foreign trade for developing countries (Kinal and Lahiri 1993).

Mundlak (1978) considered the case where the endogeneity is solely attributed to the individual effects. In this case, Mundlak showed that if these individual effects are a linear function of the averages of *all* the explanatory variables across time, then the GLS estimator of this model coincides with the FE estimator. Mundlak's (1978) formulation assumes that *all* the explanatory variables are related to the individual effects. The random-effects model, on the other hand, assumes no correlation between the explanatory variables and the individual effects. Instead of this "all or nothing" correlation among the x_{it}'s and the μ_i's, Hausman and Taylor (1981) consider a model where *some* of the explanatory variables are related to the μ_i's. In particular, they consider

$$y_{it} = x'_{it}\beta + z'_i\gamma + \mu_i + \nu_{it} \tag{4}$$

where the z_i's are cross-sectional time-invariant variables. Hausman and Taylor (1981), hereafter HT, split the matrices X and Z into two sets of variables: $X = [X_1; X_2]$ and $Z = [Z_1; Z_2]$, where X_1 is $n \times k_1$, X_2 is $n \times k_2$, Z_1 is $n \times g_1$, Z_2 is $n \times g_2$, and $n = NT$. The terms X_1 and Z_1 are assumed exogenous in that they are not correlated with μ_i and ν_{it}, while X_2 and Z_2 are endogenous because they are correlated with the μ_i's but not the ν_{it}'s. The within transformation would sweep the μ_i's and remove the bias, but in the process it would also remove the Z_i's and hence the within estimator will not give an estimate of the γ's. To get around that, Hausman and Taylor (1981) suggest an instrumental variable estimator that uses $\tilde{X}_1$, $\tilde{X}_2$, $\bar{X}_1$, and Z_1 as instruments. Therefore, the matrix of regressors X_1 is used twice, once as averages and another time as deviations from averages. This is an advantage of panel data allowing instruments from *within* the model. The order condition for identification gives the result that the number of X_1's (k_1) must be at least as large as the number of Z_2's (g_2). With stronger exogeneity assumptions between X and the μ_i's, Amemiya and MaCurdy (1986) and Breusch, Mizon, and Schmidt (1989) suggest more efficient instrumental variable (IV) estimators. Cornwell and Rupert (1988) apply these IV methods to a returns to schooling example based on a panel of 595 individuals drawn from the PSID over the period 1976–1982. Recently, Metcalf (1996) shows that for the Hausman-Taylor model given in (4), using less instruments may lead to a more powerful Hausman specification test. Asymptotically, more instruments led to more efficient estimators. However, the asymptotic bias of the inefficient estimator will also be greater as the null hypothesis of no correlation is violated. The increase in bias more than offsets the increase in variance. Since the test statistic is linear in variance but quadratic in bias, its power will increase.

Cornwell, Schmidt, and Wyhowski (1992) consider a simultaneous equation model with error components that distinguishes between two types of exogenous variables, namely *singly exogenous* and *doubly exogenous* variables. A singly exogenous variable is correlated with the individual effects but not with the remainder noise, while a doubly exogenous variable is uncorrelated with both the effects and the remainder disturbance term. For this encompassing model with two types of exogeneity, Cornwell, Schmidt, and Wyhowski (1992) extend the three instrumental variable estimators considered above and give them a GMM interpretation. Wyhowski (1994) extend these results to the two-way error components model, while Revankar (1992) establishes conditions for exact equivalence of instrumental variables in a simultaneous-equation two-way error components model.

V. DYNAMIC PANEL DATA MODELS

The dynamic error components regression is characterized by the presence of a lagged dependent variable among the regressors, i.e.,

$$y_{it} = \delta y_{i,t-1} + x'_{it}\beta + \mu_i + \nu_{it}, \qquad i = 1, \ldots, N; \quad t = 1, \ldots, T \qquad (5)$$

where δ is a scalar, x'_{it} is $1 \times k$, and β is $k \times 1$. This model has been extensively studied by Anderson and Hsiao (1982) and Sevestre and Trognon (1985).* Since y_{it} is a function of μ_i, $y_{i,t-1}$ is also a function of μ_i. Therefore, $y_{i,t-1}$, a right-hand regressor in (5), is correlated with the error term. This renders the OLS estimator biased and inconsistent even if the ν_{it}'s are not serially correlated. For the FE estimator, the within transformation wipes out the μ_i's, but $\tilde{y}_{i,t-1}$ will still be correlated with $\tilde{\nu}_{it}$ even if the ν_{it}'s are not serially correlated. In fact, the within estimator will be biased of $O(1/T)$ and its consistency will depend upon T being large; see Nickell (1981) and Kiviet (1995), who shows that the bias of the FE estimator in a dynamic panel data model has an $O(N^{-1}T^{3/2})$ approximation error. The same problem occurs with the random effects GLS estimator. In order to apply GLS, quasi-demeaning is performed, and $y^*_{i,t-1}$ will be correlated with u^*_{it}. An alternative transformation that wipes out the individual effects yet does not create the above problem is the first-difference (FD) transformation. In fact, Anderson and Hsiao (1982) suggested first-differencing the model to get rid of the μ_i's and then using $\Delta y_{i,t-2} = y_{i,t-2} - y_{i,t-3}$ or simply $y_{i,t-2}$ as an instrument for $\Delta y_{i,t-1} = y_{i,t-1} - y_{i,t-2}$. These instruments will not be correlated with $\Delta \nu_{it} = \nu_{i,t} - \nu_{i,t-1}$, as long as the ν_{it}'s themselves are not serially correlated. This IV estimation method leads to consistent, but not necessarily efficient, estimates of the parameters in the model because it does not make use of all the available moment conditions (Ahn and Schmidt 1995) and it does not take into account the differenced structure on the residual disturbances ($\Delta \nu_{it}$). Arellano

---

*In particular, the assumptions made on the initial values are of utmost importance (Anderson and Hsiao 1982, Bhargava and Sargan 1983, Hsiao 1986). Hsiao (1986) summarizes the consistency properties of the MLE and GLS under a RE dynamic model depending on the initial values assumption and the way in which N and T tend to infinity.

(1989) finds that for simple dynamic error components models the estimator that uses differences $\Delta y_{i,t-2}$ rather than levels $y_{i,t-2}$ for instruments has a singularity point and very large variances over a significant range of parameter values. In contrast, the estimator that uses instruments in levels, i.e., $y_{i,t-2}$, has no singularities and much smaller variances and is therefore recommended. Additional instruments can be obtained in a dynamic panel data model if one utilizes the orthogonality conditions that exist between lagged values of y_{it} and the disturbances v_{it} (Holtz-Eakin 1988, Holtz-Eakin, Newey, and Rosen 1988, Arellano and Bond 1991). Based on these additional moments, Arellano and Bond (1991) suggest a GMM estimator and propose a Sargan-type test for overidentifying restrictions.* Arellano and Bover (1995) develop a unifying GMM framework for looking at efficient IV estimators for dynamic panel data models. They do that in the context of the Hausman and Taylor (1981) model given in (4). Ahn and Schmidt (1995) show that under the standard assumptions used in a dynamic panel data model, there are additional moment conditions that are ignored by the IV estimators suggested by Arellano and Bond (1991). They show how these additional restrictions can be utilized in a GMM framework. Ahn and Schmidt (1995) also consider the dynamic version of the Hausman and Taylor (1981) model and show how one can make efficient use of exogenous variables as instruments. In particular, they show that the strong exogeneity assumption implies more orthogonality conditions which lie in the deviations from mean space. These are irrelevant in the static Hausman-Taylor model but are relevant for the dynamic version of that model.

An alternative approach to estimating dynamic panel data models have been suggested by Keane and Runkle (1992). Drawing upon the forward filtering idea from the time-series literature, this method of estimation first transforms the model to eliminate the general and arbitrary serial correlation pattern in the data. By doing so, one can use the set of original predetermined instruments to obtain consistent parameter estimates of the model. First differencing is also used in dynamic panel data models to get rid of individual specific effect, and the resulting first-differenced errors are serially correlated of an MA(1) type with unit root if the original v_{it}'s are classical errors. In this case, there will be gain in efficiency in performing the Keane and Runkle filtering procedure on the FD model. Underlying this estimation procedure are two important hypotheses that are testable. The first is H_A; the set of instruments are *strictly exogenous*. In order to test H_A, Keane and Runkle propose a test based on the difference between fixed-effects 2SLS (FE-2SLS) and first-difference 2SLS (FD-2SLS). FE-2SLS is consistent only if H_A is true. In fact if the matrix of instruments contains predetermined variables then FE-2SLS would not be consistent.

---

*Bhargava (1991) gives sufficient conditions for the identification of static and dynamic panel data models with endogenous regressors.

In contrast, FD-2SLS is consistent whether H_A is true or not. If H_A is not rejected, one should check whether the individual effects are correlated with the set of instruments. In this case, the usual Hausman and Taylor (1981) test applies. However, if H_A is rejected, the instruments are predetermined and the Hausman-Taylor test is inappropriate. In this case, the test will be based upon the difference between FD-2SLS and 2SLS. Under the null, both estimators are consistent, but if the null is not true FD-2SLS remains consistent while 2SLS does not. These two tests are Hausman (1978)-type tests except that the variances are complicated because Keane and Runkle do not use the efficient estimator under the null (Schmidt, Ahn, and Wyhowski 1992). Keane and Runkle (1992) apply their testing and estimation procedures to a simple version of the rational expectations life-cycle consumption model. See also Baltagi and Griffin (1995) for another application to liquor demand.

Alternative estimation methods of a static and dynamic panel data model with arbitrary error structure are considered by Chamberlain (1982, 1984). Chamberlain (1984) considered the panel data model as a multivariate regression of T equations subject to restrictions and derives an efficient minimum distance estimator that is robust to residual autocorrelation of arbitrary form. He also first-differenced these equations to get rid of the individual effects and derived an asymptotically equivalent estimator to his efficient minimum distance estimator based on 3SLS of the $T - 2$ differenced equations. Building on Chamberlain's work, Arellano (1990) developed minimum chi-square tests for various covariance restrictions. These tests are based on 3SLS residuals of the dynamic error component model and can be calculated from a generalized linear regression involving the sample autocovariance and dummy variables. The asymptotic distribution of the unrestricted autocovariance estimates is derived without imposing the normality assumption. In particular, Arellano (1990) considered testing covariance restrictions for error components or first-difference structures with white noise, moving-average, or autoregressive schemes. If these covariance restrictions are true, 3SLS is inefficient and Arellano (1990) proposed a GLS estimator which achieves asymptotic efficiency in the sense that it has the same limiting distribution as the optimal minimum distance estimator. More recently, Li and Stengos (1996) derived a $\sqrt{N}$-consistent estimator for a semiparametric dynamic panel data model, while Li and Stengos (1995) proposed a nonnested test for parametric versus semiparametric dynamic panel data models.

VI. INCOMPLETE PANEL DATA MODELS

Incomplete panels are more likely to be the norm in typical economic empirical settings. For example, if one is collecting data on a set of countries over time, a researcher may find some countries can be traced back longer than others. Similarly, in collecting data on firms over time, a researcher may find that some firms have dropped out of the market while new entrants emerged over the sample period observed. For randomly missing observations, unbalanced panels have been dealt with in Fuller and Battese (1974), Baltagi (1985), Wansbeek and Kapteyn (1989), and Baltagi and Chang (1994).* For the unbalanced one-way error component model, GLS can still be performed as a least-squares regression. However, BQU estimators of the variance components are a function of the variance components themselves. Still, unbalanced ANOVA methods are available (Searle 1987). Baltagi and Chang (1994) performed extensive Monte Carlo experiments varying the degree of unbalancedness in the panel as well as the variance components. Some of the main results include the following: (i) As far as the estimation of regression coefficients are concerned, the simple ANOVA-type feasible GLS estimators compare well with the more complicated estimators such as MLE and MINQUE and are never more than 4% above the MSE of true GLS. (ii) For the estimation of the remainder variance component σ_ν^2, these methods show little difference in relative MSE performance. However, for the individual specific variance component estimation, σ_μ^2, the ANOVA-type estimators perform poorly relative to MLE and MINQUE methods when the variance component σ_μ^2 is large and the pattern is severely unbalanced. (iii) Better estimates of the variance components, in the MSE sense, do not necessarily imply better estimates of the regression coefficients. This echoes similar findings in the balanced panel data case. (iv) Extracting a balanced panel out of an unbalanced panel by either maximizing the number of households observed or the total number of observations lead in both cases to an enormous loss in efficiency and is not recommended.[†] For an empirical application, see Mendelsohn et al. (1992), who use panel data on repeated single-family home sales in the harbor area surrounding New Bedford, Massachusetts, over the period 1969 to 1988 to study the damage associated with proximity to a hazardous waste site. Mendelsohn et al. (1992) find a significant reduction in housing values, between \$7000 and \$10,000 (1989 dollars), as a result of these houses' proximity to hazardous waste sites. The extension of the unbalanced error components model to the two-way model including time effects is more involved. Wansbeek and Kapteyn (1989) derive the FE, MLE, and a feasible GLS estimator based on quadratic unbiased estimators of the variance components and compare their performance using Monte Carlo experiments.

---

*Other methods of dealing with missing data include (i) inputting the missing values and analyzing the filled-in data by complete panel data methods, and (ii) discarding the nonrespondents and weighting the respondents to compensate for the loss of cases; see Little (1988) and the section on nonresponse adjustments in Kasprzyk et al. (1989).

[†] Chowdhury (1991) showed that for the fixed effects error component model, the within estimator based on the entire unbalanced panel is efficient relative to any within estimator based on a sub-balanced pattern. Also, Mátyás and Lovrics (1991) performed some Monte Carlo experiments to compare the loss in efficiency of FE and GLS based on the entire incomplete panel data and complete subpanel. They find the loss in efficiency is negligible if $NT > 250$, but serious for $NT < 150$.

Rotating panels attempt to keep the same number of households in the survey by replacing the fraction of households that drop from the sample in each period by an equal number of freshly surveyed households. This is a necessity in surveys where a high rate of attrition is expected from one period to the next. For the estimation of general rotation schemes as well as maximum likelihood estimation under normality (Biorn 1981). Estimation of the consumer price index in the United States is based on a complex rotating panel survey, with 20% of the sample being replaced by rotation each year (Valliant 1991). With rotating panels, the fresh group of individuals that are added to the panel with each wave provide a means of testing for time-in-sample bias effects. This has been done for various labor force characteristics in the Current Population Survey. For example, several studies have found that the first rotation reported an unemployment rate 10% higher than that of the full sample (Bailar 1975). While the findings indicate a pervasive effect of rotation group bias in panel surveys, the survey conditions do not remain the same in practice, and hence it is hard to disentangle the effects of time-in-sample bias from other effects.

For some countries, panel data may not exist. Instead the researcher may find annual household surveys based on a large random sample of the populations. Examples of some of these cross-sectional consumer expenditure surveys include the British Family Expenditure Survey, which surveys about 7000 households annually. Examples of repeated surveys in the United States include the Current Population Survey and the National Crime Survey. For these repeated cross-sectional surveys, it may be impossible to track the same household over time as required in a genuine panel. Instead, Deaton (1985) suggests tracking cohorts and estimating economic relationships based on cohort means rather than individual observations. One cohort could be the set of all males born between 1945 and 1950. This age cohort is well defined and can be easily identified from the data. Deaton (1985) argued that these pseudo panels do not suffer the attrition problem that plagues genuine panels, and may be available over longer time periods compared to genuine panels.* For this psuedo panel with T observations on C cohorts, the fixed effects estimator $\tilde{\beta}_{FE}$, based on the within-"cohort" transformation, is a natural candidate for estimating β. However, Deaton (1985) argued that these sample-based averages of the cohort means can only estimate the unobserved population cohort means with measurement error. Therefore, one has to correct the within estimator for measurement error using estimates of the errors in measurement variance-covariance matrix obtained from the individual data. Details are given in Deaton (1985). There is an obvious trade-off in the construction of a pseudo panel. The larger the number of cohorts, the smaller is

---

*Blundell and Meghir (1990) also argue that pseudo panels allow the estimation of life-cycle models which are free from aggregation bias. In addition, Moffitt (1993) explains that a lot of researchers in the United States prefer to use pseudo panels like the Current Population Survey because it has larger more representative samples and the questions asked are more consistently defined over time than the available U.S. panels.

the number of individuals per cohort. In this case, C is large and the pseudo panel is based on a large number of observations. However, the fact that the average cohort size $n_c = N/C$ is not large implies that the sample cohort averages are not precise estimates of the population cohort means. In this case, we have a large number C of imprecise observations. In contrast, a pseudo panel constructed with a smaller number of cohorts and therefore more individuals per cohort is trading a large pseudo panel with imprecise observations for a smaller pseudo panel with more precise observations. Verbeek and Nijman (1992b) find that $n_c \rightarrow \infty$ is a crucial condition for the consistency of the within estimator and that the bias of the within estimator may be substantial even for large n_c. On the other hand, Deaton's estimator is consistent for β, for finite n_c, when either C or T tend to infinity.

Moffitt (1993) extends Deaton's (1985) analysis to the estimation of dynamic models with repeated cross sections. Moffitt illustrates his estimation method for the linear fixed-effects life-cycle model of labor supply using repeated cross sections from the U.S. Current Population Survey. The sample included white males, ages 20–59, drawn from 21 waves over the period 1968 to 1988. In order to keep the estimation problem manageable, the data was randomly subsampled to include a total of 15,500 observations. Moffitt concludes that there is a considerable amount of parsimony achieved in the specification of age and cohort effects. Also, individual characteristics are considerably more important than either age, cohort, or year effects. Blundell, Meghir, and Neves (1993) use the annual U.K. Family Expenditure Survey covering the period 1970–1984 to study the intertemporal labor supply and consumption of married women. The total number of households considered was 43,671. These were allocated to 10 different cohorts depending on the year of birth. The average number of observations per cohort was 364. Their findings indicate reasonably sized intertemporal labor supply elasticities. Collado (1995) proposed a GMM estimator corrected for measurement error to deal with a dynamic pseudopanel data model. This estimator is consistent as C tends to infinity for a fixed T and n_c.

VII. LIMITED DEPENDENT VARIABLES AND PANEL DATA

In many economic studies, the dependent variable is discrete, indicating, for example, that a household purchased a car or that an individual is unemployed or that he or she joined the union. For example, let $y_{it} = 1$ if the ith individual participates in the labor force at time t. This occurs if y_{it}^*, the difference between the ith individual's offered wage and his unobserved reservation wage is positive. This can be described more formally as follows:

$$
\begin{aligned}
y_{it} &= 1 & \text{if } y_{it}^* > 0 \\
&= 0 & \text{if } y_{it}^* \leq 0
\end{aligned}
\tag{6}
$$

where

$$y_{it}^* = x_{it}'\beta + \mu_i + v_{it} \tag{7}$$

That is, y_{it}^* can be explained by a set of regressors x_{it} and error components disturbances. In this case,

$$\Pr[y_{it} = 1] = \Pr[y_{it}^* > 0] = \Pr[v_{it} > -x_{it}'\beta - \mu_i] = F(x_{it}'\beta + \mu_i)$$

The last equality holds as long as the density function describing the cumulative distribution function F is symmetric around zero. For panel data, the presence of individual effects complicates matters significantly. For the one-way error component model with random individual effects $E(u_{it}u_{is}) = \sigma_\mu^2$ for any $t, s = 1, 2, \ldots, T$, and the joint likelihood of $(y_{it}, \ldots, y_{Nt})$ can no longer be written as the product of the marginal likelihoods of the y_{it}'s. This complicates the derivation of maximum likelihood and will now involve bivariate numerical integration. On the other hand, if there are no random individual effects, the joint likelihood will be the product of the marginals and one can proceed as in the usual cross-sectional limited dependent variable case. For the fixed effects model, with limited dependent variable, the model is nonlinear and it is not possible to get rid of the μ_i's by taking differences or performing the FE transformation, as a result β and σ_v^2 cannot be estimated consistently for T fixed, since the inconsistency in the μ_i's is transmitted to β and σ_v^2 (Hsiao 1986). The usual solution around this incidental parameters (μ_i's) problem is to find a minimal sufficient statistic for the μ_i's which does not depend on the β's. Since the maximum likelihood estimates are in general functions of these minimum sufficient statistics, one can obtain the latter by differentiating the log-likelihood function with respect to μ_i. For the logit model, this yields the result that $\sum_{t=1}^{T} y_{it}$ is a minimum sufficient statistic for μ_i. Chamberlain (1980) suggests maximizing the *conditional* likelihood function

$$L_c = \prod_{i=1}^{N} \Pr\left(y_{i1}, \ldots, y_{iT} / \sum_{t=1}^{T} y_{it}\right) \tag{8}$$

rather than the unconditional likelihood function. For the fixed-effects logit model, this approach results in a computationally convenient estimator. However, the computations rise geometrically with T and are excessive for $T > 10$.

In order to test for fixed individual effects, one can perform a Hausman-type test based on the difference between Chamberlain's conditional maximum likelihood estimator and the usual logit maximum likelihood estimator, ignoring the individual effects. The latter estimator is consistent and efficient only under the null of no individual effects and inconsistent under the alternative. Chamberlain's estimator is consistent whether H_0 is true or not, but it is inefficient under H_0 because it may not use all the data. Both estimators can be easily obtained from the usual logit maximum likelihood routines. The constant is dropped and estimates of the asymptotic variances are used to form Hausman's χ^2 statistic. This will be distributed as χ_K^2 under H_0. For an application studying the linkage between unemployment and mental health problems in Sweden using the Swedish Level of Living Surveys (Björklund 1985).

In contrast to the fixed-effects logit model, the conditional likelihood approach does not yield computational simplifications for the fixed-effects *probit* model. In particular, the fixed effects cannot be swept away and maximizing the likelihood over all the parameters including the fixed effects will in general lead to inconsistent estimates for large N and fixed T.* Heckman (1981b) performed some limited Monte Carlo experiments on a probit model with a single regressor. For $N = 100$, $T = 8$, $\sigma_\nu^2 = 1$, and $\sigma_\mu^2 = 0.5$, 1, and 3, Heckman computed the bias of the fixed-effects MLE of β using 25 replications. He found at most 10% bias for $\beta = 1$, which was always toward zero.

Although the probit model does not lend itself to a fixed effects treatment, it has been common to use it for the random-effects specification. For the random-effects probit model, maximum likelihood estimation yields a consistent and efficient estimator of β. However, MLE is computationally more involved. Essentially, one has to compute the joint probabilities of a T variate normal distribution which involves T-dimensional integrals (Hsiao 1986). This gets to be infeasible if T is big. However, by conditioning on the individual effects, this T-dimensional integral problem reduces to a single integral involving the product of a standard normal density and the difference of two normal cumulative density functions. This can be evaluated using the Gaussian quadrature procedure suggested by Butler and Moffitt (1982). This approach has the advantage of being computationally feasible even for fairly large T. For an application, see Sickles and Taubman (1986), who estimate a two-equation structural model of the health and retirement decisions of the elderly using five biennial panels of males drawn from the Retirement History Survey. For a recent Monte Carlo study on the random-effects probit model, see Guilkey and Murphy (1993). Underlying the random-effects probit model is the equicorrelation assumption between successive disturbances belonging to the same individual. In a study of labor force participation of married women, Avery, Hansen, and Hotz (1983) reject the hypothesis of equicorrelation across the disturbances and suggest a method of moments estimator that allows for a general type of serial correlation among the disturbances. Chamberlain (1984) apples both a fixed effects logit estimator and a minimum-distance random-effects probit estimator to a study of the labor force participation of 924 married women drawn from the Panel Study of Income Dynamics. Lechner (1995) suggests several specification tests for the panel data probit model.

---

*In cases where the conditional likelihood function is not feasible as in the fixed-effects probit case, Manski (1987) suggests a conditional version of his maximum score estimator which under fairly general conditions provides a strongly consistent estimator of β.

These are generalized score and Wald tests employed to detect omitted variables, neglected dynamics, heteroskedasticity, nonnormality, and random-coefficient variations. The performance of these tests in small samples is investigated using Monte Carlo experiments. Also, an empirical example on the probability of self-employment in West Germany is given which uses a random sample of 1926 working men selected from the German Socio-Economic Panel and observed over the period 1984–1989.

Heckman and MaCurdy (1980) consider a fixed-effects tobit model to estimate a life-cycle model of female labor supply. They argue that the individual effects have a specific meaning in a life-cycle model and therefore cannot be assumed independent of the x_{it}'s. Hence, a fixed effects rather than a random-effects specification is appropriate. For this fixed-effects tobit model, the model is given by (7), with $v_{it} \sim \text{IIN}(0, \sigma_v^2)$ and

$$
\begin{aligned}
y_{it} &= y_{it}^* && \text{if } y_{it}^* > 0 \\
&= 0 && \text{otherwise}
\end{aligned}
\tag{9}
$$

where y_{it} could be the expenditures on a car. This will be zero at time t, if the ith individual does not buy a car. In the latter case all we know is that $y_{it}^* \leq 0$.* As in the fixed-effects probit model, the μ_i's cannot be swept away and as a result β and σ_v^2 cannot be estimated consistently for T fixed, since the inconsistency in the μ_i's is transmitted to β and σ_v^2. Heckman and MaCurdy (1980) suggest estimating the log-likelihood using iterative methods. Recently, Honoré (1992) suggested trimmed least absolute deviations and trimmed least-squares estimators for truncated and censored regression models with fixed effects. These are semiparametric estimators with no distributional assumptions necessary on the error term. The main assumption is that the remainder error v_{it} is independent and identically distributed conditional on the x_{it}'s and the μ_i's, for $t = 1, \ldots, T$. Honoré (1992) exploits the symmetry in the distribution of the latent variables and finds that when the true values of the parameters are known, trimming can transmit the same symmetry in distribution to the observed variables. This generates orthogonality conditions which must hold at the true value of the parameters. Therefore, the resulting GMM estimator is consistent provided the orthogonality conditions are satisfied at a unique point in the parameter space. Honoré (1992) shows that these estimators are consistent and asymptotically normal. Monte Carlo results show that as long as $N \geq 200$, the asymptotic distribution is a good approximation of the small-sample distribution. However, if N is

---

*Researchers may also be interested in panel data economic relationships where the dependent variable is a count of some individual actions or events, such as the number of patents filed, the number of drugs introduced, or the number of jobs held. These models can be estimated by using Poisson panel data regressions (Hausman, Hall, and Griliches 1984).

small, the small-sample distribution of these estimators is skewed. Honoré (1993)
extends his analysis to the *dynamic* Tobit model with fixed effects; i.e.,

$$y_{it}^* = \delta y_{i,t-1} + x_{it}'\beta + \mu_i + \nu_{it} \tag{10}$$

with $y_{it} = \max\{0, y_{it}^*\}$ for $i = 1, \ldots, N, t = 1, \ldots, T$. The basic assumption is
that ν_{it} is IID$(0, \sigma_\nu^2)$ for $t = 1, \ldots, T$, conditional on y_{i0}, x_{it}, and μ_i. Honoré (1993)
shows how to trim the observations from a dynamic Tobit model so that the sym-
metry conditions are preserved for the observed variables at the true values of the
parameters. These symmetry restrictions are free of the individual effects and no
assumption is needed on the distribution of the μ_i's or their relationship with the
explanatory variables. These restrictions generate orthogonality conditions which
are satisfied at the true value of the parameters. The orthogonality conditions can
be used in turn to construct method of moments estimators. Using Monte Carlo experiments, Honoré (1993) shows that MLE for a dynamic Tobit fixed-effects model
performs poorly, whereas the GMM estimator performs quite well, when δ is the only
parameter of interest.

Recently, Keane (1994) derived a computationally practical simulation estimator for the panel data probit model. Simulation estimation methods replace intractable integrals by unbiased Monte Carlo probability simulators. This is ideal for
limited dependent variable models where for a multinominal probit model, the choice
probabilities involve multivariate integrals.* In fact, for cross-sectional data, McFadden's method of simulated moments (MSM) involves an $M - 1$ integration problem,
where M is the number of possible choices facing the individual. For panel data,
things get more complicated, because there are M choices facing any individual at
each period. This means that there are M^T possible choice sequences facing each
individual over the panel. Hence the MSM estimator becomes infeasible as T gets
large. Keane (1994) sidesteps this problem of having to simulate M^T possible choice
sequences by factorizing the method of simulated moments first-order conditions into
transition probabilities. The latter are simulated using highly accurate importance
sampling techniques. This method of simulating probabilities is referred to as the
Geweke, Hajivassiliou, and Keane (GHK) simulator because it was independently
developed by these authors. Keane (1994) performs Monte Carlo experiments and
finds that even for large T and small simulation sizes, the bias in the MSM estimator is negligible. When maximum likelihood methods are feasible, Keane (1994)
finds that the MSM estimator performs well relative to quadrature-based maximum
likelihood methods even where the latter are based on a large number of quadrature

---

*For good surveys of simulation methods, see Hajivassiliou and Ruud (1994) for limited dependent vari-
able models and Gourieroux and Monfort (1993) with special reference to panel data. The methods
surveyed include simulation of the likelihood, simulation of the moment functions, and simulation of the
score.

points. When maximum likelihood methods are not feasible, the MSM estimator out-performs the simulated maximum likelihood estimator even when the highly accurate GHK probability simulator is used. Keane (1993) applies the MSM estimator to the same data set used by Keane, Moffitt, and Runkle (1988) to study the cyclical be-havior of real wages. He finds that the Keane, Moffitt, and Runkle conclusion of a weakly procyclical movement in the real wage appears to be robust to relaxation of the equicorrelation assumption.

Heckman (1981a, 1981b, 1981c) emphasizes the importance of distinguish-ing between "true state dependence" and "spurious state dependence" in dynamic models of individual behavior. In the true case, once an individual experiences an event, his or her preferences change and he or she will behave differently in the fu-ture as compared with an identical individual that has not experienced this event in the past. In the spurious case, past experience has no effect on the probability of experiencing the event in the future. However, one cannot properly control for all the variables that distinguish one individual's decision from another. In this case, past experience which is a good proxy for these omitted variables shows up as a sig-nificant determinant of the future probability of occurrence of this event. Testing for true versus spurious state dependence is therefore important in these studies, but it is complicated by the presence of the individual effects or heterogeneity. In fact, even if there is no state dependence, $\Pr[y_{it}/x_{it}, y_{i,t-\ell}] \neq \Pr[y_{it}/x_{it}]$ as long as there are random individual effects present in the model. If in addition to the absence of the state dependence there is also no heterogeneity, then $\Pr[y_{it}/x_{it}, y_{i,t-\ell}] = \Pr[y_{it}/x_{it}]$. A test for this equality can be based on a test for $\gamma = 0$ in the model

$$\Pr[y_{it} = 1/x_{it}, y_{it-1}] = F(x'_{it}\beta + \gamma y_{i,t-1}) \tag{11}$$

by using standard maximum likelihood techniques. If $\gamma = 0$ is not rejected, we ignore the heterogeneity issue and proceed as in conventional limited dependent variable models not worrying about the panel data nature of the data. However, re-jecting the null does not necessarily imply that there is heterogeneity since γ can be different from zero due to serial correlation in the remainder error or due to state dependence. In order to test for time dependence one has to condition on the individ-ual effects, i.e., test $\Pr[Y_{it}/y_{i,t-\ell}, x_{it}, \mu_i] = \Pr[y_{it}/x_{it}, \mu_i]$. This can be implemented following the work of Lee (1987) and Maddala (1987). In fact, if $\gamma = 0$ is rejected, Hsiao (1996) suggests testing for time dependence against heterogeneity. If hetero-geneity is rejected, the model is misspecified. If heterogeneity is not rejected then one estimates the model correcting for heterogeneity. See Heckman (1981c) for an application to married women's employment decisions based on a three-year sample from the Panel Study of Income Dynamics. One of the main findings of this study is that neglecting heterogeneity in dynamic models overstate the effect of past experi-ence on labor market participation.

In many surveys, nonrandomly missing data may occur due to a variety of self-selection rules. One such self-selection rule is the problem of nonresponse of the economic agent. Nonresponse occurs, for example, when the individual refuses to participate in the survey or refuses to answer particular questions. This problem occurs in cross-sectional studies, but it gets aggravated in panel surveys. After all, panel surveys are repeated cross-sectional interviews. So, in addition to the above kinds of nonresponse, one may encounter individuals that refuse to participate in subsequent interviews or simply move or die. Individuals leaving the survey cause attrition in the panel. This distorts the random design of the survey and questions the representativeness of the observed sample in drawing inference about the population we are studying. Inference based on the balanced subpanel is inefficient even in randomly missing data since it is throwing away data. In nonrandomly missing data, this inference is misleading because it is no longer representative of the population. Verbeek and Nijman (1996) survey the reasons for nonresponse and distinguish between "ignorable" and "nonignorable" selection rules. This is important because, if the selection rule is ignorable for the parameters of interest, one can use the standard panel data methods for consistent estimation. If the selection rule is nonignorable, then one has to take into account the mechanism that causes the missing observations in order to obtain consistent estimates of the parameters of interest.

We now consider a simple model of nonresponse in panel data. Following the work of Hausman and Wise (1979), Ridder (1990), and Verbeek and Nijman (1996), we assume that y_{it} given by Eq. (1) is observed if a latent variable $r_{it}^* \geq 0$. This latent variable is given by

$$r_{it}^* = z_{it}'\gamma + \epsilon_i + \eta_{it} \tag{12}$$

where z_{it} is a set of explanatory variables possibly including some of the x_{it}'s. The one-way error components structure allows for heterogeneity in the selection process. The errors are assumed to be normally distributed $\epsilon_i \sim \text{IIN}(0, \sigma_\epsilon^2)$ and $\eta_{it} \sim \text{IIN}(0, \sigma_\eta^2)$ with the only nonzero covariances being $\text{cov}(\epsilon_i, \mu_i) = \sigma_{\mu\epsilon}$ and cov $(\eta_{it}, \nu_{it}) = \sigma_{\eta\nu}$. In order to get a consistent estimator for β, a generalization of Heckman's selectivity bias correction procedure from the cross-sectional to the panel data case can be employed. The conditional expectation of u_{it} given selection now involves two terms. Therefore, instead of one selectivity bias correction term, there are now two terms corresponding to the covariances $\sigma_{\mu\epsilon}$ and $\sigma_{\eta\nu}$. However, unlike the cross-sectional case, these correction terms cannot be computed from simple probit regressions and require numerical integration. Fortunately, this is only a one-dimensional integration problem because of the error component structure. Once the correction terms are estimated, they are included in the regression equation as in the cross-sectional case and OLS or GLS can be run on the resulting augmented model. For details, see Verbeek and Nijman (1996), who also warn about heteroskedasticity and serial correlation in the second step regression if the selection rule in nonignorable. Verbeek and Nijman (1996) also discuss MLE for this random-effects probit model with selection bias.

Before one embarks on these complicated estimation procedures one should first test whether the selection rule is ignorable. Verbeek and Nijman (1992a) consider a Lagrange multiplier test for H_0; $\sigma_{\nu\eta} = \sigma_{\mu\epsilon} = 0$. The null hypothesis is a sufficient condition for the selection rule to be ignorable for the random-effects model. Unfortunately, this also requires numerical integration over a maximum of two dimensions and is cumbersome to use in applied work. In addition, the LM test is highly dependent on the specification of the selectivity equation and the distributional assumptions. Alternatively, Verbeek and Nijman (1992a) suggest some simple Hausman-type tests based on GLS and within estimators for the unbalanced panel and the balanced subpanel. All four estimators are consistent under the null hypothesis that the selection rule is ignorable and all four estimators are inconsistent under the alternative. In practice, Verbeek and Nijman (1992a) suggest including three simple variables in the regression to check for the presence of selection bias. These are (i) the number of waves the ith individual participates in the panel, (ii) a binary variable taking the value 1 if and only if the ith individual is observed over the entire sample, and (iii) a binary variable indicating whether the individual was present in the last period. Testing the significance of these variables is recommended as a check for selection bias. Intuitively, one is testing whether the pattern of missing observations affects the underlying regression. Wooldridge (1995) derives some simple variable addition tests of selection bias as well as easy-to-apply estimation techniques that correct for selection bias in linear fixed-effects panel data models. The auxiliary regressors are either Tobit residuals or inverse Mill's ratios, and the disturbances are allowed to be arbitrarily serially correlated and unconditionally heteroskedastic.

There are many empirical applications illustrating the effects of attrition bias; see Hausman and Wise (1979) for a study of the Gary Income Maintenance experiment. For this experimental panel study of labor supply response, the treatment effect is an income guarantee/tax rate combination. People who benefit from this experiment are more likely to remain in the sample. Therefore, the selection rule is nonignorable, and attrition can overestimate the treatment effect on labor supply. For the Gary Income Maintenance Experiment, Hausman and Wise (1979) found little effect of attrition bias on the experimental labor supply response. Similar results were obtained by Robins and West (1986) for the Seattle and Denver Income Maintenance Experiments. For the latter sample, attrition was modest (11% for married men and 7% for married women and single heads during the period studied) and its effect was not serious enough to warrant extensive correction procedures. More recently, Keane, Moffitt, and Runkle (1988) studied the movement of real wages over the business cycle for a panel data drawn from the National Longitudinal Survey of Young Men (NLS) from 1966 to 1981. They showed that self-selection biased the behavior of real wage in a procyclical direction.

VIII. FURTHER READINGS

Supplementary readings on panel data include Hsiao's (1986) Econometric Society monograph, the standard reference on the subject, Maddala's (1993) two volumes of some of the classic papers in the field, a special issue of *Empirical Economics* edited by Raj and Baltagi (1992), and a special issue of the *Journal of Econometrics* edited by Baltagi (1995a). Two recent books on panel data are Baltagi (1995b) and Mátyás and Sevestre (1996).

REFERENCES

Ahn, S. C. and S. Low (1996), A Reformulation of the Hausman Test for Regression Models with Pooled Cross-Section Time-Series Data, *Journal of Econometrics*, 71, 309–319.

Ahn, S. C. and P. Schmidt (1995), Efficient Estimation of Models for Dynamic Panel Data, *Journal of Econometrics*, 68, 5–27.

Amemiya, T. (1971), The Estimation of the Variances in a Variance-Components Model, *International Economic Review*, 12, 1–13.

Amemiya, T. and T. E. MaCurdy (1986), Instrumental-Variable Estimation of an Error Components Model, *Econometrica*, 54, 869–881.

Anderson, T. W. and C. Hsiao (1982), Formulation and Estimation of Dynamic Models Using Panel Data, *Journal of Econometrics*, 18, 47–82.

Angrist, J. D. and W. K. Newey (1991), Over-identification Tests in Earnings Functions with Fixed Effects, *Journal of Business and Economic Statistics*, 9, 317–323.

Arellano, M. (1987), computing Robust Standard Errors for Within-Groups Estimators, *Oxford Bulletin of Economics and Statistics*, 49, 431–434.

Arellano, M. (1989), A Note on the Anderson-Hsiao Estimator for Panel Data, *Economics Letters*, 31, 337–341.

Arellano, M. (1990), Some Testing for Autocorrelation in Dynamic Random Effects Models, *Review of Economic Studies*, 57, 127–134.

Arellano, M. (1993), On the Testing of Correlated Effects with Panel Data, *Journal of Econometrics*, 59, 87–97.

Arellano, M. and S. Bond (1991), Some Tests of Specification for Panel Data: Monte Carlo Evidence and an Application to Employment Equations, *Review of Economic Studies*, 58, 277–97.

Arellano, M. and O. Bover (1995), Another Look at the Instrumental Variables Estimation of Error-Component Models, *Journal of Econometrics*, 68, 29–51.

Avery, R. B. (1977), Error Components and Seemingly Unrelated Regressions, *Econometrica*, 45, 199–209.

Avery, R. B., L. P. Hansen, and V. J. Hotz (1983), Multiperiod Probit Models and Orthogonality Condition Estimation, *International Economic Review*, 24, 21–35.

Bailar, B. A. (1975), The Effects of Rotation Group Bias on Estimates from Panel Survey, *Journal of the American Statistical Association*, 70, 23–30.

Baillie, R. and B. H. Baltagi (1995), Prediction from the Regression Model with One-Way Error Components, Working Paper, Department of Economics, Texas A&M University, College Station, Texas.

Baltagi, B. H. (1980), On Seemingly Unrelated Regressions with Error Components, *Econometrica*, 48, 1547–1551.

Baltagi, B. H. (1981a), Pooling: An Experimental Study of Alternative Testing and Estimation Procedures in a Two-Way Error Components Model, *Journal of Econometrics*, 17, 21–49.

Baltagi, B. H. (1981b), Simultaneous Equations with Error Components, *Journal of Econometrics*, 17, 189–200.

Baltagi, B. H. (1984), A Monte Carlo Study for Pooling Time-Series of Cross-Section Data in the Simultaneous Equations Model, *International Economic Review*, 25, 603–624.

Baltagi, B. H. (1985), Pooling Cross-Sections with Unequal Time-Series Lengths, *Economics Letters*, 18, 133–136.

Baltagi, B. H. (1995a), Editor's Introduction: Panel Data, *Journal of Econometrics*, 68, 1–4.

Baltagi, B. H. (1995b), *Econometric Analysis of Panel Data*, Wiley, Chichester.

Baltagi, B. H. (1996), Testing for Random Individual and Time Effects Using a Gauss-Newton Regression, *Economics Letters*, 50, 189–192.

Baltagi, B. H. and Y. J. Chang (1994), Incomplete Panels: A Comparative Study of Alternative Estimators for the Unbalanced One-Way Error Component Regression Model, *Journal of Econometrics*, 62, 67–89.

Baltagi, B. H. and Y. J. Chang (1996), Testing for Random Individual Effects Using Recursive Residuals, *Econometric Reviews*, 15, 331–338.

Baltagi, B. H., Y. J. Chang, and Q. Li (1992), Monte Carlo Evidence on Panel Data Regressions with AR(1) Disturbances and an Arbitrary Variance on the Initial Observations, *Journal of Econometrics*, 52, 371–380.

Baltagi, B. H. and J. M. Griffin (1988), A Generalized Error Component Model with Heteroscedastic Disturbances, *International Economic Review*, 29, 745–753.

Baltagi, B. H. and J. M. Griffin (1995), A Dynamic Demand Model for Liquor: The Case for Pooling, *Review of Economics and Statistics*, 77, 545–553.

Baltagi, B. H. and J. M. Griffin (1997), Pooled Estimators vs. Their Heterogeneous Counterparts in the Context of Dynamic Demand for Gasoline, *Journal of Econometrics*, 77, 303–327.

Baltagi, B. H., J. Hidalgo, and Q. Li (1996), A Non-parametric Test for Poolability Using Panel Data, *Journal of Econometrics*, 75, 345–367.

Baltagi, B. H. and Q. Li (1990), A Lagrange Multiplier Test for the Error Components Model with Incomplete Panels, *Econometric Reviews*, 9, 103–107.

Baltagi, B. H. and Q. Li (1991), A Transformation That Will Circumvent the Problem of Autocorrelation in an Error Component Model, *Journal of Econometrics*, 48, 385–393.

Baltagi, B. H. and Q. Li (1992), Prediction in the One-Way Error Component Model with Serial Correlation, *Journal of Forecasting*, 11, 561–567.

Baltagi, B. H. and Q. Li (1995), Testing AR(1) against MA(1) Disturbances in an Error Component Model, *Journal of Econometrics*, 68, 133–151.

Baltagi, B. H. and N. Pinnoi (1995), Public Capital Stock and State Productivity Growth: Further Evidence from an Error Components Model, *Empirical Economics*, 20, 351–359.

Becketti, S., W. Gould, L. Lillard, and F. Welch (1988), The Panel Study of Income Dynamics after Fourteen Years: An Evaluation, *Journal of Labor Economics*, 6, 472–492.

Bhargava, A. (1991), Identification and Panel Data Models with Endogenous Regressors, *Review of Economic Studies*, 58, 129–140.

Bhargava, A., L. Franzini, and W. Narendranathan (1982), Serial Correlation and Fixed Effects Model, *Review of Economic Studies*, 49, 533–549.

Bhargava, A. and J. D. Sargan (1983), Estimating Dynamic Random Effects Models from Panel Data Covering Short Time Periods, *Econometrica*, 51, 1635–1659.

Biorn, E. (1981), Estimating Economic Relations from Incomplete Cross-Section/Time-Series Data, *Journal of Econometrics*, 16, 221–236.

Biorn, E. (1992), The Bias of Some Estimators for Panel Data Models with Measurement Errors, *Empirical Economics*, 15, 221–236.

Björklund, A. (1985), Unemployment and Mental Health: Some Evidence from Panel Data, *The Journal of Human Resources*, 20, 469–483.

Blundell, R. W. and C. H. Meghir (1990), Panel Data and Life-Cycle Models, in J. Hartog, G. Ridder, and J. Theeuwes (eds.), *Panel Data and Labor Market Studies*, North-Holland, Amsterdam, 231–252.

Blundell, R., C. Meghir, and P. Neves (1993), Labor Supply and Intertemporal Substitution, *Journal of Econometrics*, 59, 137–160.

Boumahdi, R. and A. Thomas (1991), Testing for Unit Roots Using Panel Data: Application to the French Stock Market Efficiency, *Economics Letters*, 37, 77–79.

Breitung, J. and W. Meyer (1994), Testing for Unit Roots in Panel Data: Are Wages on Different Bargaining Levels Cointegrated?, *Applied Economics*, 26, 353–361.

Breusch, T. S. (1987), Maximum Likelihood Estimation of Random Effects Models, *Journal of Econometrics*, 36, 383–389.

Breusch, T. S., G. E. Mizon, and P. Schmidt (1989), Efficient Estimation Using Panel Data, *Econometrica*, 57, 695–700.

Breusch, T. S. and A. R. Pagan (1980), The Lagrange Multiplier Test and Its Applications to Model Specification in Econometrics, *Review of Econometric Studies*, 47, 239–253.

Butler, J. S. and R. Moffitt (1982), A Computationally Efficient Quadrature Procedure for the One Factor Multinominal Probit Model, *Econometrica*, 50, 761–764.

Chamberlain, G. (1980), Analysis of Covariance with Qualitative Data, *Review of Economic Studies*, 47, 225–238.

Chamberlain, G. (1982), Multivariate Regression Models for Panel Data, *Journal of Econometrics*, 18, 5–46.

Chamberlain, G. (1984), Panel Data, in Z. Griliches and M. Intrilligator (eds.), *Handbook of Econometrics*, North-Holland, Amsterdam.

Chesher, A. (1984), Testing for Neglected Heterogeneity. *Econometrica*, 52, 865–872.

Chow, G. C. (1960), Tests of Equality between Sets of Coefficients in Two Linear Regressions, *Econometrica*, 28, 591–605.

Chowdhury, G. (1991), A Comparison of Covariance Estimators for Complete and Incomplete Panel Data Models, *Oxford Bulletin of Economics and Statistics*, 53, 88–93.

Chowdhury, G. (1994), Fixed Effects with Interpersonal and Intertemporal Covariance, *Empirical Economics*, 19, 523–532.

Collado, M. D. (1995), Estimating Dynamic Models from Time-Series of Cross-Sections, University of Carlos III de Madrid, Madrid, Spain.

Cornwell, C. and P. Rupert (1988), Efficient Estimation with Panel Data: An Empirical Comparison of Instrumental Variables Estimators, *Journal of Applied Econometrics*, 3, 149–155.

Cornwell, C., P. Schmidt, and D. Wyhowski (1992), Simultaneous Equations and Panel Data, *Journal of Econometrics*, 51, 151–181.

Deaton, A. (1985), Panel Data from Time Series of Cross-Sections, *Journal of Econometrics*, 30, 109–126.

Dielman, T. E. (1989), *Pooled Cross-Sectional and Time Series Data Analysis*, Marcel Dekker, New York.

Duncan, G. J. and D. H. Hill (1985), An Investigation of the Extent and Consequences of Measurement Error in Labor Economic Survey Data, *Journal of Labor Economics*, 3, 508–532.

Fuller, W. A. and G. E. Battese (1974), Estimation of Linear Models with Cross-Error Structure, *Journal of Econometrics*, 2, 67–78.

Galbraith, J. W. and V. Zinde-Walsh (1995), Transforming the Error-Component Model for Estimation with General ARMA Disturbances, *Journal of Econometrics*, 66, 349–355.

Gourieroux, C. and A. Monfort (1993), Simulation-Based Inference: A Survey with Special Reference to Panel Data Models, *Journal of Econometrics*, 59, 5–33.

Griliches, Z. and J. A. Hausman (1986), Errors in Variables in Panel Data, *Journal of Econometrics*, 31, 93–118.

Guilkey, D. K. and J. L. Murphy (1993), Estimation and Testing in the Random Effects Probit Model, *Journal of Econometrics*, 59, 301–317.

Hajivassiliou, V. A. and P. A. Ruud (1994), Classical Estimation of LDV Models Using Simulation, in R. F. Engle and D. McFadden (eds.), *Handbook of Econometrics*, North-Holland, Amsterdam.

Hamerle, A. (1990), On a Simple Test for Neglected Heterogeneity in Panel Studies, *Biometrics*, 46, 193–199.

Han, A. K. and D. Park (1989), Testing for Structural Change in Panel Data: Application to a Study of U.S. Foreign Trade in Manufacturing Goods, *Review of Economics and Statistics*, 71, 135–142.

Haque, N. U., K. Lahiri, and P. Montiel (1993), Estimation of a Macroeconomic Model with Rational Expectations and Capital Controls for Developing Countries, *Journal of Development Economics*, 42, 337–356.

Hausman, J. A. (1978), Specification Tests in Econometrics, *Econometrica*, 46, 1251–1271.

Hausman, J. A., B. H. Hall, and Z. Griliches (1984), Econometric Models for Count Data with an Application to the Patents–R&D Relationship, *Econometrica*, 52, 909–938.

Hausman, J. A. and W. E. Taylor (1981), Panel Data and Unobservable Individual Effects, *Econometrica*, 49, 1377-1398.

Hausman, J. A. and D. Wise (1979), Attrition Bias in Experimental and Panel Data: the Gary Income Maintenance Experiment, *Econometrica*, 47, 455–473.

Heckman, J. J. (1981a), Statistical Models for Discrete Panel Data, in C. F. Manski and D. McFadden (eds.), *Structural Analysis of Discrete Data with Econometric Applications*, MIT Press, Cambridge.

Heckman, J. J. (1981b), The Incidental Parameters Problem and the Problem of Initial Conditions in Estimating a Discrete Time-Discrete Data Stochastic Process, in C. F. Manski and D. McFadden (eds.), *Structural Analysis of Discrete Data with Econometric Applications*, MIT Press, Cambridge.

Heckman, J. J. (1981c), Heterogeneity and State Dependence, in S. Rosen (ed.), *Studies in Labor Markets*, Chicago University Press, Chicago.

Heckman, J. J. and T. E. MaCurdy (1980), A Life-Cycle Model of Female Labor Supply, *Review of Economic Studies*, 47, 47–74.

Holtz-Eakin, D. (1988), Testing for Individual Effects in Autoregressive Models, *Journal of Econometrics*, 39, 297–307.

Holtz-Eakin, D., W. Newey, and H. S. Rosen (1988), Estimating Vector Autoregressions with Panel Data, *Econometrica*, 56, 1371–1395.

Honda, Y. (1985), Testing the Error Components Model with Non-normal Disturbances, *Review of Economic Studies*, 52, 681–690.

Honda, Y. (1991), A Standardized Test for the Error Components Model with the Two-Way Layout, *Economics Letters*, 37, 125–128.

Honoré, B. E. (1992), Trimmed LAD and Least Squares Estimation of Truncated and Censored Regression Models with Fixed Effects, *Econometrica*, 60, 533–565.

Honoreé, B. E. (1993), Orthogonality Conditions for Tobit Models with Fixed Effects and Lagged Dependent Variables, *Journal of Econometrics*, 59, 35–61.

Horowitz, J. L. and M. Markatou (1996), Semiparametric Estimation of Regression Models for Panel Data, *Review of Economic Studies*, 63, 145–168.

Howrey, E. P. and H. R. Varian (1984), Estimating the Distributional Impact of Time-of-Day Pricing of Electricity, *Journal of Econometrics*, 26, 65–82.

Hsiao, C. (1986), *Analysis of Panel Data*, Cambridge University Press, Cambridge.

Hsiao, C. (1991), Identification and Estimation of Dichotomous Latent Variables Models Using Panel Data, *Review of Economic Studies*, 58, 717–731.

Hsiao, C. (1996), Logit and Probit Models, in L. Mátyás and P. Sevestre (eds.), *The Econometrics of Panel Data: A Handbook of Theory and Applications*, Kluwer, Dordrecht, 410–428.

Im, K., M. H. Pesaran, and Y. Shin (1996), Testing for Unit Roots in Heterogeneous Panels, Working Paper, University of Cambridge.

Kang, S. (1985), A Note on the Equivalence of Specification Tests in the Two-Factor Multivariate Variance Components Model, *Journal of Econometrics*, 28, 193–203.

Kao, C. and J. F. Schnell (1987a), Errors in Variables in Panel Data with Binary Dependent Variable, *Economics Letters*, 24, 45–49.

Kao, C. and J. F. Schnell (1987b), Errors in Variables in a Random Effects Probit Model for Panel Data, *Economics Letters*, 24, 339–342.

Kasprzyk, D., G. J. Duncan, G. Kalton, and M. P. Singh (1989), *Panel Surveys*, Wiley, New York.

Keane, M. (1993), Simulation Estimation for Panel Data Models with Limited Dependent Variables, in G. S. Maddala, C. R. Rao, and H. D Vinod (eds.), *Handbook of Statistics*, North-Holland, Amsterdam.

Keane, M. (1994), A Computationally Practical Simulation Estimator for Panel Data, *Econometrica*, 62, 95–116.

Keane, M. P., R. Moffitt, and D. E. Runkle (1988), Real Wages over the Business Cycle: Estimating the Impact of Heterogeneity with Micro Data. *Journal of Political Economy*, 96, 1232–1266.

Keane, M. P. and D. E. Runkle (1992), On the Estimation of Panel-Data Models with Serial Correlation When Instruments Are Not Strictly Exogenous, *Journal of Business and Economics Statistics*, 10, 1–9.

Kiefer, N. M. (1980), Estimation of Fixed Effects Models for Time Series of Cross Sections with Arbitrary Intertemporal Covariance, *Journal of Econometrics*, 14, 195–202.

Kinal, T. and K. Lahiri (1990), A Computational Algorithm for Multiple Equation Models with Panel Data, *Economics Letters*, 34, 143–146.

Kinal, T. and K. Lahiri (1993), A Simplified Algorithm for Estimation of Simultaneous Equations Error Components Models with an Application to a Model of Developing Country Foreign Trade, *Journal of Applied Econometrics*, 8, 81–92.

Kiviet, J. F. (1995), On Bias, Inconsistency and Efficiency of Some Estimators in Dynamic Panel Data Models, *Journal of Econometrics*, 68, 53–78.

Krishnakumar, J. (1988), *Estimation of Simultaneous Equation Models with Error Components Structure*, Springer-Verlag, Berlin.

Lechner, M. (1995), Some Specification Tests for Probit Models Estimated on Panel Data, *Journal of Business and Economic Statistics*, 13, 475–488.

Lee, L. F. (1987), Non-parametric Testing of Discrete Panel Data Models, *Journal of Econometrics*, 34, 147–177.

Levin, A. and C. F. Lin (1996), Unit Root Tests in Panel Data: Asymptotic and Finite Sample Properties, *Journal of Econometrics*, forthcoming.

Li, Q. and T. Stengos (1992), A Hausman Specification Test Based on Root N Consistent Semiparametric Estimators, *Economics Letters*, 40, 141–146.

Li, Q. and T. Stengos (1994), Adaptive Estimation in the Panel Data Error Component Model with Heteroskedasticity of Unknown Form, *International Economic Review*, 35, 981–1000.

Li, Q. and T. Stengos (1995), A Semi-parametric Non-nested Test in a Dynamic Panel Data Model, *Economics Letters*, 49, 1–6.

Li, Q. and T. Stengos (1996), Semi-parametric Estimation of Partially Linear Panel Data Models, *Journal of Econometrics*, 71, 389–397.

Lillard, L. A. and R. J. Willis (1978), Dynamic Aspects of Earning Mobility, *Econometrica*, 46, 985–1012.

Little, R. J. A. (1988), Missing-Data Adjustments in Large Surveys, *Journal of Business and Economic Statistics*, 6, 287–297.

MacDonald, R. (1996), Panel Unit Root Tests and Real Exchange Rates, *Economics Letters*, 50, 7–11.

MaCurdy, T. A. (1982), The Use of Time Series Processes to Model the Error Structure of Earnings in a Longitudinal Data Analysis, *Journal of Econometrics*, 18, 83–114.

Maddala, G. S. (1987), Limited Dependent Variable Models Using Panel Data, *The Journal of Human Resources*, 22, 307–338.

Maddala, G. S. (ed.) (1993), *The Econometrics of Panel Data*, Vols. I, II, Edward Elgar, Cheltenham.

Maddala, G. S. and T. D. Mount (1973), A Comparative Study of Alternative Estimators for Variance Components Models Used in Econometric Applications, *Journal of the American Statistical Association*, 68, 324–328.

Maddala, G. S., V. K. Srivastava, and H. Li (1994), Shrinkage Estimators for the Estimation of Short-Run and Long-Run Parameters from Panel Data Models, Working Paper, Ohio State University, Ohio.

Magnus, J. R. (1982), Multivariate Error Components Analysis of Linear and Nonlinear Regression Models by Maximum Likelihood, *Journal of Econometrics*, 19, 239–285.

Manski, C. F. (1987), Semiparametric Analysis of Random Effects Linear Models from Binary Panel Data, *Econometrica*, 55, 357–362.

Mátyás, L. and L. Lovrics (1990), Small Sample Properties of Simultaneous Error Components Models, *Economics Letters*, 32, 25–34.

Mátyás, L. and L. Lovrics (1991), Missing Observations and Panel Data: A Monte Carlo Analysis, *Economics Letters*, 37, 39–44.

Mátyás, L. and P. Sevestre (1996), eds. *The Econometrics of Panel Data: A Handbook of Theory and Applications*, Kluwer, Dordrecht.

Mazodier, P. and A. Trognon (1978), Heteroskedasticity and Stratification in Error Components Models, *Annales de l'INSEE*, 30–31, 451–482.

Mendelsohn, R., D. Hellerstein, M. Huguenin, R. Unsworth, and R. Brazee (1992), Measuring Hazardous Waste Damages with Panel Models, *Journal of Environmental Economics and Management*, 22, 259–271.

Metcalf, G. E. (1996), Specification Testing in Panel Data with Instrumental Variables, *Journal of Econometrics*, 71, 291–307.

Moffitt, R. (1993), Identification and Estimation of Dynamic Models with a Time Series of Repeated Cross-Sections, *Journal Econometrics*, 59, 99–123.

Moulton, B. R. (1986), Random Group Effects and the Precision of Regression Estimates, *Journal of Econometrics*, 32, 385–397.

Moulton, B. R. and W. C. Randolph (1989), Alternative Tests of the Error Components Model, *Econometrica*, 57, 685–693.

Mundlak, Y. (1978), On the Pooling of Time Series and Cross-Section Data, *Econometrica*, 46, 69–85.

Nickell, S. (1981), Biases in Dynamic Models with Fixed Effects, *Econometrica*, 49, 1417–1426.

Orme, C. (1993), A Comment on "A Simple Test for Neglected Heterogeneity in Panel Studies," *Biometrics*, 49, 665–667.

Pesaran, M. H. and R. Smith (1995), Estimating Long-Run Relationships from Dynamic Heterogeneous Panels, *Journal of Econometrics*, 68, 79–113.

Prucha, I. R. (1984), On the Asymptotic Efficiency of Feasible Aitken Estimators for Seemingly Unrelated Regression Models with Error Components, *Econometrica*, 52, 203–207.

Prucha, I. R. (1985), Maximum Likelihood and Instrumental Variable Estimation in Simultaneous Equation Systems with Error Components, *International Economic Review*, 26, 491–506.

Quah, D. (1994), Exploiting Cross-Section Variation for Unit Root Inference in Dynamic Data, *Economics Letters*, 44, 9–19.

Randolph, W. C. (1988), A Transformation for Heteroscedastic Error Components Regression Models, *Economics Letters*, 27, 349–354.

Raj, B. and B. Baltagi (eds.) (1992), *Panel Data Analysis*, Physica-Verlag, Heidelberg.

Rao, C. R. (1972), Estimation Variance and Covariance Components in Linear Models, *Journal of the American Statistical Association*, 67, 112–115.

Revankar, N. S. (1979), Error Component Models with Serially Correlated Time Effects, *Journal of the Indian Statistical Association*, 17, 137–160.

Revankar, N. S. (1992), Exact Equivalence of Instrumental Variable Estimators in an Error Component Structural System, *Empirical Economics*, 17, 77–84.

Ridder, G. (1990), Attrition in Multi-wave Panel Data, in J. Hartog, G. Ridder, and J. Theeuwes (eds.), *Panel Data and Labor Market Studies*, North-Holland, Amsterdam, 45–67.

Robertson, D. and J. Symons (1992), Some Strange Properties of Panel Data Estimators, *Journal of Applied Econometrics*, 7, 175–189.

Robins, P. K. and R. W. West (1986), Sample Attrition and Labor Supply Response in Experimental Panel Data: A Study of Alternative Correction Procedures, *Journal of Business and Economics Statistics*, 4, 329–338.

Schmidt, P. (1983), A Note on a Fixed Effect Model with Arbitrary Interpersonal Covariance, *Journal of Econometrics*, 22, 391–393.

Schmidt, P., S. C. Ahn, and D. Wyhowski (1992), Comment, *Journal of Business and Economic Statistics*, 10, 10–14.

Searle, S. R., 1987, *Linear Models for Unbalanced Data*, Wiley, New York.

Sevestre, P. and A. Trognon (1985), A Note on Autoregressive Error Component Models, *Journal of Econometrics*, 28, 231–245.

Sickles, R. C. (1985), A Nonlinear Multivariate Error Components Analysis of Technology and Specific Factor Productivity Growth with an Application to U.S. Airlines, *Journal of Econometrics*, 27, 61–78.

Sickles, R. C. and P. Taubman (1986), A Multivariate Error Components Analysis of the Health and Retirement Study of the Elderly, *Econometrica*, 54, 1339–1356.

Swamy, P. A. V. B. (1971), *Statistical Inference in Random Coefficient Regression Models*, Springer-Verlag, New York.

Swamy, P. A. V. B. and S. S. Arora (1972), The Exact Finite Sample Properties of the Estimators of Coefficients in the Error Components Regression Models, *Econometrica*, 40, 261–275.

Taub, A. J. (1979), Prediction in the Context of the Variance-Components Model, *Journal of Econometrics*, 10, 103–108.

Taylor, W. E. (1980), Small Sample Considerations in Estimation from Panel Data, *Journal of Econometrics*, 13, 203–223.

Valliant, R. (1991), Variance Estimation for Price Indexes from a Two-Stage Sample with Rotating Panels, *Journal of Business and Economic Statistics*, 9, 409–422.

Verbeek, M. and Th. E. Nijman (1992a), Testing for Selectivity Bias in Panel Data Models, *International Economic Review*, 33, 681–703.

Verbeek, M. and Th. E. Nijman (1992b), Can Cohort Data Be Treated as Genuine Panel Data?, *Empirical Economics*, 17, 9–23.

Verbeek, M. and Th. E. Nijman (1996), Incomplete Panels and Selection Bias, in L. Mátyás and P. Sevestre (eds.), *The Econometrics of Panel Data: A Handbook of Theory and Applications*, Kluwer, Dordrecht, 449–490.

Wallace, T. D. (1972), Weaker Criteria and Tests for Linear Restrictions in Regression, *Econometrica*, 40, 689–698.

Wallace, T. D. and A. Hussain (1969), The Use of Error Components Models in Combining Cross-Section and Time-Series Data, *Econometrica*, 37, 55–72.

Wansbeek, T. J. and A. Kapteyn (1978), The Separation of Individual Variation and Systematic Change in the Analysis of Panel Data, *Annales de l'INSEE*, 30–31, 659–680.

Wansbeek, T. J. and A. Kapteyn (1989), Estimation of the Error Components Model with Incomplete Panels, *Journal of Econometrics*, 41, 341–361.

Wansbeek, T. J. and A. Kapteyn (1992), Simple Estimators for Dynamic Panel Data Models with Errors in Variables, in R. Bewley and T. Van Hoa (eds.), *Contributions to Consumer Demand and Econometrics: Essays in Honor of Henri Theil*, St. Martin's Press, New York, 238–251.

Wansbeek, T. J. and R. H. Koning (1989), Measurement Error and Panel Data, *Statistica Neerlandica*, 45, 85–92.

Wooldridge, J. M. (1995), Selection Corrections for Panel Data Models under Conditional Mean Independence Assumptions, *Journal of Econometrics*, 68, 115-132.

Wyhowski, D. J. (1994), Estimation of a Panel Data Model in the Presence of Correlation between Regressors and a Two-Way Error Component, *Econometric Theory*, 10, 130–139.

Ziemer, R. F. and M. E. Wetzstein (1983), A Stein-Rule Method for Pooling Data, *Economics Letters*, 11, 137–143.

9

Econometric Analysis in Complex Surveys

Aman Ullah and Robert V. Breunig
University of California at Riverside, Riverside, California

I. INTRODUCTION

In the last five decades there has been a significant growth of research in econometric methods and their application in various areas of economics. Indeed, in the last two decades, the tremendous growth in econometrics has dichotomized the subject into cross-sectional (micro) econometrics and time-series (macro) econometrics. Whereas the new cross-sectional methodology was partly due to the nature of the data and the empirical issues in microbased labor economics and industrial organization, the new time-series methodology was an outcome of the challenging empirical issues and data problems in macroeconomics and finance. Despite these developments, econometric inference methods (especially in cross-sectional econometrics) have been confined to the assumptions that data is generated as a simple random sample with replacement or that it is coming from an infinite population (Johnston 1991, Greene 1993). These assumptions are certainly not valid in the case of survey data used in development and labor economics. Surveys usually have a well-defined frame consisting of a finite population of individuals, households, or villages. Sample data for analysis is generated from this finite population using a sampling design different from random sampling with replacement (RSWR). Sampling schemes such as systematic sampling, stratified random sampling, and cluster sampling may be used alone or in combination. These have been the subject of four decades of extensive work in statistics literature (Kish 1965, Cochran 1953, Sukhatme 1984, Levy and Lemeshow 1991, Thompson 1992).

The history of survey sampling can be traced back to the early eighteenth century, and even earlier (see Hansen 1987 and Deaton 1994 for detailed references).

Engel (1895) used a nonrepresentative sample of 200 Belgian households to establish that the share of the budget allocated to food is higher for the poor. Kiaer (1897) (see Kiaer 1976 for translation) was perhaps first to collect a large-scale, representative sample and discuss the principles, uses, and limitations of various sampling designs. His paper was influential in the International Statistical Institute's 1903 resolution supporting and promoting the use of the "representative method" of sampling. A turning point in the history of survey sampling came in the 1920s with the path-breaking work in statistical estimation theory and practice of R. A. Fisher at the Rothamsted experimental station (Fisher 1925). Fisher emphasized randomization, replication, and stratification in sampling design. His work led to the calculation of statistical estimates and their precision by Yates and Zacapony (1935) and Cochran (1939). Indeed, Fisher's work paved the way for Neyman's (1934) classic paper, which, for the first time, gave a systematic discussion of inference from random samples drawn from a finite population, contained a comparison of purposive sampling and random sampling, introduced the concept of the confidence interval, established the asymptotic normality of the sample average, and provided the optimal sample sizes within strata independently of Tschuprow's (1923) work. Later Neyman (1938) developed the theory of what is known as "two-stage" sampling. In the United States, around this period, important research work on sampling design was done by the researchers at the Department of Agriculture and the statistical lab at Iowa State University, and at the U.S. Bureau of the Census regarding labor force surveys dealing with issues such as measuring employment and labor force participation. These latter surveys, using systematic sampling designs, were necessitated by the Great Depression; see Stephan (1948) for detailed references. Sampling with probability proportional to measures of size at the successive stages of sampling was introduced during the work on labor force studies.

A parallel development on the application of survey sampling took place under the leadership of P. C. Mahalanobis at the Indian Statistical Institute in the 1930s and P. V. Sukhatme at the Indian Council of Agriculture Research. Mahalanobis introduced the concepts of developing sampling designs based on cost and variance estimates and on methods of evaluating survey errors.

The work during the 1930s and the 1940s revolutionized the collection of household survey data after the war. Major developments include the first national sample survey data developed annually (1950–1970) and then every five years at the Indian Statistical Institute, the household survey data now collected in United States, United Kingdom, and Taiwan, the Living Standard Measurement Study (LSMS) survey of Peru and the Ivory Coast by the World Bank, and Malaysian family survey data by the Rand Corporation. These household data are now extensively used in development economics to study poverty, income distribution, and economic welfare.

A common feature of the survey data described above is that the ultimate sample of households selected is rarely, if ever, based on RSWR. Yet, despite three decades worth of developments in statistical inference based on survey data, econometric analysis is carried on under the false assumption of RSWR; although see the excellent works of Pudney (1989), Deaton (1994), and Howes and Lanjouw (1994) for notable exceptions. This is especially a matter of concern in development economics where measures of income inequality, poverty and elasticities are used in policymaking by governments and international agencies. We think there are perhaps two reasons for this state of affairs. One is the statistical complexity of the various sampling designs for an average development economist; the second is a complete lack of exposure to the statistical literature on survey design in econometrics texts. Given this deficiency a systematic development of the parametric and nonparametric econometric inference (estimation and testing) of various econometric models, under various practical sampling designs, is urgently needed. This is an ambitious project and it is by no means attempted here. Instead, this chapter is a modest beginning in this direction. Some new results are also presented. Essentially, our objectives are as follows. The first is to provide a unified econometric framework of the five decades of diverse statistical literature on estimating the population mean. We refer to this as the mean model. The second is to explore the implications of results from the mean model for the linear regression model and the nonparemetric kernel density estimation. The third is to explore the implications of misspecifying the sampling design on the properties of econometric estimators. It is hoped that this chapter will contribute to further development of econometric inference results in other practical econometric models and for other parameters of interest.

In Section II we present the estimation of the finite population mean, the density, and linear regression coefficients under RSWR and random sampling without replacement (RSWOR). Section III deals with stratified sampling. Section IV considers cluster sampling, systematic sampling, and two-stage sampling. In Section V we give some limited simulation results. Finally, in the Appendix we provide some technical details of the results in Section II.

II. RANDOM SAMPLING

A. Random Sampling without and with Replacement (RSWOR and RSWR)

Let us consider a finite population of size N for an economic variable Y, and write the population mean model as

$$Y_i = \beta + U_i, \qquad i = 1, \ldots, N \tag{1}$$

where Y_i is the ith population observation, U_i is the ith error, and β and σ^2 are the population mean and variance, respectively, given by

$$\beta = \frac{1}{N} \sum_{i=1}^{N} Y_i, \qquad \sigma^2 = \frac{1}{N} \sum_{i=1}^{N} (Y_i - \beta)^2 = \frac{N-1}{N} S^2 \tag{2}$$

where

$$S^2 = \sum_{i=1}^{N} \frac{(Y_i - \beta)^2}{N - 1}$$

The errors U_i are nonsampling errors which sum to zero by the definition of β in (2). Therefore, U_i and Y_i are nonstochastic variables. However, if we treat the finite population model (1) as generated from an infinite population or superpopulation model, then U_i and Y_i are stochastic. This case is not considered here.

A random sample without replacement of size n, often referred to in the literature as a simple random sample (SRS), is taken from the above finite population of size N. We denote the sample observations as y_i and write (1) for these observations as

$$y_i = \beta + u_i, \qquad i = 1, \ldots, n \tag{3}$$

where u_i is now the ith sampling error.

Since the sampling is without replacement,

$$\pi_r(i) = P[y_i = Y_r] = \frac{1}{N} \tag{4}$$

is the probability that the rth population unit is selected in the ith draw and

$$\pi_{rs}(i, j) = P[y_i - Y_r \text{ and } y_j = Y_s] = \frac{1}{N(N - 1)} \tag{5}$$

is the probability that the (r, s)th unit is selected in the (i, j)th draw where $i, j = 1, \ldots, n$ and $r, s = 1, \ldots, N$ ($i \neq j, r \neq s$). These probabilities provide

$$\pi_r = \sum_{i=1}^{N} \pi_r(i) = \frac{n}{N} \tag{6}$$

the probability of selection of the rth population unit in the sample of size n, and

$$\pi_{rs} = \sum_{\substack{i=1 \\ i \neq j}}^{n} \sum_{j=1}^{n} \pi_{rs}(i, j) = \frac{n}{N} \left(\frac{n-1}{N-1} \right) \tag{7}$$

the probability of selection of the (r, s)th population unit in the sample.

In view of (4) to (7), we get

$$Eu_i = 0, \qquad Eu_i^2 = \sigma^2 = \frac{N-1}{N} S^2 \tag{8}$$

and, for $i \neq j$,

$$Eu_i u_j = \sigma_{12} = -\frac{\sigma^2}{N-1} = -\frac{S^2}{N} = \rho \sigma^2 \tag{9}$$

where $\rho = -1/(N-1)$ is the intraclass population correlation between y_i and y_j. Further, we can verify that

$$Eu_i^3 = \gamma_1 \sigma^3 = \frac{1}{N} \sum_{i=1}^{N} (Y_i - \beta)^3 \tag{10}$$

$$Eu_i^4 = (\gamma_2 + 3)\sigma^4 = \frac{1}{N} \sum_{i=1}^{N} (Y_i - \beta)^4$$

$$Eu_i^2 u_j = \sigma_{112} = \frac{1}{N(N-1)} \sum_{\substack{i=1 \\ i \neq j}}^{N} \sum_{j=1}^{N} (Y_i - \beta)^2 (Y_j - \beta) = -\frac{\gamma_1 \sigma^3}{N-1}$$

$$Eu_i^2 u_j^2 = \sigma_{1122} = \frac{1}{N(N-1)} \sum_{\substack{i=1 \\ i \neq j}}^{N} \sum_{j=1}^{N} (Y_i - \beta)^2 (Y_j - \beta)^2$$

$$= \frac{N - (\gamma_2 + 3)}{N - 1} \sigma^4$$

$$Eu_i^3 u_j = \sigma_{1112} = \frac{1}{N(N-1)} \sum_{\substack{i=1 \\ i \neq j}}^{N} \sum_{j=1}^{N} (Y_i - \beta)^3 (Y_j - \beta) = -\frac{(\gamma_2 + 3)\sigma^4}{N - 1}$$

and for $i \neq j \neq k \neq l$,

$$Eu_i u_j u_k = \sigma_{123} \tag{11}$$

$$= \frac{1}{N(N-1)(N-2)} \sum_i \sum_{\substack{j \\ i \neq j \neq k}} \sum_k (Y_i - \beta)(Y_j - \beta)(Y_k - \beta)$$

$$= \frac{2\gamma_1}{(N-1)} \frac{\sigma^3}{(N-2)}$$

$$Eu_i u_j u_k u_l = \sigma_{1234}$$

$$= \frac{1}{N(N-1)(N-2)(N-3)} \sum_i \sum_{\substack{j \\ i \neq j \neq k \neq l}} \sum_k \sum_l$$

$$\times (Y_i - \beta)(Y_j - \beta)(Y_k - \beta)(Y_l - \beta)$$

$$= \frac{3[N - 2(\gamma_2 + 3)]}{(N-1)(N-2)(N-3)} \sigma^4$$

$$Eu_i^2 u_j u_k = \sigma_{1123}$$

$$= \frac{1}{N(N-1)(N-2)} \sum_i \sum_{\substack{j \\ i \neq j \neq k}} \sum_k (Y_i - \beta)^2 (Y_j - \beta)(Y_k - \beta)$$

$$= \frac{2(\gamma_2 + 3) - N}{(N-1)(N-2)} \sigma^4$$

where γ_1 and γ_2 are Pearson's measures of skewness and excess kurtosis. For normal distribution, $\gamma_1 = \gamma_2 = 0$. The outcomes (8) to (11) indicate that RSWOR represents a set of n identically distributed but correlated random variables y_i.

In the case of RSWR the draws are independent, so

$$Eu_i = 0, \quad Eu_i^2 = \sigma^2, \quad Eu_i^3 = \gamma_1\sigma^3, \quad Eu_i^4 = (\gamma_2 + 3)\sigma^4 \tag{12}$$

because $\sigma_{12} = \sigma_{1112} = \sigma_{123} = \sigma_{1234} = \sigma_{1123} = 0$ and $\sigma_{1122} = \sigma^4$. This also holds if we are sampling from an infinite population ($N \to \infty$), the case usually considered in econometrics. In what follows we analyze the effect of assuming RSWR when the true design is RSWOR.

B.   Estimation of Parameters

Let us write the model (3) in a more compact form as

$$y = \iota\beta + u \tag{13}$$

where y is an $n \times 1$ vector, ι is an $n \times 1$ vector of unit elements, β is a scalar parameter (finite population mean), and u is an $n \times 1$ error vector such that

$$Eu = 0, \qquad V(u) = \Sigma = \sigma^2 \begin{bmatrix} 1 & \rho & \cdots & \rho \\ \rho & 1 & & \rho \\ \vdots & & \ddots & \\ \rho & \rho & & 1 \end{bmatrix} \tag{14}$$

by using (8) and (9).

The $n \times n$ matrix Σ can be rewritten as

$$\Sigma = \sigma^2(1 - \rho)\left[I + \frac{\rho}{1 - \rho}\iota\iota'\right] \tag{15}$$

and its well-known inverse, using $(I + bb')^{-1} = [I - bb'/(1 + b; b)]$, where b is an $n \times 1$ vector, is

$$\Sigma^{-1} = \frac{1}{\sigma^2(1 - \rho)(1 + \rho(n - 1))}[(1 + \rho(n - 1))I - \rho\iota\iota'] \tag{16}$$

The least-squares (LS) estimators of β in (13) is obtained by minimizing $u'u$ with respect to β. This gives

$$b = (\iota'\iota)^{-1}\iota'y = \frac{1}{n}\sum_{i=1}^{n} y_i = \bar{y} \tag{17}$$

The estimator b is unbiased, $Eb = \beta$, and its variance is

$$V(b) = (\iota'\iota)^{-1}\iota'\Sigma\iota(\iota'\iota)^{-1} \tag{18}$$

$$= \frac{\sigma^2}{n}[1 + (n - 1)\rho]$$

$$= \frac{N - n}{nN}S^2 = \frac{1}{n}\left(1 - \frac{n}{N}\right)S^2$$

where the last equality gives the familiar expression of the variance of the sample mean under RSWOR. The term $1 - n/N$ is known as the finite population correction (fpc). For $n \to N$, $V(b) \to 0$.

The efficient generalized least-squares estimator of β is

$$\hat{\beta} = (\iota'\Sigma^{-1}\iota)^{-1}\iota'\Sigma^{-1}y = (\iota'\iota)^{-1}\iota'y = \bar{y} \tag{19}$$

where the second equality follows by using (16). Thus, under (15), the two estimators and their variances are the same.

When the sampling is RSWR or the population is infinite, $\Sigma = \sigma^2 I$ because $\rho = 0$. In this case, $Eb = \beta$ and

$$V(b) = \frac{\sigma^2}{n} \tag{20}$$

which also follows from the last equality of (18) where $n/N \to 0$ as $N \to \infty$. From (18) and (20)

$$\frac{V(b_{\text{RSWOR}})}{V(b_{\text{RSWR}})} = 1 + (n - 1)\rho = \frac{N - n}{N - 1} \leq 1 \tag{21}$$

The above results indicate that the LS estimator b is unbiased for both RSWOR and RSWR. However, if the sampling is actually without replacement, the variance formula in (20) is wrong and gives an overestimate of the correct variance (18). To obtain the correct variance one needs to deflate (20) by $(N-n)/(N-1)$. For example if $n = 20$ and $N = 40$ the correct variance will be approximately 50% smaller than the wrong variance. The smallness of the variance of b_{RSWOR} is due to the negative correlation ρ.

In order to calculate the variance of b we look into an unbiased estimator of S^2. This is given by (using (13))

$$s^2 = \frac{1}{n - 1}\sum_{i=1}^{n}(y_i - b)^2 = \frac{1}{n - 1}y'My = \frac{1}{n - 1}u'Mu \tag{22}$$

where $M = I - \iota\iota'/n$ is an idempotent matrix. From the result (116) in the Appendix it is straightforward to show that s^2 is an unbaised estimator,

$$Es^2 = (n - 1)^{-1}\sigma^2(n - 1)(1 - \rho) = \frac{N}{N - 1}\sigma^2 = S^2 \tag{23}$$

because $M\iota = 0$ and $\rho = -(N-1)^{-1}$. For ratio estimators of S^2 based on the auxiliary information, see Prasad and Singh (1992).

Proposition I.   *When the sampling is without replacement we get*

$$V(s^2) = \frac{1}{n}\left[\left\{\gamma_2 + 3 - n\left(\frac{N}{N-1}\right)^2\right\}\sigma^4\right.$$

$$\left. - 3\sigma_{1122} - 4\sigma_{1112} + 12\sigma_{1123} - 6\sigma_{1234}\right]$$

$$+ \frac{n+1}{n-1}(\sigma_{1122} - 2\sigma_{1123} + \sigma_{1234}) \tag{24}$$

Proof. $V(s^2) = Es^2 - S^4 = (n-1)^{-2}E(u'Mu)^2 - S^4$. Now using (8) to (11) and (118) in the Appendix we get (24).

Corollary I.   *When the sampling is with replacement we get*

$$V(s^2) = \sigma^4\left[\frac{\gamma_2}{n} + \frac{n+1}{n-1} - \left(\frac{N}{N-1}\right)^2\right] \tag{25}$$

This follows by substituting $\sigma_{1122} = \sigma^4$ and $\sigma_{1112} = \sigma_{1123} = \sigma_{1234} = 0$ in (24). When the population is large $V(s^2) = \sigma^4(\gamma_2/n + 2/(n-1))$ (Sukhatme 1984), which reduces to the well known result $V(s^2) = 2\sigma^4/(n-1)$ under normality ($\gamma_2 = 0$).

From the above results it is clear that, as in the case of the LS estimator b, the estimator s^2 is unbiased for both RSWOR and RSWR and the variance of s^2 under RSWOR differs from that under RSWR. But, unlike b, the variance of s^2 is affected by nonnormality through γ_2. For example when the population is large the variance of s^2, compared to the normal case, is larger for all nonnormal distributions with $\gamma_2 > 0$. Also the effect of nonnormality does not disappear for large n since the ratio of the variance under nonnormality to the variance under normality is $1 + \gamma_2(n-1)/n$, which converges to $1 + \gamma_2$ as $n \to \infty$.

Finally, we note that the estimator s^2 is used to obtain an unbiased estimator of $V(b)$:

$$\hat{V}(b) = \frac{N-n}{nN}s^2. \tag{26}$$

This, however, does not guarantee that $\sqrt{\hat{V}(b)}$ will be an unbiased estimator of $\sqrt{V(b)}$; see Ullah and Breunig (1996) for the magnitude of the bias. In the special case of sampling from an infinite, normal population, an unbiased estimator of $\sqrt{\hat{V}(b)}$ can be obtained as $s*/n$, where

$$s* = s \frac{\Gamma\left(\frac{n-1}{2}\right)\sqrt{\frac{n-1}{2}}}{\Gamma\left(\frac{n}{2}\right)}$$

An application of knowing s^2 and b is to obtain the estimator of the population coefficient of variation (cv)

$$cv = \frac{\sigma}{\beta} \tag{27}$$

This is given by

$$\widehat{cv} = \frac{s}{b} \tag{28}$$

C. The Bias and MSE of the Coefficient of Variation

The sample $\widehat{cv}$ is extensively used in applied sciences for various purposes. In economics $\widehat{cv}$ has been widely considered to compare inequalities in income of different regions or groups of individuals or households (Kakwani 1980, Sen 1992). Beach et al. (1994) consider the conditional cv for analyzing third-order stochastic dominance when Lorenz curves cross. (We note that third-order stochastic dominance tests using standardized income on income distributions with different mans can lead to test results which contradict the dominance ordering on the unstandardized distributions.) Despite widespread use of the $\widehat{cv}$, not much is known about its finite sampling properties under the nonnormal population; although see Bowman and Shenton (1981), who have considered mean and variance of $\widehat{cv}$ under the Weibull distribution, Neuts (1982), who has considered mixtures of distributions, and Singh (1993), who has analyzed the sampling properties of the inverse of $\widehat{cv}$-squared. Sampling of $\widehat{cv}$ under normality has been well analyzed going back to McKay (1932). Warren (1982) considers McKay's results and provides other references. More recently, Gupta and Ma (1996) analyze the $\widehat{cv}$ in k-normal populations. The results presented below could be extended to the case of more than one variable.

Here we present the sampling properties of $\hat{\theta} = (\widehat{cv})^2$ of $\theta = (cv)^2$ instead of $\widehat{cv}$. We analyze $(\widehat{cv})^2$ because comparing $\widehat{cv}$ across regions/groups gives similar results to comparing $\hat{\theta}$, and the approximate sampling properties of $\widehat{cv}$ require expanding both s and $1/b$, whereas $\hat{\theta}$ require the expansion of $1/b$ only.

In what follows, we present both small-σ and Nagar-type large-n approximations for the mean and mean-squared error (MSE) of $\hat{\theta}$. For details on small-σ approximation see Kadane (1971) and Srivastava et al. (1980), where they show a better performance of the small-σ approximation under certain situations.

Proposition 2. *When the sampling is without replacement the bias of $\hat{\theta}$, up to $O(\sigma^4)$, and the MSE of $\hat{\theta}$, up to $O(\sigma^4)$, respectively, are given by*

$$\text{Bias}(\hat{\theta}) = -\rho\theta - \frac{2}{n}\theta^{3/2}\xi_1 + \frac{3\theta^2}{n^2}\xi_2 \tag{29}$$

and

$$\text{MSE}(\hat{\theta}) = \frac{\theta^2}{n}\left[\xi_3 + \frac{n(n+1)}{n-1}\xi_4\right] \tag{30}$$

where

$$\xi_1 = \frac{1}{\sigma^3}[(\gamma_1\sigma^3 - 3\sigma_{112} + 2\sigma_{123}) + n(\sigma_{112} - \sigma_{123})] \tag{31}$$

$$\xi_2 = \frac{1}{\sigma^4}[(\gamma_2 + 3)\sigma^4 - (3-n)\sigma_{1122} - 2(2-n)\sigma_{1112}$$
$$+ (12 - 8n + n^2)\sigma_{1123} - (6 - 5n + n^2)\sigma_{1234}]$$

$$\xi_3 = \frac{1}{\sigma^4}[((\gamma_2 + 3) - n(1 - 2\rho))\sigma^4 - 3\sigma_{1122} - 4\sigma_{1112} + 12\sigma_{1123} - 6\sigma_{1234}]$$

$$\xi_4 = \frac{1}{\sigma^4}(\sigma_{1122} - 2\sigma_{1123} + \sigma_{1234})$$

and σ's are as given in (9) to (11).

The proof of Proposition 2 is given in Appendix B.

Corollary 2.   *When the sampling is with replacement the bias of $\hat{\theta}$, up to $O(\sigma^4)$ and the MSE of $\hat{\theta}$, up to $O(\sigma^4)$, respectively, are given by*

$$\text{Bias}(\hat{\theta}) = -\frac{2}{n}\theta^{3/2}\gamma_1 + \frac{3\theta^2}{n^2}(\gamma_2 + n) \tag{32}$$

$$\text{MSE}(\hat{\theta}) = \frac{\theta^2}{n}\left[\gamma_2 + \frac{2n}{n-1}\right] \tag{33}$$

The proof of Corollary 2 follows by substituting $\sigma_{1122} = \sigma^4$ and $\sigma_{123} = \sigma_{1112} = \sigma_{1123} = \sigma_{1234} = \rho = 0$ in Proposition 2.

From (32) we can also obtain the bias, up to $O(n^{-1})$, as

$$\text{Bias}(\hat{\theta}) = \frac{1}{n}[3\theta^2 - 2\theta^{3/2}\gamma_1) \tag{34}$$

This result compares with Breunig (1996), where the large-n expansion is used to obtain (34). Breunig (1996) also provides the MSE, up to $O(n^{-2})$, as

$$\text{MSE}(\hat{\theta}) = \frac{\theta^2}{n}\left[\gamma_2 + 4\theta - 4\gamma_1\theta^{1/2} + 2\frac{n}{n-1}\right] \tag{35}$$
$$+ \frac{\theta^2}{n^2}\left[\theta\left(24\gamma_2\frac{n}{n-1} + 20\gamma_1^2\frac{n+1}{n-1} + 20\frac{n}{n-1}\right)\right.$$
$$\left. - 4\theta^{1/2}\left(\gamma_3 + 4\gamma_1\frac{n}{n-1}\right) + 75\theta^2 - 108\theta^{3/2}\gamma_1\right]$$

From (34) we observe that the bias goes to zero as $n \to \infty$. Further, the asymptotic MSE (AMSE) is

$$\text{AMSE}(\hat{\theta}) = \theta^2 \left[\gamma_2 + 4\theta - 4\gamma_1 \theta^{1/2} + 2\left(\frac{n}{n-1}\right) \right] \tag{36}$$

Thus, for distributions with $\gamma_2 > 0$ ($\gamma_2 < 0$) the AMSE will be above (below) the AMSE under normality. Further, we note that the bias, up to $O(\sigma^3)$ and $O(n^{-1})$, is positive for negatively skewed distributions, but the bias to $O(\sigma^3)$ is negative for positively skewed distributions. However, to $O(n^{-1})$, the bias is negative for positively skewed distributions provided

$$cv = \theta^{1/2} < \frac{2}{3}\gamma_1 \tag{37}$$

Generally, $\hat{\theta}$ will give an underestimate of the true θ for positively skewed distributions. Since income distributions are generally positively skewed, it is possible that, in the past, the use of $\hat{\theta}$ in measuring income inequality gave an underestimation of inequality. In view of this, Breunig (1996) has suggested an estimator of θ which adjusts the bias of $\hat{\theta}$. This is given as $\tilde{\theta} = \hat{\theta} - \widehat{\text{Bias}}(\hat{\theta})$, where $\widehat{\text{Bias}}(\hat{\theta})$ is the bias of $\hat{\theta}$ in (34) with θ replaced by $\hat{\theta}$ and γ_1 by $\hat{\gamma}_1 = \sum(y_i - \bar{y})^3/n\hat{s}^3$. Although we do not attempt it here, it will be interesting to analyze these results for sampling without replacement. This could be done by using (29) and (30). Another possible extension involves the use of the geometric mean, useful when the data are expressed in ratio form. We could then formulate the sample cv as the ratio of the sample geometric mean to the sample standard deviation. The authors are unaware of any development of the finite-sample moments for such a statistic, even under random sampling with replacement.

D. Sampling with Unequal Probabilities

When sampling is with unequal probabilities, $\pi_r(i)$ in (4), π_r in (6), and $\pi_{r,s}$ in (7) are not constants. In this case, we first transform (13) by

$$W^{1/2}y = W^{1/2}\iota\beta + W^{1/2}u \tag{38}$$

and then obtain the LS estimator of β by minimizing the weighted squared error $u'Wu = (y - \iota\beta)'W(y - \iota\beta)$, where $W = \text{Diag}(w_1, \ldots, w_n)$ is an $n \times n$ stochastic diagonal weight matrix whose elements w_i, known as the normalized expansion factors, satisfy $\iota'w\iota = \sum_1^n w_i = 1$. This gives the weighted LS estimator of β as

$$b_w = \iota'Wy = \sum_{i=1}^n w_i y_i \tag{39}$$

The stochastic weights w_i are chosen such that (using (4) and (6)) the sample is representative of population in the sense that the sample mean, on average, is identical to the population mean. That is,

$$Eb_w = E\left(\sum_{i=1}^{n} w_i y_i\right) = \sum_{i=1}^{n}\sum_{j=1}^{N} w_j Y_j \pi_j(i) \tag{40}$$

$$= \sum_{j=1}^{N} w_j Y_j \pi_j = \beta$$

This gives

$$w_i = \frac{1}{N\pi_i} \tag{41}$$

and $b_w = N^{-1} \sum_{1}^{n}(y_i/\pi_i)$, where π_i is the probability of selection of the ith population unit in the sample. When $\pi_i = n/N$ we get $b_w = b$ as given in (17).

An alternative way to obtain (41) is to write $Eb_w = E(\sum_{1}^{n} w_i y_i) = E(\sum_{1}^{N} w_i d_i Y_i) = \sum_{1}^{N}(w_i.Ed_i)Y_i = \sum_{1}^{n} w_i \pi_i Y_i$ where d_i is a dummy random variable which is 1 when Y_i is in the sample, and 0 otherwise. Since the probability of selection of the ith population unit in the sample is π_i, we can verify that

$$Ed_i = \pi_i; \quad V(d_i) = \pi_i(1 - \pi_i) \tag{42}$$

$$\text{cov}(d_i, d_j) = \pi_{ij} - \pi_i \pi_j$$

where π_{ij} is defined in (7). Using (42),

$$V(b_w) = \sum_{i=1}^{N} w_i^2 Y_i^2 \pi_i(1 - \pi_i) + \sum_{i \neq j}^{N} w_i w_j Y_i Y_j (\pi_{ij} - \pi_i \pi_j) \tag{43}$$

When $\pi_i = n/N$ and $\pi_{ij} = n(n-1)/N(n-1)$ we get (18). We point out here again that both mean and variance of b_w contain Y_i since the error is coming from the sampling, not from the errors in the superpopulation.

E. Estimation of Regression Model

Let us consider the mean of the conditional population of Y given a vector of k variables $X_1, \ldots, X_k$ as

$$Y = X^*\beta + U \tag{44}$$

where Y is an $N \times 1$ vector and X^* is an $N \times k$ matrix. The model (44) is a conditional mean model if the conditional mean of the nonsampling errors, $E(U|X)$, is zero or

$\beta = (X \ast' X\ast)^{-1} X \ast' Y$ is the finite population regression coefficient. As before, a random sample without replacement of size n is taken from the population of size N. The sample model is

$$y = X\beta + u \tag{45}$$

where y is an $n \times 1$ vector and X is an $n \times k$ matrix. Under RSWOR, $E(u|X) = 0$, and $V(u|x) = \Sigma$, where Σ is as given in (15). The LS and GLS estimators of β and their variances, respectively, are

$$b = (X'X)^{-1}X'y, \qquad V(b) = (X'X)^{-1}X'\Sigma X(X'X)^{-1} \tag{46}$$
$$b_{\text{GLS}} = (X'\Sigma^{-1}X)^{-1}X'\Sigma^{-1}y, \qquad V(b_{\text{GLS}}) = (X'\Sigma^{-1}X)^{-1}$$

When sampling is with replacement, or as $N \to \infty$, $\rho = 0$ and we get $b = b_{\text{GLS}}$. It is easy to verify that

$$V(b) = \sigma^2(1 - \rho)(X'X)^{-1} + \sigma^2\rho(X'X)^{-1}X'\iota'X(X'X)^{-1} \tag{47}$$
$$= \sigma^2(X'X)^{-1} + \sigma^2\rho(X'X)^{-1}X'(\iota\iota' - I)X(X'X)$$

where $\iota\iota' - I$ is an indefinite matrix. Thus the sign of $V(b) - \sigma^2(X'X)^{-1}$ is not determined. This implies that if the sampling is without replacement and we consider the variance of b as though we had RSWR then this variance will be an underestimate or overestimate of the variance depending upon the design matrix X. Another point is that b_{GLS} is not identical to b numerically. These results are in contrast to the mean model (13) where $k = 1$ and $X = \iota$. However, it follows from Zinde-Walsh and Ullah (1987) and Ghosh and Sinha (1980) have shown that the F-test for testing restrictions on β using b_{GLS}, with Σ in (15), will be numerically the same as the F-test based on b. This result also holds for the likelihood ratio, Lagrange multiplier, and Wald tests.

F.   Nonparametric Estimation of Density and Regression Functions

As in the earlier sections, let $Y_i, i = 1, \ldots, N$, be the population of Y with probability density f, and $y_i = 1, \ldots, n$ be the sample drawn without replacement from f. Further, as in Section II.D, let $d_i, i = 1, \ldots, N$ be the dummy random variables such that $d_i = 1$ if Y_i is selected in the sample and $d_i = 0$ otherwise. Note that the d_i's are not independent and $Ed_i^p = n/N$ and $Ed_i^p d_j^q = n(n-1)/N(N-1)$ for any $p > 0, q > 0$.

$$f(Y) = \frac{1}{Nh_N} \sum_{i=1}^{N} K\left(\frac{Y_i - Y}{h_N}\right) = \frac{1}{N}\sum_{i=1}^{N} K_i \tag{48}$$

where $K_i = h_N^{-1} K((Y_i - Y)/h_N)$. Rosenblatt's (1956) kernel estimator, based on y_i, is then

$$\hat{f}(y) = \frac{1}{n} \sum_{i=1}^{n} k_i = \frac{1}{N\lambda} \sum_{i=1}^{N} d_i K_i \tag{49}$$

where $k_i = h_n^{-1} K((y_i - y)/h_n)$ and $\lambda = n/N$. Thus the finite population density estimation problem reduces to the problem of estimating the population mean discussed in (1) to (3). It therefore follows from the results in Section II.A that

$$E\hat{f}(y) = f(Y), \qquad V(\hat{f}(y)) = \left(\frac{1}{n} - \frac{1}{N}\right) S_k^2 \tag{50}$$

where $S_k^2 = (N-1)^{-1} \sum_1^N (K_i - \overline{K})^2$. However, if we treat the finite population Y_i, $i = 1, \dots, N$, itself as an i.i.d. random observation from an infinite (super) population with density f^*, then as $(n, N) \to \infty$ such that $\lambda \to c > 0$,

$$E\hat{f}(y) = (N\lambda)^{-1} N(Ed_1) \left(E \frac{1}{h_N} K \left(\frac{Y_1 - Y}{h_N} \right) \right) \tag{51}$$

$$\to f^*$$

and

$$(nh)^{1/2} V(\hat{f}(y)) \to f^* \int K^2(\psi) \, d\psi \tag{52}$$

The details of (51) and (52) can be worked out by following Rosenblatt (1956) and Pagan and Ullah (1995). For the asymptotic theory in the parametric, finite population models, see Fuller (1984) and the references therein.

In the regression context we can write the nonparametric versions of the finite population and sample models in (44) and (45), respectively as

$$Y_i = m(X_i) + U_i \tag{53}$$

and

$$y_i = m(x_i) + u_i \tag{54}$$

$$\simeq m(x_o) + (x_i - x_o) m'(x_o) + u_i$$

where x_o is a given point. More compactly,

$$y = Z(x_o) \delta(x_o) + u \tag{55}$$

where $Z(x_o) = [\iota \ (x - \iota x_o)]$, $\delta(x_o) = [m(x_o) \ m'(x_o)]'$, and $V(u) = \Sigma$ as given in (46). The local least-squares estimation of $\delta(x_o)$ can then be carried on by minimizing $u' \Sigma^{-1/2} K \Sigma^{-1/2} u$. The properties of $\hat{\delta}(x_o)$ so obtained can be worked out by using the procedures developed in Ruppert and Wand (1994).

III. STRATIFIED SAMPLING

A. Estimation of Mean Model

A large number of surveys use stratified random sampling, which involves dividing the total population first into various strata (subpopulations), typically by geographical regions, such as rural/urban or states, or by certain characteristics such as blue-collar and white-collar workers. Then RSWOR is used within each stratum. The sampling units are usually households or individuals. There are several advantages to this method as opposed to RSWOR or RSWR from the entire population. First, the stratification provides a more representative sample overall and so reduces the variance, especially when the variation within strata is small but the variation between strata is large. Stratification allows for different types of sampling schemes in different strata, desirable perhaps because of cost considerations. For example, one can perform SRS in urban areas, where households are closely concentrated, and cluster sampling in rural areas, where households are widely dispersed (see Section IV). Finally, stratification helps to obtain enough sample observations from small subpopulations of special interest.

The population mean model for stratified sampling can be written as

$$Y_{ij} = \beta_i + U_{ij}, \qquad i = 1, \ldots, M, \quad j = 1, \ldots, N_i \tag{56}$$

where Y_{ij} is the jth unit in the ith stratum, β_i is the mean of the ith stratum (subpopulation),

$$\beta_i = \frac{1}{N_i} \sum_{j=1}^{N_i} Y_{ij} \tag{57}$$

and U_{ij} is the error. The variance of *the* ith stratum is

$$S_i^2 = \frac{1}{N_i - 1} \sum_{j=1}^{N_i} (Y_{ij} - \overline{Y}_i)^2 = \frac{N_i}{N_i - 1} \sigma_i^2 \tag{58}$$

In a more compact form

$$Y_i = \beta_i + U_i \tag{59}$$

where Y_i is an $N_i \times 1$ vector of observations. The population size is $N = \sum_1^M N_i$.

The stratified sample observations, generated by RSWOR in each stratum, follow

$$y_{ij} = \beta_i + u_{ij}, \qquad i = 1, \ldots, M, \quad j = 1, \ldots, n_i \tag{60}$$

or, more compactly,

$$y_i = \iota_i \beta_i + u_i, \qquad V(u_i) = \Sigma_i \tag{61}$$

where ι_i is $n_i \times 1$ vector of unit elements, y_i is an $n_i \times 1$ vector of sample observations (y_{ij}) in the ith stratum, and Σ_i is Σ in (14) with $n = n_i$. The total size of the sample is $n = \sum_{i=1}^{M} n_i$. The model (60) for the ith subpopulation (stratum) is the same as that of the population model in (13). Thus, the results for Section II go through for the estimation of the ith stratum parameters. For example, the LS estimator of β_i is

$$b_i = \frac{1}{n_i} \sum_{j=1}^{n_i} y_{ij} = (\iota_i' \iota_i)^{-1} \iota_i' y_i \tag{62}$$

and

$$s_i^2 = \frac{1}{n_i - 1} \sum_{j=1}^{n_i} (y_{ij} - b_i)^2 = \frac{y_i' M_i y_i}{n_i - 1} \tag{63}$$

Here, M_i is the matrix M in (22) with $\iota = \iota_i$. The parameter of interest may, however, be the overall mean of the population. That is,

$$\beta = \frac{1}{N} \sum_{i=1}^{M} \sum_{j=1}^{N_i} Y_{ij} = \sum_{i=1}^{M} \theta_i \beta_i \tag{64}$$

where $\theta_i = N_i/N$ is the proportion of the total population in the ith stratum. An estimator of β is then

$$b_{st} = \sum_{i=1}^{M} \theta_i b_i \tag{65}$$

which is unbiased. Further

$$V(b_{st}) = \sum_{i=1}^{M} \theta_i^2 V(b_i) = \sum_{i=1}^{M} \theta_i^2 \left(1 - \frac{n_i}{N_i}\right) \frac{S_i^2}{n_i} \tag{66}$$

provided that the sample elements are chosen independently for each stratum.

To see the LS and weighted LS solutions of β we rewrite (60) as

$$y_{ij} = \beta + \beta_i^* + u_{ij} \tag{67}$$

where $\beta_i^* = \beta_i - \beta$, $Eu_{ij} = 0$, $Eu_{ij}^2 = \sigma_i^2$, $Eu_{ij}u_{i'j'} = \rho_i \sigma_i^2$ for $i = i'$, $j \neq j'$, and 0 when $i \neq i'$. In the model (67), however, β_i^* and β are not identifiable. So we impose the restriction $\sum_{1}^{M} (n_i/n)\beta_i^* = 0$. Using this we get the LS estimators of β_i and β by minimizing $\sum_{1}^{M} \sum_{i}^{N_i} (y_{ij} - \beta - \beta_i^*)^2$. This gives b_i* and hence b_i as in (62), and

$$b = \frac{1}{n} \sum_{i=1}^{M} \sum_{j=1}^{n_i} y_{ij} = \sum_{i=1}^{M} p_i b_i, \qquad p_i = \frac{n_1}{n} \tag{68}$$

which is biased because $Ey_{ij} = \beta_i$ so the $Eb = \sum_1^M p_i\beta_i$. Further,

$$V(b) = \sum_{i=1}^{n} p_i^2 V(b_i) \tag{69}$$

But n_i/N_i, the probability of selection of the jth population element in the sample of size n_i in the ith stratum, may not be constant across strata. Therefore the weighted LS described in Section II.D will be more useful here. For this we minimize $\sum_1^M \sum_1^{n_i} w_{ij}(y_{ij} - \beta - \beta_i^*)^2$ with the restriction that $\sum_1^M \sum_1^{n_i} w_{ij}\beta_i^* = 0$ and $\sum_1^M \sum_1^{n_i} w_{ij} = 1$. This gives

$$b_w = \sum_{i=1}^{M}\sum_{j=1}^{n_i} w_{ij}y_{ij} \tag{70}$$

and

$$b_{iw} = \sum_{j=1}^{n_i} w_{ij}y_{ij} \tag{71}$$

This inflation or expansion factor w_{ij} is chosen such that $Eb = \beta$ and $Eb_i = \beta_i$. This gives

$$b_w = \sum_{i=1}^{M}\sum_{j=1}^{n_i} \frac{1}{N\pi_{ij}}y_{ij}, \qquad b_{iw} = \sum_{j=1}^{n_i} \frac{1}{N\pi_{ij}}y_{ij} \tag{72}$$

If we substitute $\pi_{ij} = n_i/N_i$, we get $b_w = b_{st}$ and $b_{iw} = b_i$.

We observe that the LS estimator b will become unbiased if strata are homogenous with regard to sampling fractions, $n_i/N_i = n/N$, in which case the combined sample is SRS (this is known as proportional stratified sampling, a special case of stratification in which the data will be self-weighting); or if $\beta_i = \beta$ for all strata, in which case the population is homogeneous with respect to means. Alternatively, if we consider β_i^* to be random with mean zero and constant variance, then (67) can be treated as a one-way error component model. In this case, the LS estimator b in (68) is unbiased as well as efficient since b is also the GLS estimator in (67).

In contrast, if we assume $\beta_i = \beta$ and pool all the observations (ignoring strata and treating the data as a random sample of size n from a population of size N), then the variance of the pooled estimator $b_P = b$, known as the SRS estimator, is

$$V(b_P) = V(b_{SRS}) = \left(\frac{1}{n} - \frac{1}{N}\right) S^2 \tag{73}$$

This is the same as $V(b_{st})$ or $V(b)$ if $n_i/n = N_i/N$ and $\beta_i = \beta$. In this case the population is homogeneous and the combined sample is a simple random sample.

In general, $V(b_{SRS}) > V(b_{st})$, especially if within-strata heterogeneity is low and between-strata heterogeneity is high. To see this, consider $V(b_{st})$ for case of proportional stratified sampling, where $n_i = nN_i/N$. In this case

$$\frac{V(b_{st})}{V(b_{SRS})} = \frac{\sum_{i=1}^{M} \theta_i S_i^2}{\sum_1^M \theta_i(\beta_i - \beta)^2 + \sum_{i=1}^{M} \theta_i S_i^2} \tag{74}$$

where we use $S^2 \simeq N^{-1} \sum_1^M \sum_1^{N_i}(Y_{ij} - \beta)^2 \simeq \sum_1^M \theta_i S_i^2 + \sum_1^M \theta_i(\beta_i - \beta)^2$. Thus, if between-strata variation, $V(b_i)$, is zero, $V(b_{SRS}) = V(b_{st})$. But if $V(b_i)$ is nonzero, $V(b_{st})$ is smaller than $V(b_{SRS})$.

In order to minimize the variance of the weighted estimator, b_{st}, one must choose n_i such that $V(b_{st})$ is a minimum subject to $\sum_1^M n_i = n$. The optimal n_i will be $n_i = nN_i s_i / \sum_{k=1}^M n_k$.

In practice, the unbiased estimator of $V(b_{st})$ can be calculated by substituting S_i^2 with s_i^2 in (66). Further, if $\beta_i = \beta$, the unbiased estimate of the variance of b_{SRS} can be calculated as

$$\hat{V}(b_{SRS}) = \left(\frac{1}{n} - \frac{1}{N}\right) \frac{\sum_{i=1}^{M}(N_i - 1)s_i^2}{N - 1} \tag{75}$$

Alternatively,

$$\hat{V}(b_{SRS}) = \left(\frac{1}{n} - \frac{1}{N}\right) s^2 \tag{76}$$

where $s^2 = (n - M)^{-1} \sum_1^M (n_i - 1)s_i^2$ is the pooled estimator if $S_i^2 = S^2$. If $S_i^2 \neq S^2$ but $n_i/n = N_i/N$, one may consider

$$s_*^2 = \frac{1}{n-1} \sum_1^M \sum_1^{n_i}(y_{ij} - b)^2$$

$$= \frac{1}{n-1} \left[\sum_1^M (n_i - 1)s_i^2 + \sum_1^M n_i b_i^2 - \frac{1}{n}\left(\sum_1^M n_i b_i\right)^2 \right]$$

However, if $n_i/n \neq N_i/N$ and $\beta_i \neq \beta$, then

$$\hat{V}(b_{SRS}) = \left(\frac{1}{n} - \frac{1}{N}\right) \left[\frac{\sum_{i=1}^{M}(N_i - 1)s_i^2}{N - 1} + q \right] \tag{77}$$

where q is an unbiased estimator of $(N - 1)^{-1} \sum_1^M N_i(\beta_i - \beta)^2$ (Cochran 1953, p. 99).

From the above analysis it is clear that if the sample observations y_{ij} are generated by stratified random sampling then they should be reweighted to resemble the population by replicating (inflating) sampling units, using the inflation or expansion factor, and treating the enlarged sample as if it were the population. The inflation factor, θ_{ij}, for each sampling unit j in the ith stratum is the reciprocal of its sampling probability; that is, $\theta_{ij} = 1/\pi_{ij} = N_i/n_i$. If we multiply each sample observation by its inflation factor θ_{ij}, we obtain an unbiased estimate of the population total. Alternatively if we multiply the sample observations by their weights $w_{ij} = \theta_{ij}/\sum\sum\theta_{ij} = 1/N\pi_{ij}$, the normalized inflation factor, we get an unbiased estimate of population mean, as shown in (72) and Section II. Exactly the same procedures can be used to obtain estimates of medians, variances, and other parameters. We will examine the weighting for regression parameters.

B. Regression Model

Suppose now that the parameters of interest are no longer population means, but the parameters of a linear regression model

$$y_i = X_i\beta_i + u_i \tag{78}$$

where β_i is a $k \times 1$ vector and $i = 1, \ldots, M$ strata. As in the mean model the parameters of interest are β_i and β, where β is the overall parameter or the weighted average of β_i:

$$\beta = \sum_{i=1}^{M} \theta_i\beta_i \tag{79}$$

By the analogy of the population mean case one may consider $\theta_i = N_i/N$. If the population is stratified on some economic grounds such as rural and urban, then the estimates of β_i will be useful in their own right. However, if the population is divided into a large number of strata on administrative grounds, then studying β will be more meaningful. The estimates of β_i are

$$b_i = (X_i'X_i)^{-1}X_i'y_i \tag{80}$$

with $V(b_i) = \sigma_i^2(X_i'X_i)^{-1}$. If we have RSWOR $b_i = (X_i'\Sigma_i^{-1}X_i)^{-1}X_i'\Sigma_i^{-1}y_i$ with $V(b_i) = (X_i'\Sigma_i^{-1}X_i)^{-1}$, Σ_i is Σ with ι replaced by ι_i and n by n_i.

To estimate β we write

$$y_i = X_i\beta + X_i(\beta_i - \beta) + u_i = X_i\beta + X_i\delta_i + u_i \tag{81}$$

or, more compactly, for $i = 1, \ldots, M$,

$$y = X\beta + X^*\delta + u \tag{82}$$

where y is $n \times 1$, X is $n \times k$, and X^* is $n \times kM$, $n = \sum_1^M n_i$. Then the pooled (RSWR) estimator is

$$b_P = b_{\text{RSWR}} = (X'X)^{-1}X'y = \left(\sum_{i=1}^M X_i'X_i\right)^{-1} \sum_{i=1}^M X_i'y_i \qquad (83)$$

$$= \left(\sum_{i=1}^M p_iQ_i\right)^{-1} \sum_{i=1}^M p_iQ_ib_i = \sum_{i=1}^M w_ib_i$$

provided $\sigma_i^2 = \sigma^2$, $Q_i = X_i'X_i/n_i$, and $w_i = (\sum_{i=1}^M p_iQ_i)^{-1}p_iQ_i$. For RSWOR, $b_p = b_{\text{SRS}} = (X'\Omega^{-1}X)^{-1}X'\Omega^{-1}y$, where $\Omega = \text{Diag}((\sigma_i^2\Sigma_i))$. It is easy to see that b is a biased estimator. This bias is

$$E(b_{\text{SRS}}) - \beta = (X'X)^{-1}X'X^*\delta = \left(\sum_{i=1}^M p_iQ_i\right)^{-1} \sum_{i=1}^M p_iQ_i\delta_i \qquad (84)$$

However, if $\beta_i - \beta = \delta_i$ is random and independent of X_i then the bias is zero. Also, the bias vanishes if $\beta_i = \beta$. When X is stochastic, we are concerned with the consistency of our estimator not its bias. We thus want to consider the probability limit as $n_i \to \infty$.

So far, the two situations under which the bias vanishes mimic the results from the mean model. Also, as before, the bias of pooled, or unweighted, estimator b_p depends on the structure of the sample. This is so even when $Q_i = Q$ for all i. However, unlike in the mean model case, even if $n_i/n = N_i/N = \theta_i$ the bias still depends on the sample. When $Q_i = Q$ and $n_i/n = N_i/N$, the bias vanishes provided β is defined as in (79).

Now consider the weighted estimator of β by minimizing $u'Wu$, where $W = ((w_{ij}))$, $w_{ij} = 1/(\pi_{ij} \cdot N)$. This gives

$$b_w = (X'WX)^{-1}X'Wy \qquad (85)$$

When $k = 1$ and $X = \iota$, we get the same result as in Section II. Further, if $\pi_{ij} = \pi_i = n_i/N_i$, then

$$b_w = \left(\sum_{i=1}^M \pi_i^{-1}X_i'X_i\right)^{-1} \sum_{i=1}^M \pi_i^{-1}X_i'y_i \qquad (86)$$

$$= \left(\sum_{i=1}^M \theta_iQ_i\right)^{-1} \sum_{i=1}^M \theta_ic_i, \qquad c_i = \frac{X_i'y_i}{n_i}$$

$$= \left(\sum_{i=1}^M \theta_iQ_i\right)^{-1} \sum_{i=1}^M \theta_iQ_ib_i = \sum_{i=1}^M w_ib_i$$

and

$$Eb_w - \beta = \left(\sum_{i=1}^{M} \theta_i Q_i \right)^{-1} \sum_{i=1}^{M} \theta_i Q_i (\beta_i - \beta) \tag{87}$$

so that the bias does not depend on the sample values; w_i is the same as in (83) with p_i replaced by θ_i. Furthermore if $\beta_i - \beta$ is random, $Q_i = Q$ or $\beta_i = \beta$, then the bias vanishes. For $Q_i = Q$ the bias vanishes when $w_i = \theta_i$. Note that for $\beta_i = \beta$ both b_w and b_P are unbiased but b_P will be more efficient by the Gauss Markov theorem (DuMouchel and Duncan 1983). Kish and Frankel (1974), however, argue in favor of b_w and the parameter of interest

$$\beta = \sum_{i=1}^{M} w_i \beta_i \tag{88}$$

where w_i is as in (86). When $Q_i = Q$, $w_i = \theta_i$ and (88) reduces to (79). Note that b_w is unbiased only if the parameter of interest is β in (88). If the parameter of interest is given by (79), b_w will not be unbiased.

In practice, one may be able to make a choice between b_P and b_w by comparing b_P and b_w and performing a Hausman-type specification test. This amounts to doing an artificial regression of the type

$$y = X\beta + WX\gamma + u \tag{89}$$

and testing for $\gamma = 0$ by the standard F-test. Alternatively, one can combine the two by using Stein-type shrinkage estimators as

$$b_S = \left(1 - \frac{c}{F} \right) b_P + \frac{c}{F} b_w \tag{90}$$

The properties of b_S are not known, but they can be developed by following Judge and Bock (1978) or Vinod and Ullah (1981).

Magee et al. (1996) suggest an alternative estimator to weighted least squares when the sampling probabilities are known but the form of the sample design is unknown. They propose a conditional maximum likelihood estimator, which, under certain conditions, is superior (in the mean-squared error sense) to WLS or OLS. They treat the weights (sampling probabilities) as having been generated by a stochastic process and independently distributed throughout the population, an assumption which is violated under either stratification or clustered sampling. They suggest using their ML estimator in any case, since information about the sample is usually unknown. Magee (1996) and Magee et al. (1996) suggest a way to improve weighted least-squares regression for survey data. Including weights often injects additional heteroskedasticity into the model and the WLS estimator, though consistent, often has a high variance. Magee (1996) suggests creating new weights by multiplying the weights by a function parameterized so as to minimize variance. Again, independence across the sample is assumed, necessitating some adjusting of the procedure for use under stratification or clustered sampling.

For further discussion of regression, including inference on finite population and superpopulation parameters, Kalton (1983) provides a clear and readable introduction, Pfefferman (1993) reviews some of the recent work on regression models and weighting, Selden (1994) considers the case of weighted, generalized least squares in the mean model case, and Godambe (1995, 1997) provides a more general model.

The estimators b_P, b_w, and b_S are all weighted averages of b_i. Another weighted average of b_i follows by considering $\beta_i - \beta$ to be random with mean vector 0 and diagonal covariance matrix Δ_β. Thus, heterogeneity across strata is due to variance only. In this case, $X_i(\beta_i - \beta) + u_i$ has the variance $\sigma_i^2 + X_i \Delta X_i'$ and we can get the GLS estimator of β as

$$b_{\text{GLS}} = \sum_{i=1}^{M} w_i b_i \tag{91}$$

where $w_i = ([\sum_1^n (\Delta_\beta + V_i)^{-1}]^{-1}(\Delta_\beta + V_i)^{-1}; V_i = \sigma_1^2(X_i'X_i)^{-1})$. This is the well-known Swamy (1971) estimator. The estimators b_P and b_w will both be inefficient under this scenario. Their standard errors need to be calculated by using a White (1980) kind of adjustment.

The above procedures are useful when heterogeneity across strata is present. This is important when the number of strata are large. If the number of strata are small, one could do separate regressions, combine them by allowing stratum-specific intercepts, or use strata dummies with the same variables.

Although we do not explore questions of prediction here, they will be important in choosing the parameters of interest. The census parameter, β in (79) would be of interest when one would like to predict the change in the independent variable for a small change in the dependent variable for *every* member of society. If a particular policy would affect some strata differently than others, this might not be a parameter which aids in prediction.

For the purpose of prediction, however, nonparametric analysis will be more useful since it directly estimates the regression function and thus avoids the problem of defining the parameter of interest. This is not taken up here, but will be an interesting subject of a future study.

Also not considered here are the usual diagnostic tests used in both parametric and nonparametric ecomometric analysis. Obviously, these too must be adapted to account for complex sampling procedures.

IV. CLUSTER AND SYSTEMATIC SAMPLING

A. Cluster Sampling

Let us consider Y_{ij} to be population observations of the ith group or cluster, $i = 1, \ldots, M$ and $j = 1, \ldots, N_i$ elements in the ith cluster. Cluster sampling or single-stage sampling involves drawing a sample of m clusters out of M and then sampling all N_i, $i = 1, \ldots, m$, elements in the sample. The sampling is done, therefore, only at the first stage. However, if we further take a random sample of n_i elements out of N_i at the second stage, then the overall sampling is called subsampling or two-stage sampling. The first-stage units, clusters, could be villages or street blocks, and the second-stage units could be households. The primary advantage of cluster sampling is that it drastically reduces survey cost per second-stage unit. The disadvantage is that it usually leads to higher variance due to correlation among the elements within clusters. The standard assumption of uncorrelated observations in cross-sectional data is certainly not true for clustered samples.

It is useful to begin by first considering the problem of estimating population means β and β_i under single-stage (cluster) sampling. For this we consider the population model as $Y_{ij} = \beta + \beta_i^* + U_{ij}$ and the sample model as

$$y_{ij} = \beta + \beta_i^* + u_{ij}, \qquad i = 1, \ldots, m, \quad j = 1, \ldots, N_i \tag{92}$$

where we assume RSWR such that

$$Eu_{ij} = 0, \quad Eu_{ij}^2 = \sigma^2, \quad Eu_{ij}u_{ij'} = \rho\sigma^2, \qquad j \neq j' \tag{93}$$

$$Eu_{ij}u_{i'j'} = 0, \qquad i \neq i'$$

$\rho > 0$ is the intracluster correlation coefficient. Equation (93) implies that the elements within clusters are correlated but are uncorrelated across clusters. Thus,

$$Eu = 0 \qquad \text{and} \qquad Euu' = \text{Diag}(\Sigma_1, \ldots, \Sigma_m) = \Omega \tag{94}$$

where Ω is an $n \times n$ block diagonal matrix with $\Sigma_i = \sigma^2[(1 - \rho)I + \rho\iota_i\iota_i']$. The LS estimators of β_i and β, respectively, are

$$b_i = \frac{1}{N_i} \sum_{j=1}^{N_i} y_{ij} \qquad \text{and} \qquad b_c = \frac{1}{m} \sum_{i=1}^{m} b_i \tag{95}$$

If $\beta_i = \beta$, the estimator b_c is unbiased. Further, its variance is given by

$$V(b_c) = \frac{1}{n^2} \sum_{i=1}^{m} (\iota_i'\Sigma_i\iota_i) = \frac{\sigma_u^2}{n}[1 + (\bar{n} - 1)\rho] \tag{96}$$

where $n = \Sigma_1^m N_i$, and $\bar{n} = n^{-1}\Sigma_1^m Nn_i^2$ is the weighted mean of cluster sizes. As shown in Section II, the LS estimator b_c is the same as the GLS estimator for the structure of Σ_i in (94). An alternative expression of (96) can be written by following the results in Section II.A. This is $V(b_c) = \sigma_b^2/m$ where $\sigma_b^2 = \sum_1^M (b_i - b_c)^2/M$. For the equivalence of this expression with (96) see Kish (1965).

If we ignore the clustering and consider RSWR of size n, then

$$V(b_c) = V(b_{\mathrm{RSWR}}) = \frac{\sigma_u^2}{n} \qquad (97)$$

Thus, the usual formula of RSWR has to be inflated by $d = 1 + (\bar{n} - 1)\rho)$ to account for the intracluster correlation. For $\rho > 0$, Eq. (97) may provide serious underestimation of the true variance in (96).

It can be shown that

$$\hat{\sigma}_u^2 = \frac{1}{n-d}\hat{u}'\hat{u} = \frac{1}{n-d}\sum_{}^{m}\sum_{}^{N_i} \hat{u}_{ij}^2 \qquad (98)$$

is an unbiased estimator of σ_u^2, where $\hat{u}_{ij} = y_{ij} - b_c$. In practice, one could consider $\hat{u}'\hat{u}/n$, which is consistent. Further, for $i = 1, \ldots, m$,

$$N_i(N_i - 1)\hat{\rho}\hat{\sigma}_u^2 = \sum_{j}^{N_i}\sum_{\substack{j' \\ j \neq j'}}^{N_i} \hat{u}_{ij}\hat{u}_{ij}, \qquad (99)$$

gives

$$\hat{\rho} = \frac{\sum_{i=1}^{m} \sum_{j=1}^{N_i} \sum_{j \neq j}^{N_i} \hat{u}_{ij}\hat{u}_{ij'}}{\hat{\sigma}_u^2 n(\bar{n} - 1)} \qquad (100)$$

Deaton (1994) provides numerical examples of the effect of ignoring ρ in the calculation of standard errors by b_c. He showed that for estimated food price elasticities in Pakistani villages, ρ is between .3 and .6, leading to underestimation of $V(b_c)$ by a factor greater than 2 when the mean cluster size is 12. Now we turn to the case of two-stage sampling where, within each selected cluster, we pick a sample of $n_i < N_i$ units. The probability of selection of every element in the chosen cluster is mn_i/MN_i. It is easy to verify that the estimator of the mean β is $b_{2S} = \sum_1^m b_i/m$, where b_i is $\sum_i^{n_i} y_{ij}/n_i$. For the variance of b_{2S}, see Kish (1965).

B. Systematic Sampling

Systematic sampling is one of the most common techniques used in development economics. In systematic sampling, the sampling units are (usually) arranged in random order with respect to the variable of interest. Of the first K units, one is selected at random. Then every Kth unit is sampled in order. This sampling design is the easiest to implement because it involves drawing only one sample. Systematic sampling can be thought of as a kind of one-stage cluster sampling. The population is arranged into K clusters, each with n elements. One of these clusters is chosen and every element within that cluster is sampled. For simplicity, we assume that there are N elements in the population and that N/K is an integer. (In other words, the N elements are exhaustively and uniquely assigned to the K clusters, each of which has the same number of elements n, an integer.)

Consider the finite population model $Y = \beta + U$, where the data are randomly ordered. The population mean and variance are defined as in Section I. If the population is divided into K clusters, we can write the model as

$$Y_{kj} = \beta_k + U_{kj}, \qquad k = 1, \ldots, K, \quad j = 1, \ldots, n \tag{101}$$

Our sample would consist of one randomly chosen cluster k, written as

$$y_j = \beta + u_j, \qquad j = 1, \ldots, n \tag{102}$$

The LS estimator of the mean, when cluster k is chosen, is

$$b_k b_{\text{SYS}} = \frac{1}{n} \sum_{j=1}^{n} y_j$$

which will be unbiased when N/k is an integer. Further,

$$V(b_{\text{SYS}}) = \frac{1}{K} \sum_{k=1}^{K} (b_k - \beta)^2 \tag{103}$$

In general, we will not be able to estimate this variance. In the case where our data consists of one systematic sample, the population mean, β, is unknown as are the remaining $K - 1$ unsampled clusters. In some surveys, resampling is possible. In this case, information can be gathered about the within and across-cluster heterogeneity and an approximation for $V(b_{\text{SYS}})$ as a function of the intracluster correlation coefficient, ρ:

$$V(b_{\text{SYS}}) = \frac{\sigma^2}{n}[1 + \rho(n-1)] \tag{104}$$

In general, if the data is randomly arranged with respect to the variable of interest, systematic sampling should give broad coverage of the population, the estimate of the mean should be unbiased, and the approximation of $V(b_{\text{SYS}})$ by assuming simple random sampling will not be too unreasonable. If the data is clustered and the clusters are ordered, then a systematic sample will perform better than either SRS or clustered sampling. This follows because the systematic sample, picking every kth elements, will cover most, if not all, clusters. This very broad coverage will give a precise estimate of the mean.

C.   Regression Model

In the regression model

$$y = X\beta + u \tag{105}$$

where X is $n \times k$, $n = \sum_1^n n_i$, the LS and GLS estimators of β and their variances, respectively, are

$$b_c = b_{LS} = (X'X)^{-1}X'y, \qquad V(b_{LS}) = (X'X)^{-1}X'\Omega X(X'X)^{-1} \tag{106}$$

and

$$b_{GLS} = (X'\Omega^{-1}X)^{-1}X'\Omega^{-1}Y, \qquad V(b_{GLS}) = (X'\Omega^{-1}X)^{-1} \tag{107}$$

Ω is as defined in Section III.B. Suppose X contains observations which vary only across clusters but are constant within clusters. Further assume that the cluster sizes are the same so that $\bar{n} = n$. Then it follows from Kloek (1981) that the LS estimator is numerically the same as the GLS estimator. Further, the true variance of b_c is

$$V(b_c) = \sigma^2 (X'X)^{-1} d \tag{108}$$

which reduces to the variance in the mean model where $k = 1$ and $X = \iota$. Moulton (1990) and Deaton (1994) provide examples of potential underestimation, using the usual variance $\sigma^2(X'X)^{-1}$.

The results for the case of observations in X changing with clusters or when $n_i \neq n$ is not well developed in the literature, although see Pfefferman and Smith (1985) who provide an upper bound of $V(b_c)$. Also, the efficiency of b_{GLS} compared to b_{LS} needs to be analyzed.

An alternative to estimating Ω by first estimating ρ and σ^2 is to consider Ω to be of unknown form and then estimate $X'\Omega X$ consistently by $X'\hat{\Omega}X$, as suggested by White (1980) and Arellano (1987).

D.   Unequal Probability of Selection

Consider again the mean model

$$y_{ij} = \beta + \beta_i^* + u_{ij} \tag{109}$$

where $\sum \beta_i^* = 0$. Then the LS estimators b_i and b of β_i and β, respectively, are as given in Section IV.A. In practice, two-stage sampling is carried out by using the probability weight proportional to the size of the cluster. In this situation the estimator b will be biased. An unbiased weighted estimator is

$$b_w = \sum_{}^{m} \sum_{}^{n_i} w_{ij} y_{ij} \tag{110}$$

where w_{ij} is such that $Eb_w = \beta$. This is

$$w_{ij} = \frac{N_i}{Nn_i} \frac{1}{\pi_i} = \frac{\theta_i}{n_i \pi_i} \tag{111}$$

where π_i is the probability of selection of the ith cluster in the sample. For RSWR, $\pi_i = m/M$. We note that $w_{ij} = (P_{ij})^{-1}$, where P_{ij} is the probability of selecting $y_{ij} =$ probability of selecting the ith cluster $\times$ probability of selecting the jth unit of the population in the sample given the ith cluster is selected. This gives

$$b_w = \sum_{i=1}^{m} \left(\frac{\theta_i}{\pi_i} \right) b_i \tag{112}$$

If the cluster is chosen with probability proportional to size of the cluster (i.e., $\pi_i = mN_i/N = m\theta_i$,

$$b_w = \frac{1}{m} \sum_{i=1}^{m} b_i \tag{113}$$

In practice π_i is proportional to estimated size N_i^*. This gives

$$b_w = \frac{1}{m} \sum_{i=1}^{m} \frac{\theta_i}{\theta_i^*} b_i \tag{114}$$

where $\theta_i^* = N_i^*/N^*$, $N^* = \sum_1^m N_i^*$.

When we consider single-stage sampling or cluster sampling, the estimator b_w can be written as

$$b_w = \frac{1}{M} \sum_{i=1}^{m} \frac{b_i}{\pi_i} \tag{115}$$

For the RSWR $\pi_i = m/M$.

In the regression case

$$y_i = X_i \beta_i + u_i$$

$b_i = (X_i' X_i)^{-1} X_i' y_i$, and we can consider $b_w = \sum_1^m (\theta_i b_i/\pi_i)$ as before (Konijn 1962). For cluster sampling we can again consider $b_w = m^{-1} \sum_1^m b_i/\pi_i$.

We can also estimate β by considering β_i to be random so that $y_i = X_i \beta_i + u_i = X_i \beta + X_i \delta_i + u_i$, where $\delta_i = \beta_i - \beta$. This will involve Swamy-type random coefficient estimation of β described in Section III (also see Porter 1973).

Fuller (1975) estimates the one-way error component model

$$y_i = X_i \beta + \epsilon_i l_i + u_i$$

to capture the correlation across the elements of the ith cluster, where ϵ_i has mean zero and constant variance.

V. SIMULATION

In this section, we present a summary of the results from a detailed simulation of the mean model under complex sampling. The two primary objectives of the simulation are (1) to illustrate the effect of ignoring sample design in data analysis, and (2) to ascertain the properties of our estimators under various sample designs where analytical results do not exist.

The first step in our simulation was the creation of several finite "populations." The populations created ranged in size from 50 to 20,000, with means ranging from 1 to 2000. They were drawn from an "infinite" population of randomly distributed, normal numbers. From these finite "populations" we then drew n observations, using the sampling design in question. For the questions under consideration in this section, the shape of the distribution is irrelevant, so only normal random numbers were considered. For analyzing other variables, such as the distribution of $V(s^2)$ considered in Section I, the shape of the distribution does matter, and conclusions based on simulations using only normally distributed populations should be made with caution.

To demonstrate the first point, a sample of size n was drawn using the sample design of interest (stratified, clustered, etc.), then b and $V(b)$ were estimated using the information on how the sample was drawn. Then taking this same sample and ignoring the sample design, we calculated b_{RSWR} and $V(b_{RSWR})$—i.e., treating the sample as if it were a random sample drawn with replacement. These values are averaged over 1000 repetitions. We then compare the average bias (b) and bias (b_{RSWR}) and the ratio of the averages of the two variances, which can be interpreted as the degree of over- or underestimation arising from ignoring (or misspecifying) the sample design.

The second type of simulation we have undertaken, to answer the second question raised above, involves drawing a separate sample for each sampling design of interest. From these distinct samples, we calculate b and $V(b)$ for each of the sample designs. After r repetitions, we compute the average bias and the simulation variance of the estimator for each design.

By way of example, let us consider a simulation of both kinds comparing sampling under finite population and infinite population from Section I. Recall that

$$V(b_{SRS}) = \frac{s^2}{n}d = V(b_{RSWR})\left(1 - \frac{n}{N}\right)$$

Following the first method, we take 1000 samples of size n from our population and calculate $V(b_{SRS})$, using the fact that the sample has been drawn without replacement from a finite population. Then we calculate $V(b_{RSWR})$ and compare the ratio of the average of these two over the 1000 repetitions.

The ratio should be the inverse of the finite population correction. Indeed this is confirmed in the results in Table 1. Results from the second type of simu-

Table I   Effect of Ignoring Sample Design (Sampling with Replacement)

Population size	Sample size				
	5	10	20	50	100
50	1.11	1.25	1.67	—	—
100	1.05	1.11	1.25	2.00	—
500	1.01	1.02	1.04	1.11	1.25
1000	1.00	1.01	1.02	1.05	1.11

Table entries show the degree of overestimation of the variance of the sample mean: $\frac{var(b_{RSWR})}{var(b_{SRS})}$.

lation are presented in Table 2. Here, two separate samples are drawn from the same population—one under SRS, the other under sampling with replacement. Results for 10,000 repetitions are reported. As expected, the results closely approximate those in Table 1. One way to interpret these results is that for the same sample size, sampling without replacement is more precise than sampling with replacement. (Since the variance of the estimator b under SRS is, on average, smaller.) We can also think of the ratio as representing the "cost" of assuming that sampling is from an infinite population when in fact sampling is from a finite population.

Tables 3 through 6 present comparisons between RSWR and stratified sampling without replacement, conducted under the first method described. In Table 3, we consider two strata, each with a population of 1000, where a sample of size n_i is drawn from each stratum. Since $n_1 \neq n_2$ and both strata have a population of 1000, the sampling probability is unequal. As we saw in Section III, the unweighted estimator of β will be biased. As we can see from Table 3, the more unequal the

Table 2   Efficiency Gains from Sampling without Replacement vs. Sampling with Replacement

Population size	Sample size				
	5	10	20	50	100
50	1.08	1.22	1.63	—	—
100	1.063	1.082	1.24	1.98	—
500	.993	1.009	1.054	1.12	1.253
1000	1.00	1.00	1.005	1.055	1.108

Entries in table are $\frac{var(b_{SWR})}{var(b_{SRS})}$ averaged over 10,000 repetitions of each design.

Table 3   Sampling with Unequal Probabilities

Sample size (Stratum 1, 2)	Bias b_w	Penalty of not considering sampling structure		
		Bias b_{RSWR}	Ratio of variances $Var(b_{RSWR})/Var(b_w)$	Ratio of MSEs $MSE(b_{RSWR})/MSE(b_w)$
(5, 10)	0.94	17.97	1.46	1.81
(5, 20)	−0.13	30.11	1.15	2.99
(5, 50)	0.30	41.62	0.60	4.75
(10, 5)	0.24	−16.62	1.23	1.37
(20, 5)	1.16	−29.68	0.72	1.97
(50, 5)	−0.99	−41.37	0.27	3.07

2 Strata: $\beta_1 = 200$; $\sigma_1 = 59$; $\beta_2 = 300$; $\sigma_2 = 74.8$; $\beta = 250$.

sampling probabilities, the greater the bias in the unweighted estimator, b_{RSWR}, and the greater the ratio of mean-squared errors. Most data in labor and development economics is stratified, and the most common case is unequal sampling probabilities, either by design or because of different rates of nonresponse across strata. Thus, as the simulation shows, a potentially serious bias problem exists even in calculating a simple mean. The general intuition behind these results extends to the regression case.

In Tables 4 to 6, we present results from the stratified case, but with equal probabilities of selection in both strata. In Table 4, we see that even though b_{RSWR}

Table 4   Sampling with Equal Probabilities

Sample sizes (Stratum 1, 2)	Population means (Stratum 1, 2)	Penalty of not considering sampling structure		
		Bias b_{RSWR} = Bias b_w	Ratio of variances $Var(b_{RSWR})/$ $Var(b_w)$	Ratio of MSEs $MSE(b_{RSWR})/$ $MSE(b_w)$
(5, 5)	(200, 300)	0.33	1.59	1.30
	(200, 400)	0.29	2.29	1.64
	(200, 600)	0.12	3.53	2.24
	(200, 800)	2.22	5.46	3.19
(100, 100)	(200, 300)	−0.20	1.74	1.36
	(200, 400)	0.15	2.37	1.67
	(200, 600)	0.05	3.68	2.35
	(200, 800)	0.26	5.97	3.32

Table 5 Improved Efficiency: $\beta_1 \neq \beta_2$

Sample sizes (Stratum 1, 2)	Population means (Stratum 1, 2)	Penalty of not considering sampling structure	
		Ratio of variances $Var(b_{RSWR})/Var(b_w)$	Ratio of MSEs $MSE(b_{RSWR})/MSE(b_w)$
(5, 5)	(200, 300)	2.20	1.56
	(200, 600)	20.22	10.44
	(200, 1000)	74.95	37.82
(20, 20)	(200, 300)	2.16	1.54
	(200, 600)	18.44	9.67
	(200, 1000)	68.06	34.86
(50, 50)	(200, 300)	2.20	1.61
	(200, 600)	18.97	9.76
	(200, 1000)	69.58	34.51
(100, 100)	(200, 300)	2.32	1.66
	(200, 600)	19.81	10.05
	(200, 1000)	72.75	35.55

(unweighted) is unbiased under equal sampling probability, it is not efficient compared to b_w. In Tables 5 and 6, we consider across-strata heterogeneity in the mean and the element variances separately, since both affect the ratio between b_{RSWR} and b_W. In Table 5 we first consider the case where $\sigma_1^2 = \sigma_2^2$, but $\beta_1 \neq \beta_2$. Table 6 presents the case where $\sigma_1^2 \neq \sigma_2^2$, but $\beta_1 = \beta_2$. We note that the increase in precision as measured by the ratio of variances is increasing as the distance between the two strata means, β_1 and β_2, increases. It is not uncommon in development eco-

Table 6 Improved Efficiency: $\sigma_1^2 \neq \sigma_2^2$

Sample sizes (Stratum 1, 2)	Population σ^2 (Stratum 1, 2)	Penalty of not considering sampling structure		
		Bias b_{RSWR} = Bias b_w	Ratio of variances $Var(b_{RSWR})/Var(b_w)$	Ratio of MSEs $MSE(b_{RSWR})/MSE(b_w)$
(5, 5)	(2500, 3520)	−12.3	1.01	1.00
	(2500, 9650)	13.4	1.01	1.00
(50, 50)	(2500, 3520)	−5.5	1.05	1.03
	(2500, 9650)	−6.9	1.05	1.03

Table 7 One-Stage Clustered Sampling

# of clusters sampled (c)	Total sample size (n = c* = m)	True pop. mean	Pop ρ	$\hat{\rho}$	Ratio of variances $Var(b_{CLU})/Var(b_{RSWR})$	Expected kish design effect (d)
5	100	1000	.12	.088	3.13	3.28
		1000	.17	.13	5.00	4.23
		1000	.26	.20	6.48	5.94
		1000	.44	.344	7.24	9.36
		1000	.48	.38	10.02	10.12
10	200	1000	.12	.11	2.39	3.28
		1000	.17	.15	3.89	4.23
		1000	.26	.24	5.78	5.94
		1000	.44	.41	10.36	9.36
		1000	.48	.45	11.78	10.12
25	500	1000	.12	.119	3.69	3.28
		1000	.17	.169	4.29	4.23
		1000	.26	.256	5.42	5.94
		1000	.44	.439	9.69	9.36
		1000	.48	.48	10.09	10.12

nomics to encounter stratified samples where the urban mean income is three times that of rural mean income. In the case where $\beta_1 = 200$ and $\beta_2 = 600$, we see that this can lead to an overestimate of the variance of the population mean by a factor of 20. It increases for large sample sizes, because the finite population correction has a proportionally larger effect. From Table 6, we see that stratification does not improve efficiency when the strata have the same mean regardless of the difference in within-stratum variance. The small increases in efficiency that we see are the result of the increasing effect of the finite population correction as the sample size increases. In other simulations, not reported here, we show that assuming a stratified structure when the data does not have one leads to no gains in efficiency. This result follows intuitively from the results in Table 6.

Table 7 presents the cost of ignoring the one-stage clustered sample design and assuming that the sample is actually a RSWR. Here we have drawn a clustered sample from one stratum with a population of 1000, which is divided into 50 clusters, each of size 10. We present the ratio of variances, $V(b_c)/V(b_{RSWR})$, where we have calculated $V(b_c)$ using the estimated sample value of $\hat{\rho}$. We compare this to the expected Kish design effect $1 + (m - 1)\rho$ given our knowledge of the true value of ρ: $\hat{\rho}$ gives a slight underestimation, which disappears as $n \to N$.

Table 8   Comparisons of SRS, Clustered, and Systematic Sampling Designs

# of clusters sampled	Total sample size	True pop. mean	Population ρ	Ratio of variances		
				SRS/ clustered	SRS/ systematic	Clustered/ systematic
5	100	1000	.12	.28	1.31	4.60
		1000	.26	.15	1.88	12.4
		1000	.48	.09	2.30	25.74
		1000	.64	.06	2.92	47.37
10	200	1000	.12	.24	1.09	4.52
		1000	.26	.11	1.69	14.7
		1000	.48	.08	6.24	81.15
		1000	.64	.07	1.04	14.78

Tables 8 and 9 present results comparing RSWOR (SRS), stratified sampling, cluster sampling, and systematic sampling. We use the second method described above for simulation.

Table 8 presents results from 5000 replications comparing SRS, clustered, and systematic samples drawn from the same population. The last three columns compare the variances of the different estimators. Systematic sampling performs best in the simulation reported in Table 8. Since the data is ordered by cluster, the systematic sample gives the broadest coverage of the population, taking at least one observation from each cluster. (See Section IV.B) As expected, clustered sampling gives the highest variance for given sample size. Table 9 compares stratified and sys-

Table 9   Systematic Sampling Compared with Stratified Systematic Sampling

Sample size (for each stratum)	Strata means	Strata pop.	Bias b_w	Bias b_{sys}	Ratio of variances $Var(b_{sys})/Var(b_w)$	Ratio of MSEs $MSE(b_{RSWR})/MSE(b_w)$
5	(2, 900)	100	3.21	−0.09	1.06	1.03
	(1000, 1000)	500	0.09	−0.54	1.05	1.02
25	(2, 900)	100	4.11	−3.11	2.55	1.78
	(1000, 1000)	500	−0.17	−0.04	1.46	1.23
50	(2, 900)	100	1.89	0.30	1.89	1.44
	(1000, 1000)	500	−0.16	0.77	1.74	1.37

tematic sampling in a stratified population. Results are for 10,000 repetitions. The systematic sample does not perform as well as the stratified random sample since stratification will more evenly cover the population over repeated sampling.

As the results of Sections I through IV demonstrate, problems of inference and estimation arise when data is gathered under a complex sampling design. The simulation helps to demonstrate that these are of more than trivial interest. Unequal sampling probabilities are the rule, not the exception, and treating such data as having been drawn under RSWR will lead to biased estimators. Even where the disproportion is 2 to 1, this leads to large bias as shown in Table 3.

As different strata will usually have different parameter means, ignoring stratification will lead to large overestimates of the true variance of our estimate of β. A recent survey of income in Kenya showed that average rural income was one-third that of average urban income. Ignoring stratification when calculating a population mean in this case will lead to confidence intervals which are 20 times too wide. The exact opposite problem occurs in clustering. Intraclass correlation coefficients of .5 are common in developing country studies. The simulation shows that ignoring the sample design leads to an underestimate of the variance by a factor of 10, more if the average cluster size is greater than 20.

Bias problems and misestimates of standard errors are exacerbated in more complex sample designs which combine different aspects of stratification, clustering, and systematic sampling. Clearly, the same problems will arise in the regression case. The simulation demonstrates that assuming away sample design effects as trivial is unjustified. Instead, more careful attention should be paid to using available methods of analysis and information on sampling to construct unbiased and more precise estimates.

VI.   APPENDIX

A.   Some Useful Expectations

Suppose the elements of an $n \times 1$ vector u satisfy (8) to (11). Let A and B be $n \times n$ symmetric matrices of known constants, b an $n \times 1$ vector of known constants, ι an $n \times 1$ vector unit of elements, and $A * B$ the Hadamard product of A and B. Then we can verify the following expectations:

$$E(u'Au) = E\left[\sum_i^n \sum_j^n u_i u_j a_{ij}\right] = E\left[\sum_i^n u_i^2 a_{ii} + \sum_{i \neq j}^n u_i u_j a_{ij}\right] \quad (116)$$

$$= \sigma^2 \text{tr } A + \rho \sigma^2 (\iota' A \iota - \text{tr } A)$$

$$= \sigma^2 (1 - \rho) \text{tr } A + \rho \sigma^2 \iota' A \iota$$

$$E(u'Au)(b'u) = E\left[\sum_i^n \sum_j^n \sum_k^n u_i u_j u_k a_{ij} b_k\right] \tag{117}$$

$$= \gamma_1 \sigma^3 \sum_i^n a_{ii} b_i + \sigma_{112}\left(\sum_{i \neq j}^n a_{ii} b_j + 2\sum_{i \neq j}^n a_{ij} b_j\right)$$

$$+ \sigma_{123} \sum_{i \neq j \neq k}^n a_{ij} b_k$$

$$= (\gamma_1 \sigma^3 - 3\sigma_{112} + 2\sigma_{123})\iota'(I * A)b$$

$$+ (\sigma_{112} - \sigma_{123})((\text{tr } A)\iota'b + 2\iota'Ab) + \sigma_{123}(\iota'A\iota)\iota'b$$

$$E(u'Au)(u'Bu) \tag{118}$$

$$= E\left[\sum_i^n \sum_j^n \sum_k^n \sum_l^n u_i u_j u_k u_l a_{ij} b_{kl}\right]$$

$$= [(\gamma_2 + 3)\sigma^4 + 3\sigma_{1122} - 4\sigma_{1112} + 12\sigma_{1123} - 6\sigma_{1234}]\text{tr}(A * B)$$

$$+ (\sigma_{1122} - 2\sigma_{1123} + \sigma_{1234})((\text{tr } A)(\text{tr } B) + 2 \text{ tr } AB)$$

$$+ (2\sigma_{1112} - 6\sigma_{1123} + 4\sigma_{1234})\iota'((I * B)A + (I * A)B)\iota$$

$$+ (\sigma_{1123} - \sigma_{1234})(\text{tr } A(\iota'B\iota) + \text{tr } B(\iota'A\iota)) + 4(\sigma_{1123} - \sigma_{1234})\iota'BA\iota$$

$$+ \sigma_{1234}(\iota'A\iota)(\iota'B\iota)$$

When sampling is with replacement, $\sigma_{1122} = \sigma^4$ and $\rho = 0 = \sigma_{112} = \sigma_{1112} = \sigma_{123} = \sigma_{1123} = \sigma_{1234}$. In this case (116) to (118) reduce to the standard results in the literature:

$$E(u'Au) = \sigma^2 \text{ tr } A \tag{119}$$

$$E(u'Au)(b'u) = \gamma_1 \sigma^3 \iota'(I * A)b \tag{120}$$

$$E(u'Au)(u'Bu) = \gamma_2 \sigma^4 \text{ tr}(A * B) + \sigma^4(\text{tr } A \text{ tr } B + 2 \text{ tr } AB) \tag{121}$$

B. Proof of Proposition 2

Let us use (2) and (17) in (28) and write the estimator $\hat{\theta} = (\widehat{cv})^2$ of $\theta = (cv)^2$ as

$$\hat{\theta} = \frac{s^2}{b^2} = \frac{n}{n-1}\frac{y'My}{y'\overline{M}y} \tag{122}$$

where $\overline{M} = I - M = \iota\iota'/n$. Further we write the model (13) as

$$y = \iota\overline{\beta}\sigma v, \qquad u = \sigma v \tag{123}$$

where v is the error vector where moments are determined by the moments of u in (8) to (11). Then $y'My = \sigma^2 v'Mv$ and $y'\overline{M}y = n\beta^2 + \sigma(2\beta v'\iota) + \sigma^2 v'\overline{M}v$, and these give, up to $O(\sigma^4)$,

$$\frac{n-1}{n}\hat{\theta} = \sigma^2 \frac{v'Mv}{n\beta^2}\left[1 + \frac{2\sigma\beta v'\iota + \sigma^2 v'\overline{M}v}{n\beta^2}\right]^{-1} \tag{124}$$

$$= \sigma^2 \frac{v'mv}{n\beta^2}\left[1 - \frac{2\sigma\beta v'\iota + \sigma^2 v'\overline{M}v}{n\beta^2} + \left(\frac{2\sigma\beta v'\iota + \sigma^2 v'\overline{M}v}{n\beta^2}\right)^2 - \cdots\right]$$

$$= \sigma^2 a_2 + \sigma^3 a_3 + \sigma^4 a_4$$

where

$$\sigma^2 a_2 = \frac{1}{n\beta^2}(u'Mu), \qquad \sigma^3 a_3 = \frac{-2}{n^2\beta^3}(u'Mu)u'\iota \tag{125}$$

$$\sigma^4 a_4 = \frac{3}{n^2\beta^4}(u'Mu)(u'\overline{M}u)$$

Now, using (116) to (118) we can easily verify that

$$\sigma^2 Ea_2 = \frac{1}{n}\theta(1-\rho)(n-1)$$

$$\sigma^3 Ea_3 = \frac{-2(n-1)}{n^2\beta^3}[\gamma_1\sigma^3 - 3\sigma_{112} + 2\sigma_{123} + n(\sigma_{112} - \sigma_{123})]$$

$$\sigma^4 Ea_4 = \frac{3(n-1)}{n^3\beta^4}[(\gamma_2 + 3)\sigma^4 - 3\sigma_{1122} - 4\sigma_{1112} + 12\sigma_{1123}$$

$$- 6\sigma_{1234} + n(\sigma_{1122} + 2\sigma_{1112} - 8\sigma_{1123} + 5\sigma_{1234})$$

$$+ n^2(\sigma_{1123} - \sigma_{1234})]$$

ACKNOWLEDGMENTS

The research on this topic is an outcome of the first author's visit to the Policy Research Division of the World Bank and helpful discussions with C. Howes, E. Jimenez, and M. Ravallion. The authors are thankful for comments by participants of seminars at York University, University of Windsor, UCR, and University of Guelph and participants in the Canadian Econometric Studies Group 1996 meeting. Special thanks to Gordon Anderson, David Giles, V. P. Godambe, and Lonnie Magee. Financial support from the Academic Senate, UCR, is gratefully acknowledged by the first author.

REFERENCES

Arellano, M. (1987), Computing Robust Standard Errors for Within-Group Estimators, *Oxford Bulletin of Economics and Statistics*, 49, 431–434.

Beach, C. M., R. Davidson, and G. A. Slotsve (1994), Distribution Free Statistical Inference for Inequality Dominance with Crossing Lorenz Curves, manuscript, Vanderbilt University, Nashville.

Bowman, K. O. and L. R. Shenton (1981), Moment Series for the Coefficient of Variation in Weibull Sampling, *American Statistical Association—Proceedings of the Statistical Computing Section*, 148–153.

Breunig, R. (1996), Almost Unbiased Estimation of an Inequality Measure, manuscript, University of California, Riverside.

Cochran, W. G. (1953), *Sampling Techniques*, Wiley, New York.

Cochran, W. G. (1939), The Use of the Analysis of Variance in Enumeration by Sampling, *Journal of the American Statistical Association*, 34, 492–510.

Deaton, A. (1994), The Analysis of Household Surveys: Microeconometric Analysis for Development Policy, manuscript, Princeton University.

DuMouchel, W. H. and G. J. Duncan (1983), Using Sample Survey Weights in Multiple Regression Analysis of Stratified Samples, *Journal American Statistical Association*, 78, 535–543.

Engel, E. (1895), Die Productions-Und Consummations-Erhaltnisse des Konigreichs Sachsen, in *Die Lebenkosten belgischer Arbielter Fa* (E. Engel, ed.), Dresden.

Fisher, R. A. (1925), *Statistical Methods for Research Workers*, Oliver and Boyd, Edinburgh.

Fuller, W. A. (1975), Regression Analysis for Sample Survey, *Sankhya: The Indian Journal of Statistics, Series C*, 87, 117–132.

Fuller W. A. (1984), Least Squares and Related Analyses for Complex Survey Designs, *Survey Methodology*, 10, 97–118.

Ghosh, M. and B. K. Sinha (1980), On the Robustness of Least Squares Procedures in Regression Models, *Journal of Multivariate Analysis*, 10, 332–342.

Godambe, V. P. (1995), Estimation of Parameters in Survey Sampling: Optimality, *Canadian Journal of Statistics*, 23, 227–243.

Godambe, V. P. (1997), Estimation of Parameters in Survey Sampling, *Journal of the American Statistical Association*, forthcoming.

Greene, W. M. (1993), *Econometric Analysis*, 2nd ed., Macmillan, New York.

Gupta, R. and S. Ma (1996), Testing the Equality of Coefficients of Variation in K Normal Populations, *Communications in Statistics—Theory and Methods*, 25, 115–132.

Hansen, M. (1987), History of Survey Sampling, *Statistical Science*, 2, 180–190.

Howes, S. and J. O. Lanjouw (1994), Making Poverty Comparisons Taking into Account Survey Design: How and Why, manuscript, World Bank.

Johnston, J. (1991), *Econometric Methods*, 3rd ed., McGraw-Hill, London.

Judge, G. G. and M. E. Bock (1978), *The Statistical Implications of Pre-Test and Stein Rule Estimators in Econometrics*, North-Holland, Amsterdam.

Kadane, J. B. (1971), Comparisons of k-Class Estimators When the Disturbances are Small, *Econometrica*, 39, 723–737.

Kakwani, N. (1980), *Income Inequality and Poverty: Methods of Estimation and Policy Applications*, World Bank, Oxford University Press, Oxford.

Kalton, G. (1983), Introduction to Survey Sampling, Sage University Paper #35.

Kiaer, A. N. (1897), The Representative Method of Statistical Surveys (translation 1976, original in Norwegian), *Kristiania Videnskabsselskabets Skrifter, Historisk-filosofiske Klasse*, 4: (34–56), Statistisk Sentralbyra, Oslo.

Kish, L. (1965), *Survey Sampling*, Wiley, New York.

Kish, L. and M. R. Frankel (1974) Inference from Complex Samples, *Journal of Royal Statistical Society, Series B*, 1, 1–35.

Kloek, T. (1981), OLS Estimation in a Model Where a Microvariable Is Explained by Aggregates and Contemporaneous Disturbances are Equi-correlated, *Econometric*, 49, 205–207.

Konjin, M. S. (1962), Regression Analysis in Sample Surveys, *Journal of the American Statistical Association*, 57, 590–606.

Levy, P. S. and S. Lemeshow (1991), *Sampling of Populations: Methods and Applications*, 2nd ed., Wiley, New York.

Magee, L. (1996), Improving Survey-Weighted Least Squares Regression, manuscript, McMaster University.

Magee, L., A. L. Robb, and J. B. Burbidge (1996), On the Use of Sampling Weights When Estimating Regression Models with Survey Data, manuscript, McMaster University.

McKay, A. T. (1932), Distributions of the Coefficient of Variation and the Extended t-Distribution, *Journal of the Royal Statistical Society*, 95, 695–698.

Moulton, B. R. (1990), An Illustration of a Pitfall in Estimating the Effects of Aggregate Variables on Microunits, *Review of Economics and Statistics*, 72, 334–338.

Neuts, M. (1982), On the Coefficient of Variation of Mixtures of Probability Distributions, *Communications in Statistics—Simulation and Computation*, 11, 649–657.

Neyman, J. (1938), Contribution to the Theory of Sampling Human Populations, *Journal of the American Statistical Association*, 33, 101–116.

Neyman, J. (1934), On the Two Different Aspects of the Representative Method: The Method of Stratified Sampling and the Method of Purposive Selection, *Journal of Royal Statistical Society*, 97, 558–606.

Pagan, A. and A. Ullah (1995), Nonparametric Econometrics, manuscript, Australian National University, Australia.

Pfefferman, D. (1993), The Role of Sampling Weights When Modeling Survey Data, *International Statistical Review*, 61, 317–337.

Pfefferman, D. and T. M. F. Smith (1985), Regression Models for Grouped Populations in Cross-Sectional Surveys, *International Statistical Review*, 53, 37–59.

Porter, R. D. (1973), On the Use of Survey Sample Weights in the Linear Model, *Annals of Economics and Social Measurement*, 2/2, 141–158.

Prasad, B. and M. P. Singh (1992), Unbiased Estimators of Finite Population Variance Using Auxiliary Information in Sample Surveys, *Communications in Statistics—Theory and Methods*, 21, 1367–1376.

Pudney, S. (1989), *Modeling Individual Choice: The Econometrics of Corners, Kinks, and Holes*, Basil Blackwell, Oxford.

Rosenblatt, M. (1956), Remarks on Some Nonparametric Estimates of Density Function, *Annals of Mathematical Statistics*, 27, 832–837.

Ruppert, D. and M. P. Wand (1994), Multivariate Locally Weighted Least Squares Regression, *The Annals of Statistics*, 22, 1346–1370.

Selden, T. (1994), Weighted Generalized Least Squares Estimation for Complex Survey Data, *Economics Letters*, 46, 1–6.

Sen, A. (1992), *Inequality Reexamined*, Russell Sage Foundation, Oxford University Press, New York.

Singh, M. (1993), Behaviour of Sample Coefficient of Variation Drawn from Several Distributions, *Sankhya: The Indian Journal of Statistics*, 55, 65–76.

Srivastava, V. K., T. D. Dwivedi, M. Beluisky, and R. Tiwari (1980), A Numerical Comparison of Exact Large Sample and Small Disturbance Approximations of Properties of K-Class Estimators, *International Economic Review*, 21, 249–252.

Stephan, F. F. (1948), History of the Uses of Modern Sampling Procedures, *Journal of American Statistical Association*, 43.

Sukhatme, P. V. (1984), *Sampling Theory of Surveys with Applications*, Iowa State University Press, Ames, IA.

Swamy, P. A. V. B. (1971), *Statistical Inference in Random Coefficient Regression Models*, Springer-Verlag, Berlin.

Thompson, S. (1992), *Sampling*, Wiley, New York.

Tschuprow, A. (1923), On the Mathematical Expectation of the Moments of Frequency Distributions in the Case of Correlated Observation, *Metron*, 2, 461–493, 646–680.

Ullah, A. and R. Breunig (1996), On the Bias of the Standard Errors of the LS Residual and the Regression Coefficients under Normal and Nonnormal Errors, *Econometric Theory, Problems and Solutions*, 12, 868.

Vinod, H. D. and A. Ullah (1981), *Recent Advances in Regression Methods*, Marcel Dekker, New York.

Warren, W. G. (1982), On the Adequacy of the Chi-Squared Approximation for the Coefficient of Variation, *Communications in Statistics—Simulation and Computation*, 11, 659–666.

White, H. (1980). A Heteroskedasticity-Consistent Covariance Matrix Estimator and a Direct Test for Heteroskedasticity, *Econometrica*, 48, 817–838.

Yates, F. and I. Zacapony (1935), The Estimation of the Efficiency of Sampling with Special Reference to Sampling for Yield in Cereal Experiments, *Journal of Agricultural Science*, 25, 543–577.

Zinde-Walsh, V. and A. Ullah (1987), On Robustness of Tests of Linear Restrictions in Regression Models with Elliptical Error Distributions, in *Time-Series Econometric Modeling* (I. B. Macneil and G. J. Umphrey, eds.), Reidel, Holland.

10

Information Recovery
In Simultaneous-Equations Statistical Models

Amos Golan
University of California at Berkeley, Berkeley, California and
American University, Washington, D.C.

George Judge
University of California at Berkeley, Berkeley, California

Douglas Miller
Iowa State University, Ames, Iowa

I. INTRODUCTION

For the last five decades a significant portion of econometric effort has been directed to

1.   Develop ways for analyzing multiple-equation statistical models that describe a range of underlying economic data generation processes
2.   Develop a corresponding basis for estimation and inference to be used in data reduction and information recovery

Despite the productive efforts of many, questions remain concerning the insecure assumptions underlying the sampling theory, likelihood, and asymptotic approaches and the usefulness of traditional multiple-equation estimation and inference procedures in helping us find order when using the partial-incomplete underlying economic data that is normally found in practice. Against this backdrop, we propose a new method of estimation in multiple-equation statistical models that is widely applicable because it does not require the specification of a parametric family for the likelihood function. The estimation rule is robust with respect to likelihood, is flexible with respect to the dynamic, stochastic, and feedback nature of economic data as well as to the introduction of prior information, and is computationally simple. Using linear and quadratic risk measures, we compare the finite-sample performance of this method to other widely used traditional estimation rules.

The organization of the chapter is as follows: In Section II the simultaneous-equations statistical model is introduced, traditional estimation rules are identified, and corresponding asymptotic and finite-sample performances are noted. In Section III the maximum-entropy approach to recovering information in the case of inverse problems with noise is formulated and corresponding asymptotic sampling properties are developed. In Section IV sampling experiments are proposed as a basis for comparing finite-sample performance of the alternative estimation rules, and the resulting empirical sampling-risk results are evaluated. Section V contains summary comments and recommendations.

II. THE SIMULTANEOUS-EQUATIONS STATISTICAL MODEL (SESM) AND TRADITIONAL ESTIMATORS

To provide a format for analyzing the SESM that reflects an instantaneous feedback mechanism between some of the variables in the stochastic system of equations, consider a statistical model consistent with the data generation process for a system of G simultaneous equations:

$$Y\Gamma + XB + E = 0 \tag{1}$$

where Y is a $T \times G$ matrix of observations on G endogenous variables, X is a $T \times K$ matrix of observations on the K exogenous predetermined variables, Γ and B are comformable matrices of unknown parameters, and E is a $T \times G$ matrix of unobservable equation errors. The rows of E are assumed to be independently distributed, with joint density function that has a zero mean vector and a $G \times G$ covariance matrix Σ. The reduced form counterpart of (1) is

$$Y = -XB\Gamma^{-1} - E\Gamma^{-1} = X\Pi + D \tag{2}$$

where $\Pi\Gamma \equiv -B$ or

$$\Pi\Gamma + B = [\Pi \quad I_k] \begin{bmatrix} \Gamma \\ B \end{bmatrix} = 0 \tag{3}$$

D has a joint density function $f([0], \Omega)$ and Y has some corresponding density function $f(Y|\Pi, \Omega)$.

For single-equation-analysis purposes, it is conventional to assume that Γ contains -1's on the diagonal and to rewrite the ith equation as

$$\mathbf{y}_i = [Y_i, Y_i^* : X_i, X_i^*] \begin{bmatrix} \gamma_1 \\ \beta_i \end{bmatrix} + \mathbf{e}_i = [Y_i, Y_i^* : X_i, X_i^*] \begin{bmatrix} \mathbf{g}_i \\ \mathbf{0} \\ \mathbf{b}_i \\ \mathbf{0} \end{bmatrix} + \mathbf{e}_i$$

(4)

$$= [Y_i, X_i] \begin{bmatrix} \mathbf{g}_i \\ \mathbf{b}_i \end{bmatrix} + \mathbf{e}_i = Z_i \delta_i + \mathbf{e}_i$$

where $\mathbf{y}_i$ and Y_i represent the endogenous jointly determined variables in the ith equation and X_i represents the exogenous predetermined variables in the ith equation, X_i^* represents the exogenous predetermined variables appearing in the system but not included in equation i, and X represents the K exogenous predetermined variables in the *system* of equations. Let Z_i be a $T \times (G_i + K_i)$ matrix representing the G_i endogenous Y_i and K_i exogenous predetermined X_i variables that appear in the ith equation with nonzero coefficients. Further, $\delta_i = (\mathbf{g}_i', \mathbf{b}_i')'$ is a $(G_i + K_i)$-dimensional vector of unknown and unobservable parameters corresponding to the endogenous and exogenous variables in the ith equation, and $\mathbf{e}_i$ is a T-dimensional random vector for the ith equation that is traditionally assumed to have mean $\mathbf{0}$ and scale parameter σ_{ii}. The variables $\mathbf{y}_i, Y_i, X_i$ are observed, and δ_i and $\mathbf{e}_i$ are unobserved and unobservable.

Given the ith equation (4), the *complete* system of G equations may be written as

$$\begin{bmatrix} \mathbf{y}_1 \\ \mathbf{y}_2 \\ \vdots \\ \mathbf{y}_G \end{bmatrix} = \begin{bmatrix} Z_1 & & & \\ & Z_2 & & \\ & & \ddots & \\ & & & Z_G \end{bmatrix} \begin{bmatrix} \delta_1 \\ \delta_2 \\ \vdots \\ \delta_G \end{bmatrix} + \begin{bmatrix} \mathbf{e}_1 \\ \mathbf{e}_2 \\ \vdots \\ \mathbf{e}_G \end{bmatrix}$$

(5)

or compactly as

$$\mathbf{y} = Z\delta + \mathbf{e}$$

(6)

where, in the context of the traditional SESM, $\mathbf{y}$ and $\mathbf{e}$ are GT-dimensional random vectors and δ is an unknown and unobservable $\sum_i (G_i + K_i)$-dimensional vector. Traditionally $\mathbf{e}$ is assumed to have mean $\mathbf{0}$ and $\text{cov}(\mathbf{e}) = \Sigma \otimes I_T$, where Σ is a $G \times G$ unknown covariance matrix. Given (5) and (6), the corresponding system of reduced form equations may be written as

$$\mathbf{y} = (I_G \otimes X)\pi + \mathbf{d}$$

(7)

where π is a GK-dimensional vector of unknown parameters and the random vector $\mathbf{d}$ has mean vector $\mathbf{0}$ and $\text{cov}(\mathbf{d}) = \Omega \otimes I_T$.

From a sampling-theory point of view, besides direct least squares (DLS), the following estimators for δ that seek the consistency property are relevant: two-stage least squares (2SLS), k-class (KC), limited-information maximum likelihood (LIML), three-stage least squares (3SLS), and full-information maximum likelihood (FIML). These estimators are widely reported and reviewed in econometrics books such as Judge et al. (1985) and Davidson and MacKinnon (1993). In terms of sampling properties, FIML estimation of the SESM produces consistent and asymptotically efficient structural parameter estimates *given that the distribution function is correctly specified*. A nice review of finite-sample results for sampling-theory SESM estimators is given by Zellner (1996), Tsurumi (1990), and Park (1982). These authors also review Bayesian approaches to the SESM and report conditional and unconditional performance results for the Bayesian estimators, and White (1994) develops properties for the quasi-ML approach. The experimental design underlying the finite-sample results reported in these papers and the papers by Cragg (1966, 1967) form the basis for the maximum-entropy estimator sampling performance results given in Section IV.

III. A MAXIMUM-ENTROPY FORMULATION AND ESTIMATOR

As an alternative to traditional frequentist and Bayes estimators for the unknown structural parameters δ_i and δ, we consider a variation of the generalized maximum entropy (GME) estimation rule, proposed by Golan, Judge, and Miller (1996). This estimation rule is based on the entropy measure of Shannon (1948), the maximum entropy (ME) formalism of Jaynes (1957a, 1957b, 1984), Levine (1980), Shore and Johnson (1980), Skilling (1989), and Csiszár (1991), and the cross-entropy principle of Gokhale and Kullback (1978), Good (1963), and Kullback (1959).

In formulating the GME estimator for the SESM, the possible outcomes for the unknown and unobservable parameters and equation errors are viewed probabilistically. To reflect this specification, the statistical model (7), involving a single equation from the system of reduced-form equations, is reparameterized as

$$\mathbf{y}_i = X\boldsymbol{\pi}_i + \mathbf{d}_i \equiv XS_i^{\pi}\mathbf{p}_i^{\pi} + V_i\mathbf{w}_i \tag{8}$$

where S_i^{π} is a block-diagonal matrix reflecting transformation of the possible outcomes of each π_{ik} to the interval [0, 1] by defining a set of $M \geq 2$ discrete support points $\mathbf{s}_{ik}^{\pi} = [s_{ik1}^{\pi}, s_{ik2}^{\pi}, \ldots, s_{ikM}^{\pi}]'$ and a vector of corresponding unknown weights (probabilities) $\mathbf{p}_{ik}^{\pi} = [p_{ik1}^{\pi}, p_{ik2}^{\pi}, \ldots, p_{ikM}^{\pi}]'$, such that

$$\mathbf{s}_{ik}^{\pi}{}'\mathbf{p}_{ik}^{\pi} \equiv \sum_m p_{ikm}^{\pi} s_{ikm}^{\pi} = \pi_{ik} \tag{9}$$

Similarly, V_i is a matrix that provides a transformation of the possible outcomes for d_{it} to the interval [0, 1] by defining a set of $J \geq 2$ discrete support points

$\mathbf{v}_{it} = [v_{it1}, v_{it2}, \ldots, v_{itJ}]'$, *distributed uniformly and evenly around zero*, and a vector of corresponding unknown weights (probabilities) $\mathbf{w}_{it} = [w_{it1}, w_{it2}, \ldots, w_{itJ}]'$ such that

$$\mathbf{v}'_{it}\mathbf{w}_{it} \equiv \sum_j v_{itj}w_{itj} = d_{it} \tag{10}$$

When prior information is available, consistent with the set of discrete points $\mathbf{s}^{\pi}_{ik}$ in (9), this may be specified by corresponding *prior* probabilities $\mathbf{q}^{\pi}_{ik} = [q^{\pi}_{ik1}, q^{\pi}_{ik2}, \ldots, q^{\pi}_{ikM}]'$. Also, consistent with the set of discrete points $\mathbf{v}_{it}$, we may specify corresponding *prior* probabilities $\mathbf{u}_{it} = [u_{it1}, u_{it2}, \ldots, u_{itJ}]'$. For the complete system of reduced-form equations, the statistical model (7) is reparameterized as

$$\mathbf{y} = (I_G \otimes X)\boldsymbol{\pi} + \mathbf{d} = (I_G \otimes X)S^{\pi}\mathbf{p}^{\pi} + V\mathbf{w} \tag{11}$$

where S^{π} and V are block-diagonal matrices with S^{π}_i and V_i as the blocks.

Further, even though we do not make use of the structural equations directly, we reparameterize B and Γ, or similarly $\boldsymbol{\delta}$, in the following way. Let

$$\delta_{ik} \equiv \sum_m p^{\delta}_{ikm}s^{\delta}_{ikm} = \mathbf{s}^{\delta}_{ik}{}'\mathbf{p}^{\delta}_{ik} \tag{12}$$

with $m = 1, 2, \ldots, M$ and $M \geq 2$.

To simplify notation, we also use the definitions

$$\beta_{ik} \equiv \sum_m p^{\beta}_{ikm}s^{\beta}_{ikm}, \qquad k = 1, 2, \ldots, G_i, \quad i = 1, 2, \ldots, G \tag{13}$$

and

$$\gamma_{in} \equiv \sum_m p^{\gamma}_{inm}s^{\gamma}_{inm}, \qquad i = 1, 2, \ldots, G, \quad n = 1, 2, \ldots, K_i \tag{14}$$

which are the two subsets of $\boldsymbol{\delta}_i$.

Using these definitions and the corresponding reparameterizations, the generalized maximum-entropy estimation problem for estimating a system of G equations is formulated as

$$\underset{\mathbf{p}^{\pi}, \mathbf{w}, \mathbf{p}^{\delta}}{\text{Max}} H(\mathbf{p}^{\pi}, \mathbf{w}, \mathbf{p}^{\delta}) = -\mathbf{p}^{\pi\prime} \ln \mathbf{p}^{\pi} - \mathbf{w}' \ln \mathbf{w} - \mathbf{p}^{\delta\prime} \ln \mathbf{p}^{\delta} \tag{15}$$

subject to the GT reparameterized reduced-form equations

$$\mathbf{y} = (I_G \otimes X)S^{\pi}\mathbf{p}^{\pi} + V\mathbf{w} \tag{16}$$

the structural equation to reduced-form restrictions (3), reflecting (13) and (14),

$$-\beta_{ik} \equiv -\sum_m p^{\beta}_{ikm}s^{\beta}_{ikm} = \sum_n \pi_{nk}\gamma_{in} \equiv \sum_n \sum_m \pi_{nk}p^{\gamma}_{inm}s^{\gamma}_{inm} \tag{17}$$

and the adding-up conditions

$$(I_{G_i} \otimes \mathbf{1}'_M)\mathbf{p}_i^\delta = \mathbf{1}_{iG_i} \tag{18}$$

$$(I_{K_i} \otimes \mathbf{1}'_M)\mathbf{p}_i^\pi = \mathbf{1}_{iK_i} \tag{19}$$

$$(I_T \otimes \mathbf{1}'_J)\mathbf{w}_i = \mathbf{1}_{iT} \tag{20}$$

where $\mathbf{1}_{i\ell}$ is an $\ell \times 1$ vector of unit values.

Carrying through the first-order conditions and solving the resulting system yields the solution

$$\hat{p}^\pi_{ikm} = \frac{\exp(-S_i^\pi \hat{\lambda}'_i X[\cdot, i])}{\sum_m \exp(-S_i^\pi \hat{\lambda}'_i X[\cdot, i])} \equiv \frac{\exp(-S_i^\pi \hat{\lambda}'_i X[\cdot, i])}{\Omega^\pi_{ik}(\hat{\lambda}_i)} \tag{21}$$

$$\hat{w}_{itj} = \frac{\exp(-V'_i \hat{\lambda}'_i)}{\sum_j \exp(-V'_i \hat{\lambda}'_i)} \equiv \frac{\exp(-V'_i \hat{\lambda}'_i)}{\psi_{it}(\hat{\lambda}_i)} \tag{22}$$

where $X[\cdot, i]$ is the ith column of X and λ_i is a $T \times 1$ vector of Lagrange multipliers for equation i's constraints in (16):

$$\hat{p}^\gamma_{inm} = \frac{\exp\left(-\sum_k \hat{\rho}_{ik}\hat{\pi}_{nk}s^\gamma_{inm}\right)}{\sum_m \exp\left(-\sum_k \hat{\rho}_{ik}\hat{\pi}_{nk}s^\gamma_{inm}\right)} \equiv \frac{\exp\left(-\sum_k \hat{\rho}_{ik}\hat{\pi}_{nk}s^\gamma_{inm}\right)}{\Omega^\gamma_{in}(\hat{\rho}_i)} \tag{23}$$

and

$$\hat{p}^\beta_{ikm} = \frac{\exp(-\hat{\rho}_{ik}s^\beta_{ikm})}{\sum_m \exp(-\hat{\rho}_{ik}s^\beta_{ikm})} \equiv \frac{\exp(-\hat{\rho}_{ik}s^\beta_{ikm})}{\Omega^\beta_{ik}(\hat{\rho}_i)} \tag{24}$$

where ρ_i is the vector of Lagrange multipliers associated with (17) that refers to equation i, and $\Omega(\cdot)$ and $\psi(\cdot)$ are the partition (normalization) functions for the probabilities.

Finally, using (9), (10), and (12) yields

$$\hat{\pi}_i = \mathbf{S}_i^{\pi'}\hat{\mathbf{p}}_i^\pi \tag{25}$$

$$\hat{\mathbf{d}}_i = \mathbf{V}'_i\mathbf{w}_i \tag{26}$$

$$\hat{\delta}_i = S_i^{\delta'}\hat{\mathbf{p}}_i^\delta \tag{27}$$

or $\hat{\gamma}_i = \mathbf{S}_i^{\gamma'}\mathbf{p}_i^\gamma$ and $\hat{\beta}_i = \mathbf{S}_i^{\beta'}\hat{\mathbf{p}}_i^\beta$.

To obtain the GME solution it is sufficient to have two points in the support spaces that convert the elements from the real line into a $[0, 1]$ space. The larger the number of support points, the more moments can be captured in the optimization-estimation process.

If prior information exists concerning the structural or reduced-form parameters or the noise components, this information can be introduced through the prior probabilities $\mathbf{q}_i^\pi$, $\mathbf{q}_i^\delta$, and $\mathbf{u}_i$. This information can be used in the following generalized cross-entropy (GCE) formulation:

$$
\begin{aligned}
\operatorname*{Min}_{\mathbf{p}^{\pi},\mathbf{w},\mathbf{p}^{\delta}} I(\mathbf{p}^{\pi},\mathbf{w},\mathbf{p}^{\delta};\mathbf{q}^{\pi},\mathbf{u},\mathbf{q}^{\delta}) &\equiv \sum_i \sum_k \sum_m p_{ikm}^{\pi} \ln\left(\frac{p_{ikm}^{\pi}}{q_{ikm}^{\pi}}\right) \\
&+ \sum_i \sum_t \sum_j w_{itj} \ln\left(\frac{w_{itj}}{u_{itj}}\right) + \sum_i \sum_k \sum_m p_{ikm}^{\delta} \ln\left(\frac{p_{ikm}^{\delta}}{q_{ikm}^{\delta}}\right)
\end{aligned}
\tag{28}
$$

subject to (16)–(20). Carrying through the first-order conditions and solving the resulting system yields the estimated probabilities

$$
\tilde{p}_{ikm}^{\pi} = \frac{q_{ikm}^{\pi} \exp(S_i^{\pi} \tilde{\lambda}_i' X[\cdot, i])}{\sum_m q_{ikm}^{\pi} \exp(S_i^{\pi} \tilde{\lambda}_i' X[\cdot, i])} \equiv \frac{q_{ikm}^{\pi} \exp(S_i^{\pi} \tilde{\lambda}_i' X[\cdot, i])}{\Omega_{ik}^{\pi}(\tilde{\lambda}_i)}
\tag{29}
$$

and

$$
\tilde{w}_{itj} = \frac{u_{itj} \exp(V_i \tilde{\lambda}_i')}{\sum_j u_{itj} \exp(V_i \tilde{\lambda}')} \equiv \frac{u_{itj} \exp(V_i \tilde{\lambda}')}{\Psi_{it}(\tilde{\lambda}_i)}
\tag{30}
$$

In a similar way, estimates are provided for the structural parameters $\tilde{\mathbf{p}}^{\gamma}$ and $\tilde{\mathbf{p}}^{\beta}$. Finally, the point estimates $\tilde{\pi}$, $\tilde{\mathbf{d}}$, and $\tilde{\delta}$ (or $\tilde{\gamma}$ and $\tilde{\beta}$) are recovered as in (25)–(27).

A. Discussion and Large Sample Properties

As shown in Chapter 6 of Golan, Judge, and Miller (1996), for the traditional linear statistical (regression) model, the Hessian matrix of the GCE problem is positive definite for $\mathbf{p}_i$, $\mathbf{w}_i \gg 0$, and thus satisfies the sufficient condition for a unique global minimum. When prior information does not exist, both $\mathbf{q}_{ik}$ and $\mathbf{u}_{it}$ become uniform (e.g., $q_{ikm} = 1/M$ for all m and k) and the GCE solution is equivalent to the GME solution. Although the GME-GCE solutions do not have a closed form, the dual unconstrained formulation proposed by Miller (1994) and Golan, Judge, and Miller (1996) may be used to evaluate the sampling behavior of the solutions within the extremum of M-estimation framework (Huber, 1981). In general, the GME-GCE solutions may be viewed as discrete members of the exponential family of probability distributions, and these functional forms may be used to relate the original parameter vector, $\mathbf{p}_i$, to the dual parameters, λ_i and ρ_i. The large-sample properties of the GME-GCE estimators are based on the asymptotic behavior of the dual parameters, and the relationship follows the corresponding results in the exponential family literature (Brown 1986, Johansen 1979).

Following Golan, Judge, and Miller (1996, Chap. 6), we develop the dual unconstrained GME-SESM. Given the Lagrangian for the optimization problem (15)–(20), we substitute the maximum-entropy posteriors probabilities (21)–(24) into the Lagrangian where, for simplicity, we use $\mathbf{p}^{\beta}$ and $\mathbf{p}^{\gamma}$ instead of $\mathbf{p}^{\delta}$. Further, since these posteriors already satisfy the adding-up requirements, Eqs. (18)–(20) are omitted from the Lagrangian. Using some simple algebra, one gets the dual unconstrained problem. Specifically,

$$\mathcal{L}(\lambda, \rho) = -\mathbf{p}^{\pi\prime} \ln \mathbf{p}^{\pi} - \mathbf{p}^{\beta\prime} \ln \mathbf{p}^{\beta} - \mathbf{p}^{\gamma\prime} \ln \mathbf{p}^{\gamma} - \mathbf{w}' \ln \mathbf{w}$$

$$+ \sum_t \sum_i \lambda_{it} \left[y_{it} - \sum_k \sum_m x_{tk} p_{ikm}^{\pi} s_{ikm}^{\pi} - \sum_j w_{tij} v_{ij} \right]$$

$$+ \sum_i \sum_k \rho_{ik} \left[-\sum_m p_{ikm}^{\beta} s_{ikm}^{\beta} - \sum_n \sum_m \pi_{nk} p_{inm}^{\gamma} s_{inm}^{\gamma} \right]$$

$$= \sum_i \sum_k \sum_m p_{ikm}^{\pi} \left[\sum_t \lambda_{ti} x_{tk} s_{ikm}^{\pi} - \ln(\Omega_{ik}^{\pi}(\lambda_i)) \right]$$

$$+ \sum_i \sum_t \sum_j w_{tij} [\lambda_{ti} v_{ij} - \ln(\psi_{it}(\lambda_i))]$$

$$+ \sum_i \sum_k \sum_m p_{ikm}^{\beta} [\delta_{ik} s_{ikm}^{\beta} - \ln(\Omega_{ik}^{\beta}(\rho_i))] \qquad (31)$$

$$+ \sum_i \sum_n \sum_m p_{inm}^{\gamma} \left[\sum_k \delta_{ik} \pi_{nk} s_{inm}^{\gamma} - \ln(\Omega_{in}^{\gamma}(\rho_i)) \right]$$

$$+ \sum_t \sum_i \lambda_{it} \left[y_{it} - \sum_k \sum_m x_{tk} p_{ikm}^{\pi} s_{ikm}^{\pi} - \sum_j w_{tij} v_{ij} \right]$$

$$+ \sum_i \sum_k \rho_{ik} \left[-\sum_m p_{ikm}^{\beta} s_{ikm}^{\beta} - \sum_n \sum_m \pi_{nk} p_{inm}^{\gamma} s_{inm}^{\gamma} \right]$$

$$= \sum_t \sum_i \lambda_{it} y_{it} - \sum_i \sum_k \ln[\Omega_{ik}^{\pi}(\lambda_i)] - \sum_i \sum_t \ln[\psi_{it}(\lambda_i)]$$

$$- \sum_i \sum_k \ln[\Omega_{ik}^{\beta}(\rho_i)] - \sum_i \sum_n \ln[\Omega_{in}^{\gamma}(\rho_i)]$$

Minimizing the dual unconstrained GME model with respect to λ and ρ yields $\hat{\lambda}$ and $\hat{\rho}$, which, in turn, yield $\hat{\mathbf{p}}^{\pi}$, $\hat{\mathbf{p}}^{\gamma}$, $\hat{\mathbf{p}}^{\beta}$, and $\hat{\mathbf{w}}$. Investigating the concentrated, or dual, objective function (31) reveals the following properties. First, letting $G = 1$, the system reduces to the simple (one equation) linear statistical model where the last two terms disappear and the summation and indices involving i are deleted from the first three. Thus, we have the GME estimator for the linear statistical model. Second, the first three terms correspond to the reduced-form system of equations and involve the data $\mathbf{y}'\lambda$ and the sum of the partition functions for π and $\mathbf{w}$, respectively. This part can be viewed as an empirical likelihood function (Golan and Judge, 1996a) for the reduced-form equations. Third, the last two terms correspond to the definition (3) or its reparameterized form (17). There is no noise component involved in these two terms, so they are related to the classical (pure) maximum-entropy formulation.

What remains is to show that (i) $\hat{\pi}$ is a statistically consistent estimator of π, and (ii) $\hat{\gamma}$ and $\hat{\beta}$ are consistent internally estimates of γ and β. Part (i) is a trivial generalization of Golan, Judge, and Miller (1996, Proposition 6.3, p. 104). Part (ii) follows the principle of classical (pure) maximum entropy. That is, given the estimated π, which serves as the data for the (pure) ME problem, the entropies of γ and β are maximized. This ensures estimates that can be realized in the greatest number of ways consistent with the data (Jaynes 1957a,b, Levine 1980, Golan, Judge, and Miller 1996, Chap. 3). Furthermore, for those equations that are exactly identified, the maximum-entropy approach yields the exact mathematical inversion.

An alternative consistency motivation may be based on the following heuristic argument: Given the value of π, the problem (15)–(20) can be viewed as equivalent to maximizing (15) subject to

$$\mathbf{y} - (I_G \otimes X)\pi = V\mathbf{w} \tag{32}$$

$$\pi = S^\pi \mathbf{p}^\pi \tag{33}$$

and (17)–(20). The solution is then one of choosing $\mathbf{p}^\pi$ with maximum entropy $-\mathbf{p}^{\pi\prime} \ln \mathbf{p}^\pi$ to satisfy (32), with the remainder of the problem ($\mathbf{w}$ and $\mathbf{p}^\delta$) being separable from the choice of $\mathbf{p}^\pi$. Given the continuity of the constraint functions in the original problem, and since $\hat{\pi} \xrightarrow{p} \pi$, it is reasonable to view the problem of deriving $\mathbf{w}$ and $\mathbf{p}^\delta$ through (32), (33), and (15)–(20) as leading to the informational consistency of the separable maximum-entropy problem relating to β and γ.

B. Remarks

In applied work, many times emphasis is focused on one structural equation in the system of equations. For this situation it is traditional to use the 2SLS method of moments (Hansen 1982) or LIML estimators. Within a GME context several possibilities exist, and a range of these is discussed in Chapter 12 of Golan, Judge, and Miller (1996). One straightforward GME possibility is just to make use of the information, $\mathbf{y}_i$, $\mathbf{z}_i$, in the structural equation of interest. Although this formulation ignores X_i^*, the exogenous variables in the remainder of the system, the sampling results provided by Golan, Judge, and Miller (1996) suggest it performs well relative to traditional sampling-theory competitors.

A GME single-equation formulation and estimation rule consistent with the objectives of LIML may be developed as a special case of the complete system of equations formulation (15)–(27). In this single-structural-equation formulation, the objective function identifies the ith equation and it is maximized subject to (6), the zero-restriction condition for the ith equation (17), and appropriate adding-up conditions. Although this format is compatible with the LIML formulation, the distributional (and other) specification requirements and the solution basis differ significantly.

Finally, we note the relationship between the general method of moments (GMM) estimator (Hansen 1982) and the GME estimation rule. To develop this relationship, we consider a single overidentified structural equation and premultiply by the idempotent matrix $X(X'X)^{-1}X'$, as proposed by Basmann (1957), to rewrite the statistical model as

$$Z_i'X(X'X)^{-1}X'y_i = Z_i'X(X'X)^{-1}X'Z_iS_i\mathbf{p}_i + Z_i'X(X'X)^{-1}X'V_i\mathbf{w}_i \qquad (34)$$

Letting $V_{ij} = 0$ for all j yields the pure moment condition

$$Z_iX(X'X)^{-1}X'y_i = Z_iX(X'X)^{-1}X'Z_iS_i\mathbf{p}_i \qquad (35)$$

which, with the addition of a reparameterization for δ_i, is identical to the usual first-order conditions for the GMM estimator. If (35) replaces the relaxed moment condition (34) as the consistency relation in the GME formulation, then traditional GMM estimates for δ_i result. If the relaxed moment relation (34) is used in the GME-GCE formulation, and the bounded parameter space S contains the true parameter vector δ_i, then asymptotically the resulting estimates have, under standard regularity conditions, the same large-sample properties as the GMM estimators (Judge et al 1988, pp. 641–643). In finite samples, sampling results presented in Golan, Judge, and Miller (1996, Chap. 12) suggest that, under a squared error measure, GME is a superior performing estimator.

IV. SAMPLING EXPERIMENTS

Although analytic small-sample results are available in a few cases, much of the information that we have about the finite-sample properties of simultaneous-equations estimation rules comes from sampling experiments conducted over a four-decade period. Despite the usefulness of these studies, many questions are unresolved. For example, when considered in a loss-risk context, the rankings of traditional estimation rules remain somewhat in doubt. To obtain some experience with the MaxEnt estimator specified in Section III, and to gauge how they compare performance-wise with traditional sampling-theory rules, we conducted a limited range of sampling experiments. These experiments focused on some of the special characteristics of nonexperimental economic data such as small samples, collinear relations among variables, and the lack of independence between some of the right-hand-side variables and the equation errors. As a basis for judging estimator performance, we use the quadratic loss measures $\|\hat{\theta} - \theta\|^2$ for some unknown θ.

A. Sampling Design

In the sampling experiments, we work with a linear simultaneous-equations model involving three structural equations. The model follows Tsurumi (1990) and is a modification of the model structure employed by Cragg (1967). In the context of (1),

$$
\Gamma = \begin{pmatrix} -1.0 & .267 & .087 \\ .222 & -1.0 & 0 \\ 0 & .046 & -1.0 \end{pmatrix} \qquad
B = \begin{pmatrix} 6.2 & 4.4 & 4.0 \\ 0 & .74 & 0 \\ .7 & 0 & .53 \\ 0 & 0 & .11 \\ .96 & .13 & 0 \\ 0 & 0 & .56 \\ .06 & 0 & 0 \end{pmatrix} \tag{36}
$$

Using zero restrictions, all three equations are overidentified.

The exogenous variables $x_{t2}, \ldots, x_{t7}$ are drawn from a normal $(0, 1)$ distribution. To reflect the structure of the correlation among the exogenous variables, we use the condition number $\kappa(X'X)$, which is the ratio of the largest and smallest singular values of X with columns scaled to unit length. We use $\kappa(X'X) = 1$ for orthonormal X's and the condition number $\kappa(X'X) = 90$ for moderately high collinearity. Each experiment to be reported involved 1000 samples of size $T = 20$. The errors for the structural equations were drawings from a multivariate normal distribution with mean zero and covariance $\Sigma \otimes I_T$ with

$$
\Sigma = \begin{pmatrix} 1 & -1 & -.125 \\ -1 & 4 & .0625 \\ -.125 & 0.625 & 8 \end{pmatrix} \tag{37}
$$

For comparison purposes, the results for an alternative covariance and drawings from a $t_{(3)}$ distribution are also reported. In addition to the zero and normalization restrictions, the support spaces specified for the structural and reduced-form parameters and equation errors are $s_{ik}^{\pi} = s_{ik}^{\beta} = [-5, -2.5, 0, 2.5, 5]'$ for $k = 2, 3, \ldots, 7$; $s_{i1}^{\pi} = s_{i2}^{\beta} = [-20, -10, 0, 10, 20]'$ for $k = 1$; and $s_{in}^{\gamma} = [-2, -1, 0, 1, 2]'$ for $i, n = 1, 2, 3$. The errors' support V is specified as $v_{it} = [-3\sigma_{y_i}, 0, 3\sigma_{y_1}]'$, where σ_{y_i} is the empirical standard deviation of $\mathbf{y}_i$. For all experiments, the sampling performances (empirical risks) are reported for the whole system of structural equations and for a single parameter γ_{12} in (1). For comparison purposes, 3SLS results that use the *cor-*

rect covariance matrix (37) are reported. To provide results when the analysis focuses only on one equation, a GME formulation using only the information in $(\mathbf{y}_i, Z_i)$ is reported.

B. Sampling Results

The sampling results for a range of experiments are summarized in Table 1. The results for the base experiment $(T = 20, \kappa(X'X) = 1$, and normal errors) are given in the first row of Table 1. Focusing on the MSE results for the reduced-form parameter vector π, the unrestricted LS empirical MSE is 125, and thus close to its theoretical value. In contrast, the GME estimator of π that takes account of zero restrictions in the system yields MSE($\hat{\pi}$) = 4.11. This reduction in MSE relative to the *traditional* unrestricted π estimator is impressive. In terms of the structural parameters, note, relative to the 3SLS estimator with *known* error covariance, the superior empirical risk performance of the GME estimator. The empirical sampling variability of the GME estimator is given in parentheses for the structural parameters Γ and B. These results reflect the relative stability of the GME estimator, even under conditions of nonnormal errors or a high condition number. Intuitively speaking, significant improvement of the GME relative to the 3SLS is due to (i) shrinkage possibilities for both the signal and noise components, (ii) use of a dual loss function, and (iii) avoiding distributional assumptions (restrictions).

To reflect the sampling characteristics of the GME and 3SLS (with *known* covariance) estimators, we follow Tsurumi (1990) and focus on the γ_{12} parameter in (1) and give in Figure 1 a frequency plot for the two estimators. Relative to the 3SLS *known* error covariance estimator, the high concentration of γ_{12} and the restricted range variability of the GME estimates are nicely reflected in the empirical histogram. The empirical bias of the 3SLS (known covariance) estimator is slightly smaller than that of the GME estimator.

In terms of results for (1), the GME results indicate the empirical-risk gain when information from the whole system is used (column 4) relative to using only information from $(\mathbf{y}_i, Z_i)$ in column 7. In contrast, note the empirical-risk superiority of the GME $(\mathbf{y}_i, Z_i)$ estimation rule over the 3SLS (known covariance) estimation rule for (1). Also note the GME $(\mathbf{y}_i, Z_i)$ estimation rule remains stable and is concentrated near γ_{12} as the condition number or error distribution changes. Finally, repeating the experiment involving $T = 20$ and $\kappa(X'X) = 1$), but using relaxed moment constraints

$$\sum_t x_{tk} y_{ti} = \sum_t \sum_k \sum_m x_{tk} x_{tk} s^\pi_{ikm} p^\pi_{ikm} + \sum_t \sum_j x_{tk} v_{ij} w_{tij} \tag{38}$$

yields estimates similar to those reported in the first row of Table 1.

Although we have chosen as a basis of comparison the unattainable traditional 3SLS estimator with *known error covariance*, the empirical-risk superiority under a

Table 1   Empirical-Risk Results from 1000 Experiments Using the SESM, GME, and 3SLS Estimators with MSE Performance Measures[a]

Experiment description	Results from an analysis of the complete system						Using only y_1, Y_1, X_1	
	MSE($\hat{\pi}$)	MSE($\hat{B}$)	MSE($\hat{\Gamma}$)	MSE($\hat{B}+\hat{\Gamma}$)	MSE($\hat{\beta}_1+\hat{\gamma}_1$)	$\bar{\gamma}_{12}$	MSE($\hat{\beta}_1+\hat{\gamma}_1$)	$\bar{\gamma}_{12}$
GME, $T=20$ normal errors $\kappa(X'X)=1$	4.11	7.10 (2.95)	0.09 (0.03)	7.20	3.31 (1.30)	0.39 (0.01)	24.17 (0.06)	0.31
3SLS (known Cov) normal errors $\kappa(X'X)=1$				699.70		0.16 (0.13)		
GME, $T=20$ $t(3)$ errors $\kappa(X'X)=1$	8.77	11.39 (7.81)	0.21 (0.15)	11.59	4.74 (3.09)	0.36 (0.05)	24.60 (0.11)	0.31
GME, $T=20$ normal errors $\kappa(X'X)=90$	4.09	6.88 (2.83)	0.08 (0.03)	6.97	3.02 (1.19)	0.38 (0.01)	23.88 (0.05)	0.31
GME, $T=20$[b] normal errors $\kappa(X'X)=1$	4.21	6.41 (2.64)	0.08 (0.02)	6.50	2.50 (1.23)	0.33 (0.01)	24.05 (0.07)	0.31

[a]Numbers in parentheses are empirical variances.

[b]$\text{Cov} = \begin{pmatrix} 1.4 & -2.3 & 0.9 \\ & 4.1 & 0.6 \\ & & 8 \end{pmatrix}$

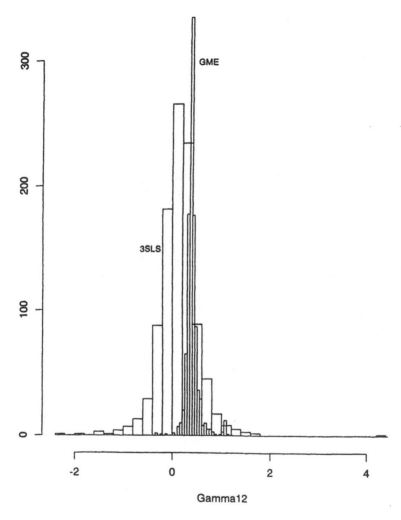

Figure I   Empirical histogram of the GME and 3SLS for Gamma 12.

MSE measure of the GME estimation rule appears to hold over a range of conditions normally found in practice.

V. CONCLUDING REMARKS

We propose a new GME method for recovering the unknown parameters in a simultaneous-equations statistical model that is (i) robust in respect to likelihood; (ii) flexible in respect to introducing sample and nonsample information; (iii) works well in both ill-posed (e.g., collinear X's) and well-posed problems and with small samples of data; (iv) has the usual asymptotic sampling properties; (v) in finite samples, under a squared error loss measure and relative to traditional estimators, is a high-performing estimation rule; and (vi) is computationally simple.

In contrast to traditional estimation for simultaneous-equations models, it permits the sample information to be introduced in either a data or moment form. It permits information recovery in case of nonlinear and/or nonstationary expectational models (Golan, Judge, and Karp 1996) and with discrete and/or limited endogenous regressors (Golan, Judge, and Perloff 1995, 1996). Using the normalized entropy concept provides a basis for selecting among alternative competing statistical models (Golan and Judge 1996b). By employing the entropy measure for each of the unknown endogenous and exogenous variables, when all the support spaces s are defined to be symmetric about zero, it is possible to identify the extraneous variables in each of the G equations. This problem will be further developed in future work.

The finite-sample results reported suggest, relative to the 3SLS rule with *known* covariance, the superior performance of the GME rule under selected experimental designs. What is needed at this point is extensive sampling experiments that make a sharp comparison with traditional sampling theory and Bayes' estimators for the SESM.

ACKNOWLEDGMENTS

It is a pleasure to acknowledge the helpful suggestions of Ron Mittelhammer.

REFERENCES

Basmann, R. L. (1957), A General Classical Method of Linear Estimation of Coefficients in a Structural Equation, *Econometrica*, 25, 77–83.

Brown, L. D. (1986), *Fundamentals of Statistical Exponential Families*, Institute of Mathematical Statistics, Hayward, CA.

Cragg, J. G. (1966), On the Sensitivity of Simultaneous Equations Estimators to the Stochastic Assumptions of the Models, *Journal of the American Statistical Association*, 61, 136–151.

Cragg, J. G. (1967), On the Relative Small-Sample Properties of Several Structural-Equation Estimators, *Econometrica*, 35, 89–110.

Csiszár, I. (1991), Why Least Squares and Maximum Entropy? An Axiomatic Approach to Inference for Linear Inverse Problems, *Annals of Statistics*, 19, 2032–2066.

Davidson, R. and J. G. MacKinnon (1993), *Estimation and Inference in Econometrics*, Oxford University Press, New York.

Gokhale, D. V. and S. Kullback (1978), *The Information in Contingency Tables*, Marcel Dekker, New York.

Golan, A. and G. Judge (1996a), A Maximum Entropy Approach to Empirical Likelihood Estimation and Inference, University of California, Berkeley, unpublished paper.

Golan, A. and G. Judge (1996b), A Simultaneous Estimation and Variable Selection Rule, University of California, Berkeley, unpublished paper.

Golan, A., G. G. Judge, and L. Karp (1996), A Maximum Entropy Approach of Estimation and Inference in Dynamic Models or Counting Fish in the Sea Using Maximum Entropy, *Journal of Economic Dynamics and Control*, 20, 559–582.

Golan, A., G. G. Judge, and D. Miller (1996), *Maximum Entropy Econometrics: Robust Estimation with Limited Data*, Wiley, New York.

Golan, A., G. Judge, and J. Perloff (1995), Estimation and Inference with Censored and Ordered Multinomial Response Data, University of California, Berkeley, unpublished paper.

Golan, A., G. Judge, and J. Perloff (1996), A Maximum Entropy Approach to Recovering Information from Multinomial Response Data, *Journal of the American Statistical Association*, 91, 841–853.

Good, I. J. (1963), Maximum Entropy for Hypothesis Formulation, Especially for Multidimensional Contingency Tables, *Annals of Mathematical Statistics*, 34, 911-934.

Hansen, L. P. (1982), Large Sample Properties of Generalized Method of Moments Estimators, *Econometrica*, 50, 1029–1054.

Huber, P. (1981), *Robust Statistics*, Wiley, New York.

Jaynes, E. T. (1957a), Information Theory and Statistical Mechanics, *Physics Review*, 106, 620–630.

Jaynes, E. T. (1957b), Information Theory and Statistical Mechanics II, *Physics Review*, 108, 171–190.

Jaynes, E. T. (1984), Prior Information and Ambiguity in Inverse Problems, in D. W. McLaughlin (ed.), *Inverse Problems*, SIAM Proceedings, American Mathematical Society, Providence, RI, 151–166.

Johansen, S. (1979), *An Introduction to the Theory of Regular Exponential Families*, Institute of Mathematical Statistics, Hayward, CA.

Judge, G. G., W. E. Griffiths, R. C. Hill, H. Lütkepohl, and T.-C. Lee (1985), *The Theory and Practice of Econometrics*, 2nd ed., Wiley, New York.

Judge, G. G., R. C. Hill, W. E. Griffiths, H. Lütkepohl, and T.-C. Lee (1988), *Introduction to the Theory and Practice of Econometrics*, 2nd ed., Wiley, New York.

Kullback, J. (1959), *Information Theory and Statistics*, Wiley, New York.

Levine, R. D. (1980), An Information Theoretical Approach to Inversion Problems, *Journal of Physics A*, 13, 91–108.

Miller, D. J. (1994), Solving Generalized Maximum Entropy Problems with Unconstrained Numerical Techniques, University of California, unpublished paper.

Park, S.-B. (1982), Some Sampling Properties of Minimum Expected Loss (MELO) Estimators of Structural Coefficients, *Journal of Econometrics*, 18, 295–311.

Shannon, C. E. (1948), A Mathematical Theory of Communications, *Bell System Technical Journal*, 27, 379–423.

Shore, J. E. and R. W. Johnson (1980), Axiomatic Derivation of the Principle of Maximum Entropy and the Principle of Minimum Cross-Entropy, *IEEE Transactions on Information Theory*, IT-26(1), 26–37.

Skilling, J. (1989), The Axioms of Maximum Entropy, in J. Skilling (ed.), *Maximum Entropy and Bayesian Methods in Science and Engineering*, Kluwer, Dordrecht, 173–187.

Tsurumi, H. (1990), Comparing Bayesian and Non-Bayesian Limited Information Estimators, in S. Geisser, J. S. Hodges, S. J. Press, and A. Zellner (eds.), *Bayesian and Likelihood Methods in Statistics and Econometrics*, North-Holland, Amsterdam, 179–207.

White, H. (1994), *Estimation, Inference and Specification Analysis*, Cambridge University Press, Cambridge.

Zellner, A. (1997), The Finite Sample Properties of Simultaneous Equations' Estimates and Estimators: Bayesian and Non-Bayesian Approaches, *Journal of Econometrics*, Annals Issue in Honor of Carl F. Christ (L. R. Klein, ed.) (in press).

11

Diagnostic Testing in Econometrics
Variable Addition, RESET, and Fourier Approximations

Linda F. DeBenedictis
Ministry of Human Resources, Victoria, British Columbia, Canada

David E. A. Giles
University of Victoria, Victoria, British Columbia, Canada

I. INTRODUCTION

The consequences of model misspecification in regression analysis can be severe in terms of the adverse effects on the sampling properties of both estimators and tests. There are also commensurate implications for forecasts and for other inferences that may be drawn from the fitted model. Accordingly, the econometrics literature places a good deal of emphasis on procedures for interrogating the quality of a model's specification. These procedures address the assumptions that may have been made about the distribution of the model's error term, and they also focus on the structural specification of the model, in terms of its functional form, the choice of regressors, and possible measurement errors.

Much has been written about "diagnostic tests" for model misspecification in econometrics in recent years. The last two decades, in particular, have seen a surge of interest in this topic which has, to a large degree, redressed what was previously an imbalance between the intellectual effort directed toward pure estimation issues and that directed toward testing issues of various sorts. There is no doubt that diagnostic testing is now firmly established as a central topic in both econometric theory and practice, in sympathy with Hendry (1980, p. 403), urging that we should "test, test, and test." Some useful general references in this field include Krämer and Sonnberger (1986), Godfrey (1988, 1996), and White (1994), among others.

As discussed by Pagan (1984), the majority of the statistical procedures proposed for measuring the inadequacy of econometric models can be allocated into one of two categories—"variable-addition" methods and "variable-transformation" methods. More than a decade later, it remains the case that the first of these categories still provides a useful basis for discussing and evaluating a wide range of the diagnostic tests that econometricians use. The equivalence between variable-addition tests and tests based on "Gauss-Newton regressions" is noted, for instance, by Davidson and MacKinnon (1993, p. 194) and essentially exploited by MacKinnon and Magee (1990). Indeed it is the case that many diagnostic tests can be viewed and categorized in more than one way.

In this chapter we limit our attention to diagnostic tests in econometrics which can be classified as "variable-addition" tests. This will serve to focus the discussion in a manageable way. Pagan (1984) and Pagan and Hall (1983) provide an excellent discussion of this topic. Our purpose here is to summarize some of the salient features of that literature and then to use it as a vehicle for proposing a new variant of what is perhaps the best-known variable addition test—Ramsey's (1969) "regression specification error (RESET) test."

The layout of the chapter is as follows. In the next section we discuss some general issues relating to the use of variable-addition tests for model misspecification. Section III discusses the formulation of the standard RESET test and the extent to which the distribution of its statistic can be evaluated analytically. In Section IV we introduce a modification of this test, which we call the FRESET test (as it is based on a Fourier approximation), and we consider some practical issues associated with its implementation. A comparative Monte Carlo experiment, designed to explore the power of the FRESET test under (otherwise) standard conditions, is described in Section V. Section VI summarizes the associated results. The last section contains some conclusions and recommendations which strongly favor the new FRESET test over existing alternatives; and we note some work in progress which extends the present study by considering the robustness of tests of this type to nonspherical errors in the data-generating process.

II.   VARIABLE-ADDITION TESTS IN ECONOMETRICS

A.   Preliminaries

One important theme that underlies many specification tests in econometrics is the idea that if a model is correctly specified, then (typically) there are many weakly consistent estimators of the model's parameters, and so the associated estimates should differ very little if the sample size is sufficiently large. A substantial divergence of these estimates may be taken as a signal of some sort of model misspecification (e.g., White 1994, Chap. 9). Depending on the estimates which are being compared, tests for various types of model misspecification may be constructed. Indeed, this basic idea underlies the well-known family of Hausman (1978) tests and the information matrix tests of White (1982, 1987). This approach to specification testing is based on the stance that, in practice, there is generally little information about the precise *form* of any misspecification in the model. Accordingly, no specific alternative specification is postulated, and a pure significance test is used. This stands in contrast with testing procedures in which an explicit alternative hypothesis is stated, and used in the construction and implementation of the test (even though a rejection of the null hypothesis need not lead one to accept the stated alternative). In the latter case, we frequently "nest" the null within the alternative specification and then test whether the associated parametric restrictions are consistent with the evidence in the data. The use of likelihood ratio, Wald, and Lagrange multiplier tests, for example, in this situation are common and well understood.

As noted, specification tests which do not involve the formulation of a specific alternative hypothesis are pure significance tests. They require the construction of a sample statistic whose null distribution is known, at least approximately or asymptotically. This statistic is then used to test the consistency of the null with the sample evidence. In the following discussion we will encounter tests which involve a specific alternative hypothesis, although the latter may involve the use of proxy variables to allow for uncertainties in the alternative specification. Our subsequent focus on the RESET test involves a procedure which really falls somewhere between these two categories, in that although a specific alternative hypothesis is formulated, it is largely a device to facilitate a test of a null specification. Accordingly, it should be kept in mind that the test is essentially a "destructive" one, rather than a "constructive" one, in the sense that a rejection of the null hypothesis (and hence of the model's specification) generally will not suggest any specific way of reformulating the model in a satisfactory form. This is certainly a limitation on its usefulness, so it is all the more important that it should have good power properties. If the null specification is to be rejected, with minimal direction as to how the model should be respecified, then at least one would hope that we are rejecting for the right reason(s). Accordingly, in our reconsideration of the RESET test in Sections III and IV we emphasize power properties in a range of circumstances.

Variable-addition tests are based on the idea that if the model specification is "complete," then additions to the model should have an insignificant impact, in some sense. As noted by Pagan and Hall (1983) and Pagan (1984), there are many forms that such additions can take. For instance, consider a standard linear multiple regression model, with k fixed regressors and T observations:

$$y = X\beta + u \tag{1}$$

where it may be assumed that $(y \mid X) \sim N[X\beta, \sigma^2 I_T]$. One could test this specification in terms of the adequacy of the assumed conditional mean of y, namely $X\beta$; or one might test the adequacy of the assumed conditional covariance matrix, $\sigma^2 I_T$. The assumed normality could be tested with reference to higher-order moments, as in Jarque and Bera (1980). In most of these cases, tests can be constructed by fitting auxiliary regressions which include suitable augmentation terms, and then testing the significance of the latter.

B.   Variable Addition and the Conditional Mean

For example, if it is suspected that the conditional mean of the model may be misspecified, then one could fit an "augmented" model

$$y = X\beta + W\gamma + u \tag{2}$$

and test the hypothesis that $\gamma = 0$. This assumes, of course, that W is known and observable. In the event that it is *not*, a matrix of corresponding proxy variables, W^*, may be substituted for W, and (2) may be written as

$$y = X\beta + W^*\gamma + (W - W^*)\gamma + u = X\beta + W^*\gamma + e \tag{3}$$

and we could again test if $\gamma = 0$. As Pagan (1984, p. 106) notes, the effect of this substitution will show up in terms of the power of the test that is being performed. An alternative way of viewing (2) (or (3) if the appropriate substitution of the proxy variables is made below) is by way of an auxiliary regression with residuals from (1) as the dependent variable:

$$y - Xb = X(\beta - b) + W\gamma + u \tag{4}$$

where $b = (X'X)^{-1}X'y$ is the least-squares estimator of β in (1). This last model is identical to (2), and the test of $\gamma = 0$ will yield an identical answer in each case. However, (4) emphasises the role of diagnostic tests in terms of explaining residual values.

The choice of W (or W^*) will be determined by the particular way in which the researcher suspects that the conditional mean of the basic model may be misspecified. Obvious situations that will be of interest include a wrongly specified functional form for the model or terms that have been omitted wrongly from the set of explanatory variables. There is, of course, a natural connection between these two types of model misspecification, as we discuss further. In addition, tests of serial independence of the errors, structural stability, exogeneity of the regressors, and those which discriminate between nonnested models, can all be expressed as variable-addition tests which focus on the conditional mean of the data-generating process. All of these situations are discussed in some detail by Pagan (1984). Accordingly, we will simply summarize some of the salient points here, and we will focus particularly on functional form and omitted effects, as these are associated most directly with the RESET test and hence with the primary focus of this chapter.

The way in which tests for serial independence can be cast in terms of variable-addition tests is easily illustrated. Consider model (1), but take as the maintained hypothesis an AR(1) representation for the disturbances. That is, assume that $u_t =$

$\rho u_{t-1} + \varepsilon_t$, where $|\rho| < 1$. We wish to test the hypothesis that $\rho = 0$. The mean of y in (1) is then conditional on the past history of y and X so it is conditional on previous values of the errors. Accordingly, the natural variable-addition test would involve setting W in (2) to be just the lagged value of u. Of course, the latter is unobservable, so the proxy variable approach of (3) would be used in practice, with W^* comprising just the lagged OLS residual series (u^*_{-1}) from the basic specification, (1). Of course, in the case of a higher-order AR process, extra lags of u^* would be used in the construction of W^*, and we would again test if $\gamma = 0$. It is important to note that the same form of variable-addition tests would be used if the alternative hypothesis is that the errors follow a moving-average process, and such tests are generally powerful against both alternatives. The standard Durbin-Watson test can be linked to this approach to testing for model misspecification, and various other standard tests for serial independence in the context of dynamic models, such as those of Godfrey (1978), Breusch (1978), and Durbin (1970), can all be derived in this general manner. Tests for structural stability which can be given a variable-addition interpretation include those of Salkever (1976), where the variables that are used to augment the basic model are suitably defined "zero-one" dummy variables. Further, the well-known tests for regressor exogeneity proposed by Durbin (1954), Wu (1973), and Hausman (1978) can also be reexpressed as variable-addition tests which use appropriate instrumental variables in the construction of the proxy matrix W^* (e.g., Pagan 1984, pp. 114–115).

The problem of testing between (nonnested) models is one which has attracted considerable attention during the last 20 years (e.g., McAleer 1987, 1995). Such tests frequently can be interpreted as variable-addition tests which focus on the specification of the conditional mean of the model. By way of illustration, recall that in model (1) the conditional mean of y (given X, and the past history of the regressors and of y) is $X\beta$. Suppose that there is a competing model for explaining y, with a conditional mean of $X^+\mu$, where X and X^+ are nonnested, and μ is a conformable vector of unknown parameters. To test one specification of the model against the other, there are various ways of applying the variable-addition principle. One obvious possibility (assuming an adequate number of degrees of freedom) would be to assign $W^* = X^+$ in (3), and then apply a conventional F-test. This is the approach suggested by Pesaran (1974) in one of the earlier contributions to this aspect of the econometrics literature. Another possibility, which is less demanding on degrees of freedom, is to set $W^* = X^+(X^{+\prime}X^+)^{-1}X^{+\prime}y$ (that is, using the ordinary least-squares (OLS) estimate of the conditional mean from the second model as the proxy variable), which gives us the J-test of Davidson and MacKinnon (1981). There have been numerous variants on the latter theme, as discussed by McAleer (1987), largely with the intention of improving the small-sample powers of the associated variable-addition tests.

Our main concern is with variable-addition tests which address possible misspecification of the functional form of the model or the omission of relevant explanatory effects. The treatment of the latter issue fits naturally into the framework of Eqs. (1)–(4). The key decision that has to be made in order to implement a variable-addition test in this case is the choice of W (or, more likely, W^*). If we have some idea what effects may have been omitted wrongly, then this determines the choice of the "additional" variables, and if we were to make a perfect choice then the usual F-test of $\gamma = 0$ would be exact and uniformly most powerful (UMP). Of course, this really misses the entire point of our present discussion, which is based on the premise that we have specified the model to the best of our knowledge and ability, but are still concerned that there may be some further, unknown, omitted effects. In this case, some ingenuity may be required in the construction of W or W^*, which is what makes the RESET test (and our modification of this procedure in this chapter) of particular interest. We leave a more detailed discussion of the RESET test to Section III.

In many cases, testing the basic model for a possible misspecification of its functional form can be considered in terms of testing for omitted effects in the conditional mean. This is trivially clear if, for example, the fitted model includes simply a regressor, x_t, but the correct specification involves a polynomial in x_t. Constructing W^* with columns made up of powers of x_t would provide an optimal test in this case. Similarly, if the fitted model included x_t as a regressor, but the correct specification involved some (observable) transformation of x_t, such as $\log(x_t)$, then (2) could be constructed so as to include *both* the regressor and its transformation, and the significance of the latter could be tested in the usual way. Again, of course, this would be feasible only if one had some prior information about the likely nature of the misspecification of the functional form. (See also, Godfrey, McAleer, and McKenzie 1988).

More generally, suppose that model (1) is being considered, but in fact the correct specification is

$$y = f(X, \beta_1, \beta_2) + u \tag{5}$$

where f is a nonlinear function which is continuously differentiable with respect to the parameter subvector, β_2. If (1) is nested within (5) by setting $\beta_2 = \beta_2^0$, then by taking a Taylor series expansion of f about the (vector) point β_2^0, we can (after some rearrangement) represent (5) by a model which is of the form (3) and then proceed in the usual way to test (1) against (5) by testing if $\gamma = 0$ via a RESET test. More specifically, this can be achieved, to a first-order approximation, by setting

$$W^* = \frac{\partial f[X, b_1, \beta_2^0]}{\partial \beta_2}, \qquad \gamma = (\beta_2 - \beta_2^0) \tag{6}$$

where b_1 is the least-squares estimator of the subvector β_1, obtained subject to the restriction that $\beta_2 = \beta_2^0$. As we will see again in Section III, although the presence of b_1 makes W^* random, the usual F-test of $\gamma = 0$ is still valid, as a consequence of the Milliken-Graybill (1970) theorem.

As we have seen, many tests of model misspecification can be formulated as variable-addition tests in which attention focuses on the conditional mean of the underlying data-generating process. This provides a very useful basis for assessing the sampling properties of such tests. Model (3) forms the basis of the particular specification test (the RESET test) that we will be considering in detail later. Essentially, we will be concerned with obtaining a matrix of proxy variables, W^*, that better represents an arbitrary form of misspecification than do the usual choices of this matrix. In this manner, we hope to improve on the ability of such tests to reject models which are falsely specified in the form of Eq. (1).

C. Variable-Addition and Higher Moments

Although our primary interest is with variable-addition tests which focus on misspecification relating to the conditional mean of the model, some brief comments are in order with respect to related tests which focus on the conditional variance and on higher-order moments. Of the latter, only the third and fourth moments have been considered traditionally in the context of variable-addition tests, as the basis of testing the assumption of normally distributed disturbances (e.g., Jarque and Bera 1987). Tests which deal with the conditional variance of the underlying process have been considered in a variable-addition format by a number of authors.

Essentially, the assumption that the variance, σ^2, of the error term in (1) is a constant is addressed by considering alternative formulations, such as

$$\sigma_t^2 = \sigma^2 + z_t \phi \tag{7}$$

where z_t is an observation on a vector of r known variables, and ϕ is $r \times 1$. We then test the hypothesis that $\phi = 0$. To make this test operational, (7) needs to be reformulated as a "regression relationship" with an observable "dependent variable" and a stochastic "error term." The squared tth element of u in (1) gives us σ_t^2, on average, so it is natural to use the corresponding squared OLS residuals on the left side of (7). Then

$$(u_t^*)^2 = \sigma^2 + z_t \phi + (u_t^*)^2 - \sigma_t^2 = \sigma^2 + z_t \phi + v_t \tag{8}$$

where $v_t = (u_t^*)^2 - \sigma_t^2$. Equation (8) can be estimated by OLS to give estimates of σ^2 and ϕ and to provide a natural test of $\phi = 0$. The (asymptotic) legitimacy of the usual t-test (or F-test) in this capacity is established, for example, by Amemiya (1977).

So the approach in (8) is essentially analogous to the variable-addition approach in the case of (2) for the conditional mean of the model. As was the situation there, in practice we might not be able to measure the z_t vector, and a replacement proxy vector, z_t^*, might be used instead. Then the counterpart to (3) would be

$$(u_t^*)^2 = \sigma^2 + z_t \phi + (z_t - z_t^*)\phi + v_t = \sigma^2 + z_t^* \phi + v_t^* \tag{9}$$

where we again test if $\phi = 0$, and the choice of z_t^* determines the particular form of heteroskedasticity against which we are testing.

For example, if z^* is an appropriately defined scalar dummy variable then we can test against a single break in the value of the error variance at a known point. This same idea also relates to the more general homoskedasticity tests of Harrison and McCabe (1979) and Breusch and Pagan (1979). Similarly, Garbade's (1977) test for systematic structural change can be expressed as the above type of variable-addition test, with $z_t^* = tx_t^2$; and Engle's (1982) test against ARCH(1) errors amounts to a variable addition test with $z_t^* = (u_{t-1}^*)^2$. Higher-order ARCH and GARCH processes* can be accommodated by including additional lags of $(u_t^*)^2$ in the definition of z^*. Pagan (1984, pp. 115–118) provides further details as well as other examples of specification tests which can be given a variable-addition interpretation with respect to the conditional variance of the errors.

D. Multiple Testing

Variable-addition tests have an important distributional characteristic which we have not yet discussed. To see this, first note that under the assumptions of model (1), the UMP test of $\gamma = 0$ will be a standard F-test if X and W (or W^*) are both non-stochastic and of full column rank. In the event that either the original or "additional" regressors are stochastic (and correlated with the errors), and/or the errors are nonnormal, the usual F-statistic for testing if $\gamma = 0$ can be scaled to form a statistic which will be asymptotically chi-square. More specifically, if there are T observations and if $\mathrm{rank}(X) = k$ and $\mathrm{rank}(W) = p$, then the usual F-statistic will be $F_{p,\nu}$, under the null (where $\nu = T - k - p$). Then pF will be asymptotically χ_p^2 under the null.[†] Now, suppose that we test the model's specification by means of a variable addition test based on (2) and denote the usual test statistic by F^w. Then, suppose we consider a second test for misspecification by fitting the "augmented" model

$$y = X\beta + Z\delta + v \tag{10}$$

where $\mathrm{rank}(Z) = q$, say. In the latter case, denote the statistic for testing if $\delta = 0$ by F^z. Asymptotically, pF^w is χ_p^2, and qF^z is χ_q^2, under the respective null hypotheses.

Now, from the usual properties of independent chi-square statistics, we know that if the above two tests are independent, then $pF^w + qF^z$ is asymptotically χ_{p+q}^2 under the null that $\gamma = \delta = 0$. As discussed by Bera and McKenzie (1987) and Eastwood and Godfrey (1992, p. 120), independence of the tests requires that plim

---

*Lee (1991) shows the equivalence of ARCH(p) and GARCH(p, q) tests under the null, for a constant q, where p is any natural number.
[†] Its distribution under the alternative is discussed in Section III.

$(Z'W/T) = 0$ and that either $\text{plim}(Z'X/T) = 0$ or $\text{plim}(W'X/T) = 0$. The advantages of this are the following. First, if these variable addition specification tests are applied in the context of (2) and (10) sequentially, then the overall significance level can be controlled. Specifically, if the (asymptotic) significance levels for these two tests are set at α_w and α_z respectively, then[*] the overall joint significance level will be $1 - (1 - \alpha_w)(1 - \alpha_z)$. Second, the need to fit a "supermodel," which includes all of the (distinct) columns of X, W, and Z as regressors, can be avoided. The extension of these ideas to testing subsets of the regressors is discussed in detail by Eastwood and Godfrey (1992, pp. 120–122).

A somewhat related independence issue also arises in the context of certain variable-addition tests, especially when the two tests in question focus on the different moments of the underlying stochastic process. Consider, for example, tests of the type discussed in Section II.B. These deal with the first moment of the distribution and are simply tests of exact restrictions on a certain regression coefficient vector. Now, suppose that we also wanted to test for a shift in the variance of the error term in the model, perhaps at some known point(s) in the sample. It is well known[†] that many tests of the latter type (e.g., those of Goldfeld and Quandt 1965, Harrison and McCabe 1979, and Breusch and Pagan 1979) are independent (at least asymptotically in some cases) of the usual test for exact restrictions on the regression coefficients. Once again, this eases the task of controlling the overall significance level, if we are testing for two types of model misspecification concurrently but do not wish to construct "omnibus tests."

The discussion assumes, implicitly, that the two (or more) variable-addition tests in question are not only independent but that they are applied separately but "concurrently." By the latter, we mean that each test is applied regardless of the outcome(s) of the other test(s). That is, we are abstracting from any "pretest testing" issues. Of course, in practice, this may be too strong an assumption. A particular variable-addition test (such as the RESET test) which relates to the specification of the conditional mean of the model might be applied only if the model "passes" another (separate and perhaps independent) test of the specification of the conditional variance of the model. If it fails the latter test, then a *different* variable addition test for the specification of the mean may be appropriate. That is, the choice of the *form* of one test may be contingent on the outcome of a prior test of a different feature of the model's specification. Even if the first-stage and second-stage tests are independent, there remains a "pretesting problem" of substance.[‡] The *true* significance level of

---

[*]Essentially, this follows from Basu's (1955) independence theorem. For example, see Anderson (1971, pp. 34–43, 116–134, 270–276) and the asymptotic extensions discussed by Mizon (1977a, 1977b).

[†]For some general discussion of this point, see Phillips and McCabe (1983), Pagan and Hall (1983), and Pagan (1984, pp. 116–117, 125–127). Phillips and McCabe (1989) also provide extensions to other tests where the statistics can be expressed as ratios of quadratic forms in a normal random vector.

[‡]For a more comprehensive discussion of this point, see Giles and Giles (1993, pp. 176–180).

the two-part test for the specification of the conditional mean of the model will differ from the sizes nominally assigned to either of its component parts, because of the randomization of the choice of second-stage test which results from the application of the first-stage test (for the specification of the conditional variance of the model).

More generally, specification tests of the variable-addition type may *not* be independent of each other. As noted by Pagan (1984, pp. 125–127), it is unusual for tests which focus on the same conditional moment to be mutually independent. One is more likely to encounter such independence between tests which relate to *different* moments of the underlying process (as in the discussion above). In such cases there are essentially two options open. The first is to construct joint variable-addition tests of the various forms of misspecification that are of interest. This may be a somewhat daunting task, and although some progress along these lines has been made (e.g., Bera and Jarque 1982), there is still little allowance for this in the standard econometrics computer packages. The second option is to apply separate variable-addition tests for the individual types of model misspecification, and then adopt an "induced testing strategy" by rejecting the model if at least one of the individual test statistics is significant. Generally, in view of the associated nonindependence and the likely complexity of the joint distribution of the individual test statistics, the best that one can do is to compute bounds on the overall significance level for the "induced test." The standard approach in this case would be to use Bonferroni inequalities (e.g., David 1981, Schwager 1984), though generally such bounds may be quite wide and, hence, relatively uninformative. A brief discussion of some related issues is given by Krämer and Sonnberger (1986, pp. 147–155), and Savin (1984) deals specifically with the relationship between multiple t-tests and the F-test. This, of course, is directly relevant to the case of certain variable-addition tests for the specification of the model's conditional mean.

E.   Other Distributional Issues

There are several further distributional issues which are important in the context of variable-addition tests. In view of our subsequent emphasis on the RESET test in this chapter, it is convenient and appropriate to explore these issues briefly in the context of tests which focus on the conditional mean of the model. However, it should be recognized that the general points that are made in the rest of this section also apply to variable-addition tests relating to other moments of the data-generating process. Under our earlier assumptions, the basic form of the test in which we are now interested is an F-test of $\gamma = 0$ in the context of model (2). In that model, if W is truly the precise representation of the omitted effects, then the F-test will be UMP. Working, instead, with the matrix of proxy variables W^*, as in (3), does not affect the null distribution of the test statistic in general, but it does affect the power of the test, of course. Indeed, the reduction in power associated with the use of the proxy variables increases as the correlations between the columns of W and those of W^* decrease. Ohtani and Giles (1993) provide some exact results relating to this phenomenon under very general distributional assumptions, and find the reduction in power to be more pronounced as the error distribution departs from normality. They also show that, regardless of the degree of nonnormality, the test can be biased* as the hypothesis error grows, and they prove that the usual null distribution for the F-statistic for testing if $\gamma = 0$ still holds even under these more general conditions.

Of course, in practice, the whole point of the analysis is that the existence, form, and degree of model misspecification are unknown. Although the general form of W^* will be chosen to reflect the type of misspecification against which one is testing, the extent to which W^* is a "good" proxy for W (and hence for the omitted effect) will not be able to be determined exactly. This being the case, in general it is difficult to make specific statements about the power of such variable-addition tests. As long as W^* is correlated with W (asymptotically), a variable-addition test based on model (3) will be consistent. That is, for a given degree of specification error, as the sample size grows the power of the test will approach unity. In view of the immediately preceding comments, the convergence path will depend on the forms of W and W^*.

The essential consistency of a basic variable-addition test of $\gamma = 0$ in (2) is readily established. Following Eastwood and Godfrey (1992, pp. 123–125), and assuming that $X'X$, $W'W$, and $X'W$ are each $O_p(T)$, the consistency of a test based on F^w (as defined in Section II.D, and assuming independent and homoskedastic disturbances) is ensured if $\text{plim}(F^w/T) \neq 0$ under the alternative. Now, as is well known, we can write

$$F^w = \frac{(\text{RSS} - \text{USS})/p}{\text{USS}/(T - k - p)} \tag{11}$$

where RSS denotes the sum of the squared residuals when (2) is estimated by OLS subject to the restriction that $\gamma = 0$, and USS denotes the corresponding sum of squares when (2) is estimated by unrestricted OLS. Under quite weak conditions the denominator in (11) converges in probability to σ^2 (by Khintchine's theorem). So, by Slutsky's theorem, in order to show that the test is consistent, it is sufficient to establish that plim (RSS - USS)/$T \neq 0$ under the alternative.[†] Now, for our problem we can write

$$\frac{\text{RSS} - \text{USS}}{T} = \gamma' \left[R \left(\frac{X^{*'}X^*}{T} \right)^{-1} R' \right]^{-1} \gamma \tag{12}$$

---

*A biased test is one whose power can fall below its significance level in some region of the relevant parameter space.

[†] Clearly, this plim is zero if the null is true, because then both RSS/T and USS/T are consistent estimators of σ^2.

where $R = [I_k, O_p]$, and $X^* = (X : W)$. Given our assumption about the orders in probability of the data matrices, we can write $\text{plim}(X^{*\prime}X^*/T) = Q^*$, say, where Q^* is finite and nonsingular. Then it follows immediately from (12) that $\text{plim}[(\text{RSS} - \text{USS})/T] > 0$, if $\gamma \neq 0$, so the test is consistent. It is now clear why consistency is retained if W^* is substituted for W, as long as these two matrices are asymptotically correlated. It is also clear that this result will still hold even if W is random or if W^* is random (as in the case of a RESET test involving some function of the OLS prediction vector from (1) in the construction of W^*).

Godfrey (1988, pp. 102–106) discusses another important issue that arises in this context, and which is highly pertinent for our own analysis of the RESET test in this chapter. In general, if we test one of the specifications in (2)–(4) against model (1), by testing if $\gamma = 0$, it is likely that in fact the true data-generating process differs from both the null and maintained hypotheses. That is, we will generally be testing against an incorrect alternative specification. In such cases, the determination of even the asymptotic power of the usual F-test (or its large-sample chi-square counterpart) is not straightforward, and is best approached by considering a sequence of local alternatives. Not surprisingly, it turns out that the asymptotic power of the test depends on the (unknown) extent to which the maintained model differs from the true data-generating process.

Of course, in practice the errors in the model may be serially correlated and/or heteroskedastic, in which case variable-addition tests of this type generally will be *inconsistent*, and their power properties need to be considered afresh in either large or small samples. Early work by Thursby (1979, 1982) suggested that the RESET test might be robust to autocorrelated errors, but as noted by Pagan (1984, p. 127) and explored by Porter and Kashyap (1984), this is clearly not the case. We abstract from this situation in the development of a new version of the RESET test later in this chapter, but it is a topic that is being dealt with in some detail in our current research. Finally, we should keep in mind that a consistent test need not necessarily have high power in small samples, so this remains an issue of substance when considering specific variable-addition tests.

Finally, it is worth commenting on the problem of discriminating between two or more variable-addition tests, each of which is consistent in the sense described above. If the significance level for the tests is fixed (as opposed to being allowed to decrease as the sample size increases), then there are at least two fairly standard ways of dealing with this issue. These involve bounding the powers of the tests away from unity as the sample size grows without limit. The first approach is to use the "approximate slope" analysis of Bahadur (1960, 1967). This amounts to determining how the *asymptotic* significance level of the test must be reduced if the power of the test is to be held constant under some fixed alternative. The test statistics are $O_p(T)$, and they are compared against critical values which increase with T. The approximate slope for a test statistic, S, which is asymptotically chi-square, is just plim(S/T).* This is the same quantity that we considered in the determination of the consistency of such a test. A choice between two consistent tests which test the same null against the same alternative can be made by selecting so as to maximize the approximate slope. Of course, once again this does not guarantee good power properties in small samples.

A second approach which may be used to discriminate between consistent tests is to consider a sequence of local alternatives. In this approach, the alternative hypothesis is adjusted so that it approaches the null hypothesis in a manner that ensures that the test statistics are $O_p(1)$. Then a fixed critical value can be used, and the asymptotic powers of the tests that can be compared as they will each be less than unity. In our discussion of the traditional RESET test in the next section and of a new variant of this test in Section IV, neither of these approaches to discriminating between consistent tests is particularly helpful. There, the tests that are being compared have the same algebraic construction, but they differ in terms of the number of columns in W^* and the way in which these columns are constructed from the OLS prediction vector associated with the basic model. In general, the "approximate slope" and "local alternatives" approaches do not provide tractable expressions upon which to base clear comparisons in this case. However, in practical terms our interest lies in the small-sample properties of the tests, and it is on this characteristic that we focus.

III. THE RESET TEST

Among the many "diagnostic tests" that econometricians routinely use, some variant or other of the RESET test is widely employed to test for a nonzero mean of the error term. That is, it tests implicitly whether a regression model is correctly specified in terms of the regressors that have been included. Among the reasons for the popularity of this test are that it is easily implemented and that it is an *exact* test whose statistic follows an F-distribution under the null. The construction of the test does, however, require a choice to be made over the nature of certain "augmenting regressors" that are employed to model the misspecification, as we saw in Section II.B. Depending on this choice, the RESET test statistic has a nonnull distribution which may be doubly noncentral F or totally nonstandard. Although this has no bearing on the size of the test, it has obvious implications for its power.

The most common construction of the RESET test involves augmenting the regression of interest with powers of the prediction vector from a least-squares regression of the original specification and testing their joint significance. As a result

---

*For example, see Geweke (1981), Magee (1987), and Eastwood and Godfrey (1992, p. 132).

of the Monte Carlo evidence provided by Ramsey and Gilbert (1972) and Thursby (1989), for example, it is common for the second, third, and fourth powers of the prediction vector to be used in this way.* Essentially, Ramsey's original suggestion, following earlier work by Anscombe (1961), involves approximating the unknown nonzero mean of the errors, which reflects the extent of the model misspecification, by some analytic function of the conditional mean of the model. The specific construction of the RESET test noted above then invokes a polynomial approximation, with the least-squares estimator of the conditional mean replacing its true counterpart.

Other possibilities include using powers and/or cross products of the individual regressors, rather than powers of the prediction vector, to form the augmenting terms. Thursby and Schmidt (1977) provide simulation results which appear to favor this approach. However, all of the variants of the RESET test that have been proposed to date appear to rely on the use of *local* approximations, essentially of a Taylor series type, of the conditional mean of the regression. Intuitively, there may be gains in terms of the test's performance if a *global* approximation were used instead. This chapter pursues this intuition by suggesting the use of an (essentially unbiased) Fourier flexible approximation. This suggestion captures the spirit of the development of cost and production function modeling, and the associated transition from polynomial functions (e.g., Johnston 1960) to Translog functions (e.g., Christensen et al. 1971, 1973) and then to Fourier functional forms (e.g., Gallant 1981, Mitchell and Onvural 1995, 1996).

Although Ramsey (1969) proposed a battery of specification tests for the linear regression model, with the passage of time and the associated development of the testing literature, the RESET test is the one which has survived. Ramsey's original discussion was based on the use of Theil's (1965, 1968) "BLUS" residuals, but the analysis was subsequently recast in terms of the usual OLS residuals (e.g., Ramsey and Schmidt 1976, Ramsey 1983), and we will follow the latter convention in this chapter. As Godfrey (1988, p. 106) emphasizes, one of the principles which underlies the RESET test is that the researcher has only the same amount of information available when testing the specification of a regression model as was available when the model was originally formulated and estimated. Accordingly direct tests against new theories, perhaps embodying additional variables, are ruled out.

A convenient way of discussing and implementing the standard RESET test is as follows. Suppose that the regression model under consideration is (1), which we reproduce here:

$$y = X\beta + u \tag{13}$$

---

*For instance, the SHAZAM (1993) package adopts this approach. Clearly, the first power cannot be used as an extra regressor in the "augmented" equation as the design matrix would then be perfectly collinear.

where X is $T \times k$, of rank k, and nonstochastic; and it is *assumed* that $u \sim N[0, \sigma^2 I_T]$. Such a model would be "correctly specified." Now, suppose we allow for the possibility that the model is misspecified through the omission of relevant regressors or the wrong choice of functional form. In this case, $E[u \mid X] = \xi \neq 0$. The basis of the RESET test is to approximate ξ by $Z\theta$, which corresponds to $W^*\gamma$ in (3), and fit an augmented model,

$$y = X\beta + Z\theta + \varepsilon \tag{14}$$

We then test if $\xi = 0$ by testing if $\theta = 0$, using the usual F-test for restrictions on a subset of the regression coefficients. Different choices of the $T \times p$ matrix Z lead to different variants of the RESET test. As noted, the most common choice is to construct Z to have tth row vector

$$Z_t = [(Xb)_t^2, (Xb)_t^3, \ldots, (Xb)_t^{p+1}] \tag{15}$$

where often $p = 3$; $b = (X'X)^{-1}X'y$ is the OLS estimator of β from (1); and $(Xb)_t$ is therefore the tth element of the associated prediction vector, $\hat{y} = Xb$.

If Z is chosen in this way (or if it is nonrandom, or if it depends on y only through b), then, as a consequence of the Milliken-Graybill (1970) theorem, the RESET test statistic is F-distributed with p and $T - k - p$ degrees of freedom under the null that $\theta = 0$,[*] provided that the disturbances are NID$(0, \sigma^2)$. If Z is nonstochastic (as would be the case if Thursby and Schmidt's 1977 proposal of using powers of the regressors to augment the model were followed) then the test statistic's nonnull distribution is doubly noncentral F with the same degrees of freedom[†] and with numerator and denominator noncentrality parameters:

$$\lambda_1 = \frac{\xi' M_x Z (Z' M_x Z)^{-1} Z' M_x \xi}{\sigma^2} \tag{16}$$

$$\lambda_2 = \frac{\xi'[M_x - M_x Z (Z' M_x Z)^{-1} Z' M_x]\xi}{\sigma^2} \tag{17}$$

In this case, the power of the RESET test can be computed exactly for any given degree of misspecification, ξ, by recognizing that

$$\Pr[F(p, T - k - p; \lambda_1, \lambda_2) < c]$$

$$= \Pr\left[\frac{1}{p}\chi^2(p; \lambda_1) - \left(\frac{c}{T - k - p}\right)\chi^2(T - k - p; \lambda_2) < 0\right] \tag{18}$$

---

[*]Note that ξ is being *approximated* by $Z\theta$.
[†]For example, see Ramsey (1969) and Thursby and Schmidt (1977).

where F denotes a doubly noncentral F-statistic, and χ^2 denotes a noncentral chi-square statistic, each with degrees of freedom and noncentrality parameters as shown. The algorithm of Davies (1980), which is conveniently coded in the SHAZAM (1993) package, provides an efficient and accurate means of computing such probabilities, although we are not aware of any studies which do this in the context of the RESET test.

If, as is usual in practice, Z is constructed from powers of Xb or is random for any other reason, then λ_1 and λ_2 will be random and the nonnull distribution of the RESET test statistic will no longer be doubly noncentral F. The power characteristics associated with its nonstandard distribution will depend on the specific choice of Z (and hence on the number of powers of Xb that may be used, if this choice is adopted), and are then best explored via Monte Carlo simulation. This is the approach that we adopt in this chapter.

IV. FOURIER APPROXIMATIONS AND THE FRESET TEST

As noted, the essence of Ramsey's RESET test is to approximate the unknown ξ by some analytic function of $X\beta$ (or, more precisely, of the observable $\hat{y} = Xb$). A power series approximation is one obvious possibility, but there are other approximations which may be more attractive. In particular, as Gallant (1981) has argued in a different context, one weakness of Taylor series approximations (which include the power series and Translog approximations, for example) is that they have only *local* validity. Taylor's theorem is valid only in a neighborhood of some unspecified size containing a specific value of the argument of the function to be approximated. On the other hand, a Fourier approximation has *global* validity. Such approximations can take the form of a conventional sine/cosine expansion or the less conventional Jacobi, Laguerre, or Hermite expansions. We consider using a sine/cosine expansion of Xb to approximate ξ.* Although Gallant (1981) has suggested that the precision of a truncated Fourier series can generally be improved by adding a second-order Taylor series approximation (see also Mitchell and Onvural 1995), we do not pursue this refinement here.

In order to obtain a Fourier representation of $Z\theta$ in (14), and hence a Fourier approximation of the unknown ξ, by using Xb, we need to first transform the elements of this vector to lie in an interval of length 2π, such as $[-\pi, +\pi]$. This is because a Fourier representation is defined only if the domain of the function is in an inter-

---

*In view of the Monte Carlo evidence provided by Thursby and Schmidt (1977), in principle it would also be interesting to consider multivariate Fourier expansions in terms of the original regressors.

val of length 2π. Mitchell and Onvural (1995, 1996) and other authors use a linear transformation.* In our case, this amounts to constructing

$$w_t = \frac{\pi\{2(Xb)_t - [(Xb)_{max} + (Xb)_{min}]\}}{(Xb)_{max} - (Xb)_{min}} \tag{19}$$

where $(Xb)_{min}$ and $(Xb)_{max}$ are respectively the smallest and largest elements of the prediction vector. We also consider an alternative sinusoidal transformation, based on Box (1966):

$$w_t = 2\pi \, \sin^2[(Xb)_t] - \pi \tag{20}$$

Then the Z matrix for (2) is constructed to have tth row vector

$$Z_t = [\sin(w_t), \cos(w_t), \sin(2w_t), \cos(2w_t), \sin(3w_t), \\ \cos(3w_t), \ldots, \sin(p'w_t), \cos(p'w_t)] \tag{21}$$

for some arbitrary truncation level p'. This recognizes that the Fourier approximation, $g(x)$, of a continuously differentiable function, $f(x)$, is

$$g(x) = u_0 + \sum [u_j \cos(jx) + v_j \sin(jx)] \tag{22}$$

where[†]

$$u_0 = \frac{1}{2\pi} \int f(x) \, dx,$$

$$u_j = \frac{1}{\pi} \int f(x) \cos(jx) dx, \quad v_j = \frac{1}{\pi} \int f(x) \sin(jx) dx \tag{23}$$

Equation (22) gives an exact representation of $f(x)$ by $g(x)$, except near $x = -\pi$ and $x = +\pi$. An approximation is obtained by truncating the range of summation in (22) to a finite number of terms, p', and this approximation will be *globally* valid. The failure of the representation at the exact endpoints of the $[-\pi, +\pi]$ interval can generate an approximation error which is often referred to as "Gibbs' phenomenon." Accordingly, some authors modify transformations such as (19) or (20) to achieve a range which is just inside this interval. We have experimented with this refinement to the above transformations and have found that it makes negligible difference in the case considered. The results reported in Section VI do not incorporate this refinement.

---

*If the data have to be positive, as in the case of cost functions, then a positive interval such as $[c, c+2\pi]$ would be appropriate.

†The range of summation in (22) is from $j = 1$ to $j = \infty$; the ranges of integration in (23) are each from $-\pi$ to $+\pi$.

In our case, the u_j's and v_j's in (23) correspond to elements of θ in(14), so they are estimated as part of the testing procedure.* Note that Z is random here, but only through b. Our FRESET test involves constructing Z in this way and then testing if $\theta = 0$ in (14). Under this null, the FRESET test statistic is still central F with $2p'$ and $t - k - 2p'$ degrees of freedom. Its nonnull distribution will depend upon the form of the model misspecification, the nature of the regressors, and the choice of p'. This distribution will differ from that of the RESET test statistic based on a (truncated) Taylor series of Xb. In the following discussion we use the titles FRESETL and FRESETS to refer to the FRESET tests based on the linear transformation (19) and the sinusoidal transformation (20) respectively.

V. A MONTE CARLO EXPERIMENT

We have undertaken a Monte Carlo experiment to compare the properties of the new FRESET tests with those of the conventional RESET test for some different types of model misspecification. Table 1 shows the different formulations of the tests that we have considered in all parts of the experiment. Effectively, we have considered choices[†] of $p = 1, 2,$ or 3 and $p' = 2$ or 3. In the case of the RESET test, the variables whose significance is tested (i.e., the "extra" Z variables which are added to the basic model) comprise powers of the prediction vector from the *original* model under test, as in (15). For the FRESET test they comprise sines and cosines of multiples of this vector, as in (21), once the linear transformation (19) or the sinusoidal transformation (20) has been used for the FRESETL and FRESETS variants respectively.

Three models form the basis of our experiment, and these are summarized in Table 2. In each case we show a particular data-generating process (DGP), or "true" model specification, together with the model that is actually fitted to the data. The latter "null" model is the one whose specification is being tested. Our model 1 allows for misspecification through static variable omission, and corresponds to models 6–8 (depending on the value of γ) of Thursby and Schmidt (1977, p. 638). Our model 2 allows for a static misspecification of the functional form, and our model 3 involves the omission of a dynamic effect. In each case, x_2, x_3, and x_4 are as in Ramsey and

---

*The parameter u_0 gets "absorbed" into the coefficient of the intercept in the model (14).

[†] We found that setting $p' > 3$ resulted in a singular matrix when constructing the FRESET tests. Eastwood and Gallant (1991) suggest that setting the number of parameters equal to the sample size raised to the two-thirds power will ensure consistency and asymptotic normality when estimating a Fourier function. Setting $p' = 2$ or $p' = 3$ is broadly in keeping with this for our sample sizes. As Mitchell and Onvural (1996) note, increasing p' will increase the variance of test statistics. In the context of the FRESET test it seems wise to limit the value of p'.

Table 1   Test Variables

Test	p	Test variables
RESET	1	1) $\hat{y}_t^2$
	2	2) $\hat{y}_t^2, \hat{y}_t^3$
	3	3) $\hat{y}_t^2, \hat{y}_t^3$, and $\hat{y}_t^4$
	p'	
FRESETL	2	1) $\sin(w_t), \cos(w_t), \sin(2w_t), \cos(2w_t)$
(linear transformation)	3	2) $\sin(w_t), \cos(w_t), \sin(2w_t), \cos(2w_t),$ $\sin(3w_t), \cos(3w_t)$
FRESETS	2	3) $\sin(w_t), \cos(w_t), \sin(2w_t), \cos(2w_t)$
(sinusoidal transformation)	3	4) $\sin(w_t), \cos(w_t), \sin(2w_t), \cos(2w_t),$ $\sin(3w_t), \cos(3w_t)$

Table 2   Models

Model	Specification	Problem
1	DGP: $y_t = 1.0 - 0.4x_{3t} + x_{4t} + \gamma x_{2t} + v_t$	Omitted variable
	Null: $y_t = \beta_0 + \beta_3 x_{3t} + \beta_4 x_{4t} + v_t$	(omitted static effect)
2	DGP: $y_t = 1.0 - 0.4x_{3t} + x_{4t}(1 + \gamma x_{2t}) + v_t$	Incorrect functional form.
	Null: $y_t = \beta_0 + \beta_3 x_{3t} + \beta_4 x_{4t} + v_t$	(omitted multiplicative effect)
3	DGP: $y_t = 1.0 - 0.4x_{3t} + x_{4t} + \gamma y_{t-1} + v_t$	Incorrect functional form
	Null: $y_t = \beta_0 + \beta_3 x_{3t} + \beta_4 x_{4t} + v_t$	(omitted dynamic effect)

Gilbert (1972) and Thursby and Schmidt (1977),* and sample sizes of $T = 20$ and $T = 50$ have been considered.

Various values of γ were considered in the range $[-8.0, +8.0]$ in models 1 and 2, though in the latter the graphs and tables reported relate to a "narrower" range as the results "stabilize" quite quickly. Values of γ in the (stationary) range $[-0.9, +0.9]$ were considered in model 3. If $\gamma = 0$ the fitted (null) model is correctly specified. Other values of γ generate varying degrees of model misspecification, and we are interested in the probability that each test rejects the null model (by rejecting the null hypothesis that $\theta = 0$ in (14)) when $\gamma \neq 0$. For convenience, we will term these rejection rates "powers" in the ensuing discussion. However, care should be

---

*Ramsey and Gilbert (1972, p. 185) provide data for two series, x_1 and x_2, for a sample size of $T = 10$. (We follow them in "repeating" these values to generate regressors which are "fixed in repeated samples" when considering samples of size $T = 20, 50$.) As in Thursby and Schmidt (1977), $x_3 = x_1 + x_2$; and $x_4 = x_1^2/10$.

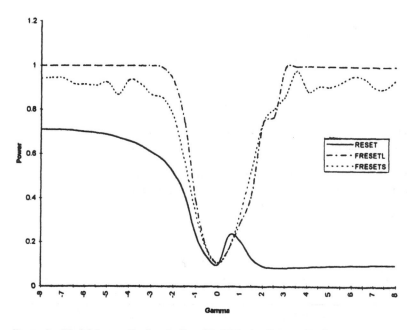

Figure 1   Model 1, $p = 3$, $p' = 3$, $T = 20$, 10% significance level.

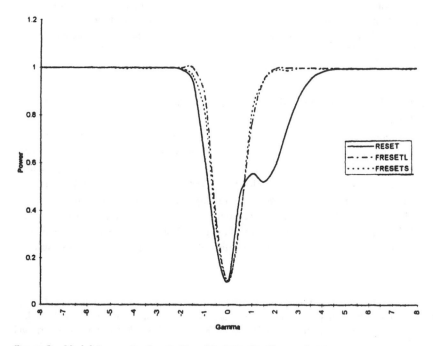

Figure 2   Model 1, $p = 3$, $p' = 3$, $T = 50$, 10% significance level.

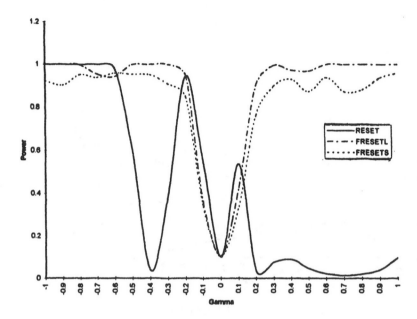

Figure 3 Model 2, $p = 3$, $p' = 3$, $T = 20$, 10% significance level.

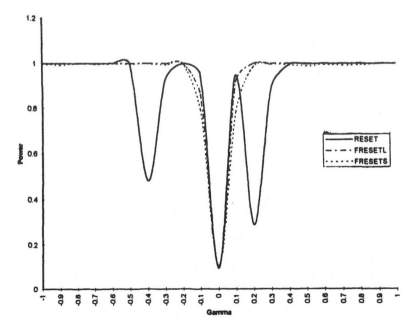

Figure 4 Model 2, $p = 3$, $p' = 3$, $T = 50$, 10% significance level.

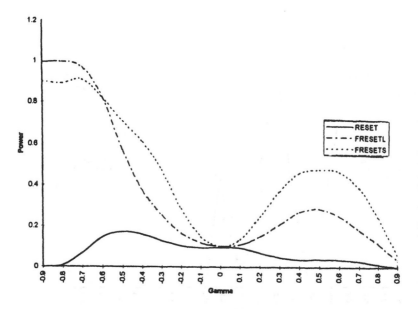

Figure 5   Model 3, $p = 3$, $p' = 3$, $T = 20$, 10% significance level.

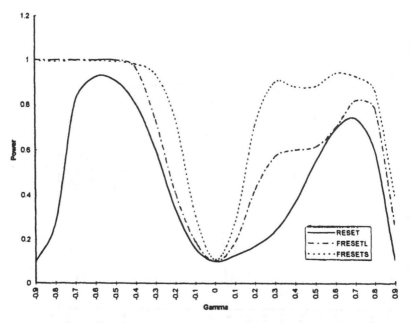

Figure 6   Model 3, $p = 3$, $p' = 3$, $T = 50$, 10% significance level.

taken over their interpretation in the present context. Strictly, of course, the power of the RESET or FRESET test is the rejection probability when $\theta \neq 0$. As noted in Section II, this power can be determined (in principle), as the test statistics are typically doubly noncentral F when $\theta \neq 0$. Only in the very special case where $Z\theta$ in (14) exactly coincides with the specification error, ξ, would "powers" of the sort that we are computing and reporting actually correspond to the *formal* powers of the tests. (In model 1, for example, this would require that the RESET test be applied by using just x_2, fortuitously, as the only "augmenting" variable rather than augmenting the model with powers of the prediction vector.) Accordingly, it is not surprising that, in general, the shapes of the various graphs reported in the next section do not accord with that for the *true* power curve for an F-test.

The error term, ν_t was generated to be standard normal, though of course the tests are scale invariant and so the results are invariant to the value of the true error variance. The tests were conducted at the 5% and 10% significance levels. As the RESET and FRESET test statistics are exactly F-distributed if $\gamma = 0$, there is no "size distortion" if the appropriate F critical values are used—the nominal and true significance levels coincide. Precise critical values were generated using the Davies (1980) algorithm as coded in the DISTRIB command in the SHAZAM (1993) package. Each component of the experiment is based on 5000 replications. Accordingly, from the properties of a binomial proportion, the standard error associated with a rejection probability, π (in Tables 3–5), takes the value $[\pi(1 - \pi)/5000]^{1/2}$, which takes its maximum value of 0.0071 when $\gamma = 0.5$. The simulations were undertaken using SHAZAM code on both a PC and a DEC Alpha 3000/400.

VI. MONTE CARLO RESULTS

In this section, we present the results of Monte Carlo experiments designed to gather evidence on the power of the various RESET and the above FRESET tests in Table 1 for each model in Table 2. The experimental results and the graphs of the rejection rates are given below. For convenience we will refer to the graphs below as "power" curves. As discussed, it is important to note that these are not conventional power curves. Only results for case 3 for the RESET and case 2 for the FRESET tests in Table 1 are presented in detail, for reasons which will become evident. The entries in Tables 3–5 are the proportions of rejections of the null hypothesis. Not surprisingly, for all models considered, the RESET, FRESETL, and FRESETS tests exhibit higher rejection rates at the 10% significance level than at the 5% level. However, the "pattern" of the power curves is insensitive to the choice of significance level, hence we will focus on the 10% significance level in the remainder of the discussion.

Generally, the patterns of the power curves differ only marginally when the sample size is increased from 20 to 50. The results for a sample size of 50 display higher power than the comparable sample-size-20 results, reflecting the consistency

Table 3 Model 1, $p = 3$, $p' = 3$

| | $T = 20$ | | | | | | $T = 50$ | | | | | |
| | 5% | | | 10% | | | 5% | | | 10% | | |
γ	RESET	FRESETS	FRESETL	RESET	FRESETS	FRESETL	RESET	FRESETS	FRESETL	RESET	FRESETS	FRESETL
-8.0	0.188	0.893	1.000	0.710	1.000	0.942	1.000	0.998	1.000	1.000	1.000	0.998
-7.5	0.209	0.909	1.000	0.709	1.000	0.946	1.000	0.999	1.000	1.000	1.000	0.999
-7.0	0.233	0.913	1.000	0.708	1.000	0.944	1.000	0.999	1.000	1.000	1.000	1.000
-6.5	0.258	0.874	1.000	0.705	1.000	0.918	1.000	0.998	1.000	1.000	1.000	0.999
-6.0	0.284	0.868	1.000	0.700	1.000	0.915	1.000	0.996	1.000	1.000	1.000	0.999
-5.5	0.309	0.870	1.000	0.695	1.000	0.911	1.000	0.995	1.000	1.000	1.000	0.998
-5.0	0.331	0.879	1.000	0.687	1.000	0.925	1.000	0.998	1.000	1.000	1.000	0.999
-4.5	0.350	0.801	1.000	0.674	1.000	0.869	1.000	0.992	1.000	1.000	1.000	0.996
-4.0	0.366	0.896	1.000	0.658	1.000	0.937	1.000	0.998	1.000	1.000	1.000	0.998
-3.5	0.378	0.872	1.000	0.639	1.000	0.918	1.000	0.994	1.000	1.000	1.000	0.996
-3.0	0.384	0.795	0.997	0.609	1.000	0.870	0.999	0.998	1.000	1.000	1.000	0.998
-2.5	0.369	0.767	0.965	0.576	0.992	0.854	0.996	0.995	1.000	0.999	1.000	0.998
-2.0	0.337	0.646	0.850	0.516	0.930	0.766	0.975	0.991	1.000	0.992	1.000	0.995
-1.5	0.266	0.414	0.580	0.419	0.738	0.565	0.866	0.941	0.997	0.941	0.998	0.965

−1.0	0.137	0.216	0.254	0.235	0.392	0.334	0.479	0.706	0.791	0.617	0.882	0.806
−0.5	0.064	0.092	0.091	0.133	0.168	0.161	0.152	0.207	0.233	0.247	0.352	0.315
0.0	0.046	0.050	0.054	0.101	0.106	0.105	0.052	0.054	0.053	0.104	0.102	0.101
0.5	0.141	0.095	0.096	0.233	0.173	0.173	0.339	0.237	0.249	0.471	0.366	0.355
1.0	0.117	0.204	0.176	0.206	0.289	0.332	0.409	0.713	0.643	0.556	0.766	0.817
1.5	0.061	0.381	0.267	0.126	0.413	0.528	0.365	0.916	0.907	0.524	0.954	0.953
2.0	0.037	0.617	0.608	0.091	0.745	0.749	0.406	0.987	0.996	0.593	0.999	0.992
2.5	0.027	0.720	0.679	0.083	0.771	0.807	0.559	0.980	0.997	0.748	1.000	0.989
3.0	0.021	0.778	0.983	0.085	0.994	0.866	0.727	0.998	1.000	0.881	1.000	0.999
3.5	0.019	0.960	0.992	0.087	0.999	0.979	0.854	1.000	1.000	0.956	1.000	1.000
4.0	0.017	0.829	0.999	0.089	1.000	0.387	0.935	0.995	1.000	0.987	1.000	0.997
4.5	0.014	0.853	1.000	0.091	1.000	0.910	0.976	0.998	1.000	0.998	1.000	0.999
5.0	0.011	0.857	1.000	0.093	1.000	0.908	0.993	0.995	1.000	0.999	1.000	0.996
5.5	0.009	0.879	1.000	0.094	1.000	0.929	0.998	0.996	1.000	1.000	1.000	0.997
6.0	0.007	0.915	1.000	0.095	1.000	0.956	0.999	1.000	1.000	1.000	1.000	1.000
6.5	0.005	0.920	1.000	0.096	1.000	0.949	1.000	0.999	1.000	1.000	1.000	0.999
7.0	0.005	0.846	1.000	0.097	1.000	0.907	1.000	0.999	1.000	1.000	1.000	1.000
7.5	0.003	0.811	1.000	0.098	1.000	0.903	1.000	0.994	1.000	1.000	1.000	0.997
8.0	0.002	0.913	1.000	0.096	1.000	0.946	1.000	0.999	1.000	1.000	1.000	0.999

of the tests. This is also in accord with the fact that the larger sample size yields "smoother" power curves in the FRESETL and FRESETS cases for models 1 and 2, as in Figs. 1 to 4.

The probability of rejecting a true null hypothesis when specification error is present depends, in part, on the number of variables included in Z_t. In general, our results indicate, regardless of the type of misspecification, that the use of $p = 3$ in the construction of Z_t as in (15) yields the most powerful RESET test. However, this does not always hold, such as in model 3 where the RESET test with only the term $\hat{y}_t^2$ included in the auxiliary regression yields higher power for relatively large positive misspecification ($\gamma > 0.3$) and large negative misspecification ($\gamma < -0.7$).

The FRESETL and FRESETS tests with $p' = 3$ terms are generally the most powerful of these tests. The pattern of the power curves tends to fluctuate less and the results indicate higher rejection rates than in the comparable $p' = 2$ case. This is not surprising, as we would expect a better degree of approximation to the omitted effect as more terms are included. However, the ability to increase the number of test variables included in the auxiliary regression is constrained by the degrees of freedom. We focus primarily on the RESET test with $p = 3$ and the FRESET tests with $p' = 3$.

In all cases, the FRESET tests perform equally as well as, and in many cases yield higher powers than, the comparable RESET tests. A comparison of the rejection rates of the various tests for the three models considered indicates FRESETL is the most powerful test for models 1 and 2. The FRESETS test yields higher power for model 3 than the FRESETL and RESET tests, with the exception of high levels of misspecification, where FRESETL exhibits higher rejection rates. The FRESETS test yields higher rejection rates than the comparable RESET test for models 1 and 2, with two exceptions. First, model 1 in the presence of a high degree of misspecification; second, model 2 in the presence of positive levels of misspecification ($\gamma > 0$). However, FRESETL yields higher rejection rates than the RESET test for the two exceptions. The FRESETL test dominates the RESET test for model 3, as in Figs. 5 and 6.

The power of the RESET test is excellent for models 1 and 2, and $p = 3$, with sample size 50. Then for larger coefficients of the omitted variable, the proportion of rejections increases to 100%. For model 1, the use of the squares, cubes, and fourth powers of the predicted values as the test variables for the RESET test results in power which generally increases as the coefficient of the omitted variable becomes increasingly negative. In the presence of positive coefficients of the omitted variable, the rejection rate generally increases initially as the level of misspecification increases but decreases as the coefficient of the omitted variable continues to increase. However, power begins to marginally increase again at moderate levels of misspecification ($\gamma = 3$).

Our results for model 2 indicate the power of the RESET test increases as the coefficient of the omitted variable increases for lower and higher levels of misspec-

Table 4 Model 2, $p = 3$, $p' = 3$

	T = 20						T = 50					
	5%			10%			5%			10%		
γ	RESET	FRESETS	FRESETL	RESET	FRESETS	FRESETL	RESET	FRESETS	FRESETL	RESET	FRESETS	FRESETL
-1.0	1.000	0.891	1.000	1.000	1.000	0.922	1.000	0.995	1.000	1.000	1.000	0.998
-0.9	1.000	0.865	1.000	1.000	1.000	0.902	1.000	0.989	1.000	1.000	1.000	0.992
-0.8	1.000	0.924	0.994	1.000	0.999	0.950	1.000	0.998	1.000	1.000	1.000	0.999
-0.7	0.996	0.910	0.832	1.000	0.954	0.937	1.000	0.994	1.000	1.000	1.000	0.996
-0.6	0.883	0.938	0.843	0.980	0.942	0.958	1.000	0.998	1.000	1.000	1.000	0.998
-0.5	0.339	0.931	0.996	0.604	0.999	0.953	1.000	0.996	1.000	1.000	1.000	0.997
-0.4	0.013	0.926	1.000	0.037	1.000	0.950	0.309	0.998	1.000	0.481	1.000	0.999
-0.3	0.243	0.866	0.999	0.391	1.000	0.909	0.870	0.989	1.000	0.926	1.000	0.994
-0.2	0.880	0.735	0.874	0.945	0.947	0.847	1.000	0.994	1.000	1.000	1.000	0.997
-0.1	0.404	0.221	0.231	0.552	0.363	0.340	0.899	0.697	0.776	0.952	0.862	0.792
0.0	0.046	0.050	0.054	0.101	0.105	0.106	0.043	0.049	0.048	0.094	0.101	0.105
0.1	0.387	0.216	0.290	0.539	0.428	0.339	0.898	0.694	0.847	0.947	0.917	0.785
0.2	0.014	0.647	0.809	0.032	0.902	0.766	0.148	0.975	1.000	0.290	1.000	0.985
0.3	0.022	0.851	0.980	0.076	0.995	0.899	0.729	0.991	1.000	0.888	1.000	0.996
0.4	0.015	0.892	0.945	0.090	0.973	0.930	0.955	0.993	1.000	0.994	1.000	0.995
0.5	0.005	0.825	0.940	0.047	0.969	0.873	0.984	0.981	1.000	0.998	1.000	0.988
0.6	0.001	0.912	1.000	0.021	1.000	0.938	0.994	0.993	1.000	1.000	1.000	0.995
0.7	0.001	0.831	1.000	0.014	1.000	0.873	0.999	0.988	1.000	1.000	1.000	0.992
0.8	0.000	0.840	1.000	0.020	1.000	0.882	1.000	0.991	1.000	1.000	1.000	0.994
0.9	0.001	0.910	1.000	0.041	1.000	0.939	1.000	0.998	1.000	1.000	1.000	0.999
1.0	0.001	0.936	1.000	0.099	1.000	0.960	1.000	0.997	1.000	1.000	1.000	0.999

Table 5 Model 3, $p = 3$, $p' = 3$

| | $T = 20$ | | | | | | $T = 50$ | | | | | |
| | 5% | | | 10% | | | 5% | | | 10% | | |
γ	RESET	FRESETS	FRESETL	RESET	FRESETS	FRESETL	RESET	FRESETS	FRESETL	RESET	FRESETS	FRESETL
-0.9	0.000	0.858	0.989	0.000	0.995	0.899	0.000	0.997	1.000	0.099	1.000	0.998
-0.8	0.001	0.849	0.987	0.009	0.998	0.891	0.065	0.990	1.000	0.278	1.000	0.993
-0.7	0.020	0.851	0.913	0.069	0.970	0.909	0.586	0.995	1.000	0.815	1.000	0.998
-0.6	0.055	0.709	0.672	0.147	0.825	0.819	0.816	0.994	1.000	0.924	1.000	0.996
-0.5	0.073	0.573	0.402	0.173	0.572	0.708	0.801	0.986	0.998	0.908	1.000	0.993
-0.4	0.068	0.482	0.252	0.160	0.376	0.609	0.674	0.964	0.934	0.805	0.967	0.982
-0.3	0.060	0.337	0.159	0.129	0.254	0.474	0.449	0.896	0.631	0.603	0.734	0.936
-0.2	0.052	0.174	0.094	0.104	0.167	0.286	0.222	0.617	0.293	0.338	0.410	0.740
-0.1	0.046	0.077	0.064	0.095	0.120	0.150	0.090	0.200	0.108	0.160	0.191	0.309
0.0	0.045	0.053	0.052	0.096	0.102	0.101	0.059	0.055	0.051	0.101	0.103	0.109
0.1	0.042	0.074	0.057	0.094	0.110	0.136	0.069	0.179	0.110	0.130	0.192	0.285
0.2	0.031	0.156	0.085	0.069	0.150	0.249	0.099	0.591	0.329	0.172	0.427	0.712
0.3	0.020	0.255	0.132	0.047	0.211	0.373	0.142	0.840	0.496	0.241	0.575	0.900
0.4	0.014	0.327	0.181	0.037	0.272	0.462	0.233	0.836	0.527	0.372	0.604	0.882
0.5	0.011	0.360	0.197	0.039	0.287	0.478	0.382	0.839	0.518	0.555	0.618	0.883
0.6	0.011	0.357	0.172	0.037	0.251	0.469	0.524	0.909	0.588	0.695	0.700	0.941
0.7	0.007	0.285	0.122	0.029	0.184	0.388	0.550	0.883	0.722	0.741	0.820	0.928
0.8	0.003	0.162	0.076	0.013	0.115	0.242	0.371	0.786	0.659	0.593	0.781	0.861
0.9	0.000	0.037	0.019	0.002	0.037	0.059	0.034	0.277	0.157	0.100	0.250	0.381

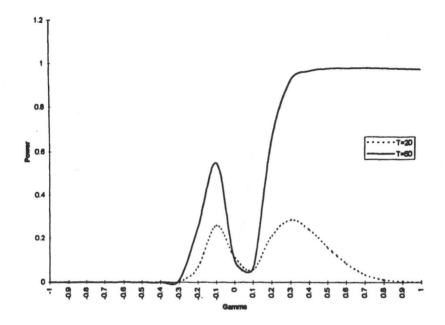

Figure 7   Model 2, RESET test ($p = 1$), 10% significance level.

ification. This result generally holds, but as can be seen by Figure 7, it is possible for power to decrease as the level of misspecification increases. The test yields low power at positive levels of misspecification for a sample size of 20 when there is an omitted multiplicative variable. For model 3 and both sample sizes, the rejection rate initially increases as the coefficient of the omitted variable increases and then falls as the degree of misspecification continues to increase.

The powers of the FRESETL and FRESETS tests are excellent for models 1 and 2, when $p' = 3$. The proportion of rejections increases to 100% as the coefficient of the omitted variable increases, with the exception of the FRESETS test* for model 1. The inclusion of three sine and three cosine terms of the predicted values as the test variables for the FRESETL and FRESETS tests results in power generally increasing as the coefficient of the omitted variable increases for models 1 and 2 with both sample sizes. However, as can be seen by Figure 8, it is possible for the rejection rate to decrease as the coefficient of the omitted variable increases in the $p' = 2$ case. For models 1 and 2 the "power" curve increases at a faster rate and is "smoother" for sample size 50.

---

*For FRESETS, the number of rejections is greater than 90% for higher levels of misspecification.

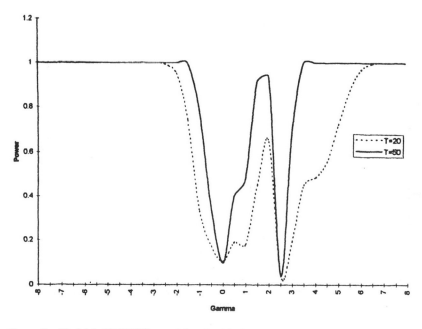

Figure 8   Model 1, FRESETL test ($p' = 2$), 10% significance level.

For model 3, our results indicate that the rejection rate increases initially as the coefficient of the omitted variable increases and then decreases as the level of misspecification continues to increase for positive omitted coefficients of the lagged variable. However, the rejection rate increases as misspecification increases for negative coefficients of the omitted lagged variable.

Finally, we have also considered the inclusion of an irrelevant explanatory variable in order to examine the robustness of the RESET and FRESET tests to an overspecification of the model. We consider model 1 where the DGP now becomes

$$y_t = \beta_0 + \beta_3 x_{3t} + \beta_4 x_{4t} + v_t \tag{24}$$

and the "null" becomes

$$y_t = 1.0 - 0.4x_{3t} + x_{4t} + \gamma x_{2t} + v_t \tag{25}$$

In this case, the coefficient (γ) of the redundant regressor is freely estimated and therefore we cannot consider a range of preassigned γ values. Our results indicate that the power results differ negligibly from the true significance levels, as the rejection rates fall within two maximum standard deviations of the size.* That is, the tests appear to be quite robust to a simple overspecification of the model.

---

*As calculated by binomial proportions.

VII.   CONCLUSIONS

We have considered the problem of testing a regression relationship for possible misspecification in terms of its functional form and/or selection of regressors, in the spirit of the family of variable-addition tests. We have proposed a new variant of Ramsey's (1969) RESET test, which uses a global (rather than local) approximation to the misspecified part of the conditional mean of the model. Rather than basing the variable-addition procedure on polynomial terms of a prediction vector from the basic model, we suggest the use of a Fourier series approximation. Two ways of transforming the predicted values are considered so that they lie in an admissible range for such an approximation—a linear transformation gives rise to what we have termed the FRESETL test, while a sinusoidal transformation results in what we have called the FRESETS test.

These two new test statistics share the property of the usual RESET statistic of being exactly F-distributed under the null hypothesis of no conditional mean misspecification, at least in the context of a model with nonstochastic regressors and spherical disturbances. We have undertaken a Monte Carlo experiment to determine the power properties of the two variants of the FRESET test, and to compare these with the power of the RESET test for different forms of model misspecification, and under varying choices of the number of "augmentation" terms in all cases. Our simulation results suggest that using the global Fourier approximation may have advantages over using the more traditional (local) Taylor's series approximation in terms of the tests' abilities to detect misspecification of the model's conditional mean.

Our results also suggest that using a Fourier approximation with three sine and cosine terms results in a test which performs well in terms of "power." The empirical rates of rejection for false model specifications exhibited by the FRESET tests are at least as great as (and generally greater than) those shown by the RESET test. The FRESETL test is generally the best overall when the misspecified model is static, while the FRESETS test is best overall when the model is misspecified through the omission of a dynamic effect. In practical terms, this favors a recommendation for using the latter variant of the test.

Although the proposals and results in this chapter seem to offer some improvements on what is, arguably, the most commonly applied variable-addition test in empirical econometric analysis, there is still a good deal of research to be undertaken in order to explore the features of both the RESET and the FRESET tests in more general situations. Contrary to earlier apparent "evidence," such as that of Thursby (1979, 1982), it is now well recognized (e.g., Pagan 1984, p. 127; Godfrey 1988, p. 107; Porter and Kashyap 1984) that the RESET test is *not* robust to the presence of autocorrelation and/or heteroskedasticity in the model's errors. The same is clearly true of the FRESET tests. Work in progress by the authors investigates the degree of size distortion of these tests and their ability to reject falsely specified models in these more general situations. As well as considering the same forms of the RESET and FRESET tests that are investigated here, we are also exploring the properties of the corresponding tests when the robust error covariance matrix estimators of White (1980) and Newey and West (1987) are used in their construction. In these cases, as well as in the case of models in which the null specification is dynamic and/or nonlinear in the parameters, asymptotically valid (chi-square) counterparts to the RESET and FRESET tests are readily constructed, as in Section II.D. The finite-sample qualities of these variants of the tests are also under investigation by the authors.

ACKNOWLEDGMENTS

We are grateful to Karlyn Mitchell for helpful correspondence in relation to the use of Fourier functional forms and for supplying unpublished material, to Lindsay Tedds for her excellent research assistance, and to Peter Kennedy, John Small, and Michael Veall for their helpful comments. This research was supported by University of Victoria Internal Research Grant 1-41566.

REFERENCES

Amemiya, T. (1977), A Note on a Heteroscedastic Model, *Journal of Econometrics*, 6, 365–370.

Anderson, T. W. (1971), *The Statistical Analysis of Time Series*, Wiley, New York.

Anscombe, F. J. (1961), Examination of Residuals, *Proceedings of the Fourth Berkeley Symposium on Mathematical Statistics and Probability*, 1, 1–36.

Bahadur (196), Stochastic Comparison of Tests, *Annals of Mathematical Statistics*, 31, 276–295.

Bahadur (1967), Rates of Convergence of Estimates and Tests Statistics, *Annals of Mathematical Statistics*, 38, 303–324.

Basu, D. (1955), On Statistics Independent of a Complete Sufficient Statistic, *Sankhyā*, 15, 377–380.

Bera, A. K. and C. M. Jarque (1982), Model Specification Tests: A Simultaneous Approach, *Journal of Econometrics*, 20, 59–82.

Bera, A. K. and C. R. McKenzie (1987), Additivity and Separability of the Lagrange Multiplier, Likelihood Ratio and Wald Tests, *Journal of Quantitative Economics*, 3, 53–63.

Box, M. J. (1966), A Comparison of Several Current Optimization Methods, and the Use of Transformations in Constrained Problems, *Computer Journal*, 9, 67–77.

Breusch, T. S. (1978), Testing for Autocorrelation in Dynamic Linear Models, *Australian Economic Papers*, 17, 334–355.

Breusch, T. S. and A. R. Pagan (1979), A Simple Test for Heteroscedasticity and Random Coefficient Variation, *Econometrica*, 47, 1287–1294.

Christensen, L. R., D. W. Jorgensen, and L. J. Lau (1971), Conjugate Duality and the Transcendental Logarithmic Production Function (abstract), *Econometrica*, 39, 255–256.

Christensen, L. R., D. W. Jorgensen, and L. J. Lau (1973), Transcendental Logarithmic Production Frontiers, *Review of Economics and Statistics*, 55, 28–45.

David, H. A. (1981), *Order Statistics*, Wiley, New York.

Davidson, R. and J. G. MacKinnon (1981), Several Tests for Model Specification in the Presence of Alternative Hypotheses, *Econometrica*, 49, 781–793.

Davidson, R. and J. G. MacKinnon (1993). *Estimation and Inference in Econometrics*, Oxford University Press, Oxford.

Davies, R. B. (1980), The Distribution of a Linear Combination of χ^2 Random Variables: Algorithm AS 155, *Applied Statistics*, 29, 323–333.

Durbin, J. (1954), Errors in Variables, *Review of the International Statistical Institute*, 22, 23–32.

Durbin, J. (1970), Testing for Serial Correlation in Least Squares Regression When Some of the Regressors Are Lagged Dependent Variables, *Econometrica*, 38, 410–421.

Eastwood, A. and L. G. Godfrey (1992), The Properties and Constructive Use of Misspecification Tests for Multiple Regression Models, in L. G. Godfrey (ed.), *The Implementation and Constructive Use of Mis-specification Tests in Econometrics*, Manchester University Press, Manchester, 109–175.

Eastwood, B. J. and A. R. Gallant (1991), Adaptive Rules for Semiparametric Estimators that Achieve Asymptotic Normality, *Econometric Theory*, 7, 307–340.

Engle, R. F. (1982), Autoregressive Conditional Heteroscedasticity with Estimates of the Variance of U.K. Inflation, *Econometrica*, 50, 987–1007.

Gallant, A. R. (1981), On the Bias in Functional Forms and an Essentially Unbiased Form: The Fourier Flexible Form, *Journal of Econometrics*, 15, 211–245.

Garbade, K. (1977), Two Methods of Examining the Stability of Regression Coefficients, *Journal of the American Statistical Association*, 72, 54–63.

Geweke, J. (1981), The Approximate Slopes of Econometric Tests, *Econometrica*, 49, 1427–1442.

Giles, J. A. and D. E. A. Giles (1993), Pre-test Estimation and Testing in Econometrics: Recent Developments, *Journal of Economic Surveys*, 7, 145–197.

Godfrey, L. G. (1978), Testing against General Autoregressive and Moving Average Error Models When the Regressors Include Lagged Dependent Variables, *Econometrica*, 46, 1293–1302.

Godfrey, L. G. (1988), *Misspecification Tests in Econometrics: The Lagrange Multiplier Principle and Other Approaches*, Cambridge University Press, Cambridge.

Godfrey, L. G. (1996), Misspecification Tests and Their Uses in Econometrics, *Journal of Statistical Planning and Inference*, 49, 241–260.

Godfrey, L. G., M. McAleer, and C. R. McKenzie (1988), Variable Addition and Lagrange Multiplier Tests for Linear and Logarithmic Regression Models, *Review of Economic Studies*, 70, 492–503.

Goldfeld, S. M. and R. E. Quandt (1965), Some Tests for Homoscedasticity, *Journal of the American Statistical Association*, 60, 539–547.

Harrison, M. J. and B. P. M. McCabe (1979), A Test for Heteroscedasticity Based on Ordinary Least Squares Residuals, *Journal of the American Statistical Association*, 74, 494–499.

Hausman, J. A. (1978), Specification Tests in Econometrics, *Econometrica*, 46, 1251–1272.

Hendry, D. F. (1980), Econometrics: Alchemy or Science?, *Economica*, 47, 387–406.

Jarque, C. J. and A. K. Bera (1980), Efficient Tests for Normality, Homoscedasticity and Serial Independence of Regression Residuals, *Economics Letters*, 6, 255–259.

Jarque, C. J. and A. K. Bera (1987), A Test for Normality of Observations and Regression Residuals, *International Statistical Review*, 55, 163–172.

Johnston, J. (1960), *Statistical Cost Analysis*, McGraw-Hill, New York.

Krämer, W. and H. Sonnberger (1986), *The Linear Regression Model under Test*, Physica-Verlag, Heidelberg.

Lee, J. H. H. (1991), A Lagrange Multiplier Test for GARCH Models, *Economics Letters*, 37, 265–271.

MacKinnon, J.G. and L. Magee (1990), Transforming the Dependent Variable in Regression Models, *International Economic Review*, 31, 315–339.

Magee, L. (1987), Approximating the Approximate Slopes of LR, W and LM Test Statistics, *Econometric Theory*, 3, 247–271.

McAleer, M. (1987), Specification Tests for Separate Models: A Survey, in M. L. King and D. E. A. Giles (eds.), *Specification Analysis in the Linear Model*, Routledge and Kegan Paul, London, 146–196.

McAleer, M. (1995), Sherlock Holmes and the Search for Truth: A Diagnostic Tale, reprinted in L. Oxley et al. (eds.), *Surveys in Econometrics*, Blackwell, Oxford, 91–138.

Milliken, G. A. and F. A Graybill (1970), Extensions of the General Linear Hypothesis Model, *Journal of the American Statistical Association*, 65, 797–807.

Mitchell, K. and N. M. Onvural (1995), Fourier Flexible Cost Functions: An Exposition and Illustration Using North Carolina S & L's, mimeo., College of Management, North Carolina State University.

Mitchell, K. and N. M. Onvural (1996), Economies of Scale and Scope at Large Commercial Banks: Evidence from the Fourier Flexible Functional Form, *Journal of Money, Credit and Banking*, 28, 178–199.

Mizon, G. E. (1997a), Inferential Procedures in Nonlinear Models; An Application in a U.K. Cross Section Study of Factor Substitution and Returns to Scale, *Econometrica*, 45, 1221–1242.

Mizon, G. E. (1977b), Model Selection Procedures, in M. J. Artis and A. R. Nobay (eds.), *Studies in Modern Economic Analysis*, Blackwell, Oxford, 97–120.

Newey, W. and K. West (1987), A Simple, Positive-Definite, Heteroskedasticity and Autocorrelation Consistent Covariance Matrix, *Econometrica*, 55, 703–708.

Ohtani, K. and J. A. Giles (1993), Testing Linear Restrictions on Coefficients in a Linear Model with Proxy Variables and Spherically Symmetric Disturbances, *Journal of Econometrics*, 57, 393–406.

Pagan, A. R. (1984), Model Evaluation by Variable Addition, in D. F. Hendry and K. F. Wallis (eds.), *Econometrics and Quantitative Economics*, Blackwell, Oxford, 103–133.

Pagan, A. R. and A. D. Hall (1983), Diagnostic Tests as Residual Analysis, *Econometric Reviews*, 2, 159–218.

Pesaran, M. H. (1974), On the Problem of Model Selection, *Review of Economic Studies*, 41, 153–171.

Phillips, G. D. A. and B. P. M. McCabe (1983), The Independence of Tests for Structural Change in Econometric Models, *Economics Letters*, 12, 283–287.

Phillips, G. D. A. and B. P. M. McCabe (1989), A Sequential Approach to Testing for Structural Change in Econometric Models, in W. Krämer (ed.), *Econometrics of Structural Change*, Physica-Verlag, Heidelberg, 87–101.

Porter, R. D. and A. K. Kashyap (1984), Autocorrelation and the Sensitivity of RESET, *Economics Letters*, 14, 229–233.

Ramsey, J. B. (1969), Tests for Specification Errors in Classical Linear Least-Squares Regression Analysis, *Journal of the Royal Statistical Society B*, 31, 350–371.

Ramsey, J. B. (1983), Comment on "Diagnostic Tests as Residual Analysis," *Econometric Reviews*, 2, 241–248.

Ramsey, J. B. and R. Gilbert (1972), A Monte Carlo Study of Some Small Sample Properties of Tests for Specification Error, *Journal of the American Statistical Association*, 67, 180–186.

Ramsey, J. B. and P. Schmidt (1976), Some Further Results in the Use of OLS and BLUS Residuals in Specification Error Tests, *Journal of the American Statistical Association*, 71, 389–390.

Salkever, D. S. (1976), The Use of Dummy Variables to Compute Prediction Errors and Confidence Intervals, *Journal of Econometrics*, 4, 393–397.

Savin, N. E. (1984), Multiple Testing, in Z. Griliches and M. D. Intriligator (eds.), *Handbook of Econometrics*, Vol. 2, North-Holland, Amsterdam, 827–879.

Schwager, S. J. (1984), Bonferroni Sometimes Loses, *American Statistician*, 38, 192–197.

SHAZAM (1993), *SHAZAM Econometrics Program: User's Reference Manual, Version 7.0*, McGraw-Hill, New York.

Theil, H. (1965), The Analysis of Disturbances in Regression Analysis, *Journal of the American Statistical Association*, 60, 1067–1079.

Theil, H. (1968), A Simplification of the BLUS Procedure for Analyzing Regression Disturbances, *Journal of the American Statistical Association*, 63, 242–251.

Thursby, J. G. (1979), Alternative Specification Error Tests: A Comparative Study, *Journal of the American Statistical Association*, 74, 222–223.

Thursby, J. G. (1982), Misspecification, Heteroskedasticity, and the Chow and Goldfeld-Quandt Tests, *Review of Economics and Statistics*, 64, 314–321.

Thursby, J. G. (1989), A Comparison of Several Specification Error Tests for a General Alternative, *International Economic Review*, 30, 217–230.

Thursby, J. G. and P. Schmidt (1977), Some Properties of Tests for Specification Error in a Linear Regression Model, *Journal of the American Statistical Association*, 72, 635–641.

White, H. (1980), A Heteroskedastic-Consistent Covariance Matrix and a Direct Test for Heteroskedasticity, *Econometrica*, 48, 421–448.

White, H. (1982), Maximum Likelihood Estimation of Misspecified Models, *Econometrica*, 50, 1–25.

White, H. (1987), Specification Tests in Dynamic Models, in T. Bewley (ed.), *Advances in Econometrics—Fifth World Congress*, Vol. 1, Cambridge University Press, New York, 1–58.

White, H. (1994), *Estimation, Inference and Specification Analysis*, Cambridge University Press, Cambridge.

Wu, D. M. (1973), Alternative Tests of Independence between Stochastic Regressors and Disturbances, *Econometrica*, 41, 733–750.

12

Applications of the Bootstrap in Econometrics and Economic Statistics

Michael R. Veall
McMaster University, Hamilton, Ontario, Canada

I. INTRODUCTION

Empirical economists and econometricians have long looked to simulation studies for guidance concerning the choice and performance of estimators whose theoretical justification is often only asymptotic. For example, in any econometrics class that describes a method based on an asymptotic distribution, an instructor will almost always encounter the question, "How large a sample is large enough?" It is a good question, particularly when in most applications the character of the data (e.g., trending vs. nontrending) matters as much as the actual number of observations. Most instructors will give an answer based on simulation studies, but in an application there is often a considerable gap between cases that have been studied by Monte Carlo methods and the case at hand. The bootstrap method may be regarded as a simulation study that is tailored to the actual data being studied, with the results used either to fill in statistical gaps that do not yield easily to analytic methods (such as providing standard errors or confidence intervals when they are otherwise unavailable) or to adjust the original statistical estimates in an attempt to improve finite-sample accuracy. It is therefore not surprising that the bootstrap has proven useful to many empirical researchers in economics, especially as the approach replaces difficult or intractable theoretical calculations with computer calculations that are becoming cheaper and cheaper over time. While bootstrap-like notions had existed previously, even within econometrics, the seminal work for these developments is Efron (1979), the classic paper in statistics that named the bootstrap, developed it as a unified technique, and demonstrated how computer power could widen the scope of its implementation.

We shall suggest that there are two principal stages in the development of the bootstrap. In the first stage, the bootstrap was used as a replacement for analytic methods of calculating standard errors, confidence intervals, etc. This makes obvious sense when there are no available analytic methods (the "better than nothing" case). This first-stage bootstrap has also sometimes been used by researchers under the impression that its finite-sample properties would prove superior to the analytic alternatives. While this is sometimes true, there is little general theoretical support for this position. Section II focuses on this first stage of development. In Section III a number of applications are described, emphasizing cases such as confidence intervals for forecasts and inference after specification search, where the bootstrap may be used because there is no good alternative. Section IV emphasizes the second stage of the bootstrap literature concerning cases where asymptotic, analytic tools are available but in which bootstrap refinements are used to improve finite-sample performance. These are most valuable when the original estimation is plagued by substantial bias. Section V considers more applications, sometimes comparing the results of different approaches. The cases of system estimation and nonlinearities are emphasized. Section VI summarizes briefly and concludes. A principal conclusion is that the bootstrap, while no panacea, may be an important step toward a style of econometric practice that routinely checks the applicability of inferential tools that are not exact in the statistical sense and hence depend on some form of potentially unreliable asymptotic approximation. This argument for simulation is in addition to that of McCloskey and Ziliak (1996), who suggest simulation as a tool for elucidating the economic meaning of econometric results.

Finally a caveat: no paper could now cover all the published applications of the bootstrap in empirical economics, let alone the theoretical developments which may prove relevant to future practice. This chapter instead tries to describe the development of the bootstrap along the lines indicated, using the vocabulary most economists will know from their experience with regression methods, with emphasis on the standard techniques involving confidence intervals or hypothesis tests. The focus is on what an economic statistician or econometrician can reasonably expect from the bootstrap at its current stage of development. Hence, only a few areas of bootstrap application are considered, including inference problems when a system of equations has been estimated or when the estimate of interest is a nonlinear function of the parameter estimates. These are chosen to illustrate the strengths and weaknesses of the bootstrap approach. There are a number of surveys that cover many applications, including Veall (1989), Jeong and Maddala (1993), Vinod (1993), and Li and Maddala (1996). The last survey is particularly recommended for researchers interested in bootstrapping in time-series contexts, with potential problems due to unit roots, lagged dependent variables, and serial correlation. There are also books that treat the statistical background of the bootstrap in more detail, with Efron's early monograph (1982) still a useful reference. Efron and Tibshirani (1993) is probably the most straightforward reference for econometric practitioners, while some of the discussion in Hall (1992) is at a more difficult theoretical level and includes more emphasis on the nature of bootstrap approximation. (LePage and Billard 1992 also include theoretical papers on the limits of the bootstrap: See, for example, Arcones and Giné 1992, who prove that the bootstrap distribution of any M-estimator converges weakly to the true limiting distribution in probability. This provides the basis for a similar result in Hahn 1996 for generalized method of moments estimators.) It should be noted that the Hall and Efron/Tibshirani books represent different approaches on a key aspect of bootstrap practice, as will be discussed further in Sections IV, V, and VI.

II. THE SIMPLE BOOTSTRAP

A. The Basic Method

Because it is so familiar to economists, we use the linear regression model as our example, and, where possible, we describe the bootstrap using the language of a Monte Carlo experiment. The linear regression model is

$$Y_i = X_i\beta + u_i$$
$$= X_i\hat{\beta} + e_i, \qquad i = 1, \ldots, n \tag{1}$$

Y_i and the row k-vector X_i are observations on the dependent and independent variables respectively with the first element of X_i a 1, β and $\hat{\beta}$ are column vectors of parameters and their estimates respectively, u_i are random errors which will be assumed to be independently and identically distributed (i. i. d) with mean zero and variance σ^2, and e_i are regression residuals. The bootstrap estimate of the variance matrix in this case is calculated by the following simulation experiment.

Step 1:   Create an artificial sample:

$$Y_i^{*1} = X_i\hat{\beta} + e_i^{*1}, \qquad i = 1, \ldots, n \tag{2}$$

The elements e_i^{*1} are created by resampling from e_i, that is drawing from e_i randomly but with replacement. (We have assumed there is an intercept so that the e_i have mean zero; otherwise we would have to work with deviations in e_i by subtracting off the mean.) The artificial sample can be thought of as the result of one trial of a Monte Carlo experiment where the independent variables are set as the actual values of X_i, the parameters are set as the actual data estimates $\hat{\beta}$ and its disturbances are the draws from the empirical distribution function of the residuals e_i, that is the discrete probability distribution function that attaches probability $1/n$ to each value e_i.

Step 2:   Estimate $\hat{\beta}^{*1}$ on artificial sample 1.

Step 3:   Repeat steps 1 and 2 B times.

Step 4:   The sampling distribution of the $\hat{\beta}^{*j} - \hat{\beta}$ over the B Monte Carlo bootstrap trials is an estimate of the sampling distribution of $\hat{\beta} - \beta$. In particular, the variance of $\hat{\beta}$ is estimated by $\hat{v}^*(\hat{\beta}^{*j})$, the sample variance matrix of the $\hat{\beta}^{*j}$ over the B bootstrap samples.

An alternative, known as the parametric bootstrap, is identical to the above except the disturbances are drawn from a particular distribution with its parameters set as estimates from the residuals, most commonly a normal random number generator with mean zero and variance equal to the residual variance. It is also possible to generate the disturbances as weighted averages of draws from the empirical distribution function of the residuals and the parametric distribution; this method may be called the smoothed bootstrap with the weights determining the degree of smoothing. Other techniques in Monte Carlo analysis, such as importance sampling and antithetic variates, can also be applied to the bootstrap. (See Efron and Tibshirani 1993 for some discussion and references.) These alternatives have the same essential properties as the ordinary bootstrap, which will remain our focus as it is the method of resampling that has almost exclusively been employed in econometric applications.

B.   Why the Bootstrap Works

To consider why this approach may work, consider the sampling distribution of a $\hat{\beta}^{*j}$, where j denotes the jth bootstrap sample drawn from a population based on the real sample in the manner described in the previous section. Call $v(\hat{\beta}^{*j})$ the variance matrix of $\hat{\beta}^{*j}$ based on this distribution function. Just as in any simulation experiment, as B grows large $\hat{v}^*(\hat{\beta}^{*j})$ approaches $v(\hat{\beta}^{*j})$. Note that $\hat{\beta}^{*j} = (X^T X)^{-1} X^T Y^{*j} = \hat{\beta} + (X^T X)^{-1} X^T e^{*j}$, where X is the $n \times k$ matrix of the row vectors X_i, and Y^{*j} and e^{*j} are the N-vectors corresponding to the jth artificial sample. Treating e^{*j} as a random variable drawn from the empirical distribution function, it is straightforward to show that, because the e_i's sum to zero and have equal probability weights $1/n$, $E(e^{*j}) = 0$ and $v(e^{*j}) = (e^T e/n)I_n$, where e is the vector of e_i's and I_n is the $n \times n$ identity matrix. Therefore, $E(\hat{\beta}^{*j}) = \hat{\beta}$ and $v(\hat{\beta}^{*j}) = v((X^T X)^{-1} X^T e^{*j}) = (X^T X)^{-1} X^T v(e_1^{*j}) X (X^T X)^{-1} = (e^T e/n)(X^T X)^{-1}$. But note this last expression is just the usual estimate of $v(\hat{\beta})$ as calculated by a regression package without bootstrapping, with the exception that the division is by n rather than $n - k$. (This is sometimes corrected in practice by scaling the original residuals by $(n/(n-k))^{1/2}$.) We have therefore shown that we can compute the ordinary least-squares (OLS) estimate of the covariance matrix of the coefficients if we use as our population a probability model based on the sample. As it is sometimes put for bootstrapping statistics more generally, as the sample is drawn from an underlying population, a resample from that sample can be thought of as being drawn from something similar to the underlying population. Hence we may infer the population variance of a statistic by observing its behavior in resampling. A similar approach can be used to calculate other statistics such as the limit points of confidence intervals.

Why have we bothered to show this simple result, particularly as it seems the bootstrap is redundant in this case, in that it is not essentially different from the analytic estimate? First, the fact that the bootstrap is so close to the right answer in a case where we know the right answer is reassuring given that the bootstrap is also proposed in other contexts in which analytic formulas are not known. Second, that the bootstrap estimate has divisor n when we know that the exact test requires (among other things) that the divisor be $n - k$ emphasizes that the justification for the use of the simple bootstrap is typically only asymptotic, although in this case a simple rescaling eliminates the discrepancy. (We shall refer to this issue later.) Third, it shows that the bootstrap estimate of the variance, just like the standard analytic variance estimate, will only be "correct" for inference if the initial model is correctly specified. For example, the disturbances must be identically and independently distributed; hence practitioners should check these assumptions by suitable diagnostics before bootstrapping. Finally, the exercise emphasizes that the bootstrap does not create any additional information. It is simply a computational device to utilize information already in the original sample.

The type of bootstrap just described in the regression context was called "residual resampling" by Kiviet (1984), because it kept the fixed structure of the independent variables and only resampled on the residuals. Alternatively we could use what Kiviet called "complete resampling," where resampling is from the row $(k + 1)$-vectors (Y_i, X_i) and then the bootstrap algorithm proceeds as usual. As discussed in Jeong and Maddala (1993, pp. 577–578), this method should give a consistent estimate of the variance-covariance matrix of the estimated coefficients even in the presence of random X's and/or heteroskedasticity. The discussion in Efron and Tibshirani (1993, Chap. 9) goes in a different direction, pointing out that the bootstrap will yield an estimate of the sampling distribution of $\hat{\beta}$ provided only that the original observation vectors were chosen from the underlying $(k + 1)$-variate probability distribution. This does not even depend on the existence of a "true" linear model as in the first line of (1). Naturally, however, if there is no true model, interpretation will be difficult: we shall have an estimate of the sampling distribution of $\hat{\beta}$ without knowing its relationship, if any, to the parameter being estimated, if any. Even granting knowledge of the true model, the assumption of random draws in the complete resample precludes most time-series estimation; it, however, may be a natural approach to use in cross-sectional estimation.

C. An Example

As an example consider the Theil (1971) textile data where Y_t, the log of Netherlands annual textile consumption for 1923–1939, is regressed on X_{t1}, the log of real income per capita, and X_{t2}, the log of the relative price of textiles:

$$\hat{Y}_t = \hat{\beta}_0 + \hat{\beta}_1 X_{t1} + \hat{\beta}_2 X_{t2} = 1.37 + 1.14 X_{t1} - .83 X_{t2} \tag{3}$$

Table I   Various Standard Error Estimates for the Theil Data

Technique	B^a	s.e.$(\hat{\beta}_0)^a$	s.e.$(\hat{\beta}_1)$	s.e.$(\hat{\beta}_2)$
Analytic	n.a.	.3061	.1560	.0361
Bootstrap (unscaled)	50	.2893	.1460	.0321
	100	.2616	.1352	.0310
	1000	.2824	.1449	.0332
Bootstrap (scaled)	50	.3188	.1609	.0353
	100	.2883	.1489	.0342
	1000	.3120	.1608	.0362
Bootstrap (parametric)	50	.3272	.1627	.0348
	100	.3029	.1526	.0366
	1000	.3003	.1531	.0347
Bootstrap (complete resampling)	50	.2926	.1641	.0379
	100	.2617	.1476	.0430
	1000	.2613	.1375	.0391

^{a}s.e. = standard error, and B = number of bootstraps.

Residual bootstrapping is based on the assumption that the underlying disturbance terms are independently and identically distributed, so these assumptions should be checked (as has been done for this example extensively in the numerous exercises in the manual for the computer package SHAZAM 1993). Table 1 reports a variety of standard error estimates, where the bootstrap standard errors are calculated in the usual manner as the square root of the diagonal elements of the bootstrap estimate of the variance matrix.

The first thing to notice from the table is that all the residual bootstrap methods yield answers close to the analytic case; the complete resampling bootstrap results are not as close, nor should they be, for as described above they are more like estimates of the heteroskedasticity-consistent standard errors rather than the ordinary standard errors. It is also clear that the number of bootstraps is not that important, although more seems to be better. Scaling the residuals (by $(n/(n-k))^{1/2} = (17/14)^{1/2}$ in this case) increases the bootstrap standard errors somewhat as would be expected, from being a bit smaller than their analytic counterparts to being a bit larger. The parametric bootstrap, which uses the normal distribution with mean zero and variance $e^T e/(n-k)$ instead of the empirical distribution function of the residuals in order to generate the artificial samples, yields very similar answers to the other residual bootstrap techniques.

D.   Bootstrapping Confidence Intervals

The main reason to calculate standard errors is for inference, by means of hypothesis tests or confidence regions. Hypothesis tests involving the coefficients are executable by determining whether corresponding confidence regions cover the null hypothesis parameter values. Noting this equivalence, we shall concentrate our discussion on bootstrapping confidence regions. Moreover for simplicity and because the theory is better developed in this area, we focus on confidence regions involving one parameter, which of course correspond to t-statistics in a hypothesis-testing framework.

One way to use the bootstrap in inference would be to use the bootstrap standard errors instead of analytic standard errors in the basic confidence interval formula. We call these bootstrap standard error confidence intervals, and they are of the form $\hat{\beta}_i \pm t_{n-k}^{.95} \cdot \text{s.e.}^*(\hat{\beta}_i)$. As the bootstrapped standard errors were similar to the analytic standard errors in the linear regression context, therefore bootstrap confidence intervals of this form will not be much different from their analytic counterparts. Another more popular bootstrap confidence interval has been the percentile method. In this, the artificial sample values of the statistic of interest, such as a regression coefficient, are sorted and the $1 - 2\alpha$ confidence intervals are set as the αth and $(1 - \alpha)$th percentiles. While some find it intuitively more pleasing, in terms of regression coefficient estimates these intervals tend to be similar to analytic or bootstrap standard error confidence intervals, even when the disturbances are nonnormal (although the combination of severe nonnormality, few degrees of freedom, and very small α can lead to substantial differences). Regression estimates are weighted averages, and very commonly this ensures (by way of a central limit theorem), that the distribution function of $\hat{\beta}^*$ is approximately normal, as it is for $\hat{\beta}$. Hence for constructing confidence intervals of linear regression coefficients, there is relatively little to be gained by bootstrap methods.

More generally, Hall (1992) demonstrates that in most cases (essentially for any root-n-consistent statistic that may be expanded in Edgeworth form), the endpoints of the bootstrap percentile confidence intervals and of the bootstrap standard error confidence intervals are accurate only to $O(n^{-1/2})$, which in general is the accuracy that any analytic asymptotic method can be expected to achieve. Similar results hold for the accuracy of the tail coverage—that is, the degree of approximation of the area in each tail outside the confidence interval to its putative value of α. Given this, the best case for using these types of bootstrap methods is when there is really no alternative. We shall discuss two cases: (1) forecasting from the linear regression model and (2) inference after a specification search.

III.   APPLICATIONS OF FIRST-STAGE BOOTSTRAPPING METHODS

A.   Confidence Intervals for Forecasts

One context where the bootstrap may be very useful is in the confidence intervals for forecasts. If we stay with the linear regression context, the forecast is $x_{n+1}\hat{\beta}$, where

x_{n+1} is the row vector of observations on the k right-hand-side variables for period $n + 1$. The forecast error e^f_{n+1} is

$$e^f_{n+1} = u_{n+1} + x_{n+1}(\hat{\beta} - \beta) \qquad (4)$$

While the variance of this expression can be estimated analytically, note that in constructing confidence intervals, a central limit theorem does not ensure normality: the second term of (4) may tend to the normal distribution as n grows large, but the first will only be normal if the disturbances themselves are normal. To deal with this nonnormality is difficult analytically but is straightforward using the bootstrap. Following Freedman and Peters (1984a, 1984b), we bootstrap just as above only we focus on the forecast $x_{n+1}\hat{\beta}^{*j}$ as the statistic of interest. We then calculate a "simulated actual" by adding an additional single bootstrapped residual to the actual forecast $x_{n+1}\hat{\beta}$. The difference between the forecast and the simulated actual is the simulated forecast error; we obtain an estimate of the probability distribution of the forecast error by repeating the process B times.

Moreover, unlike the standard OLS case, in bootstrap simulation it is easy to incorporate uncertainty in the x_{n+1} in the forecast confidence intervals, the importance of which is stressed by Feldstein (1971). Early contributions to this approach are the stochastic simulation methods of, for example, Brown and Mariano (1984) and especially Fair (1979, 1980), who, independently of the bootstrap literature, proposed the same bootstrap method of evaluating forecast uncertainty. In addition, Fair proposes a modification to the basic bootstrap uncertainty measures to make allowance for specification error and applies the method to macroeconomic forecasting in the United States. Freedman and Peters (1984a) use the bootstrap technique to develop forecast standard errors in a generalized least-squares application involving United States electricity consumption by region. Veall (1987a) applies the method, with emphasis on the percentile method and the uncertainty in the independent variable forecasts, to forecasting the demand for electricity in Ontario in a time-series context; Veall (1987b) is a Monte Carlo study that confirms the reliability of the approach for this problem. Bernard and Veall (1987) extend the same exercise for Quebec emphasizing the dynamics still more. Prescott and Stengos (1987) use the same approach for studying the United States supply of pork.

B.   The Bootstrap and Specification Search

Much of the focus in econometric theory is on sampling error, yet, in practice, specification error is the more vexing question in many econometric applications. We typically cannot create more data by laboratory-type experiments, and it is relatively rare to have such a large data set that we can select the specification on a training set and then estimate on different data. Hence, specification search and estimation are typically on the same data. While the process of changing a specification in response to initial results may be an important part of empirical modeling, the resulting "pretesting" may lead to the coefficients and standard errors estimated from the eventual model being seriously biased.

There have been a few attempts to use simulation methods as a way of treating the specification error problem. The work of Fair (1979, 1980), which began to deal with specification uncertainty, has already been cited in the prediction context. Efron and Gong (1983) attempted to grapple with the problem of specification search by studying the sampling distribution for estimates from an entire data-mining procedure by bootstrap simulation and provide an example relating to hepatitis diagnosis. The entire decision tree of the investigator is laid out (e.g., step 1: estimate whole model; step 2: drop all variables whose coefficients have t-statistics less than 1, etc.) and then applied on the data. Then the same entire decision tree is applied to each of the bootstrap samples, generated either by the complete resampling method or the residual resampling method using either the first-stage or the final model.

Brownstone (1990) and Veall (1992) apply this technique to econometric examples, the former also estimating the standard errors of Stein-James shrinkage estimators in this context. (See Vinod and Raj 1988 for an application involving ridge regression.) Freedman, Navidi, and Peters (1988) and Dijkstra and Veldkamp (1988) study this kind of method for stylized data-mining procedures and, in general, find that the simulation method does not yield accurate standard errors for the data-mining estimator. The results of Freedman, Navidi, and Peters are particularly discouraging, although it should be noted that their example is based on an initial stage of estimating 75 coefficients from 100 observations. It must also be remembered that the method is only valid if the estimation procedure can be modeled as if it were a prespecified decision tree: if new hypotheses and approaches were entertained only after seeing the first set of results, then, strictly speaking, even bootstrapping the entire procedure as run does not solve the fundamental problem. Nonetheless this method does seem to be the only feasible possibility at this time for dealing with the pretest issue in any real estimation context. It does seem a minimal requirement for any estimated econometric model that if Monte Carlo samples are generated from it and the entire data-mining procedure is applied to those samples, the results over the Monte Carlo samples should be consistent in all important respects with the results from the original data.

Finally, in a Bayesian context, Geweke (1986) proposes a useful method of implementing strong priors on coefficient signs in regression methods, essentially by bootstrapping an unrestricted regression model but basing all estimates on the artificial samples for which the estimates meet the sign restrictions. Hence, if the only prior is that the income coefficient is positive, the estimate of the income coefficient will be the average of all the positive estimates over the bootstrap samples and the standard error will be the standard deviation of these estimates. Some researchers may wonder why they should ignore the negative estimates in this context, but this is the consequence of the prior they purport to hold. Chalfant, Gray, and White (1991) is an application of the same technique.

IV. BOOTSTRAP REFINEMENTS

A.   Simple Bias Correction

As we have discussed, application of bootstrap confidence intervals or test procedures based upon bootstrap standard errors or on the percentile method does not lead to a more accurate (Edgeworth) approximation than asymptotic methods, and we know of no other theoretical method of comparison. Hence there seems to be little justification for using the bootstrap when asymptotic methods are available. There is, however, an informal argument that if one can regard the bootstrap as a Monte Carlo experiment, the bootstrap distribution of an estimator should in some sense agree with the analytic distribution. If, for example, the analytic standard errors are smaller than the bootstrap standard errors, it suggests that there is a downward bias in the analytic standard errors, and hence perhaps the "true" standard errors are bigger than both the analytic and the bootstrap standard errors, with the bootstrap standard errors therefore a better choice as they are closer to the truth. Such a view seems to be behind the early bootstrap application of Korajczyk (1985), for example, who found for a system of econometric equations modeling the foreign exchange market, that bootstrap standard errors exceeded analytic standard errors and that switching to bootstrap standard errors meant that a null hypothesis associated with rational behavior could no longer be rejected. Given a common tendency for bootstrap standard errors to be larger than their analytic counterparts, we hope it will not become the practice to invoke the bootstrap approach only when the economist in one thinks that the null hypothesis should be protected.

We should differentiate at least two possible bias problems that come up in standard regression modeling and its extensions. There can be a bias in coefficient estimation, perhaps best captured by the well-known bias toward zero of estimated serial correlation coefficients. There can also be bias in standard error estimation. Both of these can affect inference.

A simple method of calculating bias for coefficient estimates (Efron, 1982, p. 33) is to subtract the actual value from the average of the bootstrap values over the B bootstrap samples. Hence, if the actual data coefficient estimate is .5 and the average of the bootstrap values is .7, the bias estimate is .2 and one could even propose a bias-adjusted estimate of $.5 - .2 = .3$. The bias-adjusted result, which is very much like a jackknife estimate, may, however, be subject to very high sampling variability.

Freedman and Peters (1984a, 1984b) follow an instructive approach along these same lines for the standard error, which is to calculate for each coefficient a bias factor as the ratio of the bootstrap standard error to some average of the analytic standard errors as calculated on each of the bootstrap samples. This gives a better idea of how the analytic standard errors are biased, and it therefore seems natural to come up with a "bias-adjusted" standard error by multiplying the original analytic standard error estimate by the bias factor. We note that, in the linear regression context, if standard errors have been based on a variance estimate in which n rather than $n - k$ has been used as the divisor, the Freedman and Peters approach will "automatically" adjust the standard error for degrees of freedom. Marais (1986) refines this approach and finds the method can be quite accurate in the context of estimating systems of regression equation. However, because the purpose of standard errors is for inference, the Freedman and Peters approach has been eclipsed by approaches that directly adjust the potential bias in the test procedures or in the confidence intervals.

B.   The Bias-Corrected Percentile Method

Efron (1987) proposes one attempt to solve the problem of finite-sample bias. Assume we have a scalar β, which is not necessarily a single-equation regression model coefficient and is more likely a coefficient from a system of equations or perhaps a nonlinear function of coefficients that can be estimated by linear methods. Consider the interval

$$(\hat{\beta}^{*(\alpha 1)}, \hat{\beta}^{*(\alpha 2)}) \tag{5}$$

where $\hat{\beta}^{*(\alpha)}$ is the 100αth percentile of the $\hat{\beta}^*$'s over the B bootstrap runs. If we set $\alpha_1 = \alpha$ and $\alpha_2 = 1 - \alpha$, we have the ordinary bootstrap percentile confidence intervals with intended coverage $1 - 2\alpha$ but with no adjustment for bias. Efron proposes the BCa confidence intervals where instead we implement (5) with

$$\alpha_1 = \Phi(\hat{z}_0 + \hat{w}_0(\alpha)) \qquad \alpha_2 = \Phi(\hat{z}_0 + \hat{w}_0(1 - \alpha)) \tag{6}$$

where

$$w_0(\alpha) = \frac{\hat{z}_0 + z^{(\alpha)}}{1 - \hat{a}(\hat{z}_0 + z^{(\alpha)})} \tag{7}$$

and

$$\hat{z}_0 = \Phi^{-1}(\hat{F}^*(\hat{\beta})) \qquad z^{(\alpha)} = \Phi^{-1}(\alpha) \tag{8}$$

where Φ is the standard normal cumulative distribution function, $\hat{F}^*$ is the empirical cumulative distribution function of the $\hat{\beta}^*$'s, $w_0(1 - \alpha)$ is defined analogously to $w_0(\alpha)$, and $\hat{a}$ is called the acceleration constant and will be discussed.

Consider first the case where $\hat{a} = 0$, which yields what have been called the bias-corrected (BC) percentile confidence intervals. This method primarily deals with bias in the coefficient estimates. If there were no tendency toward bias in the coefficient estimate of the $\hat{\beta}$, we should expect that in the empirical distribution function of the $\hat{\beta}^*$'s, half should be above $\hat{\beta}$ and half below; i.e., that $\hat{F}^*(\hat{\beta}) = .5$ and hence that $\hat{z}_0 = \Phi^{-1}(.5) = 0$. This in turn implies that $\alpha_1 = \alpha$ and $\alpha_2 = 1 - \alpha$. Hence, with no coefficient bias, the BC percentile confidence intervals are just the ordinary bootstrap percentile confidence intervals. Now suppose $\hat{F}^*(\hat{\beta})$ exceeded .5, suggesting a negative bias in the coefficient estimates. Therefore $\hat{z}_0$ will be positive and $\alpha_1 = \Phi(2\hat{z}_0 + z^{(\alpha)})$ and $\alpha_2 = \Phi(2\hat{z}_0 + z^{(1-\alpha)})$ will exceed α and $1 - \alpha$ respectively. Hence, the BC method adjusts to the downward bias in the coefficient estimate by shifting the entire confidence interval up. (One option to adjust the coefficient estimate itself is to use $\alpha = .5$ in the above formula.)

The role of the acceleration constant $\hat{a}$ is less obvious, but it is partially related to bias in the estimation of the standard error. For example, if $\hat{z}_0 = 0$ and given that $z^{(\alpha)}$ is negative and $z^{1-\alpha}$ is positive, it can be shown that changing $\hat{a}$ from zero to a small positive value will widen the BCa bootstrap percentile confidence intervals. More generally, we could argue that the usual normal approximation

$$\frac{\hat{\beta} - \beta}{\text{se}(\hat{\beta})} \sim N(0, 1) \tag{9}$$

(where $\text{se}(\hat{\beta})$ is the standard error of $\hat{\beta}$) may be generalized to

$$\frac{m(\hat{\beta}) - m(\beta)}{\text{se}_0(m(\hat{\beta}))(1 + a(\beta - \beta_0))} \sim N(-z_0, 1) \tag{10}$$

for some increasing transformation m, where $\text{se}_0(m(\hat{\beta}))$ denotes the standard error of $m(\hat{\beta})$ when the true value β equals any conveniently chosen β_0: recall the point of the exercise is that in finite samples, $\text{se}(m(\hat{\beta}))$ will depend on the value of β and the approximation in the denominator in (10) attempts to capture this. Efron (1987) or Efron and Tibshirani (1993, pp. 326–327) show that if we use the normalizing transform m, calculate confidence intervals based on the normal distribution, and then transform back using m^{-1}, we obtain the BCa intervals except that a and z_0 need to be estimated. Note that m does not need to be known. These papers also argue that in one-parameter families, a good approximation for $\hat{a}$ is one-sixth of the skewness coefficient of the score function of β, evaluated at $\hat{\beta}$; for multiparameter families, they offer a formula based on the infinitesimal jackknife. However, most econometricians will prefer, at least computationally, the simpler jackknife formula (Efron and Tibshirani, 1993, p. 186):

$$\hat{a} = \frac{\sum(\hat{\beta}_J - \hat{\beta}_{(i)})^3}{6\{\sum(\hat{\beta}_J - \hat{\beta}_{(i)})^2\}^{3/2}} \tag{11}$$

where the summations run from 1 to n, $\hat{\beta}_{(i)}$ is $\hat{\beta}$ calculated on a sample with the ith observation deleted, and $\hat{\beta}_J$, the jackknife estimator of β, is the average of the $\hat{\beta}_{(i)}$.

C.   Percentile-t Methods

An alternative refinement to BCa methods is the percentile-t bootstrap confidence intervals. Suppose after estimating $\hat{\beta}$ and se($\hat{\beta}$) on the original sample, $\hat{\beta}^*$ and se($\hat{\beta}^*$) are estimated on each bootstrap sample. (A key requirement for this method is that some form of standard error estimate is available for both the original and bootstrapped data.) If one thinks of the bootstrap process as a Monte Carlo experiment, it is natural to think of $\hat{\beta}^* = \hat{\beta}$ as the null hypothesis to be tested in each trial and hence natural to calculate the t-ratio $t^* = (\hat{\beta}^* - \hat{\beta})/\text{se}(\hat{\beta}^*)$ on each trial. The bootstrap procedure therefore essentially generates a distribution for this t-ratio under a particular null hypothesis and the $1 - 2\alpha$ percentile-t confidence intervals become

$$(\hat{\beta} - t^{*1-\alpha}\text{se}(\hat{\beta}), \hat{\beta} - t^{*\alpha}\text{se}(\hat{\beta})) \tag{12}$$

where $t^{*\alpha}$ is the αth percentile of the t^*'s. Essentially this technique uses the bootstrap to create its own critical values instead of using those supplied by the usual t-distribution.

D.   Comparison of the BCa and Percentile-t Methods

Both the BCa and the percentile-t corrections are a refinement of the standard asymptotic confidence intervals. As discussed, if we consider either the endpoints themselves or the tail-area coverage, analytic confidence intervals, bootstrap standard error confidence intervals and percentile confidence intervals are all only first-order accurate, meaning that the approximation error is a term of $O(n^{-1/2})$. The BCa and percentile-t confidence intervals are both second-order accurate in that the approximation error is a term of $O(n^{-1})$; that is, the approximation error goes to zero significantly faster as n increases (Hall 1988, 1992; Efron and Tibshirani 1993).

There has been considerable discussion of the relative merits of the BCa method versus the percentile-t method. Hall (1988, 1992), one of the leading proponents of the percentile-t method, notes that it avoids the calculations discussed above. Although the jackknife method using (11) is not excessively difficult in many contexts, certainly the concept of the acceleration constant is not a comfortable or familiar one to many practitioners. There is indeed some reason to be uncomfortable. While possible, the generalization in BCa to the multivariate confidence regions case is somewhat difficult. Moreover, even in the single-coefficient case above, it can be seen that the bounds (5) based on (6)–(8) are not necessarily monotonic in α: i.e., it is possible, because of the jump discontinuity in (7) when the denominator is zero, for increases in α to widen rather than narrow the confidence bounds. The investigator needs to be careful if $\hat{a}$ has sufficient magnitude for this to occur in the relevant range of α.

The percentile-t method is not without its flaws. It is not transformation invariant: we cannot obtain the confidence interval endpoints for $h(\beta)$ by simply plugging the endpoints of the β confidence intervals into the function h. This is a familiar problem in statistics (described in an econometrics context by Gregory and Veall 1985), but leaves open the possibility that different ways of specifying restrictions or parameters can lead to markedly different substantive results. A solution proposed by Tibshirani (1988) is to implement the percentile-t method with a variance-stabilizing transform.

While the BCa method is transformation invariant in the single-parameter case, it is not in the multiparameter case if the transformation h involves the parameters: the confidence interval for $\beta_1\beta_2 - 1$ is not the same as the confidence interval for $\beta_1 - 1/\beta_2$. While this seems quite natural, if the idea is to perform a test for the null hypothesis that $\beta_1\beta_2 - 1 = 0$ (equivalent to seeing whether the confidence interval of $\beta_1\beta_2 - 1$ covers zero), it is disconcerting that it matters if we instead test the algebraically equivalent null hypothesis $\beta_1 - 1/\beta_2 = 0$.

E. The Barnard Method

A final method we shall discuss was developed before the bootstrap but can be thought of as a type of percentile-t bootstrap. Barnard (1963) proposed simulating the distribution of a test statistic under the null hypothesis. In the context of confidence intervals, this means that rather than calculating the actual confidence interval around the estimated parameter and determining whether it covers the parameter value corresponding to the null hypothesis, we simulate the confidence interval under the null hypothesis and see whether it covers the estimated value. These confidence intervals will be

$$(\beta_0 - t^{**1-\alpha}\text{se}(\hat{\beta}), \beta_0 - t^{**\alpha}\text{se}(\hat{\beta})) \tag{13}$$

where β_0 is the value of β under the null hypothesis and t^{**} are the t-values generated by a bootstrap simulation using the estimates that are generated with the null hypothesis. In the case, for example, when there are no other parameters to estimate, a confidence interval calculated in this manner (or the corresponding test) is exact in finite samples. This result extends if there are parameters to estimate even under the null hypothesis, provided those other parameters do not enter the null distribution of the parameter under test. (An example would be in the (static) linear regression case, where the null distribution of each coefficient does not depend on the other coefficients.) However, if this property does not hold, such as in many dynamic or nonlinear regression contexts, the property of exactness is lost due to the estimation of the other "nuisance" parameters: see Dufour (1995), who incidentally attributes the initial idea behind this procedure to Dwass (1957) rather than to Barnard. Theil and his associates provide examples of the Barnard method (Rosalsky, Finke, and Theil 1984, Theil, Shonkwiler, and Taylor 1985, Taylor, Shonkwiler, and Theil 1986,

Shonkwiler and Theil 1986, Theil and Shonkwiler 1986), some of which will be discussed subsequently. Van Giersbergen and Kiviet (1993) promote the approach in the context of a dynamic regression model. The method of "calibration" now commonly used in macroeconomics, in which the distribution of statistics is generated under a null hypothesis imposed under an elaborate computer model called an "artificial economy," can be seen as an extension of this basic idea. See Gregory and Smith (1993) for a survey.

V. APPLICATIONS OF BOOTSTRAP REFINEMENTS

A. Size Correction

While it is sometimes called a bootstrap correction, what we have called here the Barnard approach can be used as a size correction for almost any test. Horowitz (1994), for example, uses this approach to correct the size of the information matrix test for heteroskedasticity, Fan and Li (1995) use it to correct the J-test for testing a nonnested hypothesis, Theil and Shonkwiler (1986) use it to study tests for serial correlation, and Davidson and MacKinnon (1996) use both the J-test and serial correlation test examples in a more general discussion of this type of bootstrap testing. Horowitz argues that 100 bootstraps are sufficient for many cases where α is not too small (so that $t^{*.05}$ would be the fifth smallest value of t^* and $t^{*.95}$ would be the 96th smallest, that is the fifth largest) (see also Marriot 1979, Hall 1986). While Horowitz's results are supported by his Monte Carlo simulation results which check for empirical size, size is not the only criterion of interest, and it is possible that the use of such a small number of bootstraps may affect power.

B. Estimation of Systems of Equations

Various bootstrap methods have been applied to tests on systems of equations, in what is often called the SUR model. Laitinen (1978) illustrated that finite-sample inference for this model could be a problem in the context of an expenditure system for m goods:

$$w_{it} = x_t^T \beta_i + u_{it} \tag{14}$$

where w_{it} is the expenditure share of good i at time t, x_t^T is a row vector consisting of a 1 (corresponding to a constant), total expenditure, and all m prices, and u_{it} represents a disturbance term which has no correlation with any disturbance term at any other time but which may have a contemporaneous correlation across commodities. The homogeneity property from consumer theory suggests that the last $m + 1$ elements of each β_i should sum to zero, implying that if all prices and total expenditure were changed in the same proportion there would be no change in expenditure shares and, hence, in physical quantities purchased. Because the right-hand-side variables are the same in each equation, if the disturbances are normally distributed, OLS is maximum likelihood.

Laitinen noted that before his paper this kind of proposition was commonly tested with a Wald test based on the OLS estimates of each share equation and a cross-equation variance-covariance matrix estimated from the OLS residuals. The resulting test is asymptotically distributed as chi-square, with $m - 1$ degrees of freedom. Checking the asymptotic distribution using a simulation experiment based on data from Theil (1975), when the number of commodities is 14, a true (by construction) null hypothesis is rejected by the Wald test based on a nominal 5% level, 87 times out of 100 rather than the expected 5. Laitinen argued that this is one reason there are typically so many rejections of homogeneity in actual applications. One intuition is that the variance-covariance matrix is badly misestimated by maximum likelihood methods because there is no adjustment for degrees of freedom lost due to parameter estimation, particularly as the number of estimated parameters increases directly with the number of equations.

As it turns out, for the special case of the homogeneity test, Laitinen finds an exact test using the Hotelling T^2 distribution. But his exact solution does not apply in other contexts, for example the closely related problem of testing for the property of symmetry in demand systems. Meisner (1979) and Bera, Byron, and Jarque (1981) use simulation methods to examine the test for symmetry and find very inaccurate test sizes and poor power as well.

Theil, Shonkwiler, and Taylor (1985) and Taylor, Shonkwiler, and Theil (1986) apply the Barnard method directly to the demand homogeneity and symmetry Wald tests; Shonkwiler and Theil (1986) use the same method to develop critical values for alternative, non-Wald tests that they show can have superior power. Raj and Taylor (1989) apply a Barnard-type bootstrap to testing within-equation restrictions in demand systems.

Other researchers generate ordinary bootstrap standard errors and base tests on these. Korajczyk (1985) and Freedman and Peters (1984a, 1984b) have been discussed. Atkinson and Wilson (1992) have a different reading of Freedman and Peters than ours and believe Freedman and Peters are arguing for direct application of the bootstrap standard errors with no adjustment for bias. Given our discussion above that ordinary bootstrap quantities are not theoretically more accurate than their analytic counterparts, it is not surprising that Atkinson and Wilson find in Monte Carlo analysis based on relatively small systems that SUR standard errors (calculated entirely using OLS residuals) may be no more accurate than ordinary bootstrap standard errors. Rilstone and Veall (1996a), emphasizing that the purpose of estimating the variance-covariance matrix and standard errors is for inference, find that percentile-t confidence intervals are considerably more accurate than those based on OLS/SUR, although the BCa confidence intervals performance is only fair. Rilstone (1993) has similar findings for percentile-t (see also Rayner 1991) and BCa intervals in a single-equation regression context with AR(1) errors.

C.   Nonlinearities

While discussing the lack of invariance of percentile-t methods, it was pointed out that sometimes test results and confidence intervals are sensitive to nonlinearities. In demand systems of the kind just described, for example, even if we ignore the SUR problem there are potential difficulties in inferences because the desired estimates of the price and income elasticities are nonlinear in the estimated parameters. Green, Rocke, and Hahn (1987) find that the bootstrap estimates of the standard errors of the estimated price elasticities are not much different from their asymptotic counterparts, but there are substantial differences for the income elasticity. Krinsky and Robb (1986, 1990, 1991) find no large differences between analytic and bootstrap alternatives for a different data set and a translog system of equations, nor do McManus and Rosalsky (1985) in a nonlinear earnings equation. George, Oksanen, and Veall (1995) find some differences in a context where desired stock is estimated as a nonlinear quantity, as do Veall and Zimmermann (1993) in another dynamic, nonlinear context where they also use simulation to estimate power.

The point of these examples is that sometimes nonlinearities seem to matter and sometimes they do not. If we return to the simple example of Gregory and Veall (1985), it is easy to see what can cause the problem. If we compare the quantities $\beta_1\beta_2 - 1$ and $\beta_1 - 1/\beta_2$, for example, it is immediately clear that the second is much more nonlinear as β_2 approaches zero. Rilstone and Veall (1996b) examine the use of percentile-t methods and BCa methods and find that neither works well at all in this simple nonlinear example. While it may be simply that any approximation will break down with enough nonlinearity (e.g., if small enough β_2 above), nonetheless we must conclude that the bootstrap is not yet a complete answer to the problems associated with finite-sample inference involving nonlinear quantities.

VI.   SUMMARY AND CONCLUSIONS

An important role of the bootstrap is to provide standard errors and other tools of inference when there are no other available methods. As discussed, such methods are no more accurate, at least in a theoretical sense, than analytic methods based on asymptotic approximations, but in the typical case where the use of the bootstrap can be justified asymptotically, bootstrap standard errors cautiously used may be valuable as "better than nothing" when analytic alternatives are not available. (This is particularly true when the appropriateness of the bootstrap is itself checked by simulation as discussed in the final paragraph.)

However, this survey has raised some questions with respect to whether bootstrap methods are necessarily more accurate than analytic, asymptotic methods. Only bootstrap refinements, such as the BCa method with its analytic bias correction based on an estimated "acceleration" coefficient or the percentile-t method, are more accurate in the Edgeworth sense. Yet even these have their flaws: it is possible for BCa $(1 - 2\alpha)$ confidence intervals to narrow as α increases; percentile-t confidence intervals are not transformation invariant. These flaws seem to be reflected in actual performance problems, for example the severe shortcomings of the bootstrap in the simple nonlinear case just discussed. Only some applications of the Barnard method, involving simulation under a null hypothesis which does not depend on nuisance parameters, yield bootstrap-based tests which are exact in finite samples.

Fortunately, while simulation reveals the problem it may also provide the answer. The first point to emphasize is that analytic, asymptotic methods often have bad finite-sample performance. However, the quality of finite-sample performance of such methods is usually based on speculation unless a simulation experiment is done. If a simulation is done which tends to verify the accuracy of asymptotic methods, further simulations are not a priority. But based on current understanding of the bootstrap, if the bootstrap and analytic, asymptotic methods differ it may be that the bootstrap results are slightly to be preferred (especially refined bootstrap methods), but what really needs to be done is further simulation study of the bootstrap itself. (See Beran 1988, 1990 or Beran and Ducharme 1991 for approaches along these lines, in which, for example, the percentile-t bootstrap is performed using bootstrap standard errors.) While this "bootstrapping the bootstrap" approach seems computationally tedious, computer time is the one thing that is getting cheaper and, except in the very rare case of exact tests, we now see that many kinds of results cannot be relied upon in finite samples unless they can be confirmed in a simulation experiment. In some sense, our answer to the student's question in the introduction as to how big a sample is big enough to use asymptotics reliably is, "It depends, but I can tell you a way that may help to find out for any given problem." Hence, while adding layers of simulation to our standard econometric practice may seem difficult, it is comforting to know that there is at least one feasible method to check the asymptotic approximations that are so widespread in econometrics.

ACKNOWLEDGMENTS

The author acknowledges the research assistance of Deb Fretz, the useful comments of a referee, and the financial support of the Social Sciences and Humanities Research Council of Canada.

REFERENCES

Arcones, M. and E. Giné (1992), On the Bootstrap of M-estimators and Other Statistical Functionals, in P. E. LePage and L. Billard (eds.), *Exploring the Limits of the Bootstrap*, Wiley, New York, 13–47.

Atkinson, S. E. and P. W. Wilson (1992), The Bias of Bootstrapped versus Conventional Standard Errors in the General Linear and SUR Models, *Econometric Theory*, 8, 258–275.

Barnard, G. A. (1963), Contribution to Discussion, *Journal of the Royal Statistical Society B*, 25, 294.

Bera, A., R. P. Byron, and C. M. Jarque (1981), Further Evidence on Asymptotic Tests for Homogeneity and Symmetry in Large Demand Systems, *Economics Letters*, 8, 101–105.

Beran, R. (1988), Prepivoting Test Statistics: A Bootstrap View of Asymptotic Refinements, *Journal of the American Statistical Association*, 83, 687–697.

Beran, R. (1990), Refining Bootstrap Simultaneous Confidence Sets, *Journal of the American Statistical Association*, 85, 417–426.

Beran, R. and G. Ducharme (1991), *Asymptotic Theory for Bootstrap Methods in Statistics*, Centre de Recherches Mathematiques, Montreal.

Bernard, J.-T. and M. R. Veall (1987), The Probability Distribution of Future Demand, *Journal of Business and Economics Statistics*, 5, 417–424.

Brown, B. and R. Mariano (1984), Residual-Based Stochastic Prediction and Estimation in a Nonlinear Simultaneous System, *Econometrica*, 52, 321–343.

Brownstone, D. (1990), Bootstrapping Improved Estimators for Linear Regression Models, *Journal of Econometrics*, 44, 171–187.

Chalfant, J., R. S. Gray, and K. J. White (1991), Evaluating Prior Beliefs in a Demand System: The Case of Meat Demand in Canada, *American Journal of Agricultural Economics*, 73, 476–490.

Davidson, R. and J. G. MacKinnon (1996), The Size Distortion of Bootstrap Tests, paper presented at the meetings of the Canadian Econometric Study Group, Waterloo, September.

Dijkstra, T. K. and J. H. Veldkamp (1988), Data-Driven Selection of Regressors and the Bootstrap, in T. K. Dijkstra (ed.), *On Model Uncertainty and Its Statistical Implications*, Springer, Berlin, 17–38.

Dufour, J.-M. (1995), Monte Carlo Tests with Nuisance Parameters: A General Approach to Finite-Sample Inference and Nonstandard Asymptotics in Econometrics, paper presented at the meetings of the Canadian Econometric Study Group, Montreal, September.

Dwass, M. (1957), Modified Randomization Tests for Nonparametric Hypotheses, *Annals of Mathematics and Statistics*, 28, 181–187.

Efron, B. (1979), Bootstrap Methods: Another Look at the Jackknife, *Annals of Statistics*, 7, 1–26.

Efron, B. (1982), *The Jackknife, the Bootstrap and Other Resampling Plans*, Society for Industrial and Applied Mathematics, Philadelphia, Pa.

Efron, B. (1987), Better Bootstrap Confidence Intervals (with discussion), *Journal of the American Statistical Association*, 82, 171–200.

Efron, B. and G. Gong (1983), A Leisurely Look at the Bootstrap, the Jackknife and Cross-Validation, *American Statistician*, 37, 36–48.

Efron, B. and R. J. Tibshirani (1993), *An Introduction to the Bootstrap*, Chapman and Hall, New York.

Fair, R. C. (1979), An Analysis of the Accuracy of Four Macroeconometric Models, *Journal of Political Economy*, 87, 701–718.

Fair, R. C. (1980), Estimating the Expected Predictive Accuracy of Econometric Models, *International Economic Review*, 21, 355-378.

Fan, Y. and Q. Li (1995), Bootstrapping J-type Tests for Non-nested Regression Models, *Economics Letters*, 48, 107–112.

Feldstein, M. (1971), The Error and Forecast in Econometric Models When the Forecast-Period Exogenous Variables Are Stochastic, *Econometrica*, 39, 55–60.

Freedman, D. A., W. Navidi, and S. C. Peters (1988), On the Impact of Variable Selection in Fitting Regression Equations, in T. K. Dijkstra (ed.), *On Model Uncertainty and Its Statistical Implications*, Springer, Berlin, 1–16.

Freedman, D. A. and S. C. Peters (1984a), Bootstrapping a Regression Equation: Some Empirical Results, *Journal of the American Statistical Association*, 79, 97–106.

Freedman, D. A. and S. C. Peters (1984b), Bootstrapping an Econometric Model: Some Empirical Results, *Journal of Business and Economics Statistics*, 2, 150–158.

George, P. J., E. H. Oksanen, and M. R. Veall (1995), Analytic and Bootstrap Approaches to Testing a Market Saturation Hypothesis, *Mathematics and Computers in Statistics*, 29, 311–315.

Geweke, J. (1986), Exact Inference in the Inequality Constrained Normal Linear Regression Model, *Journal of Applied Econometrics*, 1, 127–141.

Green, R., D. Rocke, and W. Hahn (1987), Standard Errors for Elasticities: A Comparison of Bootstrap and Asymptotic Standard Errors, *Journal of Business and Economics Statistics*, 4, 145–149.

Gregory, A. W. and G. W. Smith (1993), Statistical Aspects of Calibration in Macroeconomics, in G. S. Maddala, C. R. Rao, and H. D. Vinod (eds.), *Handbook of Statistics*, Vol. 11, Elsevier, Amsterdam, 703–719.

Gregory, A. W. and M. R. Veall (1985), On Formulating Wald Tests of Nonlinear Restrictions, *Econometrica*, 53, 1465–1468.

Hahn, J. (1996), A Note on Bootstrapping Generalized Method of Moments Estimators, *Econometric Theory*, 12, 187–197.

Hall, P. (1986), On the Number of Bootstrap Simulations Required to Construct a Confidence Interval, *Annals of Statistics*, 14, 1453–1462.

Hall, P. (1988), Theoretical Comparison of Bootstrap Confidence Intervals (with discussion), *Annals of Statistics*, 16, 927–952.

Hall, P. (1992), *The Bootstrap and Edgeworth Expansions*, Springer, Berlin.

Horowitz, J. L. (1994), Bootstrap-Based Critical Values for the Information Matrix Test, *Journal of Econometrics*, 61, 395–411.

Jeong, J. and G. S. Maddala (1993), A Perspective on Application of Bootstrap Methods, in G. S. Maddala, C. R. Rao, and H. D. Vinod (eds.), *Handbook of Statistics*, Vol. 11, Elsevier, Amsterdam, 573–610.

Kiviet, J. F. (1984), Bootstrap Inference in Lagged Dependent Variable Models, University of Amsterdam Working Paper.

Korajczyk, R. A. (1985), The Pricing of Forward Contracts for Foreign Exchange, *Journal of Political Economy*, 93, 346–368.

Krinsky, I. and A. L. Robb (1986), On Approximating the Statistical Properties of Elasticities, *Review of Economics and Statistics*, 68, 715–719.

Krinsky, I. and A. L. Robb (1990), On Approximating the Statistical Properties of Elasticities: A Correction, *Review of Economics and Statistics*, 72, 189–190.

Krinsky, I. and A. L. Robb (1991), Three Methods for Calculating the Statistical Properties of Elasticities: A Comparison, *Empirical Economics*, 16, 199–209.

Laitinen, K. (1978), Why Is Demand Homogeneity So Often Rejected?, *Economics Letters*, 1, 187–191.

LePage, P. E. and L. Billard (eds.) (1992), *Exploring the Limits of the Bootstrap*, Wiley, New York.

Li, H. and G. S. Maddala (1996), Bootstrapping Time Series Models (with discussion), *Econometric Reviews*, 15, 115–195.

Marais, M. L. (1986), On the Finite Sample Performance of Estimated Generalized Least Squares in Seemingly Unrelated Regression: Nonnormal Disturbances and Alternative Standard Error Estimators, Graduate School of Business, University of Chicago Working Paper.

Marriot, F. H. C. (1979), Barnard's Monte Carlo Test, How Many Simulations?, *Applied Statistics*, 28, 75–77.

McCloskey, D. N. and S. T. Ziliak (1996), The Standard Error of Regressions, *Journal of Economic Literature*, 34, 97–114.

McManus, W. S. and M. C. Rosalsky (1985), Are All Asymptotic Standard Errors Awful?, *Economics Letters*, 17, 243–245.

Meisner, J. F. (1979), The Sad Fate of the Asymptotic Slutsky Symmetry Test for Large Systems, *Economics Letters*, 2, 231–233.

Prescott, D. M. and T. Stengos (1987), Bootstrapping Confidence Intervals: An Application to Forecasting the Supply of Pork, *American Journal of Agricultural Economics*, 69, 266–273.

Raj, B. and T. G. Taylor (1989), Do "Bootstrap Tests" Provide Significance Levels Equivalent to the Exact Test? Empirical Evidence from Testing Linear Within-Equation Restrictions in Large Demand Systems, *Journal of Quantitative Economics*, 5, 73–89.

Rayner, R. K. (1991), Resampling Methods for Tests in Regression Models with Autocorrelated Errors, *Economics Letters*, 36, 281–284.

Rilstone, P. (1993), Some Improvements for Bootstrapping Regression Estimators under First-Order Serial Correlation, *Economics Letters*, 42, 335–339.

Rilstone, P. and M. R. Veall (1996a), Using Bootstrapped Confidence Intervals for Improved Inferences with Seemingly Unrelated Regression Equations, *Econometric Theory*, 12, 569–580.

Rilstone, P. and M. R. Veall (1996b), Comment on Bootstrapping Time Series Models by H. Li and G. S. Maddala, *Econometric Reviews*, 15, 177–181.

Rosalsky, M. C., R. Finke, and H. Theil (1984), The Downward Bias of Asymptotic Standard Errors of Maximum Likelihood Estimates of Nonlinear Systems, *Economics Letters*, 14, 207–211.

SHAZAM User's Reference Manual Version 7.0 (1993), McGraw-Hill, New York.

Shonkwiler, J. S. and H. Theil (1986), Some Evidence on the Power of Monte Carlo Tests in Systems of Equations, *Economics Letters*, 20, 53–54.

Taylor, R. G., J. S. Shonkwiler, and H. Theil (1986), Monte Carlo and Bootstrap Testing of Demand Homogeneity, *Economics Letters*, 20, 55–57.

Theil, H. (1971), *Principles of Econometrics*, Wiley, New York.

Theil, H. (1975), *Theory and Measurement of Consumer Demand*, Vol. 1, North-Holland, Amsterdam.

Theil, H. and J. S. Shonkwiler (1986), Monte Carlo Tests of Autocorrelation, *Economics Letters*, 20, 157–160.

Theil, H., J. S. Shonkwiler, and T. G. Taylor (1985), A Monte Carlo Test of Slutsky Symmetry, *Economics Letters*, 19, 331–332.

Tibshirani, R. (1988), Variance Stabilization and the Bootstrap, *Biometrika*, 75, 433-444.

van Giersbergen, N. P. A. and J. F. Kiviet (1993), How to Implement Bootstrap Hypothesis Testing in Regression Models, University of Amsterdam Working Paper.

Veall, M. R. (1987a), Bootstrapping the Probability Distribution of Peak Electricity Demand, *International Economic Review*, 28, 203–212.

Veall, M. R. (1987b), Bootstrapping and Forecast Uncertainty: A Monte Carlo Analysis, in I. B. MacNeill and G. J. Umphry (eds.), *Time Series and Econometric Modelling*, Reidel, Dordrecht, 373–384.

Veall, M. R. (1989), Applications of Computationally-Intensive Methods to Econometrics, *Bulletin of the International Statistical Institute, Proceedings of the 47th Session*, 3, 75–78.

Veall, M. R. (1992), Bootstrapping the Process of Model Selection: An Econometric Example, *Journal of Applied Econometrics*, 7, 93–99.

Veall, M. R. and K. F. Zimmermann (1993), The Size and Power of Integrability Tests in Dynamic Demand Systems, *Computational Statistics*, 8, 127–139.

Vinod, H. D. (1993), Bootstrap Methods: Applications in Econometrics, in G. S. Maddala, C. R. Rao, and H. D. Vinod (eds.), *Handbook of Statistics*, Vol. 11, Elsevier, Amsterdam, 629–661.

Vinod, H. D. and B. Raj (1988), Economic Issues in Bell System Divestiture: A Bootstrap Application, *Applied Statistics*, 37, 251–261.

13
Detection of Unusual Observations in Regression and Multivariate Data

Ali S. Hadi
Cornell University, Ithaca, New York

Mun S. Son
University of Vermont, Burlington, Vermont

I. INTRODUCTION AND MOTIVATING EXAMPLES

Regression and multivariate analysis techniques are commonly used to analyze data from many fields of study including economic and social sciences. These data often contain unusual observations. Unusual observations, usually referred to as outliers, are observations that do not conform to the pattern (model) suggested by the majority of the observations in a data set. If they exist in the data, outliers can distort the analysis of data and the conclusions based on the analysis. For example, outliers can distort parameter estimation, invalidate test statistics, and lead to incorrect statistical inference. We illustrate this point and the methods presented in this chapter by the following data set.

Example 1: Financial Data. In this chapter we make a repeated use of the following data set, which we refer to as the financial data. The data set was collected and thoroughly analyzed by Jeff M. Semanscin (a student in one of the authors' applied regression methods class) using Standard & Poor's Compustat PC Plus. The purpose of using the data here is only to illustrate the methods presented in this chapter. There are several variables in the data set, but for illustrative purposes we consider only a subset consisting of the following three variables:

X_1: Book value in dollars per share at the end of 1992

X_2: Net sales in millions of dollars in 1992

X_3: Sales-to-assets ratio in 1992

This subset of the data is shown in Table 1. It consists of 26 observations (financial companies).

Let us first think of this data set as a trivariate data. Figure 1 shows the trivariate scatter plot after it has been rotated to show the outliers in the data. The four outlying points marked on the graph are detected by the multivariate outliers detection method presented here. The mean and covariance matrix of the data are

$$\begin{pmatrix} 22.06 \\ 5.86 \\ 0.24 \end{pmatrix} \quad \text{and} \quad \begin{pmatrix} 168.14 & 23.44 & -1.51 \\ 23.44 & 35.24 & -0.62 \\ -1.51 & -0.62 & 0.18 \end{pmatrix} \quad (1)$$

respectively. When the outliers are deleted, mean and covariance matrix become

$$\begin{pmatrix} 21.65 \\ 5.36 \\ 0.12 \end{pmatrix} \quad \text{and} \quad \begin{pmatrix} 125.24 & 32.59 & -0.45 \\ 32.59 & 18.47 & -0.11 \\ -0.45 & -0.11 & 0.00 \end{pmatrix} \quad (2)$$

respectively. Note the dramatic effects of outliers on the estimated variances and covariances. To illustrate how the confidence regions can change substantially because of outliers, let us examine the bivariate scatter plot of X_1 versus X_2. The scatter plot,

Table I   Financial Data for 29 Financial Companies

Number	X_1	X_2	X_3	Number	X_1	X_2	X_3
1	14.58	26.961	0.17	14	2.46	0.247	0.26
2	21.15	4.816	0.15	15	15.75	2.213	0.09
3	19.26	3.394	0.08	16	25.19	2.825	0.09
4	39.93	5.455	0.08	17	34.30	7.281	0.07
5	6.12	1.495	1.52	18	39.26	7.382	0.10
6	32.25	9.112	0.09	19	30.80	9.228	0.08
7	32.43	11.078	0.08	20	13.51	3.364	0.14
8	8.30	0.806	0.28	21	15.96	3.840	0.08
9	16.68	4.461	0.08	22	6.75	0.196	0.16
10	24.79	14.559	0.09	23	31.84	15.372	0.12
11	19.26	1.114	1.82	24	1.52	0.909	0.17
12	19.42	4.190	0.09	25	17.70	5.099	0.23
13	27.02	2.009	0.08	26	57.44	5.067	0.10

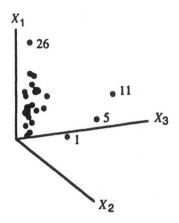

Figure I   Financial data: Trivariate Scatter plot of X_1, X_2, and X_3 with the outliers indicated by their numbers.

together with two ellipses expected to contain 95% of the observations, are shown in Figure 2. The larger ellipse is based on the mean and covariance matrix of the full data (all 26 observations) and the smaller ellipse is based on the mean and covariance matrix of the data without the outliers (indicated by their numbers on the scatter plot). Observe the huge difference between the two ellipses in terms of their

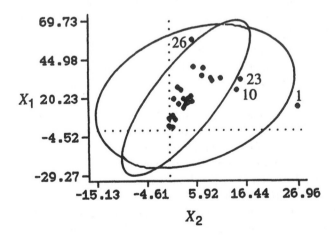

Figure 2   Financial data: Bivariate scatter plot of X_1 versus X_2 with two ellipses (expected to contain 95% of the observations). The larger ellipse is based on the mean and covariance matrix of the full data (all 26 observations), and the smaller ellipse is based on the mean and covariance matrix of the data without the outliers (indicated by their numbers).

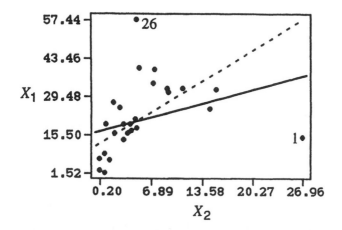

Figure 3   Financial data: Scatter plot of X_1 versus X_2 with two superimposed regression lines. The solid line is the least-squares regression line obtained using the full data. The dotted line is obtained when the outliers (1 and 26) are deleted.

sizes, orientations, and shapes. Note also how the larger ellipse is affected by the outliers. The larger ellipse detects only one observation as an outlier, whereas the smaller ellipse declares four observations as outliers.

Let us now think of the data as regression data, where distinction between dependent and independent variables has to be made. Consider, for example, the simple regression of X_1 on X_2. The scatter plot of X_1 versus X_2, with two super-imposed least-squares regression lines, is shown in Figure 3. The solid line is the least-squares regression line obtained using the full data. The dotted line is obtained when the two outlying observations 1 and 26 are deleted. Again, we obtain two sub-stantially different lines.

The above example shows that outliers can lead to misleading or erroneous conclusions. It is therefore important for data analysts to first identify and examine outliers if they exist in the data, before making conclusions based on data.

Before we proceed any further, we wish to make an important point. After reading the literature on outlier detection, some people are left with the incorrect impression that once outliers are identified, they should be deleted from the data and the analysis continues. We do not advocate automatic deletion (or even automatic down-weighing) of outliers because outliers are not necessarily bad observations. On the contrary, if they are correct, they may be the most informative points in the data. For example, they may indicate that the data did not come from a normally distributed population, as commonly assumed by almost all multivariate analysis techniques; or they may indicate that the model is not linear. For this reason the outliers should be

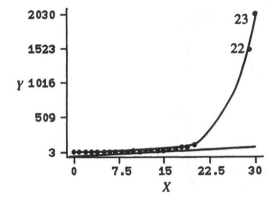

Figure 4   A scatter plot of population size, Y, versus time, X. The curve is obtained by fitting an exponential function to the full data. The straight line is the least-squares line when observations 22 and 23 are deleted.

called the *unusual* or even the *interesting* observations. To emphasize that outliers can be the most informative points in the data, we use the exponential growth data described in the following example.

Example 2: Exponential Growth Data.   Figure 4 is the scatter plot of two variables, the size of a certain population, Y, and time, X. As can be seen from the scatter of points, the majority of the points resemble a linear relationship between population size and time as indicated by the straight line in Figure 4. According to this model, points 22 and 23 in the upper right corner are outliers. If these points, however, are correct, they are the only observations in the data set that indicate that the data follow a nonlinear (e.g., exponential) model, such as the one shown in the graph. Think of this as a population of bacteria which increases very slowly over a period of time, then once somebody sneezes the population size explodes.

What do we do with outliers once they are identified? Because outliers can be the most informative observations in the data set, they should not be automatically discarded without justification. Instead, they should be examined to determine why they are outlying. Based on this examination, appropriate corrective actions can then be taken. These corrective actions include correction of errors in the data, deletion or downweighing outliers, transforming the data, considering a different model, redesigning the experiment or the sample survey, collecting more data, etc.

Outliers in multivariate data are intrinsically more difficult to detect than outliers in univariate and bivariate data. For example, in simple regression and in univariate, bivariate, and trivariate data, outliers can be detected easily by graphing the data. In higher than three dimensions and in the presence of multiple outliers, it is difficult to detect outliers because

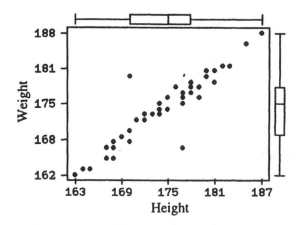

Figure 5   A scatter plot of weight versus height with the box plots on the margin of each variable. The two bivariate outliers cannot be detected by examining only univariate graphs such as box plots.

1. It is difficult to graph high-dimensional data. Furthermore, examining lower-dimensional plots of the data does not necessarily detect multivariate outliers. This point is illustrated using the following example.

Example 3: The Weight-Height Data.   The two bivariate outliers which appear in the scatter plot of weight versus height in Figure 5 cannot be detected by examining only univariate graphs (e.g., box plots) of weight and height separately. Note, however, that methods that work for higher-dimensional data will continue to work for lower-dimensional data, but the converse is not generally true.

2. If the data contain a single outlier, the problem of identifying the outlier is simple, but if the data contain more than one outlier the problem of identifying them becomes difficult due to the masking and swamping problems. Masking occurs when a method fails to detect outlying observations (false negative decisions). Swamping occurs when a method declares *good* points as outliers (false positive decisions). Masking and swamping are serious problems and are the cause of the failure of many outlier detection methods. Note that methods that work in the presence of multiple outliers will continue to work if the data contain a single outlier or no outliers at all, but the converse is not generally true.

In this chapter we discuss some methods for the detection of outliers in regression and multivariate data. We concentrate our attention on methods that have been recently developed and that require reasonable computational effort. Relevant outlier detection methods are found, for example, in Rohlf (1975), Hawkins

(1980), Schwager and Margolin (1982), Beckman and Cook (1983), Barnett and Lewis (1984), Hampel et al. (1986), Bacon-Shone and Fung (1987), Rousseeuw and Leroy (1987), Fung (1988), Rasmussen (1988), and Caroni and Prescott (1992). The rest of the chapter is organized as follows. Section II describes a unified framework for the detection of outliers in both multivariate and regression data. Sections III and IV discuss the specifics of this unified framework for the detection of outliers in multivariate and regression data, respectively. Section V deals with the problem of outliers detection in very large data sets.

II. A UNIFIED FRAMEWORK FOR OUTLIER DETECTION

In regression analysis situations the data set contains a response variable y, consisting of n observations, and a matrix X, consisting of n rows (observations) and p columns (covariates). In multivariate analysis situations there is no y. The data set contains only X, but we think of X as a random sample generated from a multivariate elliptically symmetric distribution such as a multivariate normal or a multivariate t-distribution. Our objective is to identify outliers if they exist in the data set in each of these two situations.

It has been long recognized that classical methods, such as least-squares residuals or Mahalanobis distances, are not effective in the detection of outliers because they are not robust; that is, they are affected by the outliers that they are supposed to detect. One way out of this problem is to replace classical methods by robust methods, which produce estimates that are resistant to the presence of outliers and/or to violations of distributional assumptions. Indeed, several books have been devoted either entirely or in large part to robust methods and/or outlier detection techniques; see, for example, Barnett and Lewis (1984), Hawkins (1980), Huber (1981), Hampel et al. (1986), Rousseeuw and Leroy (1987), and Chatterjee and Hadi (1988). Other relevant articles include Maronna (1976), Campbell (1980), Rousseeuw and Yohai (1984), and Lopuha (1989).

Robust methods have been suggested for many years now, but they have not yet been widely used in practice because they involve extensive computations. The most widely known of the robust methods are the least median of squares (for regression data) and minimum-volume ellipsoids (for multivariate data) estimators. These methods are highly effective because they are not affected by outliers, but they are computationally prohibitive, like other robust estimation methods.

Another way out of the problem has been recently developed by Hadi (1992a, 1994) and Hadi and Simonoff (1993, 1994, 1997). The main idea of these methods is to first form a *basic* subset of about half of the data which is presumably free of outliers, then add observations that are consistent with the basic subset. If all the observations are added to the basic subset, the data set is declared to be free of outliers; otherwise the observations that are not consistent with the basic subset are declared to be outliers. To determine whether the observations are consistent with the basic subset, a suitable metric is chosen to measure the distance between each observation in the data and the center of the observations in the basic subset (multivariate data) or the least-squares line based on the observations in the basic subset (regression data). Thus, the method is carried out in two stages; the first is to find the basic subset, and the second is to test for the outliers. To implement this method the following questions have to be addressed:

1. How do we divide the data set into basic and nonbasic subsets?
2. What distance measure should we use?
3. How large must the distances be for the corresponding observations to be declared as outliers?

The answers to these questions depend on whether we are dealing with multivariate or regression data.

III. OUTLIERS IN MULTIVARIATE DATA

In multivariate analysis situations the data matrix X is viewed as a sample from a multivariate elliptically symmetric distribution. Thus, an elliptically symmetric distribution is assumed to be the underlying model. To detect the outliers in X we need to measure the distance between the ith observation x_i and the fitted model. If a good distance measure is chosen, then observations with large values would indicate that they are outliers. Since the model is elliptically symmetric, an appropriate measure of distance would be the elliptical distance

$$d_i(c, V) = \sqrt{(x_i - c)^T V^{-1}(x_i - c)}, \qquad i = 1, \ldots, n \qquad (3)$$

The elliptical distance $d(c, V)$ measures the distance between the ith observation, x_i, and a location (center) estimator c, relative to a measure of dispersion, V. The classical choices of c and V are $c = \bar{x}$ (the sample mean) and $V = S$ (the sample covariance matrix), respectively. This choice of c and V gives

$$d_i(\bar{x}, S) = \sqrt{(x_i - \bar{x})^T S^{-1}(x_i - \bar{x})}, \qquad i = 1, \ldots, n \qquad (4)$$

which is known as the Mahalanobis distance. If the data come from a p-variate normal distribution, the $d_i^2(\bar{x}, S)$ follows approximately a χ^2-distribution with p degrees of freedom. Thus, using an α-level of significance, values of $d_i(\bar{x}, S)$ larger than $\sqrt{\chi^2(p, \alpha/n)}$ are declared to be outliers.

Unfortunately, $d_i(\bar{x}, S)$ is affected by outliers. For example, some of the outliers may still have small values of $d_i(\bar{x}, S)$ (masking) and some of the observations which are not outliers may have large $d_i(\bar{x}, S)$ (swamping). This is illustrated by the following example.

Table 2 Financial Data: The Mahalanobis Distances for the Trivariate and Bivariate Data Sets Graphed in Figures 1 and 2

| Number | Data set | | Number | Data set | |
	Trivariate	Bivariate		Trivariate	Bivariate
1	3.97	3.96	14	1.67	1.60
2	0.33	0.18	15	0.92	0.69
3	0.68	0.43	16	0.78	0.66
4	1.47	1.47	17	0.96	0.95
5	3.03	1.29	18	1.34	1.34
6	0.85	0.85	19	0.79	0.77
7	1.04	1.04	20	0.85	0.70
8	1.24	1.20	21	0.77	0.51
9	0.69	0.43	22	1.49	1.34
10	1.49	1.49	23	1.63	1.63
11	3.79	0.80	24	1.77	1.63
12	0.57	0.31	25	0.36	0.34
13	0.99	0.89	26	2.92	2.91

Example 4: Mahalanobis Distance. Consider the financial data described in Example 1. The Mahalanobis distances for the trivariate and bivariate data sets graphed in Figures 1 and 2 are given in Table 2. The corresponding index plots are shown in Figure 6a and b, respectively. In both cases, the Mahalanobis distance declares only observation 1 as an outlier (the distances for observation 1 of 3.97 and 3.96 are slightly larger than the cutoff point of 3.86). By comparison with the

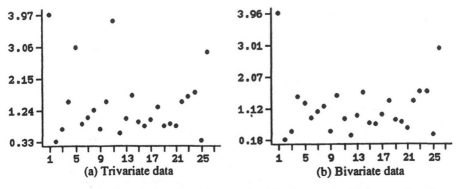

Figure 6 Financial data: Index plot of the Mahalanobis distance for (a) the trivariate data $\{X_1, X_2, X_3\}$ and (b) the bivariate data $\{X_1$ and $X_2\}$.

trivariate and bivariate scatter plots in Figures 1 and 2, we see that the Mahalanobis distance failed to identify observations 5, 11, and 26 in the trivariate case and observations 10, 23 and 26 in the bivariate case.

The Mahalanobis distance fails to detect the outliers because it depends on the sample mean and covariance matrix, which are known to be sensitive to the presence of outliers. One way to solve this problem is to replace the mean and the covariance matrix by more robust estimators of the location and scale. There are many robust estimators for multivariate data. One problem with robust methods, however, is that they are computationally extensive and at times practically infeasible. This may explain why robust statistics have not been widely implemented in statistical packages, although they have appeared in the literature for so many years. Alternatively, using the mean and the covariance matrix of the basic subset, we obtain

$$d_i(\overline{\mathbf{x}}_b, \mathbf{S}_b) = \sqrt{(\mathbf{x}_i - \overline{\mathbf{x}}_b)^T \mathbf{S}_b^{-1} (\mathbf{x}_i - \overline{\mathbf{x}}_b)}, \qquad i = 1, \ldots, n \qquad (5)$$

where $\overline{\mathbf{x}}_b$ and $\mathbf{S}_b$ are the mean and the covariance matrix of the basic subset, respectively. Thus, the distance $d_i(\overline{\mathbf{x}}_b, \mathbf{S}_b)$ depends on the observations contained in the basic subset. If the data come from a p-variate normal distribution, then $d_i^2(\overline{\mathbf{x}}_b, \mathbf{S}_b)$ follows approximately a χ^2-distribution with p degrees of freedom. If the basic subset is free from outliers, then $\overline{\mathbf{x}}_b$ and $\mathbf{S}_b$ will not be affected by the outliers and the $d_i(\overline{\mathbf{x}}_b, \mathbf{S}_b)$ would be effective in the detection of the outliers. Thus, an important task remaining here is how to obtain a basic subset which is likely to be outlier-free. Let h be the integer part of $(n + p + 1)/2$, $\mathbf{m}$ the vector containing the coordinatewise medians, and

$$\mathbf{A} = (n - 1)^{-1} \sum_{i=1}^{n} (\mathbf{x}_i - \mathbf{m})(\mathbf{x}_i - \mathbf{m})^T \qquad (6)$$

The two stages of the method in the multivariate case are given in Algorithms 1 and 2 (Hadi 1994). In these algorithms $\overline{\mathbf{x}}_b$ and $\mathbf{S}_b$ are the mean and covariance matrix of the observations in the current basic subset $\mathbf{X}_b$.

Algorithm 1: Finding the Basic Subset

Input: An $n \times p$ matrix of multivariate data.

Output: A basic subset of size h observations that are likely to be free from outliers.

> *Step 0:*   Compute $d_i(\mathbf{m}, \mathbf{A})$. Let $\overline{\mathbf{x}}_b$ and $\mathbf{S}_b$ be the mean and the covariance matrix of the observations with the h smallest values of $d_i(\overline{\mathbf{m}}, \mathbf{A})$. Compute $d_i(\overline{\mathbf{x}}_b, \mathbf{S}_b)$. Rearrange the observations in ascending order according to $d_i(\overline{\mathbf{x}}_b, \mathbf{S}_b)$. Divide the observations into two initial subsets: a *basic* subset containing the first $p + 1$ observations and a *nonbasic* subset containing the remaining $n - p - 1$ observations.

Step 1:   If the basic subset is of full column rank, compute $d_i(\overline{\mathbf{x}}_b, \mathbf{S}_b)$, where $\overline{\mathbf{x}}_b$ and $\mathbf{S}_b$ are the mean and covariance matrix of the observations in the current basic subset. If the basic subset is not of full rank, increase the basic subsets by as many observations as needed for the basic subset to become full rank. If needed, the observations are added according to their ranked order.

Step 2:   Rearrange the n observations in ascending order according to $d_i(\overline{\mathbf{x}}_b, \mathbf{S}_b)$. Let s be the number of observations in the current basic subset. Divide the observations into two initial subsets: a *basic* subset containing the first $s + 1$ observations and a *nonbasic* subset containing the remaining $n - s - 1$ observations.

Step 3:   Repeat Steps 1 and 2 until the basic subset contains h observations.

Algorithm 2: Testing for Outliers

Input: An $n \times p$ matrix of multivariate data and a basic subset $\mathbf{X}_b$ of size h observations obtained by Algorithm 1.

Output: The set of observations identified as outliers.

Step 0:   Let $s = h$ and $\overline{\mathbf{x}}_b$ and $\mathbf{S}_b$ be the mean and the covariance matrix of the observations in the current basic subset.

Step 1:   Compute $d_i(\overline{\mathbf{x}}_b, \mathbf{S}_b)$. Rearrange the observations in ascending order according to $d_i(\overline{\mathbf{x}}_b, \mathbf{S}_b)$.

Step 2:   Let

$$c_\alpha = \left(1 + \frac{2}{n - 1 - 3p} + \frac{p+1}{n - p}\right)\sqrt{\chi^2\left(p, \frac{\alpha}{n}\right)} \qquad (7)$$

be the Bonferroni-adjusted critical value based on a χ^2-distribution with p degrees of freedom multiplied by a correction factor. If min $\{d_i(\overline{\mathbf{x}}_b, \mathbf{S}_b); i \in nonbasicsubset\} \geq c_\alpha$, stop and declare all observations with $d_i(\overline{\mathbf{x}}_b, \mathbf{S}_b)$ to be outliers. Otherwise, go to Step 3.

Step 3:   For a new basic subset by taking the first $s + 1$ observations ordered according to $d_i(\overline{\mathbf{x}}_b, \mathbf{S}_b)$. If $n = s + 1$, then declare no outliers in the data set and stop; otherwise go to Step 1.

If desired, the final basic subset obtained in Algorithm 2 can be used to compute the final distance for each of the observations in the data set. The method proposed here is easy to compute. It is also effective in identifying the outliers when tried on real as well as simulated data. This method has been implemented in some commercially available statistical packages (e.g., *Stata*; see Gould and Hadi 1993).

Example 5: Financial Data.   Consider again the financial data described in Example 1. The $d_i(\overline{\mathbf{x}}_b, \mathbf{S}_b)$ for the trivariate and bivariate data sets graphed in Figures 1

Table 3   Financial Data: The $d_i(\bar{\mathbf{x}}_b, \mathbf{S}_b)$ for the Trivariate and Bivariate Data Sets Graphed in Figures 1 and 2

	Data set			Data set	
Number	Trivariate	Bivariate	Number	Trivariate	Bivariate
1	5.73	10.27	14	1.73	1.27
2	0.44	0.22	15	1.03	0.51
3	0.90	0.23	16	0.92	1.07
4	1.75	1.87	17	0.94	1.01
5	21.93	1.02	18	1.53	1.40
6	0.77	1.34	19	0.74	1.43
7	1.03	2.05	20	0.59	0.56
8	2.04	0.91	21	1.10	0.41
9	1.05	0.54	22	1.10	1.06
10	2.10	4.05	23	1.99	3.83
11	27.63	1.19	24	1.46	1.38
12	0.69	0.12	25	1.55	0.68
13	1.26	1.60	26	3.86	4.01

and 2 are given Table 3. The corresponding index plots are shown in Figure 7a and b, respectively. In both cases, the Mahalanobis distance declares only observation 1 as an outlier (the distances for observation 1 of 3.97 and 3.96 are slightly larger than the cutoff point of 3.86). By comparison with the trivariate and bivariate scatter plots in Figures 1 and 2, we see that $d_i(\bar{\mathbf{x}}_b, \mathbf{S}_b)$ identifies all outliers in both data sets.

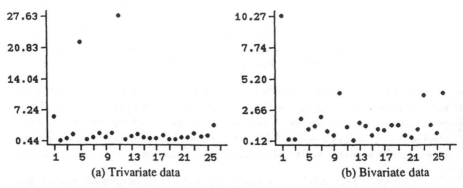

Figure 7   Financial data: Index plot of $d_i(\bar{\mathbf{x}}_b, \mathbf{S}_b)$ for (a) the trivariate data $\{X_1, X_2, X_3\}$ and (b) the bivariate data $\{X_1, X_2\}$.

IV. OUTLIERS IN REGRESSION DATA

In regression analysis situations the data consist of an n-response vector $\mathbf{y}$ and an $n \times p$ covariate matrix $\mathbf{X}$ assumed to be of full-column rank. A model that relates $\mathbf{y}$ and $\mathbf{X}$ is usually assumed to be of a linear form, $\mathbf{y} = \mathbf{X}\beta + \epsilon$, where β is the vector of regression coefficients, ϵ is a vector of random errors assumed to have a multivariate normal distribution with mean $\mathbf{0}$ and covariance matrix $\sigma^2 \mathbf{I}_n$, σ^2 is an unknown scalar, and $\mathbf{I}_n$ is an identity matrix of order n.

Unlike the case of multivariate analysis data, in which all unusual observations are labeled as outliers, in the regression analysis case unusual observations are classified into three classes: *outliers*, *high-leverage points*, and *influential observations* (see, e.g., Chatterjee and Hadi 1986). We deal with the detection of each of these types of observations separately.

A. Detection of Outliers

As we mentioned, outliers are observations that do not conform to the pattern (model) suggested by the majority of the observations in a data set. To detect the outliers in $\mathbf{X}$ we need to measure the distance between the ith observation $\mathbf{x}_i$ and the fitted model. The classical choice of a distance here is the least-squares standardized residual

$$r_i(\sigma) = \frac{e_i}{\sigma\sqrt{1 - p_{ii}}}, \qquad i = 1, \ldots, n \tag{8}$$

where e_i is the ith element of the residual vector $\mathbf{e} = (\mathbf{I}_n - \mathbf{P})\mathbf{y}$ and p_{ii} is the ith diagonal element of the projection matrix

$$\mathbf{P} = \mathbf{X}(\mathbf{X}^T\mathbf{X})^{-1}\mathbf{X}^T \tag{9}$$

Replacing σ by $\hat{\sigma} = \sqrt{\mathbf{e}^T\mathbf{e}/(n - p)}$, we obtain

$$r_i(\hat{\sigma}) = \frac{e_i}{\hat{\sigma}\sqrt{1 - p_{ii}}}, \qquad i = 1, \ldots, n \tag{10}$$

which is known as the *internally Studentized residual*. Replacing σ or $\hat{\sigma}_{(i)}$, we obtain the *externally Studentized residual*,

$$r_i(\hat{\sigma}_{(i)}) = \frac{e_i}{\hat{\sigma}_{(i)}\sqrt{1 - p_{ii}}}, \qquad i = 1, \ldots, n \tag{11}$$

where $\hat{\sigma}_{(i)}$ is the estimate of σ when the ith observation is deleted. For simplicity of notation, we write r_i and r_i^* instead of $r_i(\hat{\sigma})$ and $r_i(\hat{\sigma}_{(i)})$, respectively. Note that r_i and r_i^* are related by

$$r_i^* = r_i\sqrt{\frac{n - p - 1}{n - p - r_i^2}} \tag{12}$$

hence r_i and r_i^* are equivalent.

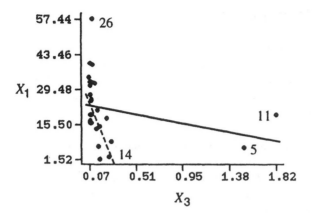

Figure 8   Financial data: Scatter plot of X_1 versus X_3. The solid line is the regression line based on all 26 observations and the dotted line is obtained when the observations 5, 11, and 26 are deleted.

Outliers tend to have large absolute values of r_i or r_i^*, but unfortunately this is not always the case. Outliers can have small residuals (masking), and observations which are not outliers can have large residuals (swamping). The reason for this, again, is that r_i and r_i^* are affected by outliers. For example, the two points in the lower right corner of the scatter plot of X_1 versus X_3 in Figure 8 have very small residuals because they are close to the solid regression line, which is based on all 26 observations, yet they are very far from the dotted regression line, which is obtained when the outliers 5, 11, and 26 are deleted. Thus, using the r_i or r_i^*, observations 5 and 11 will be masked. On the other hand, observation 14, which is close to the dotted regression line, is far from the solid regression line (swamping).

To deal with masking and swamping problems, we replace r_i and r_i^* by a more robust residual. One alternative is to use the least median of squares (LMS) residual. However, the LMS is computationally intensive. Another alternative is to define the residual with respect to the fitted model based only on a basic subset which is likely to be free from the outliers. Let y_b and X_b be the observations in the basic subset. Let $\hat{\beta}_b$ and $\hat{\sigma}_b^2$ be the least-squares estimate of β and σ^2 based on the observations in the basic subset, respectively. A robust version of the residual (which can actually be thought of as the scaled prediction error) can be defined as

$$
d_i = \begin{cases} \dfrac{|y_i - \mathbf{x}_i^T \hat{\beta}_b|}{\hat{\sigma}_b \sqrt{1 - \mathbf{x}_i^T (\mathbf{X}_b^T \mathbf{X}_b)^{-1} \mathbf{x}_i}} & \text{if } i \in \text{basic subset} \\[4mm] \dfrac{|y_i - \mathbf{x}_i^T \hat{\beta}_b|}{\hat{\sigma}_b \sqrt{1 + \mathbf{x}_i^T (\mathbf{X}_b^T \mathbf{X}_b)^{-1} \mathbf{x}_i}} & \text{if } i \in \text{basic subset} \end{cases} \tag{13}
$$

The two stages of the method in the regression case are given in Algorithms 3 and 4 (Hadi and Simonoff 1993). In these algorithms $\hat{\beta}_b$ and $\hat{\sigma}_b^2$ are the least-squares estimates of β and σ^2 based on the current basic subset y_b and X_b.

Algorithm 3: Finding the Basic Subset

Input: An n-vector y and an $n \times p$ matrix X.

Output: A basic subset of size h observations that is likely to be free from outliers.

Step 0:   Compute r_i. Rearrange the observations in ascending order according to $|r_i|$. Divide the observations into two initial subsets: a *basic* subset containing the first $p + 1$ observations and a *nonbasic* subset containing the remaining $n - p - 1$ observations.

Step 1:   If X_b is of full-column rank, compute d_i in (13). If X_b is not of full rank, increase the basic subsets by as many observations as needed for X_b to become full rank. If needed, the observations are added according to their ranked order.

Step 2:   Rearrange the n observations in ascending order according to d_i. Let s be the number of observations in the current basic subset. Divide the observations into two initial subsets: a *basic* subset containing the first $s + 1$ observations and a *nonbasic* subset containing the remaining $n - s - 1$ observations.

Step 3:   Repeat Steps 1 and 2 until the basic subset contains h observations.

Algorithm 4: Testing for Outliers

Input: An n-vector y, an $n \times p$ matrix X, and a basic subset y_b and X_b of size h observations obtained by Algorithm 3.

Output: The set of observations identified as outliers.

Step 0:   Let $s = h$ and let $\hat{\beta}_b$ and $\hat{\sigma}_b^2$ be the least-squares estimates of β_b and σ_b^2 based on the current basic subset y_b and X_b.

Step 1:   Compute d_i as in (13) but use the current y_b and X_b. Rearrange the observations in ascending order according to d_i.

Step 2:   Let

$$c_\alpha = t\left(s - p, \frac{\alpha}{2(s + 1)}\right) \tag{14}$$

be the Bonferroni-adjusted critical value based on a t-distribution with $s - p$ degrees of freedom. If $\min\{d_i; i \in \text{nonbasic subset}\} \geq c_\alpha$, stop and declare all observations with $d_i \geq c_\alpha$ to be outliers. Otherwise, go to Step 3.

Step 3:   Form a new basic subset by taking the first $s+1$ observations ordered according to d_i. if $n = s + 1$, then declare no outliers in the data set and stop; otherwise go to Step 1.

Table 4   Financial Data: The $d_i(\bar{x}_b, S_b)$ for the Trivariate
and Bivariate Data Sets Graphed in Figures 1 and 2

Number	d_i	Number	d_i
1	−3.70	14	−1.25
2	0.07	15	−0.02
3	0.16	16	1.01
4	2.24	17	1.15
5	−1.06	18	1.75
6	0.50	19	0.30
7	0.09	20	−0.55
8	−0.64	21	−0.35
9	−0.40	22	−0.70
10	−1.84	23	−1.09
11	0.67	24	−1.51
12	0.00	25	−0.41
13	1.43	26	4.29

If desired, the final basic subset obtained in Algorithm 4 can be used to compute the final residual for each of the observations in the data set.

Notice the similarity between Algorithms 1 and 2 for the detection of outliers
in multivariate data and Algorithms 3 and 4 for the detection of outliers in regression
data. They are special cases of the unified framework discussed in Section 2.

Example 6: Financial Data.   Consider again the financial data described in
Example 1. The final d_i obtained by Algorithm 4 for the simple regression of X_1 on
X_2 is given in Table 4. The corresponding index plot is shown in Figure 9. Comparing
these results with the scatter plot in Figure 3, we see that the method identified the
two outliers marked on the graph.

B.   Detection of High-Leverage Points

Observations in the **X**-space that exert undue leverage in determining the fitted line
are called *high-leverage* points. This can be seen by writing the ith fitted value as

$$\hat{y}_i = \sum_{j=1}^{n} p_{ij}y_j = p_{ii} + \sum_{j \neq i} p_{ij}y_j \tag{15}$$

where p_{ij} is the ijth element of the matrix **P** in (9), and observing the p_{ii} is the
weight or leverage attached to y_i in determining the ith fitted values. Two important
and interesting properties of p_{ii} are

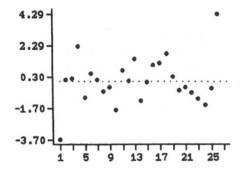

Figure 9   Financial data: Index plot of the final d_i obtained by Algorithm 4 for the simple regression of X_1 on X_2. Two observations (1 and 26) are declared outliers.

$$0 \leq p_{ii} \leq 1 \tag{16}$$

and

$$p_{ii} + \frac{e_i^2}{e^T e} \leq 1 \tag{17}$$

For proofs of these and many other properties of the matrix $\mathbf{P}$ and its elements, see Chatterjee and Hadi (1988, Chapter 2). Consequently, the larger the p_{ii} the higher the leverage of the y_i in determining $\hat{y}_i$. For example, in the extreme case where $p_{ii} = 1$, we have $\hat{y}_i = y_i$ and $e_i = 0$; that is y_i completely determines $\hat{y}_i$. Thus, observations with large values of p_{ii} are called *high-leverage points* and the p_{ii}'s are called *leverage values*.

The presence of high-leverage points, individually or in groups, makes it very difficult to identify the outliers. Therefore, the $\mathbf{X}$ data should be examined for the presence of high-leverage points. High-leverage points tend to have large values of p_{ii}. Unfortunately, high-leverage points may not always have large leverage values because a group of points can collaborate together and collectively induce high leverage, although their individual leverage values are not high. In other words, while all points with large leverage values are high-leverage points, some observations with small leverage values may be collectively a high-leverage group. Such a group of high-leverage points can be identified by exploiting the relationship between the concept of high-leverage and outliers in the multivariate $\mathbf{X}$-space. One can think of high-leverage points simply as outliers in the $\mathbf{X}$-space. Thus, to identify high-leverage points, one can think of $\mathbf{X}$ as multivariate data and apply Algorithms 1 and 2 to identify the outliers in $\mathbf{X}$ (see also Rousseeuw and van Zomeren 1990, 1991). In the context of regression, these outliers are called high-leverage points.

C. Detection of Influential Observations

Observations, individually or collectively, that excessively influence the fitted regression equation as compared to the other observations are called *influential* observations. A bewilderingly large number of statistical quantities have been proposed to study outliers and influential observations in regression analysis (Chatterjee and Hadi 1986). One of the commonly used measures of influence is known as Cook's distance (Cook 1977), which is defined as

$$C_i = \frac{(\hat{\beta} - \hat{\beta}_{(i)})^T \mathbf{X}^T \mathbf{X}(\hat{\beta} - \hat{\beta}_{(i)})}{p\hat{\sigma}^2} \tag{18}$$

where $\hat{\beta}_{(i)}$ is the least-squares estimate of β when the ith observation is deleted. A comparison with (3) shows that C_i is the squared elliptical distance between $\hat{\beta}$ and $\hat{\beta}_{(i)}$. Thus, a large value of C_i indicates that the ith observation is influential on $\hat{\beta}$. After some algebraic manipulations, one can show that

$$C_i = \frac{r_i^2}{p} \frac{p_{ii}}{1 - p_{ii}} \tag{19}$$

from which it follows that C_i is a multiplicative function of the residual and leverage values. Although a large value of C_i indicates that the ith observation is influential on $\hat{\beta}$, a small value of C_i does not necessarily indicate that the ith observation is not influential. This can be seen from (17) because a high-leverage point tends to have a small residual, hence a small value of C_i. From (19) it can be seen that an observation will be influential on $\hat{\beta}$ if it is an outlier (large value of $|r_i|$), a high-leverage point (large value of p_{ii}), or both. Hadi (1992b) utilizes this idea and develops the additive influence measure

$$H_i = \frac{p_{ii}}{1 - p_{ii}} + \frac{p}{1 - p_{ii}} \frac{d_i^2}{1 - d_i^2} \tag{20}$$

where $d_i^2 = e_i^2/\mathbf{e}^T\mathbf{e}$ is the square of the ith normalized residual. The first term is a leverage term which measures outlyingness in the $\mathbf{X}$-space. The function $p_{ii}/(1-p_{ii})$ is known as the potential function. The second term in H_i is a residual term which measures outlyingness of the observation in the $\mathbf{y}$-direction. Since H_i is an additive function of the residual and potential functions, it will be large if the observation is an outlier in either the $\mathbf{X}$-space, the $\mathbf{y}$-space, or both. To determine which is the case, Hadi (1992b) suggests plotting the potential versus the residual function, that is, the scatter plot of

$$\frac{p_{ii}}{1 - p_{ii}} \quad \text{versus} \quad \frac{p}{1 - p_{ii}} \frac{d_i^2}{1 - d_i^2} \tag{21}$$

This plot is referred to as the potential-residual (P-R) plot. In the P-R plot, high-leverage points are located in the upper area of the plot and observations with large

Table 5 Financial Data: Measures of Outlyingness, Leverage, and Influence Obtained When X_1 is Regressed on X_2

Number	r_i^*	p_{ii}	C_i	H_i	Number	r_i^*	p_{ii}	C_i	H_i
1	−2.89	0.54	3.80	1.80	14	−1.33	0.07	0.07	0.23
2	−0.02	0.04	0.00	0.04	15	−0.31	0.05	0.00	0.07
3	−0.09	0.05	0.00	0.05	16	0.41	0.05	0.00	0.07
4	1.51	0.04	0.04	0.24	17	0.91	0.04	0.02	0.11
5	−1.07	0.06	0.04	0.16	18	1.33	0.04	0.04	0.20
6	0.65	0.05	0.01	0.09	19	0.52	0.05	0.01	0.08
7	0.56	0.07	0.01	0.10	20	−0.55	0.05	0.01	0.07
8	−0.85	0.07	0.03	0.13	21	−0.38	0.04	0.00	0.06
9	−0.35	0.04	0.00	0.05	22	−0.95	0.07	0.04	0.16
10	−0.25	0.12	0.00	0.15	23	0.29	0.14	0.01	0.17
11	0.03	0.06	0.00	0.07	24	−1.45	0.07	0.07	0.25
12	−0.12	0.04	0.00	0.04	25	−0.31	0.04	0.00	0.05
13	0.61	0.06	0.01	0.09	26	3.53	0.04	0.17	1.10

prediction error are located in the area to the right. Both H_i and the P-R plot have been implemented in commercially available statistics packages such as *Data Desk* and *Stata*.

Example 7: Financial Data.   Consider again the financial data in Example 1. The r_i^*, p_{ii}, C_i, and H_i obtained when X_1 is regressed on X_2 are given in Table 5. The corresponding index plots are shown in Figure 10. The P-R plot is given in Figure 11. Observation 26 has the largest value of r_i^*, which is the only observation that exceeds the cutoff value of $t(n - p, \alpha/2n) = 3.50$. Observation 1 is the only high-leverage point in the data. Observation 1 is also identified by C_i as the only influential observation in the data. Two observations (1 and 26) are identified by H_i as influential. These two observations are separated from the rest of other points in the P-R plot. Observation 26 is an outlier because it is located in the lower-right corner of the plot. Observation 1 is both an outlier and a leverage point.

V. DEALING WITH VERY LARGE DATA SETS

The methods presented in Sections III and IV have been shown to perform well in many real-life and simulated data. They produce results in a reasonable amount of time for small to medium data sets. But for large data sets, increasing the basic subset one observation at a time can be time consuming. Hadi and Velleman (1997) adapt these methods to large data sets as follows:

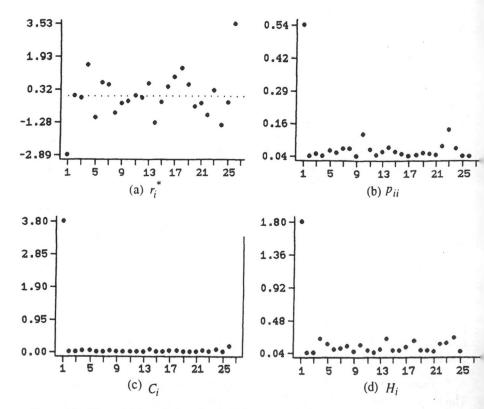

Figure 10   Financial data: Index plot of (a) r_i^*, (b) p_{ii}, (c) C_i, and (d) H_i obtained when X_1 is regressed on X_2.

1. In some applications the data analyst may have some reasons to believe that a certain subset of the data is free from outliers. In these applications, this subset can be used as the first basic subset instead of the one obtained in Step 0 of Algorithms 1 and 3. Actually, this suggestion is applicable to data sets of all sizes, but the computational savings that result from eliminating Step 0 of Algorithms 1 and 3 increase as the size of the data set increases. Additional computational savings can also be achieved if the size of the chosen basic subset is larger than the initial size of $p + 1$.

2. In Step 2 of Algorithms 1 and 3, the basic subset size is increased by one observation at a time. Computational savings can be realized here by adding to the basic subset all observations that are below a certain cutoff point. In this way the subset size grows more rapidly than in the original algorithms.

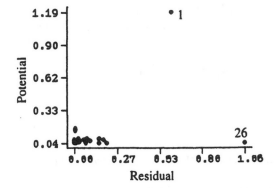

Figure 11   Financial data: The Potential-Residual plot obtained when X_1 is regressed on X_2.

3. In Algorithms 2 and 4, testing starts only after the size of the basic subset reaches h. Computational time can be saved by starting testing as soon as the basic subset stabilizes or includes a certain prespecified number of observations per parameter (e.g., six or more observations per parameter).

Note that in Algorithms 1–4, the nonbasic subset size can be at most $n - h$ observations. These observations, which constitute less than 50% of the data, are interpreted as outliers. The above modifications imply that the nonbasic subset can be as large as $n - s$, where s is the size of the initial basic subset chosen by the data analyst. In this case, the basic and nonbasic subsets are regarded as two distinct subgroups in the data set. Each of these subgroups can be further divided into two smaller subsets by applying the above modified algorithms as many times as desired. In this way, the modified algorithms can be thought of as methods for finding homogeneous groups, rather than finding outliers, in data sets.

REFERENCES

Bacon-Shone, J. and W. K. Fung (1987), A New Graphical Method for Detecting Single and Multiple Outliers in Univariate and Multivariate Data, *Journal of the Royal Statistical Society* (C) 36, 153–162.

Barnett, V. and T. Lewis (1984), *Outliers in Statistical Data*, 2nd ed., Wiley, New York.

Beckman, R. and R. D. Cook (1983), Outlier. . . s, *Technometrics*, 25, 119–149.

Campbell, N. A. (1980), Robust Procedures in Multivariate Analysis. I: Robust Covariance Estimation, *Applied Statistics*, 29, 231–237.

Caroni, C. and P. Prescott (1992), Sequential Application of Wilks's Multivariate Outlier Test, *Applied Statistics*, 41, 355-364.

Chatterjee, S. and A. S. Hadi (1986), Influential Observations, High Leverage Points, and Outliers in Linear Regression (with discussions), *Statistical Science*, 1, 379–416.

Chatterjee, S. and A. S. Hadi (1988), *Sensitivity Analysis in Linear Regression*, Wiley, New York.

Cook. R. D. (1977), Detection of Influential Observations in Linear Regression, *Technometrics*, 19, 15–18.

Fung, W. K. (1988), Critical Values for Testing in Multivariate Statistical Outliers, *Journal of Statistical Computation and Simulation*, 30, 195–212.

Gould, W. and A. S. Hadi (1993), Identifying Multivariate Outliers, *Stata Technical Bulletin*, 11, 2–5.

Hadi, A. S. (1992a), Identifying Multiple Outliers in Multivariate Data, *Journal of the Royal Statistical Society, Series (B)*, 54, 761–771.

Hadi, A. S. (1992b), A New Measure of Overall Potential Influence in Linear Regression, *Computational Statistics and Data Analysis*, 14, 1–27.

Hadi, A. S. (1994), A Modification of a Method for the Detection of Outliers in Multivariate Samples, *Journal of the Royal Statistical Society, Series (B)*, 56, No. 2, 393–396.

Hadi, A. S. and J. S. Simonoff (1993), Procedures for the Identification of Multiple Outliers in Linear Models, *Journal of the American Statistical Association*, 88, 424, 1264–1272.

Hadi, A. S. and J. S. Simonoff (1994), Improving the Estimation and Outlier Identification Properties of the Least Median of Squares and Minimum Volume Ellipsoid Estimators, *Parisankhyan Sammikkha*, 1, 61–70.

Hadi, A. S. and J. S. Simonoff (1997), A More Robust Outlier Identifier for Regression Data, *Bulletin of the International Statistical Institute*, 281–282.

Hadi, A. S. and P. F. Velleman (1997), Computationally Efficient Adaptive Methods for the Identification of Outliers and Homogeneous Groups in Large Data Sets, *Proceedings of the Statistical Computing Section*, American Statistical Association (in press).

Hampel, F. R., E. M. Ronchetti, P. J. Rousseeuw, and W. A. Stahel (1986), *Robust Statistics: The Approach Based on Influence Functions*, Wiley, New York.

Hawkins, D. M. (1980), *Identification of Outliers*, Chapman and Hall, London.

Huber, P. (1981), *Robust Statistics*, Wiley, New York.

Lopuha, H. P. (1989), On the Relation between S-Estimators and M-Estimators of Multivariate Location and Covariance, *Annals of Statistics*, 17, 1662–1683.

Maronna, R. A. (1976), Robust M-Estimators of Multivariate Location and Scatter, *Annals of Statistics*, 4, 51–67.

Rasmussen, J. L. (1988), Evaluating Outlier Identification Test: Mahalanobis D Squared and Comrey Dk, *Multivariate Behavioral Research*, 23, 189–202.

Rohlf, F. J. (1975), Generalization of the Gap Test for the Detection of Multivariate Outliers, *Biometrics*, 31, 93–101.

Rousseeuw, P. J. and A. M. Leroy (1987), *Robust Regression and Outlier Detection*, Wiley, New York.

Rousseeuw, P. J. and B. C. van Zomeren (1990), Unmasking Multivariate Outliers and Leverage Points (with discussion), *Journal of the American Statistical Association*, 85, 633–651.

Rousseeuw, P. J. and B. C. van Zomeren (1991), Robust Distances: Simulations and Cutoff Values, in *Directions in Robust Statistics and Diagnostics: Part II* (W. Stahel and S. Weisberg, eds.), Springer-Verlag, New York, 195–203.

Rousseeuw, P. J. and V. J. Yohai (1984), Robust Regression by Means of S Estimators, in *Robust and Nonlinear Time Series Analysis, Lecture Notes in Statistics*, Springer-Verlag, New York, Vol. 26, 256–272.

Schwager, S. J. and B. H. Margolin (1982), Detection of Multivariate Normal Outliers, *Annals of Statistics*, 10, 943–954.

14

Union-Intersection and Sample-Split Methods in Econometrics with Applications to MA and SURE Models

Jean-Marie Dufour
University of Montreal, Montreal, Quebec, Canada

Olivier Torrès
Université de Lille, Villeneuve d'Ascq, France

I. INTRODUCTION

Straightforward application of usual inference procedures (tests and confidence regions) in econometrics is often impossible. The problem usually comes from an insufficient specification of the probability distribution generating the data, as occurs for example when one makes assumptions only on the first few moments of an error distribution. However, the problem also arises in parametric models that specify the data generating process up to a finite vector of parameters. This is typically the case when the assumptions made on the distribution depart from those made in the standard linear regression framework, such as the absence of serial independence and homoskedasticity.

This chapter treats in a unified way two apparently distinct categories of problems where distributional results are difficult to establish. The first one consists of comparing and pooling information about parameter estimates from samples whose stochastic relationship is totally unspecified. In such cases, it is not possible to write a usable joint likelihood function and standard finite-sample or asymptotic methods are not applicable. The second one consists of making inferences in models for which the distributions of standard test and confidence set procedures are difficult to establish, e.g., because of the presence of nuisance parameters, but for which relevant test statistics based on appropriately selected subsamples are distributionally more tractable.

To illustrate the problems we will study, consider the case where we have $m \geq 1$ regression equations of the form

$$
\begin{aligned}
y_{i,t} &= x'_{i,t}\beta_i + u_{i,t}, \qquad t = t_i + 1, \ldots, t_i + N_i \\
u_i &= (u_{i,t_i+1}, \ldots, u_{i,t_i+N_i})' \sim N(\mathbf{0}, \Omega_i), \qquad i = 1, 2, \ldots, m
\end{aligned}
\tag{1}
$$

where β_i is an unknown $k_i \times 1$ vector $(k_i < N_i)$, $x_{i,t}$ is a $k_i \times 1$ vector of fixed (or strictly exogenous) regressors, and Ω_i is an unknown positive-definite nonsingular $N_i \times N_i$ matrix, $i \in I = \{1, 2, \ldots, m\}$. This setup describes situations frequently met in econometrics. Special cases of interest include the following.

1. Models in which each equation expresses a similar relationship between analogous variables (i.e., the coefficients β_i have the same economic interpretation), but each one corresponds to a different sample and the different samples may be dependent in a way that is difficult to specify (e.g., this includes many panel data models).
2. Models with structural change: this situation is a special case of the previous one, where the different samples correspond to different subperiods.
3. Stacked regressions where each equation represents a different economic relation, possibly with different numbers of regressors which have different economic interpretations.
4. Time-series models where the dependence between the m equations in (1) is induced by serial dependence.

For example, take $m = 2$. A model of type 1 could express the relation between the log of the wage and a variable measuring the level of education for two individuals. The coefficient β is then interpreted as the return to education (Ashenfelter and Krueger 1992), and we may wish to test whether this return is the same for individuals 1 and 2. In models of type 2 we may wish to know whether the parameter linking variable y to variable x is the same over the whole period of observation. An example of a type 3 model could be two equations where $y_{1,t}$ and $y_{2,t}$ represent the consumption of two different goods and $x_{1,t}$ and $x_{2,t}$ are different vectors of explanatory variables. Model 3 is composed of two distinct relationships, but for some reason we want to test the equality of the two coefficients. An important example of a type 4 model is a linear regression model with errors that follow a moving-average (MA) process of order 1, where the first equation contains the odd-numbered observations and the second equation the even-numbered observations.

The most common practice in such situations is to rely on asymptotic inference procedures. The lack of reliability of such methods is well documented in the literature. This feature of asymptotic tests has been established by Park and Mitchell (1980), Miyazaki and Griffiths (1984), Nankervis and Savin (1987), and DeJong et al.

(1992) in the context of AR(1) models. Burnside and Eichenbaum (1994) provide evidence on the poor performance of GMM-based Wald test statistics. For more general theoretical results on the inaccuracy of asymptotic methods, the reader may consult Dufour (1997); see also Nelson, Startz, and Zivot (1996), Savin and Würtz (1996), and Wang and Zivot (1996). Furthermore, there are situations where usual asymptotic procedures do not apply. For instance, consider a model for panel data with time-dependent errors: if no assumption is made on the dependence structure, it is not at all clear what should be done.

The main characteristic of model (1) is that the vector of dependent variables $y = (y'_1, \ldots, y'_m)'$ is in some way divided into m subsamples (different individuals and/or different subperiods), whose relationship is unknown. Because the joint distribution of the vector of errors $u = (u'_1, u'_2, \ldots, u'_m)'$ is not specified, usual inference methods based on the whole sample y are not applicable. This chapter develops inference procedures which are valid in such contexts.

The general issues we consider can be described as follows. Given several data sets whose stochastic relationship is not specified (or difficult to model), but on which we can make inferences separately, we study the following problems: (I) how to combine separate tests for an hypothesis of interest bearing on the different data sets (more precisely, how to test the *intersection* of several related hypotheses pertaining to different data sets); for example, in model (1), we may wish to test whether the linear restrictions $C_i \beta_i = \gamma_{i0}, i = 1, 2, \ldots, m$, hold *jointly*; (II) how to test cross-restrictions between the separated models (such as $\beta_1 = \beta_2 = \cdots = \beta_m$, when $k_i = k, i = 1, 2, \ldots, m$), which involves testing the *union* of a large (possibly infinite) number of hypotheses of the preceding type (e.g., $\beta_i = \beta_0, i = 1, 2, \ldots, m$, for some β_0); (III) how to combine confidence sets (e.g., confidence intervals or confidence ellipsoids) for a common parameter of interest and based on different data sets in order to obtain more accurate confidence sets. All these problems require procedures for *pooling* information obtained from separate, possibly nonindependent, samples and for making comparisons between them.

Besides being applicable to situations where the stochastic relationship between the different samples is completely unknown, the methods proposed will also be useful for inference on various models in which the distribution of a standard statistic based on the complete sample is quite difficult to establish (e.g., because of nuisance parameters), while the distributional properties of test statistics can be considerably simplified by looking at properly chosen subsamples. This is the case, for example, in seemingly unrelated regressions (SURE) and linear regressions with MA errors.

The methods proposed here rely on a systematic exploitation of Boole-Bonferroni inequalities (Alt 1982) which allow one to bound the probability of the union (or intersection) of a finite set of events from their marginal probabilities, without any knowledge of their dependence structure. Although such techniques have been used in the simultaneous inference literature to build simultaneous confidence intervals, especially in standard linear regressions (Miller 1981, Savin 1984), it does not appear they have been exploited for the class of problems studied here. In particular, for general problems of type I, we discuss the use of *induced tests* based on rejecting the null hypothesis when at least one of the several separate hypotheses is rejected by one of several separate tests, with the overall level of the procedure being controlled by Boole-Bonferroni inequalities. For problems of type II, we propose using *empty intersection tests* which reject the null hypothesis when the intersection of a number of separate confidence sets (or intervals) is empty. In the case of confidence intervals, this leads to simple rules that reject the null hypothesis when the distance between two parameter estimates based on separate samples is greater than the *sum* of the corresponding critical points. We also discuss how one can perform empty intersection tests based on confidence ellipsoids and confidence boxes. For problems of type III, we propose using the intersection of several separate confidence sets as a way of *pooling* the information in the different samples to gain efficiency. These common characteristics have led us to use the terminology *union-intersection* (UI) methods.

The techniques discussed in this chapter for type I problems are akin to procedures proposed for combining test statistics (Folks 1984) and for *meta-analysis* (Hedges and Olkin 1985). Meta-analysis tries to combine the evidence reported in different studies and articles on particular scientific questions; it has often been used to synthesize medical studies. However, these studies have concentrated on situations where the separate samples can be treated as independent and do not deal with econometric problems. Conversely, these methods are practically ignored in the econometric literature. Note also that the techniques we propose for problems of types II and III can be viewed as extensions of the union-intersection method proposed by Roy (1953) (Arnold 1981, pp. 363–364) for testing linear hypotheses in multivariate linear regressions, in the sense that an *infinity* of relatively simple hypothesis tests are explicitly considered and combined. A central difference here comes from the fact that the "simple" null hypotheses we consider are themselves tested via induced tests (because we study quite distinct setups) and from the different nature of the models studied.

As pointed out, our methods have the advantage of being versatile and straightforward to implement, even when important pieces of information are missing. These also turn out to be easily applicable in various problems where the distributional properties of test statistics can be considerably simplified by looking at appropriately selected subsamples. We show in particular that this is the case for several inference problems in SURE models and linear regressions with MA errors. This provides original and rather striking examples of "sample-split techniques" for simplifying distributional properties. For other recent illustrations of this general idea, the reader may consult Angrist and Krueger (1994), Dufour and Jasiak (1995), and Staiger and Stock (1997). In the first reference, the authors propose a sample-split technique to obtain IV estimators with improved properties, while the two other papers suggest similar methods to obtain more reliable tests in structural models.

The paper is organized as follows. Section II presents the general theory: in the context of a general statistical model, we derive procedures for testing null hypotheses of types I and II. In Section III, we consider the problem of pooling confidence sets obtained from different data sets (type III problems). In Section IV, we apply our results to test the equality of linear combinations of parameters of different equations in a SURE model, an interesting setup where standard tests and confidence sets only have an asymptotic justification (for a review, see Srivastava and Giles 1987). In particular, we impose no restrictions on the contemporaneous covariance matrix, allowing for different variances and instantaneous cross-correlation. In Section V, we study inference for linear regression models with MA(q) errors. We show that our inference technique is very well suited for testing hypotheses on regression coefficients in the presence of MA errors. We study in detail the case of an MA(1) process and consider the problem of testing an hypothesis about the mean. We compare our procedure with some alternative tests. It appears much easier to implement than other commonly used procedures, since it does not require estimation of MA parameters. We also study the performance of our method by simulation. The results show that sample-split combined test procedures are reliable from the point of view of level control and enjoy surprisingly good power properties. We conclude in Section VI.

II. HYPOTHESIS TESTING: GENERAL THEORY

In this section, we consider a general statistical model characterized by a sample space $\mathcal{Y}$ and a family $\mathcal{L} = \{P_\theta : \theta \in \Theta\}$ of probability distributions parameterized by θ, where Θ is the set of admissible values of θ. Let $\mathcal{L}_0$ be a subset of $\mathcal{L}$ and suppose we wish to test $H_0 : P_\theta \in \mathcal{L}_0$ against $H_1 : P_\theta \in \mathcal{L} \backslash \mathcal{L}_0$. If the model is identified, which will be assumed, this amounts to testing $H_0 : \theta \in \Theta_0$ against $H_1 : \theta \in \Theta_1$, where $\theta \in \Theta_0 \Leftrightarrow P_\theta \in \mathcal{L}_0$.

We consider here three classes of inference problems concerning θ. First, we study situations where Θ_0 can be expressed as a finite intersection of subsets of Θ; i.e., $\Theta_0 = \bigcap_{\gamma \in \Gamma} \Theta_{0\gamma}$, where Γ is an index set of the form $\Gamma = \{1, 2, \ldots, r\}$ and $\Theta_{0,\gamma} \subset \Theta$, $\gamma \in \Gamma$. Second, we examine null hypotheses which restrict θ to a subset Θ_0 of Θ, where Θ_0 can be written as $\Theta_0 = \bigcup_{\gamma \in \Gamma} \Theta_0(\gamma)$, $\Theta_0(\gamma) \subset \Theta$, $\gamma \in \Gamma$. In this case, Γ is not constrained to be a finite set. Third, we consider situations where the information about θ is available from different subsamples whose joint distribution is unknown. We then try to pool these pieces of information by combining inferences based on each subsample.

A. H_0 as the Finite Intersection of Subhypotheses

The test procedure in this section is based on the fact that although H_0 may not be easily testable it can be expressed as the intersection of subhypotheses, $H_{0\gamma} : \theta \in \Theta_{0\gamma}$, each one of which can be tested by usual procedures. The decision rule is built from the logical equivalence that H_0 is wrong if and only if *any* of its components $H_{0\gamma}$ is wrong.

Assume that we can test $H_{0\gamma}$ using a statistic T_γ such that, for any $\theta \in \Theta_{0\gamma}$, $P_\theta(\{y \in \mathcal{Y} : T_\gamma(y) \geq x\})$ is known, for all $x \in \mathbb{R}$, $\gamma \in \Gamma = \{1, 2, \ldots, r\}$. The relation between these statistics is unknown or difficult to establish [as it is the case in model (1)]. We wish to combine the information on the true probability distribution of the model, brought by each of the r statistics. Since H_0 is true if and only if all the $H_{0\gamma}$'s are individually true, a natural way of testing H_0 is to proceed as follows. Using the r statistics T_γ, we build r critical regions $W_\gamma(\alpha_\gamma) = T_\gamma^{-1}([t_\gamma(\alpha_\gamma), \infty))$, where $t_\gamma(\alpha_\gamma)$ is chosen so that $P_\theta[W_\gamma(\alpha_\gamma)] = \alpha_\gamma$ under $H_{0\gamma}$. We reject the null hypothesis H_0 when the vector of observations y lies in at least one of the $W_\gamma(\alpha_\gamma)$ regions, or equivalently if $T_\gamma(y) \geq t_\gamma(\alpha_\gamma)$ for at least one γ. The rejection region corresponding to this decision rule is $\bigcup_{\gamma \in \Gamma} W_\gamma(\alpha_\gamma)$. Such a test is called an *induced* test of H_0 (Savin 1984). Its size is impossible or difficult to determine since the joint distribution of the statistics T_γ is generally unknown or intractable. It is, however, possible to choose the individual levels α_γ so that the induced test has level $\alpha \in (0, 1)$, for by subadditivity

$$P_\theta \left(\bigcup_{\gamma \in \Gamma} W_\gamma(\alpha_\gamma) \right) \leq \sum_{\gamma \in \Gamma} P_\theta[W_\gamma(\alpha_\gamma)] = \sum_{\gamma \in \Gamma} \alpha_\gamma$$

for any $\theta \in \bigcap_{\gamma \in \Gamma} \Theta_{0\gamma} = \Theta_0$. Therefore, if we want the induced test to have level α, we only need to choose the α_γ's so that they sum to α (or less).

To our knowledge, there is no criterion for choosing the α_γ's in an optimal manner. Without such a rule, in most of our applications we will give the null hypotheses $H_{0\gamma}$ the same degree of protection against an erroneous rejection by taking $\alpha_\gamma = \alpha_0 = \alpha/r$, $\forall \gamma \in \Gamma$. However, there may exist situations where we wish to weigh the $H_{0\gamma}$'s in a different way. In particular, if for some reason we know that one of the decisions $d_{\gamma'}$ (say, accepting or rejecting $H_{0\gamma'}$) is less reliable than the other decisions, we are naturally led to give $d_{\gamma'}$ less impact on the final decision concerning the acceptance or rejection of H_0. In other words, we will choose $\alpha_{\gamma'} < \alpha_\gamma$, $\forall \gamma \neq \gamma'$.

In the case where we choose $\alpha_\gamma = \alpha_0 = \alpha/r$, $\forall \gamma \in \Gamma$, we reject $H_{0\gamma}$ at level α_0 when y is in $W_\gamma(\alpha_0)$. Assuming $F_{\gamma,\theta}(x)$ is continuous in x, this region of $\mathcal{Y}$ can be reexpressed as

$$\begin{aligned} W_\gamma(\alpha_0) &= \{y \in \mathcal{Y} : 1 - F_{\gamma,\theta}[T_\gamma(y)] \leq 1 - F_{\gamma,\theta}[t_\gamma(\alpha_0)]\} \\ &= \{y \in \mathcal{Y} : 1 - F_{\gamma,\theta}[T_\gamma(y)] \leq \alpha_0\} \end{aligned}$$

for all $\theta \in \Theta_0$, where $F_{\gamma,\theta}(z) \equiv P_\theta[T_\gamma^{-1}((-\infty, z])]$ is the probability distribution function of $T_\gamma(Y)$. Then an equivalent alternative rejection criterion is to reject $H_{0\gamma}$ when the p-value $\Lambda_\gamma(y) \equiv 1 - F_{\gamma,\theta}[t_\gamma(y)]$ is less than α_0. So according to this rule, a rejection of H_0 occurs whenever at least one of the r p-values Λ_γ is not greater than α_γ, and the critical region of this test procedure is

$$W(\alpha) = \{y \in \mathcal{Y} \min_{\gamma \in \Gamma} \Lambda_\gamma(y) \le \alpha_0\}$$

If we assume that the statistics T_γ are identically distributed under the null hypothesis, then $F_{\gamma,\theta} = F_\theta, \forall \theta \in \Theta_0$ and $t_\gamma(\alpha_0) = t(\alpha_0), \forall \gamma \in \Gamma$, hence (with probability 1)

$$W(\alpha) = \{y \in \mathcal{Y} : \max_{\gamma \in \Gamma} T_\gamma(y) \ge t(\alpha_0)\}$$

This criterion is derived heuristically from the logical equivalence that H_0 is true if and only if all the $H_{0\gamma}$'s are true. It is similar to Tippett's (1931) procedure for combining inferences obtained from independent studies: using the fact that the r p-values are i.i.d. $\mathcal{U}_{[0,1]}$ under H_0, Tippett (1931) suggested the rule:

reject H_0 at level α if $\min\{\Lambda_\gamma : \gamma \in \Gamma\} \le 1 - (1 - \alpha)^{1/r}$

Such procedures were also proposed for meta-analysis (Hedges and Olkin 1985), but have seldom been used in econometrics. We show that an extension of Tippett's procedure to the case where the p-values are not independent can be fruitfully applied to several econometric problems. The analogy with meta-analysis comes from the fact that inference on the ith equation in (1) is made "independently" from inference on any other equation, although the test statistics may not be stochastically independent. Since dependence of the test statistics is a common situation in econometrics, we do not assume the independence of the p-values, which leads one to use α/r instead of $1 - (1 - \alpha)^{1/r}$. When α is small, the difference between α/r and $1 - (1 - \alpha)^{1/r}$ is typically quite small. For some applications of this approach to independent test statistics, see Dufour (1990), McCabe (1988), and Phillips and McCabe (1988, 1989).

It is possible to demonstrate optimality properties for induced tests. For example, consider a test procedure which combines r p-values $\Lambda_1, \Lambda_2, \ldots, \Lambda_r$ so that it rejects H_0 when $S(\Lambda_1, \Lambda_2, \ldots, \Lambda_r) \le s$, where S is some function from $\mathbb{R}^r$ into $\mathbb{R}$ and s is a critical value such that $P[S(\Lambda_1, \Lambda_2, \ldots, \Lambda_r) \le s] \le \alpha$ under H_0. Birnbaum (1954) showed that every monotone combined test procedure is admissible in the class of all combined test procedures.* A combined test procedure S is monotone if S is a nondecreasing function, i.e., if $x_i^* \le x_i, i = 1, 2, \ldots,$

---

*On admissibility of decision rules, see Lehmann (1986, Section 1.8, p. 17).

$r \Rightarrow S(x_1^*, x_2^*, \ldots, x_r^*) \leq S(x_1, x_2, \ldots, x_r)$. In our case, $S(\Lambda_1, \Lambda_2, \ldots, \Lambda_r) = \min\{\Lambda_1, \Lambda_2, \ldots, \Lambda_r\}$, is clearly nondecreasing. For further discussion of admissibility issues in such contexts, see Folks (1984).

B. H_0 as the Union of Subhypotheses

Let us consider a null hypothesis of the form $H_0 : \theta \in \Theta_0$, where $\Theta_0 = \bigcup_{\gamma \in \Gamma} \Theta_0(\gamma)$. The solution to this testing problem is very similar to the one suggested above in the case of an intersection of subhypotheses. Once again it is based on the fact that H_0 is wrong if and only if each of its components is wrong. If each hypothesis $H_0(\gamma) : \theta \in \Theta_0(\gamma)$ can be tested using the rejection region $W_\gamma(\alpha_\gamma) = T^{-1}([t_\gamma(\alpha_\gamma), \infty))$, satisfying $P_\theta[W_\gamma(\alpha_\gamma)] = \alpha_\gamma, \forall \theta \in \Theta_0(\gamma)$, it would appear natural to consider the overall rejection region $W(\alpha_\gamma, \gamma \in \Gamma) = \bigcap_{\gamma \in \Gamma} W_\gamma(\alpha_\gamma)$ for a test of H_0.

However, difficult problems arise when one wants to implement this procedure as described above. First, if Γ contains a finite number p of elements, we have

$$P_\theta[W(\alpha_\gamma, \gamma \in \Gamma)] \geq 1 - \sum_{\gamma=1}^{p} [1 - P_\theta[W_\gamma(\alpha_\gamma)]]$$

which provides a *lower* bound for the probability of making a type I error. Of course, this type of bound is of no use since we try to bound from above the probability of an erroneous rejection of H_0. Appropriate upper bounds for the probability of an intersection are difficult to obtain. Second, when Γ is infinite, it is impossible to build $W_\gamma(\alpha_\gamma)$ for every $\gamma \in \Gamma$.

It is however interesting to note that some null hypotheses can be written as the union of several hypotheses (possibly an infinite number of such hypotheses). It is then natural to construct an overall rejection region which is equivalent to the infinite intersection $\bigcap_{\gamma \in \Gamma} W_\gamma(\alpha_\gamma)$. For example, consider the hypothesis $H_0 : \theta_1 = \theta_2 = \cdots = \theta_m$, where θ_i is a $q \times 1$ subvector of the initial parameter vector θ. We note that H_0 is true if and only if $\exists \theta_0 \in \mathbb{R}^q$ such that $\theta_1 = \theta_2 = \cdots = \theta_m = \theta_0$, where θ_0 is the unknown true value of θ_i under the null. Defining $\Theta_0(\theta_0) \equiv \{\theta \in \Theta : \theta_1 = \theta_2 = \cdots = \theta_m = \theta_0\}$, we have $\Theta_0 = \bigcup_{\theta_0 \in \mathbb{R}^q} \Theta_0(\theta_0)$. H_0 can be expressed as an infinite union of subhypotheses $H_0(\theta_0) : \theta \in \Theta_0(\theta_0)$. Therefore H_0 is true if and only if anyone of the $H_0(\theta_0)$'s is true.

Obviously, it is impossible to test every $H_0(\theta_0)$. Instead, we propose the following procedure. For each $i \in \{1, 2, \ldots, m\}$, we build a confidence region $C_i(y_i, \alpha_i)$ for θ_i with level $1 - \alpha_i$ using the sample y_i, where the α_i's are chosen so that $\sum_{i=1}^{m} \alpha_i = \alpha$. This region satisfies

$$P_\theta[A_i(\theta_i, \alpha_i)] \geq 1 - \alpha_i, \qquad \forall \theta \in \Theta$$

where $A_i(\theta_i, \alpha_i) = \{y \in \mathcal{Y} : C_i(y_i, \alpha_i) \ni \theta_i\}$, $i = 1, 2, \ldots, m$, and $G \ni x$ means that the set "G contains x." In particular, if θ_0 is the true value of θ_i, we have

$$P_\theta[A_i(\theta_0, \alpha_i)] \geq 1 - \alpha_i, \qquad \forall \theta \in \Theta_0$$

Proposition I. A (conservative) α-level test of $H_0 : \theta_1 = \theta_2 = \cdots = \theta_m$ is given by the rejection region

$$W(\alpha, m) = \left\{ y \in \mathcal{Y} : \bigcap_{i=1}^{m} C_i(y_i, \alpha_i) = \emptyset \right\}$$

where α_i, $i = 1, 2, \ldots, m$ satisfy $\sum_{i=1}^{m} \alpha_i \leq \alpha$.

Proof. We need to show that $P_\theta[W(\alpha, m)] \leq \alpha, \forall \theta \in \Theta_0, \forall \alpha \in (0, 1)$. Note that $\forall \theta_0 \in \mathbb{R}^q$,

$$\bigcap_{i=1}^{m} C_i(y_i, \alpha_i) = \emptyset \Rightarrow \exists j \in \{1, 2, \ldots, m\} : C_j(y_j, \alpha_j) \not\ni \theta_0$$

Hence, using the Boole-Bonferroni inequality

$$P_\theta[W(\alpha, m)] \leq P_\theta \left[\bigcup_{i=1}^{m} \{y \in \mathcal{Y} : C_i(y_i, \alpha_i) \not\ni \theta_0 \} \right]$$

$$\leq \sum_{i=1}^{m} P_\theta[\mathcal{Y} \backslash A_i(\theta_0, \alpha_i)] \leq \sum_{i=1}^{m} \alpha_i \leq \alpha, \qquad \forall \theta \in \Theta$$

We shall call a critical region of the form of $W(\alpha, m)$ an *empty intersection* test. In our notation, $W(\alpha, m)$ does not depend directly upon α, but on how the α_i's are chosen to satisfy the constraint $\sum_{i=1}^{m} \alpha_i \leq \alpha$. For this procedure to be applicable, we need to have confidence regions $C_i(y_i, \alpha_i)$ with levels $1 - \alpha_i$. This is of course possible in model (1) as long as $\Omega_i = \sigma_i^2 I_{N_i}, i \in \{1, 2, \ldots, m\}$. We describe three interesting special cases for which the procedure takes a simple and appealing form.

I. Intersection of Confidence Intervals: The Sum of Critical Points Rule

Consider a situation where $q = 1$. Typically, $C_i(y_i, \alpha_i)$ has the form

$$C_i(y_i, \alpha_i) = [\hat{\theta}_i - c_{iL}(y_i, \alpha_i), \hat{\theta}_i + c_{iU}(y_i, \alpha_i)]$$

where $\hat{\theta}_i$ is some estimator of θ_i, $\sum_{i=1}^{m} \alpha_i \leq \alpha \in (0, 1)$, and $c_{iL}(y_i, \alpha_i) > 0$, $c_{iU}(y_i, \alpha_i) > 0$ for all possible values of y_i. Furthermore, it is usually the case that $c_{iL}(y_i, \alpha_i) = c_{iU}(y_i, \alpha_i)$ but we shall not need this restriction here. It is easy to see that the following lemma holds.

Lemma I. The intersection of a finite number m of intervals $I_i \equiv [L_i, U_i] \subset \mathbb{R}, i = 1, 2, \ldots, m$, with nonempty interiors is empty if and only if

$$\min\{U_i : i = 1, 2, \ldots, m\} < \max\{L_i : i = 1, 2, \ldots, m\}.$$

Proof. Define $U_M \equiv \min\{U_i : i = 1, 2, \ldots, m\}$, $L_M \equiv \max\{L_i : i = 1, 2, \ldots, m\}$, and $\check{I} \equiv \{x \in \mathbb{R} : L_M \leq x \leq U_M\}$. Then

$$\check{I} \neq \emptyset \Longleftrightarrow \exists x \text{ such that } L_i \leq x \leq U_i \text{ for } i = 1, 2, \ldots, m$$

$$\Longleftrightarrow \exists x \text{ such that } L_M \leq x \leq U_M$$

or equivalently,

$$\check{I} = \emptyset \Longleftrightarrow \forall x, x \notin [L_M, U_M] \Longleftrightarrow U_M < L_M$$

From Lemma 1 and Proposition 1, we reject H_0 if and only if

$$\min\{\hat{\theta}_i + c_{iU}(y_i, \alpha_i) : i = 1, 2, \ldots, m\}$$
$$< \max\{\hat{\theta}_i - c_{iL}(y_i, \alpha_i) : i = 1, 2, \ldots, m\}$$

But this condition is equivalent to

$$\exists j, k \in \{1, 2, \ldots, m\} \text{ such that } \hat{\theta}_j + c_{jU}(y_j, \alpha_j) < \hat{\theta}_k - c_{kL}(y_k, \alpha_k)$$

or

$$\exists j, k \in \{1, 2, \ldots, m\} \text{ such that } \frac{|\hat{\theta}_k - \hat{\theta}_j|}{c_{jU}(y_j, \alpha_j) + c_{kL}(y_k, \alpha_k)} > 1$$

Finally, we reject H_0 if and only if

$$\max_{j,k\in\{1,2,\ldots,m\}} \left[\frac{|\hat{\theta}_k - \hat{\theta}_j|}{c_{jU}(y_j, \alpha_j) + c_{kL}(y_k, \alpha_k)} \right] > 1$$

In the case where $m = 2$, with $c_{jU}(y_j, \alpha_j) = c_{jL}(y_j, \alpha_j) = c_j(y_j, \alpha_j)$, $j = 1, 2$, this criterion takes the simple form: *reject the null hypothesis when the distance between the two estimates is larger than the sum of the two corresponding "critical point."* The rejection region is then

$$W(\alpha, 2) = \{y \in \mathcal{Y} : |\hat{\theta}_1 - \hat{\theta}_2| > c_1(y_1, \alpha_1) + c_2(y_2, \alpha_2)\}$$

For $m > 2$, we reject the null hypothesis when at least one of the distances $|\hat{\theta}_k - \hat{\theta}_j|$ is larger than the sum $c_j(y_j, \alpha_j) + c_k(y_k, \alpha_k)$. We will now extend this procedure to multidimensional parameters and consider confidence ellipsoids.

2. Intersection of Two Confidence Ellipsoids

Consider the null hypothesis $H_0 : \theta_1 = \theta_2$, where θ_i is a $q \times 1$ vector. As before, H_0 can be restated as $H_0 : \theta \in \{\theta \in \Theta : \exists \theta_0 \in \mathbb{R}^q : \theta_1 = \theta_2 = \theta_0\}$. Suppose that for $i = 1, 2$, we have a confidence ellipsoid $C_i(y_i, \alpha_i)$ for θ_i, such that

$$C_i(y_i, \alpha_i) \ni \theta_0 \Leftrightarrow (\hat{\theta}_i - \theta_0)'A_i(\hat{\theta}_i - \theta_0) \leq c_i(\alpha_i)$$

where A_i is a $q \times q$ positive definite matrix whose elements depend on y_i, $\hat{\theta}_i$ is an estimator of θ_i, and $c_i(\alpha_i)$ is a constant such that

$$\mathbf{P}_\theta[\{y \in \mathcal{Y} : C_i(y_i, \alpha_i) \ni \theta\}] \geq 1 - \alpha_i, \qquad \forall \theta \in \Theta$$

Then there exists two $q \times q$ matrices P_1 and P_2 such that $P_1' A_1 P_1 = I_q$, $P_2'(P_1' A_2 P_1) P_2 = D$, where D is a diagonal nonnegative definite $q \times q$ matrix, $|P_1| \neq 0$ and $P_2 P_2' = I_q$. It is easy to show that

$$(\hat{\theta}_1 - \theta_0)' A_1 (\hat{\theta}_1 - \theta_0) \leq c_1(\alpha_1) \Leftrightarrow (\hat{\gamma}_1 - \gamma)'(\hat{\gamma}_1 - \gamma) \leq c_1(\alpha_1)$$

$$(\hat{\theta}_2 - \theta_0)' A_2 (\hat{\theta}_2 - \theta_0) \leq c_2(\alpha_2) \Leftrightarrow (\hat{\gamma}_2 - \gamma)' D (\hat{\gamma}_2 - \gamma) \leq c_2(\alpha_2)$$

where $\gamma = P_2' P_1^{-1} \theta_0$ and $\hat{\gamma}_i = P_2' P_1^{-1} \hat{\theta}_i$, $i = 1, 2$. Setting

$$E_1(\alpha_1) = \{\gamma \in \mathbb{R}^q : (\hat{\gamma}_1 - \gamma)'(\hat{\gamma}_1 - \gamma) \leq c_1(\alpha_1)\}$$

$$E_2(\alpha_2) = \{\gamma \in \mathbb{R}^q : (\hat{\gamma}_2 - \gamma)' D (\hat{\gamma}_2 - \gamma) \leq c_2(\alpha_2)\}$$

the rejection criterion $C_1(y_1, \alpha_1) \cap C_2(y_2, \alpha_2) = \emptyset$ of Proposition 1 is seen to be equivalent to $E_1(\alpha_1) \cap E_2(\alpha_2) = \emptyset$.

To determine whether the intersection of the two ellipsoids is empty, it is sufficient to find the set $E_2^*(\alpha_2)$ of solutions of the problem

$$\min_{\gamma \in E_2(\alpha_2)} \|\gamma - \hat{\gamma}_1\|^2 \tag{2}$$

and check whether there is at least one element of $E_2^*(\alpha_2)$ lying in $E_1(\alpha_1)$, in which case the two confidence ellipsoids have a non empty intersection and H_0 is accepted at level $\alpha_1 + \alpha_2$. This is justified by the following lemma.

Lemma 2.   Let $E_2^*(\alpha_2) \subset E_2(\alpha_2)$ be the set of the solutions of (2). Then

$$E_1(\alpha_1) \cap E_2(\alpha_2) \neq \emptyset \Longleftrightarrow E_1(\alpha_1) \cap E_2^*(\alpha_2) \neq \emptyset$$

Proof.   ($\Leftarrow$) Let $E_1(\alpha_1) \cap E_2^*(\alpha_2) \neq \emptyset$. Since $E_2^*(\alpha_2) \subseteq E_2(\alpha_2)$, it follows trivially that $E_1(\alpha_1) \cap E_2(\alpha_2) \neq \emptyset$.

($\Rightarrow$) Let $E_1(\alpha_1) \cap E_2(\alpha_2) \neq \emptyset$. Then we can find $\tilde{\gamma}$ such that $\tilde{\gamma} \in E_1(\alpha_1)$ and $\tilde{\gamma} \in E_2(\alpha_2)$. In other words, $\tilde{\gamma}$ is an element of $E_2(\alpha_2)$ that satisfies the condition $\|\tilde{\gamma} - \hat{\gamma}_1\|^2 \leq c_1(\alpha_1)$, which entails that $\min_{\gamma \in E_2(\alpha_2)} \|\gamma - \hat{\gamma}_1\|^2 \leq c_1(\alpha_1)$. Now suppose $E_1(\alpha_1) \cap E_2^*(\alpha_2) = \emptyset$. This means that the following implication must hold:

$$\|\gamma_0 - \hat{\gamma}_1\|^2 \equiv \min_{\gamma \in E_2(\alpha_2)} \|\gamma - \hat{\gamma}_1\|^2 \Rightarrow \gamma_0 \notin E_1(\alpha_1) \Rightarrow \|\gamma_0 - \hat{\gamma}_1\|^2 > c_1(\alpha_1)$$

Since $E_2^*(\alpha_2)$ is not empty, it follows that $\min_{\gamma \in E_2(\alpha_2)} \|\gamma - \hat{\gamma}_1\|^2 > c_1(\alpha_1)$, a contradiction. Thus we must have $E_1(\alpha_1) \cap E_2^*(\alpha_2) \neq \emptyset$.

Although any numerical calculus computer package is able to solve (2), we propose a simple two-step procedure for deciding whether the intersection of two ellipsoids is empty. This method can prove useful for high-dimensional problems. The two steps are the following:

1. Check whether $\hat{\theta}_1 \in C_2(y_2, \alpha_2)$ or $\hat{\theta}_2 \in C_1(y_1, \alpha_1)$. If one of these events is realized, then $C_1(y_1, \alpha_1) \cap C_2(y_2, \alpha_2)$ is not empty, and H_0 is accepted. Otherwise, go to the second stage of the procedure.
2. Since $\hat{\theta}_1 \notin C_2(y_2, \alpha_2)$, it follows (by convexity) that $E_2^*(\alpha_2)$ is a subset of the boundary $\partial E_2(\alpha_2)$ and

$$\min_{\gamma \in \partial E_2(\alpha_2)} \|\gamma - \hat{\gamma}_1\|^2 = \min_{\gamma \in E_2(\alpha_2)} \|\gamma - \hat{\gamma}_1\|^2 > 0 \tag{3}$$

so that we can check whether $E_1(\alpha_1) \cap E_2(\alpha_2) \neq \emptyset$ by checking if $E_1(\alpha_1) \cap \partial E_2(\alpha_2) \neq \emptyset$. If the latter condition is satisfied, H_0 is accepted; otherwise H_0 is rejected.

To be more specific, step 2 simply requires one to find the vector $\tilde{\gamma}$ which minimizes $\|\gamma - \hat{\gamma}_1\|^2$ subject to the restriction $(\gamma - \hat{\gamma}_2)' D (\gamma - \hat{\gamma}_2) = c_2(\alpha_2)$, and then to reject H_0 when $\|\tilde{\gamma} - \hat{\gamma}_1\|^2 > c_1(\alpha_1)$.

3.   Intersection of Two Confidence Boxes

When the covariance matrices of the estimator $\hat{\theta}_i$ are unknown, one cannot typically build confidence ellipsoids. To illustrate such situations, consider the following example. Two published papers investigate econometric relationships of the form

$$y_i = X_i \beta_i + u_i, \qquad i = 1, 2 \tag{4}$$

where $u_i \sim N(0, \sigma_i^2 I_{N_i})$, and β_i is a $k \times 1$ vector of unknown parameters.

We wish to compare the two parameter vectors β_1 and β_2. However, only the standard errors of the coefficients are known (or reported), not their covariance matrices. Then it is not possible to use the previous procedure. But it is possible to use simultaneous inference techniques and build simultaneous confidence boxes (hyperrectangles). Various methods for building such confidence sets are described in Miller (1981) and Savin (1984). More precisely, let us build for each of the two regressions in (4) k simultaneous confidence intervals, denoted by $C_i^j(y_i, \alpha_i^j)$ for the component β_i^j of β_i, $j = 1, 2, \ldots, k, i = 1, 2$, such that

$$P_\theta \left[\bigcap_{j=1}^{k} \{y \in \mathcal{Y} : C_i^j(y_i, \alpha_i^j) \ni \beta_i^j\} \right] \geq 1 - \alpha_i, \qquad \forall \theta \in \Theta, \quad i = 1, 2$$

Then choosing the α_i's so that $\alpha_1 + \alpha_2 = \alpha$, and applying the results of Proposition 1, we reject $H_0 : \boldsymbol{\beta}_1 = \boldsymbol{\beta}_2$ at level α when the intersection of the two hyperrectangles is empty.

Checking whether the intersection of the two boxes is empty is especially simple because one simply needs to see whether the confidence intervals for each component of $\boldsymbol{\beta}_i$ have an empty intersection (as in Section II.B.1). Furthermore it is straightforward to extend this technique in order to compare more than two regressions. Similarly, although we proposed a test for the null hypothesis that all the parameter vectors $\boldsymbol{\beta}_i$ are equal (then imposing that in each equation has the same number of parameters), it is easy to extend this procedure in order to test the equality of *linear* transformations of $\boldsymbol{\beta}_i, i = 1, 2, \ldots, m$. Indeed, the method relies only on the ability to derive confidence regions for parameters which are restricted to be equal under the null. This is clearly possible whenever the parameters of interest are of the form $R_i\boldsymbol{\beta}_i$. The procedure is actually applicable to any function $h(\theta)$ of the parameter, provided we are able to build a confidence region for $h(\theta)$.

III. CONFIDENCE SET ESTIMATION

In the previous section, we described a general method for testing hypotheses in several contexts. The main feature of the procedure is that a single final decision concerning a family of probability distributions is taken by combining several individual (partial) decisions on that family.

In many situations, we may wish to go a step further. For instance, consider again model (1) and the null hypothesis $H_0 : \boldsymbol{\beta}_1 = \boldsymbol{\beta}_2 = \cdots = \boldsymbol{\beta}_m$. The results of Section II show how to test such an hypothesis. Suppose H_0 is taken for granted. It is then natural to ask what could be a valid confidence region for $\boldsymbol{\beta}$, the unknown common value of $\boldsymbol{\beta}_i, i = 1, 2, \ldots, m$. The main difficulty here comes from the fact that only the marginal distributions of the separate samples are specified, not their joint distribution. Suppose each of these marginal distributions can be used to build a confidence region for $\boldsymbol{\beta}$. The problem is then to find a way of pooling these pieces of information on the true value of $\boldsymbol{\beta}$ and derive a single confidence region which is based on the whole sample. This can be done as follows. Suppose each one of the separate observations vectors $y_1, y_2, \ldots, y_m$ has a distribution which depends on $\theta \in \Theta$. Although the joint distribution of $y = (y_1', y_2', \ldots, y_m')'$ is unknown, we assume it is possible to build m separate confidence regions $C_i(\alpha_i, y_i)$ for θ such that

$$P_\theta[C_i(\alpha_i, y_i) \ni \theta] \geq 1 - \alpha_i, \qquad \forall \theta \in \Theta, \quad i = 1, 2, \ldots, m$$

Then a natural way to exploit simultaneously these different pieces of information consists of taking the intersection $\bigcap_{i=1}^{m} C_i(\alpha_i, y_i)$ of the different confidence regions. It is easy to see that

$$P_\theta \left[\bigcap_{i=1}^{m} C_i(\alpha_i, y_i) \ni \theta \right] \geq 1 - \sum_{i=1}^{m} P_\theta[C_i(\alpha_i, y_i) \not\ni \theta]$$

$$\geq 1 - \sum_{i=1}^{m} \alpha_i, \qquad \forall \theta \in \Theta$$

Thus selecting $\alpha_1, \alpha_2, \ldots, \alpha_m$ so that $\sum_{i=1}^{m} \alpha_i = \alpha$, we can get any desired confidence level. This procedure can be especially useful when one of the subsamples yields a particularly accurate confidence set for θ.

In the next sections, we show how the procedures described in Sections II and III can be used to make inference in SURE and regressions with MA errors.

IV. EXACT INFERENCE IN SURE MODELS

A. The Model and the Procedures

In this section we consider the SURE-type model:

$$
\begin{aligned}
y_i &= X_i \beta_i + u_i \\
u_i &\sim N(0, \sigma_{ii} I_{N_i}), \qquad i = 1, 2, \ldots, m
\end{aligned}
\tag{5}
$$

where X_i is a $N_i \times k_i$ fixed matrix of rank $k_i < N_i$, β_i is a $k_i \times 1$ vector of unknown parameters, $u_i = (u_{i1}, u_{i2}, \ldots, u_{iN_i})'$, and $E(u_{it} u_{is}) = 0, \forall t \neq s$. Note we do not impose any restriction on the relationship between the m equations, so that the above model is more general than the standard SURE model. The null hypotheses of interest are $H_0^{(1)} : \lambda_i = \lambda_{0i}, i = 1, 2, \ldots, m$, and $H_0^{(2)} : \lambda_1 = \lambda_2 = \cdots = \lambda_m$, where $\lambda_i = R_i \beta_i$, R_i is a known $q_i \times k_i$ matrix with rank $q_i \leq k_i, i = 1, 2, \ldots, m$, and λ_{0i} is a known $q_i \times 1$ vector, $i = 1, 2, \ldots, m$.[*] An interesting special case of $H_0^{(1)}$ is $\beta_1 = \beta_2 = \cdots = \beta_m = \beta_0$, which is obtained by choosing $k_i = k$, $R_i = I_k$, $\lambda_{0i} = \beta_0$, a known $k \times 1$ vector, in the above setup.

We will consider two versions of (5), depending on whether we make the assumption $(A1): u = (u_1', u_2', \ldots, u_m')' \sim N(0, \sigma^2 I_N)$, where $N = \sum_{i=1}^{m} N_i$. Under $A1$, there exists an optimal test of $H_0^{(1)}$ given be the critical region associ-

---

[*]For $H_0^{(2)}$ we must have $q = q_i, \forall i = 1, 2, \ldots, m$ and $q \leq \min\{k_i : i = 1, 2, \ldots, m\}$.

ated with the Fisher F-statistic, based on the stacked model $y = X\beta + u$, where $y = (y_1', y_2', \ldots, y_m')'$, $\beta = (\beta_1', \ldots, \beta_m')'$, $X = \text{diag}(X_i)_{i=1,2,\ldots,m}$, and

$$F = \frac{(\hat{\Lambda} - \Lambda_0)'[s^2 R(X'X)^{-1}R']^{-1}(\hat{\Lambda} - \Lambda_0)}{Q}$$

with $\Lambda = (\lambda_1', \lambda_2', \ldots, \lambda_m')'$, $\hat{\Lambda} = (\hat{\lambda}_1', \hat{\lambda}_2', \ldots, \hat{\lambda}_m')'$, $\Lambda_0 = (\lambda_{01}', \lambda_{02}', \ldots, \lambda_{0m}')'$, $\hat{\lambda}_i = R_i \hat{\beta}_i$, $\hat{\beta}_i = (X_i'X_i)^{-1}X_i'y_i$, $i = 1, 2, \ldots, m$, $R = \text{diag}(R_i)_{i=1,2,\ldots,m}$, $s^2 = \|(I_N - X(X'X)^{-1}X')y\|^2/(N - K)$, $Q = \sum_{i=1}^m q_i$ and $K = \sum_{i=1}^m k_i$.

When we introduce heteroskedasticity in the model by allowing the variances to differ across equations, our procedure is still valid, but the Fisher procedure is not. As an alternative, one would typically use an asymptotic method based on a generalized least-squares estimation and a critical region defined by a Wald, a Lagrange multiplier or a likelihood ratio statistic. But, as we already mentioned in the introduction, it is well known that these approximations are not reliable.

An induced test of $H_0^{(1)}$ consists in testing $H_{0i}^{(1)} : \lambda_i = \lambda_{0i}$ at level α_i using the critical region $W_i(\alpha_i) = \{y \in \mathcal{Y} : F_i > F(\alpha_i; q_i, N_i - k_i)\}$, where

$$F_i = \frac{(\hat{\lambda}_i - \lambda_{0i})'(s_i^2 R_i(X_i'X_i)^{-1}R_i')^{-1}(\hat{\lambda}_i - \lambda_{0i})}{q_i}$$

with $\hat{\lambda}_i = R_i \hat{\beta}_i$, $s_i^2 = \|(I_{N_i} - X_i(X_i'X_i)^{-1}X_i')y_i\|^2/(N_i - k_i)$, and $F(\alpha_i; q_i, N_i - k_i)$ is the $1 - \alpha_i$ percentile of the Fisher distribution with $(q_i, N_i - k_i)$ degrees of freedom. The α_i's can be chosen so that $\sum_{i=1}^m \alpha_i = \alpha$. Then the level α critical region for an induced test of $H_0^{(1)}$ is $\bigcup_{i=1}^m W_i(\alpha_i)$.

If we wish to test $H_0^{(2)}$ at level α, we simply have to build m confidence regions at level $1 - \alpha_i$ for λ_i which are defined by

$$C_i(y_i, \alpha_i) = \{x \in \mathbb{R}^q : (\hat{\lambda}_i - x)'(s_i^2 R_i(X_i'X_i)^{-1}R_i')^{-1}(\hat{\lambda}_i - x)$$

$$\leq q_i F(\alpha_i; q_i, N_i - k_i)\}$$

in the λ_i space, and reject $H_0^{(2)}$ whenever $\bigcap_{i=1}^m C_i(y_i, \alpha_i) = \emptyset$, with $\sum_{i=1}^m \alpha_i = \alpha$.

Note that, under assumption $(A1)$, the induced procedure for a test of $H_0^{(1)}$ can be improved by taking into account the independence of the regressions. In Section II, we showed that the rejection region associated with an induced test of $H_0^{(1)}$ is $\bigcup_{i=1}^m W_i(\alpha_i)$, where $W_i(\alpha_i)$ is the critical region for a test of $\beta_i = \beta_0$ at level α_i. Under $(A1)$, we have

$$P_\theta\left[\bigcup_{i=1}^m W_i(\alpha_i)\right] = 1 - P_\theta\left[\bigcap_{i=1}^m \mathcal{Y}\backslash W_i(\alpha_i)\right] = 1 - \prod_{i=1}^m P_\theta[\mathcal{Y}\backslash W_i(\alpha_i)]$$

Under $H_0^{(1)}$ we have $P_\theta[\mathcal{Y}\backslash W_i(\alpha_i)] = 1 - \alpha_i$, Thus by choosing the α_i's so that $\prod_{i=1}^m (1 - \alpha_i) = \alpha$, we get a test of $H_0^{(1)}$ which has level α. If $\alpha_i = \alpha_0$, $i = 1, 2, \ldots, m$, we must have $\alpha_0 = 1 - (1 - \alpha)^{1/m}$.

Unfortunately, the independence assumption ($A1$) is not helpful when we turn to the test of $H_0^{(2)}$. But the procedure of Section II.B remains valid. To see why it is difficult to exploit the independence of the regressions, consider the case where $m = 2$ and $k = 1$. We can build two confidence intervals $C_i(y_i, \alpha_0) = [\hat{\beta}_i - c_i(y_i, \alpha_0), \hat{\beta}_i + c_i(y_i, \alpha_0)]$, with $\alpha_0 = \alpha/2$. According to our rejection criterion (see Section II.B.1), we reject $H_0^{(2)}$ at level α when $|\hat{\beta}_1 - \hat{\beta}_2| > c_1(y_1, \alpha_0) + c_2(y_2, \alpha_0)$. It is quite difficult to find the size of this critical region.

Consider now model (5) where assumption ($A1$) is not imposed. This model has $m(m+1)/2 + \sum_{i=1}^{m} k_i$ parameters and $\sum_{i=1}^{m} N_i$ observations. In this case, no usual finite-sample or asymptotic test procedure appears to be available for comparing the coefficients of the different regressions. But the induced test method allows one to test $H_0^{(1)}$ and $H_0^{(2)}$ relatively easily.

B. Some Examples

We now present some illustrations of the procedure described in the previous sections.

I.   Testing Restrictions in a System of Demands for Inputs

The first example we consider is taken from Berndt (1991, pp. 460–462). We consider the problem of testing restrictions on the parameters of a generalized Leontieff cost function. We assume that the production technology has constant returns to scale and incorporates only two inputs, capital (K) and labor (L), whose prices are P_K and P_L respectively. If we denote the output by Y and the total cost by C, the generalized Leontieff cost function is

$$C = Y \cdot (d_{KK}P_K + 2d_{KL}(P_K P_L)^{1/2} + d_{LL}P_L)$$

If the producer has a cost-minimizing strategy, it can be shown that the demands for factors K and L are given by

$$\frac{K}{Y} = d_{KK} + d_{KL}\left(\frac{P_L}{P_K}\right)^{1/2}, \qquad \frac{L}{Y} = d_{LL} + d_{KL}\left(\frac{P_K}{P_L}\right)^{1/2}$$

A stochastic version of this model would consist in the two-equation SURE model

$$k_t = a_k + b_k p_t^k + u_t^k, \qquad l_t = a_l + b_l p_t^l + u_t^l$$

where u^k and u^l are two Gaussian random vectors with zero mean and covariance matrices $\sigma_k^2 I_N$ and $\sigma_l^2 I_N$, respectively, and $N = 25$ is the sample size for each variable of the model. A restriction imposed by the theory is $b_k = b_l$, which will be our null hypothesis. To test H_0, the procedure described in Section II.B.1 is particularly well suited since we have no a priori information on the relation between the

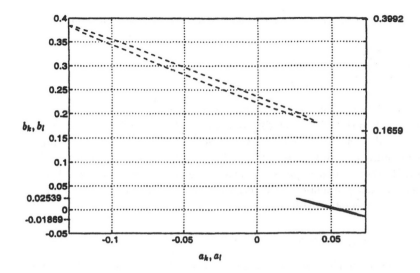

Figure I 97.5% confidence ellipsoids and intervals in the Berndt example. Confidence ellipsoid for $(a_k, b_k)'$: —; confidence ellipsoid for $(a_l, b_l)'$: ---; confidence intervals for b_k and u_l appear on the left and right vertical axes, respectively.

random variables u_t^k and u_s^l. Using the data provided in Berndt (1991), which are described in Berndt and Wood (1975), we performed separate tests of the following null hypotheses:

$$H_0 : b_k = b_l, \qquad H_0^* : \begin{pmatrix} a_k \\ b_k \end{pmatrix} = \begin{pmatrix} a_l \\ b_l \end{pmatrix}$$

The results of the estimation are

$$k_t = 0.0490 \quad + 0.00342 \; p_t^k + \hat{u}_t^k$$
$$\quad (.000125) \quad (.000084)$$

$$l_t = -0.04464 + 0.28295 \; p_t^l + \hat{u}_t^l$$
$$\quad (.001621) \quad (.002350)$$

where the standard errors are given in parentheses. In Figure 1, we show the two 97.5% level confidence ellipsoids required for testing H_0^*. It is straightforward to see that we can reject both null hypotheses at level 5% because none of the regions intersect. Similarly, the 97.5% level confidence intervals for b_k and b_l are respectively $(-0.01869, 0.02539)$ and $(0.1659, 0.3992)$, and so do not intersect.

Since no information on the joint distribution of u_t^k and u_t^l is available, usual GLS procedures cannot be applied in this context. However, suppose that we assume that $(u_1^k, u_2^k, \ldots, u_{25}^k, u_1^l, u_2^l, \ldots, u_{25}^l)'$ is a Gaussian random vector with variance matrix

$$\Omega = \begin{pmatrix} \sigma_1^2 I_{25} & \sigma_{12} I_{25} \\ \sigma_{12} I_{25} & \sigma_2^2 I_{25} \end{pmatrix}$$

as is usually done in SURE models. Using standard GLS techniques, the estimate of $(a_k, b_k, a_l, b_l)'$ is $(0.05100, 0.00235, -0.04886, 0.28804)'$ and the F-statistics for testing H_0 and H_0^* are 27.61 and 938.37, respectively. Since the corresponding 5% asymptotic critical values are 4.05175 and 3.19958, the null hypotheses are both rejected. However, one may prefer the empty intersection test procedure, because it makes a weaker assumption on the error distribution. Moreover, GLS-based tests only have an asymptotic justification.

2. Testing Restrictions on Returns to Schooling

This example is taken from Ashenfelter and Zimmerman (1993). The study considers the following SURE model:

$$Y_{1j} = \theta_1 X_{1j} + \lambda_2 X_{2j} + w_{1j}, \qquad Y_{2j} = \lambda_1 X_{1j} + \theta_2 X_{2j} + w_{2j}$$

where Y_{ij} and X_{ij} represent the log wage and the schooling of the ith brother in the jth family. These equations are the reduced form of a structural model which expresses the relationship between the wage and years of schooling:

$$Y_{1j} = \beta_1 X_{1j} + F_J + v_{1j}, \qquad Y_{2j} + \beta_2 X_{2j} + F_j + v_{2j}$$
$$F_j = \lambda_1 X_{1j} + \lambda_2 X_{2j} + \xi_j$$

where F is a family specific component. We must have $\theta_i = \beta_i + \lambda_i, i = 1, 2$.

The structural model has been estimated over a sample of 143 pairs of brothers. The estimates reported by Ashenfelter and Zimmerman (1993, Table 3) are given below, with standard errors in parentheses:

$$\hat{\theta}_1 = 0.052, \quad \hat{\lambda}_1 = 0.018, \quad \hat{\theta}_2 = 0.068, \quad \hat{\lambda}_2 = 0.006$$
$$(0.015) \qquad\quad (0.020) \qquad\quad (0.019) \qquad\quad (0.015)$$

A natural hypothesis to test here is $H_0 : (\beta_1, \lambda_1)' = (\beta_2, \lambda_2)'$. This can easily be tested from the estimated structural model, since H_0 is equivalent to H_0^* : $(\theta_1, \lambda_1)' = (\theta_2, \lambda_2)'$. Here, we will use the hyperrectangle technique, because Ashenfelter and Zimmerman (1993) do not provide the full estimated covariance matrix for each regression. We first find a confidence interval with level $1 - \alpha/4$ for each one of the mean parameters in the structural model, and check whether the two rectangles so obtained overlap, in which case we accept the null hypothesis. This is done for $\alpha = 5\%$. Each event $[0.0140, 0.0900] \ni \theta_1, [-0.0326, 0.0686] \ni \lambda_1, [0.0199, 0.1161] \ni \theta_2, [-0.0320, 0.0440] \ni \lambda_2$ occurs with probability 0.9875.

We accept the null hypothesis at level 5%, since the two boxes $[0.0140, 0.0900] \times [-0.0326, 0.0686]$ and $[0.0199, 0.1161] \times [-0.0320, 0.0440]$ have a nonempty intersection, which is $[0.0199, 0.0900] \times [-0.0320, 0.0440]$.

V. EXACT INFERENCE IN LINEAR REGRESSION MODELS WITH MA(q) ERRORS

In this section, we show that the procedures developed in Section II can be useful for inference in some dynamic models.

A. A Test on the Mean of a General MA(q) Model

In this section, we consider models of the form

$$
\begin{aligned}
y_t &= m_t + u_t, \quad u_t = \Psi(B)\varepsilon_t, \quad t \in \mathbf{T} = \{1, 2, \ldots, T\} \\
\varepsilon &\equiv (\varepsilon_{1-q}, \varepsilon_{2-q}, \ldots, \varepsilon_0, \varepsilon_1, \ldots, \varepsilon_T)' \sim N(0, \sigma^2 I_{T+q})
\end{aligned}
\tag{6}
$$

where $\Psi(z) = \psi_0 + \psi_1 z + \psi_2 z^2 + \cdots + \psi_q z^q$, $\psi_0 \equiv 1$, $m_t = \sum_{k=1}^{K} x_{tk} b_k = x_t' b$, $b \equiv (b_1, b_2, \ldots, b_K)'$ is a vector of unknown coefficients, and $x_t \equiv (x_{t1}, x_{t2}, \ldots, x_{tK})'$, $t = 1, 2, \ldots, T$, are vectors of fixed (or stricly exogenous) variables. In model (6), $y \sim N(m, \Omega)$, where $m = (Em_1, Em_2, \ldots, Em_T)'$ and $\Omega = (\omega_{t,s})_{t,s=1,2,\ldots,T}$, with

$$
\omega_{t,s} = \begin{cases} \sigma^2 \displaystyle\sum_{i=|t-s|}^{q} \psi_i \psi_{i-|t-s|}, & \text{if } |t - s| \le q \\ 0, & \text{if } |t - s| > q \end{cases}
\tag{7}
$$

(7) shows the key feature of model (6): observations distant by more than q periods from each other are mutually independent. Then, we are naturally led to consider model (6) for subsamples obtained as follows. Define subsets of $\mathbf{T}$, $J_i \equiv \{i, i + (q + 1), i + 2(q + 1), \ldots, i + n_i(q + 1)\}$, where $n_i \equiv I[(T - i)/(q + 1)]$ ($I[x]$ denotes the integer part of x), $i = 1, 2, \ldots, q + 1$, and consider the $q + 1$ equations

$$
\begin{aligned}
y_t &= m_t + u_t, \quad t \in J_i, \\
u &= (u_t : t \in J_i)' \sim N(0, \sigma_u^2 I_{n_i+1}), \quad i = 1, 2, \ldots, q + 1
\end{aligned}
\tag{8}
$$

Equation (8) belongs to the class of model (1). In each equation, the error term satisfies the assumptions of the linear regression model, so that it is possible to apply usual inference procedures to test restriction on b, $H_0 : b \in \Phi$. This null hypothesis can be seen as the intersection of $q + 1$ hypotheses $H_{0,i}$, each of which restricts the mean of the ith subsample to be in Φ, $i = 1, 2, \ldots, q + 1$. The methods presented in Sections II and III are perfectly suited to such situations. We build $q + 1$ critical regions with level $\alpha/(q + 1)$ to test each one of the hypotheses $H_{0,i}$, and reject the null hypothesis at level α if the vector of observations belongs to the union of these regions. Note we did not make any assumption on the roots of $\Psi(z)$. In particular, we did not restrict the MA process $\{\Psi(B)\varepsilon_t : t \in \mathbf{T}\}$ to be invertible.

In the next subsection we apply the procedure to a MA(1) process with a constant and provide comparisons with some alternative procedures such as asymptotic tests and bounds tests.

B. Exact Inference in the Context of a MA(1) Process

1. An Induced Test on the Mean

Consider the model described by (6), with $q = 1$, $K = 1$ and $x_t = 1$, $\forall t \in \mathbf{T}$:

$$y_t = \beta + \varepsilon_t + \psi \varepsilon_{t-1}, \qquad \varepsilon_t \overset{\text{ind}}{\sim} N(0, \sigma^2), \qquad t \in \mathbf{T} \tag{9}$$

The vector of parameters is $\theta = (\beta, \psi, \sigma^2)'$. The null hypothesis we consider is $H_0 : \theta \in \Theta_0$, $\Theta_0 = \{\theta \in \Theta : \beta = 0\}$. According to our procedure, assuming T is even, we form two subsamples of size $T/2$, $(y_t, t \in J_i)$, where $J_1 = \{1, 3, 5, \ldots, T-1\}$ and $J_2 = \{2, 4, 6, \ldots, T\}$. For each subsample, we make inference on β from the regression equation

$$y_t = \beta + u_t, \qquad t \in J_i, \quad \mathbf{u}_i = (u_t : t \in J_i)' \sim N(\mathbf{0}, \sigma_u^2 I_{T/2}), i = 1, 2 \tag{10}$$

A natural critical region with level $\alpha/2$ for testing $\beta = 0$ is then given by

$$W_i(\alpha/2) = \left\{ y \in \mathcal{Y} : \frac{|\hat{\beta}_i|}{[\hat{V}(\hat{\beta}_i)]^{1/2}} > t\left(\frac{T}{2} - 1; \frac{\alpha}{4}\right) \right\}$$

where $\hat{\beta}_i$ is the OLS estimator of β and $\hat{V}(\hat{\beta}_i)$ the usual unbiased estimator of the variance of $\hat{\beta}_i$ from regression (10) using sample $(y_t : t \in J_i)$; $t(T/2 - 1; \alpha/4)$ is the upper $1 - \alpha/4$ percentile of Student's t distribution with $T/2 - 1$ degrees of freedom. We reject $H_0 : \beta = 0$ at level α if $y \in W_1(\alpha/2) \cup W_2(\alpha/2)$.

2. Alternative Procedures

We compared this procedure with two alternatives. The first one consists in testing H_0 using bounds proposed by Hillier and King (1987), Zinde-Walsh and Ullah (1987), Vinod (1976), and Kiviet (1980); see also Vinod and Ullah (1981, Chap. 4). The latter are based on standard least-squares-based tests statistics for testing $\beta = 0$ obtained from the complete sample, such as the t-statistic or its absolute value. Since the distributions of the latter depend on the unknown value of the moving average parameter ψ, one finds instead bounds $t^l(\alpha)$ and $t^u(\alpha)$ which do not depend on the parameter vector θ and such that

$$\mathsf{P}_\theta[T(y) > t^l(\alpha)] \geq \alpha \geq \mathsf{P}_\theta[T(y) > t^u(\alpha)]$$

for all $\theta \in \Theta_0, \alpha \in (0, 1)$. Then the decision rule that consists in rejecting H_0 when $T(y) > t^u(\alpha)$ and accepting H_0 when $T(y) < t^l(\alpha)$ has level α. An inconvenient feature of such procedures is that they may be unconclusive (when $T(y) \in [t^l(\alpha), t^u(\alpha)]$). Obviously, to avoid losses of power, the bounds should be as tight as possible.

In all the above references on bounds tests, the bounds are derived assuming that the MA parameter is known, so that they depend on it, even under the null hypothesis. Therefore we will denote by $t^l_\theta(\alpha)$ and $t^u_\theta(\alpha)$ the lower and upper bounds on $t_\theta(\alpha)$. But as ψ is unknown, we have to find the supremum, $t^u(\alpha)$, over the set $\{t^u_\theta(\alpha) : t^u_\theta(\alpha) \geq t_\theta(\alpha), \forall \theta \in \Theta_0\}$, to make sure that the test based on the rejection region

$$W(\alpha) = \{y \in \mathcal{Y} : T(y) > t^u(\alpha)\}$$

satisfies the level constraint

$$\sup_{\theta \in \Theta_0} P_\theta[W(\alpha)] \leq \alpha$$

Since the moving-average parameter is not restricted by H_0, the set of admissible values for ψ is $\mathbb{R}$. The upper bound is then likely to be quite large.

In the context of model (9), $T(y)$ is typically the usual t-statistic, its square or its absolute value. Since under H_0, its distribution only depends on ψ (and the sample size), we write t_ψ, t^u_ψ, and t^l_ψ instead of t_θ, t^u_θ, and t^l_θ, respectively.

Here, we only use the bounds of Zinde-Walsh and Ullah (1987) and Kiviet (1980), denoted by $t^u_{Z,\psi}(\alpha)$ and $t^u_{K,\psi}(\alpha)$, because they are respectively tighter than those of Hillier and King (1987) and Vinod (1976). The supremum $t^u_K(\alpha)$ of $t^u_{K,\psi}(\alpha)$ for $\psi \in \mathbb{R}$ is difficult to establish, but Kiviet (1980, Table 6, p. 357), gives the values of the bounds for $\psi \in \{.2, .3, .5, .9\}$, and it can be seen that $t^u_{K,.9}(\alpha) \geq t^u_{K,\psi}(\alpha)$, for $\psi \in \{.2, .3, .5, .9\}$. We note that these bounds increase with ψ, and we suspect that the supremum is arbitrarily large, possibly infinite when $\psi \neq 1$. Nevertheless, we will use $t^u_{K,.9}(\alpha)$ as the relevant upper bound in our simulations. Zinde-Walsh and Ullah (1987) derived bounds on the Fisher statistic (or on the square of the t-statistic in our case). $t^u_{Z,\psi}(\alpha)$ is proportional to the ratio $\lambda_{\max}(\psi)/\lambda_{\min}(\psi)$ of the highest and lowers eigenvalues of the covariance matrix of y:

$$t^u_{Z,\psi}(\alpha) = [t_0(\alpha)]^2 \frac{\lambda_{\max}(\psi)}{\lambda_{\min}(\psi)}$$

We need to make here a remark about the accuracy of Zinde-Walsh and Ullah's bound. Their test rejects H_0 at level α when $[T(y)]^2 > \sup_{\psi \in \mathbb{R}} t^u_{Z,\psi}(\alpha) \equiv t^u_Z(\alpha)$. The critical value $t^u_Z(\alpha)$ is not easy to determine analytically, so instead of finding the maximum of $t^u_{Z,\psi}(\alpha)$ on $\mathbb{R}$, we reckoned $t^u_{Z,\psi}(0.05)$ for some values of ψ in the interval $[-1, 2]$. We found a maximum at $\psi = 1$, and a minimum at $\psi = -1$, for every sample size we considered. Although $t^u_{Z,1}(0.05) \leq t^u_Z(0.05)$, we used this

Table 1 Zinde-Walsh and Ullah's Bounds

Sample Size T	25	50	75	100
$t^{U}_{Z,1}(0.05)$	1 164.1972	4 254.3396	9 291.4222	16 274.6855

value as the upper bound. Doing so gives more power to the Zinde-Walsh–Ullah test than it really has, because it may reject H_0 more often than it would do if we used $t^{u}_{Z}(0.05)$. Despite this fact, $t^{u}_{Z,1}(0.05)$ is so large (see Table 1) that the power of the test is zero everywhere on the set of alternatives we considered, for any sample size and for any ψ (see Section V.B.3).

The second alternative consists of using asymptotic tests. In this category, we considered three commonly used test. The first category includes tests based on a GLS estimation of (9). In the first step, one finds a consistent estimator $\hat{\Omega}$ of Ω and $\hat{P}$ such that $\hat{P}'\hat{P} = \hat{\Omega}^{-1}$. In the second step, we multiply both sides of (9) by $\hat{P}$ and apply ordinary least squares (OLS) to that transformed model. In the last step, we test H_0 using the standard F-test. We examine two estimation procedures that lead to a consistent estimator of β, resulting in two test statistics. The first one is detailed in Fomby, Hill, and Johnson (1984, pp. 220–221). We denote it by GLS-MM because in the first step of GLS, we estimate the MA parameter ψ by the method of moments. ψ is estimated by minimizing the distance (in the sense of the Euclidean norm on $\mathbb{R}$) between the sample and true first-order autocorrelations. The second estimation procedure uses exact maximum likelihood in the first step of GLS and will be denoted by GLS-ML.*

The third test we consider is motivated by a central limit theorem (Brockwell and Davis 1991, p. 219) which establishes the following property: if a process, with mean β, has an infinite-order MA representation with IID error terms and MA coefficients ψ_i, $i = \ldots, -2, -1, 0, 1, 2, \ldots$, satisfying the conditions

$$\sum_{i=-\infty}^{\infty} |\psi_i| < \infty, \qquad \sum_{i=-\infty}^{\infty} \psi_i \neq 0$$

then the sample mean of the process is asymptotically normally distributed, with mean β and variance $T^{-1}\sum_{k=-\infty}^{\infty}\gamma(k)$, where $\gamma(k)$ is the autocovariance at lag k. Note that the last condition on the ψ_i's is not satisfied for the MA(1) process (9) with $\psi = -1$, but as ψ is unknown, we might not be aware of this fact or ignore it. Then a

*For further discussion of ML estimation in this context, see Tunnicliffe Wilson (1989) and Laskar and King (1995).

Table 2   Size and Critical Values of 5% Level Asymptotic Tests

| Sample size | T = 25 | | | | T = 50 | | | |
	ψ^*	Size (%)	ACV	CCV	ψ^*	Size (%)	ACV	CCV
GLS-MM	−0.5	19.22	4.25968	30.664	−0.5	18.59	4.0384	59.555
GLS-ML	−0.5	27.87	4.25968	37.979	−0.5	15.06	4.0384	14.615
NW	1	15.03	3.840	8.459	1	10.25	3.840	5.789
Sample size	T = 75				T = 100			
	ψ^*	Size (%)	ACV	CCV	ψ^*	Size (%)	ACV	CCV
GLS-MM	−0.5	16.98	3.97024	64.502	−0.5	14.98	3.9371	38.789
GLS-ML	−0.5	10.13	3.97024	6.396	−0.5	7.84	3.9371	4.983
NW	1	8.82	3.840	5.243	1	8.08	3.840	4.907

natural way of testing H_0 is to estimate β by the sample mean $\overline{Y}_T$ and the asymptotic variance by the consistent estimator proposed in Newey and West (1987):

$$\hat{\phi}_T(p) = \frac{1}{T}\left[r_T(0) + 2 \sum_{k=1}^{p} \left(1 - \frac{K}{p+1}\right) r_T(k)\right]$$

where $r_T(k)$ is the sample autocovariance at lag k. Then, if H_0 is true, the statistic

$$\xi_T^{NW} = \frac{T\overline{Y}_T^2}{\hat{\phi}_T(p)}$$

has an asymptotic χ^2 distribution with 1 degree of freedom. We will denote this procedure by NW.*

Before presenting the results of our simulations, we wish to insist on a very important condition one has to impose when comparing the relative performance of two tests. In the Neyman-Person approach to the problem of testing a null hypothesis H_0 against an alternative H_1, it is meaningless to say that a test A has a higher power than a test B, if the two tests do not have the same level. A test of $H_0 : \theta \in \Theta$ with critical region W has *level* α if $\sup_{\theta \in \Theta_0} P_\theta(W) \le \alpha$, and it has *size*

---

*Of course, the list of the methods considered in the present simulation is not exhaustive. For example, possible variants of the NW method include the covariance matrix estimators proposed by Wooldridge (1989). Bayesian methods (Kennedy and Simons 1991) and marginal likelihood methods (King 1996) could also be used in this context. But space and time limitations have precluded us from including all proposed methods in our simulations.

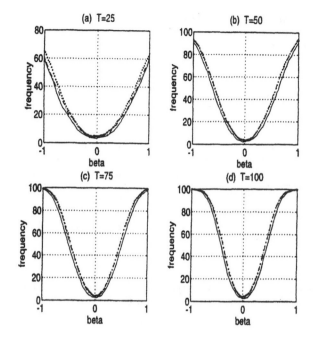

Figure 2   Rejection frequencies of $H_0 : \beta = 0$ in model (4) with $\psi = 1, T = 25, 50, 75, 100$; induced test (—), tests based on subsamples $(y_t : t \in J_1)$ and $(y_t : t \in J_2)$ ($\cdots$ and $-\cdot-$).

α if $\sup_{\theta \in \Theta_0} \mathsf{P}_\theta(W) = \alpha$; see Lehmann (1986, Section 3.1). When the probability of rejection $\mathsf{P}_\theta(W)$ under the null hypothesis ($\theta \in \Theta_0$) is not constant, *controlling the level* of a test requires one to ensure that $\mathsf{P}_\theta(W) \leq \alpha$ for all $\theta \in \Theta_0$, and *controlling its size* involves ensuring that the maximum (or supremum) of the rejection probabilities over $\theta \in \Theta_0$ is equal to α. Of course, this may lead to a difficult search over the parameter space. When the distribution of a test statistic depends on a nuisance parameter (the unknown value of ψ, in the present case), correcting the size of a test requires one to find a critical value such that the maximum probability of rejection under the null hypothesis (irrespective of ψ) is equal to α. A way of doing this is to detect the value of ψ, ψ^* say, for which the discrepancy between the level and the size is maximum. For that value, we simulate S times the test statistic, $T_{\psi^*}(\mathbf{y})$.[*] We then take the observation of rank $(95 \times S)/100 + 1$ of the statistic as our corrected 5% level critical value: we reject H_0 at level 5% when $T_{\psi^*}(\mathbf{y})$ is larger than or equal to that value. Table 2 reports ψ^*, the size (in %), the 5% asymptotic critical value

---

[*]For all asymptotic test procedures, we set $S = 10,000$.

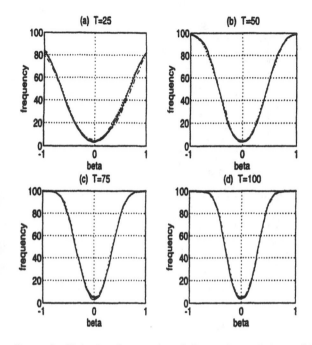

Figure 3   Rejection frequencies of $H_0 : \beta = 0$ in model (4) with $\psi = 0.5, T = 25, 50, 75, 100$; induced test (—), tests based on subsamples $(y_t : t \in J_1)$ and $(y_t : t \in J_2)$ $(\cdots$ and $- \cdot -)$.

(ACV), and the 5% corrected critical value (CCV), for each sample size T, and each of the three asymptotic procedures.

3.   Simulations

In our simulations, we proceeded as follows. For $\psi \in \{-1, -.5, 0, .5, 1\}$ and $T \in \{25, 50, 75, 100\}$, we considered a grid of β values around $\beta_0 = 0$. In each case, 1000 independent samples $(y_t, t = 1, 2, \ldots, T)$ were generated and the following statistics were computed: (1) the t-statistic based on the whole sample; (2) the t-statistic based on the two subsamples $(y_t : t \in J_t)$ and $(y_t : t \in J_2)$ containing the odd and even observations respectively; (3) the GLS-MM and GLS-ML based F-statistics; (4) the ξ_n^{NW}-statistic. Using these statistics, the following tests were implemented at level 5% and the corresponding rejection frequencies were computed: (a) Zinde-Walsh and Ullah's bounds test; (b) Kiviet's bounds test;[*] (c) GLS-

---

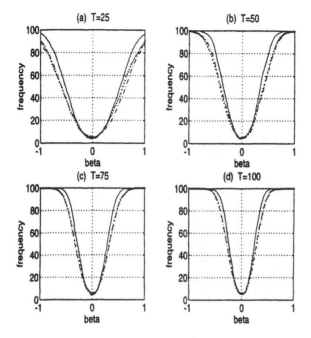

Figure 4 Rejection frequencies of H_0 : $\beta = 0$ in model (4) with $\psi = 0, T = 25, 50, 75, 100$; induced test (—), tests based on subsamples $(y_t : t \in J_1)$ and $(y_t : t \in J_2)$ ($\cdots$ and $-\cdot-$).

MM asymptotic test (corrected and uncorrected for size); (d) GLS-ML asymptotic test (corrected and uncorrected for size); (e) NW asymptotic test (corrected and uncorrected for size); (f) the induced test which combines the standard t-tests based on the two subsamples $(y_t : t \in J_t)$ and $(y_t : t \in J_2)$; (g) the separate tests based on the subsamples $(y_t : t \in J_1)$ and $(y_t : t \in J_2)$. The results are presented in Figures 2 to 6 and Tables 3 to 7.

As it became clear in the description of the induced test, when applying such a procedure to model (9), one is led to split the sample in two, and make two tests at level $\alpha/2$. At first sight, the procedure displays features that may seem quite unattractive. First, it splits the available sample in two, and second it combines two tests whose levels are only $\alpha/2$ (instead of α). From these two remarks, one may expect the procedure to lack power. But we should keep in mind that, since the two "subtests" have level $\alpha/2$, the resulting induced test has level certainly greater than $\alpha/2$ (although not greater than α). Furthermore, this test actually uses the information contained in the whole sample. Then it becomes less clear whether the induced test procedure automatically leads to a loss of power relatively to other alternative

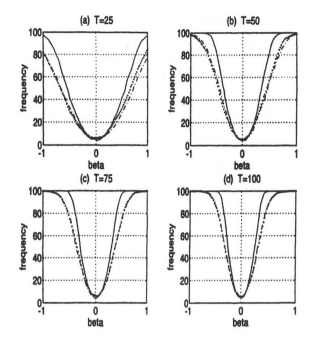

Figure 5   Rejection frequencies of $H_0 : \beta = 0$ in model (4) with $\psi = -.5, T = $ 25, 50, 75, 100; induced test (—), tests based on subsamples $(y_t : t \in J_1)$ and $(y_t : t \in J_2)$ ($\cdots$ and $-\cdot-$).

tests. Two questions arises from these remarks: (1) Is combining preferable to not combining? i.e., should our decision at level α rely on an induced test procedure or on a test based on one of the subsample only? (2) How does the induced test compare with the procedures mentioned in Section V.B.2?

Figures 2 to 6 answer the first question. They show that the power of the induced test (solid line) is generally higher than that of an α-level test based on one of the two subsamples (dashed lines). In other words, combining is preferable to not combining. When it is not the case (when the true value of the MA parameter is unity, $\psi = 1$, see Figures 2a to 2d), the power loss from using the induced test is very small, so that one would usually prefer the sample-split procedure that uses all the observations.

Tables 3 to 7 report the estimated probability of a rejection of $H_0 : \beta = 0$ for different sample sizes ($T \in \{25, 50, 75, 100\}$) and true values for $\beta(\beta \in \{-1, -.8, -.5, -.2, 0, .2, .5, .8, 1\})$, for each one of the test procedures of Section V.B.2. If we first consider bounds tests, we note that the Kiviet test is dominated by the induced test, except for $\psi = .5$ and $\psi = 1$. We already mentioned in Section

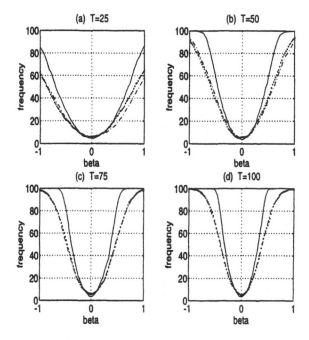

Figure 6 Rejection frequencies of H_0 : $\beta = 0$ in model (4) with $\psi = -1, T = 25, 50, 75, 100$; induced test (—), tests based on subsamples $(y_t : t \in J_1)$ and $(y_t : t \in J_2)$ $(\cdots$ and $- \cdot -)$.

V.B.2 that the bound which has been used here, namely $t^u_{K,.9}(0.05)$, is not appropriate because we do not know whether this value satisfies the level constraint:

$$\sup_{\theta \in \Theta_0} \mathsf{P}_\theta[\{y \in \mathcal{Y} : T(y) > t^u_{K,.9}(0.05)\}] \leq 0.05$$

In other words, a critical region based on Kiviet's bounds has an unknown level. Moreover, what makes the induced test more attractive relatively to Kiviet's test is that it avoids the calculation of a bound that changes with the sample size. Finally, because Zinde-Walsh and Ullah's upper bounds are so large (see Table 1), the power of their test is zero for all ψ. These are not reported in Tables 3–7.

The most surprising result which emerges from our Monte Carlo study can be seen in Tables 3, 4, and 5. Once the asymptotic critical values used for the GLS-MM and GLS-ML tests have been corrected so that the corresponding critical regions have the desired level, our procedure becomes more powerful than these alternatives for many plausible values of ψ. The difference between estimated power functions grows as ψ increases, but diminishes when the sample size T gets larger. The GLS-MM method seems to be the worst of all the asymptotic procedures studied here,

Table 3 Rejection Frequencies of $H_0 : \beta = 0$ in Model (9) with $\psi = 1$

β	Sample size $T = 25$					Sample size $T = 75$			
	GLS-ML	GLS-MM	NW	Kiv.	Ind. test	GLS-ML	GLS-MM	NW	Ind. test
−1.0	0.80	3.40	56.00	66.70	59.80	97.70	0.70	99.10	98.80
−0.8	0.20	1.10	37.80	46.60	40.00	84.90	0.10	91.50	89.70
−0.5	0.00	0.20	17.00	20.10	16.60	40.00	0.00	52.00	48.00
−0.2	0.00	0.00	6.10	6.20	5.20	6.60	0.00	14.30	9.90
−0.1	0.00	0.00	4.50	5.00	3.50	2.20	0.00	7.00	4.40
0.0	0.00	0.00	4.30	4.30	3.50	1.10	0.00	4.30	2.70
0.1	0.00	0.00	5.20	5.10	4.30	2.10	0.00	6.50	3.80
0.2	0.00	0.00	6.50	7.60	5.30	5.80	0.00	11.80	9.80
0.5	0.10	0.60	18.10	21.20	18.10	38.60	0.00	55.20	48.90
0.8	0.10	1.80	38.80	46.40	40.50	83.60	0.20	91.10	89.40
1.0	0.30	3.90	55.00	64.90	59.60	96.70	0.80	98.90	98.80

β	Sample size $T = 50$					Sample size $T = 100$			
	GLS-ML	GLS-MM	NW	Kiv.	Ind. test	GLS-ML	GLS-MM	NW	Ind. test
−1.0	46.00	0.20	92.40	93.30	90.70	99.50	25.60	99.80	99.70
−0.8	21.50	0.00	77.30	78.30	72.30	97.00	6.70	97.70	96.90
−0.5	3.50	0.00	40.10	40.20	32.90	61.30	0.20	69.10	63.20
−0.2	0.20	0.00	9.60	10.00	6.80	11.90	0.00	17.40	12.20
−0.1	0.00	0.00	4.90	4.50	3.50	4.60	0.00	7.90	5.00
0.0	0.10	0.00	3.90	3.90	2.40	3.40	0.00	4.40	2.80
0.1	0.10	0.00	4.30	4.70	3.50	5.60	0.00	7.40	3.90
0.2	0.20	0.00	8.20	9.20	6.60	11.60	0.00	15.10	10.30
0.5	3.70	0.00	34.90	40.60	33.70	62.80	0.30	67.00	60.40
0.8	21.00	0.10	74.00	78.60	72.70	96.50	4.60	97.30	96.60
1.0	45.10	0.40	89.70	95.00	90.30	99.90	22.90	99.90	99.80

Table 4 Rejection Frequencies of $H_0 : \beta = 0$ in Model (9) with $\psi = .5$

β	Sample size $T = 25$					Sample size $T = 75$			
	GLS-ML	GLS-MM	NW	Kiv.	Ind. test	GLS-ML	GLS-MM	NW	Ind. test
-1.0	4.80	10.70	79.40	85.30	83.90	100.00	8.50	100.00	100.00
-0.8	2.50	4.70	61.20	67.30	66.30	98.80	1.00	99.50	99.40
-0.5	0.60	1.00	27.20	30.30	29.00	65.50	0.10	77.50	76.70
-0.2	0.00	0.00	6.70	6.80	7.00	11.30	0.00	19.40	18.10
-0.1	0.00	0.00	4.60	3.90	4.60	2.80	0.00	8.40	7.00
0.0	0.00	0.00	4.30	3.50	4.20	0.80	0.00	4.30	2.90
0.1	0.00	0.00	5.60	4.70	4.70	3.10	0.00	7.80	6.40
0.2	0.10	0.10	7.80	8.10	8.30	10.40	0.00	19.30	17.10
0.5	0.50	0.60	27.80	30.90	30.20	64.70	0.10	77.70	75.60
0.8	1.60	4.60	60.50	66.80	64.70	97.80	1.10	99.50	99.40
1.0	5.00	12.10	78.30	84.80	82.80	100.00	7.20	100.00	100.00
	Sample size $T = 50$					Sample size $T = 100$			
	GLS-ML	GLS-MM	NW	Kiv.	Ind. test	GLS-ML	GLS-MM	NW	Ind. test
-1.0	83.00	2.40	99.30	99.80	99.40	100.00	65.10	100.00	100.00
-0.8	50.80	0.20	93.80	94.30	94.00	99.80	26.80	99.90	99.90
-0.5	10.80	0.00	58.30	59.20	55.80	86.80	1.10	90.90	89.00
-0.2	0.40	0.00	12.70	11.50	11.90	20.50	0.00	24.70	22.50
-0.1	0.10	0.00	6.50	4.10	5.60	6.00	0.00	10.90	7.90
0.0	0.10	0.00	3.30	2.90	3.60	3.40	0.00	4.30	3.50
0.1	0.20	0.00	5.70	4.30	5.20	7.40	0.00	8.70	6.70
0.2	0.40	0.00	13.00	12.20	11.90	19.10	0.00	22.30	20.70
0.5	10.40	0.00	59.70	59.30	57.40	88.10	1.00	89.70	88.10
0.8	50.60	0.50	94.80	96.00	94.70	100.00	25.20	100.00	99.90
1.0	82.10	3.50	99.40	99.70	99.50	100.00	63.30	100.00	100.00

Table 5 Rejection Frequencies of $H_0 : \beta = 0$ in Model (9) with $\psi = 0$

	Sample size $T = 25$					Sample size $T = 75$			
β	GLS-ML	GLS-MM	NW	Kiv.	Ind. test	GLS-ML	GLS-MM	NW	Ind. test
-1.0	35.30	45.80	97.80	96.20	97.00	100.00	70.30	100.00	100.00
-0.8	19.30	23.10	89.30	81.70	85.20	100.00	28.20	100.00	100.00
-0.5	6.20	5.10	48.40	31.70	45.30	96.90	0.80	98.50	94.30
-0.2	2.50	0.90	8.90	2.80	9.90	25.80	0.00	33.70	28.20
-0.1	1.20	0.30	3.30	0.90	6.60	7.20	0.00	11.80	10.40
0.0	1.00	0.20	3.10	0.40	4.70	1.40	0.00	2.70	4.50
0.1	1.40	0.60	5.10	0.80	7.00	6.90	0.00	10.60	9.50
0.2	1.90	0.70	9.40	3.00	11.00	23.40	0.00	32.10	27.10
0.5	7.00	5.40	48.30	31.80	45.70	96.10	0.80	98.00	94.30
0.8	20.80	23.80	87.50	80.50	85.90	100.00	26.70	100.00	100.00
1.0	37.70	45.30	98.00	97.20	97.00	100.00	70.70	100.00	100.00

	Sample size $T = 50$					Sample size $T = 100$			
β	GLS-ML	GLS-MM	NW	Kiv.	Ind. test	GLS-ML	GLS-MM	NW	Ind. test
-1.0	99.70	40.80	100.00	100.00	100.00	100.00	99.80	100.00	100.00
-0.8	95.00	13.20	100.00	99.90	99.60	100.00	93.20	100.00	100.00
-0.5	45.10	0.50	88.60	73.30	80.20	99.80	20.80	99.60	98.30
-0.2	2.70	0.00	21.40	6.90	19.40	42.50	0.00	47.50	36.30
-0.1	0.50	0.00	7.50	1.40	6.50	12.30	0.00	14.60	12.50
0.0	0.30	0.00	2.40	0.10	4.70	3.90	0.00	3.30	5.20
0.1	0.90	0.00	6.90	1.60	8.00	12.60	0.00	12.70	11.50
0.2	2.60	0.00	20.80	7.10	18.30	43.50	0.20	43.60	35.40
0.5	43.80	0.60	89.00	72.90	80.20	100.00	17.40	99.60	98.60
0.8	95.10	13.30	99.90	99.70	99.60	100.00	94.00	100.00	100.00
1.0	99.60	41.30	100.00	100.00	100.00	100.00	99.90	100.00	100.00

Table 6 Rejection Frequencies of $H_0 : \beta = 0$ in Model (9) with $\psi = -0.5$

Sample size $T = 25$

β	GLS-ML	GLS-MM	NW	Kiv.	Ind. test
-1.0	93.50	91.70	100.00	97.50	97.20
-0.8	82.80	76.30	99.80	77.30	84.90
-0.5	49.80	41.70	82.70	10.10	40.30
-0.2	18.00	18.20	6.30	0.00	10.80
-0.1	8.00	8.60	1.40	0.00	6.40
0.0	4.50	4.30	0.10	0.00	4.40
0.1	10.60	9.70	1.90	0.00	6.20
0.2	18.90	19.50	8.70	0.00	11.10
0.5	50.00	41.80	81.40	10.50	42.80
0.8	82.10	76.00	99.80	76.60	84.40
1.0	93.10	92.40	100.00	97.90	97.90

Sample size $T = 50$

β	GLS-ML	GLS-MM	NW	Kiv.	Ind. test
-1.0	100.00	97.90	100.00	100.00	100.00
-0.8	100.00	86.90	100.00	100.00	100.00
-0.5	98.60	38.90	100.00	66.10	81.80
-0.2	33.50	16.80	39.70	0.10	18.30
-0.1	10.30	10.30	4.60	0.00	7.00
0.0	2.90	5.00	0.10	0.00	4.40
0.1	9.50	9.10	5.60	0.00	7.10
0.2	33.90	15.80	37.20	0.00	17.30
0.5	99.00	41.30	99.90	67.00	80.20
0.8	100.00	87.70	100.00	99.90	99.90
1.0	100.00	98.10	100.00	100.00	100.00

Sample size $T = 75$

β	GLS-ML	GLS-MM	NW	Ind. test
-1.0	100.00	100.00	100.00	100.00
-0.8	100.00	98.50	100.00	100.00
-0.5	100.00	56.70	100.00	96.90
-0.2	82.90	14.70	65.10	24.60
-0.1	29.60	9.30	10.80	9.90
0.0	3.60	5.30	0.40	4.00
0.1	29.20	9.60	12.20	9.40
0.2	81.50	15.30	67.10	23.80
0.5	100.00	58.00	100.00	97.60
0.8	100.00	98.10	100.00	100.00
1.0	100.00	100.00	100.00	100.00

Sample size $T = 100$

β	GLS-ML	GLS-MM	NW	Ind. test
-1.0	100.00	100.00	100.00	100.00
-0.8	100.00	100.00	100.00	100.00
-0.5	100.00	97.00	100.00	99.90
-0.2	94.90	20.60	84.60	32.30
-0.1	45.50	9.40	18.60	11.00
0.0	5.00	4.40	0.40	4.30
0.1	44.40	9.30	16.80	10.10
0.2	95.30	19.70	82.70	32.00
0.5	100.00	97.40	100.00	99.80
0.8	100.00	100.00	100.00	100.00
1.0	100.00	100.00	100.00	100.00

Table 7 Rejection Frequencies of $H_0 : \beta = 0$ in Model (9) with $\psi = -1$

	Sample size $T = 25$					Sample size $T = 75$			
β	GLS-ML	GLS-MM	NW	Kiv.	Ind. test	GLS-ML	GLS-MM	NW	Ind. test
-1.0	99.50	93.10	100.00	79.90	85.20	100.00	100.00	100.00	100.00
-0.8	98.20	80.10	100.00	34.80	64.00	100.00	99.20	100.00	100.00
-0.5	89.80	58.30	89.10	0.00	28.70	100.00	82.30	100.00	83.60
-0.2	49.90	40.80	0.10	0.00	9.00	100.00	51.60	84.60	17.10
-0.1	2.20	4.30	0.00	0.00	5.90	98.80	47.00	0.10	8.10
0.0	0.00	0.00	0.00	0.00	4.80	0.70	0.00	0.00	3.80
0.1	2.30	2.80	0.00	0.00	6.10	99.10	47.00	0.10	8.10
0.2	52.40	41.40	0.40	0.00	9.30	100.00	51.40	87.90	17.90
0.5	90.70	59.40	88.10	0.40	28.50	100.00	82.00	100.00	85.50
0.8	98.10	81.30	100.00	34.30	64.70	100.00	99.30	100.00	100.00
1.0	99.60	93.10	100.00	79.40	87.20	100.00	100.00	100.00	100.00

	Sample size $T = 50$					Sample size $T = 100$			
β	GLS-ML	GLS-MM	NW	Kiv.	Ind. test	GLS-ML	GLS-MM	NW	Ind. test
-1.0	100.00	98.30	100.00	100.00	100.00	100.00	100.00	100.00	100.00
-0.8	100.00	92.60	100.00	99.20	99.00	100.00	100.00	100.00	100.00
-0.5	100.00	69.30	100.00	13.10	58.20	100.00	99.20	100.00	98.30
-0.2	96.60	46.00	29.10	0.00	12.10	100.00	61.20	99.70	21.70
-0.1	75.00	42.10	0.00	0.00	6.20	100.00	48.60	0.50	8.50
0.0	0.00	0.00	0.00	0.00	3.80	1.10	0.00	0.00	3.80
0.1	76.00	41.90	0.00	0.00	6.50	100.00	48.90	0.40	8.70
0.2	96.40	45.80	28.20	0.00	10.80	100.00	61.90	99.50	20.70
0.5	100.00	69.30	100.00	12.10	59.20	100.00	99.00	100.00	98.50
0.8	100.00	92.30	100.00	99.50	98.70	100.00	100.00	100.00	100.00
1.0	100.00	98.30	100.00	100.00	100.00	100.00	100.00	100.00	100.00

whereas GLS-ML appears to benefit from the asymptotic efficiency property of maximum likelihood estimators. But for nonnegative values of ψ, the sample size has to be $T = 100$ for the GLS-ML test to have a probability of correctly rejecting the null as high as the induced test. The GLS-MM test is still dominated for some negative values of $\psi(\psi = -.5)$, irrespective of the sample size. Only when ψ is close to -1 does this procedure become admissible.

While the two commonly used asymptotic inference procedures, GLS-MM and GLS-ML, cannot be recommended on the ground of our Monte Carlo study, the conclusion is less negative for the NW method. Except for small sample sizes ($T = 25$) and large values of the MA parameter ($\psi = 1, .5$), it does better than the induced test procedure. This result is somewhat unexpected because the Newey-West estimator of $V(\overline{Y}_T)$ does not take into account the autocovariance structure of the process. However, although the induced test is conservative, it is more powerful than NW test for alternatives close to the null hypothesis when ψ is negative. Furthermore, it is important to remember that the NW test suffers from level distortions (see Table 2) that are not easy to correct in practice.

4.   An Example: An Induced Test on the Mean of the Canadian Per Capita GDP Series

We now apply our procedure to test the nullity of the mean of a process that has a MA(1) representation. Our series is the first difference of the Canadian per capita GDP, denominated in real 1980 Purchasing Power Parity-adjusted US dollars, observed yearly from 1901 to 1987. It is taken from Bernard and Durlauf (1995). Figure 7 plots the series. Using standard Box-Jenkins procedure (autocorrelation and partial autocorrelation functions), we identified a MA(1) process for the series (see Table 8).

We then consider a model like (9). ML estimation of (9) gives $\hat{\beta} = 136.1810$ and $\hat{\psi} = 0.4211$ with estimated variances 945.1919 and 0.0095, respectively. The estimated $\text{Cov}(\hat{\beta}, \hat{\psi})$ is 0.0834 and the sample variance of the residuals is 40117.5725.

To implement an induced test for the nullity of the mean parameters, β, at level 5%, we split the sample in two parts, $\{y_t : t \in J_i\}, i = 1, 2$, which contain respectively the odd and the even observations, and make two 2.5% tests of $\beta = 0$, using the statistics $t_i = \sqrt{n_i}|\overline{y}_i|/s_i$, where $\overline{y}_i = \sum_{j \in J_i} y_j/n_i$, $s_i^2 = (\sum_{j \in J_i}(y_j - \overline{y}_i)^2)/(n_i - 1)$, and n_i is the size of subsample $i, i = 1, 2$. We reject the null hypothesis when $t_1 > t(\alpha/4, \nu_1)$ or $t_2 > t(\alpha/4, \nu_2)$, where $t(\alpha/4, \nu)$ is the $1 - (\alpha/4)$ percentile of Student's t distribution with ν degrees of freedom. We also perform both GLS-MM and GLS-ML asymptotic tests. Our results are reported in Table 9. $\hat{\beta}$ is the two step estimator of β, $\hat{\psi}$ the estimator of ψ that has been obtained in the first step to estimate the error covariance matrix, and t the test statistic, whose distribution will be approximated by a student's t-distribution with 86 degrees of freedom. Both subtests

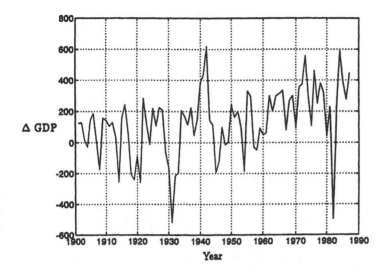

Figure 7 First differences of the Canadian per capita GDP. (From Bernard and Durlauf 1994.)

reject the null hypothesis at level 2.5%. Hence the induced test rejects the nullity of the mean at level 5%. The two asymptotic tests also reject the null hypothesis, if we admit that the asymptotic critical value is a good approximation when the sample size is 87. Our findings are consistent with the results of the Monte Carlo study of Section V.B.3. For similar sample sizes ($T = 75$ or $T = 100$) we found that the GLS-MM test produces larger values of the test statistic than the GLS-ML test does. This is what we have here with $T = 87$.

If we decide to include a linear trend in the mean of the MA(1) process, our induced test procedure still applies. The per capita GDP series now admits the representation

$$y_t = \beta_0 + \beta_1 t + \varepsilon_t + \psi \varepsilon_{t-1}, \qquad \varepsilon_t \overset{\text{ind}}{\sim} N(0, \sigma^2)$$

Table 8 Sample Autocorrelations of the Canadian Per Capita GDP Series

ag	1	2	3	4	5	6	7	8	9	10	11	12
utocorrelation	.41	.19	.10	−.04	.05	.07	.12	.04	−.04	.09	.08	.20
tandard error	.11	.12	.13	.13	.13	.13	.13	.13	.13	.13	.13	.13
jung-box Q-statistic	15.4	18.8	19.8	19.9	20.2	20.6	22.1	22.3	22.5	23.3	24.0	28.3

Table 9 Induced and Asymptotic Tests. Model $y_t = \beta + \varepsilon_t + \psi \varepsilon_{t-1}$

	$i = 1$	$i = 2$	GLS-MM	GLS-ML
$\hat\beta$	127.6836	125.4406	122.2522	123.6574
$t(\nu)$	4.1892 (43)	3.5076 (42)	7.5112 (86)	6.9345 (86)
p-value	0.00014	0.00109	0.00000	0.00000
$\hat\psi$	—	—	05298	04221

for $t = 1, 2, \ldots, T$. We consider the three hypotheses

$$H_0^{(0)} : \beta_0 = 0, \qquad H_0^{(1)} : \beta_1 = 0, \qquad H_0 : \begin{pmatrix} \beta_0 \\ \beta_1 \end{pmatrix} = \begin{pmatrix} 0 \\ 0 \end{pmatrix}$$

For each hypothesis, we perform the induced test as well as the asymptotic test. Results appear in Table 10. We note that only one of the subtests rejects the presence of a linear trend. However, according to our decision rule, this is enough to reject $H_0^{(1)}$. Both GLS-MM and GLS-ML unambiguously reject this hypothesis. But we know from our simulations that the asymptotic tests tend to reject the null too often when it is true. For the parameters β_j, $j = 1, 2$, we also report two confidence intervals I_1^j and I_2^j, each with level 97.5%, based on the two subsamples $(y_t : t \in J_1)$ and $(y_t : t \in J_2)$. The intersection $I_1^j \cap I_2^j$ gives the set of values $\gamma \in \mathbb{R}$ such that the hypothesis $H_0^j(\gamma) : \beta_j = \gamma$ is not rejected at level 5% by the induced test. These intervals are

Table 10 Induced and Asymptotic Tests. Model $y_t = \beta_0 + \beta_1 t + \varepsilon_t + \psi \varepsilon_{t-1}$

	$i = 1$	$i = 2$	GLS-MM	GLS-ML
$\hat\beta_0$	−37.0695	−6.2981	−22.5258	−22.5578
$\hat\beta_1$	3.7444	2.9941	3.3507	3.3554
$t_0(\nu)$	−0.6832 (42)	−0.0903 (41)	−0.6726 (85)	−0.6577 (85)
p-value ($H_0^{(0)}$)	0.49823	0.92849	0.50303	0.51251
$t_1(\nu)$	3.5058 (42)	2.1674 (41)	5.0392 (85)	4.9319 (85)
p-value ($H_0^{(1)}$)	0.00110	0.03606	0.00000	0.00000
$F(\nu_1, \nu_2)$	17.2244 (2, 42)	9.0421 (2, 41)	37.9250 (2, 85)	39.3875 (2, 85)
p-value (H_0)	0.00000	0.00056	0.00000	0.00000
$\hat\psi$	—	—	0.3536	0.3253

t_j, $j = 1, 2$, and F denote Student's t- and Fisher's F-statistics used for testing $H_0^{(j)}$, $j = 1, 2$, and H_0, respectively.

Table 11   Induced and Asymptotic Tests. Model $y_t = \beta + \varepsilon_t + \psi_1\varepsilon_{t-1} + \psi_2\varepsilon_{t-2}$

	$i = 1$	$i = 2$	$i = 3$	GLS-MM	GLS-ML
$\hat{\beta}$	122.4644	130.7687	126.4918	128.9405	129.7032
$t(\nu)$	2.6573 (28)	4.1328 (28)	2.9150 (28)	3.6828 (86)	3.2812 (86)
p-value	0.01286	0.00029	0.00692	0.00040	0.00149
$\hat{\psi}_1$	—	—	—	0.4096	0.3931
$\hat{\psi}_2$	—	—	—	0.2354	0.1037

t denotes the statistic used for testing $H_0 : \beta = 0$.

$$I_1^0 = [-163.2002, 89.0612] \qquad I_2^0 = [-168.5787, 155.9825]$$
$$I_1^1 = [1.2616, 6.2272], \qquad I_2^1 = [-0.2201, 6.2083]$$

yielding the following 95% confidence intervals for β_0 and β_1:

$$\beta_0 \in [-163.2002, 89.0612], \qquad \beta_1 \in [1.2616, 6.2083]$$

These entail that β_0 is not significantly different from 0, while β_1 is.

When we apply the induced test procedure, we implicitly assume that we have correctly identified a MA(1) process. An interesting issue is to look at what we get if, instead of the true MA(1) representation, we use a MA(2) model to build our test statistics. In this case, we split the sample in three parts, make three tests at level 5/3%, and reject the null $H_0 : \beta = 0$ when our sample falls in one of the three critical regions. The results are in Table 11. We first note that as two of the subsample based tests reject H_0 at level 5/3%, we reject the null at level 5%. We also note that both asymptotic tests reject the null hypothesis. But we know that using 5% critical values obtained from asymptotic distributions leads to a probability of making a type I error larger than 5%. Therefore, although asymptotic tests and tests based on the sample split yield the same decision of rejecting H_0, we put more confidence in the sample split procedure.

VI. CONCLUDING REMARKS

In this chapter we proposed a set of inference methods for comparing and pooling information obtained from different data sets, which simply use separate tests (or confidence sets) based on the different data sets. The methods described are based on a systematic exploitation of Boole-Bonferroni inequalities and can yield exact tests and confidence sets without the need to specify at all the relationship between the data sets, even with small sample sizes. As a result, they are quite versatile and usually easy to implement. The general problems studied include (1) combining separate tests based on different data sets for an hypothesis of interest (more precisely, for the intersection of similar hypotheses), to obtain more powerful tests; (2) comparing parameters estimated from the different data sets (e.g., to test their equality); (3) combining confidence regions based on different samples to obtain a more accurate confidence set. For problem 1, we were led to consider Bonferroni-type induced tests; for problem 2, we proposed *empty intersection* tests; and for problem 3, we suggested taking the intersection of separate confidence sets with appropriate levels.

We also showed that the methods proposed can be quite useful in various models where usual inference procedures based on a complete sample involve difficult distributional problems (e.g., because of nuisance parameters), but for which distributional properties of test statistics computed on appropriately selected subsamples are simpler. This leads to an interesting form of *sample-split* (SS) method. One first splits the sample into several subsets of observations from which separate inferences (tests of confidence sets) are obtained. Then these results are recombined, using the general union-intersection (UI) methods already described, to obtain a single inference which uses the full sample. The way the data is split depends on the model considered. In some situations the structure naturally suggests the division. This is for example true when the model contains several equations. In other cases, the division is based on more elaborate arguments, as in moving average models.

The UI/SS methods proposed can be applied to a wide spectrum of econometric situations and models. We discussed and illustrated their applications in two cases where only asymptotic methods are typically available, namely inference in SURE models and linear regressions with MA errors. In the latter case, we also presented an extensive Monte Carlo study comparing the UI-SS method for testing an hypothesis about a mean with other available approaches. Two main conclusions emerged from these results: first, they provided further evidence on the size distortions associated with usual asymptotic procedures; second, they showed that UI-SS tests not only have the predicted levels, but enjoy good power properties. In view of the fact that these methods involve splitting the sample and lead to conservative procedures, hence leading one to expect a power loss, this is indeed quite remarkable. Our results show that the Bonferroni-based recombination of the evidence obtained from the different subsamples apparently makes up for the loss. For another application of UI-SS approach to autoregressive and other types of dynamic models, the reader may consult Dufour and Torrès (1995).

Before closing, it is worthwhile noting a few other points. First, the UI (or UI-SS) procedures are often simpler to implement than usual asymptotic procedures. For SURE models and linear regressions with MA errors, they only require critical values from standard distributions. For MA models, they avoid the task of estimating MA coefficients. Second, they offer some extra robustness to model specification, as illustrated by SURE where no assumption on the relationship between the different equations is needed. Third, although we stressed here the derivation of finite sample methods, there is nothing that forbids the application of UI (or UI-SS) methods to situations where only asymptotically justified tests or confidence sets are available from the separate data sets. In such cases, the methods are applied in exactly the same way. This feature may be especially attractive for gaining robustness to model specification. We think all these properties make this UI-SS approach an attractive and potentially quite useful addition to the methods available to applied econometricians.

ACKNOWLEDGMENTS

We thank Eugene Savin, the editor, David Giles, and an anonymous referee for several useful comments. This work was supported by grants from the Social Sciences Research Council of Canada, the Natural Sciences and Engineering Council of Canada, and the Government of Québec (Fonds FCAR).

REFERENCES

Alt, F. B. (1982), Bonferroni Inequalities and Intervals, in *Encyclopedia of Statistical Sciences*, Vol. 1 (N. L. Johnson and S. Kotz, eds.), Wiley, New York, 294–301.

Angrist, J. D. and A. B. Krueger (1994), Split Sample Instrumental Variables, *Technical Working Paper 150*, NBER, Cambridge, MA.

Arnold, S. F. (1981), *The Theory of Linear Models and Multivariate Analysis*, Wiley, New York.

Ashenfelter, O. and A. Krueger (1992), Estimates of the Economic Return to Schooling from a New Sample of Twins, *Working Paper 304*, Industrial Relations Sections, Princeton University.

Ashenfelter, O. and D. J. Zimmerman (1993), Estimates of the Returns to Schooling from Sibling Data: Fathers, Sons, and Brothers, *Working Paper 318*, Industrial Relations Section, Princeton University.

Bernard, A. B. and S. N. Durlauf (1995), Convergence in International Output, *Journal of Applied Econometrics*, 10, 97–108.

Berndt, E. R. (1991), *The Practice of Econometrics: Classic and Contemporary*, Addison-Wesley, Reading (MA).

Berndt, E. R. and D. O. Wood (1975), Technology, Prices and the Derived Demand for Energy, *Review of Economics and Statistics*, 53, 259–268.

Birnbaum, A. (1954), Combining Independent Tests of Significance, *Journal of the American Statistical Association*, 49, 559–575.

Brockwell, P. J. and R. A. Davis (1991), *Time Series: Theory and Methods*, Springer-Verlag, New York.

Burnside, C. and M. Eichenbaum (1994), Small Sample Properties of Generalized Method of Moments Based Wald Tests, *NBER Technical Working Paper*, No 155.

DeJong, D. N., J. C. Nankervis, N. E. Savin, and C. H. Whiteman (1992), The Power of Unit Root Tests in Time Series with Autoregressive Errors, *Journal of Econometrics*, 53, 323–343.

Dufour, J.-M. (1990), Exact Tests and Confidence Sets in Linear Regressions with Autocorrelated Errors, *Econometrica*, 58, 479–494.

Dufour, J.-M. (1997), Some Impossibility Theorems in Econometrics, with Applications to Structural and Dynamic Models, *Econometrica*, forthcoming.

Dufour, J.-M. and J. Jasiak (1995), Finite Sample Inference Methods for Simultaneous Equations and Models with Unobserved and Generated Regressors, *Discussion Paper*, CRDE, Universitéde Montréal.

Dufour, J.-M. and O. Torrès (1995), Two-Sided Autoregressions and Exact Inference for Stationary and Nonstationary Autoregressive Processes, *Discussion Paper*, CRDE, Universitéde Montréal.

Folks, J. L. (1984), Combination of Independent Tests, in *Handbook of Statistics, Volume 4, Nonparametric Methods* (P. R. Krishnaiah and P. K. Sen, eds.), North Holland, Amsterdam, pp. 113–121.

Fomby, T. B., R. C. Hill and S. R. Johnson (1984), *Advanced Econometric Methods*, Springer-Verlag, New York.

Gouriéroux, C. and A. Monfort (1989), *Statistique et Modèles Économétriques*, vol. 2, Economica, Paris.

Hedges, L. V. and I. Olkin (1985), *Statistical Methods for Meta-Analysis*, Academic Press, San Diego.

Hillier, G. H. and M. L. King (1987), Linear Regression with Correlated Errors: Bounds on Coefficient Estimates and t-values, in *Specification Analysis in the Linear Model* (M. L. King and D. E. A. Giles, ed.), Routledge & Kegan Paul, London, Ch. 4.

Kennedy, P. and D. Simons (1991), Fighting the Teflon Factor: Comparing Classical and Bayesian Estimators for Autocorrelated Errors, *Journal of Econometrics*, 48, 15–27.

King, M. L. (1996), Hypothesis Testing in the Presence of Nuisance Parameters, *Journal of Statistical Planning and Inference*, 50, 103–120.

Kiviet, J. F. (1980), Effects of ARMA Errors on the Significance Tests for Regression Coefficients: Comments on Vinod's Article; Improved and Additional Results, *Journal of the American Statistical Association*, 75, 353–358.

Laskar, M. R. and M. L. King (1995), Parameter Orthogonality and Likelihood Functions, in *Proceedings of the 1995 Econometrics Conference at Monash*, Melbourne (Australia), (C. S. Forbes, P. Kofman, and T. R. L. Fry, eds.), Monash University, 253–289.

Lehmann, E. L. (1986), *Testing Statistical Hypotheses*, Wiley, New York.

McCabe, B. P. M. (1988), A Multiple Decision Theory Analysis of Structural Change in Regression, *Econometric Theory*, 4, 499–508.

Miller, R. G. Jr (1981), *Simultaneous Statistical Inference*, Springer-Verlag, New York.

Miyazaki, S. and W. E. Griffiths (1984), The Properties of Some Covariance Matrix Estimators in Linear Models with AR(1) Errors, *Economics Letters*, 14, 351–356.

Nankervis, J. C. and N. E. Savin (1987), Finite Sample Distributions of t and F Statistics in an AR(1) Model with an Exogenous Variable, *Econometric Theory*, 3, 387–408.

Nelson, C. R., R. Startz, and E. Zivot (1996), Valid Confidence Intervals and Inference in the Presence of Weak Instruments, *Technical Report*, Department of Economics, University of Washington.

Newey, W. K. and K. D. West (1987), A Simple, Positive Semi-Definite, Heteroskedasticity and Autocorrelation Consistent Covariance Matrix, *Econometrica*, 55, 703–708.

Park, R. E. and B. M. Mitchell (1980), Estimating the Autocorrelated Error Model with Trended Data, *Journal of Econometrics*, 13, 185–201.

Phillips, G. D. A. and B. P. M. McCabe (1988), Some Applications of Basu's Independence Theorems in Econometrics, *Statistica Neerlandica*, 42, 37–46.

Phillips, G. D. A. and B. P. M. McCabe (1989), A Sequential Approach to Testing Econometric Models, *Empirical Economics*, 14, 151–165.

Roy, S. N. (1953), On a Heuristic Method of Test Construction and Its Use in Multivariate Analysis, *Annals of Mathematical Statistics*, 24, 220–238.

Savin, N. E. (1984), Multiple Hypothesis Testing, in *Handbook of Econometrics* (Z. Griliches and M. D. Intriligator, eds.), North-Holland, Amsterdam, Chap. 14.

Savin, N. E. and A. Würtz (1996), The Effect of Nuisance Parameters on the Power of LM Tests in Logit and Probit Models, *Technical Report*, Department of Economics, University of Iowa, Iowa City, IA.

Srivastava, V. K. and D. E. Giles (1987), *Seemingly Unrelated Regression Models*, Marcel Dekker, New York.

Staiger, D. and J. H. Stock (1997), Instrumental Variable Regression with Weak Instruments, *Econometrica*, 65, 557–586.

Tippett, L. H. C. (1931), *The Methods of Statistics*, Williams & Norgate, London.

Tunnicliffe Wilson, G. (1989), On the Use of Marginal Likelihood in Time Series Model Estimation, *Journal of the Royal Statistical Society, Series B*, 51, 15–27.

Vinod, H. D. (1976), Effects of ARMA Errors on the Significance Tests for Regression Coefficients, *Journal of the American Statistical Association*, 71, 929–933.

Vinod, H. D. and A. Ullah (1981), *Recent Advances in Regression Methods*, Marcel Dekker, New York.

Wang, J. and E. Zivot (1996), Inference on a Structural Parameter in Instrumental Variables Regression with Correlated Instruments, *Technical Report*, Department of Economics, University of Washington.

Wooldridge, J. M. (1989), A Computationally Simple Heteroskedasticity and Serial Correlation Robust Standard Error for the Linear Regression Model, *Economics Letters*, 31, 239–243.

Zinde-Walsh, V. and A. Ullah (1987), On the Robustness of Tests of Linear Restrictions in Regression Models with Elliptical Error Distributions, in *Time Series and Econometric Modelling* (I. B. MacNeill and G. J. Umphrey, eds.), Reidel, Dordrecht.

15
Modeling Economic Relationships with Smooth Transition Regressions

Timo Teräsvirta
Stockholm School of Economics, Stockholm, Sweden

I. INTRODUCTION

The assumption of linearity has long dominated macroeconometric model building. If linear approximations to economic relationships had not been successful in empirical work they would no doubt have been abandoned long ago, but this has not been the case. On the contrary, very powerful concepts such as that of cointegrated variables (Granger 1981) have been built upon the idea of linear relationships between variables. The statistical theory of linear models with cointegrated variables is well established and enables the analysis of linear systems of cointegrated variables; for recent overviews see Banerjee et al. (1993), Hatanaka (1996), and Johansen (1995). A coherent modeling strategy built upon the multivariate normal distribution, its conditional means, and the idea of moving from general to more specific models has been proposed and successfully put into practice in a number of applications. The idea has been to merge economic theory and statistical considerations in search of a congruent model. Hendry (1995) contains a detailed account of these principles.

Nevertheless, there exist situations in which the underlying economic theory is strongly nonlinear, and the econometric methods have reflected that fact. Modeling economies with rationing constitutes an example. Maddala (1986) surveyed the resulting switching regression models that were disequilibrium models with the "minimum condition." The class of switching regression models also includes models that assume a finite number of linear regimes but without the minimum condition. Which regime generates the observations at any given point of time in these models depends instead on a so-called switching variable that can either be observable or unobservable. In the former case, the switchpoint itself is often assumed unknown;

see Goldfeld and Quandt (1972, Section 9.2). Piecewise regression models, see for example Ertel and Fowlkes (1976) and the references therein, belong to this category. The general idea forms a basis for testing parameter constancy against a structural break at an unknown point as in Quandt (1960) and Andrews and Ploberger (1994). Goldfeld and Quandt (1973) first discussed the case in which the sequence of unobserved switching variables is an irreducible first-order Markov chain, and Lindgren (1978) derived the maximum likelihood estimators of parameters of such hidden Markov models.

Switching regression models may be generalized in such a way that the transition from one extreme regime to the other is not discrete but smooth. Such generalizations are the topic of this chapter. Instead of a usually (small) finite number of regimes there exists in these generalizations a continuum of them. Bacon and Watts (1971) first suggested such a model and coined the term "smooth transition" to illustrate how a locally linear equation changes from the one extreme linear parameterization to the other as a function of the continuous transition variable. In the econometrics literature, Goldfeld and Quandt (1972, pp. 263–264) presented a similar idea. Suppose one has the following switching regression model of two variables y_t and x_t:

$$
\begin{aligned}
y_t = \alpha_1 \{1 - D(z_t)\} + \alpha_2 D(z_t) + \{\beta_1 (1 - D(z_t)) + \beta_2 D(z_t)\} x_t \\
+ \{1 - D(z_t)\} u_{1t} + D(z_t) u_{2t}
\end{aligned}
\tag{1}
$$

where $u_{it} \sim \text{nid}(0, \sigma_i^2)$, $i = 1, 2$, are the error terms of the two regimes, z_t is the transition variable, and $D(z_t)$ is the Heaviside function:

$$
D(z_t) = \begin{cases} 1, & z_t \geq c \\ 0, & z_t < c \end{cases}
\tag{2}
$$

Goldfeld and Quandt pointed out that the estimation of the parameters of (1) including c in (2) is complicated and that the problem can be simplified by using a feasible approximation of (2). Their suggestion was to define

$$
D(z_t) = (2\pi\sigma^2)^{-1/2} \int_{-\infty}^{z_t} \exp\{-(1/2\sigma^2)(z - c)^2\} \, dz
\tag{3}
$$

i.e., to assume that the transition function is the cumulative distribution function of the normal (c, σ^2) variable and that $\sigma_1^2 = \sigma_2^2$ in (1). Replacing (2) by (3) in the switching regression (1) defines a smooth transition regression. In the time series literature, Chan and Tong (1986) made a similar suggestion in order to generalize a univariate switching autoregressive or "threshold autoregressive" model, albeit not for computational reasons. Tong (1990) contained a detailed account of threshold autoregressive models. Maddala (1977, p. 396) proposed that (3) be replaced by the logistic function

$$
D(z_t) = (1 + \exp\{\delta_1 + \delta_2 z_t\})^{-1}
\tag{4}
$$

which is one of the alternatives to be considered in this chapter; see also Granger and Teräsvirta (1993) and Teräsvirta (1994).

A smooth transition between two extreme regimes may be an attractive parameterization because the resulting smooth transition regression (STR) model is locally linear and thus often allows easy interpretation. Also from the point of view of theory, the assumption of a small number (usually two) of regimes may sometimes be too restrictive compared to the STR alternative. For instance, instead of assuming that an economy just has two discrete states, expansion and contraction, say, it may be more convenient and realistic to assume a continuum of states between the two extremes. Another argument is that economic agents may not all act promptly and uniformly at the same moment; their response to news requiring action may contain delays. Nevertheless, these two viewpoints are not competitors. As is already obvious from Goldfeld and Quandt (1972, pp. 263–264), the two-regime switching regression model is a special case of an STR model and can therefore be treated in that framework.

Furthermore, an STR model may be used in the same way as the switching regression to serve as an alternative against which to test parameter constancy in a linear model. The alternative to parameter constancy in this framework is a continuous change in parameters, which often is statistically a more convenient case to handle than just a single structural break. Lin and Teräsvirta (1994) discussed this possibility which requires defining $z_t = t$ in (4) while the STR model has the form

$$y_t = (\varphi + \theta D(t))' x_t + u_t \tag{5}$$

where x_t is a vector of explanatory variables and φ and θ are parameter vectors. If the null of parameter constancy, $D(t) \equiv 0$ in (5), is rejected one can estimate the alternative including the parameters in $D(t)$ and find out how the parameter vector $\varphi + \theta D(t)$ is changing over time. For multivariate applications, see for example Albæk and Hansen (1995), Coutts, Mills and Roberts (1995), Heinesen (1996), Jansen and Teräsvirta (1996), Lütkepohl, Teräsvirta, and Wolters (1995) and Wolters, Teräsvirta, and Lütkepohl (1997).

Generally, the transition variable of an STR model is an observable economic variable in the tradition of Fair and Jaffee (1972) who applied a switching regression model to modeling the demand for and supply of housing. Models with an unobserved transition variable will not be discussed any further here. The application in Section V is also related to the housing market. Jansen and Teräsvirta (1996) applied the general STR model in various ways as a vehicle for testing super exogeneity; see, for example, Hendry (1995, Chap. 5) or Ericsson (1992) for definitions of this and other types of exogeneity. Other applications of the STR model to-date include Granger, Teräsvirta, and Anderson (1993), Semmler and Koçkesen (1995), and Chiarella, Semmler, and Koçkesen (1996).

The approach in this chapter is classical. However, Bacon and Watts (1971) used the Bayesian approach for estimating the parameters of their smooth transition model. They first obtained the joint marginal posterior distribution for the parameters of the transition function; for example in (3) those are c and $1/\sigma^2$. Then they estimated the remaining parameters conditionally on the mean of this joint posterior distribution. Later, Tsurumi (1982) applied the same idea to his "gradual switching" simultaneous-equation model. More recent Bayesian treatments have been mainly concerned with switching regression or threshold autoregressive models. Péguin-Feissolle (1994) who discussed the smooth transition autoregressive model constitutes an exception. Published work on threshold autoregressive models includes Pole and Smith (1985), Geweke and Terui (1993), Pfann, Schotman, and Tschernig (1996), and Cook and Broemeling (1996).

This chapter is organized as follows. The STR model is defined and its properties and potential in applications are discussed in Section II. Section III considers statistical inference in STR models. This includes testing linearity which should precede any nonlinear modeling. The modeling cycle consisting of specification, estimation, and evaluation of STR models is described in Section IV. Section V contains an application to a U.K. house price equation considered in Hendry (1984), and Section VI concludes.

II. SMOOTH TRANSITION REGRESSION MODEL

Consider the nonlinear regression model

$$y_t = x_t'\varphi + (x_t'\theta)G(\gamma, c; s_t) + u_t \qquad (t = 1, \ldots, T) \tag{6}$$

where $x_t = (1, x_{1t}, \ldots, x_{pt})' = (1, y_{t-1}, \ldots, y_{t-k}; z_{1t}, \ldots, z_{mt})'$ with $p = k + m$ is the vector of explanatory variables, $\varphi = (\varphi_0, \varphi_1, \ldots, \varphi_p)'$ and $\theta = (\theta_0, \theta_1, \ldots, \theta_p)'$ are parameter vectors, and $\{u_t\}$ is a sequence of independent, identically distributed errors. Some of the parameters φ_i and θ_j may be zero a priori or the restriction $\varphi_i = -\theta_i$ may hold for some i. In (6), G is a bounded continuous transition function; it is customary to bound G between zero and unity, and s_t is the transition variable. It may be a single stochastic variable, for example, an element of x_t, a linear combination of stochastic variables or a deterministic variable such as a linear time trend. By writing (6) as

$$y_t = x_t'(\varphi + \theta G) + u_t \tag{7}$$

it is seen that the model is locally linear in x_t and that the combined parameter vector $\varphi + \theta G$ is a function of the transition variable s_t. If G is bounded between 0 and 1, the combined parameters fluctuate between φ and $\varphi + \theta$. In dynamic modeling, this property makes it possible, for example, to characterize an economy with dynamic properties in expansion being different from those in contraction. Teräsvirta and

Anderson (1992) contained univariate examples of this kind. Model (6) is an example of the smooth transition regression (STR) model discussed in the Introduction.

The practical applicability of (6) depends on how G is defined. A few definitions have been suggested in the literature; see, for example, Granger and Teräsvirta (1993, Chap. 7). If G has the form

$$G_1(\gamma; c, s_t) = (1 + \exp\{-\gamma(s_t - c)\})^{-1}, \qquad \gamma > 0 \tag{8}$$

then the STR model (6) is called the logistic STR of LSTR1 model. The transition function (8) is a monotonically increasing function of s_t. The restriction $\gamma > 0$ is an identifying restriction. The slope parameter γ indicates how rapid the transition from zero to unity is as a function of s_t and the location parameter c determines where the transition occurs. If $\gamma \to \infty$ in (8), (6) becomes a two-regime switching regression model with the switching variable s_t. In this special case, $s_t = c$ is the switchpoint between the regimes $y_t = x_t'\varphi + u_t$ and $y_t = x_t'(\varphi + \theta) + u_t$.

Monotonic transition may not always be a satisfactory alternative in applications. A simple nonmonotonic alternative is

$$G_2(\gamma, c; s_t) = (1 + \exp\{-\gamma(s_t - c_1)(s_t - c_2)\})^{-1}, \\ \gamma > 0, \quad c_1 \leq c_2 \tag{9}$$

where the restrictions on γ, c_1, and c_2 are identifying restrictions. This transition function is symmetric about $(c_1 + c_2)/2$, and $G(\gamma, c; s_t) \to 1$ for $s_t \to \pm\infty$. The minimum value of G remains between 0 and 1/2, the upper limit holds for $c_1 = c_2$. On the other hand, when $\gamma \to \infty$, $G(\gamma, c; s_t) \to 0$ for $c_1 \leq s_t \leq c_2$; for other values $G(\gamma, c; s_t) \to 1$. This is a special case of a three-regime switching regressions model in which the two outer regimes are equal. The STR model (6) with transition function (9) is called the LSTR2 model.

Jansen and Teräsvirta (1996) suggested (9) as a generalization of the exponential STR (ESTR) model discussed in the literature; see Granger and Teräsvirta (1993, Chap. 7). The transition function of an ESTR model is defined as

$$G(\gamma, c; s_t) = 1 - \exp\{-\gamma(s_t - c)^2\}, \qquad \gamma > 0 \tag{10}$$

which is closely related to the case $c_1 = c_2$ in (9). The transition function is symmetric about c and $G(\gamma, c; s_t) \to 1$ for $s_t \to \pm\infty$. However, when $\gamma \to \infty$, the transition function $G(\gamma, c; s_t) \to 1$ except that for $s_t = c$ the limit is 0. Thus for large values of γ it is difficult in practice to distinguish an ESTR model from a linear model. By introducing another parameter as in (9) one obtains an STR model with more useful limiting properties as $\gamma \to \infty$. An LSTR2 model with three extreme regimes and fairly sharp transitions between them achieved by (9) may in fact be quite an interesting alternative in practice; see for example the case discussed in Anderson (1997). In all theoretical derivations of the chapter, however, it is assumed that γ is finite.

Defining $s_t = t$ yields an important special case of the STR model. Then (7) becomes

$$y_t = x_t'(\varphi + \theta G(\gamma, c; t)) + u_t \tag{11}$$

Model (11) can be interpreted as a linear model whose parameters change over time. It contains as a special case the presence of a single structural break which has been the most popular alternative to parameter constancy in econometric work. This special case is obtained by completing (6) by (8) with $s_t = t$ and letting $\gamma \to \infty$ in (8). Lin and Teräsvirta (1994) defined another nonmonotonic transition function (see also Jansen and Teräsvirta 1996)

$$G(\gamma, c; s_t) = (1 + \exp\{-\gamma(t - c_1)(t - c_2)(t - c_3)\})^{-1} \tag{12}$$

where $\gamma > 0$, $c_1 \le c_2 \le c_3$. In fact, Lin and Teräsvirta (1994) defined the exponent of (12) directly as a third-order polynomial without requiring the roots to be real. As we shall see, this does not make any difference as far as testing parameter constancy is concerned. On the other hand, if an STR model with (12) is to be estimated, restricting the roots to be real alleviates the potential problem of very high correlation between the estimator of θ on the one hand and that of γ and possibly c_1, c_2, and c_3 on the other. At the same time one does not give up too much generality in the sense that (12) still allows quite a lot of flexibility in the transition function.

Many parameter constancy tests explicitly or implicitly assume the alternative to parameter constancy to be a single structural break. If this null hypothesis is rejected it is often not obvious what to do next. More information can be obtained by using recursive tests: see, for example, Hendry (1995, Chap. 16). An advantage of testing parameter constancy in the STR framework is that any rejection of the null hypothesis is a rejection against a parametric alternative. In case of a rejection the parameters of the alternative can be estimated, which helps obtain information about where and how parameter constancy breaks down if it does. This information in turn is helpful in deciding how the specification of the model should be improved in order to obtain a model with constant parameters.

It is of course possible to define (12) also for other transition variables than $s_t = t$. However, macroeconomic time series are usually not overly long. When modeling with such series it may be advisable to restrict the order of the exponent of the transition function to two to avoid excessive difficulties in parameter estimation unless there is economic theory suggesting a higher order. This is done here excepting the case $s_t = t$. In that case the experience has shown that there are less problems: a heuristic explanation to that is that for $s_t = t$, superconsistency makes the estimation of the parameters in the exponent easier in small samples than if s_t is stationary.

III. INFERENCE IN SMOOTH TRANSITION REGRESSION MODELS

A. Testing Linearity Against STR

The first question an applied econometrician is facing in considering a nonlinear model is: is it worthwhile? If the economic relationship to be modeled can be adequately characterized by a linear model, then working with a nonlinear model is waste of time. Besides, as will be seen, fitting certain types of nonlinear models such as STR models to data is not a statistically feasible undertaking if the data-generating process is linear. Econometric modeling with STR models thus has to begin with testing linearity against STR. In order to discuss testing statistical hypotheses within the STR model, additional assumptions about (6) are necessary. The stochastic variables among $z_{1t}, \ldots, z_{kt}$ are assumed stationary whereas the nonstochastic ones are dummy variables. Furthermore, s_t is assumed to be a stationary variable (if it is not a time trend), and all cross-moments $E z_{it} z_{jt}$, $E z_{it} s_t^k$, $E y_{t-i} s_t^k$, and $E y_{t-1} z_{jt}$, $k \le 3$, are assumed to exist. Note, however, that some of the variables z_{jt} may be stationary linear combinations of I(1) variables. The parameters of such combinations may be treated as known because their least squares estimates are super consistent, and the inference to be discussed is still valid. Finally, the errors are assumed uncorrelated with x_t and s_t.

A minor redefinition of the transition functions will allow convenient notation in this section. Let $G^* = G - 1/2$ where G is any of the transition functions defined in the previous section. Rewrite (6) as

$$y_t = x_t' \varphi + (x_t' \theta) G^*(\gamma, c; s_t) + u_t, \qquad \gamma > 0 \tag{13}$$

(although φ and θ are changed the previous notation is retained). In order to derive the test statistic, assume that $u_t \sim \text{nid}(0, \sigma^2)$. The conditional log-likelihood function of the model is

$$\sum_{t=1}^{T} \ell(\varphi, \theta, \gamma, c; y_t \mid x_t, s_t) = \alpha - \frac{T}{2} \log \sigma^2 - \frac{1}{2\sigma^2} \sum_{t=1}^{T} u_t^2 \tag{14}$$

The null hypothesis of linearity in (13) is $H_0: \gamma = 0$ against $H_1: \gamma > 0$. It is assumed that the roots of the lag polynomial $1 - \sum_{j=1}^{k} \varphi_j L^j$ are outside the unit circle when the null hypothesis holds. It is seen from (13) that the linearity hypothesis can equally well be expressed as $H_0^\theta: \theta = 0$. This is an indication of an identification problem in (13): the model is identified under the alternative but not under the null hypothesis. This poses the following statistical problem. Suppose that the investigator wants to apply a likelihood ratio test to testing H_0. In order to do that (13) has to be estimated both under H_0 and H_1. However, the parameters of (13) cannot be consistently estimated under H_0 because θ and c are then nuisance parameters whose values do not affect the value of the likelihood. For this reason, the likelihood ratio statistic does not have its standard asymptotic χ^2 distribution under the null hypothesis. The two other classical tests, the Lagrange multiplier and the Wald test share the same property.

Davies (1977, 1987) first discussed solutions to this problem; for later contributions in the econometrics literature, see for example Shively (1988), King and Shively (1993), Lee, White, and Granger (1993), Andrews and Ploberger (1994), and Hansen (1996). It occurs in connection with many nonlinear models which nest a linear model such as the STR model, the switching regression model with an unknown switchpoint or the Hidden Markov model of Goldfeld and Quandt (1973) and Lindgren (1978). A common feature in much of the literature is that under the null hypothesis there is a single nuisance parameter in the model. When the STR model is concerned there are at least two such parameters, whichever way one formulates the null hypothesis. A way of solving the identification problem by circumventing it is discussed for example in Granger and Teräsvirta (1993, Chap. 6) and is also considered here. It is based on the work of Saikkonen and Luukkonen (1988) and Luukkonen, Saikkonen, and Teräsvirta (1988a).

To discuss this idea, take the logistic transition function $G_1^* = G_1 - 1/2$ and its Taylor series approximation with the null hypothesis $\gamma = 0$ as the expansion point. The latter can be written as

$$T_1 = \delta_0 + \delta_1 s_t + R_1(\gamma, c; s_t) \tag{15}$$

where R_1 is the remainder and δ_0 and δ_1 are constants. Substituting T_1 for G_1^* in (13) yields

$$y_t = x_t'\beta_0 + (x_t s_t)'\beta_1 + u_t^* \tag{16}$$

where $u_t^* = u_t + (x_t'\theta)R_1(\gamma, c; s_t)$ and β_1 is a $(p+1) \times 1$ parameter vector. The zero- and first-order terms emerging due to the substitution merge with the corresponding terms in (13) leading to (16). This approximation may be viewed either as an approximation to the STR model (the conditional mean of y_t) or to the log-likelihood (14). Use of this Taylor expansion in (13) or (14) amounts to giving up information about the structure of the alternative in order to circumvent the identification problem and obtain a simple test of the null hypothesis. Parameter vector β_1 has the property $\beta_1 = \gamma \tilde{\beta}_1$; for details see Luukkonen, Saikkonen, and Teräsvirta (1988a). Thus the null hypothesis $H_0: \gamma = 0$ in (13) implies $H_0': \beta_1 = 0$ and $H_1': \beta_1 \neq 0$ within (16). Since (16) is linear in parameters and $u_t^* = u_t$ when H_0' holds, one can test this null hypothesis by a straightforward LM-type test. Under H_0,

$$\chi_{\text{LM}}^2(p+1) = \hat{\sigma}^{-2} \left(\sum_{t=1}^{T} \hat{u}_t w_t \right)' (\hat{M}_{11} - \hat{M}_{10}\hat{M}_{00}^{-1}\hat{M}_{01})^{-1} \left(\sum_{t=1}^{T} w_t \hat{u}_t \right) \tag{17}$$

where $\hat{M}_{00} = \sum_{t=1}^{T} z_t z_t'$, $\hat{M}_{01} = \hat{M}_{10}' = \sum_{t=1}^{T} z_t w_t'$, $\hat{M}_{11} = \sum_{t=1}^{T} w_t w_t'$, $\hat{\sigma}^2 = $ $(1/T) \sum_{t=1}^{T} \hat{u}_t^2$, $\hat{u}_t$ is the residual estimated under the null hypothesis, $z_t = x_t$ and $w_t = x_t s_t$, has an asymptotic χ^2 distribution with $p + 1$ degrees of freedom when the moments and cross-moments implied by (17) exist. For detailed derivations, see, for example Luukkonen, Saikkonen, and Teräsvirta (1988a, 1988b) or Granger and Teräsvirta (1993, Chap. 6).

The above notation requires that s_t is not an element of x_t. If it is then the auxiliary regression becomes

$$y_t = x_t' \beta_0 + (\tilde{x}_t s_t)' \beta_1 + u_t^* \tag{18}$$

where $\tilde{x}_t = (x_{1t}, \ldots, x_{pt})'$ and β_1 is a $p \times 1$ vector. Furthermore, $w_t = \tilde{x}_t s_t$ in (17) and the asymptotic null distribution of the test statistic thus has p degrees of freedom. This is probably the most common case in practice except when $s_t = t$, and in the following the notation will conform to it.

The LM-type statistic (17) seems to have good power already in small samples (see for example Luukkonen, Saikkonen, and Teräsvirta 1988a and Petruccelli 1990), but in a special occasion it only has trivial power against H_1'. This is the case when s_t is an element of x_t and $\theta = (\theta_0, 0, \ldots, 0)'$, $\theta_0 \neq 0$. In other words, the only nonlinear element in (13) is the intercept. Then $\beta_1 = 0$ in (17) even under H_1' and the test thus has no power. To remedy this situation, Luukkonen, Saikkonen, and Teräsvirta (1988a) suggested a third-order Taylor approximation to G_1^*. This can be written as

$$T_3 = \delta_0 + \delta_1 s_t + \delta_2 s_t^2 + \delta_3 s_t^3 + R_3(\gamma, c; s_t) \tag{19}$$

where R_3 is a remainder and δ_j, $j = 0, 1, 2, 3$, are constants. When (19) is substituted for G_1^* in (13) one obtains

$$y_t = x_t' \beta_0 + (\tilde{x}_t s_t)' \beta_1 + (\tilde{x}_t s_t^2)' \beta_2 + (\tilde{x}_t s_t^3)' \beta_3 + u_t^* \tag{20}$$

where $u_t^* = u_t + (x_t' \theta) R_3(\gamma, c; s_t)$ and $\beta_j = \gamma \tilde{\beta}_j$, $j = 1, 2, 3$. The LM-type test of H_0': $\beta_j = 0$, $j = 1, 2, 3$, against H_1': "at least one $\beta_j \neq 0$" can be constructed as before. The test statistic is (17) with $w_t = (\tilde{x}_t' s_t, \tilde{x}_t' s_t^2, \tilde{x}_t' s_t^3)'$ and the number of degrees of freedom in the asymptotic χ^2 distribution under H_0 is $3p$. This result requires the existence of all the moments implied by w_t and (17).

When x_t has a large number of elements, the auxiliary null hypothesis will sometimes be large compared to the sample size. In that case the asymptotic χ^2 distribution is likely to be a poor approximation to the actual small sample distribution. It has been found out (see Granger and Teräsvirta 1993, Chap. 7) that an F-approximation to (17) works much better (the empirical size of the test remains close to the nominal size while power is good). The test can be carried out in stages:

1. Regress y_t on x_t and compute the residual sum of squares $SSR_0 = (1/T)$ $\sum_{t=1}^{T} \hat{u}_t^2$.

2. Regress $\hat{u}_t$ (or y_t) on x_t, $\tilde{x}_t s_t$, $\tilde{x}_t s_t^2$, and $\tilde{x}_t s_t^3$ and compute the residual sum of squares $SSR_1 = (1/T) \sum_{t=1}^{T} \hat{v}_t^2$.

3. Compute

$$F = \frac{(SSR_0 - SSR_1)/3p}{SSR_1/(T - 4p - 1)} \tag{21}$$

Under H_0': $\beta_1 = \beta_2 = \beta_3 = 0$, F has approximately an F-distribution with $3p$ and $T - 4p - 1$ degrees of freedom.

The above theory works when $\{s_t\}$ is stationary. It continues to work when $s_t = t$. In that case, t is not an element of x_t. The auxiliary regression when testing parameter constancy of a linear model against STR with transition function (12) (first-order Taylor approximation), is

$$y_t = x_t'\beta_0 + (x_t t)'\beta_1 + (x_t t^2)'\beta_2 + (x_t t^3)'\beta_3 + u_t^* \tag{22}$$

(Lin and Teräsvirta 1994). The F statistic corresponding to (21) thus has $3(p + 1)$ and $T - 4p - 4$ degrees of freedom.

The parameter constancy test can also be carried out for any subset of parameters. In many modeling situations considering various subsets is advisable when one wants to obtain a clear idea of which parameters may actually be nonconstant if the model appears to have unstable parameters. This is done by assuming that the appropriate elements $\theta_i = 0$ in (13) in which case the corresponding elements of β_j, $j = 1, 2, 3$, also equal zero a priori and are not included in the null hypothesis. The empirical example in Section V will illustrate the possibilities of this approach. The same also applies to testing linearity against STR. In some occasions, economic theory behind the model or statistical considerations may suggest such a priori parameter restrictions. A dummy variable with all values equal to zero except one constitutes an example of a variable whose coefficient must always be assumed constant.

The above theory covers the LSTR1 model, whereas testing linearity against the LSTR2 model has not been discussed separately. The first-order Taylor approximation of G_2^* at the expansion point $\gamma = 0$ eventually leads to the auxiliary regression (20) with $\beta_3 = 0$. Thus the null hypothesis $\gamma = 0$ is transformed into H_0': $\beta_1 = \beta_2 = 0$. This indicates that the LM-type test of H_0': $\beta_1 = \beta_2 = \beta_3 = 0$ within (20) also has power against the LSTR2 model. If the alternative to the linear model is an ESTR model defined by (10) the auxiliary regression based on the first-order Taylor approximation to the transition function is identical to that of the LSTR2 model. A similar remark holds for (22). Note in particular, that (22) can also be regarded as a first-order Taylor approximation to (13) with (12).

Sometimes it may not be clear from economic theory which variable should be taken to be the transition variable under the alternative. Suppose, however, that the choice is between the elements of $\tilde{x}_t$. Then one can define the linear combination $a'\tilde{x}_t$ where $a = (0, \ldots, 1, 0, 0, \ldots, 0)'$ is a $p \times 1$ vector with the only unit element corresponding to the true but unknown transition variable and substitute it for s_t in (8). Proceeding as in Luukkonen, Saikkonen, and Teräsvirta (1988a) leads to the auxiliary regression

$$
\begin{aligned}
y_t = x'_t \beta_0 &+ \sum_{i=1}^{p} \sum_{j=i}^{p} \beta_{1ij} x_{it} x_{jt} + \sum_{i=1}^{p} \sum_{j=1}^{p} \beta_{2ij} \tilde{x}_{it} \tilde{x}_{jt}^2 \\
&+ \sum_{i=1}^{p} \sum_{j=1}^{p} \beta_{3ij} \tilde{x}_{it} \tilde{x}_{jt}^3 + u_t^*
\end{aligned}
\tag{23}
$$

with the linearity hypothesis H'_0: $\beta_{1ij} = 0, i = 1, 2, \ldots, p; j = i, i+1, \ldots, p$; $\beta_{2ij} = \beta_{3ij} = 0, i, j = 1, 2, \ldots, p$. The LM-type test statistic is formed accordingly and the number of degrees of freedom in the asymptotic χ^2 distribution under H'_0 is $p(p+1)/2 + 2p^2$. This quickly grows large with increasing p and the authors therefore suggested an "economy version" based on

$$
y_t = x'_t \beta_0 + \sum_{i=1}^{p} \sum_{j=1}^{p} \beta_{1ij} \tilde{x}_{it} \tilde{x}_{jt} + \sum_{j=1}^{p} \beta_{3ij} \tilde{x}_{jt}^3 + u_t^*
\tag{24}
$$

The number of degrees of freedom in the test based on (24) is $p(p+1)/2 + p$, the null hypothesis of linearity being H'_0: $\beta_{1ij} = 0, i = 1, 2, \ldots, j; j = i, i + 1, \ldots, p$; $\beta_{3j} = 0, j = 1, 2, \ldots, p$. Of course, the choice of potential transition variables may be restricted to only a subset of variables in $\tilde{x}_t$. In that case, the null hypothesis must be modified accordingly and the relevant coefficients in (23) or (24) set equal to zero a priori.

The above statistical theory has the advantage that the asymptotic null distributions are standard and the tests can be carried out just by using ordinary least squares. Although they are designed against STR they are also sensitive to other types of nonlinearity. After rejecting linearity it may therefore not be clear what to do next. However, one may have decided to consider STR models if linearity is rejected. In that case, tests based on the above auxiliary regression may be used only for testing linearity but also, in case of a rejection, for the specification of an STR model. This argument is elaborated in Section IV.B.

The LM-type test statistics continue to have reasonable power in small samples when $\gamma \to \infty$, at least when the alternative is an LSTR1 model; see Luukkonen, Saikkonen, and Teräsvirta (1988a) and Hansen (1996). But then, if the alternative is a switching regression model (it is assumed a priori that γ is infinite) the above theory does not work. Hansen (1996) recently considered a general framework for hypothesis testing when the model is only identified under the alternative. His results have bearing for switching regression models as well. Let $\nu \in N$ be the vector of nuisance parameters and $S_T(\nu)$ a test statistic which is a function of the nuisance parameters. Hansen showed how to obtain critical values for statistics such as $\sup_{\nu \in N} S_T(\nu)$ (Davies 1977, 1987) or $\text{ave}_{\nu \in N} S_T(\nu) = \int_N S_T(\nu) \, dW(\nu)$ (Andrews and Ploberger 1994) by simulation. Hansen's technique is not restricted to the case where there only exists a single nuisance parameter, but the amount of computations required to perform any of the possible tests increases with the number of nuisance parameters. He applied the test to testing linearity against a two-regime threshold autoregressive model with an unknown transition variable (delay). In general, the supremum and average tests should be preferred to the LM-type tests discussed above if one is able to exclude smooth transition as an alternative to linearity in favor of a discrete switch. This is the case, for example, if one is investigating structural change and knows beforehand that the only alternative to constant parameters during the observation period must be a single structural break.

B.   Misspecification Testing in STR Models

In the previous section the focus was on testing linearity against STR. This section deals with the situation in which the parameters of the STR model have been estimated and the validity of the assumptions of the model is checked in the light of the results. The misspecification tests to be discussed here are those Eitrheim and Teräsvirta (1996) recently derived in a univariate setting. As the authors remarked, generalizing them to STR models is straightforward. An assumption one has to make is that the parameters of the STR model have been estimated consistently and that the estimates are asymptotically normal; see Wooldridge (1994) and Escribano and Mira (1995) for discussions of conditions for this.

Estimation of an STR model is carried out, among other things, under the assumption of no error autocorrelation and that of parameter constancy. These two assumptions thus have to be tested, and procedures to that effect are discussed here. Furthermore, it is of interest to try and find out whether or not the estimated STR model captures all nonlinear features present in the data. Statistical inference for treating this problem is another topic of this section. To retain the same order of presentation as in Section III.A it is considered before tests of parameter stability.

1.   Test of No Error Autocorrelation

Consider first the test of no error autocorrelation in the STR model (assume for simplicity that the transition variable s_t is an element of $\tilde{x}_t$). More generally,

$$y_t = M(x_t; \psi) + u_t$$

$$u_t = a'v_t + \varepsilon_t \qquad (t = 1, \ldots, T) \tag{25}$$

where M is at least twice continuously differentiable with respect to the parameters, $a = (a_1, \ldots, a_q)'$ is a parameter vector, $v_t = (u_{t-1}, \ldots, u_{t-q})'$, and $\varepsilon_t \sim IN(0, \sigma^2)$.

The null hypothesis is H_0: $a = 0$, and H_1: $a \neq 0$. The conditional log-likelihood, given the fixed starting values $y_0, y_{-1}, \ldots, y_{-q+1}$ and $x_0, x_{-1}, \ldots, x_{-q+1}$, has the form

$$L = c - \frac{T}{2} \ln \sigma^2 - \frac{1}{2\sigma^2} \sum_{t=1}^{T}$$

$$\times \left\{ y_t - \sum_{j=1}^{q} a_j y_{t-j} = M(x_t; \psi) + \sum_{j=1}^{q} a_j M(x_{t-j}; \psi) \right\}^2 \tag{26}$$

The information matrix related to (26) is block diagonal such that the element corresponding to the second derivative of (26) with respect to σ^2 forms its own block. The variance σ^2 can thus be treated as a fixed constant in (26) when deriving the test statistic. The first partial derivatives of the log-likelihood with respect to a and ψ are

$$\frac{\partial L}{\partial a_j} = \sum_{t=1}^{T} \frac{\varepsilon_t}{\sigma^2} \{y_{t-j} - M(x_{t-j}; \psi)\}, \qquad j = 1, \ldots, q$$

$$\frac{\partial L}{\partial \psi} = -\sum_{t=1}^{T} \frac{\varepsilon_t}{\sigma^2} \left\{ \frac{\partial M(x_t; \psi)}{\partial \psi} - \sum_{j=1}^{q} a_j \frac{\partial M(x_{t-j}; \psi)}{\partial \psi} \right\}$$

Assume now that (25) is an LSTR1 model so that

$$M(x_t; \psi) = \varphi' x_t + (\theta' x_t) G_1(\gamma, c; s_t) \tag{27}$$

with $\psi = (\varphi', \theta', \gamma, c)'$ and

$$\frac{\partial M(x_t; \psi)}{\partial \psi} = \left(\frac{\partial M}{\partial \varphi'} \frac{\partial M}{\partial \theta'} \frac{\partial M}{\partial \gamma} \frac{\partial M}{\partial c} \right)' = (x_t' \; x_t' G_1(\gamma, c; s_t) \; g_\gamma(t) \; g_c(t))' \tag{28}$$

Furthermore, in (28)

$$g_\gamma(t) = \left[\exp\left\{ \frac{\gamma}{2}(s_t - c) \right\} + \exp\left\{ -\frac{\gamma}{2}(s_t - c) \right\} \right]^{-2} (s_t - c)\theta' x_t \tag{29}$$

and

$$g_c(t) = \gamma \left[\exp\left\{ \frac{\gamma}{2}(s_t - c) \right\} + \exp\left\{ -\frac{\gamma}{2}(s_t - c) \right\} \right]^{-2} \theta' x_t \tag{30}$$

The exponential terms in (29) and (30) are bounded for $\gamma < \infty$. It follows that the existence of the necessary fourth moments is required for consistent estimation of parameters.

The test statistic is (17) with $w_t = \hat{v}_t = (\hat{u}_{t-1}, \ldots, \hat{u}_{t-q})'$ and $z_t = \hat{z}_t = \partial M(x_t; \hat{\psi})/\partial\psi$, and its asymptotic null distribution under the null hypothesis is χ^2 with q degrees of freedom. The "^" indicates consistent estimates under H_0. The approximate F-version of the test may be computed in three stages (Eitrheim and Teräsvirta 1996):

1. Estimate the LSTR1 model by NLS under the assumption of uncorrelated errors and compute the residual sum of squares $SSR_0 = \sum_{t=1}^{T} \hat{u}_t^2$.
2. Regress $\hat{u}_t$ on $\hat{v}_t$ and $\hat{z}_t$ and compute the residual sum of squares, SSR_1.
3. Compute the test statistic $F_{LM} = \{(SSR_0 - SSR_1)/q\}/\{SSR_1/(T - n - q)\}$, where n is the dimension of the gradient vector $\hat{z}_t$.

In small samples, the F-statistic with q and $T-n-q$ degrees of freedom is preferable to the asymptotic χ^2 variant because its superior size properties. To carry out the corresponding test when $G = G_2$ (LSTR2 model) requires an obvious modification of z_t and the assumption of the existence of the appropriate sixth moments in (17).

A computational detail is worth mentioning here. The auxiliary regression in Stage 2 contains q lags of $\hat{u}_t$. A standard way of proceeding in such a situation is to trim all the time series by omitting the first q observations in them. However, if this is done, $\sum_{t=q+1}^{T} \hat{u}_t \hat{z}_t \neq 0$ so that the trimming affects the empirical size of the test. In small samples the effect may not be negligible. A better solution is not to shorten the series but rather replace the missing observations in the beginning of the series of lagged residuals by zeros. Also, if the STR model in question happens to be a very difficult one to estimate it may be that the NLS estimation algorithm is not able to do a perfect job; that is, $\sum_{t=1}^{T} \hat{u}_t \hat{z}_t \neq 0$. To remedy that, step 1 may be replaced by

1'. Estimate the LSTR1 model by NLS under the assumption of uncorrelated errors. Regress the residuals $\hat{u}_t$ on $\hat{z}_t$ and compute the residual sum of squares $SSR_0 = \sum_{t=1}^{T} \hat{u}_t^{*2}$.

Step 1 guarantees $\sum_{i=1}^{T} \hat{u}_t^* \hat{z}_t = 0$ and prevents size distortion of the test. This extension of step 1 can also be recommended for the other two tests discussed in this section.

2. Test of No Remaining Nonlinearity

Since a purpose of the STR model is to give an adequate characterization to the nonlinear features in the data it is of particular interest to find out how successful the estimated model is in this respect. In order to test the adequacy of the STR model, one may extend the model in some suitable way. Eitrheim and Teräsvirta (1996) suggested such an extension for STAR models. Following their approach, define the *additive* STR model as

$$y_t = x_t'\varphi + (x_t'\theta)G(\gamma_1, c_1; s_t) + (x_t'\psi)H(\gamma_2, c_2; r_t) + u_t$$
$$(t = 1, \ldots, T) \tag{31}$$

This STR model has two additive nonlinear components, and the transition function H where r_t is assumed an element of x_t may be defined analogously to (8) or (9). Since testing the presence of the additional component is discussed, assume $H(0, c_2, r_t) \equiv 0$ for notational simplicity. When adequacy of the standard STR model is the issue it can be investigated by testing H_0: $\gamma_2 = 0$ in (31). Because of the parameterization, the additive STR model (31) is not identified under this null hypothesis. This problem may be solved (Eitrheim and Teräsvirta 1996) in the same way as in Section III.A. This means that the transition function H is replaced by its third-order Taylor approximation

$$T_3(\gamma_2, c_2; r_t) = \delta_0 + \delta_1 r_t + \delta_2 r_t^2 + \delta_3 r_t^3 + R_3(\gamma_2, c_2; r_t)$$

in (31). Doing that and rearranging terms yields

$$\begin{aligned} y_t = x_t'\beta_0 + (x_t'\theta)G(\gamma_1, c_1; s_t) + (\tilde{x}_t r_t)'\beta_1 + (\tilde{x}_t r_t^2)'\beta_2 \\ + (\tilde{x}_t r_t^3)'\beta_3 + u_t^* \end{aligned} \tag{32}$$

where $\beta_j = \gamma_2 \tilde{\beta}_j$, $j = 1, 2, 3$, and $u_t^* = u_t + (x_t'\psi)R_3(\gamma_2, c_2; r_t)$. The null hypothesis is H_0': $\beta_1 = \beta_2 = \beta_3 = 0$ and when it holds, $u_t^* = u_t$. Deriving the appropriate test statistic with the asymptotic χ^2 null distribution is straightforward. In practice, the test can be carried out in the three stages described above. In the present case, $\hat{v}_t = (\tilde{x}_t' r_t, \tilde{x}_t' r_t^2, \tilde{x}_t' r_t^3)'$ in stage 2, whereas $\hat{z}_t$ is the same as before. The degrees of freedom in the F-statistic are $3p$ and $T - 4p - 1$, respectively. If $G \equiv 0$ in (32), the test collapses into the linearity test discussed in Section III.A.

In the above it is assumed that all elements of ψ in (30) are nonzero. This is not necessary, and the elements of $\tilde{x}_t$ included in the second nonlinear component may be selected freely. On the other hand, write $x_t'\varphi = x_{1t}'\varphi_1 + x_{2t}'\varphi_2$ in (13) and consider the case in which the parameters have been estimated under the restriction $\varphi_2 = 0$. Then (32) can be written as

$$\begin{aligned} y_t = x_{1t}'\beta_{01} + x_{2t}'\beta_{02} + (x_t'\theta)G(\gamma_1, c_1; s_t) + (\tilde{x}_t r_t)'\beta_1 \\ + (\tilde{x}_t r_t^2)'\beta_2 + (\tilde{x}_t r_t^3)'\beta_3 + u_t^* \end{aligned} \tag{33}$$

and the null hypothesis of no remaining nonlinearity H_0': $\beta_{02} = 0, \beta_1 = \beta_2 = \beta_3 = 0$. One can also separately test H_0'': $\beta_{02} = 0$ in (33) assuming $\beta_1 = \beta_2 = \beta_3 = 0$. This is simply an LM test of validity of the restrictions $\varphi_2 = 0$ in (13). Furthermore, one may not want to be specific about the transition variable in the second nonlinear component. In that case one can generalize the relevant linearity tests to this situation as discussed in the previous section. Such a generalization based on the Taylor expansion of the appropriately parameterized transition function is straightforward.

Sometimes a model builder may not want to be specific about the parametric form of the remaining nonlinearity except that it is of additive type. In that case the maintained model can be written as

$$y_t = \varphi'x_t + (\theta'x_t)G(\gamma, c; s_t) + K(x_t) + u_t \tag{34}$$

If the STR does not adequately characterize the nonlinearity in the data then $K(x_t)$ is a nonlinear function. To investigate that possibility assume that $K(x_t)$ is at least three times continuously differentiable with respect to x_t and expand $K(x_t)$ into a Taylor series about the expansion point $x_t = x_t^0$. The third-order expansion has the form

$$K(x_t) = \kappa_0' x_t + \sum_{i=1}^{p}\sum_{j=i}^{p} \kappa_{ij} x_{it} x_{jt} + \sum_{i=1}^{p}\sum_{j=i}^{p}\sum_{\ell=j}^{p} \kappa_{ij\ell} x_{it} x_{jt} x_{\ell t} + R_K(x_t) \quad (35)$$

where $R_K(x_t)$ is the remainder term. Approximating $K(x_t)$ in (34) by (35) yields

$$y_t = \beta_0' x_t + (\theta' x_t) G(\gamma, c; s_t) + \sum_{i=1}^{p}\sum_{j=i}^{p} \kappa_{ij} x_{it} x_{jt}$$

$$+ \sum_{i=1}^{p}\sum_{j=i}^{p}\sum_{\ell=j}^{p} \kappa_{ij\ell} x_{it} x_{jt} x_{\ell t} + u_t^*$$

where $u_t^* = u_t + R_K(x_t)$. The null hypothesis of no remaining nonlinearity is H_0': $\kappa_{ij} = 0, i = 1,\ldots,p; j = i,\ldots,p; \kappa_{ij\ell} = 0, i = 1,\ldots,p; j = i,\ldots,p; \ell = j,\ldots,p$. The test can be carried out as before as an F-test if all the necessary moments (sixth) for x_t exist. Because the maintained model is very general, the null hypothesis is large. As a result, the test is likely not to have very good power in small samples if p is not small. Note that if $p = 1$, the test is equivalent to the corresponding test against STR based on the auxiliary regression (33) with $r_t = x_{1t}$ and $\beta_3 = 0$.

3. Parameter Constancy

Parameter constancy is one of the key assumptions of an STR model. Testing it is therefore as important as it is in linear models. In this chapter the alternative to parameter constancy is a set of smoothly changing parameters. Following Eitrheim and Teräsvirta (1996), the definition of parameter change is based on the idea of smooth transition, and the developments in this section are just a generalization of results in Lin and Teräsvirta (1994). Consider the STR model

$$y_t = (x_t^0)' \varphi^0(t) + (x_t^1)' \theta^0(t) G(\gamma, c; s_t) + u_t \quad (36)$$

where x_t^0 contains those $p_0 \leq p + 1$ elements of x_t whose coefficients are not assumed zero a priori, and the $p_1 \times 1$ vector x_t^1, $p_1 \leq p + 1$, is defined in the same way for the nonlinear part of the model. G is defined by (8) or (9). Let $\varphi^0(t) = \varphi^0 + \lambda_1 H(t; \gamma_1, \mathbf{c}_1)$ where φ^0 and λ_1 are $p_0 \times 1$ vectors, and $\theta^0(t) = \theta^0 + \lambda_2 H(\gamma_1, \mathbf{c}_1; s_t)$ where θ^0 and λ_2 both are $p_1 \times 1$. As in Section III.A, let H be either

$$H_1(\gamma_1, c_1; s_t) = (1 + \exp\{-\gamma_1(t - c_1)\})^{-1} - \frac{1}{2} \tag{37}$$

$$H_2(\gamma_1, c_1; s_t) = (1 + \exp\{-\gamma_1(t - c_{11})(t - c_{12})\})^{-1} - \frac{1}{2} \tag{38}$$

$$H_3(\gamma_1, c_1; s_t) = (1 + \exp\{-\gamma_1(t - c_{11})(t - c_{12})(t - c_{13})\})^{-1} - \frac{1}{2} \tag{39}$$

where $\gamma_1 > 0$ and $c_{11} \le c_{12} \le c_{13}$. These definitions accord with those in Jansen and Teräsvirta (1996). Using them, the null hypothesis of parameter constancy in (36) is $H_0: \gamma_1 = 0$. Note that the parameters in the exponent of G in (36) are assumed constant a priori. The three transition functions H_j, $j = 1, 2, 3$, allow a lot of flexibility in characterizing parameter instability. For instance, choosing H_1 and allowing $\gamma_1 \to \infty$ yields a single structural break. Doing the same in H_2 if $c_{11} < c_{12}$ yields two structural breaks: the second one restores the original parametrization. Transition function H_3 is the most general one and allows nonsymmetric and nonmonotonic parameter nonconstancy.

To discuss testing parameter constancy against $\varphi^0(t)$ and $\theta^0(t)$ in (36), let the transition function be H_3. The other two functions, H_1 and H_2, are special cases of H_3 and need not be discussed separately. The null hypothesis of parameter constancy is $H_0: \gamma_1 = 0$ which is tested against $H_1: \gamma_1 > 0$. The parameterizations of $\varphi^0(t)$ and $\theta^0(t)$ have the property that (39) is unidentified when $\gamma_1 = 0$. This by now familiar complication which precludes the use of standard tests is dealt with in the same way as before. In order to construct a viable test, approximate H_3 by its first-order Taylor expansion about $\gamma_1 = 0$. This yields

$$H_3(\gamma_1, c_1; t) = \delta_0 + \delta_1(t - c_{11})(t - c_{12})(t - c_{13}) + R_3(\gamma_1, c_1; t)$$
$$= \delta_0^* + \delta_1^* t + \delta_2^* t^2 + \delta_3^* t^3 + R_3(\gamma_1, c_1; t)$$
$$= T_3(\gamma_1, c_1; t) + R_3(\gamma_1, c_1; t) \tag{40}$$

Substituting $T_3(\gamma_1, c_1; t)$ for (39) in (36) and rearranging terms leads to the approximation

$$y_t = (x_t^0)'\beta_0 + (x_t^0 t)'\beta_1 + (x_t^0 t^2)'\beta_2 + (x_t^0 t^3)'\beta_3$$
$$+ \{(x_t^1)'\beta_4 + (x_t^1 t)'\beta_5 + (x_t^1 t^2)'\beta_6 + (x_t^1 t^3)'\beta_7\}G(\gamma, c; s_t) + u_t^* \tag{41}$$

where $\beta_j = \gamma_1\tilde{\beta}_j$, $j = 1, 2, 3, 5, 6, 7$, and $u_t^* = u_t + \{(x_t^0)'\varphi^0 + (x_t^1)'\theta^0 G(\gamma_1, c; s_t)\}$ $R_3(\gamma_1 c_1; t)$. The null hypothesis of parameter constancy thus becomes $H_0': \beta_1 = \beta_2 = \beta_3 = 0, \beta_5 = \beta_6 = \beta_7 = 0$. Since $\partial G/\partial\gamma = g_\gamma(t)$ and $\partial G/\partial c = g_c(t)$ are bounded everywhere, the results in Lin and Teräsvirta (1994) generalize to this situation (Eitrheim and Teräsvirta 1996). From this fact it follows that the test statistic (17) has an asymptotic χ^2 distribution under H_0'; in this case

$$\hat{z}_t = \{(x_t^0)', (x_t^1 G(\hat{\gamma}, \hat{c}; s_t))', \hat{g}_\gamma(t), \hat{g}_c(t)\}'$$

and

$$w_t = \{(x_t^0 t)', (x_t^0 t^2)', (x_t^0 t^3)', (x_t^1 tG(\hat{\gamma}, \hat{c}; s_t))', (x_t^1 t^2 G(\hat{\gamma}, \hat{c}; s_t))',$$
$$(x_t^1 t^3 G(\hat{\gamma}, \hat{c}; s_t))'\}$$

Again, an F-version of the test is preferred to the asymptotic theory. The degrees of freedom of the F-statistic are $3(p_0 + p_1)$ and $T - 4(p_0 + p_1)$, respectively.

If the alternative to parameter constancy is characterized by H_2 then $\beta_3 = 0$ and $\beta_7 = 0$ in (41), and the null hypothesis is reduced accordingly. If only monotonic change is considered to be the alternative to stable parameters, then the transition function is the logistic function (37) and one has $\beta_2 = \beta_3 = 0$ and $\beta_6 = \beta_7 = 0$ in (41). Even this test may be restricted to cover only certain parameters of the model while the remaining ones are constant. This is done by rewriting (36) as

$$y_t = (x_t^{01})'\varphi_1 + (x_t^{02})'\varphi_2^0(t) + \{(x_t^{11})'\theta_1 + (x_t^{12})'\theta_2^0(t)\}G(\gamma, c; s_t) + u_t \quad (42)$$

in obvious notation. The maintained nonconstancy is restricted to the coefficients of x_t^{02} and x_t^{12}. Carrying out tests corresponding to appropriate block divisions of x_t^0 and x_t^1 one can gain information about which parameters in (36) may be nonconstant if not all of them are. The empirical example in Section V illustrates the usefulness of testing the stability of subsets of parameters in STR models; see also, for example, Lütkepohl, Teräsvirta, and Wolters (1995) and Wolters, Teräsvirta, and Lütkepohl (1997).

Finally, the LM test of no autoregressive conditional heteroskedasticity (Engle 1982, McLeod and Li 1983) in the error process of an STR model can be carried out in the standard fashion exactly as in linear models. The same is true for the Lomnicki-Jarque-Bera test (Lomnicki 1961, Jarque and Bera 1980) of normality of the errors.

4. Additional Remarks

The tests discussed in this and the preceding subsection as well as the linearity tests of Section III.A have been designed against parametric alternatives. There exists a wide selection of other tests that are used in connection with nonlinear modeling and testing for structural change; see for instance Granger and Teräsvirta (1993, Chap. 6) for an account. Many of those tests do not have a specific alternative hypothesis although some of them, such as RESET (Ramsey 1969) may be interpreted as LM tests of linearity against a parametric nonlinear alternative. They have rather been intended as general tests of either linearity or parameter constancy in linear models. Nonparametric tests surveyed in Tjøstheim (1994) form a large subset of this class of tests. Because the focus in this chapter is on econometric modeling with STR models these other tests have not received the attention that they otherwise would deserve.

C. Testing the Granger Noncausality Hypothesis

Testing the null hypothesis of Granger noncausality between economic variables is normally carried out in the linear framework; for the original causality definition, see Granger (1969), and for a survey of the literature see Geweke (1984). The STR model offers a possibility to do that in a parametric nonlinear framework. For a nonparametric test, see, for example, Bell, Kay, and Malley (1996). This is a potentially useful extension, because the functional form assumed for the test may affect the outcome as Hendry (1995, p. 176) argued. It is also a straightforward one because the bivariate test is just another variant of the test of no additive nonlinearity discussed in Section III.B.2. Following Skalin and Teräsvirta (1996), who considered this extension, modify (31) slightly such that

$$y_t = w_t'\varphi + (w_t'\theta)G(\gamma_1, c_1; y_{t-d}) + v_t'\vartheta + (v_t'\psi)H(\gamma_2, c_2; x_{t-e}) + u_t \quad (43)$$

where $w_t = (1, y_{t-1}, \ldots, y_{t-p})'$ and $v_t = (x_{t-1}, \ldots, x_{t-q})'$ and ϑ and ψ being two $q \times 1$ parameter vectors, $q \geq 1$. Testing the hypothesis that the stationary variable x_t does not cause the other stationary variable y_t is equivalent to testing $H_0: \vartheta = 0$ and $\gamma_2 = 0$ in (43). Note that here it cannot be assumed that the delay e in H is known. Luukkonen, Saikkonen, and Teräsvirta (1988a) discussed this situation, and an LM-type test of the present null hypothesis may be obtained following their ideas; see Skalin and Teräsvirta (1996) for details. The restriction that the contribution of lagged x to the predictability of y be of STR-type may seem a strong one for some. It is possible, however, to adopt a general functional form for x as in (34) and proceed from there. A drawback of that approach is that the dimension of the null hypothesis increases quickly with q. If it is assumed a priori that $\gamma_1 = \gamma_2 = 0$ the test collapses into the customary linear bivariate single-equation test. The application in Skalin and Teräsvirta (1996) shows that the results from the STR-based test and the bivariate linear test applied to the same data set may indeed be very different.

IV. THE MODELING CYCLE

A. Introduction

When STR models form an alternative to linear models in econometric modeling there are several practical questions to be answered before one can fit any STR model to data and more to come after the estimation. First, as mentioned, one has to find out whether or not a linear model provides an adequate description of data. If it does, nonlinear models are not needed. Second, if nonlinearity seems to be present in the data, economic theory may not be explicit about the parametric form of the STR model, and the dynamic structure of the model may not be completely specified a priori either. For instance, it may not be obvious which of the independent variables should be the transition variable. Third, after estimating an STR model, the adequacy of the model has to be checked. In order to deal with these issues, Teräsvirta (1994) proposed a modeling cycle for univariate STAR models consisting of specification, estimation, and evaluation stages. This was an application of ideas in Box and Jenkins (1970), who developed such an approach to constructing ARIMA models. Granger and Teräsvirta (1993, Chap. 7) extended the STAR modeling cycle to STR models. This cycle is the topic of the present section. The misspecification tests of estimated STR models introduced in Eitrheim and Teräsvirta (1996) and discussed in the previous section have not become available until recently and are thus new compared to previous presentations of the STAR or STR modeling cycle. The three main stages of the cycle will be considered separately. The use of the encompassing principle to compare an STR model with its rivals explaining the same phenomenon is another addition not discussed before.

B. Specification of STR Models

The specification of STR models consists of finding answers to the following question: (i) Is a linear model adequate? (ii) If it is not and STR models are considered as an alternative, which variable should be selected to be the transition variable? (iii) Should one choose an LSTR1, LSTR2 or possibly an ESTR model? To answer the first question, linearity is tested against STR assuming each of the potential transition variables in turn to be the transition variable in the test. If linearity is rejected against an STR with the transition variable x_{jt}, this alternative is tentatively accepted. Otherwise linearity is accepted in which case no further nonlinear modeling is necessary. If linearity is rejected against STR for more than one transition variable, one selects the variable giving the strongest rejection (lowest p-value) to be the transition variable. Teräsvirta (1994) provided a heuristic justification for this decision rule. In STR modeling, the linearity testing thus has a dual purpose in the sense that if linearity is rejected the test results are also used for finding the right transition variable for the STR model. A false rejection of linearity is likely to be discovered at some later stage of the modeling cycle anyway. It may be pointed out that by performing several individual tests one is not in control of the overall significance level of the linearity test. This is not crucial if the main purpose of testing is to help model building. On the other hand, the control of the overall significance level is important if the main purpose is to test an economic theory. In that case one should use such a variant of the test that does not assume that the transition variable is known a priori. These have been discussed in Section III.A.

After making a decision about the transition variable the next step is to choose the type of the model. The decision rule can be based on the auxiliary regression (20) with the appropriate transition variable as follows. Define the following sequence of null hypotheses within (20):

$H_{04}: \beta_3 = 0$

$H_{03}: \beta_2 = 0 \mid \beta_3 = 0$

$H_{02}: \beta_1 = 0 \mid \beta_2 = \beta_3 = 0$

carry out the tests and apply the following decision rule. If the rejection of H_{03} is the strongest one, choose an LSTR2 model (or an ESTR) model, otherwise select an LSTR1 model. The coefficient vectors β_j, $j = 1, 2, 3$, are functions of the parameters of the original STR model and they depend on the type of the model. The selection rule is based on this fact; for details see Granger and Teräsvirta (1993, Chap. 7) or Teräsvirta (1994).

There exists another, computationally slightly heavier but still very practicable strategy. Consider the original STR model (6). Giving fixed values to the parameters in the transition function makes (6) linear in parameters. These parameters can be estimated by OLS. Construct a two-dimensional (LSTR1) and three-dimensional (LSTR2) grid of γ and c and estimate the other parameters for these combinations of γ and c. In order to be able to choose a meaningful set of values of γ, the exponent of the transition function should be standardized; see the discussion in the next subsection. A reasonable set of values of c may be selected between the observed minimum and maximum values of the transition variable. Estimate the models for both LSTR1 and LSTR2 and select between these alternatives after comparing the fit of the best-fitting LSTR1 and LSTR2 models. This procedure can also be used to obtain initial estimates for the NLS estimation and to reduce the size of the model by imposing exclusion restrictions. An illustration can be found in the empirical example of Section V. Finally, the choice between LSTR2 and ESTR can be made after estimating an LSTR2 model by testing $c_1 = c_2$ within that model.

C. Estimation of STR Models

Estimation of the STR model (6) with (7), (8), or (9) is carried out by NLS which is equivalent to the maximum likelihood estimation in the case of normal errors. Wooldridge (1994) and Escribano and Mira (1995) recently discussed conditions for obtaining consistent and asymptotically normal estimates. The grid estimation mentioned above can be used to obtain sensible initial values. Hendry (1995, Appendix 5.5) contains a useful updated overview of numerical optimization techniques for maximizing the likelihood; for other accounts see, for instance, Judge et al. (1985, Appendix B) and Quandt (1984). One should mention that there are sometimes numerical problems in the estimation of LSTR1 models and that they are related to estimating the slope parameter γ of the transition function. First, γ is not a scale-free parameter as its value depends on the magnitude of the values of the transition variable s_t. In order to reduce this dependence it is advisable to standardize the exponent of the transition function by dividing it by the sample standard deviation

(LSTR1 model) or the sample variance (LSTR2 model) of s_t. This standardization may bring γ close to the other parameters of the model in magnitude unless γ is very large, which is numerically an advantage in the estimation. It also makes it easier to find sensible initial estimates for this parameter.

Another problem is that if γ is large the STR model is very close to a switching regression model. This makes the estimation of γ difficult in small samples because accurate estimation requires a sufficient number of observations of the transition variable s_t in a small neighborhood of c (LSTR1 model) or c_1 and c_2 (LSTR2 model). Because it is unlikely in small samples that there exists a large cluster of observations of s_t sufficiently close to these parameters, estimating γ with satisfactory precision is a problem; see, for example, Bates and Watts (1988, p. 87), Seber and Wild (1989, pp. 480–481) and Teräsvirta (1994). This in turn may cause problems in carrying out the tests discussed in Section III.B. The partial derivatives (29) and (30) reflect this difficulty. If γ and then most likely also the NLS estimate $\hat{\gamma}$ are large and there exist no observations of s_t that are quite close but not extremely close to c, (29) will be practically zero for all t. Thus the moment matrix of any of the test statistics in Section III.B will be near-singular. In (30), observations very close to c (LSTR1) will cause a blip in the series whose observations otherwise are practically zero. If none of the observations is sufficiently close to zero then the whole vector will be practically zero and the moment matrix of the test statistic again near-singular. Rescaling (29) and (30) does not help. A feasible solution is to omit these variables from $\hat{z}_t$ when carrying out the test so that the results are numerically stable. This hardly affects the results one would obtain if one had enough precision to compute them. When the omission is accompanied by stage 1' in the testing sequence any potential size distortion (usually negligible) due to the omission is eliminated.

The estimation of the STR models in the example of Section V is carried out by applying a variant with numerical derivatives of the Broyden-Fletcher-Goldfarb-Shanno (BFGS) algorithm as implemented in the OPTMUM routine of GAUSS 3.1. While the exponent of the transition function is standardized as discussed above, it will be seen that the size of γ does not cause any difficulties in that example.

D. Evaluation of STR Models

After estimating the parameters of an STR model it is necessary to test the basic assumptions underlying the estimation. Tests of no error autocorrelation, no remaining nonlinearity and parameter constancy have been considered in Section III.B. There are also other, informal ways of checking the adequacy of the model. For instance, inflated standard errors for the parameter estimates of φ_i and θ_i at certain lags are usually an indication of overspecification. As a remedy, one should try imposing one of the restrictions $\varphi_i = 0, \theta_i = 0$ and $\varphi_i = -\theta_i$. Which one to choose is most often an empirical matter and is best settled by imposing each restriction in turn, reestimating the model and comparing the results. An estimated $\hat{c}$ (LSTR1), or $\hat{c}_1$ or

$\hat{c}_2$ (LSTR2), far outside the observed range of the transition variable is often a sign of convergence to an infeasible local minimum and thus an indication of an inadequate model. In that case, however, the model usually also fails at least some of the misspecification tests discussed above. However, in the case of an LSTR2 model, if either $\hat{c}_1$ or $\hat{c}_2$ lies far outside the observed range of the transition variable while $\hat{\gamma}$ is not small, it may also indicate that an LSTR1 model is a better choice than an LSTR2 one.

If the model fails the test of no error autocorrelation, respecification seems the only feasible solution. This is probably the most common route to follow also when the model badly fails the tests of no additional nonlinearity. When the STR model does not have constant parameters, respecification is an obvious solution as well. But there exists at least one special case in which one might actually want to have another STR component to accommodate and parameterize such nonconstancy. This is when the model contains seasonal dummies and their coefficients seem to change over time. There may not be economic reasons for such a change but seasonality may change slowly anyway because of changing institutions. This is not an uncommon situation in macroeconometric models. It has not been frequently accounted for in practice, perhaps because the econometricians have generally seen a single structural break as the most interesting alternative to parameter constancy. The paper by Farley, Hinich, and McGuire (1975) was an early exception to this rule. The STR framework offers a possibility of parameterizing such a continuous change; see the example of the next section. For other macroeconometric applications of this idea, see for example Jansen and Teräsvirta (1996) and Lütkepohl, Teräsvirta, and Wolters (1995). Gradual changes in institutions affecting other things than seasonality may also be modelled using the STR approach as in Heinesen (1996).

E. Encompassing Other Models

STR models may be constructed in areas where there already exist rival models estimated for the same time period and explaining the same economic phenomenon. In such a case, even when the estimated STR model passes the above misspecification tests it is useful to find out whether the model is an improvement over the previous quantitative explanations or not. This is the case if the STR model explains the results obtained by the rival models while the converse is not true. The STR model is then said to encompass its competitors; see, for example, Hendry (1995, Chap. 14) for a formal definition and a thorough discussion of this concept.

The encompassing property may be investigated by statistical tests. Assume for simplicity that there exists a single rival model to the STR one and that it is a linear single-equation model. Then it may be possible to construct a minimal nesting model (MNM) nesting the two competitors. This is done by extending the STR model by an additive linear component that contains those variables in the rival model that do not linearly enter the STR model. The MNM is thus an STR model. If the rival model is also an STR model then the MNM may be an additive STR model. The MNM trivially encompasses the two original models because they are nested in it; see Hendry (1995, p. 511). If an MNM can be constructed, one can apply a simplification encompassing test (Mizon and Richard 1986) to see if the STR model parsimoniously encompasses the MNM. If the rival model is linear then the test consists of testing the null hypothesis that the corresponding linear component does not additively enter the nonlinear MNM model. Such a test was discussed in Section III.B.2 when the idea was to test exclusion restrictions on coefficients of the linear component of an STR model. Accepting this hypothesis is equivalent to accepting that the STR model encompasses the MNM. Finding out if the linear model parsimoniously encompasses the MNM is tantamount to testing linearity within an MNM of STR type. This test has to be carried out using the techniques discussed in Section III.A because an STR type MNM is not identified under the null hypothesis. Suppose that this is done and the null hypothesis rejected. Then the conclusion is that the linear model does not encompass the MNM. If the STR model does, then the transitivity property of encompassing implies that the STR model encompasses its linear rival. The use of this testing procedure requires that both the STR equation and its single-equation rival are valid models in that they can be analyzed without knowledge of the rest of the system. An example of simplification encompassing tests can be found in Section V.

V. APPLICATION

A. Background

This section contains an example of the modeling cycle consisting of specification, estimation and evaluation of STR models. It is based on the data set and results in Hendry (1984), who modeled house prices in the United Kingdom in 1960–1981 using error-correction models. For another econometric analysis of this data set, see Richard and Zhang (1996). The purpose here is not to present a new model for UK house prices: in order to do that the first thing would be to extend the time series as close to the present time as possible. The main objective is instead to use the equation for house price expectations Hendry (1984) specified and estimated as a benchmark and see if the STR approach based on that equation and the same observation period leads to any new insight or yields an improved specification. This provides an opportunity to show how the STR modeling strategy works in practice.

The period Hendry (1984) considered was eventful. The nominal house prices increased 12-fold and the real prices by over 50%. The nominal prices were also clearly more volatile than the ordinary retail price index. These and other features of the observed time series are discussed in Hendry (1984). The paper also contains a description of the way the housing market functioned in the United Kingdom during the observation period. The theoretical model in the paper is based on the assumption that the housing stock H_t evolves intertemporally according to

$$H_t = (1 - \delta_t)H_{t-1} + C_t \tag{44}$$

where δ_t is the depreciation rate and C_t denotes net additions. At any time, C_t is very small relative to H_{t-1} so that H_{t-1} is taken to be the fixed supply of housing in the short run. As a result, the fluctuations in demand translate into fluctuations in the price of housing, Ph_t. Thus (see Hendry 1984, pp. 224–225 for a more complete argument), the demand equation is the one determining the price of housing. Its general form is postulated as

$$H^D = f(Ph/P, Y, \rho, R, Rm, M, T, N, F)$$
$$\quad\quad - \quad + \; - \; - \quad - \quad + \;\; ? \;\; + \;\; ?$$

where P is the general price level, Y the real income, ρ the real rental rate, R the market interest rate, and Rm is the mortgage rate of interest, which for institutional reasons may differ substantially from the market interest rate. Furthermore, M is the stock of mortgages, T the tax rate, N the size of the population, and F the average family size. Changes in the real price of housing, the real rental rate and the interest rates have a negative effect on demand. Changes in the real income, the stock of mortgage, and the size of the population have the opposite effect. The tax rate and the average family size also affect the demand for housing, but the sign is indeterminate.

In the following the focus will be solely on modeling price expectations which Hendry (1984) discussed in detail. The expectations are assumed unbiased:

$$\Delta ph_t = \Delta ph_t^e + u_t \tag{45}$$

where Δ is the difference operator, ph_t the logarithmic nominal price of housing (lowercase letters denote logarithms), ph_t^e the corresponding expectation, and u_t an innovation with respect to the information used in predicting Δph_t: $Eu_t = 0$, $\text{var}(u_t) = \sigma_u^2$, $\text{cov}(u_t, u_s) = 0$, $s \neq t$. Equation (11) in Hendry (1984, p. 228) gives the expectations the following parametric form:

$$\Delta ph_t^e = \sum_{i=1}^{2} c_i \Delta ph_{t-1} - c_3(R^0 - \Delta_4 p)_{t-1} + c_4 \Delta m_{t-1}$$
$$- c_5(ph + h - p - y - c_a)_{t-1} + c_6(m - ph - h - c_b)_{t-1}$$

where R_t^0 is the after-tax interest rate, c_a and c_b are constants, and $c_i, i = 1, \ldots, 6$, are unknown parameters. The expectation equation contains two error correction terms. The first one requires the nominal value of housing to stand in constant proportion to nominal income in the long run. The second error correction term implies a constant long-run ratio of mortgage (m) to own equity ($ph + h$); H is the housing stock.

The estimated counterpart of (45) considered here is somewhat different from (45) itself. It is Eq. (18) in Hendry (1984, p. 237) and contains more short-run dynamics than (45). The equation has been estimated for the period 1959(1)–1982(2) and has the following estimated form*

$$
\begin{aligned}
\Delta ph_t = \quad & 0.23 \ \Delta ph_{t-2} + 13.1 \ (\Delta ph_{t-1})^3 + \ 0.51 \ A_2(\Delta y_t) \\
& (0.085) \qquad\quad (4.3) \qquad\qquad\quad (0.13) \\[4pt]
+ \ & 0.17 \ (m - ph - h)_{t-1} + \ 0.43 \ (y - h)_{t-1} \\
& (0.028) \qquad\qquad\qquad\quad (0.072) \\[4pt]
+ \ & 0.82 \ F_{13}(p) + \ 0.53 \ F_{13}(m - p) - \ 0.21 \ \overline{R}^0_{t-3} \\
& (0.14) \qquad\quad (0.13) \qquad\qquad\quad (0.10) \\[4pt]
- \ & 0.51 \ \Delta R^0_{t-1} - \ 0.53 \ - \ 0.0006 \ Q_{1t} + \ 0.025 \ Q_{2t} \\
& (0.20) \qquad\quad (0.12) \quad (0.0057) \qquad (0.0049) \\[4pt]
+ \ & 0.019 \ Q_{3t} - \ 3.7 \ D^0_{1t} - \ 2.2 \ D^0_{2t} + \hat{u}_t \qquad\qquad (46) \\
& (0.0047) \qquad (1.5) \qquad (0.63)
\end{aligned}
$$

$T = 94$, $AIC = -3.80$, $R^2 = 0.78$, $\hat{\sigma} = 0.0143$, $\hat{\sigma}^2/\hat{\sigma}^2_L = 0.94$, $sk = 1.3$, $ek = 4.8$, $LJB = 117(4 \times 10^{-26})$

where the figures in parentheses below the coefficient estimates are estimated standard deviations, $\hat{\sigma}^2 = (T - k)^{-1} \sum_{t=1}^{T} \hat{u}_t^2$ is the residual variance (k is the number of estimated regression coefficients), $\hat{\sigma}^2_L$ is the residual variance of the corresponding linear model (Δph_{t-1} not cubed), sk is skewness and ek excess kurtosis of the residuals, T is the number of observations, R^2 is the coefficient of determination, LJB is the Lomnicki-Jarque-Bera normality test and the value in parentheses is the p-value of the test statistic. The small p-value is due to a few large residuals. In (46), $A_2(x) = (1/3)(3x_t + 2x_{t-1} + x_{t-2})$, $F_{13}(x) = \Delta(x_{t-1} + x_{t-3})$, $F_{13}(m - p) = \Delta\{(m - p)_{t-1} + (m - p)_{t-3}\}$, $\overline{R}^0_{t-3} = (1/2)(R^0_{t-3} + R^0_{t-4})$, Q_{it}, $i = 1, 2, 3$, are the three seasonal dummy variables, and D^0_{1t} and D^0_{2t} are two dummy variables discussed in Hendry (1984, pp. 241–242). The errors of (46) are not autocorrelated, see Table 1. Tests of no autoregressive conditional heteroskedasticity (Engle 1982) in Table 1 do not indicate any problems either.

Equation (46) is a result of a specification search. Ratio "real value of housing to income" in (45) has become "real income to stock of housing." The lag structure has been simplified by omitting lags and imposing coefficient restrictions defined by A_2, F_{13}, and $\overline{R}$. An interesting feature is the cubic price difference $(\Delta ph_{t-1})^3$ whose

---

*The estimated equation is not exactly the same as Eq. (18) in Hendry (1984). The differences are due to the fact that the data set Hendry (1984) used was no longer available in its original form. The data set used in this chapter is as close to the original data as possible.

Table I p-Values of the LM Test of No Error Autocorrelation against an AR(q) and MA(q) Error Process and the LM Test of No Autoregressive Conditional Heteroskedasticity against ARCH(q) in model (46)

	Maximum lag q				
Test	1	2	3	4	6
No error autocorrelation	0.55	0.83	0.75	0.88	0.38
No ARCH	0.99	0.99	1.00	1.00	0.99

coefficient estimate is significant and which Hendry (1984, p. 228) describes as a local approximation to a more complicated lag structure. When it is introduced, a linear first-order lag Δph_{t-1} is no longer needed in the equation.

B. Specification and Estimation of the STR Model

The presence of the cubic price lag in (46) may be interpreted as an indication of possible nonlinearity. It can be viewed as a first-order Taylor approximation to a certain ESTR equation with Δph_{t-1} as a transition variable. Instead of directly estimating such an equation a more general approach is adopted. First, one has to find out whether the expectations model is nonlinear, the alternative to linearity being the STR. If the null hypothesis of linearity is rejected, the next step is to build an adequate STR model to characterize the nonlinearity. When this is attempted, the linear parameter restrictions imposed in (46) are retained because the focus is on potential nonlinearity of the price expectations. Omitting the restrictions and starting with a richer parameterization than (46) has not been considered.

To test linearity against STR, the cubic lag $(\Delta ph_{t-1})^3$ in (46) is replaced by Δph_{t-1} so that the basic model is linear not only in parameters but also in variables. This model forms the null hypothesis in testing linearity as discussed in Section III.A. It is assumed that the dummy variables only enter the model linearly and do not thus under any circumstances appear in the nonlinear part of the model. For D_{1t}^0 and D_{2t}^0 with very few nonzero values this is the only feasible assumption, but it also covers the seasonal dummies. All the nondeterministic variables are a priori regarded as potential transition variables. Therefore, a separate test sequence is carried out, assuming each of them in turn being the transition variable under the alternative hypothesis. The results appear in Table 2. The null of linearity is strongly rejected when Δph_{t-1} is the transition variable. This accords with the specification of (46), although the test sequence (H_{04} and H_{02} rejected more strongly than H_{03}) suggests an LSTR1 rather than an LSTR2 or an ESTR model. The other strong rejection occurs when $F_{13}(m-p)$ involving changes in the real value of mortgage is assumed to be the transition variable. Since the first rejection is the strongest one

Table 2 p-Values of Tests of Linearity of the Linear (Δph_{t-1} not cubed) U.K. Housing Price Equation against STR, Transition Variable Assumed Known

Linearity test	Transition variable[a]								
	Δph_{t-1}	Δph_{t-2}	$A_2(\Delta y_t)$	$(m - ph - h)_{t-1}$	$(y - h)_{t-1}$	$F_{13}(p)$	$F_{13}(m - p)$	$\overline{R}^0_{t-1}$	ΔR^0_{t-1}
F	0.00014	0.057	0.22	0.47	0.41	0.82	0.00037	0.76	0.80
F_4	0.017	0.34					0.0036		
F_3	0.066	0.74					0.069		
F_2	0.0013	0.0025					0.037		

[a]The p-values for the whole sequence of tests are given only if the p-value of the general test (F) lies below 0.1.

Table 3 p-Values of Parameter Constancy Tests of the Linear (Δph_{t-1} not cubed) U.K. House Price Equation against STR-Type Nonconstancy

Parameter constancy test	Null hypothesis			
	(1)	(2)	(3)	(4)
F_1	0.090	0.026	0.85	0.42
F_2	0.19	0.056	0.54	0.50
F_3	0.37	0.24	0.78	0.53

(1): H_0: "All parameters except the coefficients of D_1^0 and D_2^0 are constant."
(2): H_0: "Intercept and the coefficients of seasonal dummy variables are constant."
(3): H_0: "Coefficients of Δph_{t-1} and Δph_{t-2} are constant."
(4): H_0: "Coefficients of 'exogenous' variables are constant."
Notes: (1) The parameters not under test are assumed constant also under the alternative. (2) Test F_j is a test against an STR model with transition function H_j, $j = 1, 2, 3$; see Section III.B.3 and definitions (23)–(25).

and as (46) may be interpreted as an approximation to an STR model with Δph_{t-1} as the transition variable it is tentatively assumed that the data have been generated by such an LSTR1 model. Furthermore, the constancy of parameters in the linear equation was tested applying the tests discussed in Section III.A. The results can be found in Table 3. The overall test vaguely indicates that the equation may not have constant parameters. The main reason for this seems to be that seasonality has been changing over time. Nevertheless, the evidence against parameter constancy is clearly weaker than that against linearity, and the nonlinearity is therefore dealt with first. It can be mentioned that if the parameter constancy of (46) is investigated by the same tests (results not reported here), the rejection is stronger than in the case of the linear equation.

After choosing an LSTR1 model for Δph_t one has to specify the parameter structure of the model. In order to do that one first assumes a fully parameterized model and estimates the parameters using ordinary least squares and the two-dimensional grid for γ and c as proposed in Section IV.B. The results appear in Table 4. As discussed, they can be used to specify the parameter structure of the STR model. Noting that the t-values are conditional on $\hat{\gamma}$ and $\hat{c}$, it is seen that only few coefficients in the nonlinear part of the equation may be nonzero. Somewhat arbitrarily first retaining those with $|t| \geq 1.6$ estimating the corresponding STR model and removing the redundant variables leads to the result that only the intercept and the error-correcting variable $(m - ph - h)_{t-1}$ have significant nonlinear coefficients (they had the highest t-ratios already in Table 4). One thus obtains the following STR model:

Table 4 Grid Estimation of the Fully Parameterized LSTR1 Model for the U.K. Housing Prices, 1959(1)–1981(2)

Linear coefficient of	Estimate	Standard deviation	t-Value
Δph_{t-1}	−0.065	0.092	−0.71
Δph_{t-2}	0.21	0.073	2.8
$A_2(\Delta y_t)$	0.53	0.10	5.1
$(m - ph - h)_{t-1}$	0.13	0.023	5.5
$(y - h)_{t-1}$	0.37	0.061	6.1
$F_{13}(p)$	0.82	0.13	6.5
$F_{13}(m - p)$	0.52	0.11	4.9
$\bar{R}^0_{t-3}$	−0.26	0.082	−3.2
ΔR^0_{t-1}	−0.54	0.21	−2.6
Intercept	0.38	0.092	4.1
Q_{1t}	−0.00098	0.0045	−0.22
Q_{2t}	0.022	0.0037	5.8
Q_{3t}	0.014	0.0036	3.9
D^0_1	−2.16	0.47	−4.6
D^0_2	−3.78	1.08	−3.5

Nonlinear coefficient of			
Δph_{t-1}	−0.17	0.43	−0.41
Δph_{t-2}	1.39	0.83	1.7
$A_2(\Delta y_t)$	−1.01	2.18	−0.46
$(m - ph - h)_{t-1}$	0.37	0.12	3.1
$(y - h)_{t-1}$	−2.18	1.42	−1.5
$F_{13}(p)$	−2.95	4.26	−0.69
$F_{13}(m - p)$	−0.87	1.81	−0.48
$\bar{R}^0_{t-3}$	4.83	3.19	1.5
ΔR^0_{t-1}	−7.95	3.87	−2.1
Intercept	3.09	1.23	2.5
$\hat{\gamma}$	9.5		
$\hat{c}$	0.07		
R^2	0.87		
$\hat{\sigma}$	0.104		

Table 5 p-Values of the LM Test of No Error Autocorrelation against an AR(q) and MA(q) Error Process and the LM Test of No Autoregressive Conditional Heteroskedasticity against ARCH(q) in (47)

	Maximum lag q				
Test	1	2	3	4	6
No error autocorrelation	0.34	0.30	0.38	0.51	0.48
No ARCH	0.65	0.85	0.53	0.69	0.80

$$
\begin{aligned}
\Delta ph_t =\ & 0.22\ \Delta ph_{t-2} + 0.48\ A_2(\Delta y_t) + 0.13\ (m - ph - h)_{t-1} \\
& (0.074) \qquad\quad (0.11) \qquad\qquad (0.027)
\end{aligned}
$$

$$
\begin{aligned}
+\ & 0.34\ (y - h)_{t-1} + 0.70\ F_{13}(p) + 0.43\ F_{13}(m - p) \\
& (0.060) \qquad\qquad (0.12) \qquad\qquad (0.10)
\end{aligned}
$$

$$
\begin{aligned}
-\ & 0.21\ \overline{R}^0_{t-3} - 0.51\ \Delta R^0_{t-1} - 0.38\ -\ 0.0012\ Q_{1t} \\
& (0.084) \qquad (0.21) \qquad\quad (0.11) \quad (0.0047)
\end{aligned}
$$

$$
\begin{aligned}
+\ & 0.024\ Q_{2t} + 0.0174\ Q_{3t} - 3.6\ D^0_{1t} - 2.2\ D^0_{2t} \\
& (0.0040) \qquad (0.0040) \qquad (0.24) \qquad (0.53)
\end{aligned}
$$

$$
\begin{aligned}
+\ \{\ & 3.0\ + 0.60\ (m - ph - h)_{t-1}\} \\
& (1.2) \quad (0.25)
\end{aligned}
$$

$$
\begin{aligned}
\times\ & [1 + \exp\{\ -2.5\ (\Delta ph_{t-1} - 0.088)\ /\hat{\sigma}(\Delta ph_{t-1})\}]^{-1} \\
& \qquad\quad (0.79) \qquad\qquad (0.0085)
\end{aligned}
$$

$$
+\ \hat{u}_t \tag{47}
$$

$T = 94$, $AIC = -8.63$, $R^2 = 0.85$, $\hat{\sigma} = 0.0123$, $\hat{\sigma}^2/\hat{\sigma}^2_L = 0.69$, $sk = 0.93$, $ek = 4.0$, $LJB = 76(4 \times 10^{-17})$

where $\hat{\sigma}(\Delta ph_{t-1})$ is the sample standard deviation of Δph_{t-1} and $\hat{\sigma}^2_L$ the residual variance of the corresponding linear model. The residual variance of (47) is only about 70% of that of the corresponding linear model. Results of the LM test of no error autocorrelation in Table 5 do not indicate autocorrelation, nor is there any evidence of ARCH. Large skewness and excess kurtosis estimates are mainly due to a large positive residual in 1964(1); see Figure 1.

C. Interpretation and Evaluation of the STR Model

The residuals of (47) are graphed in Figure 1 together with the residuals of (46). The main difference in the fits of the two models is due to different characterizations of the

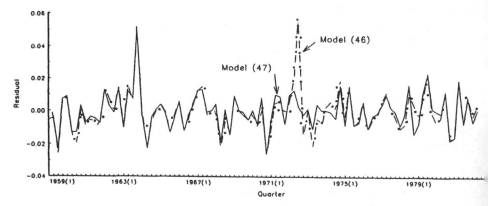

Figure I   Residuals from the linear model with a cubic lag (46) and from the STR model (47) for the first differences of the logarithmic UK housing price index, 1959(1)–1982(2).

price turbulence in 1973. The linear equation strengthened by the cubic lag does not explain the features of the housing price boom and its immediate aftermath as well as (47). Apart from that period both models have almost identical fits. It seems obvious that the nonlinear specification is mainly required to model the exceptional increase in house prices in 1973. This is also seen from Figure 2. It shows that the transition function obtains values close to zero most of the time. A comparison between (46) and the linear part of (47) indicates that they are quite similar. For those periods (46) and (47) thus may be expected to have rather similar residuals.

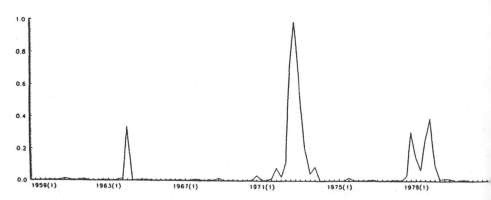

Figure 2   Values of the transition function of the STR model (47), 1959(1)–1982(2).

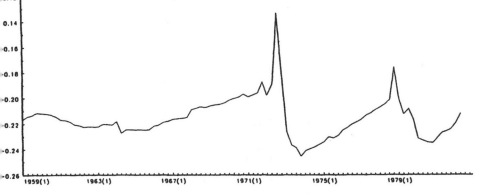

Figure 3   Values of the error-correction term (48) in equation (47), 1959(1)–1982(2).

It is instructive to find out how the nonlinear error correction (nec_t) works in 1973. From (47),

$$nec_t = 0.38 + 3.0G_1(\hat{\gamma}, \hat{c}; \Delta ph_{t-1})$$
$$+ \{0.13 + 0.60G_1(\hat{\gamma}, \hat{c}; \Delta ph_{t-1})\}(m - ph - h)_{t-1} \tag{48}$$

where

$$G_1(\hat{\gamma}, \hat{c}; \Delta ph_{t-1}) = \left[1 + \exp\left\{-\frac{2.5(\Delta ph_{t-1} - 0.088)}{\hat{\sigma}(\Delta ph_{t-1})}\right\}\right]^{-1}$$

The graph of (48) over the observation period appears in Figure 3. When a large price shock arrives (Δph_t obtains a large value) it causes a sharp increase in (48) one period later through the combined intercept $0.38 + 3.0G_1$. The error-correcting combination $(m - ph - h)$ has negative values throughout so that an increase in the value of the transition function initially weakens the error correction. This initial effect is soon offset by a large change in $(m - ph - h)_{t-1}$ as the value of the mortgage stock does not follow the rapid increase in the value of the housing stock. Because of the large positive nonlinear coefficient (0.60) of $(m - ph - h)_{t-1}$ in (48) the pull toward the equilibrium increases dramatically and eventually suppresses the price boom.

The STR model (47) seems to explain the dynamics of the unusually large increase in housing prices but does it explain all nonlinearity in the data? Table 6 contains the results of the tests of no additive nonlinearity against an additive STR model considered in Section III.B.2. Note that Δph_{t-1} is included in the second nonlinear component although it only appears in the transition function of (47). When the test is carried out with Δph_{t-1} as the transition variable, the p-value of the test equals 0.11. This indicates that the nonlinearity causing a very low p-value for the

Table 6 p-Values of Tests of No Additive Nonlinearity in the LSTR1 Model (47) for a Set of Transition Variables

Linearity test	Transition variable								
	Δph_{t-1}	Δph_{t-2}	$A_2(y_t)$	$(m - ph - h)_{t-1}$	$(y - h)_{t-1}$	$F_{13}(p)$	$F_{13}(m - p)$	$\bar{R}^0_{t-3}$	ΔR^0_{t-1}
F	0.11	0.90	0.011	**[a]	**[a]	0.65	0.17	0.72	0.89
F_4			0.011						
F_3			0.056						
F_2			0.57	0.43	0.79				

[a]Test not computed due to near-singularity of the moment matrix. In that case, results of F_2 (test against additive nonlinearity of LSTR1 type) are shown.

Table 7 *p*-Values of Parameter Constancy Tests of the LSTR1 Model (47) against STR-Type Constancy

Parameter constancy test	Null hypothesis				
	(1)	(2)	(3)	(4)	(5)
F_1	0.022	0.017	0.61	0.0078	0.71
F_2	0.19	0.15	0.84	0.038	0.87
F_3	0.30	0.19	0.96	0.12	0.71

(1): H_0: "All parameters except the coefficients of D_1^0 and D_2^0 are constant."
(2): H_0: "All parameters in the linear part of the model except the coefficients of D_1^0 and D_2^0 are constant."
(3): H_0: "All parameters in the nonlinear part of the model are constant."
(4): H_0: "Intercepts and coefficients of the seasonal dummy variables are constant."
(5): H_0: "All parameters in the linear part of the model except the dummy variables are constant."
Notes: (1) The parameters not under test are assumed constant also under the alternative. (2) Test F_j is a test against an STR model with transition function H_j, $j = 1, 2, 3$; see Section III.B.3 and definitions (23)–(25).

corresponding linearity test (Table 3) has been dealt with in a satisfactory manner. Another result pointing at the same direction is that the test with $F_{13}(m - p)$ as the transition variable has *p*-value 0.17, whereas the corresponding linearity test had a low value. On the other hand, the test with $A_2(\Delta y_t)$ as the transition variable now has a *p*-value close to 0.01, but this order of magnitude is considerably higher than that of the lowest *p*-values in the linearity tests. As at the same time all the other tests have *p*-values exceeding 0.1, this result does not cause too much concern. As a whole, it can be concluded that the STR model (47) explains most of the nonlinearity present in the data.

Parameter constancy tests of the linear model indicated some nonconstancy although the result could also have been interpreted as an effect of neglected nonlinearity on these tests. Table 7 contains results of the parameter constancy tests described in Section III.B. The results suggest that despite careful parametrization of nonlinearity the parameter nonconstancy is still a problem. It seems obvious that seasonality in U.K. house prices has been changing over time. Furthermore, the change seems to have been monotonic during the observation period because F_1 is the test with the strongest rejection of the null hypothesis just as it was in the linear model.

D. Respecification and Reestimation of the Model

To capture this parameter change one can estimate the parametric alternative to parameter constancy, which in this case means estimating an additive STR model. In this model the second transition function has time as the transition variable. A specification search indicated that the second and the third quarters have to be included in the additional nonlinear component of the model. The estimated equation has the form (note that the second transition variable t/T is standardized between 0 and 1)

$$
\begin{aligned}
\Delta ph_t = \quad & 0.23 \ \Delta ph_{t-2} + \ 0.58 \ A_2(\Delta y_t) + \ 0.12 \ (m - ph - h)_{t-1} \\
& (0.070) \qquad\qquad (0.11) \qquad\qquad (0.025) \\[4pt]
+ \ & 0.33 \ (y - h)_{t-1} + \ 0.55 \ F_{13}(p) + \ 0.31 \ F_{13}(m - p) \\
& (0.061) \qquad\qquad (0.12) \qquad\qquad (0.10) \\[4pt]
- \ & 0.32 \ \overline{R}^0_{t-3} - \ 0.48 \ \Delta R^0_{t-1} + \ 0.38 \ - \ 0.0035 \ Q_{1t} \\
& (0.082) \qquad\quad (0.19) \qquad\quad (0.098) \quad (0.0045) \\[4pt]
- \ & 0.24 \ Q_{2t} + \ 0.23 \ Q_{3t} - \ 3.9 \ D^0_{1t} - \ 1.6 \ D^0_{2t} \\
& (0.57) \qquad\; (0.54) \qquad\; (1.1) \qquad\; (0.48) \\[4pt]
+ \ & \{ \ 2.9 \ + \ 0.59 \ (m - ph - h)_{t-1} \} \\
& \ \ (1.2) \quad\ (0.24) \\[4pt]
\times \ & [1 + \exp\{- \ 2.8 \ (\Delta ph_{t-1} - \ 0.088 \ )/\hat{\sigma}(\Delta ph_{t-1})\}]^{-1} \\
& \qquad\qquad\; (0.94) \qquad\qquad (0.0075) \\[4pt]
+ \ & \{ \ 0.53 \ Q_{2t} + \ 0.50 \ Q_{3t} \} \\
& \ \ (1.1) \qquad\; (1.1) \\[4pt]
\times \ & [1 + \exp\{- \ 0.13 \ (t/T - \ 0.49 \ )/\hat{\sigma}(t/T)\}]^{-1} \\
& \qquad\qquad\; (0.28) \qquad\quad (0.16) \\[4pt]
+ \ & \hat{u}_t \hspace{8cm} (49)
\end{aligned}
$$

$T = 94$, $AIC = -8.75$, $R^2 = 0.88$, $\hat{\sigma} = 0.0114$, $\hat{\sigma}^2/\hat{\sigma}_L^2 = 0.59$, $sk = 0.86$, $ek = 3.4$, $LJB = 56 \ (7 \times 10^{-13})$

where $\hat{\sigma}(t/T) = \{(1/T) \sum_{t=1}^{T}(t/T - 1/2)^2\}^{1/2}$. The residual variance of (49) is about 60% of that of the corresponding linear model. Testing the residuals against error autocorrelation and ARCH does not indicate any model misspecification (the test results are not shown). The fit of (49) is very similar to that of (47). AIC has decreased compared to (47).

Figures 4 and 5 illustrate the effect of the second transition function. Figure 4 shows that during the observation period, the estimated transition function is practically a straight line. This explains the large standard deviations of the estimates of the coefficients of both the linear and the nonlinear seasonal dummy variables and the slope parameter of the transition function γ. Information about where the logistic transition function is bending is necessary for accurate estimation of the pa-

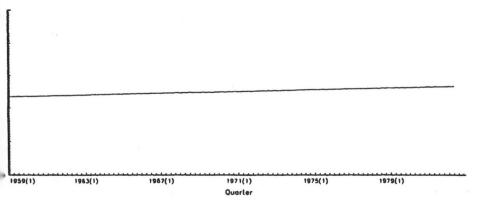

Figure 4   Values of the transition function of seasonal dummy variables in the STR model (49), 1959(1)–1982(2).

rameters involved in the characterization of seasonal effects. That this information is scarce in the sample is seen from Figure 4. A similar situation is described in Jansen and Teräsvirta (1996). Figure 5 shows the values of the time-varying seasonals $\{-0.24 + 0.53H(\hat{\gamma}, \hat{c}; t/T)\}Q_{2t}$ and $\{-0.23 + 0.50H(\hat{\gamma}, \hat{c}; t/T)\}Q_{3t}$. The figure shows that seasonality in the house prices increases from 1959(1) to 1982(2) and that the effect can be ascribed to the second and the third quarters of the year.

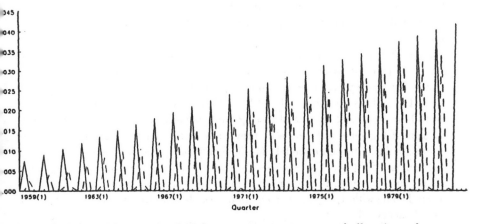

Figure 5   Values of the second and third quarter time-varying seasonal effects (second quarter = solid line, third quarter = dashed line) according to Eq. (49), 1959(1)–1982(2).

In order to find out whether or not the large uncertainty in the estimates of seasonal parameters is mainly due to overparameterization the second logistic transition function in (49) is replaced by the linear approximation (t/T) and the parameters reestimated. This yields

$$\Delta ph_t = \underset{(0.069)}{0.23} \ \Delta ph_{t-2} + \underset{(0.11)}{0.59} \ A_2(\Delta y_t) + \underset{(0.025)}{0.12} \ (m - ph - h)_{t-1}$$

$$+ \underset{(0.060)}{0.33} \ (y - h)_{t-1} + \underset{(0.12)}{0.55} \ F_{13}(p) + \underset{(0.10)}{0.31} \ F_{13}(m - p)$$

$$- \underset{(0.080)}{0.32} \ \overline{R}^0_{t-3} - \underset{(0.20)}{0.48} \ \Delta R^0_{t-1} + \underset{(0.096)}{0.38} - \underset{(0.0045)}{0.0035} \ Q_{1t}$$

$$+ \underset{(0.0063)}{0.0067} \ Q_{2t} + \underset{(0.0060)}{0.0016} \ Q_{3t} - \underset{(1.1)}{3.9} \ D^0_{1t} - \underset{(0.51)}{1.6} \ D^0_{2t}$$

$$+ \{ \underset{(1.2)}{3.0} + \underset{(0.24)}{0.59} \ (m - ph - h)_{t-1} \}$$

$$\times [1 + \exp\{ - \underset{(0.93)}{2.8} \ (\Delta ph_{t-1} - \underset{(0.0077)}{0.088} \ )\}]^{-1}$$

$$+ \underset{(0.010)}{0.035} \ (t/T)Q_{2t} + \underset{(0.0094)}{0.033} \ (t/T)Q_{3t} + \hat{u}_t \qquad (50)$$

$T = 94$, $AIC = -8.79$, $R^2 = 0.88$, $\hat{\sigma} = 0.0113$, $\hat{\sigma}^2/\hat{\sigma}^2_L = 0.58$, $sk = 0.86$, $ek = 3.4$, $LJB = 56(7 \times 10^{-13})$.

The estimates of other parameters than the seasonal dummies remain practically unchanged. The seasonality seems indeed to be changing over time. The amount of uncertainty in the parameter estimates is considerably less than in (49) and AIC smaller. The seasonals in the linear part of (50) could even be removed altogether; that is, in the beginning of the period there has been little or no seasonality in house prices. For illustration, however, those variables have been retained in the model.

The test results for models (49) and (50) are very close to each other. Those for (50) are reported. Results of testing the hypothesis of no remaining nonlinearity against the same alternative as previously can be found in Table 8. They are rather similar to those for model (47). The only conspicuous difference is that the test against LSTR with $(m - ph - h)_{t-1}$ as the transition variable now has a relatively small p-value. Because most of the other tests have clearly higher p-values (even the one with $A_2(y_t)$ as the transition variable) the current specification is tentatively accepted.

Results of the parameter constancy tests can be found in Table 9. The null of constancy is not rejected in any of the tests. The parameterization of changing

Table 8 p-Values of Tests of No Additive Nonlinearity in the LSTR1 Model (50) for a Set of Transition Variables

Linearity test	Transition variable								
	Δph_{t-1}	Δph_{t-2}	$A_2(y_t)$	$(m - ph - h)_{t-1}$	$(y - h)_{t-1}$	$F_{13}(p)$	$F_{13}(m - p)$	$\bar{R}^0_{t-3}$	ΔR^0_{t-1}
F	0.27	0.91	0.031	**[a]	**[a]	0.13	0.16	0.25	0.44
F_4			0.015						
F_3			0.37						
F_2			0.24	0.016	0.65				

[a]Test not computed due to near-singularity of the moment matrix. In that case, result of F_2 (tests against additive nonlinearity of LSTR1 type) are shown.

Table 9 p-Values of Parameter Constancy Tests of the LSTR1 Model (50) against STR-Type Nonconstancy

Parameter constancy test	Null hypothesis					
	(1)	(2)	(3)	(4)	(4a)	(5)
F_1	0.70	0.61	0.58	$**a$	0.97	0.50
F_2	$**a$	0.81	0.60	$**a$	0.85	0.73
F_3	$**a$	0.71	0.81	$**a$	0.96	0.32

(1): H_0: "All parameters except the coefficients of D_1^0 and D_2^0 are constant."
(2): H_0: "All parameters in the linear part of the model except the coefficients of D_1^0 and D_2^0 are constant."
(3): H_0: "All parameters in the nonlinear part of model are constant."
(4): H_0: "Intercepts and coefficients of the seasonal dummy variables are constant."
(4a): H_0: "Linear intercept and coefficients of the seasonal dummy variables are constant."
(5): H_0: "All parameters in the linear part of the model except the coefficients of the dummy variables are constant."
[a]Test not computed due to near-singularity of the moment matrix.
Notes: (1) The parameters not under test are assumed constant also under the alternative.
(2) Test F_j is a test against an STR model with transition function H_j, $j = 1, 2, 3$; see Section III.B.3 and definitions (23)–(25).

seasonality has removed the variation in the coefficients of seasonal dummy variables. According to (50), U.K. housing price changes now have a trending seasonal component. However, it is not very realistic to extend the conclusion far outside the sampling period. In fact, model (49) should be preferred to (50) as far as the interpretation of the "trend" is concerned. The change one observed just happens to lie within an interval for which the logistic transition function is almost linear. Of course, even the interpretation that Eq. (49) offers may turn out to be incorrect in the light of any new data, but at any rate, the assumption of a long-run linear trend in seasonality is hardly a plausible one.

E. Encompassing Tests

An objective of this application is to find out if the specification of the price expectations equation (46) can be improved by applying STR models. Equation (50) is an STR model, and the question is whether or not it can be viewed as an improvement over (46). This can be investigated by encompassing tests as discussed in Section IV.E. The task is to investigate whether the STR model (50) encompasses (46) or not and vice versa. To find out if (50) encompasses (46) the first step is to construct an MNM. This is done by augmenting (50) linearly by the cubic lag of the price change $(\Delta ph_{t-1})^3$. The augmented model trivially encompasses (50) because the latter model is nested in it. In order to see whether (50) parsimoniously encompasses the MNM one estimates the parameters of the MNM and computes the likelihood ratio statistic

$$LR = -2\{\log(L_0) - \log(L_1)\} \tag{51}$$

where L_0 is the likelihood of (50) and L_1 that of the MNM. Under H_0: "Coefficient of $(\Delta ph_{t-1})^3$ equals zero in the MNM," (51) has an asymptotic $\chi^2(1)$ distribution. In this case one obtains $LR = 0.0299$ with p-value 0.86. The conclusion is that (50) parsimoniously encompasses the MNM.

The following computational remark is in order. Equation (46) and the STR model (50) are competitors, of which the latter is nonlinear even in parameters. Nonlinear least-squares estimation of the MNM required to carry out the test of parsimonious encompassing may sometimes be difficult because the parameterization of the MNM may turn out to be very rich. In the present case the convergence was achieved in the estimation when the initial estimates of the parameters of the MNM (save the coefficient of $(\Delta ph_{t-1})^3$) were the same as the final estimates in (50). If the initial estimates were chosen further away from those values, the NLS algorithm (BFGS) either converged to a local minimum or did not converge at all.

Testing the null hypothesis that (46) parsimoniously encompasses the MNM is not equally straightforward because the latter equation is nonlinear and is not identified if linearity in parameters (the null hypothesis) holds. This situation was discussed in Section IV.E and it was pointed out the hypothesis can be tested by applying linearity tests described in Section III.A. The transition variable of the alternative needed in the test is Δph_{t-1}. The linearity test statistic is the usual one and the variables in the auxiliary regression based on the MNM include the ordinary ones in (50) but also $(\Delta ph_{t-1})^3$. Note that in accordance with (46), Δph_{t-1} only appears as the transition variable and is not present in the null model. The F-approximation to the likelihood ratio test results in $F(27, 52) = 2.42(0.0031)$, whereas $F_2(9, 70) = 2.51(0.0051)$ (F_2 is a test against LSTR1). This implies a rejection of the null hypothesis, and the conclusion is that Eq. (46) does not parsimoniously encompass the MNM. As (50) does, one has the result that (50) also encompasses (46) and thus contributes to the general understanding of the mechanism generating U.K. house price expectations during 1960–1981.

VI. FINAL REMARKS

In this chapter the emphasis is on showing how STR models can be applied to modeling problems in time series econometrics. It is demonstrated how the actual modeling is carried out in a systematic fashion through a modeling cycle. This cycle can be repeated until an adequate model passing the available diagnostics is found. Alternatively, the cycle may be terminated by concluding that the family of STR models is not an appropriate one for empirical modeling of the economic relationship in question. The central role of hypothesis testing in STR modeling becomes clear from the text. First, it is important to single out the linear cases from nonlinear ones. But testing is also an essential part of model specification and evaluation. The validity of assumptions underlying the STR model are investigated by tests after estimating the parameters of the model as well as the question whether or not an estimated STR model is an improvement over previous models in the literature.

The STR can be used, as in the example of Section V, for modeling economic relationships between variables. Another role of the STR model is that it constitutes a feasible alternative to important null hypotheses concerning linear models. The null hypothesis of parameter constancy is one: in that case the alternative to constant parameters are continuously changing parameters. As Section V shows, this is a useful alternative, for example, in testing the constancy of the pattern of seasonal fluctuations. Furthermore, although this possibility has not been discussed in this chapter in any detail, the STR model provides a convenient framework for joint testing of weak exogeneity and a restricted form of invariance, that is, for testing superexogeneity. Finally, the additive STR model may be used for testing for Granger causality in the presence of STAR-type nonlinearity.

The STR model considered in this chapter is a single-equation model. In theory, the idea of smooth transition can be extended to systems of equations. This can be done in many ways of which the ones with practical significance still need to be sorted out. Anderson and Vahid (1995) recently searched for common nonlinearities between variables, using a vector STAR model. In general, however, there is as yet little empirical experience available of such systems and more is needed. In the meantime, additional applications of the single-equation STR model are also necessary to learn more about how the proposed modeling strategy works in practice and to find out ways of improving it and developing it further.

ACKNOWLEDGMENTS

Research supported in part by the Swedish Council for Research in Humanities and Social Sciences and the Copenhagen Business School. Material from the chapter, including the empirical example, has been discussed at courses at the Copenhagen Business School, Universidad Carlos III de Madrid, University of Helsinki, and GREQAM–Université d' Aix-Marseille II & III. Comments from participants as well as those from Rolf Tschernig are gratefully appreciated. I am particularly indebted to David Hendry for making his U.K. housing market data set available to me. The responsibility for any errors and shortcomings remains mine.

REFERENCES

Albæk, K. and H. Hansen (1995), Estimating Aggregate Labour Market Relations, mimeo., Department of Economics, University of Copenhagen.

Anderson, H. M. (1997), Transaction costs and Nonlinear Adjustment towards Equilibrium in the US Treasury Bill Market, *Oxford Bulletin of Economics and Statistics*, forthcoming.

Anderson, H. M. and F. Vahid (1995), Testing Multiple Equation Systems for Common Nonlinear Components, mimeo., Texas A & M University.

Andrews, D. W. K. and W. Ploberger (1994), Optimal Tests When a Nuisance Parameter Is Present Only under the Alternative, *Econometrica*, 62, 1383–1414.

Bacon, D. W. and D. G. Watts (1971), Estimating the Transition Between Two Intersecting Straight Lines, *Biometrika*, 58, 525–534.

Banerjee, A., J. Dolado, J. W. Galbraith, and D. F. Hendry (1993), *Co-integration, Error-Correction, and the Econometric Analysis of Non-stationary Data*, Oxford University Press, Oxford.

Bates, D. M. and D. G. Watts (1988), *Nonlinear Regression Analysis and Its Applications*, Wiley, New York.

Bell, D., J. Kay, and J. Malley (1996), A Nonparametric Approach to Nonlinear Causality Testing, *Economics Letters*, 51, 7–18.

Box, G. E. P. and G. M. Jenkins (1970), *Time Series Analysis, Forecasting and Control*, Holden-Day, San Francisco.

Chan, K. S. and H. Tong (1986), On Estimating Thresholds in Autoregressive Models, *Journal of Time Series Analysis*, 7, 178–190.

Chiarella, C., W. Semmler, and L. Koçkesen (1996), The Specification and Estimation of a Nonlinear Model of Real and Stock Market Interaction, mimeo., Department of Economics, New School for Social Research, New York.

Cook, P. and L. D. Broemeling (1996), Analyzing Threshold Autoregressions with a Bayesian Approach, in T. Fomby (ed.), *Advances in Econometrics*, Vol. 11, Part B: *Bayesian Methods Applied to Time Series Data*, JAI Press, Greenwich CT, 89–107.

Coutts, J. A., T. C. Mills, and J. Roberts (1995), Parameter Stability in the Market Model: Tests and Time Varying Parameter Estimation with U.K. Data, mimeo., Sheffield University Management School.

Davies, R. B. (1977), Hypothesis Testing When a Nuisance Parameter Is Present Only under the Alternative, *Biometrika*, 64, 247–254.

Davies, R. B. (1987), Hypothesis Testing When a Nuisance Parameter Is Present Only under the Alternative, *Biometrika*, 74, 33–44.

Eitrheim, Ø. and T. Teräsvirta (1996), Testing the Adequacy of Smooth Transition Autoregressive Models, *Journal of Econometrics*, 74, 59–75.

Engle, R. F. (1982), Autoregressive Conditional Heteroskedasticity with Estimates of the Variance of the U.K. Inflation, *Econometrica*, 50, 987–1008.

Ericsson, N. R. (1992), Cointegration, Exogeneity and Policy Analysis: An Overview, *Journal of Policy Modeling*, 14, 251–280.

Ertel, J. E. and E. B. Fowlkes (1976), Some Algorithms for Linear Spline and Piecewise Multiple Linear Regression, *Journal of the American Statistical Association*, 71, 640–648.

Escribano, A. and S. Mira (1995), Nonlinear Time Series Models: Consistency and Asymptotic Normality of NLS under New Conditions, Universidad Carlos III de Madrid, Statistics and Econometrics Series 14, Working Paper 95-42.

Fair, R. C. and D. M. Jaffee (1972), Methods of Estimation for Markets in Disequilibrium, *Econometrica*, 40, 497-514.

Farley, J. U., M. Hinich, and T. W. McGuire (1975), Some Comparisons of Tests for a Shift in the Slopes of Multivariate Linear Time Series Model, *Journal of Econometrics*, 3, 297-318.

Geweke, J. (1984), Inference and Causality in Economic Time Series Models, in Z. Griliches and M. D. Intriligator (eds.), *Handbook of Econometrics*, Vol. 2, North-Holland, Amsterdam, 1101-1144.

Geweke, J. and N. Terui (1993), Bayesian Threshold Autoregressive Models of Nonlinear Time Series, *Journal of Time Series Analysis*, 14, 441-454.

Goldfeld, S. M. and R. E. Quandt (1972), *Nonlinear Methods in Econometrics*, North-Holland, Amsterdam.

Goldfeld, S. M. and R. E. Quandt (1973), A Markov Model for Switching Regression, *Journal of Econometrics*, 1, 3-16.

Granger, C. W. J. (1969), Investigating Causal Relations by Econometric Models and Cross-Spectral Methods, *Econometrica*, 37, 424-438.

Granger, C. W. J. (1981), Some Properties of Time Series Data and Their Use in Econometric Model Specification, *Journal of Econometrics*, 16, 121-130.

Granger, C. W. J. and T. Teräsvirta (1993), *Modelling Nonlinear Economic Relationships*, Oxford University Press, Oxford.

Granger, C. W. J., T. Teräsvirta, and H. Anderson (1993), Modeling Nonlinearity over the Business Cycle, in J. H. Stock and M. W. Watson (eds.), *Business Cycles, Indicators, and Forecasting*, University of Chicago Press, Chicago, 311-325.

Hansen, B. E. (1996), Inference When a Nuisance Parameter Is Not Identified under the Null Hypothesis, *Econometrica*, 64, 413-430.

Hatanaka, M. (1996), *Time-Series-Based Econometrics. Unit Roots and Co-integration*, Oxford University Press, Oxford.

Heinesen, E. (1996), The Tax Wedge and the Household Demand for Services, mimeo., Institute of Local Government Studies, Copenhagen.

Hendry, D. F. (1984), Econometric Modelling of House Prices in the United Kingdom, in D. F. Hendry and K. F. Wallis (eds.), *Econometrics and Quantitative Economics*, Blackwell, Oxford, 211-252.

Hendry, D. F. (1995), *Dynamic Econometrics*, Oxford University Press, Oxford.

Jansen, E. S. and T. Teräsvirta (1996), Testing Parameter Constancy and Super Exogeneity in Econometric Equations, *Oxford Bulletin of Economics and Statistics*, 58, 735-763.

Jarque, C. M. and A. K. Bera (1980), Efficient Tests for Normality, Homoscedasticity, and Serial Independence of Regression Residuals, *Economics Letters*, 6, 255-259.

Johansen, S. (1995), *Likelihood-Based Inference in Cointegrated Vector Autoregressive Models*, Oxford University Press, Oxford.

Judge, G. G., W. E. Griffiths, R. C. Hill, H. Lütkepohl, and T.-C. Lee (1985), *The Theory and Practice of Econometrics*, 2nd ed., Wiley, New York.

King, M. L. and T. S. Shively (1993), Locally Optimal Testing When a Nuisance Parameter Is Present Only under the Alternative, *Review of Economics and Statistics*, 75, 1-7.

Lee, T.-H., H. White and C. W. J. Granger (1993), Testing for Neglected Nonlinearity in Time Series Models: A Comparison of Neural Network Methods and Alternative Tests, *Journal of Econometrics*, 56, 269–290.

Lin, C.-F. and T. Teräsvirta (1994), Testing the Constancy of Regression Parameters Against Continuous Structural Change, *Journal of Econometrics*, 62, 211–228.

Lindgren, G. (1978), Markov Regime Models for Mixed Distributions and Switching Regressions, *Scandinavian Journal of Statistics*, 5, 81–91.

Lomnicki, Z. A. (1961), Test for Departure from Normality in the Case of Linear Stochastic Processes, *Metrika*, 4, 37–62.

Lütkepohl, H., T. Teräsvirta, and J. Wolters (1995), Investigating Stability and Linearity of a German M1 Money Demand Function, Stockholm School of Economics, Working Paper Series in Economics and Finance No. 64.

Luukkonen, R., P. Saikkonen, and T. Teräsvirta (1988a), Testing Linearity against Smooth Transition Autoregression, *Biometrika*, 75, 491–499.

Luukkonen, R., P. Saikkonen, and T. Teräsvirta (1988b), Testing Linearity in Univariate Time Series, *Scandinavian Journal of Statistics*, 15, 161–175.

Maddala, D. S. (1977), *Econometrics*, McGraw-Hill, New York.

Maddala, D. S. (1986), Disequilibrium, Self-Selection, and Switching Models, in Z. Griliches and M. D. Intriligator (eds.), *Handbook of Econometrics*, Vol. 3, North-Holland, Amsterdam, 1633–1688.

McLeod, A. I. and W. K. Li (1983), Diagnostic Checking ARMA Time Series Models Using Squared Residuals, *Journal of Time Series Analysis*, 4, 269–273.

Mizon, G. E. and J.-F. Richard (1986), The Encompassing Principle and Its Application to Non-nested Hypothesis Tests, *Econometrica*, 54, 657–678.

Péguin-Feissolle, A. (1994), Bayesian Estimation and Forecasting in Nonlinear Models. Application to an LSTAR model. *Economics Letters*, 46, 187–194.

Petruccelli, J. D. (1990), A Comparison of Tests for SETAR-Type Non-linearity in Time Series, *Journal of Forecasting*, 9, 25–36.

Pfann, G. A., P. C. Schotman, and R. Tschernig (1996), Nonlinear Interest Rate Dynamics and Implications for the Term Structure, *Journal of Econometrics*, 74, 149–176.

Pole, A. M. and A. F. M. Smith (1985), A Bayesian Analysis of Some Threshold Switching Models, *Journal of Econometrics*, 29, 97–119.

Quandt, R. E. (1960), Tests of the Hypothesis That a Linear Regression System Obeys Two Separate Regimes, *Journal of the American Statistical Association*, 55, 324–330.

Quandt, R. E. (1984), Computational Problems and Methods, in Z. Griliches and M. D. Intriligator (eds.), *Handbook of Econometrics*, Vol. 1, North-Holland, Amsterdam, 699–764.

Ramsey, J. B (1969), Test for Specification Errors in Classical Linear Least-Squares Regression Analysis, *Journal of the Royal Statistical Society B*, 31, 350–371.

Richard, J.-F. and W. Zhang (1996), Econometric Modeling of UK House Prices Using Accelerated Importance Sampling, *Oxford Bulletin of Economics and Statistics*, 58, 601–613.

Saikkonen, P. and R. Luukkonen (1988), Lagrange Multiplier Tests for Testing Non-linearities in Time Series Models, *Scandinavian Journal of Statistics*, 15, 55–68.

Seber, G. A. F. and C. J. Wild (1989), *Nonlinear Regression*, Wiley, New York.

Semmler, W. and L. Koçkesen (1995), Testing for Nonlinear Financial-Real Interaction Using a Smooth Transition Regression Model, mimeo., Department of Economics, New York School for Social Research, New York.

Shively, T. S. (1988), An Analysis of Tests for Regression Coefficient Stability, *Journal of Econometrics*, 39, 367–386.

Skalin, J. and T. Teräsvirta (1996), Another Look at Swedish Business Cycles, 1861–1988, Stockholm School of Economics, Working Paper Series in Economics and Finance, No. 130.

Teräsvirta, T. (1994), Specification, Estimation, and Evaluation of Smooth Transition Autoregressive Models, *Journal of the American Statistical Association*, 89, 208–218.

Teräsvirta, T. and H. M. Anderson (1992), Characterizing Nonlinearities in Business Cycles Using Smooth Transition Autoregressive Models, *Journal of Applied Econometrics*, 7, S119–S136.

Tjøstheim, D. (1994), Non-linear Time Series: A Selective Review, *Scandinavian Journal of Statistics*, 21, 97–130.

Tong, H. (1990), *Non-linear Time Series. A Dynamical System Approach*, Oxford University Press, Oxford.

Tsurumi, H. (1982), A Bayesian and Maximum Likelihood Analysis of a Gradual Switching Regression in a Simultaneous-equation Framework, *Journal of Econometrics*, 19, 165–182.

Wolters, J., T. Teräsvirta, and H. Lütkepohl (1997), Modeling the Demand for M3 in the Unified Germany, *Review of Economics and Statistics*, forthcoming.

Wooldridge, J. M. (1994), Estimation and Inference for Dependent Processes, in R. F. Engle and D. L. McFadden (eds.), *Handbook of Econometrics*, Vol. 4, North-Holland, Amsterdam, 2649–2738.

16

Modeling Seasonality in Economic Time Series

Philip Hans Franses
Erasmus University Rotterdam, Rotterdam, The Netherlands

I. INTRODUCTION

This chapter surveys issues concerning seasonality in economic time series. An elaborate discussion on a formal definition of seasonality is given in, e.g., Hylleberg (1986, 1992). Here I loosely refer to seasonality as the variation in time-series data that displays a certain regularity corresponding with the measurement interval. For example, for quarterly data one may consider the annually recurring positive or negative peaks in certain quarters as seasonal fluctuations. Furthermore, the observation that stock returns on Mondays seem more volatile than those on other weekdays concerns seasonality too, that is, seasonality in variance.

In many cases, seasonality in economic time series is due to weather or institutional factors. An example of the latter is that school holidays are fixed by local governments, and hence one may expect tourism spending to be high in the corresponding season. Another example is that the deadline for companies to publish their annual reports can be dictated by law. Right after that deadline, one may expect more volatility in stock markets in case the news differs from the expectations, and one may also observe changes in key macroeconomic figures such as consumer confidence indicators. Hence, part of the seasonal variation may be roughly constant over time, since for example it is unlikely that Christmas will move to other months, but another part of seasonality may change because of changes in institutional factors. Finally, seasonal patterns can also change because economic agents start to behave in a different way. In fact, if Mondays would always display high volatility, one would be able to make money through derivatives. Hence, because of the so-called weekend effect, one may expect a high volatility on Mondays, but this feature is unlikely to be constant over time, see, e.g., Franses and Paap (1995). Another example is that seasonal labor supply may make the unemployment rate to display more seasonality in the expansion stage.

In this chapter I will focus on statistical models that can describe and forecast economic time series with seasonal variation which changes over time. The running examples in this chapter to be used for illustration are taken from macroeconomics, tourism, marketing, and finance. A dominant approach in macroeconomics is to seasonally adjust the data prior to analysis; that is, one assumes that seasonality is not an interesting data feature and should be removed. In most fields of economics, however, seasonality is considered important since it can convey information on, for example, the behavior of economic agents (Ghysels 1994a). In this chapter I concur with this view and I will therefore not consider seasonal adjustment, and confine myself to models that explicitly incorporate descriptions of seasonality. I refer the reader interested in seasonal adjustment to the surveys in Hylleberg (1992), Bell and Hillmer (1984), and Maravall (1995) inter alia.

The outline of this chapter is as follows. Section II gives some summary statistics for four sample series. These statistics mainly show that seasonality does not appear constant over time, and that any changes do not occur quickly. Hence, seasonality seems to change slowly over time. Section III reviews the two approaches which are nowadays commonly used in many applications, i.e., univariate and multivariate models that incorporate seasonal unit roots and seasonal parameter variation. Due to space limitations, I only highlight some of the key features of these models and refer the interested reader to the surveys in Hylleberg (1994), Franses (1996a, b), and to the specific studies mentioned here. In Section IV, I discuss further topics of research. Section V concludes with some remarks.

II. SOME EXAMPLE SERIES

In this section I discuss some time series features of four sample series. The first time series is real consumption nondurables in the United Kingdom, which is observed quarterly for 1955.1–1988.4. The source of these data is described in Osborn (1990). The graph of this series is displayed in Figure 1. It is clear that the data display an upward-moving trend, which seems sometimes hampered by shocks, especially those around 1974 and 1979. Furthermore, seasonal variation seems a dominant source of variation. As is usual for economic time series, these data are transformed by taking natural logarithms.

The second quarterly time series is depicted in Figure 2 and it concerns the unemployment rate in Federal Germany for 1962.1 to 1991.4. The source of these data is the OECD Main Economic Indicators. The graph in Figure 2 shows that unemployment seems to increase rapidly around the recession periods 1967, 1974, and 1980–1982. The decline in unemployment occurs more slowly, and hence this series

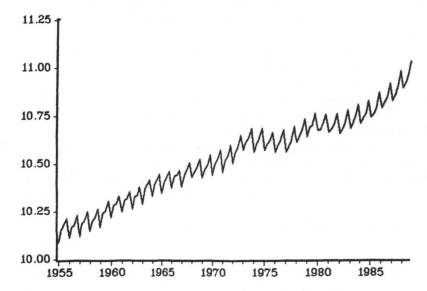

Figure 1   Nondurable consumption in the United Kingdom.

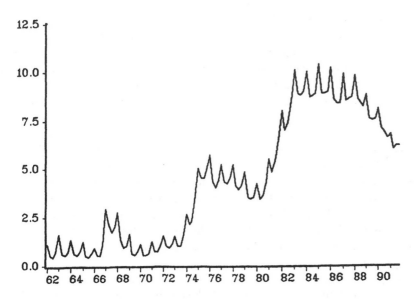

Figure 2   Unemployment rate in Germany.

displays nonlinear features. Again, seasonal fluctuations seem important sources of variation. Since this unemployment series is a percentage, the log transformation is not applied.

A sample series for the tourism sector is taken from Gonzalez and Moral (1995). It concerns the monthly real tourism and travel receipts for Spain (in billions pesetas), for the sample 1979.01–1993.12. From Figure 3 one can observe that (as expected) seasonal variation is quite pronounced, where the peaks correspond with August and the dips with December, approximately. There is an upward trend until 1989 and a slightly different trend after 1990.

Finally, the fourth sample series concerns marketing data; i.e., these are expenditures on radio advertising in The Netherlands, measured four-weekly from 1984.01 to 1994.13. The data source is the Bureau for Budget Control, Amsterdam. Its graph in Figure 4 shows that the second half of the sample displays strong seasonal variation.

For convenience, I denote all four time series as y_t. The index t runs from 1 to n. The four series are observed S times per year, where S equals 4 for consumption nondurables and unemployment, 12 for tourism in Spain, and 13 for the radio advertising data. The variables concern N years of data, where the index T is used to indicate years (with $T = 1, 2, \ldots, N$). Hence, n equals SN. The seasonal index is denoted by s, with $s = 1, 2, \ldots, S$. Finally, I denote the conventional seasonal dummy variables by $D_{s,t}$, which take a value 1 in season s for which $t = S(T-1)+s$ for $T = 1, 2, \ldots, N$, and zero otherwise.

A useful first step in analyzing the properties of seasonal time series is to calculate the autocorrelation function (ACF) of (transformations of) the data. These transformations concern the application of certain differencing filters to remove any nonstationary features. For example, when a time series has a stochastic trend, its ACF when estimated along standard lines cannot be interpreted since the variance of such a process depends on time. A stochastic trend can be removed by taking first differences Δ_1, where Δ_1 is defined by $\Delta_k y_t \equiv (1 - B^k)y_t \equiv y_t - y_{t-k}$ for $k = 1, 2, \ldots$. An analogous version of Δ_1 for seasonal time series is $\Delta_S = 1 - B^S$. In that case one says that y_t has seasonal stochastic trends. When seasonality is approximately deterministic, one may consider the residuals from the regression of $\Delta_1 y_t$ on S seasonal dummies, since in that regression one removes the seasonal constants from $\Delta_1 y_t$. Finally, it may be that seasonal and nonseasonal stochastic trends somehow interact, and hence that one considers the $\Delta_1 \Delta_S$ filter, as is advocated by Box and Jenkins (1970). Table 1 reports on the ACFs for the various transformed time series. To save space, only the autocorrelations at lags, 1, 2, $S - 1$, S, $S + 1$, and $2S$ are given.

The results in Table 1 can be said to be typical for economic time series with seasonality. For all four sample series, the ACF of y_t dies out only very slowly. The ACFs of the Δ_1 transformed data all show peaks at lags S and $2S$. When the Δ_1

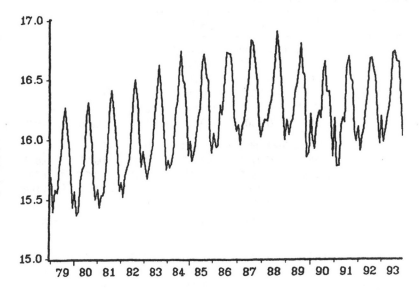

Figure 3 Tourism and travel receipts in Spain.

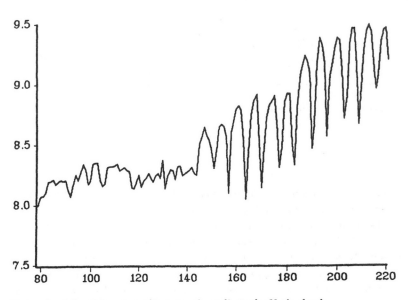

Figure 4 Advertising expenditures on the radio in the Netherlands.

Table I   Autocorrelations of Sample Series[a]

Variable	Lag	1	2	$S-1$	S	$S+1$	$2S$
		Consumption nondurables ($n=136$, $S=4$)					
y_t		0.928*	0.900*	0.876*	0.891*	0.823*	0.788*
$\Delta_1 y_t$		−0.463*	−0.014	−0.481*	0.947*	−0.438*	0.910*
$\Delta_1 y_t - \sum_{s=1}^{S} \hat{\mu}_s D_{s,t}$		−0.074	−0.359*	−0.034	0.554*	0.023	0.491*
$\Delta_S y_t$		0.779*	0.625*	0.449*	0.248*	0.238*	−0.014
$\Delta_1 \Delta_S y_t$		−0.164	0.050	0.048	−0.444*	0.236*	0.023
		Unemployment rate ($n=120$, $S=4$)					
y_t		0.969*	0.946*	0.934*	0.934*	0.891*	0.831*
$\Delta_1 y_t$		−0.195*	−0.234*	−0.224*	0.874*	−0.254*	0.798*
$\Delta_1 y_t - \sum_{s=1}^{S} \hat{\mu}_s D_{s,t}$		0.220*	0.290*	0.078	0.457*	−0.077	0.188*
$\Delta_S y_t$		0.919*	0.772*	0.593*	0.401*	0.250*	−0.002
$\Delta_1 \Delta_S y_t$		0.422*	0.189*	0.072	−0.239*	−0.239*	−0.089
		Tourism and travel receipts ($n=180$, $S=12$)					
y_t		0.833*	0.590*	0.756*	0.860*	0.728*	0.721*
$\Delta_1 y_t$		0.219*	0.091	0.330*	0.733*	0.276*	0.658*
$\Delta_1 y_t - \sum_{s=1}^{S} \hat{\mu}_s D_{s,t}$		−0.433*	−0.164*	0.169*	0.056	−0.134*	−0.042*
$\Delta_S y_t$		0.304*	0.278*	0.234*	−0.106	0.141	0.067
$\Delta_1 \Delta_S y_t$		−0.481*	−0.051	0.307*	−0.428*	0.135*	0.002
		Advertising expenditures, radio ($n=143$, $S=13$)					
y_t		0.876*	0.719*	0.662*	0.745*	0.649*	0.483*
$\Delta_1 y_t$		0.131	−0.332*	0.155	0.753*	0.187*	0.588*
$\Delta_1 y_t - \sum_{s=1}^{S} \hat{\mu}_s D_{s,t}$		0.081	−0.216*	0.121	0.506*	0.226*	0.298*
$\Delta_S y_t$		0.631*	0.465*	0.093	−0.037	0.059	−0.162
$\Delta_1 \Delta_S y_t$		−0.277*	0.026	0.180*	−0.311*	0.262*	0.004

*Significant at the 5% level.
[a] All variables are in logs, except the unemployment rate.

data are regressed on seasonal dummies ($\Delta_1 y_t - \sum_{s=1}^{S} \hat{\mu}_s D_{s,t}$), some of the large values at S and $2S$ reduce to small and insignificant values. This shows that one should always account for seasonal constants, since neglecting these may lead to spuriously large ACF values at seasonal lags. The ACFs of the $\Delta_S y_t$ series seem to die out reasonably quickly, and one may want to estimate low-order autoregressive (AR) models for these transformed series. In the last row of each panel, one can observe (and this is common to many economic series) that the ACF of $\Delta_1 \Delta_S y_t$ has significant values at lags 1, $S-1$, S, and $S+1$.

The latter feature for the $\Delta_1\Delta_S$ transformed series is also recognized in Box and Jenkins (1970), who propose to analyze the so-called airline model

$$\Delta_1\Delta_S y_t = (1 - \theta_1 B)(1 - \theta_S B^S)\varepsilon_t \tag{1}$$

where ε_t is a standard white noise series. Clearly, the theoretical ACF of (1) corresponds with nonzero values for the ACF at lags 1, $S - 1$, S, and $S + 1$. In the case where $\theta_1 = \theta_S = 1$, which amounts to overdifferencing since then $(1 - B)(1 - B^S)$ cancels from both sides of (1), the corresponding ACF values are -0.5, 0.25, -0.5, and 0.25, respectively. Most textbooks in time series advocate the use of (1). Furthermore, for many different example series model (1) fits well. Its key advantage is that it only contains two moving-average (MA) parameters to be estimated.

As an example for the consumption nondurables data it appears that the following model passes the usual LM-type diagnostic checks for residual autocorrelation at lag 1 and at lags 1-to-S:

$$\Delta_1\Delta_4 y_t = \underset{(0.0010)}{0.0002} + \hat{\varepsilon}_t - \underset{(0.069)}{0.634\,\hat{\varepsilon}_{t-4}} \tag{2}$$

where the parameters are estimated using Micro TSP routines, and where standard errors are reported in parentheses. Notice that (2) effectively contains only one parameter, which (together with the double-differencing filter) appears sufficient to remove the autocorrelation in y_t.

As another example, for the four-weekly advertising data, the following model is found to pass the diagnostic checks for residual autocorrelation:

$$\Delta_1\Delta_{13} y_t = \underset{(0.0277)}{0.0012} + \hat{\varepsilon}_t - \underset{(0.077)}{0.394\,\hat{\varepsilon}_{t-1}} - \underset{(0.072)}{0.388\,\hat{\varepsilon}_{t-2}}$$
$$- \underset{(0.069)}{0.496\,\hat{\varepsilon}_{t-13}} - \underset{(0.076)}{0.304\,\hat{\varepsilon}_{t-14}} \tag{3}$$

Due to the disaggregation level of these data, one should expect the need for several parameters to whiten the errors. For the MA part of the model, the solutions to its characteristic polynomial are six pairs of complex roots with absolute values 0.981, 0.963, 0.943, 0.930, 0.928, and 0.921, and two real roots 0.993 and 0.612. Hence, given that 13 of the 14 solutions are close to the unit circle, it may be that the Δ_{13} filter on the left-hand side of (3) is redundant. Given the fact that MA parameters are typically estimated away from the unity boundary, one may now be tempted to conclude that the Δ_1 filter for advertising should be sufficient. In fact, if so, with seasonality being approximately constant, Bell (1987) shows that the solutions to the MA polynomial for (1) approach the unit circle.

It is worthwhile to note that the solutions to $1 - z^S = 0$, or equivalently to $\exp(Si\phi) = 1$ (where $i^2 \equiv -1$) are $\{1, \cos(2\pi k/S) + i\sin(2\pi k/S)\}$ for $k =$

$1, 2, \ldots$, yielding S different solutions which all lie on the unit circle. In other words, the $\Delta_1 \Delta_S$ filter in (1) to (3) assumes $S+1$ unit roots, i.e., $S+1$ independent sources of stochastic and nonstationary variation. Even though the MA component in models as (1) seems to "repair" in some sense the possibly overestimated number of unit roots, one may question the notion that economic time series are governed by such a large number of trends.

Consider for example the so-called seasonal random walk process $\Delta_S y_t = \varepsilon_t$ and also consider the S time series $Y_{s,T}$, which are the annually observed data on y_t in season $s = 1, 2, \ldots, S$. Given the seasonal random walk process, it is clear that the individual $Y_{s,T}$ series are annually observed random walks which are independent. The observations in the various seasons are not tied together somehow, and they wander through time without any restriction. Strictly speaking, this means that "summer" can become "winter." Of course, the MA component in (1) will prevent such changes to occur rapidly, but in principle it is possible. Given the graphs in Figures 1 through 4, it seems that for economic data the seasons are somehow

Table 2 Changing Seasonality

Sample	Lags	Min $\hat{\delta}_s$	Max $\hat{\delta}_s$	SD
	Consumption nondurables ($S = 4$)			
1957.3–1973.4	2, 4, 5, 8, 9	$-0.054\ (\hat{\delta}_1)$	$0.041\ (\hat{\delta}_2)$	0.042
1974.1–1988.4	2, 4, 5, 8, 9	$-0.078\ (\hat{\delta}_1)$	$0.039\ (\hat{\delta}_4)$	0.055
	Unemployment rate ($S = 4$)			
1963.3–1976.4	1, 4, 5	$-0.679\ (\hat{\delta}_2)$	$0.590\ (\hat{\delta}_1)$	0.548
1977.1–1991.4	1, 4, 5	$-0.451\ (\hat{\delta}_2)$	$0.288\ (\hat{\delta}_1)$	0.322
	Tourism and travel receipts ($S = 12$)			
1979.07–1986.12	1, 2, 3, 4, 5	$-0.459\ (\hat{\delta}_{12})$	$0.569\ (\hat{\delta}_8)$	0.375
1987.01–1993.12	1, 2, 3, 4, 5	$-0.438\ (\hat{\delta}_{12})$	$0.434\ (\hat{\delta}_7)$	0.299
	Advertising expenditures, radio ($S = 13$)			
1980.02–1988.13	1, 2, 3, 4, 5, 13	$-0.114\ (\hat{\delta}_1)$	$0.086\ (\hat{\delta}_6)$	0.059
1989.01–1994.13	1, 2, 3, 4, 5, 13	$-0.355\ (\hat{\delta}_8)$	$0.297\ (\hat{\delta}_2)$	0.179

The auxiliary regression is

$$\Delta_1 y_t = \delta_1 D_{1,t} + \delta_2 D_{2,t} + \cdots + \delta_S D_{S,t} + \alpha_1 \Delta_1 y_{t-1} + \cdots + \alpha_i \Delta_1 y_{t-i} + \varepsilon_t$$

where the number of lags is based on LM tests for residual autocorrelation at lags 1 and 1-to-S. SD is the standard deviation of the estimated $\hat{\delta}_s$. Note that a formal test for the equality of the $\hat{\delta}_s$ across the two subsamples is only valid when this regression model is valid. For example, unemployment may not be linear, and perhaps filters as Δ_S are needed instead of Δ_1, see Sections III and IV.

tied together. In other words, although the seasonal fluctuations seem to change over time, they do so not that quickly.

To obtain a tentative impression of how seasonality can change over time, I consider the estimates of δ_s, $s = 1, 2, \ldots, S$, from the auxiliary regression

$$
\Delta_1 y_t = \delta_1 D_{1,t} + \delta_2 D_{2,t} + \cdots + \delta_S D_{S,t} + \alpha_1 \Delta_1 y_{t-1} + \cdots \\
+ \alpha_i \Delta_1 y_{t-i} + \varepsilon_t
\tag{4}
$$

where the values of i are set such that there is no autocorrelation in $\hat{\varepsilon}_t$. In Table 2, I report minimum and maximum values of $\hat{\delta}_s$ and the standard deviation of the S $\hat{\delta}_s$ parameters for two subsamples which are roughly similarly large. For consumption nondurables one can observe that the maximum value for $\hat{\delta}_s$ is obtained for the second quarter in the first sample, while it corresponds with the fourth quarter in the second sample. For unemployment, these extreme $\hat{\delta}_s$ values correspond to the same seasons, although the standard deviation decreases with about 60%. For the Spanish tourism data, August loses its importance to July toward the end of the sample. For illustrative purposes, I depict the estimates for the twelve δ_s parameters in the two subsamples in Figure 5. One can observe that tourism in Spain seems to shift slightly from the summer months more toward the winter months January to March, suggesting structural shifts in the behavior of tourists.

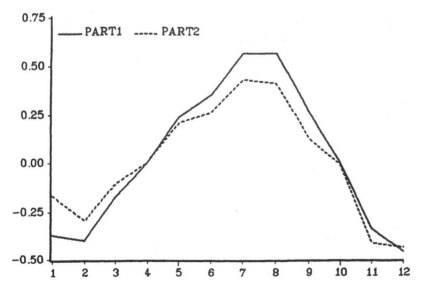

Figure 5 Estimates of monthly dummy parameters for 1979.01–1986.12 (Part 1) and for 1987.01–1993.12 (Part 2).

Finally, for radio advertising expenditures in Table 2 it seems that indeed "winter" becomes "summer"; i.e., the dip in January moves to about July/August, while the peak changes from June/July to about February. Furthermore, as could be expected from Figure 4, the variance of seasonality increases from 0.06 to 0.18, i.e., with 300%.

Notwithstanding the results for the advertising series, it seems that the seasonal fluctuations in economic time series change over time only at a slow pace. This implies that the airline model in (1) may assume too many changes in seasonality since it assumes a large number of unit roots. In the next section I will discuss two important approaches that may overcome the conceptual problems with (1), and which can yield descriptions of economic data that allow for slowly changing seasonality.

III.   MODELING STRATEGIES

Broadly speaking, there are two general strategies to model seasonal economic time series which display slowly changing seasonality. The first is to impose (depending on formal test results) one or a few seasonal unit roots on the AR part of a model, and the second is to allow the AR parameters to vary across the seasons. For brevity, I call the first approach "seasonality in dynamics," and discuss this in Section III.A and call the second approach "seasonality in parameters" (Section III.B). Recent studies in Boswijk, Franses, and Haldrup (1997) and Ghysels, Hall, and Lee (1996) seem to suggest that combinations of the two approaches do not obtain much empirical support. This section only highlights the main issues, and for further more detailed overviews the reader is referred to, for example, Franses (1996a, b).

A.   Seasonality in Dynamics

Consider the AR(p) model for a time series with seasonal frequency S:

$$y_t = \phi_1 y_{t-1} + \phi_2 y_{t-2} + \cdots + \phi_p y_{t-p} + \varepsilon_t \tag{5}$$

where typically p exceeds S. First, I focus on univariate y_t processes, while the second part of this section allows y_t to be an $m \times 1$ vector process.

1.   Univariate Time Series

When the AR polynomial in (5) denoted as $\phi_p(B)$ can be decomposed as

$$\phi_p(B) = (1 - B)(1 - B^S)\phi_{p-(S+1)} \tag{6}$$

as in (1), the y_t process is said to be both integrated (because of $1 - B$) and seasonally integrated (because of $1 - B^S$). In this case, y_t contains $S + 1$ unit roots.

The polynomial $1 - B^S$ can be decomposed as $(1 - B)(1 + B + \cdots + B^{S-1})$, and y_t is an I(2) process since its AR model contains the component $(1 - B)^2$ and it has $S - 1$ so-called seasonal unit roots. The latter unit roots correspond with the stochastic trends at the seasonal frequency, and hence can be relevant in case the seasonal pattern in y_t changes over time. Notice that these changes are permanent. For example, consider the simple process

$$y_t = -y_{t-1} + \varepsilon_t \tag{7}$$

which in case $S = 4$ is called a process with a seasonal unit root at the biannual frequency, see Hylleberg et al. (1990). Substituting lagged y_t in (7) results in

$$y_t = \varepsilon_t - \varepsilon_{t-1} + \varepsilon_{t-2} - \varepsilon_{t-3} + \cdots \tag{8}$$

and hence the variance of y_t at t equals $t\sigma^2$, which is the same as that for a standard random walk. Shocks ε_t in (7) cause permanent changes to the pattern of y_t because their effect does not die out as seen from (8), and because of the seasonal unit root, the seasonal pattern in y_t changes permanently.

Given the conceptual impact of the assumption of seasonal unit roots, and given that unit roots lead to wider forecasting intervals and to more involved modeling strategies for multivariate time series in the next step, it is important to test for the number of seasonal and nonseasonal unit roots in a univariate time series. The two commonly applied test procedures are proposed in Osborn et al. (1988) and in Hylleberg et al. (1990) [HEGY]. The OCSB method investigates whether the $\phi_p(B)$ polynomial contains the components $1 - B$, $1 - B^S$, both, or none of these. The test regression is

$$\eta(B)\Delta_1\Delta_S y_t = \mu_t + \gamma_1\Delta_S y_{t-1} + \gamma_2\Delta_1 y_{t-S} + \varepsilon_t \tag{9}$$

where $\eta(B)$ is an AR polynomial, and

$$\mu_t = \alpha_0 + \sum_{s=1}^{S-1} \alpha_s D_{s,t} + \beta_0 t + \sum_{s=1}^{S-1} \beta_s D_{s,t} t \tag{10}$$

where t reflects a deterministic trend. OCSB consider the case with $S = 4$ and where β_0 to β_{S-1} are set equal to zero. Critical values for other cases are given in Franses and Hobijn (1997). When $\gamma_1 = \gamma_2 = 0$, the $\Delta_1\Delta_S$ filter is appropriate, when $\gamma_1 = 0$ and γ_2 is not, the Δ_1 filter is needed, when $\gamma_2 = 0$ and γ_1 is not the Δ_S filter, and no filter is required when both γ_1 and γ_2 are unequal to zero. When the β's in (10) are not set equal to zero, and when these also appear useful in a model for Δ_1 or Δ_S filtered y_t, this y_t series in fact shows increasing seasonal variation. Such models may be useful when one does not want to use the logarithmic data transformation (e.g., Bowerman, Koehler, and Pack 1990).

As an example, an application of this OCSB method to the unemployment data, where $S = 4$, $\eta(B)$ contains one lag, and all β's in μ_t are set to zero, yields the t-values $t(\hat{\gamma}_1) = 0.006$ $[-2.11]$ and $t(\hat{\gamma}_2) = -4.384$ $[-3.75]$, where 5% critical values are given in brackets. Other choices for μ_t do not change the conclusion that y_t appears to contain the nonseasonal unit root 1 since $\hat{\gamma}_1$ is insignificant and that unemployment does not need the double-differencing filter.

An application to radio advertising expenditures, where now a trend is included in (9) and $\eta(B) = 1$, results in $t(\hat{\gamma}_1) = -4.913 [-2.78]$ and $t(\hat{\gamma}_2) = -5.506 [-5.81]$. Hence, for this four-weekly series it is found that the Δ_S filter may be needed since $t(\hat{\gamma}_2)$ does not appear significant. In sum, for both series the $\Delta_1 \Delta_S$ filter assumes to many unit roots. It should be mentioned that the empirical work in Osborn (1990) convincingly shows that there are only very few quarterly time series for which the $\Delta_1 \Delta_4$ filter is required to obtain stationarity.

Since the Δ_S filter assumes $S - 1$ seasonal unit roots, it is now relevant to test how many of these roots are present for a certain time series y_t. A commonly applied method for this purpose is (a variant of) the HEGY approach. For $S = 4$ this method concerns the auxiliary regression

$$\eta(B)\Delta_4 y_t = \mu_t + \pi_1 M(B)y_{t-1} + \pi_2 A(B)y_{t-1} \\ + (\pi_3 B + \pi_4)(1 - B^2)y_{t-1} + \varepsilon_t \tag{11}$$

where $M(B) = 1 + B + B^2 + B^3$ and $A(B) = -(1 - B)(1 + B^2)$. HEGY give asymptotics and critical values for $S = 4$ and μ_t with $\beta_s = 0(s = 1, \ldots, S - 1)$, Smith and Taylor (1995) allow for seasonally varying trends in case $S = 4$, and Franses and Hobijn (1997) consider additional cases. The key focus is on the π_1 to π_4 parameters in (11) since $\pi_1 = 0$ implies $1 - B$, $\pi_2 = 0$ implies $1 + B$, and $\pi_3 = \pi_4 = 0$ implies $1 + B^2$, where the latter two seasonal unit roots $\pm i$ correspond with the annual frequency. Typically, one considers t-tests for π_1 and π_2 and a joint F-test for π_3 and π_4. One may also use the joint F-tests for π_1 to π_4 or π_2 to π_4, where the latter concerns all seasonal unit roots (Ghysels, Lee, and Noh 1994).

An application of the HEGY method to the German unemployment data (where $\eta(B)$ in (11) includes two lags, and μ_t does not include any trends) results in $t(\hat{\pi}_1) = 1.292 [-2.83]$, $t(\hat{\pi}_2) = -1.914 [-2.83]$, and $F(\hat{\pi}_3, \hat{\pi}_4) = 8.380 [6.62]$, where again 5% critical values are given in brackets. Hence, to remove the nonstationarities in this series, one needs the $(1 - B)(1 + B) = 1 - B^2$ filter. In other words, one may describe the changing seasonal variation in unemployment using a seasonal unit root at the biannual frequency.

For the Spanish tourism data, the application of an extension of the HEGY regression (11), with β_0 in (10) unequal to zero and $\eta(B) = 1$, results in $t(\hat{\pi}_1) = -1.161 [-3.29]$, $t(\hat{\pi}_2) = -4.799 [-2.76]$, and $F(\hat{\pi}_1, ., \hat{\pi}_{12}) = 18.835 [4.46]$. Hence, for this monthly series only the Δ_1 filter is required. This implies that the results in Table 1 and Figure 5 can be interpreted safely since the auxiliary regression model cannot be rejected by the data.

An application of the HEGY method to many quarterly U.K. time series (also including consumption nondurables) in Osborn (1990) yields that often only one or two seasonal unit roots are present. This finding, which appears typical across empirical applications of the HEGY method (see, for example, Hylleberg, Jørgensen, and Sørensen 1993, Ghysels, Lee, and Siklos 1993, and Lee and Siklos 1993), implies that the double-differencing filter $\Delta_1 \Delta_S$ as assumed in (1) imposes too many unit roots on the AR polynomial in (5). However, as shown in Ghysels, Lee, and Noh (1994), in case (1) is the true data-generating process in simulation exercises, the empirical size of the HEGY test largely exceeds the nominal size, implying that one is inclined to find not enough unit roots. Furthermore, it can be shown that inappropriate lag augmentation in regressions as (11) can have a large impact on the empirical outcomes. Finally, if p in (5) is smaller than S, the HEGY and OCSB approaches may be difficult to apply. Additional research is needed to fully understand the theoretical and empirical properties of both test methods. For example, Ghysels, Lee, and Siklos (1993) examine the effect of lag selection of HEGY test results.

2. Multivariate Time Series

In case the y_t process in (5) is an $m \times 1$ vector time series, and hence when the ϕ_1 to ϕ_p are $m \times m$ matrices, and the components, say, $y_{1,t}$ to $y_{m,t}$ contain one or more seasonal or nonseasonal unit roots, one may wish to decrease the number of parameters in a vector AR (VAR) by investigating common properties across time series. For example, if two quarterly series $y_{1,t}$ and $y_{2,t}$ have a common seasonal unit root -1, i.e. when they both have a $1 + B$ component in their univariate AR models, these two series are seasonally cointegrated at the biannual frequency if a linear combination $G_1(B)y_{1,t} - \alpha G_2(B)y_{2,t}$ (where the $G_i(B)$ do not contain $1 + B$) does not have this seasonal unit root (Engle et al. 1993). When both variables require the Δ_4 filter, the polynomials $G_1(B)$ and $G_2(B)$ equal $1 - B + B^2 - B^3$. When for example $y_{1,t}$ only requires the $1 - B^2$ filter, $G_1(B)$ equals $1 - B$. Hence, the transformations $G_i(B)$ ($i = 1, 2$) can differ across time series, and they usually depend on the HEGY test results for univariate series.

In principle, the test method for seasonal cointegration in EGHL concerns regressions of $G_1(B)y_{1,t}$ on μ_t and $G_2(B)y_{2,t}$, and an analysis of the seasonal and nonseasonal unit root properties of the estimated residuals. Lee (1992) extends the Johansen method to test for cointegration to the seasonal case ($S = 4$) by investigating the ranks of the $m \times m$ matrices π_1 to π_4 in a VAR model similar to (11). An application to several data sets for various countries is given in Kunst (1993). In Franses and Kunst (1995) the impact of seasonal dummies on the properties of seasonally cointegrated variables is discussed. It can be shown that such seasonal dummies can imply the data to have seasonally varying trends. Since this feature may not be present in many log-transformed time series, Franses and Kunst (1995) propose to check the cointegration properties at each of the frequencies separately. Finally, a Granger representation theorem for seasonal cointegration is given in Johansen and Schaumburg (1996).

B. Seasonality in Parameters

The second currently dominant approach in modeling and forecasting economic time series with seasonality concerns so-called periodic models. Since the studies in Osborn (1988) and Osborn and Smith (1989), these models have become increasingly popular in economics. This section surveys some key aspects of periodic models for univariate and multivariate time series. A full account of the literature and of recent developments in periodic models for nonstationary seasonal time series is given in Franses (1996b). For ease of notation, I confine most of the discussion to quarterly time series ($S = 4$).

1. Univariate Time Series

Consider the simple first-order AR model with seasonal varying parameters

$$y_t = \sum_{s=1}^{4} \mu_s D_{s,t} + \sum_{s=1}^{4} \phi_s D_{s,t} y_{t-1} + \varepsilon_t \tag{12}$$

where μ_s are seasonal intercepts and the ϕ_s denote AR parameters that are allowed to vary with the season. Model (12) represents a periodic AR process (PAR) of order 1. This PAR(1) is the simplest case of the more general PAR(p) class of models, but for the present discussion expression (12) suffices to highlight some of the properties of periodic AR models. Franses (1996b) focuses at great length on more elaborate models.

For the periodic process in (12) with $\phi_s \neq \phi$ for all $s = 1, 2, 3, 4$, it is clear that the observations in each of the four seasons are described by a different model. Denoting $Y_{s,T}$ as the observation on y_t in quarter s in year T, (12) implies that $Y_{4,T} = \mu_4 + \phi_4 Y_{3,T} + \varepsilon_{4,T}$ and for example $Y_{1,T} = \mu_1 + \phi_1 Y_{4,T-1} + \varepsilon_{1,T}$. These expressions show that the models for the annual time series $Y_{s,T}$ have constant parameters. In fact, model (12) can be written as

$$\begin{bmatrix} 1 & 0 & 0 & -\phi_1 B \\ -\phi_2 & 1 & 0 & 0 \\ 0 & -\phi_3 & 1 & 0 \\ 0 & 0 & -\phi_4 & 1 \end{bmatrix} \begin{bmatrix} Y_{1,T} \\ Y_{2,T} \\ Y_{3,T} \\ Y_{4,T} \end{bmatrix} = \begin{bmatrix} \mu_1 \\ \mu_2 \\ \mu_3 \\ \mu_4 \end{bmatrix} + \begin{bmatrix} \varepsilon_{1,T} \\ \varepsilon_{2,T} \\ \varepsilon_{3,T} \\ \varepsilon_{4,T} \end{bmatrix} \tag{13}$$

where $BY_{4,T} = Y_{4,T-1}$. The model for the 4×1 vector process Y_T containing $Y_{1,T}$ to $Y_{4,T}$ is convenient to analyze the unit root properties of y_t since (13) represents the same time-series data as (12) does. The characteristic equation of (13) is

$$1 - \phi_1 \phi_2 \phi_3 \phi_4 z = 0 \tag{14}$$

implying that y_t can contain a single unit root when $\phi_1 \phi_2 \phi_3 \phi_4 = 1$. Obviously, when all $\phi_s = 1$, which is the standard random walk case, y_t has the unit root 1, and also the Y_T process has one unit root. When all $\phi_s = -1$, y_t has the seasonal unit root -1, and the Y_T process again has one unit root. Hence, seasonal unit roots in y_t correspond with regular unit roots in the Y_T process. Given this correspondence, it seems natural for a PAR process first to test whether $\phi_1\phi_2\phi_3\phi_4 = 1$ and next to test restrictions on the ϕ_s parameters. Boswijk and Franses (1996) show that the first test follows a Dickey-Fuller distribution, and that the second step involves just χ^2 asymptotics. Franses and Paap (1996a) show through simulation and through forecasting empirical series that this two-step method yields useful results. Boswijk, Franses, and Haldrup (1997) extend this method to allow for more general structures involving I(2) processes and more seasonal unit roots. Ghysels, Hall, and Lee (1996) focus on testing for restrictions as $\phi_s = -1$ in (12) in one step.

In case $\phi_1\phi_2\phi_3\phi_4 = 1$ in (12) and $\phi_s \neq 1$ or -1 for all $s = 1, 2, 3, 4$, the y_t process is said to be periodically integrated (Osborn 1988, Franses 1996b). Otherwise formulated, y_t requires a periodic differencing filter $1 - \phi_s B$ with $\phi_1\phi_2\phi_3\phi_4 = 1$ to remove the stochastic trend. Notice that these ϕ_s parameters have to be estimated from the data, and that some ϕ_s values will exceed 1. Typical values for $\hat{\phi}_s$ for quarterly data are within the range of 0.8 and 1.2. For example, the estimation results for a PAR(1) with the restriction $\phi_1\phi_2\phi_3\phi_4 = 1$ for the U.K. consumption durables sample series are $\hat{\phi}_1 = 1.003$ (0.008), $\hat{\phi}_2 = 0.932$ (0.007), $\hat{\phi}_3 = 1.030$ (0.008), and $\hat{\phi}_4 = 1.039$ (0.008), with estimated standard errors in parentheses. Note that these standard errors underestimate the true standard errors since the ϕ_s parameters are estimated superconsistently, see Boswijk and Franses (1996). An F-test for the restriction $\phi_s = 1$ obtains the value of 31.617, and this is clearly significant at the 5% level. Hence, U.K. consumption nondurables appears to be a periodically integrated process; see also Osborn (1988) where it is shown that this finding is consistent with a modified economic theory and Franses (1996b) for some results on forecasting such series.

For many sample series considered in Franses (1996b) it is found that the periodic differencing filter $1 - \phi_s B$ is appropriate to remove the unit root. This finding appears robust to data transformations and structural breaks. One of the main implications of periodic integration is that seasonality changes, which can be illustrated by writing (12) without the μ_s as

$$y_t = y_{t-4} + \varepsilon_t + \sum_{s=1}^{4} \phi_s D_{s,t} \varepsilon_{t-1} + \sum_{s=1}^{4} \phi_s \phi_{s-1} D_{s,t} \varepsilon_{t-2}$$

$$+ \sum_{s=1}^{4} \phi_s \phi_{s-1} \phi_{s-2} D_{s,t} \varepsilon_{t-3}$$

which after taking first differences and with ϕ_s values close to unity can be approximated by

$$\Delta_1 \Delta_4 y_t = \varepsilon_t - \sum_{s=1}^{4} \eta_s D_{s,t} \varepsilon_{t-4} \tag{15}$$

When model (15) is estimated under the assumption that $\eta_s = \eta$ for all s, it is very similar to the airline model in (1). Hence, periodically integrated time series may seem adequately described by the airline model. Note that this does not hold the other way around. In fact, for the consumption series, such a model is estimated in (2). In general, it can be shown that neglecting periodic parameter variation increases the lag order (Osborn 1991, Tiao and Grupe 1980); i.e., there appears to be a trade-off between lags in nonperiodic models and the number of intrayear parameters in periodic time-series models. Furthermore, neglecting periodicity reduces the power of nonseasonal unit root tests (Franses 1996b), and it may lead to the finding of spurious seasonal unit roots (Boswijk and Franses 1996).

The application of periodic integration is not necessarily restricted to periodic models as (12). In fact, Bollerslev and Ghysels (1996) introduce the stationary periodic GARCH model to describe seasonally observed financial time series. Franses and Paap (1995) extend this model to allow for persistence of volatility shocks in order to describe the stylized fact that for many daily financial time series the variance appears to change slowly over time. Franses and Paap (1995) fit the following PAR(p)-PIGARCH(1, 1) model to daily returns on the Dow-Jones index (for about 4000 daily observations):

$$y_t = \mu_s + \sum_{i=1}^{p} \phi_{is} y_{t-i} + \varepsilon_t \qquad (s = 1, 2, 3, 4, 5) \tag{16}$$

$$\varepsilon_t \sim N(0, \sigma_t^2) \tag{17}$$

$$\sigma_t^2 = \omega_s + \alpha_s \varepsilon_{t-1}^2 + \beta \sigma_{t-1}^2 \qquad (s = 1, 2, 3, 4, 5) \tag{18}$$

under the restriction that

$$\prod_{s=1}^{5} (\alpha_s + \beta) = 1 \qquad \text{with } \alpha_s \neq \alpha \quad (s = 1, 2, 3, 4, 5) \tag{19}$$

It is found that for example $\hat{\alpha}_1 + \hat{\beta} = 1.008$ and $\hat{\alpha}_2 + \hat{\beta} = 0.956$, reflecting the well-documented importance of Monday stock returns over Tuesday returns.

A second important aspect of periodic integration is that the trend/cycle component and the seasonal component in an economic time series interact. This is obvious from the fact that the ϕ_s parameters in the differencing filter $1 - \phi_s B$ differ across the seasons; see Franses (1996c) for more details. Hence, seasonal adjustment methods as Census X-11, which treat all seasons in an equal fashion, do not remove the intrinsic periodicity in a time series. As shown in Franses (1996b) it is possible to fit periodic models to seasonally adjusted data if the underlying time series shows dynamic periodicity. Strictly speaking, it does not make sense to seasonally adjust periodic time series since the key assumption for seasonal correction is that one can isolate the seasonal from the nonseasonal component. As a possible consequence of this conceptual problem, there is evidence that the NBER peaks and troughs display seasonality, even though these dates are set using seasonally adjusted data (Ghysels 1994b). In fact, Franses (1996b) shows that seasonally adjusted periodically integrated time series can generate such features. Additional evidence for the apparent link between seasonal fluctuations and the trend/cycle is presented in Barsky and Miron (1989), Beaulieu, MacKie-Mason, and Miron (1992), Canova and Ghysels (1994), Franses (1995), and Miron (1996).

Finally, a feature of periodic integration in connection with the intercept parameters μ_s in (12) is that this allows a description of variables with increasing seasonal variation without taking logs. This may be useful for such data as the trade balance, and other economic data that may take negative values. A key drawback of periodic AR models, however, is that the number of parameters increases quite rapidly if S and p increase. For example, a PAR(2) model as (12) for monthly data involves the estimation of 24 parameters. Hence it may be useful to impose certain parameter restrictions (Anderson and Vecchia 1993).

2.   Multivariate Time Series

Obviously, the number of parameters in multivariate periodic models increases quite rapidly, see Lütkepohl (1991) and Ula (1993) for periodic VARs for time series without unit roots. For example, when y_t in (12) is $m \times 1$, and all parameters are allowed to vary with the season, this periodic VAR(1) contains $4m^2$ parameters. An additional drawback of this representation is that it is quite complicated to investigate the unit root properties of the vector y_t process since the expressions for the characteristic equations similar to (14) are not as simple.

One possibility is to resort to an alternative representation of periodic VARs, as is done in Kleibergen and Franses (1995). This representation assumes that long-run cointegration relations across variables obey certain restrictions. If one wants to allow for more flexibility, one may consider the second possibility, which is to impose more structure on the dynamics and also to focus on only a single equation (Birchenhall et al. 1989, Boswijk and Franses 1995). A simple example of a resultant model for a bivariate time series is

$$\Delta_4 y_{1,t} = \alpha_s(y_{1,t-4} - \beta_s y_{2,t-4}) + \varepsilon_t \tag{20}$$

where the equilibrium and adjustment parameters are allowed to vary with the season $s = 1, 2, 3, 4$. Boswijk and Franses (1995) propose an empirical strategy for this so-called periodic cointegration model. Franses (1996b) illustrates that models such as (20) can also generate time series for which the trend/cycle and the seasonal fluctuations are dependent.

At present, the literature on multivariate periodic models is not extensive, and much more research is needed into the properties of such models and into the design of useful empirical methods for more general cases than (20).

IV. FURTHER RESEARCH TOPICS

The issue of investigating seasonal variation in economic data has gained much interest in the last few years, and this has resulted in the two approaches discussed in the previous section. There are also several studies in which seasonality is incorporated explicitly into economic theory, e.g., Osborn (1988), Hansen and Sargent (1993), Todd (1990), and Miron and Zeldes (1988). There are, however, many issues for current and future research, and in this section I will highlight only a few of these.

A.   Structural Breaks

The time-series models in the previous section assume that seasonality changes because of shocks; i.e., the changes are stochastic. It may, however, be that the changes are deterministic. Institutional changes may make economic agents to start behaving differently in certain seasons. For example, allowing country regions to fix their own school holiday periods may reduce variation in tourism spending. Another example is the introduction of a new TV and radio broadcasting channel that can change the structure of the market for advertising expenditures. Such a new radio channel was introduced in The Netherlands in 1989, around observation 154 in Figure 4. It is clear from this graph that seasonality in advertising expenditures changes dramatically. The source of this change appears to be a new pricing policy, which in turn is due to changing pricing policies for TV. This change is seasonality can be said to be deterministic.

Following the arguments in Perron and Vogelsang (1992), it can be expected that neglecting changing parameters in deterministic seasonal dummies biases seasonal unit root tests toward nonrejection; see Franses and Vogelsang (1997) for formal results. Furthermore, such changes bias tests for periodicity toward the alternative; i.e., too much periodicity is found, as shown in Franses (1996b). As an example, for the advertising series, when the OCSB regression in (9) is enlarged with 13 seasonal dummies for the period from observation 154 onward, the t-tests obtain the values $t(\hat{\gamma}_1) = -2.497\,[-2.86]$ and $t(\hat{\gamma}_2) = -11.358\,[-7.50]$, where the 5% critical values are from Franses and Hobijn (1997). Hence, the previous conclusion that radio advertising needs a Δ_{13} filter changes to the necessity of only the Δ_1 filter. All 12 seasonal unit roots disappear when one allows for deterministic shifts.

For many economic time series the location of a break that affects tests for seasonal unit roots is unknown. If one suspects such breaks, one may then use the extreme values of the various t- and F-tests to search for a break data. Asymptotic theory for this approach is presented in Franses and Vogelsang (1997). If one does not like the idea of searching for possibly inconveniently located breaks, one may wish to use this method in order to investigate the robustness of the outcomes of seasonal unit root tests. In fact, Paap, Franses, and Hoek (1997) show through simulation experiments that making mistakes in either direction yields highly imprecise forecasts.

Further research is needed to investigate the robustness of empirical findings on periodicity and seasonal unit roots to structural shifts in seasonal means. Since Ghysels and Perron (1996) document that seasonal adjustment tends to obscure structural breaks, it may be that the approach in Franses and Vogelsang (1997) can also be usefully applied for unadjusted variables to test for nonseasonal unit roots in the presence of seasonal unit roots and mean shifts. Finally, in case deterministic seasonal mean shifts occur quite frequently, it seems relevant to investigate the robustness of the empirical regularities summarized in Miron (1996).

B. Time-Varying Parameters

Allowing for structural mean shifts can be helpful to detect exactly when seasonality changes, if it does. Seasonal unit root and periodic integration models may not be very helpful to decide in which part of the sample the seasonal variation changes. Hence, in order to be able to understand more clearly what economic behavior causes such changes, one may consider models that are in between models with seasonal unit roots and deterministic shifts. For example, it may be useful to consider

$$y_t = \sum_{s=1}^{4} \delta_{s,t} D_{s,t} + u_t \tag{21}$$

where u_t is some ARMA process and $\delta_{s,t}$ are time-varying parameters for the seasonal dummies. The $\delta_{s,t}$ can be made functions of time, of lagged $\delta_{s,t}$ or of economic variables. Consequently, one can extend the HEGY approach to allow for more flexible structures in the μ_t term in (10).

Recent examples of flexible structures for seasonal variation are given in Andersen and Bollerslev (1994), Harvey and Scott (1994), and Canova and Hansen (1995). Hylleberg and Pagan (1997) put forward the so-called evolving seasonals model that amounts to an intermediate case between the models in (4) and the seasonal unit root models. This model appears useful to explain the simulation results in Hylleberg (1995), where the HEGY test appears better in some cases and the Canova-Hansen test in others. The two null models in these tests are both special cases of the evolving seasonals model. Further research is needed to investigate the practical usefulness of more flexible structures with respect to those in Section III.

C. Nonlinear Modeling

It may be possible and important to introduce even more flexible structures by allowing y_t to be described by nonlinear time-series models, while taking care of seasonality. For example, Ghysels (1994b) finds that regime shifts in the business cycle tend to occur more frequently in some seasons than in others. Additionally, Canova and Ghysels (1994) and Franses (1995) find that some macroeconomic variables show

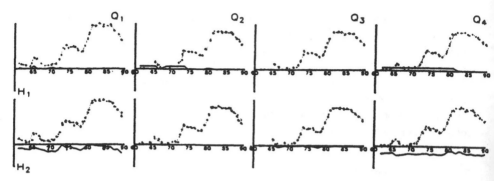

Figure 6 Quarterly output of hidden layer units.

different seasonality in expansions than in recessions, which in turn may depend on when the regime switches occur. In these studies the business cycle is defined by the NBER indicators, but one may want to estimate these dates where seasonality changes using the same data. Given the periodic structure of the business cycle, it is proposed in Ghysels (1993) to extend the Hamilton-type Markov switching model to allow for seasonality in the transition probabilities.

Such nonlinear models may also be helpful to decide when seasonality is changing. For example, Franses and Draisma (1997) propagate the linear neural network model with q hidden layers as

$$y_t = \mu_t + \phi_1 y_{t-1} + \cdots + \phi_p y_{t-p}$$

$$+ \sum_{j=1}^{q} \beta_j G(\mu_{tj}^* + \phi_{1j}^* y_{t-1} + \cdots + \phi_{pj}^* y_{t-p}) + \varepsilon_t \tag{22}$$

where G is the logistic activation function; see Kuan and White (1994) for an overview of neural network models. Although the parameters in the nonlinear component cannot be interpreted, Franses and Draisma (1997) use the $\beta_j G(\cdot)$ components (for each of the seasons) to investigate the contribution of the hidden layer components. For example, consider again the unemployment data for Germany. When μ_t in (22) is as in (10) with β_2 to β_4 set to zero and p in (22) equals 4, the Schwarz criteria for $q = 0, 1, 2, 3$, are $-756.3, -790.8, -809.5$, and -789.0, respectively. Hence the neural network model with $q = 2$ hidden layer units is selected, and German unemployment shows nonlinear features.

In Figure 6, I depict the impact of the two hidden layer units (H_1 and H_2) in each of the four quarters. The dotted line is the time series in each of the seasons ($Y_{s,T}$), and the straight line is the contribution of the relevant hidden layer output to the final output y_t. It is clear that H_1 is active only in quarters 2 and 4 (Q_2 and Q_4), where for Q_2 this activity ends around 1975 and for Q_4 it ends around 1985. Hidden layer 2 appears active only in quarters 1 and 4. These results show that the time series in quarter 3 seems linear and that changes in certain seasons occur around 1975 and 1985.

It is now interesting to see whether the nonlinear structures in these German unemployment data are robust to seasonal adjustment. For this purpose, I estimate the neural network model in (22) for the official adjusted data. For $q = 0, 1, 2, 3$, I obtain Schwarz criteria values of -918.7, -892.1, -867.7, and -855.9, respectively. Hence, seasonal adjustment appears to affect the apparent nonlinear features of this time series, at least when represented by the highly parameterized neural network model in (22). See Granger and Teräsvirta (1993) for more parsimonious nonlinear models. Ghysels, Granger, and Siklos (1996) document that Census X-11 seasonal adjustment appears to introduce nonlinear features in otherwise linear data. Finally, Franses and Paap (1996b) show that seasonal adjustment may leave nonlinearity intact but that it changes some key parameters. Obviously, it seems useful to study the properties of seasonal adjustment with respect to nonlinearity. Also, it seems of great importance to consider models that explicitly describe nonlinearity and seasonality at the same time, possibly along the lines of Lewis and Ray (1996).

D. Economics

A fourth important topic for further research considers how one can design economic models that can describe the behavior of economic agents which cause macroeconomic aggregates to show slowly changing seasonality over time. As mentioned, several studies incorporate seasonality somehow, but to my knowledge there are no studies that deal with the question of why seasonality changes; i.e., why do economic agents endogenously change their seasonal behavior? Furthermore, can (seasonal or periodic) unit root models generate time series that really mimic economic behavior?

One possible route to follow may be to focus on consumer expectations. An empirical analysis of confidence indicators in Ghysels and Nerlove (1988) and Franses (1996b) shows that, even when agents are asked to remove seasonality by focusing on annual trends, the indicators display marked seasonality. In fact, for most countries one can nowadays obtain only seasonally adjusted consumer confidence indicators, which somehow may seem counterintuitive.

Another research strategy, which may involve game-theoretic aspects, is to investigate if seasonality changes because economic agents mistake a set of large shocks as a precursor for "new" seasonality, and hence start to behave like that. Many important changes for, e.g., Germany, occurred in the fourth quarter (1966.4: economic crisis, 1973.4 and 1979.4 dramatic increase in oil prices and 1989.4: unified Germany). Hence, theoretical models where the arrival rate of shocks can have an impact on economic behavior may be useful.

V. CONCLUDING REMARKS

This chapter surveys recent and possible future research issues in modeling economic time series with seasonality. There is a growing interest in modeling seasonality and not in removing it through some seasonal adjustment methods. In areas as marketing and tourism, seasonality itself is the focus of investigation. In macroeconomics there is a tendency to apply Census X-11 type methods to remove seasonality. However, recent empirical studies have shown the large number of drawbacks of seasonally adjusted data. Furthermore, other studies have shown that it is not that difficult to model seasonality explicitly.

From a statistical point of view, there appears a consensus that not too many stochastic trends can be found in economic data. In other words, although seasonal variation changes over time, it changes quite slowly. The tools for analyzing seasonal time series can be refined in the direction of nonlinear models or alternative flexible structures. From an economic point of view, and especially in case of macroeconomics, there is a need to understand why such seasonal patterns change over time.

ACKNOWLEDGMENTS

I thank the Royal Netherlands Academy of Arts and Sciences for its financial support. Dick van Dijk, Richard Paap, and an anonymous referee provided helpful comments.

REFERENCES

Andersen, T. G. and T. Bollerslev (1994), Intraday Seasonality and Volatility Persistence in Foreign Exchange and Equity Markets, Working Paper 186, Northwestern University, Evanston, IL.

Anderson, P. L. and A. V. Vecchia (1993), Asymptotic Results for Periodic Autoregressive Moving-Average Models, *Journal of Time Series Analysis*, 14, 1–18.

Barsky, R. B. and J. A. Miron (1989), The Seasonal Cycle and the Business Cycle, *Journal of Political Economy*, 97, 503–535.

Beaulieu, J. J., J. K. MacKie-Mason, and J. A. Miron (1992), Why Do Countries and Industries with Large Seasonal Cycles Also Have Large Business Cycles?, *Quarterly Journal of Economics*, 107, 621–656.

Bell, W. R. (1987), A Note on Overdifferencing and the Equivalence of Seasonal Time Series Models with Monthly Means and Models with $(0, 1, 1)_{12}$ Seasonal Parts When $\theta = 1$, *Journal of Business and Economic Statistics*, 5, 383–387.

Bell, W. R. and S. C. Hillmer (1994), Issues Involved with the Seasonal Adjustment of Economic Time Series (with discussion), *Journal of Business and Economic Statistics*, 2, 291–320, 1984.

Birchenhall, C. R., R. C. Bladen-Hovell, A. P. L. Chui, D. R. Osborn, and J. P. Smith (1989), A Seasonal Model of Consumption, *Economic Journal*, 99, 837–843.

Bollerslev, T. and E. Ghysels (1996), Periodic Autoregressive Conditional Heteroskedasticity, *Journal of Business and Economic Statistics*, 14, 139–151.

Boswijk, H. P. and P. H. Franses (1995), Periodic Cointegration: Representation and Inference, *Review of Economics and Statistics*, 77, 436–454.

Boswijk, H. P. and P. H. Franses (1996), Unit Roots in Periodic Autoregressions, *Journal of Time Series Analysis*, 17, 221–245.

Boswijk, H. P., P. H. Franses, and N. Haldrup (1997), Multiple Unit Roots in Periodic Autoregression, *Journal of Econometrics*, 80, 167–193.

Bowerman, B. L., A. B. Koehler, and D. J. Pack (1990), Forecasting Time Series with Increasing Seasonal Variation, *Journal of Forecasting*, 9, 419–436.

Box, G. E. P. and G. M. Jenkins (1970), *Time Series Analysis; Forecasting and Control*, Holden-Day, San Francisco.

Canova, F. and E. Ghysels (1994),Changes in Seasonal Patterns: Are They Cyclical?, *Journal of Economic Dynamics and Control*, 18, 1143–1171.

Canova, F. and B. E. Hansen (1995), Are Seasonal Patterns Constant over Time?, A Test for Season Stability, *Journal of Business and Economic Statistics*, 13, 237–252.

Engle, R. F., C. W. J. Granger, S. Hylleberg, and H. S. Lee (1993), Seasonal Cointegration: The Japanese Consumption Function, *Journal of Econometrics*, 55, 275–298.

Franses, P. H. (1995), Quarterly USA Unemployment: Cycles, Seasons, and Asymmetries, *Empirical Economics*, 20, 717–725.

Franses, P. H. (1996a), Recent Advances in Modelling Seasonality, *Journal of Economic Surveys*, 10, 299–345.

Franses, P. H. (1996b), *Periodicity and Stochastic Trends in Economic Time Series*, Oxford University Press, Oxford.

Franses, P. H. (1996c), Multi-step Forecast Error Variances for Periodically Integrated Time Series, *Journal of Forecasting*, 15, 83–95.

Franses, P. H. and G. Draisma (1997), Recognizing Changing Seasonal Patterns Using Artificial Neural Networks, *Journal of Econometrics* (to appear).

Franses, P. H. and B. Hobijn (1997), Critical Values for Unit Root Tests in Seasonal Time Series, *Journal of Applied Statistics*, 24, 25–47.

Franses, P. H. and R. Kunst (1995), On the Role of Seasonal Intercepts in Seasonal Cointegration, Institute for Advanced Studies, Vienna.

Franses, P. H. and R. Paap (1995), Modeling Changing Day-of-the-Week Seasonality in Stock Returns and Volatility, Econometric Institute Report 9556, Erasmus University Rotterdam.

Franses, P. H. and R. Paap (1996a), Periodic Integration: Further Results on Model Selection and Forecasting, *Statistical Papers*, 37, 33–52.

Franses, P. H. and R. Paap (1996b), Does Seasonal Adjustment Change Inference from Markov Switching Models?, Econometric Institute Report 9615, Erasmus University Rotterdam.

Franses, P. H. and T. J. Vogelsang (1997), On Seasonal Cycles, Unit Roots and Mean Shifts, *Review of Economics and Statistics* (to appear).

Ghysels, E. (1993), A Time Series Model with Periodic Stochastic Regime Switching, Unpublished manuscript, University of Montreal.

Ghysels, E. (1994a), On the Economics and Econometrics of Seasonality, in C. A. Sims (ed.), *Advances in Econometrics, Sixth World Congress of the Econometric Society*, Cambridge University Press, Cambridge, UK.

Ghysels, E. (1994b), On the Periodic Structure of the Business Cycle, *Journal of Business and Economic Statistics*, 12, 289–298.

Ghysels, E., C. W. J. Granger, and P. L. Siklos (1996), Is Seasonal Adjustment a Linear or Nonlinear Data-Filtering Process?, *Journal of Business and Economic Statistics*, 14, 374–386.

Ghysels, E., A. Hall, and H. S. Lee (1996), On Periodic Structures and Testing for Seasonal Unit Roots, *Journal of The American Statistical Association*, 91, 1551–1559.

Ghysels, E., H. S. Lee, and J. Noh (1994), Testing for Unit Roots in Seasonal Time Series, *Journal of Econometrics*, 62, 415–442.

Ghysels, E., H. S. Lee, and P. L. Siklos (1993), On the (Mis)Specification of Seasonality and Its Consequences: An Empirical Investigation with US Data, *Empirical Economics*, 18, 747–760.

Ghysels, E. and M. Nerlove (1988), Seasonality in Surveys; A Comparison of Belgian, French and German Business Tests, *European Economic Review*, 32, 81–99.

Ghysels, E. and P. Perron (1996), The Effects of Linear Filters on Dynamic Time Series with Structural Change, *Journal of Econometrics*, 70, 69–98.

González, P. and P. Moral (1995), An Analysis of the International Tourism Demand in Spain, *International Journal of Forecasting*, 11, 233–251.

Granger, C. W. J. and T. Teräsvirta (1993), *Modelling Nonlinear Economic Relationships*, Oxford University Press, Oxford.

Hansen, L. P. and T. J. Sargent (1993), Seasonality and Approximation Errors in Rational Expectations Models, *Journal of Econometrics*, 55, 21–56.

Harvey, A. and A. Scott (1994), Seasonality in Dynamic Regression Models, *Economic Journal*, 104, 1324–1345.

Hylleberg, S. (1986), *Seasonality in Regression*, Academic Press, Orlando.

Hylleberg, S. (1992), *Modelling Seasonality*, Oxford University Press, Oxford.

Hylleberg, S. (1994), Modelling Seasonal Variation, in Hargreaves, C. P. (ed.), *Nonstationary Time Series Analysis and Cointegration*, Oxford: Oxford University Press.

Hylleberg, S. (1995), Tests for Seasonal Unit Roots: General to Specific or Specific to General?, *Journal of Econometrics*, 69, 5–26.

Hylleberg, S., R. F. Engle, C. W. J. Granger, and B. S. Yoo (1990), Seasonal Integration and Cointegration, *Journal of Econometrics*, 44, 215–238.

Hylleberg, S., C. Jørgensen, and N. K. Sørensen (1993), Seasonality in Macroeconomic Time Series, *Empirical Economics*, 18, 321–335.

Hylleberg, S. and A. R. Pagan (1997), Seasonal Integration and the Evolving Seasonals Model, *International Journal of Forecasting* (to appear).

Johansen, S. and E. Schaumburg (1996), Inference for Seasonally Cointegrated Autoregressive Processes, Mimeo, University of Copenhagen.

Kleibergen, F. and P. H. Franses (1995), Direct Cointegration Testing in Periodic Vector Autoregression, Econometric Institute Report 9524, Erasmus University Rotterdam.

Kuan, C.-M. and H. White (1994), Artificial Neural Networks: An Econometric Perspective, *Econometric Reviews*, 13, 1–91.

Kunst, R. M. (1993), Seasonal Cointegration in Macroeconomic Systems: Case Studies for Small and Large European Countries, *Review of Economics and Statistics*, 75, 325–330.

Lee, H. S. (1992), Maximum Likelihood Inference on Cointegration and Seasonal Cointegration, *Journal of Econometrics*, 54, 351–365.

Lee, H. S. and P. L Siklos (1993), The Influence of Seasonal Adjustment on the Canadian Consumption Function, 1947–1991, *Canadian Journal of Economics*, 26, 575–589.

Lewis, P. A. and B. K. Ray (1996), Modeling Periodic Threshold Autoregressions Using TSMARS, Unpublished manuscript, Department of Mathematics, New Jersey Institute of Technology.

Lütkepohl, H. (1991), *Introduction to Multiple Time Series Analysis*, Springer-Verlag, Berlin.

Maravall, A. (1995), Unobserved Components in Economic Time Series, in H. Pesaran, P. Schmidt, and M. Wickens (eds.), *Handbook of Applied Econometrics*, Basil Blackwell, Oxford.

Miron, J. A. (1996), *The Economics of Seasonal Cycles*, MIT Press, Cambridge, MA.

Miron, J. A. and S. P. Zeldes (1988), Seasonality, Cost Shocks, and the Production Smoothing Model of Inventories, *Econometrica*, 56, 877–908.

Osborn, D. R. (1988), Seasonality and Habit Persistence in a Life-Cycle Model of Consumption, *Journal of Applied Econometrics*, 3, 255–266.

Osborn, D. R. (1990), A Survey of Seasonality in UK Macroeconomic Variables, *International Journal of Forecasting*, 6, 327–336.

Osborn, D. R. (1991), The Implications of Periodically Varying Coefficients for Seasonal Time-Series Processes, *Journal of Econometrics*, 48, 373–384.

Osborn, D. R., A. P. L. Chui, J. P. Smith, and C. R. Birchenhall (1988), Seasonality and the Order of Integration of Consumption, *Oxford Bulletin of Economics and Statistics*, 50, 361–377.

Osborn, D. R. and J. P. Smith (1989), The Performance of Periodic Autoregressive Models in Forecasting Seasonal UK Consumption, *Journal of Business and Economic Statistics*, 7, 117–127.

Paap, R., P. H. Franses, and H. Hoek (1997), Mean Shifts, Unit Roots and Forecasting Seasonal Time Series, *International Journal of Forecasting* (to appear).

Perron, P. and T. J. Vogelsang (1992), Nonstationarity and Level Shifts with an Application to Purchasing Power Parity, *Journal of Business and Economic Statistics*, 10, 301–320.

Smith, R. J. and A. M. R. Taylor (1995), Additional Critical Values and Asymptotic Representations for Seasonal Unit Root Tests, Discussion Paper 95/43, Department of Economics, University of York, 1995.

Tiao, G. C. and M. R. Grupe (1980), Hidden Periodic Autoregressive-Moving Average Models in Time Series Data, *Biometrika*, 67, 365–373.

Todd, R. (1990), Periodic Linear-Quadratic Methods for Modeling Seasonality, *Journal of Economic Dynamics and Control*, 14, 763–795.

Ula, T. A. (1993), Forecasting of Multivariate Periodic Autoregressive Moving-Average Processes, *Journal of Time Series Analysis*, 14, 645–657.

17

Nonparametric and Semiparametric Econometrics of Panel Data

Aman Ullah
University of California at Riverside, Riverside, California

Nilanjana Roy
University of Victoria, Victoria, British Columbia, Canada

I. INTRODUCTION

Longitudinal or panel data refers to data where we have observations on the same cross section of individuals, households, industries, etc., over multiple periods of time. Often the panel data is short in the sense that the cross-sectional units are available for a period of 2 to 10 years. Some panel data sets are rotating panels where a proportion of cross-sectional units is kept for revisits, while the remaining is replaced by new cross-sectional units. Among the most analyzed panel data are the Michigan Panel Study of Income Dynamics (PSID) and the National Longitudinal Surveys of Labor Market Experience (NLS) in the United States, the International Crops Research Institute for the Semi-Arid Tropics Village Level Studies (ICRISAT VLS) in Hyderabad, India, and the Living Standards Surveys (LSS) in Côte d'Ivoire. Recent years have witnessed a significant growth in the availability of the panel data (Borus 1982, Ashenfelter and Solon 1982, Deaton 1994). The drive behind this growth stems from the fact that the panel data helps to study dynamics of the individual cross-sectional units. It is useful for studying intertemporal and intergenerational behavior of the cross-sectional units. From the inference point of view panel data leads to efficiency gains in the econometric estimators. For details on advantages and problems with the panel data, see Hsiao (1985), Klevmarken (1989), and Deaton (1994).

An important difference between the panel data econometric models with either cross-sectional or time-series models is that it allows for the cross-sectional and/or time heterogeneity. Within this framework two types of model are mostly estimated, one is the fixed-effects model where one makes inferences conditional on the cross-sectional units in the sample while the other is the random-effects model which is used if we want to make inferences about the population generating the cross-sectional units. Both the models control for heterogeneity. There is no agreement in the literature as to which one should be used in the applied work; see Maddala (1987) for a good set of arguments on the fixed- versus random-effects models. The econometrics of the fixed-effects model was initiated in the works of Mundlak (1961) and Hoch (1962), while the work on the random effects model was introduced by Balestra and Nerlove (1966) and developed further by Wallace and Hussain (1969), Maddala (1971), Nerlove (1971), and Fuller and Battese (1973). Since then a voluminous econometric literature has developed; for details see the excellent monographs by Heckman and Singer (1985), Hsiao (1986), Dielman (1989), and Baltagi (1995); surveys by Chamberlain (1984), Maddala (1987), and Baltagi (1996); a journal volume by Raj and Baltagi (1992); a recent handbook of approximately 900 pages by Mátyás and Sevestre (1996); and the recent work on dynamic panel data models by Pesaran and Smith (1995) and Harris and Mátyás (1996).

The motivation for this chapter is based on a simple observation that this voluminous literature has been largely confined to the linear parametric models; although see Mátyás and Sevestre (1996) and Baltagi (1996) for the references on some of the recent works on the nonlinear parametric regression and the latent variable models. It is, however, well known that the misspecified linear or nonlinear parametric models may lead to inconsistent and inefficient estimates and suboptimal test statistics. With this in view, the modest aim of this chapter is to systematically develop the nonparametric estimation of both the fixed- and random-effects panel models which are robust to the misspecification in the functional forms. Some new estimators are proposed. Further, the estimation of semiparametric models are also considered. For the nonparametric regression analysis based on either cross-sectional or time-series data and the usefulness of their application, see Härdle (1990) and Pagan and Ullah (1996). Muller (1988) has considered the nonparametric longitudinal models with fixed-design regressors and has discussed its applications to growth and other biomedical models. Generally the econometric regressors do not have fixed-design structure, and his work does not consider the random- and fixed-effects models. Only the static models are considered here, and it is hoped that in the future the results could be extended to various other nonparametric econometric models such as the dynamic models, limited dependent variable models, and duration models.

Another objective of this chapter is to explore the application of the nonparametric panel models to study the calorie-income relationship based on the ICRISAT VLS panel data. The usefulness of this application stems from the issue of possible nonlinearity in the calorie-income relation raised in Ravallion (1990). There has been an ongoing debate in the calorie-income literature in the context of developing countries on the magnitude of the income elasticity of calorie intake; see Strauss and Thomas (1990), Bhargava (1991), Bouis and Haddad (1992), and Subramanian and Deaton (1996), among others. All these authors have considered the per capita household level data on calorie intake. In an important contribution Behrman and Deolalikar (1990) considered the individual level panel data on calorie intake and estimated the income elasticity by the parametric fixed-effects method to control for the heterogeneity. We use the same individual level data as used by Behrman and De-olalikar (1990) and compare our nonparametric and parametric elasticity estimates with their parametric estimate. For this data set it has been found that the parametric and nonparametric elasticities are quite similar except that the nonparametric one gives us the added information at the various income levels. We note that Subramanian and Deaton also calculated the nonparametric estimate, but their analysis is based on a single cross section of households, and hence they could not control for the heterogeneity.

The plan of this chapter is as follows. In Section II we first review the estimation of pooled nonparametric panel data model and then present our proposed nonparametric and semiparametric estimators for the fixed-effects and random-effects models. The proposed nonparametric estimators are then used to analyze the calorie-income relationship in Section III.

II. NONPARAMETRIC PANEL DATA MODELS

A. Pooled Nadaraya-Watson Kernel Estimation

Let us consider the nonparametric regression model as

$$y_{it} = m(x_{it}) + u_{it}; \qquad i = 1, \ldots, n, \quad t = 1, \ldots, T \qquad (1)$$

where x_{it} is a vector of q regressors, $m(x_{it}) = E(y_{it} \mid x_{it})$, $E(u_{it} \mid x_{it}) = 0$, $V(u_{it} \mid x_{it}) = \sigma^2(x_{it})$, and (y_{it}, x_{it}) are assumed to be i.i.d. We consider the usual panel data situation where n is large and T is small. The nonparametric estimate of $m(x)$, the conditional mean at a point x, is the smoothed average of y values which correspond to the x_{it} values in a small interval of x such that $x_{it} - x = O(h)$, where h is the width of the interval known as window width and it tends to zero as $n \to \infty$. In a least-squares sense, this amounts to fitting a constant to the data around x. More explicitly, we use Taylor expansion of $m(x_{it})$ around x as $m(x_{it}) = m(x) + O(x_{it} - x)$, where $O(x_{it} - x)$ represents the remainder term of order at most $x_{it} - x$ or $O(h)$. The term $O(h)$ can be added with u_{it} so that for large n it is still the case that the expectation of the combined error is zero. With this in mind, but not explicitly writing it, one can write (1) as

$$y_{it} = m(x) + u_{it} \qquad (2)$$

or, more compactly, as

$$y = \iota_{nT} m(x) + u$$

and minimize $\sum^n \sum^T (y_{it} - m(x))^2 K((x_{it} - x)/h) = u'K(x)u$ with respect to $m(x)$; ι_{nT} is an $nT \times 1$ vector of unit elements and $K(x)$ is the diagonal matrix with the diagonal elements $K((x_{it} - x)/h) = K_{it}$, which is the kernel or weight function taking low values for x_{it} far away from x but high values for x_{it} close to x. The least-squares (LS) solution of $m(x)$ is

$$\hat{m}(x) = (\iota'_{nT} K(x) \iota_{nT})^{-1} \iota'_{nT} K(x) y \tag{3}$$

which is the Nadaraya (1964) and Watson (1964) kernel regression estimator. This estimator provides local averaging or smoothing, and it is essentially a local constant least-squares estimator.

The pointwise estimator of the partial derivative of $m(x)$, $\beta(x) = \partial m(x)/\partial x = [m(x + h/2) - m(x - h/2)]/h$, is then

$$\hat{\beta}(x) = \frac{\hat{m}(x + h/2) - \hat{m}(x - h/2)}{h} = \frac{\partial}{\partial x} \hat{m}(x) \tag{4}$$

as given in Rilstone and Ullah (1989). Expression (4) is a numerical derivative. Vinod and Ullah (1988) have considered the analytical derivative of $\hat{m}(x)$ in (3), which is approximately identical to (4).

The estimation of the average derivative, $\beta = E\beta(x)$, can be done by using Rilstone (1991) and Härdle and Stoker (1989). Following Rilstone (1991) a direct estimator for β is

$$\hat{\beta}_R = \sum_{i=1}^n \sum_{t=1}^T \frac{\hat{\beta}(x_{it})}{nT} \tag{5}$$

where $\hat{\beta}(x_{it})$ is obtained from (4). For the Härdle and Stoker estimator we assume the density $f(x)$ vanishes at the boundary of its support, and using integration by parts we write $\beta = E\beta(x) = -E[y(f^{(1)}(x)/f(x))]$, where $f^{(1)}(x)$ is the first derivative of $f(x)$. Then the estimator β is

$$\hat{\beta}_{HS} = -\frac{1}{nT} \sum_{i=1}^n \sum_{t=1}^T y_{it} \frac{\hat{f}^{(1)}(x_{it})}{\hat{f}(x_{it})} \tag{6}$$

where

$$\hat{f}(x) = (nTh)^{-1} \sum_{i=1}^n \sum_{t=1}^T K\left(\frac{x_{it} - x}{h}\right)$$

is a kernel density estimator and $\hat{f}^{(1)}$ is its first derivative.

The details on the asymptotic properties of the above estimators and the choice of kernel and h can be found in Härdle (1990) and Pagan and Ullah (1996). Specifically we note that, under the smoothness conditions on the kernel K, $E|u_{it}|^{2+\delta} < \infty$ for some $\delta > 0$, and assuming $h \to 0$, $nh^{q+2s} \to \infty$ and $nh^{q+4} \to 0$ as $n \to \infty$:

$$(nh^{q+2s})^{1/2} (\hat{m}^{(s)}(x) - m^{(s)}(x)) \sim N\left(0, \frac{\sigma_u^2(x)}{Tf(x)} \int (K^{(s)}(\Psi))^2 \, d\Psi\right) \tag{7}$$

where s represents the sth derivative and $\sigma_u^2(x) = V(u \mid x)$; for $s = 0$, $\hat{m}^{(0)}(x) = \hat{m}(x)$, and for $s = 1$, $\hat{m}^{(1)}(x) = \hat{\beta}(x)$. In practice $\sigma_u^2(x)$ can be replaced by its consistent estimator $\sigma_u^2(x) = (\iota_{nT}' K(x)\iota_{nT})^{-1}\iota_{nT}' K(x)\hat{u}$, where $\hat{u}$ is the vector of the squared nonparametric residuals $\hat{u}_{it}^2 = (y_{it} - \hat{m}(x_{it}))^2$. We can then use (7) to calculate the confidence intervals. The conditions for the asymptotic normality of $\hat{\beta}_R$ and $\hat{\beta}_{HS}$ are very similar to the conditions of the asymptotic normality of the pointwise $\hat{\beta}(x)$. It follows from Rilstone (1991) and Härdle and Stoker (1989) that, as $n \to \infty$,

$$\sqrt{n}(\hat{\beta}_R - \beta) \to \sqrt{n}(\hat{\beta}_{HS} - \beta) \to N\left(0, \frac{\Sigma}{T}\right) \tag{8}$$

where

$$\Sigma = E\left[\sigma_u^2(x)\left(\frac{f^{(1)}(x)}{f(x)}\right)^2\right] + V(\beta(x)) \tag{9}$$

In practice, for the confidence intervals and the hypothesis testing, an estimator of Σ can be obtained by replacing $\sigma_u^2(x)$, $f^{(1)}(x)$, and $\beta(x)$ by their kernel estimators $\hat{\sigma}_u^2(x)$, $\hat{f}^{(1)}(x)$, and $\hat{\beta}(x)$ and then taking the sample averages. Then for $q = 1$,

$$\hat{\Sigma} = \frac{1}{nT}\sum_{i=1}^{n}\sum_{t=1}^{T}\left\{\hat{\sigma}_u^2(x_{it})\left(\frac{\hat{f}^{(1)}(x_{it})}{\hat{f}(x_{it})}\right)^2 + (\hat{\beta}(x_{it}) - \hat{\bar{\beta}})^2\right\}$$

The second component of Σ is essentially the variance of the limiting distribution of the sample average $\sum^n \sum^T \beta(x_{it})/nT$, and it will be zero if the true $m(x)$ is linear in x. We also observe that the asymptotic variance does not depend on the window width h, kernel K, and number of regressors q. This is quite different compared to the pointwise asymptotic variance result in (7).

Now we turn to the small sample behavior of the bias and mean-square error (MSE) of the above estimators. First, considering $\hat{m}(x)$ and taking its expectation conditional on x_{it} we get

$$E(\hat{m}(x) \mid x_{it}) = (\iota' K(x)\iota)^{-1}\iota' K(x)m^* \tag{10}$$

where $\iota = \iota_{nT}$ and $m^* = [m(x_{11}), \ldots, m(x_{nT})]$. Further

$$V(\hat{m}(x) \mid x_{it}) = (\iota' K(x)\iota)^{-1}\iota'\Omega(x)\iota(\iota' K(x)\iota)^{-1} \tag{11}$$

where $\Omega(x) = K(x)\Sigma_1 K(x)$ and Σ_1 is a diagonal matrix with the diagonal elements $\sigma_u^2(x_{it}) = E(u_{it}^2 \mid x_{it})$. If $\Sigma_1 = \sigma_u^2 I$, $\Omega(x) = \sigma_u^2 K^2(x)$. In practice $\sigma_u^2(x)$ can be consistently estimated by $\hat{\sigma}_u^2(x)$. If we expand m^* by the Taylor series and consider n to be large then it can be shown that (Ruppert and Wand 1994, Pagan and Ullah 1996), up to $O(h^2)$,

$$E(\hat{m}(x) \mid x_{it}) - m(x) = \frac{h^2}{2}\mu_2\left[m^{(2)}(x) + 2m^{(1)}(x)\frac{f^{(1)}(x)}{f(x)}\right] \tag{12}$$

and, up to $O(1/nh^q)$,

$$V(\hat{m}(x) \mid x_{it}) = \frac{1}{nh^q} \frac{\sigma_u^2(x)}{Tf(x)} \int K^2(\Psi) \, d\Psi \tag{13}$$

where $\mu_2 = \int \Psi^2 K(\Psi) \, d\Psi < \infty$. We note that the expressions for the conditional bias and the conditional variance do not depend on x_{it}. Thus the unconditional bias, $E(\hat{m}(x)) - m(x)$, up to $O(h^2)$ and the unconditional variance, $V(\hat{m}(x))$, up to $O(1/nh^q)$ also are the same as given in (12) and (13), respectively. Similarly, we can show that (Pagan and Ullah 1996)

$$E\hat{\beta}(x) - \beta(x) = \frac{h^2}{2}\mu_2 \left[m^{(4)}(x) - 2 \left\{ \beta(x) \left(\left(\frac{f^{(1)}(x)}{f(x)} \right)^2 \right. \right. \right.$$
$$\left. \left. \left. - \frac{f^{(2)}(x)}{f(x)} \right) - m^{(3)}(x) \frac{f^{(1)}(x)}{f(x)} \right\} \right] \tag{14}$$

and

$$V(\hat{\beta}(x)) = \frac{1}{nh^{q+2}} \frac{\sigma_u^2(x)}{Tf(x)} \int (K^{(1)}(\Psi))^2 \, d\Psi \tag{15}$$

Essentially, (14) is the derivative of (12) with respect to x.

It follows from (12) and (13) that the optimal h values which minimize the integrated MSE of $\hat{m}(x)$ and $\hat{\beta}(x)$, respectively, are

$$h_0 \propto n^{-1/(q+4)} \qquad \text{and} \qquad h_1 \propto n^{-1/(q+6)} \tag{16}$$

For $q = 1$, $h_1 \propto n^{-1/7}$. This optimal h is much bigger than the optimal $h \propto n^{-2/7}$ for the average derivative estimator $\hat{\beta}_{HS}$, see Härdle and Stoker (1989), where they give approximate MSE of $\hat{\beta}_{HS}$.

B. Pooled Local Linear Kernel Estimation

An alternative to (2) is to fit a line locally—that is, to write (1) as a linear approximation

$$y_{it} = m(x) + (x_{it} - x)\beta(x) + u_{it} \tag{17}$$

or

$$y = Z(x)\delta(x) + u \tag{18}$$

where $Z(x)$ is an $nT \times (q + 1)$ matrix with itth element $[1 \ x_{it} - x]$ and $\delta(x) = [m(x) \ \beta'(x)]'$ is a $(q + 1) \times 1$ parameter vector. Again minimizing $u'K(x)u$ we get

$$\tilde{\delta}(x) = (Z'(x)K(x)Z(x))^{-1}Z'(x)K(x)y \tag{19}$$

which amounts to doing LS of $\sqrt{K_{it}}y_{it}$ on $\sqrt{K_{it}}$ and $\sqrt{K_{it}}(x_{it} - x)$ where $K_{it} = K((x_{it} - x)/h)$. The local least-squares estimators of $m(x)$ and $\beta(x)$ are then

$$\tilde{m}(x) = (1\ 0)\tilde{\delta}(x), \qquad \tilde{\beta}(x) = (0\ \iota_q)\tilde{\delta}(x) \tag{20}$$

Note that $Z(x) = \iota_{nT}$ for the Nadaraya-Watson estimator, which amounts to doing LS of $\sqrt{K_{it}}y_{it}$ on $\sqrt{K_{it}}$.

Estimator $\tilde{m}(x)$ was introduced by Stone (1977) in the time-series context; also see Cleveland (1979). Stone (1980, 1982) generalizes $\tilde{m}(x)$ and its derivatives for the higher-order polynomials and provides their optimal rates of convergence. Cleveland and Devlin (1988) discuss practical implementation of $\tilde{m}(x)$. Müller (1987) develops asymptotic properties of $\tilde{m}(x)$ when $q = 1$ and x_{it} are nonstochastic and follow a "regular" grid design. Müller shows that at interior points $\tilde{m}(x)$ is asymptotically equivalent to $\hat{m}(x)$. Fan (1992, 1993) studied the asymptotic bias and variance of $\tilde{m}(x)$ when $q = 1$ but x_{it} are stochastic and demonstrated the superior behavior of $\tilde{m}(x)$ compared to the Nadaraya-Watson estimator $\hat{m}(x)$. In particular, he shows that $\tilde{m}(x)$ has an important minimax property. Further, unlike the estimator $\hat{m}(x)$, the bias and variance of $\tilde{m}(x)$ near the boundary of the support of $f(x)$ are of the same order of magnitude as in the interior; also see Fan and Gijbels (1992). This is an important property since many economic data may be thin in the tails. In an important work Ruppert and Wand (1994) extended Fan's results for the q-regressors and local quadratic fits. For $q = 1$, they also analysed the bias and variance of $\tilde{\beta}(x)$, and investigated general polynomial fits. We note here that $\tilde{\beta}(x)$ may not be identical to

$$\tilde{\beta}_1(x) = \frac{\partial \tilde{m}(x)}{\partial x} = \frac{\tilde{m}(x + h/2) - \tilde{m}(x - h/2)}{h} \tag{21}$$

This proposed estimator is analogous to Rilstone and Ullah $\tilde{\beta}(x)$ in (4). An alternative is to calculate analytical derivative of $\tilde{m}(x)$ in (20). The calculation of $\tilde{\beta}(x)$ may be simpler as it turns out directly from the local least-squares regression (19).

The asymptotic normality of $\tilde{\delta}(x) = [\tilde{m}(x)\ \tilde{\beta}'(x)]'$ has been studied by Kneisner and Li (1996). Li, Lu, and Ullah (1996) have considered the estimator of average of $\delta(x)$, $\delta = E\delta(x)$, as

$$\tilde{\delta} = \frac{1}{nT}\sum_{i=1}^{n}\sum_{t=1}^{T}\tilde{\delta}(x_{it})$$

and established its asymptotic normality. This result also provides the asymptotic normality of the average derivative

$$\tilde{\beta} = \frac{1}{nT}\sum_{i=1}^{n}\sum_{t=1}^{T}\tilde{\beta}(x_{it}) \tag{22}$$

where $\tilde{\beta}(x_{it}) = (0\ \iota_q)\tilde{\delta}(x_{it})$.

To see the behavior of the above estimators more explicitly, we consider first the asymptotic results. From Kneisner and Li (1996), as $n \to \infty$,

$$(nh^q)^{1/2}(\tilde{m}(x) - m(x)) \sim N \left(0, \frac{\sigma_u^2(x)}{Tf(x)} \int K^2(\Psi) \, d\Psi \right) \tag{23}$$

$$(nh^{q+2})^{1/2}(\tilde{\beta}(x) - (\beta(x)) \sim N \left(0, \frac{\sigma_u^2(x)}{Tf(x)} \frac{\int K^2(\Psi)\Psi^2 \, d\Psi}{\mu_2^2} \right) \tag{24}$$

If x_{it} contains p lag dependent variables then T is replaced by $T - p$ in (23) and (24). The above results hold under some smoothness conditions on the kernel K, i.i.d. assumption of $\{y_{it}, x_{it}\}$, existence of fourth moments of x_{it} and u_{it}, and that $h \to 0, nh^{q+2} \to \infty$ and $nh^{q+4} \to 0$ as $n \to \infty$. Further, following Rilstone and Ullah (1989),

$$(nh^{q+2})^{1/2}(\tilde{\beta}_1(x) - (\beta(x)) \sim N \left(0, \frac{\sigma_u^2(x)}{Tf(x)} \int (K^{(1)}(\Psi))^2 \, d\Psi \right) \tag{25}$$

In practice one can replace $\sigma_u^2(x)$ and $f(x)$ by their consistent estimators $\hat{\sigma}_u^2(x)$ and $\hat{f}(x)$.

The above results indicate that while the asymptotic variances of $\tilde{m}(x)$ and $\tilde{\beta}_1(x)$ are the same as those of the Nadaraya-Watson estimators $\hat{m}(x)$ and the Rilstone-Ullah estimator $\hat{\beta}(x)$, respectively, the asymptotic variance of estimator $\tilde{\beta}(x)$ is different from $\tilde{\beta}_1(x)$ and $\hat{\beta}(x)$. For the standard normal kernel, however, the asymptotic variance of $\tilde{\beta}(x)$ is the same as that of $\tilde{\beta}_1(x)$ and $\hat{\beta}(x)$. This is because $\mu_2^{-2} \int \Psi^2 K^2(\Psi) \, d\Psi = \int (K^{(1)}(\Psi))^2 \, d\Psi = 1/4\sqrt{\pi}$.

Regarding the average derivative in (22) we note from Li, Lu, and Ullah (1996) that

$$n^{1/2}(\tilde{\beta} - \beta) \sim N \left(0, \frac{\Sigma}{T} \right) \tag{26}$$

where Σ is as in (9). Thus, the local linear average derivative estimator has the same asymptotic variance as $\hat{\beta}_R$ and $\hat{\beta}_{HS}$ (8). The Monte Carlo analysis in Li, Lu, and Ullah (1996), however, indicates that in small samples $\tilde{\beta}$ performs better than both $\hat{\beta}_R$ and $\hat{\beta}_{HS}$ in terms of the MSE. Further the MSE of $\tilde{\beta}$ is minimum when $h \propto n^{-2/7}$.

We now turn to the small-sample behavior of $\tilde{m}(x)$ and $\tilde{\beta}(x)$ compared to $\hat{m}(x)$ and $\hat{\beta}(x)$. Conditional on x_{it},

$$E(\tilde{\delta}(x) \mid x_{it}) = (Z'(x)K(x)Z(x))^{-1}Z'(x)K(x)m^* \tag{27}$$

$$V(\tilde{\delta}(x) \mid x_{it}) = (Z'(x)K(x)Z(x))^{-1}Z'(x)\Omega(x)Z(x)(Z'(x)K(x)Z(x))^{-1} \tag{28}$$

where m^* and $\Omega(x) = K(x)\Sigma K(x)$ are as given in (10) and (11). In practice, Σ can be consistently estimated by using $\hat{\sigma}_u^2(x)$ for $\sigma_u^2(x)$, where $\hat{\sigma}_u^2(x)$ can either be obtained by the Nadaraya-Watson kernel estimator described above or by the local linear estimator $(1\ 0)\ Z'(x)K(x)Z(x))^{-1}Z'(x)K(x)\tilde{u}$; $\tilde{u}$ is the vector of local linear squared residuals.

If we consider the Taylor expansion of m^* and take n to be large, then it can be shown that (Ruppert and Wand 1994), up to $O(h^2)$,

$$E(\tilde{m}(x) \mid x_{it}) - m(x) = \frac{h^2}{2}\mu_2 m^{(2)}(x) \tag{29}$$

and

$$E(\tilde{\beta}(x) \mid x_{it}) - \beta(x)$$
$$= \frac{h^2}{2\mu_2}\left[\frac{1}{3}\mu_4 m^{(3)}(x) + (\mu_4 - \mu_2^2)\frac{m^{(2)}(x)f^{(1)}(x)}{f(x)}\right] \tag{30}$$

where μ_4 is the fourth moment of the kernel around zero. Using (29) it can also be verified that

$$E(\tilde{\beta}_1(x) \mid x_{it}) - \beta(x) = \frac{h^2}{2}\mu_2 m^{(3)}(x) \tag{31}$$

Again, these conditional bias results are the same as the unconditional results. Further, the variance of $\tilde{m}(x)$, up to $O(1/nh^q)$, and variances of $\tilde{\beta}(x)$ and $\tilde{\beta}_1(x)$, up to $O(1/nh^{q+2})$, are the same as their asymptotic variances given above. It follows from these results that the optimal h's which minimize the integrated MSE of $\tilde{m}(x)$, $\tilde{\beta}(x)$, and $\tilde{\beta}_1(x)$ are the same as in (16) except for the proportionality constants.

Comparing the bias of Nadaraya-Watson estimator $\hat{m}(x)$ with that of the local linear estimator $\tilde{m}(x)$ we note that while the bias of $\tilde{m}(x)$ depends on the curvature behavior $m^{(2)}$, the bias of $\hat{m}(x)$ depends on $m^{(2)}$ as well as $m^{(1)}f^{(1)}/f$ due to the local constant fit. When $|m^{(1)}|$ is large or when $f^{(1)}/f$ is large, especially in highly clustered data, the bias of $\hat{m}(x)$ is large. Even when the true regression is linear $\hat{m}$ is biased but $\tilde{m}$ is unbiased. The asymptotic variances of $\hat{m}$ and $\tilde{m}$ are the same. Thus, one might expect $\tilde{m}$ to perform better than $\hat{m}$. Fan (1992, 1993) reports good finite-sample MSE performance of $\tilde{m}$ compared to $\hat{m}$. He shows that $\tilde{m}$ is the best among all linear smoothers, including orthogonal series and splines. Further $\tilde{m}$ has 100% efficiency among all linear smoothers in a minimax sense and a high minimax efficiency among all possible estimators. Fan and Gijbels (1992) have reported better performance of $\tilde{m}(x)$ near the boundary of the support of f, although see a word of caution by Ruppert and Wand (1994).

The comparison of the bias of $\tilde{\beta}(x)$ with the bias of $\hat{\beta}(x)$ is, however, not so clear. Both are seriously affected by $m^{(2)}$, $m^{(3)}$, $f^{(1)}$, and f. In contrast, the bias of our proposed estimator $\tilde{\beta}_1(x)$ is much simpler and smaller compared to $\tilde{\beta}(x)$ and $\hat{\beta}(x)$. In fact, when the true m is linear both $\tilde{\beta}(x)$ and $\hat{\beta}(x)$ are biased but $\tilde{\beta}_1(x)$ is unbiased. Though not studies here, it is conjectured that the MSE performance of $\tilde{\beta}_1$ will be much superior to both $\tilde{\beta}$ and $\hat{\beta}$.

The estimator $\bar{\delta}(x)$ in (19) can also be obtained by the local nonparametric estimation of the linear parametric model

$$y_{it} = \alpha + x_{it}\beta + u_{it} \tag{32}$$

in the sense of minimizing $\sum^n \sum^T u_{it}^2 K((x_{it} - x)/h)$ with respect to α and β. This gives the estimator of β, which will be the same as $\tilde{\beta}(x)$, and $\bar{m}(x)$ will be $\tilde{\alpha}(x) + x\tilde{\beta}(x)$. The advantage of viewing the problem in this way is that one can extend it to the local estimation of the parameters of any nonlinear parametric model $g(x_i, \beta)$. This idea stems from the recent literature on the local maximum likelihood estimation of the parameters of a parametric density (Hjort and Glad 1995). The same idea is also used in Robinson (1989b), who considered the problem of estimating β over time in the time series regression $y_t = x_t\beta + u_t$. Robinson obtained the estimator $\hat{\beta}(t^*) = (\sum^T K_t(t^*)x_t^2)^{-1} \sum^T K_t(t^*)x_t y_t$ by minimizing $\sum^T u_t^2 K_t(t^*)$, where $K_t(t^*) = K((Tt^* - t)/Th)$ for t^* in $[0, 1]$. An interpretation of (32) can be given by using the small-σ expansion. For this consider the small σ expansion of $m(x_{it})$ in (1) around μ, instead of small h around x, as $m(x_{it}) = m(\mu) + (x_{it} - \mu)\beta(\mu) + O(x_{it} - \mu)^2$ where $O(x_{it} - \mu)^2 = O(\sigma^2)$ by writing $x_{it} = \mu + \sigma\epsilon_{it}$. Thus, assuming $\sigma \to 0$, we get

$$y_{it} = m(\mu) + (x_{it} - \mu)\beta(\mu) + u_{it} \tag{33}$$
$$= \alpha + x_{it}\beta + u_{it}$$

which is (32); $\alpha = m(\mu) - \mu\beta(\mu)$. Thus while the local linear model (17) is based on the small h approximation of m, the model (32) may be interpreted as the small σ approximation of m.

We point out that when $h = \infty$ the local minimization of $\sum^n \sum^T u_{it}^2 K((x_{it} - x)/h) = K(0) \sum^n \sum^T u_{it}^2$ becomes the global minimization of $\sum^n \sum^T u_{it}^2$. This gives the usual parametric pooled LS estimators of α and β.

The idea of local linear estimation of $m(x)$, the conditional mean, can be easily extended to the local estimation of the higher central moments, $\mu_r = E[(y_{it} - m(x_{it}))^r \mid x_{it}] = E(u_{it}^r \mid x_{it})$. This will be the coefficient of $\sqrt{K_{it}}$ in the regression of $\sqrt{K_{it}}\tilde{u}_{it}^r$ on $\sqrt{K_{it}}$ and $\sqrt{K_{it}}(x_{it} - x)$, where $\tilde{u}_{it}$ is obtained from (17) or (32). It will be interesting to compare the properties of $\tilde{\mu}_r$ so obtained with those obtained by the Nadaraya-Watson-type local constant estimator $\hat{\mu}_r$, which is obtained by regressing $\sqrt{K_{it}}\hat{u}_{it}^r$ on $\sqrt{K_{it}}$, where $\hat{u}_{it}$ is the residual from (2). The comparison will especially be useful for $r = 2$, the conditional heteroskedasticity model often used in macroeconomics and finance.

C. Nonparametric Fixed-Effects (FE) Model

A nonparametric FE model can be written as

$$y_{it} = \alpha_i + m(x_{it}) + u_{it} \tag{34}$$

where α_i is the individual specific fixed parameters, and u_{it} is i.i.d. with mean zero and constant variance σ_u^2. A more general specification of (34) is discussed in Section II.D. When $m(x_{it}) = x_{it}\beta$, model (34) is the well-known linear parametric FE model studied in the literature very extensively. An important reason for the popularity of the linear parametric model is that there exists a class of transformations of (34) which eliminates α_i so that β can be estimated by either a simple LS or a generalized least-squares (GLS) estimator. It is not straightforward to get transformations which will remove α_i from (34) when $m(x_{it})$ is of unknown form. However, if we reformulate the problem of estimating m and its derivative in terms of local linear estimation of (17), then it is possible to implement some of the existing transformations. These are proposed below. First, following (17),

$$\bar{y}_{i.} = \alpha_i + m(x) + (\bar{x}_{i.} - x)\beta(x) + \bar{u}_{i.} \tag{35}$$

which gives

$$y_{it} - \bar{y}_{i.} = (x_{it} - \bar{x}_{i.})\beta(x) + u_{it} - \bar{u}_{i.} \tag{36}$$

The local FE estimator of $\beta(x)$ can then be obtained by minimizing

$$\sum_{i=1}^{n}\sum_{t=1}^{T}(y_{it} - \bar{y}_{i.} - (x_{it} - \bar{x}_{i.})\beta(x))^2 K\left(\frac{x_{it} - x}{h}\right)$$

This gives, for $q = 1$, our proposed estimator as

$$\tilde{\beta}_{FE}(x) = \sum_{i=1}^{n}\sum_{t=1}^{T}\frac{K((x_{it} - x)/h)(y_{it} - \bar{y}_{i.})(x_{it} - \bar{x}_{i.})}{\sum_{i=1}^{n}\sum_{t=1}^{T}K((x_{it} - x)/h)(x_{it} - \bar{x}_{i.})^2} \tag{37}$$

and for $q \geq 1$,

$$\tilde{\beta}_{FE}(x) = (X'M_D K(x)M_D X)^{-1}X'M_D K(x)M_D y$$

where X is an $nT \times q$ matrix, $D = I_n \otimes \iota_T$ is an $nT \times n$ matrix, and $M_D = I - DD'/T$. The estimator $\tilde{\beta}_{FE}(x)$ is essentially the LS of $\sqrt{K_{it}}(y_{it} - \bar{y}_{i.})$ on $\sqrt{K_{it}}(x_{it} - \bar{x}_{i.})$. If $h = \infty$, then $K_{it} = K(0)$ and $\tilde{\beta}_{FE}(x)$ becomes the well-known parametric FE estimator $\tilde{\beta}_{FE}$. The asymptotic properties of this can be worked out in the same way as that for $\tilde{\beta}(x)$ in (20). However, conditional on x_{it},

$$E(\tilde{\beta}_{FE}(x)) = (X'M_D K(x)M_D X)^{-1}X'M_D K(x)M_D m^* \tag{38}$$

$$V(\tilde{\beta}_{FE}(x)) = \sigma_u^2(X'M_D K(x)M_D X)^{-1}(X'M_D K^2(x)M_D X)$$
$$\times (X'M_D K(x)M_D X)^{-1}$$

provided $\sigma_{it}^2(x) = \sigma_u^2$. For a feasible version of the variance of $\tilde{\beta}_{FE}(x)$ one needs to replace σ_u^2 by $s_u^2 = \sum^n \sum^t \hat{u}_{it}^{*2}/nT$, where $\hat{u}_{it}^*$ is the residual from the regression in (36). If $\sigma_{it}^2(x) \neq \sigma_u^2$ we need to modify (38) as in (11). Also, as in Sections II.A and II.B, expanding m^* by Taylor series and taking n to be large we can obtain the asymptotic bias and variance of $\tilde{\beta}_{FE}(x)$. This, along with the asymptotic normality, will be the subject of a future study.

The FE estimator (37) can also be interpreted as the local estimation of the parametric FE model $y_{it} = \alpha_i + x_{it}\beta + u_{it}$ after it has been transformed into $y_{it} - \bar{y}_{i.} = (x_{it} - \bar{x}_{i.})\beta + u_{it} - \bar{u}_{i.}$. This will be analogous to the estimation procedure in Section II.B. One can also obtain a FE estimator of β by using first-difference transformation $y_{it} - y_{it-1} = (x_{it} - x_{it-1})\beta + u_{it} - u_{it-1}$ and then minimizing $\sum^n \sum^T (u_{it} - u_{it-1})^2 K((x_{it} - x)/h)$. This gives

$$\tilde{\beta}_{FE}(x) = (X'A'K(x)AX)^{-1}X'A'K(x)Ay \tag{39}$$

with

$$V(\tilde{\beta}_{FE}(x)) = \sigma_u^2 (X'A'KAX)^{-1}(X'A'K^2AX)(X'A'KAX)^{-1}$$

for given x_{it}; A is the matrix which transforms y into the vector of $y_{it} - y_{it-1}$. It is not clear, unlike in the linear parametric regression, whether $\tilde{\beta}_{FE}$ is the same as $\tilde{\beta}_{FE}$ for $T = 2$ or 3. This seems true even if we had estimated β by the weighted GLS procedure, which takes into account the moving-average nature of the error term.

An alternative way to estimate $\beta(x)$ is to write

$$y_{it} - y_{it-1} = m(x_{it}) - m(x_{it-1}) + u_{it} - u_{it-1} \tag{40}$$

$$= m(x) - m(x_{-1}) + (x_{it} - x)\beta(x)$$

$$-(x_{it-1} - x_{-1})\beta(x_{-1}) + u_{it} - u_{it-1}$$

and do the LS of $\sqrt{K_{it}}\Delta y_{it}$ on $\sqrt{K_{it}}$, $\sqrt{K_{it}}(x_{it} - x)$, and $\sqrt{K_{it}}(x_{it-1} - x_{-1})$, where K_{it} is now $K((x_{it} - x)/h, (x_{it-1} - x_{-1})/h)$. The relationship of the estimator so obtained with $\tilde{\beta}_{FE}$ and $\tilde{\beta}_{FE}$ remains unknown. Li and Stengos (1995) propose writing (40) as $\Delta y_{it} = m(x_{it}, x_{it-1}) + \Delta u_{it}$ and then estimating $\hat{m}$ by the nonparametric kernel method. Roy (1996) has considered the estimation of (34) by the Newey-type series estimator.

Following the results in Section II.B the average derivatives can be estimated after getting the pointwise derivatives from above.

D.   Nonparametric Random-Effects (RE) Model

A general nonparametric RE model can be considered as

$$y_{it} = m(x_{it}, \alpha_i) + u_{it} \tag{41}$$

which can be written, after expanding m around the mean values of $x_{it} = \mu$ and $\alpha_i = 0$, as

$$y_{it} = m(\mu, 0) + (x_{it} - \mu)\beta(\mu, 0) + \alpha_i\beta_1(\mu, 0) + u_{it} \qquad (42)$$

$$= \alpha + x_{it}\beta + v_i + u_{it}$$

where $m(\mu) - \mu\beta(\mu) = \alpha$, $v_i = \alpha_i\beta_1(\mu)$, and β and β_1 are derivatives with respect to x_{it} and α_i respectively. The model (41) is the RE version of the local linear model in (33) based on small-σ expansion. The RE version of the local linear model in (17) can also be written as (42) with μ replaced by x:

$$y_{it} = m(x) + (x_{it} - x)\beta(x) + v_i + u_{it} \qquad (43)$$

We consider u_{it} to be i.i.d as in Section II.C, and assume v_i also to be i.i.d with mean zero and variance σ_v^2.

The local nonparametric RE estimator of m and β in (43) can be obtained by minimizing

$$(y^* - Z^*(x)\delta(x))'K(x)(y^* - Z^*(x)\delta(x))$$

$$= \sum_{i=1}^{n}\sum_{t=1}^{T}(y_{it}^* - z_{it}^*\delta(x))^2 K\left(\frac{x_{it} - x}{h}\right)$$

where $y^* = \Omega^{-1/2}y$, $Z^*(x) = \Omega^{-1/2}Z(x)$, and $\Omega^{-1/2} = I_{nT} - (1 - \lambda^{1/2})DD'/T$; D is as in (37). $Z(x)$ and $\delta(x)$ are as defined in (18), $y_{it}^* = y_{it} - (1 - \lambda^{1/2})\bar{y}_{i.}$, $z_{it}^* = z_{it} - (1 - \lambda^{1/2})\bar{z}_{i.}$, and $\lambda = \sigma_u^2/(\sigma_u^2 + T\sigma_v^2)$. This amounts to doing the LS regression of $\sqrt{K_{it}}y_{it}^*$ on $\sqrt{K_{it}}z_{it}^* = [\sqrt{K_{it}}\lambda^{1/2} \quad \sqrt{K_{it}}(x_{it}^* - x^*)]$ which gives our proposed estimator as

$$\tilde{\delta}_{RE}(x) = (Z^{*\prime}(x)K(x)Z^*(x))^{-1}Z^{*\prime}(x)K(x)y^* \qquad (44)$$

When $h = \infty$, $K(x) = K(0)$ and we get the well-known parametric RE estimator given in econometric texts (Baltagi 1995, Hsiao 1986).

A feasible estimator of $\tilde{\delta}_{RE}(x)$ is obtained by replacing λ with its estimator $\hat{\lambda} = \hat{\sigma}_u^2/(\hat{\sigma}_u^2 + T\hat{\sigma}_v^2)$. $\hat{\sigma}_u^2$ is obtained as

$$\hat{\sigma}_u^2 = \sum_{i=1}^{n}\sum_{t=1}^{T}\frac{(y_{it} - \bar{y}_{i.} - \tilde{\beta}_{FE}(x)(x_{it} - \bar{x}_{i.}))^2}{nT} \qquad (45)$$

where $\tilde{\beta}_{FE}(x)$ is the nonparametric FE estimator given in (37). The estimator $\hat{\sigma}_v^2$ is defined as $\hat{\sigma}_v^2 = \hat{\sigma}_\eta^2 - \hat{\sigma}_u^2/T$, where $\hat{\sigma}_\eta^2 = \Sigma(\bar{y}_{i.} - \tilde{m}(\bar{x}_{i.}))^2/n$ and $\tilde{m}(\bar{x}_{i.}) = \tilde{m}(x)$ at $x = \bar{x}_{i.}$ is obtained by performing local least-squares estimation on the model $\bar{y}_{i.} = m(x) + (\bar{x}_{i.} - x)\beta(x) + v_i + \bar{u}_{i.}$.

As described in Section II.B, the estimators of α and β in (42) are also given by the estimator in (44). For given x_{it},

$$V(\tilde{\delta}_{RE}(x))$$
$$= \sigma_u^2 (Z^{*\prime}(x)K(x)Z^*(x))^{-1} Z^{*\prime}(x)K^2(x)Z^*(x)(Z^{*\prime}(x)K(x)Z^*(x))^{-1}$$

$$(46)$$

The detailed study of the asymptotic and finite-sample properties of $\tilde{\delta}(x)$ and its comparison with the FE estimators will be the subject of future research.

E. Semiparametric FE and RE Models

Here we first consider the model of the following type:

$$y_{it} = x_{it}\beta + m(z_{it}) + v_i + u_{it} \qquad (47)$$

where z_{it} is a vector of p regressors. For $v_i = 0$, the model is considered by Li and Stengos (1995), who propose the $\sqrt{n}$ consistent estimation of β by Robinson's (1988) procedure. This is given by transforming (47) with $v_i = 0$ as

$$R_{it}^{yz} = R_{it}^{xz}\beta + u_{it} \qquad (48)$$

where $R_{it}^{yz} = y_{it} - E(y_{it} \mid z_{it})$ and $R_{it}^{xz} = x_{it} - E(x_{it} \mid z_{it})$ and then applying the LS method. This gives

$$\hat{\beta}_{SP} = \left(\sum_{i=1}^n \sum_{t=1}^T R_{it}^{xz\prime} R_{it}^{xz} \right)^{-1} \sum_{i=1}^n \sum_{t=1}^T R_{it}^{xz\prime} R_{it}^{yz} \qquad (49)$$

where SP represents semiparametric pooled estimator. For a feasible version of this estimator, one needs to replace $E(y_{it} \mid z_{it})$ and $E(x_{it} \mid z_{it})$ by their respective nonparametric estimators. Li and Stengos propose the Nadaraya-Watson kernel estimator for this, but an alternative estimator can be obtained by using the local linear estimation in Section II.B. The estimators of $m(z)$ and its derivative can then be obtained by performing local linear regression of $y_{it} - \hat{y}_{it}$ on $m(z_{it})$, where $\hat{y}_{it} = x_{it}\hat{\beta}_{SP}$. Kneisner and Li (1996) look into the estimation of β (with $v_i = 0$) when z_{it} contains the lagged variables of y_{it}. They also propose the estimation of the $m(z_{it})$ and its derivative by doing the local least squares procedure for $y_{it} - \hat{y}_{it} = m(z_{it}) + u_{it}$.

In the case when v_i is nonzero but fixed, the Li and Stengos (1995) estimation of β is not affected. Once β is estimated the estimation of the derivative of $m(z)$, $\gamma(z)$, can be carried out by analyzing $y_{it}^* = y_{it} - \hat{y}_{it} = m(z_{it}) + \alpha_i + u_{it}$ with the methods described in Section II.C. From (37), our proposed semiparametric FE (SFE) estimator of the derivative of $m(z_{it})$ for $p = 1$ can be written as

$$\hat{\gamma}_{SFE}(z) = \frac{\sum_{i=1}^n \sum_{t=1}^T (z_{it} - \bar{z}_{i.})(y_{it}^* - \bar{y}_{i.}^*)K((z_{it} - z)/h)}{\sum_{i=1}^n \sum_{t=1}^T (z_{it} - \bar{z}_{i.})^2 K((z_{it} - z)/h)} \qquad (50)$$

and for $p \geq 1$,

$$\hat{\gamma}_{SFE}(z) = (Z'M_D K(z)M_D Z)^{-1} Z' M_D K(z) M_D y^*$$

It is important to be able to estimate the derivative since that is the parameter of interest in most economic applications.

When v_i is random, the model becomes the semiparametric RE (SRE) model considered in Li and Ullah (1996), who propose a $\sqrt{n}$-consistent GLS estimator of β,

$$\hat{\beta}_{SRE} = \left(\sum_{i=1}^{n} \sum_{t=1}^{T} R_{it}^{xz*\prime} R_{it}^{xz*} \right)^{-1} \left(\sum_{i=1}^{n} \sum_{t=1}^{T} R_{it}^{xz*\prime} R_{it}^{yz*} \right)$$

where $R_{it}^{yz*} = R_{it}^{yz} - (1 - \lambda^{1/2})\overline{R}_{i.}^{yz}$, $R_{it}^{xz*} = R_{it}^{xz} - (1 - \lambda^{1/2})\overline{R}_{i.}^{xz}$, and λ is as in Section II.D. Using their estimator, one can develop the estimation of $y_{it}^* = m(z_{it}) + v_i + u_{it}$ by procedures given in Section II.D; $y_{it}^* = y_{it} - x_{it}\hat{\beta}_{SRE} = y_{it} - \hat{y}_{it}$. This gives the estimators of $m(z)$ and its derivative $\gamma(z)$ from

$$\hat{\delta}_{SRE}(z) = \left(\sum_{i=1}^{n} \sum_{t=1}^{T} r_{it}^{**\prime} K_{it} r_{it}^{**} \right)^{-1} \left(\sum_{i=1}^{n} \sum_{t=1}^{T} r_{it}^{**\prime} K_{it} y_{it}^{**} \right) \tag{51}$$

as $\hat{m}(z) = (1\ 0)\hat{\delta}_{SRE}(z)$ and $\hat{\gamma}_{SRE}(z) = (0\ 1)\hat{\delta}_{SRE}(z)$, where $\hat{\delta}_{SRE}(z)$ is a vector of $\hat{m}_{SRE}(z)$ and $\hat{\gamma}_{SRE}(z)$, $r_{it}^{**} = (1\ z_{it}^{**} - z^{**})$, $z_{it}^{**} = z_{it} - (1 - \lambda^{1/2})\bar{z}_{i.}$, $z^{**} = \lambda^{1/2}z$, $y_{it}^{**} = y_{it}^* - (1 - \lambda^{1/2})\bar{y}_{i.}$, and $K_{it} = K((z_{it} - z)/h)$.

To get the feasible estimators of $\hat{\beta}_{SRE}$ and $\hat{\delta}_{SRE}(z)$ one needs to replace λ with $\check{\lambda}$ and $\tilde{\lambda}$ respectively where $\check{\lambda} = \check{\sigma}_u^2/(\check{\sigma}_u^2 + T\check{\sigma}_v^2)$ and $\tilde{\lambda} = \tilde{\sigma}_u^2/(\tilde{\sigma}_u^2 + T\tilde{\sigma}_v^2)$. Then

$$\check{\sigma}_u^2 = \frac{\sum_{i=1}^{n} \sum_{t=1}^{T} (R_{it}^{yz} - \overline{R}_{i.}^{yz} - \hat{\beta}_{FE}(R_{it}^{xz} - \overline{R}_{i.}^{xz}))^2}{nT} \tag{52}$$

where $\hat{\beta}_{FE}$ is the FE estimator from the model $R_{it}^{yz} = R_{it}^{xz}\beta + v_i + u_{it}$. The estimator $\check{\sigma}_v^2$ is given by $\check{\sigma}_\eta^2 - \check{\sigma}_u^2/T$, where $\check{\sigma}_\eta^2 = \sum^n (\overline{R}_{i.}^{yz} - \hat{\beta}_B \overline{R}_{i.}^{xz})^2/n$ and $\hat{\beta}_B$ is the between estimator obtained by performing LS estimation on the model $\overline{R}_{i.}^{yz} = \overline{R}_{i.}^{xz}\beta + v_i + \bar{u}_{i.}$. We note that our proposed feasible estimator $\hat{\beta}_{SRE}$ based on (52) is different from Li and Ullah (1996), where $\check{\sigma}_u^2$ is obtained by using the LS estimator of β instead of FE estimator of β.

Similarly, $\tilde{\lambda}$ is obtained with $\tilde{\sigma}_u^2 = \sum^n \sum^T (y_{it}^* - \bar{y}_{i.}^* - \tilde{\gamma}_{SFE}(z)(z_{it} - \bar{z}_{i.}))^2/nT$, where $\tilde{\gamma}_{SFE}(z)$ is as in (50) with $y_{it}^* = y_{it} - x_{it}\hat{\beta}_{SRE}$ and $\tilde{\sigma}_\eta^2 = \Sigma(\bar{y}_{i.}^* - \tilde{m}(z))^2/n$, where $\tilde{m}(z)$ is obtained by doing local LS on the model $\bar{y}_{i.}^* = m(z) + (\bar{z}_{i.} - z)\gamma(z) + v_i + \bar{u}_{i.}$.

In a special case of Li and Ullah, when $z_{it} = \bar{x}_{i.}$, the model becomes a semiparametric extension of the Mundlak (1978) model where $m(\bar{x}_{i.})$ is linear in $\bar{x}_{i.}$. Another semiparametric model one can consider is where $m(z_{it})$ is zero but $V(u_{it}\ |$

$x_{it}) = \sigma^2(x_{it})$. Li and Stengos (1994) suggest a two-step GLS estimator of β for such a model. One can also consider $V(\alpha_i \mid \bar{x}_{i.}) = \sigma^2(\bar{x}_{i.})$ and develop a two-step GLS estimator. The situation where both $\sigma^2(\bar{x}_{i.})$ and $\sigma^2(x_{it})$ are present has not been considered in the literature. In a recent paper Horowitz and Markatou (1996) consider the nonparametric kernel density estimation of σ_v^2 and σ_u^2 and then propose the maximum likelihood estimation of β in (47) with no $m(z)$. The issues of unit roots and serial correlation in the errors remain the subjects of future research.

One disadvantage of the semiparametric model in (47) with an unknown function of regressors, or the purely nonparametric model in Section II.A, is the "curse of dimensionality." This refers to the fact that the rate of convergence of the nonparametric estimator of $m(z)$ decreases drastically with the increase in the number of regressors. One solution explored in the literature is to use the generalized additive models of Hastie and Tibshirani (1990) and Berhame and Tibshirani (1993), which estimate the p-dimensional $m(z)$ at the convergence rate of one dimensional nonparametric estimator. Essentially the model (47) is written as (assuming β and $v_i = 0$) $y_{it} = \sum_{j=1}^{p} m_j(z_{jit}) + u_{it}$, and $m_j(z_{jit})$ is estimated by a nonparametric method; see Linton and Neilson (1996) for the kernel method of estimation.

There is an extensive semiparametric literature on the estimation of limited dependent variables models such as single-index models, censored models, and selection models (Melenberg and Soest 1993, Pagan and Ullah 1996). Essentially these semiparametric models can be considered as special cases of (47) with $v_i = 0$ where, for example, in the single-index case $x_{it}\beta + m(z_{it})$ is a function of single index, say, $z_{it}\delta$, that is, $m(z_{it}\delta)$. In the censored case $m(z_{it})$ in (47) is $m(z_{it}\delta)$, which becomes the inverse Mill's ratio under the assumption of the normality of u_{it}. When u_{it} is nonnormal, $v_i \neq 0$ but is fixed, and $m(z_{it}) = 0$, the estimation of (47) by the least absolute deviation (LAD) method has been discussed in Honoré (1992) and Keane (1993), among others.

F. Specification Testing

There is an extensive literature on various specification testing in the parametric FE and RE models (Hsiao 1986, Baltagi 1995, 1996, Greene 1993). Here we mainly look into the recent work in the context of nonparametric panel models.

Considering the nonparametric pooled model in Section II.A, we note that the pointwise hypothesis testing for the linear restrictions on the derivatives can be done by using the asymptotic normality results for the local linear or Nadaraya-Watson given there (Ullah 1988, Robinson 1989a, Lewbel 1995, and Pagan and Ullah 1996 give more details and references). In fact, the pointwise asymptotic test for various misspecifications in the local linear model $y_{it} = m(x) + (x_{it} - x)\beta(x) + u_{it}$ may follow from the corresponding tests in the linear parametric models. The global tests based on comparing the restricted residual sum of squares (RRSS) with the unrestricted RRSS or based on the conditional moments are developed, among others, in

Fan and Li (1996), Li and Wang (1994), and Bierens (1990). Their results cover the tests for linearity, exclusion of regressors and semiparametric specification, and can be extended for $y_{it} = m(x_{it}) + u_{it}$, where x_{it} and u_{it} are i.i.d. across i and t.

Frees (1995) considered the problem of testing cross-sectional correlation in the parametric panel model and noted that the Breusch and Pagan (1980) measure does not possess desirable asymptotic properties for the practical situation where n is large but T is small. In fact, he showed that the asymptotic distribution depends on the parent population even under the hypothesis of no cross-sectional correlation. In view of this, he introduced a distribution-free statistic which does not have this problem. An extension of this to nonparametric and semiparametric models will be useful. Li and Hsiao (1996) have considered the semiparametric model (47) and developed a LM-type test for the null hypothesis that u_{it} is white noise against the alternative that u_{it} has the RE specification.

An important question in the nonparametric panel data analysis is whether to pool the data. A conditional moment test for this problem, $H_0 : m_1(x_{i1}) = m_2(x_{i2})$ against $H_1 : m_1(x_{i1}) \neq m_2(x_{i2})$ assuming $T = 2$ here for simplicity, is proposed in Baltagi et al. (1995). If the H_0 is accepted then one can pool the data and use the results of Section II.A. If H_0 is rejected then the estimates for two different periods can be pooled to obtain a more precise estimate (Pinske and Robinson 1996).

III. AN APPLICATION

Here we present an empirical example based on the methodology discussed and developed in the previous sections. For a long time now there has been a debate in the nutrition-income literature in developing countries on the response of nutrition, more specifically calorie intake, resulting from a rise in income. Some of the recent articles that have engaged in this debate are by Behrman and Deolalikar (1990), Strauss and Thomas (1990), Bhargava (1991), Bouis and Haddad (1992), Subramanian and Deaton (1996), and Grimard (1995). For them estimating the income elasticity is important because it has serious policy implications on how best to reduce malnutrition. If the elasticity turns out to be close to zero, the implication is that improvement in the income of the poor will have little impact on the extent of malnutrition. Then the development policies aimed at improving nutrition will have to use policy instruments which attack malnutrition directly rather than relying solely on rising income.

Behrman and Deolalikar (1990) used individual level ICRISAT VLS panel data for two years from three villages in south central India and estimated a linear parametric FE model. In this section we consider both the standard parametric panel models and the nonparametric panel models discussed in Section II to study the calorie-income relationship based on the data set used by Behrman and Deolalikar (1990). For details on ICRISAT VLS data, see Binswanger and Jodha (1978), Ryan et al. (1984), and Walker and Ryan (1989). Our contribution to the existing literature on calorie-income relationship is that we are able to take into account both the functional form and the heterogeneity while modeling the calorie-income relationship, and this we are able to do using the results in Section II. Note that previous works have considered one or the other (i.e., the heterogeneity or the functional form) but never both together. While we recognize the fact that there are other variables which influence individual calorie intake, we choose to use income as our only regressor since it is undoubtedly the most influential factor in individuals' consumption decisions and some other authors in this literature have done the same. For example, Subramanian and Deaton (1996) studied the regression of calorie intake on expenditure. A nonparametric regression analysis of calorie intake with other variables besides income included as regressors will be the subject of a future study. We think that in the multivariate case, the semiparametric method described in Section II.E, rather than a pure nonparametric analysis, may be a better way to study the calorie-income relationship.

We consider three types of nonparametric models: constant intercept, fixed-effects, and random-effects models. These correspond to model (34) with α_i equal to a constant, α_i as an individual fixed effect, and α_i as a random effect respectively. Similarly we consider the same three types of models with linear parametric specification $m(x_{it}) = x_{it}\beta$. The dependent variable, y_{it}, in all the models represents the logarithm of individual calorie intake for the ith individual in the tth time period, the explanatory variable, x_{it}, represents the logarithm of per capita real income, and α_i represents the combined effects of unobserved individual characteristics, household characteristics, etc., which can be considered to be fixed or random, as may be the case.

The results are all based on a total number of observations of 730, that is, 365 individuals each observed over two years. For the nonparametric regression analysis the kernel used is the normal kernel given as

$$K\left(\frac{x_{it} - x}{h}\right) = \frac{1}{\sqrt{2\pi}} \exp\left\{-\frac{1}{2}\left(\frac{x_{it} - x}{h}\right)^2\right\}$$

and h, the window width, is taken as $csn^{-1/7}$, where c is a constant, s is the standard deviation of the variable x, and n is the number of observations. For details on choosing window width and kernel, see Marron (1988), Härdle (1990), and Pagan and Ullah (1996).

Our parametric estimates of the income elasticity of calorie intake from different versions of model (34) with $m(x_{it}) = x_{it}\beta$ are presented in Table 1. They are all positive and significant. Our result is in contrast to Behrman and Deolalikar's result that the elasticity estimates from the parametric FE model are zero and hence not significant. Perhaps one reason is that they had other regressors besides income in their model. Also, the sign of the current income elasticity of calorie intake was negative

Table 1   Parametric Income Elasticity of Calorie Intake

Model	Beta coefficient	t-Ratio
Constant intercept	0.138	7.104
Fixed effect	0.108	4.261
Random effect	0.126	6.673

Table 2   Nonparametric Income Elasticity of Calorie Intake

Model	Beta at the mean	t-Ratio
Constant intercept	0.136	4.789
Fixed effect	0.115	4.931
Random effect	0.117	5.906

Table 3   Nonparametric Income Elasticity of Calorie Intake

Model	Mean beta	Minimum	Maximum
Constant intercept	0.132	0.008	0.187
Fixed effect	0.121	0.062	0.134
Random effect	0.107	0.018	0.164

in their constant-intercept model, which may be due to the presence of collinearity since they had both the current and permanent income as regressors. We also calculated the t-ratios by using the cluster-corrected standard errors (Deaton 1994), but still the estimates remained significant.

Note that the magnitude of the estimated current income elasticity from the constant-intercept model is higher than that from the FE or the RE model. The result from the F-test rejects homogeneity of the intercept, and hence the coefficient estimate of the constant intercept model which fails to take into account the heterogeneity is biased. The magnitudes of elasticities from the FE and the RE model are quite similar. In fact, the Hausman specification test failed to reject the null hypothesis of no systematic difference in the two coefficients. Hence, it is up to the researcher to decide whether to use a FE model or a RE model, and this decision will depend on whether he or she wants to make inferences based on the sample or on the population (Hsiao 1986). However, in our case, the conclusion about the elasticity will not change much depending on whether one uses a FE or a RE model. From this parametric regression analysis, the overall picture is that rising income will affect calorie intake but only very slightly. What the parametric analysis does not tell us is whether the elasticity is significant at all levels of income, and if so what is its magnitude. Of course, one can do parametric regression analysis by percentile groups (for example, the elasticity for the bottom and the top deciles, say), but still one cannot get the elasticity at each income level. This question can be answered from the nonparametric regression analysis.

Table 2 gives the nonparametric elasticity estimates at the mean value of the regressor to make it somewhat comparable with the parametric elasticity estimates from Table 1. The results from the nonparametric models are similar to our parametric model results as can be seen from Table 2.

Given that the nonparametric specification gives us elasticity estimates at different income levels, we also report in Table 3 the mean, the minimum, and the maximum values of the elasticity for the different models. The mean elasticities were calculated by using $h = csn^{-2/7}$, as indicated in Section II.A. Note, however, that $h \propto n^{-2/7}$ is known to be optimal for the constant-intercept model only (Li and Ullah 1996), but the optimal h values for the FE and the RE models are not yet known. It can be seen from the table that the elasticity can be quite different for different income levels, and looking at just the mean elasticity estimate can be misleading.

We find for all three models (constant intercept, FE, and RE) and, on average, the elasticity is higher for poorer households compared to richer households (Figures 1, 2, and 3). For our constant-intercept nonparametric model the elasticity is

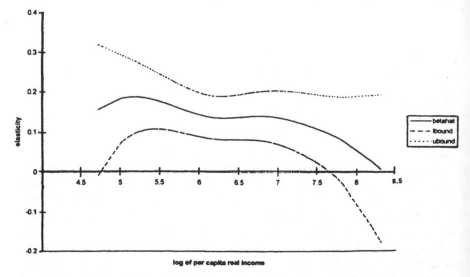

Figure I Elasticity of calorie intake with respect to per capita real income from pooled nonparametric model using local linear estimation method.

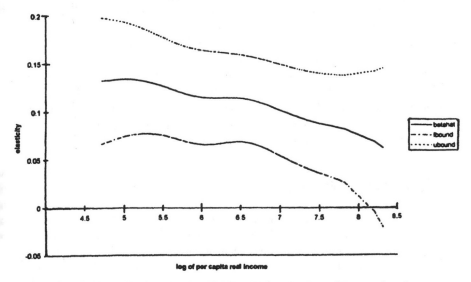

Figure 2   Elasticity of calorie intake with respect to per capita real income from a nonparametric FE model.

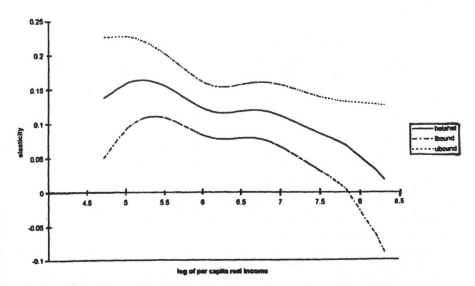

Figure 3   Elasticity of calorie intake with respect to per capita real income from a nonparametric RE model.

significant everywhere except at the tails, but the tail behavior of nonparametric estimators is generally not very good. For FE and the RE models, the elasticities are significant everywhere except at the upper tail.

Thus our results suggest that the income elasticity of calorie intake is significant but is rather low. The small magnitude of elasticity, however, does not necessarily mean that income is an ineffective policy instrument to reduce undernutrition, as has been effectively demonstrated in Ravallion (1990). Both parametric and nonparametric regression analyses give us a similar result, except that the nonparametric one gives us the added information that the elasticity gradually declines as one moves up the per capita income distribution.

ACKNOWLEDGMENTS

We are grateful to B. Baltagi, Q. Li, L. Mátyás, B. Raj, and M. Ravallion for their useful comments and suggestions. We are also thankful to the seminar participants at McGill University and UCR for their comments. We are especially grateful to Anil Deolalikar for the discussions on the subject matter of this chapter and for providing the ICRISAT VLS data. Any remaining errors are ours. The first author gratefully acknowledges research support from the Academic Senate, UCR.

REFERENCES

Ashenfelter, O. and G. Solon (1982), Longitudinal Labor Market Data-Sources, Uses and Limitations, in *What's Happening to American Labor Force and Productivity Measurements?* Proceedings of a June 17, 1982, conference sponsored by the National Council on Employment Policy, W. E. Upjohn Institute for Employment Research.

Balestra, P. and M. Nerlove (1966), Pooling Cross-Section and Time-Series Data in the Estimation of a Dynamic Model: The Demand for National Gas, *Econometrica*, 34, 585–612.

Baltagi, B. H. (1995), *Econometric Analysis of Panel Data*, Wiley, New York.

Baltagi, B. H. (1997), Panel Data Methods, in *Handbook of Applied Economics Statistics* (A. Ullah and D. E. A. Giles, eds.), Marcel Dekker, New York.

Baltagi, B. H., J. Hidalgo, and Q. Li (1996), A Nonparametric Test for Poolability Using Panel Data, *Journal of Econometrics*, 75(2), 345–367.

Behrman, J. R. and A. B Deolalikar (1990), The Intrahousehold Demand for Nutrients in Rural South India: Individual Estimates, Fixed Effects and Permanent Income, *Journal of Human Resources*, 25(4), 665–696.

Berhame, K. and R. J. Tibshirani (1993), Generalized Additive Models for Longitudinal Data, Manuscript, University of Toronto.

Bhargava, A. (1991), Estimating Short and Long Run Income Elasticities of Foods and Nutrients for Rural South India, *Journal of the Royal Statistical Society*, 154, 157–174.

Bierens, J. H. A. (1990), Consistent Conditional Moment Tests of Functional Form, *Econometrica*, 58, 1443–1458.

Binswanger, H. P. and N. S. Jodha (1978), *Manuals of Instructions for Economic Investigators in ICRISAT's Village Level Studies*, Vol. 2, Village Level Studies Series, Economics Program, ICRISAT, Patancheru.

Borus, H. E. (1982), An Inventory of Longitudinal Data Sets of Interest to Economists, *Review of Public Data Use*, 10, 113–126.

Bouis, H. E. and L. J. Haddad (1992), Are Estimates of Calorie-Income Elasticities too High? A Recalibration of the Plausible Range, *Journal of Development Economics*, 39(2), 333–364.

Breusch, T. S. and A. R. Pagan (1980), The Lagrange Multiplier Test and Its Applications to Model Specification in Econometrics, *Review of Economic Studies*, 47, 239–253.

Chamberlain, G. (1984), Panel Data, in *Handbook of Econometrics* (Z. Griliches and M. Intriligator, eds.). North-Holland, Amsterdam, 1247–1318.

Cleveland, W. S. (1979), Robust Locally Weighted Regression and Smoothing Scatter Plots, *Journal of the American Statistical Association*, 78(368), 829–836.

Cleveland, W. S.. and S. Devlin (1988), Locally Weighted Regression: An Approach to Regression Analysis by Local Fitting, *Journal of the American Statistical Association*, 83, 596–616.

Deaton, A. (1994), *The Analysis of Household Surveys: Microeconometric Analysis for Development Policy*, manuscript, Princeton University.

Dielman, T. E. (1989), *Pooled Cross-Sectional and Time Series Data Analysis*, Marcel Dekker, New York.

Fan, J. (1992), Design Adaptive Nonparametric Regression, *Journal of the American Statistical Association*, 87, 998–1024.

Fan, J. (1993), Local Linear Regression Smoothers and Their Minimax Efficiencies, *Annals of Statistics*, 21, 196–216.

Fan, J. and I. Gijbels (1992),Variable Bandwidth and Local Linear Regression Smoothers, *Annals of Statistics*, 20, 2008–2036.

Fan, Y. and Q. Li (1996), Consistent Model Specification Tests: Omitted Variables and Semiparametric Functional Forms, *Econometrica*, 64, 865–890.

Frees, E. W. (1995), Assessing Cross-Sectional Correlation in Panel Data, *Journal of Econometrics*, 69, 393–414.

Fuller, W. A. and G. E. Battese (1973), Transformations for Estimation of Linear Models with Nested Error Structure, *Journal of the American Statistical Association*, 68, 626–632.

Greene, W. H. (1993), *Econometric Analysis*, Prentice Hall, Englewood Cliffs, New Jersey.

Grimard, F. (1995), Does the Poor's Consumption of Calories Respond to Changes in Income? Evidence from Pakistan, manuscript, McGill University, Canada.

Härdle, W. (1990), *Applied Nonparametric Regression*, Cambridge University Press, Cambridge.

Härdle W. and T. M. Stoker (1989), Investigating Smooth Multiple Regression by the Method of Average Derivatives, *Journal of the American Statistical Association*, 84, 986–995.

Harris, M. R. and L. Mátyás (1996), A Comparative Analysis of Different Estimators for Dynamic Panel Data Models, Department of Economics Monash University, Australia, Working Paper.

Hastie, T. J. and R. J. Tibshirani (1990), *Generalized Additive Models*, Chapman and Hall, London.

Heckman, J. J. and B. Singer (1985), *Longitudinal Analysis of Labor Market Data*, Cambridge University Press, Cambridge.

Hjort, N. L. and I. K. Glad (1995), Nonparametric Density Estimation with a Parametric Start, *Annals of Statistics*, 23, 882–904.

Hoch, I. (1962), Estimation of Production Function Parameters Combining Time-Series and Cross-Section Data, *Econometrica*, 30, 34–53.

Honoré, B. E. (1992), Trimmed LAD and Least Squares Estimation of Truncated and Censored Regression Models with Fixed Effects, *Econometrica*, 60, 533-565.

Horowitz, J. and M. Markatou (1996), Semiparametric Estimation of Regression Models for Panel Data, *Review of Economic Studies*, 63, 145–168.

Hsiao, C. (1985), Benefits and Limitations of Panel Data, *Econometric Reviews*, 4, 121–174.

Hsiao, C. (1986), *Analysis of Panel Data*, Cambridge University Press, Cambridge.

Keane, M. P. (1993), Simulation Estimation of Panel Data Models with Limited Dependent Variables, in *Handbook of Statistics* (G. S. Maddala, C. R. Rao, and M. D. Vinod, eds.), North-Holland, Amsterdam, 545–571.

Klevmarken, N. A. (1989), Panel Studies: What Can We Learn from Them? Introduction, *European Economic Reviews*, 33, 525–529.

Kneisner, T. and Q. Li (1996), Semiparametric Panel Data Model with Dynamic Adjustment: Theoretical Considerations and an Application to Labor Supply, manuscript, University of Guelph.

Lewbel, A. (1995), Consistent Nonparametric Hypothesis Tests with an Application to Slutsky Symmetry, *Journal of Econometrics*, 67, 379–340.

Li, Q. and C. Hsiao (1996), Testing Serial Correlation in Semiparametric Panel Data Models, manuscript, University of Guelph.

Li, Q., X. Lu, and A. Ullah (1996), Estimating Average Derivatives by Local Least Squares Method, manuscript, University of Guelph.

Li, Q. and T. Stengos (1994), Adaptive Estimation in the Panel Data Error Component Model with Heteroskedasticity of Unknown Form, *International Economic Review*, 35, 981–1000.

Li, Q. and T. Stengos (1996), Semiparametric Estimation of Partially Linear Panel Data Models, *Journal of Econometrics*, 71(1–2), 389–397.

Li, Q. and A. Ullah (1996), Estimating Partially Linear Panel Data Models with One-Way Error Components Errors, manuscript, University of Guelph.

Li, Q. and S. Wang (1994), A Simple Consistent Bootstrap Test for a Parametric Regression Function, manuscript, University of Guelph.

Linton, O. and J. P. Neilson (1996), A Kernel Method of Estimating Structured Non-parametric Regression Based on Marginal Integration, *Biometrike*, 82, 93–100.

Maddala, G. S. (1971), The Use of Variance Components Models in Pooling Cross-Section and Time-Series Data, *Econometrica*, 39, 341–358.

Maddala, G. S. (1987), Recent Developments in the Econometrics of Panel Data Analysis, *Transportation Research*, 21, 303–326.

Marron, J. S. (1988), Automatic Smoothing Parameter Selection, *Empirical Economics*, 13, 66–86.

Mátyás L. and P. Sevestre (ed.) (1996), *The Econometrics of Panel Data: A Handbook of the Theory with Applications*, Kluwer, Dordrecht.

Melenberg, B. and A. V. Soest (1993), Semiparametric Estimation of the Sample Selection Model, manuscript, Tilburg University.

Müller, H. G. (1987), Weighted Local Regression and Kernel Methods for Nonparametric Curve Fitting, *Journal of the American Statistical Association*, 14, 457–471.

Müller, H. G. (1988), *Nonparametric Regression Analysis of Longitudinal Data*, Springer-Verlag, New York and Berlin.

Mundlak, Y. (1961), Empirical Production Function Free of Management Bias, *Journal of Farm Economics*, 43, 44–56.

Mundlak, Y. (1978), On the Pooling of Time Series and Cross-Section Data, *Econometrica*, 46, 69–85.

Nadaraya, E. A. (1964), On Estimating Regression, *Theory of Probability and Its Applications*, 10, 186–190.

Nerlove, M. (1971), A Note on Error Components Models, *Econometrica*, 39, 359–382.

Pagan, A. and A. Ullah (1996), *Non-parametric Econometrics*, manuscript, Australian National University, Australia.

Pesaran, M. H. and R. Smith (1995), Estimating Long-Run Relationships from Dynamic Heterogenous Panels, *Journal of Econometrics*, 68, 79–113.

Pinske, C. A. P. and P. M. Robinson (1996), Pooling Nonparametric Estimates of Regression Functions with a Similar Shape, in *The Econometrics of Panel Data: A Handbook of the Theory with Applications* (L. Mátyás and P. Sevestre, eds.), Kluwer, Dordrecht, 172.

Raj, B. and B. Baltagi (1992), *Panel Data Analysis*, Physics-Verlag, Heidelberg.

Ravallion, M. (1990), Income Effects on Undernutrition, *Economic Development and Cultural Change*, 38, 487–515.

Rilstone, P. (1991), Efficient Instrumental Variable Estimation of Nonlinear Dependent Process, manuscript, University of Laval.

Rilstone P. and A. Ullah (1989), Nonparametric Estimation of Response Coefficients, *Communications in Statistics*, 18, 2615–2627.

Robinson, P. M. (1988), Root-N Consistent Semiparametric Regression, *Econometrica*, 56, 931–954.

Robinson, P. M. (1989a), Hypothesis Testing in Semiparametric and Nonparametric Models for Economic Time Series, *Review of Economic Studies*, 56, 511–534.

Robinson, P. M. (1989b), Nonparametric Estimation of Time-Varying Parameters, in *Statistical Analysis and Forecasting of Economic Structural Change* (P. Hackl, ed.), Springer-Verlag, New York and Berlin.

Roy, N. (1996), Series Estimation of Income Elasticity of Calorie Intake Across Income Classes, *Journal of Quantitative Economics*, 12(2), 77–84.

Ruppert, D. and M. P. Wand (1994), Multivariate Locally Weighted Least Squares Regression, *Annals of Statistics*, 22, 1346–1370.

Ryan, J. G., P. D. Bidinger, P. Pushpamma, and N. Prahlad Rao (1984), *The Determinants of Individual Diets and Nutritional Status in Six Villages of Southern India*, Research Bulletin No. 7, ICRISAT, Patancheru.

Stone, C. J. (1977), Consistent Nonparametric Regression, *Annals of Statistics*, 5, 595–644. (1977).

Stone, C. J. (1980), Optimal Rates of Convergences for Nonparametric Estimation, *Annals of Statistics*, 8, 1348–1360.

Stone, C. J. (1982), Optimal Global Rates of Convergence for Nonparametric Regression, *Annals of Statistics*, 10, 1040–1053.

Strauss, S. and D. Thomas (1990), The Shape of the Calorie Expenditure Curve, *Economic Growth Centre Discussion Paper no. 595*, Yale University.

Subramanian, S. and A. Deaton (1996), The Demand for Food and Calories: Further Evidence from India, *Journal of Policital Economy*, 104(1), 133–162.

Ullah, A. (1988), *Semiparametric and Nonparametric Econometrics*, Physics-Verlag, Heidelberg.

Vinod, H. D. and A. Ullah (1988), Flexible Production Estimation by Nonparametric Kernel Estimators, in *Advances in Econometrics: Robust and Nonparametric Statistical Inference* (T. B. Fomby and C. F. Rhodes, eds.), JAI Press, Greenwich.

Walker, T. S. and J. G. Ryan (1989), *Village and Household Economies in India's Semiarid Tropics*, The Johns Hopkins University Press, Baltimore.

Wallace, T. D. and A. Hussain (1969), The Use of Error Components Model in Combining Cross-Section with Time-Series Data, *Econometrica*, 37, 55–72.

Watson, G. S. (1964), Smooth Regression Analysis, *Sankhya, Series A*, 26, 359–372.

18

On Calibration

Adrian Rodney Pagan
The Australian National University, Canberra, Australia

I. INTRODUCTION

Econometric methodologies generally develop spasmodically, frequently spurred on by advances in computing power, and it is generally the case that they are codified long after they have been in use for some time. For the past two decades macroeconometricians have tended to classify econometric methodologies as those associated with three schools of thought: the "LSE approach," VAR analysis, and Bayesian methods. Microeconometricians have tended to choose from a smaller set, and much of this research remains of the type that features quite complex statistical analysis laid over a thin veneer of economic theory. Perhaps the best examples of the latter are in a field such as labor econometrics, where one needs to be knowledgable about a myriad of estimators designed to handle the statistical issues that arise with such data. Regardless of the field, however, the passage of these decades has witnessed the growth of a different philosophy concerning the nature of modeling and how to go about it, an approach that its practitioners have increasingly given the rubric of "calibration." This literature is now very diverse. To support that contention consider the following list of studies, each of which claims to be dealing with models that are "calibrated."

1. Macroeconomic Policy Models
   Bank of Canada (1994) model (Black et al. 1994)
   New Zealand Treasury (1994) model (Murphy 1995)
   MSG model (McKibbin and Sachs 1991)
   G3 model (McKibbin and Wilcoxen 1992)
2. Macroeconomic Analysis Models
   King, Plosser, and Rebelo (1988) (real business cycles models)

3. Term Structure Models and Asset Pricing Models
    Canova and Marrinan (1993)
    Boudoukh (1993)
    Fisher (1994)
    Rudebusch (1995)
4. Macroeconomic Forecasting Models
    Ingram and Whiteman (1994) (VAR model with RBC priors)
5. Computable General Equilibrium Models
6. Monetary Models
    Nason and Cogley (1994)
    Gilles et al. (1993)
7. International Monetary Models
    Gruen and Gizycki (1993)
    Bansal et al. (1995)
8. Industry Models
    Rosen, Murphy, and Scheinkman (1994)
    Jovanovic and MacDonald (1994)

So it is clear that the practice of "calibration" is a very popular activity. It is therefore important that some investigation be made into the question of what it is and what difficulties one would face when attempting to use it when persuing an answer to a question that relies on quantitative measurement. In the following section we provide a definition and discuss what this definition means. This is followed by a more extensive treatment of some of the issues. Section IV notes that those who regard themselves as calibrators are fierce proponents of it, and this ferocity can sometimes border upon intolerance for other approaches. yet there are many unresolved questions about whether calibration really is a satisfactory answer to the quest for quantitative information. We point out some of these and ask whether they have yet been satisfactorily answered within that literature. In some instances, where the answer is in the negative, the technology exists that should enable one to be able to go at least part of the way toward a resolution of these questions.

II. A DEFINITION OF CALIBRATION

One issue that rises is how all the authors listed above can regard themselves as involved in calibration—do they mean the same thing when describing what they do in this way? To get to the heart of this we really need some definition of the term. That is not easy to come by. Definitions, such as ". . . calibration . . . is not estimation" (Kydland and Prescott 1996, p. 74) are negative rather than positive, while alternative explanations, such as Kydland and Prescott (1991), tend to be rather too diffuse. Nevertheless, I would agree with the negative statement above, in the sense that the primary focus of calibrators is not really estimation, even though they may perform estimation as part of their task. My definition is

> Calibration is the process whereby data is employed in order to measure specified characteristics of a system.

There are three key words or phrases in the definition—"process," "data employed," and "measure specified characteristics," and we need to dwell a little on the last two elements, concentrating upon how one would carry out the process distinguished in the definition. The employment of *data* distinguished calibration from exercises with models in which unknown parameters are just replaced with some values in order that they might be simulated—what King (1996) refers to as quantitative theory. The characteristics that investigators wish to measure are multifold—examples drawn from the literature include

- The moments of variables
- The density of a variable
- Derivatives of one variable with respect to another

In order to make these measurements it is necessary to describe the system which is being used. In turn this requires

- A description of relations between variables
- An assignment of values to parameters in relations
- A point at which the measurement is made

How to describe relations between variables invokes the question of the role of theory in econometric analysis. For this reason we defer discussion of it until later. This leaves the issues of how parameter values in a relation are to be assigned and what methods can be used for evaluating any model derivatives, etc., that pertain to a specific point. The last issue arises in many computable general equilibrium (CGE) models. Because the aim of experiments in such models is to perform the equivalent of comparative static analysis, it was important to have a data set that represented an equilibrium position from which the derivative could be computed. Construction of such a data set was termed "benchmarking," and the process is generally part of the "calibration" strategy of CGE researchers. Again, this points to the elusiveness of the concept. In few of the other areas in which "calibration" is practiced would we find benchmarking to be an important part of the analysis.

III. ASSIGNMENT OF PARAMETERS

This leaves us with the topic of parameter assignment. There is no one way of doing estimation that is common to all who describe themselves as calibrators. Instead, the whole gamut of estimation procedures is represented, ranging from the somewhat vague (and possibly inconsistent) prescriptions of Kydland and Prescott:

> Thus data are used to calibrate the model economy so that it mimics the world as closely as possible along a limited, but clearly specified, number of dimensions. (Kydland and Prescott 1996, p. 74)

> It is important to emphasize that the parameter values selected are not the ones that provide the best fit in some statistical sense. (Kydland and Prescott 1996, p. 74)

to maximum likelihood (e.g., McGrattan 1994), GMM (e.g., Christiano and Eichenbaum 1992, Fève and Langot 1994), and indirect estimation (e.g., Smith 1993, Bansal et al. 1995). Indirect estimation is an interesting approach in that it brings together those who are primarily interested in fitting statistical models to data with those concerned with having a theoretical model as the way of organizing the facts. In indirect estimation, as set out in Gourieroux et al. (1993) and Gallant and Tauchen (1996), the parameters of the theoretical model are derived from the estimated parameters of the statistical model. The method works from the observation that, if the theoretical model is correct, then, from the principles of encompassing, one can predict what the parameters of the statistical model should be. Hence, if we reverse the normal encompassing methodology, we can recover estimates of the parameters of the theoretical model from those of the statistical model.

It is worth asking why one does this rather than fit the theoretical model directly to the data, as would be the practice with those doing MLE, i.e., "direct" estimation. It turns out that there may be some gains to doing indirect rather than direct estimation. To see this we look at a small stochastic equilibrium model set out in Ingram (1995). In this model the system consists of equations describing the evolution of the log of the capital stock, k_t, productivity, a_t, and the real interest rate r_t, of the form (Ingram 1995, p. 20)

$$k_t = \text{const} + (\lambda_1 + \rho_a)k_{t-1} - \rho_a\lambda_1 k_{t-2}$$

$$- (\rho_a - \rho_r)\frac{\lambda_1}{\delta}\left(\frac{1}{1 - \beta\lambda_1\rho_r}\right)r_{t-1} + \text{error}$$

$$a_t = \rho_a a_{t-1} + \text{error}$$

$$r_t = \rho_r r_{t-1} + \text{error}$$

where λ_1 is a function of the discount factor β and the cost of adjustment coefficient δ, and the equation describing the evolution of the capital stock comes from solving the Euler equations.

Now let us play some games in which we describe the results from being a calibrator, who is either performing direct estimation or employing an indirect estimator as a way of measuring any unknown parameters. We will assume that the "theorist," who provides the model to be estimated, makes some errors, and we ask how robust the estimation methods are to these mistakes. The parameters to be estimated will be ρ_1, δ, β, and ρ_r. One can always get consistent estimators of ρ_a and ρ_r from the second and third equations of the system, allowing us to concentrate on the estimation of the first equation as that relevant to producing estimates of δ and β. The statistical model we choose for indirect estimation is

$$k_t = \text{const} + b_1 k_{t-1} + b_2 k_{t-2} + b_3 r_{t-1} + \text{error}$$

Case 1: Calibrationist Invalidly Assumes $\rho_1 = \rho_r$

Direct estimator of ρ_a, δ, β

The direct estimator estimates $k_t = \text{const} + (\lambda_1 + \rho_a)k_{t-1} - \rho_a \lambda_1 k_{t-2} + \text{error}$. There is clearly a specification error since r_{t-1} has been invalidly excluded. Assuming that least squares is used on each of the first two equations ρ_a will be consistently estimated, but the estimator of λ_1, and hence δ and β, will be inconsistent.*

Indirect estimator

The idea behind indirect estimation is to find what values of δ, β are implied by b_1, b_2, and b_3. In the statistical model b_1, b_2, and b_3 are all consistently estimated; i.e., $\lambda_1 + \rho_a$ and $-\rho_a \lambda_1$, are consistently estimated. Hence, the estimation of β and δ will be consistent. Thus the use of the general statistical model as the way of inferring estimates of the parameters of the theoretical model has protected us against a misspecification that comes from making an incorrect assumption about the parameters of the latter.

Case 2: Modeler Assumes $\rho_a = 0$

Direct estimators

The direct estimators of δ, β are inconsistent as the equation estimated is

$$k_t = \text{const} + (\lambda_1 + \rho_a)k_{t-1} - \rho_a \lambda_1 k_{t-2}$$
$$- \rho_r \frac{\lambda_1}{\delta}\left(\frac{1}{1 - \beta \lambda_1 \rho_r}\right) r_{t-1} + \text{error}$$

and this still involves a specification error in that there are incorrect restrictions imposed between the coefficients of r_{t-1}, k_{t-1}, and k_{t-2}.

---

*Of course we cannot recover both β and δ from a single parameter λ_1, but the variance of the error term also contains β and δ and it can be consistently estimated if λ_1 can be.

Indirect estimators

The indirect estimator will also be inconsistent for the same reason as the direct estimator, i.e., even though b_1, b_2, and b_3 are consistent, the theoretical model imposes an incorrect relation between them.

Even though indirect estimation might be more robust than direct estimation, I do not feel that one should oversell this idea, and my presumption would be that there is likely to be little gain from doing indirect estimation, at least in regards to avoiding the consequences of specification error. It does seem though that, in many instances, indirect estimation may be an *easier* way to do estimation, in that information on good statistical models of data is plentiful, and it is frequently easy to simulate from theoretical models, which is the modus operandi of indirect estimation. An example would be models of exchange rates. These can become very complex when allowance is made for intervention points, etc., and direct estimation may be very difficult. There is an extensive literature on the type of GARCH models that fit such data, so it makes sense to use these models to estimate the parameters of some underlying theoretical model of exchange rates. One might even argue that it is a philosophy that is ideally suited to calibration endeavors in that it provides the rationale for a division of labor between those designing good statistical models to fit the data and those interested in generating economic models. It is likely to be rare that any individual has skills in both of these areas and the indirect estimation principle therefore provides a way to reap the benefits of specialization when estimating the parameters of economic models.

IV.   WHAT IS THE DEBATE ABOUT?

With so much agreement one might wonder what the argument is about? I think that there are three major areas in which calibrators have a distinctive stance and these revolve around

- Weak versus strong theory
- Weak versus strong data consistency
- Role of statistics

A.   Preeminence of Theory

The following quotations provide what might be regarded as the polar cases in attitudes toward theory. At one level is the "LSE approach" to econometrics. It is not the case that such a methodology eschews theory, but it sees theory as just one element in modelling, as witnessed by the following statements.

... there is nothing that endows an economic theory with veracity a priori, and so coherence of an econometric model with an economic theory is neither necessary not sufficient for it to be a good model. (Mizon 1995, pp. 115–116)

I think it is important to emphasize that the issue is not theory versus "no theory," even though many representatives of the calibrationist approach do seem to make such a stark contrast. Empirical work in most methodological traditions today is sensitive to the need for theory. Indeed, even in the "systems of equations" approach most despised by Kydland and Prescott (1991) there are few models nowadays that do not have a strong theoretical core; see Hall (1995, pp. 980–983) for a brief review of this fact and Murphy (1988) and Powell and Murphy (1995) for a working model. The issue is more "how much theory" or "what type of theory" rather than "no theory."

In contrast to the position just advanced is a statement by Kydland and Prescott about the role of theory that seems to be shared, to different degrees, by most of those who see themselves as calibrators. It is my belief that it is this *belief in the preeminence of theory* that distinguishes a calibrator from a noncalibrator.

The degree of confidence in the answer depends on the confidence that is placed in the economic theory being used. (Kydland and Prescott 1991, p. 171)

A belief in the preeminence of theory carries with it the stance that consistency with data is of secondary importance; i.e., "strong" data consistency is not necessary when working with models. One might ask if such a stance is reasonable. I think it is if all we are doing with the model is demonstrating the *feasibility* of a particular outcome, i.e., we are doing *quantitative theory*. A good example of this would be the debate over the validity of uncovered interest parity as an essential element of many macroeconometric models. Defining the log of the spot exchange rate as S_t, its expected rate one period in the future as S_{t+1}^e, and the forward rate as F_t, covered interest parity yields

$$S_{t+1}^e - S_t = F_t - S_t.$$

Invoking rational expectations so that $S_{t+1} = S_{t+1}^e + e_{t+1}$, uncovered interest parity eventuates as

$$\Delta S_{t+1} = \beta(F_t - S_t) + e_{t+1}$$

where $\beta = 1$. Now regressions of ΔS_{t+1} on the forward discount $F_t - S_t$ frequently produce values of $\hat{\beta} < 0$, seemingly repudiating uncovered interest parity. Hence, it is interesting to see if we can construct models that would produce $\hat{\beta} \geq 0$. Gruen and Gizycki (1993) produce such a calibrated model. In this instance it does not seem relevant to ask whether the model produces other known characteristics of ΔS_t, such as GARCH and leptokurtosis; the aim of the model is just to show that it is possible to replicate a striking feature of the data with a plausible set of assumptions in a theoretical model.

B.   Consistency with Data

Of course, this raises the issue of the credibility of a model when used for a range of issues rather than just the replication of a single fact. It is rare to see a calibrated model whose originator is not aiming to say something about the real world, from statements that business cycles are largely due to supply side shocks to the idea that cycles reduce welfare by very small amounts. I really find it impossible to believe that anyone can take such conclusions or prescriptions seriously if they derive from models whose credentials have not been established by measuring them against the data. It is therefore fascinating, and troubling, to look at the first of the two polar attitudes that I discern in the literature and which are reproduced below.

> If the theory is strong and the measurements good, we have confidence that the answer for the model economy will be essentially the same as for the actual economy. (Kydland and Prescott 1996, p. 83)

The second position is certainly the opposite:

> If an econometric model is to be taken and used seriously then its credentials must be established. Two important ways to do this are to demonstrate that the model is coherent with the available relevant information, and that it is at least as good as alternative models of the same phenomenon. (Mizon 1995, p. 115)

The statement by Kydland and Prescott comes very close to blaming the data if the calibrator's model fails to fit. It is breathtaking because of our lack of strong theory. We have theory, but to think it is this "strong" is truly amazing. The idea that a model should be used just because the "theory is strong," without a demonstration that it provides a fit to an actual economy, is mind-boggling.

One might argue that few calibration exercises fail to include some evidence on their fit to data. There are however two defects currently in such presentations which hamper my acceptance of the proposition that the credibility of the maintained models has been established. One of these is the selective nature of the facts upon which fit is to be judged. In some exercises this seems to come down to a single index, e.g., the correlation between hours and productivity. In others (e.g., Burnside et al. 1993), the attempts at model evaluation are far more respectable, in that quite a number of features are examined for their correspondence with the data. Nevertheless, it is the case that few of these attempts are holistic. What is to be regarded as holistic depends upon the nature of the problem, but when the models involve restrictions upon a VAR, as is typical of most RBC and monetary models, it seems appropriate to test *all* the restrictions, and not just a subset of them.* King and Watson (1995) make

---

*Anderson (1991) makes the same point in commenting on Kydland and Prescott's (1991) paper.

this same point, although they prefer the comparison of impulse responses. As explained in Pagan (1995) I do not think impulse response comparisons are as good a method of assessing fit as a VAR, albeit it may be that any discrepancy between model and data might be usefully expressed in terms of a discrepancy between the model and data based impulse responses. In Canova et al. (1994) this philosophy was put into action with respect to the model of Burnside et al. (1993). That model looks good when judged by a limited number of features, but very poor when it is forced to address all aspects of the VAR.

Even if one abstracts from the proper way to evaluate a model, one is still left with the question of how we are to assess the magnitude of any inconsistency between data and model. Early on in calibration studies, "eyeball" tests seemed to predominate as measures of the size of the deviation between the model and reality. Consequently, the chosen metric was very fuzzy, leaving one to despair at any agreement being reached over whether a model is satisfactory. Just as beauty is in the eye of the beholder, some of the judgments rendered concerning fit seemed quite remarkable; e.g., a glance at the woeful (to me) match between model predictions and data in either Figure 5 of Hansen and Prescott (1993) or Figure 4 in Jovanovic and MacDonald (1994) leaves one with a sense of wonderment when reading that the authors described the graphical evidence as supportive of the model. Given the tendency for these same authors to pull out the measuring stick of predictive performance when judging other methodologies, e.g., "one reason for its demise was the spectacular predictive failure of the approach" (Kydland and Prescott (1991, p 166), a legitimate question would seem to be why such a benchmark should not be universal rather than particular. Fortunately, some discipline has begun to emerge in this literature, mostly through variants of statistical hypothesis testing (e.g., Burnside et al. 1993), and it is therefore time to turn to the question of the role of statistics within the calibration agenda.

C. Role of Statistics

In many papers written by calibrationists there is a clear hostility to the use of purely statistical models of data. This is most apparent in Kydland and Prescott's (1991) paper where the statistical models are identified with the "systems of equation approach." In that paper, Frisch's name is invoked as someone who heartily disapproved of this type of work. Actually, a closer reading of the paper they cite does not support that interpretation. Frisch was certainly worried about how much information there was in time-series data, and he was very much in favor of investigators "going down coal mines" in order to collect and understand the workings of the institutions they were studying, but the whole paper that is quoted so approvingly by Kydland and Prescott, is directed against the use of theoretical models that are not closely connected with modeling features seen in the real world—what he terms "playometrics."* His stance on this vividly reminds one of some parts of the calibrationist school of modelling as to be worth recording:

> In too many cases the procedure followed resembles too much the escapist procedure of the man who was facing the problem of multiplying 13 by 27. He was not very good at multiplication but very proficient in the art of adding figures, so he thought he would try to add these figures. He did and got the answer 40, which mathematically speaking was the absolutely correct answer to the problem as he had formulated it. But how well does the figure 40 tell us about the size of the figure 351?

Nevertheless, despite a suspicion about statistical analysis, calibrators do increasingly use statistics, and, when they do, I do believe there are some attendant difficulties. These stem from the observations that

> All model economies are abstractions and are by definition false. (Kydland and Prescott 1991, p. 170)

> It is pointless to test all the strong restrictions implied by this simple model: it is known to be wrong in its details, and formal statistical rejections of the null would tell us no more than we already know. The more interesting question is, How wrong is it? (Rosen et al. 1994, p. 482)

If one takes this proposition seriously then it calls into question our ability to easily decide whether a model fits the data based on some specified metric. In particular the type of analysis described below becomes problematic.

> ... first a set of statistics that summarizes relevant aspects of the behaviour of the actual economy is selected. Then the computational experiment is used to generate many independent realizations of the equilibrium process for the model economy. In this way, the sampling distribution of this set of statistics can be determined to any degree of accuracy for the model economy and compared with values of the set of statistics for the actual economy. (Kydland and Prescott 1996, p. 75)

To see what the problem is assume that the model described by the calibrator has the form

$$z_t^* = A z_{t-1}^* + \Gamma \varepsilon_t$$

where z_t^* are the "latent" (unobserved) variables of the theoretical model, ε_t are shocks that drive the theoretical model, and θ are the parameters of the model. The random variable whose realizations are the data will be z_t, and it will be definitional

---

*It is therefore rather ironic to read Kydland and Prescott's (1996) comment that "here by theory we do not mean a set of assertions about the actual economy" (p. 72).

that $z_t = z_t^* + \nu_t$, where the properties of the "observation errors" ν_t are unknown. This is simply a formal description of a misspecified model. To complete the analysis we suppose that the calibrationist computes some quantity, $g(z_t, \theta)$; e.g., this could be the sample variance of (say) output. Kydland and Prescott's proposal is to treat $g(z_t, \theta)$ as fixed and to study the distribution of $g(z_t^*, \theta)$, locating where $g(z_t, \theta)$ lies in this distribution. In order to find the distribution of $g(z_t^*, \theta)$ one only needs to be able to simulate from the theoretical economy. Despite its seductive appeal, the procedure is an invalid one, unless it is assumed that $\nu_t = -z_t^*$, as only then is it true that z_t would remain constant as different realizations of z_t^* are made. Otherwise, we need to make some assumptions about how ν_t varies with z_t^*. By far the simplest solution, used by most investigators, is to make $z_t = z_t^*$. Then the quantity of interest will be taken to be a function of θ alone and $g(z_t, \theta)$ can be compared to the value predicted by the model. However, this presumes that the model is correctly specified and contradicts Kydland and Prescott's fundamental premise about such models.*

What does one do in the face of this problem? My answer, described in more detail in Pagan (1994, pp. 8–9; 1995, p. 51), would be to take the theory of misspecification of econometric models seriously when judging the significance of a value of $g(z_t, \theta)$. The theory for completing this task is now very well developed (e.g., see White 1994), as are the computational methods. Perhaps the major obstacle to implementing the theory is to arrive at a description of how the data is generated that is independent of the theoretical model. In some cases, such as the analysis of macroeconomic data, a VAR would seem to be appropriate, but in other instance one may need to fit quite complex models, e.g., as in Bansal et al. (1995). This is how I see both statistical and theoretical models being used in a way that benefits from specialization. Unlike indirect estimation, which assumes the validity of the theoretical model and then estimates its parameters from a statistical model after generating realizations from the theoretical one, the scheme above reverses the steps, simulating data from the statistical model to be used for studying estimators and statistics that are associated with the theoretical model. Diebold et al. (1995) apply such a scheme when evaluating the quality of the "cattle cycle" model in Rosen et al. (1994).

ACKNOWLEDGMENTS

This chapter was the basis of my comments made in the Calibration Symposium at the 7th World Congress of the Econometric Society in Tokyo, August 1995. Some of the ideas are drawn from Canova et al. (1994), Kim and Pagan (1995), Pagan (1994), and Pagan (1995).

---

*This point is also relevant to those proposals for a Bayesian rather than classical assessment of the quality of the model e.g., De Jong et al. (1996).

REFERENCES

Anderson, T. M. (1991), Comment of F. E. Kydland and E. C. Prescott, The Econometrics of the General Equilibrium Approach to Business Cycles, in S. Hylleberg and M. Paldam (eds.), *New Approaches to Empirical Macroeconomics*, Blackwell, Oxford, 51–56.

Bansal, R. A., R. Gallant, R. Hussey, and G. Tauchen (1995), Nonparametric Estimation of Structural Models for High-Frequency Currency Market Data, *Journal of Econometrics*, 66, 251–287.

Black, R., D. Laxton, D. Rose, and R. Tetlow (1994), The Bank of Canada's New Quarterly Projection Models. Part 1. The Steady State Model: SSQRM, Technical Report No. 72, Bank of Canada.

Boudoukh, J. (1993), An Equilibrium Model of Nominal Bond Prices with Inflation-Output Correlation and Stochastic Volatility, *Journal of Money, Credit and Banking*, 25, 636–665.

Burnside, C., M. Eichenbaum, and S. Rebelo (1993), Labor Hoarding and the Business Cycle, *Journal of Political Economy*, 101, 245–273.

Canova, F., M. Finn, and A. R. Pagan (1994), Evaluating a Real Business Cycle Model, in C. P. Hargreaves (ed.), *Nonstationary Time Series Analysis and Cointegration*, Oxford University Press, Oxford.

Canova, F. and J. Marrinan (1996), Reconciling the Term Structure of Interest Rates with the Consumption Based ICAP Model, *Journal of Economic Dynamics and Control*, 20, 709–750.

Christiano, L. and M. Eichenbaum (1992), Current Real Business Cycles Theories and Aggregate Labor Market Fluctuations, *American Economic Review*, 82, 430–450.

DeJong, D. N., B. F. Ingram, and C. H. Whiteman (1996), Beyond Calibration, *Journal of Economic Dynamics and Control*, 14, 1–9.

Diebold, F. X., L. Ohanian, and J. Berkowitz (1995), Dynamic Equilibrium Economies: A Framework for Comparing Models and Data, NBER Technical Paper No. 174.

Fève, P. and F. Langot (1994), The RBC Model Through Statistical Inference: An Application with French Data, *Journal of Applied Econometrics*, 9, S11–S35.

Fisher, S. J. (1994), Asset Trading, Transaction Costs and the Equity Premium, *Journal of Applied Econometrics*, 9, S71–S94.

Frisch, R. (1970), Econometrics in the World Today, in W. A. Eltis, M. F. G. Scott, and J. N. Wolfe (eds.), *Induction, Growth and Trade: Essays in Honour of Sir Roy Harrod*, Clarendon Press, Oxford, 152–166.

Gallant, A. R. and G. Tauchen (1996), Which Moments to Match?, *Econometric Theory*, 12, 657–681.

Gilles, C., J. Coleman and P. Labadie (1993), Identifying Monetary Policy with a Model of the Federal Reserve, Finance and Economics Discussion Paper Series, Board of Governors of the Federal Reserve System 93–24.

Gourieroux, C., A. Monfort and E. Renault (1993), Indirect Inference, *Journal of Applied Econometrics*, 8, S85–S118.

Gruen, D. W. R. and M. C. Gizycki (1993), Explaining Forward Discounting Bias: Is It Anchoring?, Reserve Bank of Australia Research Discussion Paper, No. 9307.

Hall, S. (1995), Macroeconomics and a Bit More Reality, *Economic Journal*, 105, 974–988.

Hansen, G. D. and E. C. Prescott (1993), Did Technology Shocks Cause the 1990–1991 Recession, *American Economic Review Papers and Proceedings*, 83, 280–286.

Ingram, B. F. (1995), "Recent Advances in Solving and Estimating Dynamic Macroeconomic Models," in K. D. Hoover (ed.). Macroeconometrics: Developments, Tensions and Prospects (Kluwer Academic Publishers, Boston) 15–46.

Ingram, B. F. and C. H. Whiteman (1994), Towards a New Minnesota Prior: Forecasting Macroeconomic Series Using Real Business Cycle Model Priors, *Journal of Monetary Economics*, 47, 497–510.

Javanovic, B. and G. MacDonald (1994), The Life Cycle of a Competitive Industry, *Journal of Political Economy*, 102, 322–347.

Kim, K. and A. R. Pagan (1995), The Econometric Analysis of Calibrated Macroeconomic Models, in M. H. Pesaran and M. R. Wickens (eds.) *Handbook of Applied Econometrics*, Blackwell, Oxford, 356–390.

King, R. G. (1995), Quantitiative Theory and Econometrics, *Federal Reserve Bank of Richmond Economic Quarterly*, 81/3, 53–105.

King, R. G., C. I. Plosser, and S. T. Rebelo (1988), Production, Growth and Business Cycles. I: The Basic Neoclassical Growth Model, *Journal of Monetary Economics*, 21, 195–232.

King, R. G. and M. W. Watson (1995), On the Econometrics of Comparative Dynamics, mimeo, University of Virginia.

Kydland, F. E. and E. C. Prescott (1991), The Econometrics of the General Equilibrium Approach to Business Cycles, *Scandinavian Journal of Economics*, 93, 161–178.

Kydland, F. E. and E. C. Prescott (1996), The Computational Experiment: An Econometric Tool, *Journal of Economic Perspectives*, 10, 69–85.

McCallum, B. T. (1994), A Reconsideration of the Uncovered Interest Parity Relationship, *Journal of Monetary Economics*, 33, 105–132.

McGrattan, E. R. (1994), The Macroeconomic Effects of Distortionary Taxation, *Journal of Monetary Economics*, 33, 573–601.

McKibbin, W. J. and J. D. Sachs (1991), *Global Linkages*, Brookings Institution, Washington, DC.

McKibbin, W. J. and P. J. Wilcoxen (1993), G-Cubed: A Dynamic Multi-sector General Equilibrium Growth Model of the Global Economy, *Brookings Discussion Paper in International Economics No. 98*, Brookings Institution, Washington, DC.

Mizon, G. M. (1995), Progressive Modelling of Macroeconomic Time Series: The LSE Methodology, in K. D. Hoover (ed.), *Macroeconometrics: Developments, Tensions and Prospects*, Kluwer, Boston, 107–170.

Murphy, C. W. (1988), An Overview of the Murphy Model, in M. Burns and C. W. Murphy (eds.), *Macroeconomic Modelling in Australia* (supplementary conference issue of *Australian Economic Papers*), 61–68.

Murphy, C. W. (1995), *A Model of the New Zealand Economy*, New Zealand Treasury, Wellington.

Nason, J. M. and T. Cogley (1994), Testing the Implications of Long-Run Neutrality for Monetary Business Cycle Models, *Journal of Applied Econometrics*, S37–S70.

Pagan, A. R. (1994), Calibration and Economic Research: An Overview, *Journal of Applied Econometrics*, 9, S1–S10.

Pagan, A. R. (1995), Some Observations on the Solution, Estimation and Use of Modern Macroeconometric Models, in K. D. Hoover (ed.), *Macroeconometrics: Developments, Tensions and Prospects*, Kluwer, Boston, 47–55.

Powell, A. A. and C. W. Murphy (1995), *Inside a Modern Macroeconometric Model: A Guide to the Murphy Model*, Springer-Verlag, Berlin and New York.

Rosen, S., K. M. Murphy, and J. A. Scheinkman (1994), Cattle Cycles, *Journal of Political Economy*, 102, 468–492.

Rudebusch, G. D. (1995), Federal Reserve Interest Rate Targeting, Rational Expectations, and the Term Structure, *Journal of Monetary Economics*, 35, 245–274.

Smith, A. A. (1993), Estimating Non-linear Time-Series Models Using Simulated Vector Autoregression, *Journal of Applied Econometrics*, 8, 563–584.

White, H. (1994), *Estimation, Inference and Specification Analysis*, Cambridge University Press, Cambridge.

Index

9 780367 579371